Inventions of Opportunity:
Matching Technology with Market Needs

Inventions of Opportunity:

Matching Technology with Market Needs

Selections from the pages of the Hewlett-Packard Journal

Introduction by William R. Hewlett

Hewlett-Packard Company

Palo Alto, California, U.S.A. 1983

Inventions of Opportunity:
Matching Technology with Market Needs

Library of Congress Catalog Card Number: 83-81659
ISBN 0-9612030-0-5
Hewlett-Packard Part Number 92233B
Manufactured in the United States of America

Hewlett-Packard Company
3000 Hanover Street, Palo Alto, California 94304 U.S.A.

About this Book

Most of this book consists of reprints of articles that were published in the Hewlett-Packard Journal from its beginning in 1949 to the present time. These articles were selected because they deal with products and technologies—inventions—that represent important milestones in the technological history of the Hewlett-Packard Company. In the Introduction to this book, the term "engineering of opportunity" is given to the research and development process that inspired many of these inventions. Hence the book's title: Inventions of Opportunity.

The excerpts from the Hewlett-Packard Journal in this book were reproduced by photographing printed copies of the original issues. The original page numbers in the bottom margins have been retained. Each page has also been given a new number that identifies its position in this book. These new sequential page numbers are at the top outside corners of the pages. Page numbers in the Contents are the new numbers.

The introductory sections of this book are new material and have been given Roman-numeral page numbers.

Introduction

This book is a collection of reprints of significant articles published in the HP Journal since the magazine's inception in 1949. A panel of senior Hewlett-Packard engineers and scientists selected the articles on the basis that at the time of publication each represented a major advancement in state-of-the-art electronic technology, and in many cases led to a major product line for the company.

Before commenting on individual articles, it might be helpful if I provide some background information about the environment at HP that helped produce the equipment and the techniques described.

At the time of the HP Journal's first issue, the company had 150 employees and annual sales of about 2¼ million dollars. The engineering staff was small, but efficient. To make the most of this modest research and development program, we perfected a technique that we referred to as engineering of opportunity. Essentially, it consisted of trying to match a product need with a technology that might produce the basis for such a product.

To do this required that one have a keen understanding of the potential needs of the measurement field. Such knowledge was supplied in part from our marketing organization, and from personal contacts with friends and acquaintances in the engineering and scientific community. But perhaps to a greater extent, this knowledge was derived from our experience in generating advanced measurement techniques for our own development programs, which in many cases were in the forefront of technology. This source became known as "the next bench syndrome."

The technique of engineering of opportunity had a number of advantages. It was relatively low risk, it allowed quick reaction time, and it was ideally suited to the instrument business since each product could stand alone and not necessarily be coordinated with other developments.

One early example of engineering of opportunity that comes to mind is the case of an extended-range resistance-capacity (RC) oscillator. In the years just after World War II, Dr. Bernard M. Oliver (Barney to all) was working at Bell Labs. We had known each other while attending Stanford, and I always made it a point to visit Barney when I was in the New York area. On one such trip he asked if we had ever thought of using a three-phase ring circuit configuration to get around the problems of the parasitic capacities that plagued conventional RC circuits. It was at once obvious that this provided a solution to the problems that had bothered us from the beginning. We started a project, and within one year produced an oscillator that covered the range of frequencies from 10 Hz to 10 MHz in six decade ranges. This instrument stayed in the product line for twelve years and was an important contribution to corporate profits.

Let me say a word about the organizational structure of the company during its early years. Prior to 1957 we were basically a one-plant operation with one central research and development team. The company was small and intimate, with close ties between engineering, production, and marketing. By the mid-1950s, the centralized R&D lab had been subdivided into four sections: Audio-Video, an extension of the company's original product line; Microwave, a result of work started during World War II, primarily for the Bureau of Naval Research; Time and Frequency, an outgrowth of high-speed digital counting circuits (described in one of the following articles); and Oscilloscopes, a program that evolved more from a desire to have a full line of basic measurement products than from any then-existing innovative concepts.

Late in 1957, because of concern that the company was getting too large (1,500 people!) and fear that we might lose the personal touch of the small operating units which we believed was so important, it was decided to divisionalize the company using the four R&D product-line identifications described above. It should be mentioned that in making this decision we were somewhat worried that separating the units, particularly any future geographical dispersion, might adversely affect the valuable exchange of ideas among the engineering teams. Fortunately, this did not turn out to be the case, as time and again important technologies developed by one division were quickly picked up by the others.

An HP semiconductor operation established in the mid-1950s also deserves comment. This project designed and produced special-purpose diodes for use in our own instruments. Substantially cheaper commercial diodes were available at the time, but their performance was less satisfactory than ours. Using proprietary diodes allowed us to produce a vastly superior product that compensated many times over for the slight additional cost. Our capabilities in the rapidly advancing technology of solid-state physics were very limited, however, and recognizing the importance of the technology we were determined to accelerate entry into this very competitive field. To achieve this, we established a subsidiary called Hewlett-Packard Associates (HPA) in 1961 to conduct solid-state research and development programs. HP held an interest in the subsidiary, but gave equity interest to the top engineers and scientists with expertise in this field who joined the effort. A buy-out provision was included in this agreement. The decision to establish HPA had important downstream, long-term benefits to the company. Some of the solid-state components from this project are discussed in several of the following articles, and a very large number of our products are now dependent upon the company's integrated-circuit programs.

One final note regarding HP's R&D structure is appropriate. After several years of divisionalization, experience showed that it was very difficult for the R&D sections of the operating divisions to carry on much innovative work outside of their own fields of specialization. The engineering staffs felt that they needed all the R&D dollars available simply to maintain or enhance a competitive position in their chosen fields. Thus, in the mid-1960s we concluded the time had come to have a central research facility. As a result, HP Laboratories was established in 1966, and headed by Barney Oliver who had come to HP from Bell Labs in 1952. His charter was wide open—as a central research organization's should be—to carry on basic and applied research studies, to assist the operating divisions in finding solutions to their technical problems, and, if necessary, to develop prototype products in new and promising fields. In this latter role, HP Labs has been immensely successful. A number of major new product lines have had their genesis from this source, not the least of which is the whole field of computation, an area of activity that now represents more than 50 percent of the company's total business annually.

I'll conclude this introduction with a few observations and comments about some of the article reprints that follow.

Design Notes on the Resistance-Capacity Oscillator Circuit *(pages 1 and 5)*

The first two issues of the HP Journal in 1949 contained articles on the resistance-capacity oscillator—some ten years after the product's introduction. Through the intervening years many improvements had been made on the original design, but there were continuing problems and these articles deal with methods of overcoming them.

One persistent problem had to do with extending the frequency, both high and low. Reference was made earlier in this introduction to a three-phase oscillator suitable for generating signals up to 10 MHz. As mentioned, this required a very different configuration of circuits, and there was a need for techniques using a more conventional configuration to extend the high-frequency range. In similar fashion, there were many requests for very low-frequency oscillators, primarily for use in geophysical work. The first article discusses various techniques for achieving these range extensions.

In the second article, several other aspects of RC oscillators are described, including long- and short-term stability, synchronization, and some other special requirements. At the time of its introduction, the RC oscillator supplanted the beat-frequency oscillator. Beat-frequency oscillators had very good high-frequency stability, but were plagued with low-frequency stability problems. The RC oscillator depended upon the stability of both the capacitor and the resistor to generate frequencies as low as 20 Hz. With the capacitors then available, the resistors were in the order of 10 megohms. Fortunately, during the 1940s the technique of carbon-deposited resistors was perfected, and although they did not provide the degree of low-frequency stability that might be hoped for, they were far better than what could be obtained with beat-frequency oscillators. At the high end of the frequency range, wire-wound resistors could be used, and while the frequency stability was not comparable with a beat-frequency oscillator it was quite adequate.

These articles are included, I think, more for nostalgia than for their scientific importance. The RC oscillator was, after all, the first commercial product of Hewlett-Packard.

The High-Speed Frequency Counter *(page 9)*

The subject discussed in this article, high-speed frequency counting, is a good example of serendipity.

In the late 1940s, much emphasis was being placed on nuclear instrumentation, and better techniques were needed to handle higher count rates. Conventional counters at the time operated to about 100,000 counts per second. It was our view that there would be a market for a prescaler that would accept count rates as high as 10 million counts per second and scale them down by a factor of 100 to be fed into then-existing scalers. The prescaler was successfully developed, but the need for it never really materialized.

We soon recognized, however, that this technology could be used in the field of frequency measurement by combining this high-speed counting technique with a very accurate time-gate circuit. Again, not a new idea, but of great significance because now frequency measurement could be extended to 10 MHz versus the 100 kHz or so previously available. Basically, this technique revolutionized the measurement of frequency due to its accuracy and ease of use.

The Low-Frequency Function Generator *(page 13)*

This article is a good example of what was referred to earlier as engineering of opportunity. Although we had RC oscillators that operated as low as 1 Hz, there were needs for much lower-frequency sources. The San Francisco Section of the IRE (predecessor of the IEEE) held a student paper contest, and one of the papers described a novel method of extending the frequency range of an "oscillator" to extremely low frequencies. It was obvious that this was the method needed to meet known requirements. Arrangements were made with both the student and his professor to acquire the rights to this technique, and in a very short length of time we were able to produce an oscillator that operated very satisfactorily at frequencies as low as 0.01 Hz. Incidentally, it also gave triangular and rectangular waves. The basic technique was the use of waveform shaping, a common enough approach now, but very new at that time.

The Clip-on DC Milliammeter *(page 17)*

The instrument described in this article is an example of an old technique used for modern purposes. The principle is essentially similar to that used in a conventional flux-gate compass, i.e., a magnetic amplifier that produces a second harmonic component of the existing ac current. This second harmonic is proportional to the dc flux surrounding the conductor being measured. The use of negative feedback increases the accuracy and stability of the measurement. To illustrate some of the practical problems relating to the design of this instrument, the earth's magnetic field is about three hundred times stronger than the field measured by the instrument on its most sensitive range. The applications for this instrument had obvious advantages, for it could measure direct currents without the insertion of an external instrument or without the load on a circuit of a resistor across which a voltage could be measured.

Like many of our other products, the suggestion came from the outside. One of our senior sales engineers with aviation instrumentation experience had made the comment, "Have you ever thought of using the flux-gate compass principle to create a clip-on milliameter?" That was all that was necessary.

The Sampling Oscilloscope *(page 20)*

This article describes one of the important developments by the company in the early 1960s. There was an ever-increasing demand for methods to measure high-speed waveshapes of periodic recurring signals. Conventional techniques at the time were capable of producing oscilloscopes with a top frequency range of 50 to 100 MHz. The technology described in this article permitted the development of an oscilloscope with a 500 MHz upper limit. The capabilities of this scope depended upon a very old technique almost forgotten, that of waveform sampling. Early use of waveform sampling was in the power generation field, where the waveform of an alternator could be measured by means of a movable commutator that could sample the instantaneous voltage of an alternator at various positions of the rotor. Actual measurement of this voltage was achieved with a condenser and a simple dc voltmeter. The technique had proven to be applicable to very high-speed signals, and the sampling scope described was the first practical application of this concept.

Also discussed in this article is a concurrent development of a solid-state device designed to generate very fast, sharp pulses. One of our engineers was investigating a circuit using a solid-state diode for harmonic generation. He discovered that under certain circumstances, the harmonic spectrum suggested that a very steep wavefront must be present. Conventional oscilloscopes, however, did not show such a wavefront. Using an early prototype of the sampling scope, he discovered that indeed there was a very sharp wavefront produced at the end of a very short reverse current conductance. It was determined that this wavefront resulted from the depletion of minority carriers stored during the forward portion of the cycle. When these carriers were depleted, no further conductance was possible.

Thus, the reverse conducting diode and the sampling scope were mutually supportive. The scope detected the phenomenon, which was then incorporated into the design of the scope itself.

Time Domain Reflectometry *(page 28)*

Again, the application of an old technique using the very high-speed oscilloscope described earlier. In essence, with high-speed technology it was possible to see reflections from various discontinuities on a transmission line or circuit. The technique also drew upon previous work done at Hewlett-Packard in the field of square wave impulse testing of linear systems (B.M. Oliver, "Square Wave and Pulse Testing of Linear Systems," Hewlett-Packard Journal, Vol. 7, no. 3, November 1955).

The 50 MHz Frequency Synthesizer *(page 36)*

Today, frequency synthesizers are in common and widespread use; 20 years ago, that was not the case. Thus, this article was of considerable importance, for it described what was probably the first commercial application of frequency synthesis. Barney Oliver's preface to the article is particularly perceptive in light of subsequent developments: "It will be very interesting to watch the applications develop for frequency synthesizers. Almost certainly, they will become as common as counters."

The development of the synthesizer was of additional interest in that it marked a change in HP's product-development procedure. Most prior developments were basically a one-person, or at most a small-team, short-term effort. In contrast, development of the frequency synthesizer was distinctly a team effort, requiring almost 40-man-years of engineering spread over a period of about three years—a major commitment to a new technology. This effort resulted in an instrument that provided almost unheard of resolution, and had an added advantage in that it was electrically programmable—an important feature in light of the rapidly developing field of automated testing.

The Flying Clock *(page 44)*

Rather than discussing the technical development of a product, this article describes a unique application for it. A word of background is necessary.

For a number of years HP had had an interest in the accurate measurement of frequency and time, with most of the effort being on frequency measurement; but, measurement of frequency by counter techniques implied the ability to measure time accurately. Much work had been carried out using conventional means, such as precision quartz resonators with stability in the order of 5×10^{-10}/day. We felt, however, that the developing field of atomic standards held promise and should be pursued. Of the many choices available, the cesium-beam technique seemed to hold the greatest promise for a practical time standard of a portable nature. The HP 5060A Cesium-Beam Standard resulted.

This paper describes a test in which a pair of these instruments were first checked in the U.S. against time standards at the Naval Observatory and the National Bureau of Standards, and then flown as passengers aboard a commercial airliner to Switzerland for a comparison against Swiss standards. They were then returned to the U.S. for a recheck against the same standards that had been used initially. Much valuable information was obtained, and the accuracy and reliability of our frequency standards demonstrated.

Subsequently, the flying clock experiment was repeated by flying a standard around the world and checking national standards in 12 countries. At a later date, similar standards were flown simultaneously around the world in opposite directions, successfully demonstrating Einstein's time shift theory. Eventually, cesium-beam standards became the international standards for time.

A New Microwave Spectrum Analyzer *(page 50)*

Spectrum analyzers had been around for some time when this article was written. Prior to the HP unit described, however, these instruments had some very practical problems and limitations. For example, results were often ambiguous due to harmonic and spurious signals. Further, they were not accurate with reference to frequency, and therefore required frequent calibration. Finally, they tended to be narrowband, and thus needed an array of plug-in tuners to cover the required bands of interest. A new approach was necessary, and the tools were at hand.

The solution was an improved backward-wave oscillator (BWO), with particular attention given to a "clean" signal. A local manufacturer designed a tube to our specifications. It covered the frequency range from 2 to 4 GHz, and was voltage tunable. Thus, with a low-noise 2 GHz IF amplifier, a direct frequency of 10 MHz to 4 GHz was obtainable with image frequencies now spaced 4 GHz apart, rather than the 0.4 GHz available from conventional design. By use of a 10 MHz quartz crystal oscillator, it was possible to obtain one percent accuracy in tuning and frequency sweep. It also was possible to achieve automatic bandwidth selection for optimum resolution as the sweep width and rate were changed. With a logarithmic display, a 60 dB dynamic range was obtainable.

For its time, the instrument revolutionized the field of spectrum analysis, and opened up many new areas of application.

Microwave Harmonic Generation *(page 58)*

The article on the sampling oscilloscope (page 20) provided an introduction to a solid-state device that eventually became known as the step recovery diode. This article on harmonic generation using the step recovery diode is of interest for two reasons: one, as a discussion on the practical use of this diode as a harmonic generator, and two, because it is an example of the contributions of HP Associates to the parent company. HPA was uniquely qualified to perfect and produce step recovery diodes so as to maximize both the harmonic content and the power output available.

The Quartz Thermometer *(page 66)*

To support HP's vital interest in frequency standards, a section was established in HP Labs to perfect the use of quartz crystal resonators to stabilize the output frequency. This turned out to be a very imaginative group of people, and having solved some of the more pressing problems of stability, they looked at other possible uses for quartz resonators.

It had long been known that a simple cut of the quartz crystal had a large variation of output frequency as a function of temperature. Primary effort, therefore, had been directed towards minimizing the temperature dependence. Our group took a different tack. Perhaps, they reasoned, we could use the temperature sensitivity to measure temperature by measuring the resultant frequency shift. The problem, however, was that the frequency shift was not linear with temperature. Indeed, it had significant second and third order terms. In cutting a crystal for a given mode, there are at least two degrees of freedom depending upon the two rotations available about the central axis of the crystal. The question: Is there a set of rotations that would cause the second and third terms to disappear, but still retain a significant first order? Both theory and experiment proved this to be the case, and it became the basis for an electronic thermometer operating over the temperature range −40°C to 230°C with a resolution of about 0.0001°C. It also had the advantage that since the signal was digital in character, it could be transmitted from remote sites.

The RF Vector Voltmeter *(page 73)*

Once again, a product offshoot of sampling technology. This article describes a very sophisticated instrument for making amplitude and phase measurements in the difficult frequency range of 1 MHz to 1,000 MHz. It is not just the development of a vastly superior capability that recommends this article, but also it is an example of how a fundamental technique such as sampling can find important applications throughout the field of instrumentation.

A New 1 GHz Sampling Voltmeter *(page 84)*

A further amplification of sampling technology, with the sampling achieved this time in a manner that is noncoherent with the input signal. The price paid, however, was loss of waveform, but for a voltmeter this is not a necessity. The advantage was simplicity.

The performance speaks for itself—a range from 10 kHz to 1 GHz, and a full-scale sensitivity of 1 mV. A very novel and imaginative technique.

An Ultra Wideband Oscilloscope *(page 91)*

These are second-generation articles describing the use of very wideband sampling technology. They discuss an oscilloscope with a range from dc to 12.4 GHz, and applications to time domain reflectometry (TDR) with a rise-time resolution of 40 picoseconds as well as methods of coupling to systems operating up to 12.4 GHz without introducing serious measurement errors.

Intervening articles since the original one on sampling oscilloscopes showed how a fundamentally new measurement technology can increase measurement capability at very high frequencies in different ways with different techniques, but all using the sampling principle. These articles, which appeared almost seven years after the original, return to the basic application of oscillography and TDR with the frequency range increased by a factor of about 24 and with resolution for TDR by a factor of at least 4. The principles have not changed, just their execution.

Automatic Network Analyzers *(pages 104 and 156)*

HP has been active in the field of microwave measurement since the late 1940s—first with sources, then detectors, standing-wave indicators, directional couplers, etc., all with the aim of simplifying measurement in this very difficult field. As discussed earlier, the sampling technique was introduced into high-frequency measurement with the RF vector voltmeter in 1966. The technique, therefore, appeared at hand for the long sought solution to a more automated and error-free system of microwave measurement. Two articles were published in the February 1967 issue of the HP Journal that were of considerable significance in this regard. They were: "An Automated New Nework Analyzer for Sweep Measurement of Amplitude and Phase from 0.1 to 12.4 GHz," and "The Engineer, Automated Network Analysis and the Computer—Signs of Things to Come."

What was especially interesting about the equipment and procedures described in these articles was the degree to which developments in other divisions of the company contributed to their success. Mention has been made of sampling technology derived from the sampling oscilloscope development, both techniques and hardware to make actual connections with the circuits under test. Most important of all, however, was the recognition of the tremendous advantage to be gained from a computer designed specifically for instrumentation purposes. The combination of computer and measurement has been a dominant feature of the company from that time on.

Three years after the introduction of the network

analyzer, a fully automated version was introduced and described in the February 1970 HP Journal. This article, "A System for Automatic Network Analysis," was followed by papers on software design, accuracy, and applications.

A historical note highlights the importance of these network analyzers. At the IEEE Convention where they were first described and demonstrated, we had a number of major customers who were so interested in the product that they requested permission to test their experimental microwave transistors on the instrument to confirm past measurements and to speed up development of these devices.

Computation *(pages 115, 126, and 213)*

In contrast to the approach of article-by-article comment, it seems more logical to discuss the whole question of HP's entry into the field of computation in a unified fashion.

The story starts in 1965, with a young engineer who had recently graduated from Stanford with a PhD degree in the blossoming field of computer science. His interests in this field prompted his HP Labs supervisor to initiate a low-level project on minicomputers. The work was carried far enough to demonstrate the feasibility of the approach, and its further go-ahead was justified on the basis that there were increasing demands from our traditional customers for automated testing. To this end, a number of years earlier we had set up a separate organization, Dymec, to work on developing automated testing. A small computer would obviously be a logical adjunct to this work, so the responsibility for computer development was transferred from HP Labs to Dymec.

In November 1966 we introduced our first computer, the HP 2116A. We made it very clear, however, that we were not in the computer business, but in the field of instrument automation. The purpose of this decision was to sharpen our focus on one specific application rather than attempt to cover too broad a field. One by-product of this decision was that the computer had to be capable of operation in the same environment as our instruments, i.e., to meet class B specs. Thus, at a time when most available computers could only operate under tightly controlled conditions (temperature, line voltage, shock, etc.) the HP 2116A did not have to be so pampered. Therefore, not only did our computer prove useful in automated measurement, it also was in demand because of its reliability. Examples were uses on "Texas Tower," and in the field of timesharing where up time was of prime concern. The first two articles, "Computers for Instrument Systems" and "Successful Instrument-Computer Marriages," discuss this part of the history. It is interesting to note that a large number of subsequent articles that have appeared in the HP Journal deal, in one phase or another, with automated testing. Occasionally, they talk about developments for HP's own use, but in most cases they discuss packaged units such as that described in the article, "A System for Automatic Network Analysis" (page 156).

Within a few months after the transfer of the minicomputer project to Dymec, a new project took its place in HP Labs—to develop a powerful desktop electronic calculator. We were approached by a young inventor who had perfected a simple four-function calculator of practical design using reverse Polish notation. At about the same time, a technique for achieving single-key transcendental functions also was brought to our attention. The question: Can the two technologies be made to work together? The answer: Yes, and a formal development project was initiated. This was a formidable program, for not only were the technical problems pioneering and difficult, but as this calculator was to be used for general scientific work—often by customers unskilled in computer operation—it had to be well-styled, "friendly," and easy to use. This obviously was a major problem, and ultimate manufacturing responsibility was assigned to our division in Loveland, Colorado. Engineers from Loveland came to Palo Alto and worked hand in hand with engineers in HP Labs to complete a prototype unit, thus insuring that the transfer to Loveland would be achieved without difficulty. Within two years, the project was completed and the unit in production. There are four articles which discuss various aspects of this program. I particularly call your attention to Barney Oliver's article, "How the Model 9100A Was Developed," one of several articles about the calculator in the September 1968 HP Journal.

The third part of this trilogy on HP's entry into the computational field pertains to the HP-35, the first handheld scientific electronic calculator. Its story began in the fall of 1970, with the suggestion that such a product might be possible. Time was of the essence, for the field of handheld calculators was already blooming, albeit calculators with only four functions and mostly using discrete components. It was agreed that a concrete proposal should be readied by Groundhog Day in February 1971, at which time a final decision would be made. The day arrived, a proposal was submitted, and it was approved. By September of that year, prototype working models were available, and two months later manufacturing was underway.

This turned out to be one of the most successful products ever introduced by HP. At the time the decision was made to go ahead with the program, it was estimated that if 10,000 units could be sold in the first year of production, the program would be a success. We sold over 100,000 in the first year! Because the product was so unique to the standard HP product line, and because the potential customer base was so different, we decided to set up a special division with responsibility for research, manufacturing, and marketing in this new field. The four articles in the June 1972 HP Journal discuss in much detail this interesting project.

I think it can be said that never has a major computational device, in this case the slide rule, been so quickly and completely supplanted by the introduction of a modern technology such as that of the HP-35.

Solid-State Displays *(page 145)*

This article is in fact a report covering almost six years of work on electroluminescent diodes (LEDs) by HP and its subsidiary HP Associates, which I commented on earlier. The development effort represented a major commitment by the company to the belief that LEDs would have major applications as alpha and numeric displays. Time certainly has proved that this commitment to investigate a poorly understood phenomenon was a wise one.

The techniques described represented a significant advantage over display technology at the time, primarily gas-discharge tubes. The article discusses this application mainly in a dot-matrix display, but extensive use (particularly in early handheld calculators) also was made of the segmented display.

The Fourier Analyzer *(page 179)*

While the concept and potential of Fourier transformation had long been recognized, it wasn't until the introduction in the early 1960s of an effective algorithm for a fast Fourier transform and the advent of the digital computer that the practical use of the theory became realizable.

HP has been in the measurement business since its founding, and for many years almost all measurements were in the analog domain. An early exception to this was the introduction in 1951 of HP's somewhat crude, but accurate, digital technique for precision measurement of frequency (commented on earlier). It was the late 1960s, however, before HP engineers were able to investigate the possibility of combining the then-available low-cost HP minicomputer with the fast Fourier transform (FFT), as devised by Cooley and Tukey, for application to measurement problems. This investigation bore remarkably successful results, thus once again demonstrating the phenomenon of using modern technology with older theory to provide solutions to important measurement problems.

The first of the two articles discusses the theory and use of the Fourier transform for measurement purposes, and the second describes a versatile implementation of these concepts. An example, not included in the articles, might serve to highlight the importance and power of this technique. It has to do with the application of vibration analysis to a sports racing car. It had been discovered that at certain speeds, a vibration was set up that literally took the rear wheels off the ground. By minor relocation of the roll bar, this vibration was suppressed and a hitherto unusable configuration became practical.

A Better Laser Interferometer *(page 198)*

Sophisticated electronics finds many uses. This is particularly true in the field of instrumentation. One must be selective, however, for although many potential applications are technically interesting to work upon, not all meet the requirement of a sufficient benefit/cost ratio, or compatibility with existing or potential market channels. Distance measurement is a case in point, as will be explained later in this introduction.

Two articles relating to the electronic measurement of distance are included in this book. The first of these discusses measurement of relatively short distances (200 feet) using interferometric patterns. Although distance measurement by means of interference patterns was not new, having been pioneered in 1880 by A.A. Michelson, simple single-frequency methods were not practical since the interferometer cannot differentiate between negative and positive changes in distance. This limitation, however, can be overcome by splitting the beam and shifting the phase of one part by 90 degrees. The invention of the laser with its highly monochromatic light allowed interferometric methods to be usable up to moderate distances. There were still many practical problems, however.

As we examined these problems, it was evident that if we could obtain two frequencies offset by a megahertz or so, most of the difficulties could be overcome. The question: How to achieve this? This is an interesting example of clearly stating a problem and then, once stated, a possible solution becoming clear. In this case, it was the use of Zeemann splitting. It was well known that an axial magnetic field will split a laser's frequency into two components, the amount of frequency separation depending upon the magnetic field. These two frequencies are circularly polarized in opposite directions when a correctly designed laser plasma tube is used, and may be converted to horizontal or vertical polarization by means of a birefringent quartz plate. It is then a simple matter to heterodyne these two frequencies, and by means of a simple counter mechanism to determine relative motion to a fraction of the wavelength of the light.

This technique greatly simplified measurement of distances up to 200 feet, and several interesting applications are discussed in the supporting articles.

The Hewlett-Packard Interface Bus *(page 225)*

In the late 1960s, we found ourselves under increasing pressure from our customers to provide means for external electrical control for our instruments and for remote reading. As was previously mentioned, the requirement for remote control and data collection several years earlier had led directly to the development of the HP instrument-oriented computer. This computer, by means of special interface boards, could accommodate various instrument protocols. But as the number of instruments increased, the technique soon became cumbersome. Given the structure of HP, this new problem was approached in a somewhat ad-hoc manner, with each division or even different instrument teams providing different solutions, albeit optimized for the particular problems involved. Numerous approaches, therefore, were proposed to simplify the interface system design problem, but none seemed to be sufficiently cost/performance-effective.

The evolutionary efforts were gradual until the notion of a simple byte-serial interface with identical interface circuitry in each instrument was experimented with in early 1971. Once feasibility was proven, an explosion of interest and effort took place. In December of 1971, the decision was made among several divisions, most notably Loveland and Santa Clara, to pursue detailed specification of what is known today as the HP-IB. Pilot runs were halted in mid-course to incorporate the new concepts, and critical aspects of the interface bus were finalized (e.g., three-wire handshake for data transfer to make the interface itself independent of timing and to facilitate customer use). Within a period of about five months, instruments using the HP-IB began to roll off the production lines, and automatic test systems became much, much easier to assemble and program.

At about the same time, interest in similar interface concepts surfaced in Germany, and an international standards project was assigned to TC66 within the IEC (International Electrotechnical Commission). This indicated that the HP situation was being experienced throughout many parts of the world, as well as in the U.S.A. Customers wanted to build automatic test systems composed of products from different sources, from different HP divisions, and from different companies wherever they might be located.

Once the broad objectives were established in national and international standards bodies, it was determined that HP's concepts met the needs well enough to be used as the starting point for elaboration of what is now known as IEEE Standard 488, and its international counterpart IEC Standard 625. Today, HP uses these standard interface concepts in over 200 products, and many other manufacturers around the globe incorporate the concepts in their products as well. The basic interface concepts described in these HP Journal articles have indeed provided a common, economical, and easy-to-use interface scheme for the benefit of both customers and manufacturers.

The HP 3000 Computer System *(page 235)*

A little more than six years after the introduction of HP's instrumentation minicomputer in 1966, the company announced its first small general-purpose computer system, the HP 3000. The development of this system was a significant HP milestone for at least two reasons. First, it served as HP's entry into the growing market for adminstrative and business computers. Second, it signaled a major diversification from HP's traditional field of electronic test and measurement instrumentation.

The HP 3000 could accommodate many different software programs (including those developed by users) and computer languages. Some of the key elements of its architecture included separation of code and data, automatic relocation of programs and data, reentrancy to permit code sharing, recursion to allow a program routine to call itself, and dynamic storage—features that in the early 1970s were normally found only on larger mainframes. This architecture coupled with the HP 3000 MPE (Multiprogramming Executive) operating system provided particularly effective on-line, terminal-intensive processing, which could be done even while the system was performing concurrent batch processing. The technology of the original unit was the springboard for successive generations of the HP 3000, each of which has offered greater power and increased capabilities including, in recent years, distributed computer networking.

In effect, the first HP 3000 spawned the beginning of a hierarchy of HP computational equipment for business and industry that, in total, now represents about one half of the company's annual sales volume.

The Logic Analyzer *(page 258)*

The article on the logic analyzer is an example of an

early step in providing modern instrumentation for the digital revolution.

In the late 1960s, some of our divisions began designing instruments using a large number of integrated circuits with arithmetic sequences several thousand states long. The engineers soon discovered that although the oscilloscope had been their right arm in analog design, it was distinctly deficient for digital logic design. HP has a great way of inventing itself out of problems, and in this case the logic design development team produced a string of clever handheld products to meet this new measurement need. These included the logic probe, clip, pulser, and current tracer, all of which eventually appeared in the company's product line.

Recognizing the importance of this developing field, we established a special team for the sole purpose of developing equipment for logic troubleshooting. In analyzing the problem, it became apparent that what was needed was the digital equivalent of the storage oscilloscope. Each analog function had its digital counterpart. Vertical amplifier became level detector. Trigger, delay, storage, and display functions became gate register, memory, and state indicator. The 5000A Logic Analyzer was the first in the family of products specifically designed to help develop and troubleshoot complicated digital products.

The Fully Programmable Pocket Calculator *(page 273)*

After the great initial success of the first scientific handheld calculator and several generations of modifications and improvements, the direction of future development was not clear.

The first HP desktop electronic calculator had been capable of being programmed and of storing such programs on a magnetic card 2 inches wide and 3.6 inches long. This gave the calculator great flexibility, and saved much time in programming. It was intriguing to speculate on the possibility of developing a handheld calculator that would have both programmability and magnetic card read/write capability. In addition, the development of integrated circuits with greatly enhanced power would markedly expand both the memory and hardware capacity of such a unit. We were aware that this would not be an easy product to develop, but the benefits so outweighed the practical problems that the decision was made to attempt such a project.

The article, "The 'Personal Computer': A Fully Programmable Pocket Calculator," and the three papers that follow discuss the results of this development project that eventually led to the HP-65. It is interesting to note that this calculator, announced in the spring of 1974, had roughly the same power as the 9100A Desktop Calculator introduced only 5½ years earlier. But, in contrast, it would fit into your pocket, and the cost was only about 1/6th of the earlier model.

Gallium Arsenide Field-Effect Transistors *(page 292)*

Microwave semiconductor technology, pioneered at HP Labs and refined at our Santa Rosa Division's Microwave Technology Center, has allowed HP to make contributions in swept microwave sources.

First, silicon bipolar transistors made possible yttrium-iron-garnet-tuned oscillators and power amplifiers covering up to 8 GHz. By 1974, the gallium arsenide field-effect transistor (GaAs FET) was becoming the premier device for low-noise amplification to 12 GHz and beyond. For source amplification requirements, however, noise figure was not the primary concern; more important was the ability to provide moderate output power over a broad frequency range. HP engineers at Santa Rosa discovered that these devices were capable of efficient power amplification, and with relatively simple matching structures, could cover octave frequency ranges.

The sweeper plug-ins described in this article used such an amplifier. In fact, it was the first commercial application of the GaAs FET.

The Signature Analyzer *(page 298)*

Increased use of digital circuitry posed field repair problems that analog instruments were not suitable to solve. Early approaches to this problem involved simply locating the defective part and exchanging it with a previously tested replacement board. Whereas this worked when there was a limited number of products to be serviced, it became too complex and expensive as the number of products to be repaired increased.

The article on signature analysis, a new digital field service method, describes the genesis of a portable field instrument to solve this important problem. Once again, the dual nature of HP as a producer of digital products as well as a designer of test and service equipment is evident. This relation, both for digital and analog equipment, has been one of the traditional strengths of the company.

A Fully Integrated Total Station *(page 318)*

Several years prior to the date of this article HP had developed a method for measuring distances as great as two miles or so using the transit time of light. It was bulky and inconvenient to use, for it required that the transit or theodolite being used for angle measurement be removed from the tripod and replaced by the distance measuring equipment. It did, however, have sufficient accuracy for survey purposes and greatly simplified work in rugged terrain. Since we had demonstrated the utility of electronic distance measuring, it was a tantalizing thought to try to combine into a a single instrument both distance measuring capability and the measurement of vertical and horizontal angles. The problems, however, were formidable. For one, it would be necessary to shrink the distance measuring component by a substantial amount. This appeared straightforward, but not simple. The real problem lay in developing a theodolite with accuracies commensurate with the distance measuring component.

Theodolite development and manufacturing was an old and sophisticated technology, one in which we had no background. Thus, in a sense, we were at a great disadvantage compared with companies that had been in this field for many years. On the other hand, we had certain advantages. First, we knew the technology of electronic distance measuring. Second, and perhaps more important, was the fact that in trying to design a theodolite from scratch, we had no preconceived ideas of how this might be accomplished. We could look at all technologies, both old and new, and choose the best. Third, with our experience in handheld calculators, we should be able to greatly simplify the practical use of such a combined instrument and greatly relieve the drudgery of data reduction.

The decision was made to undertake the development of such a device, and the HP 3820A Total Station was the result. No attempt will be made to discuss the technical aspects of the development as they are well covered in the article. Suffice it to comment that the development was not easy, with many new skills having to be acquired, but the ultimate product met the desired specifications as to accuracy, reliability, and ease of use. The field results were most satisfactory, with numerous reports of excellent accuracy and significant reductions in surveying time. In one case, a U.S. Government agency completed a traverse over most difficult terrain in less than half the time that might be experienced using conventional equipment, and with a closer accuracy—about one part in 250,000 over the 27 mile traverse.

Post Script. As hinted at in an earlier comment, it is sad to note that HP found it necessary to drop out of the civil engineering field due to its limited nature and cyclical characteristic. Despite this decision, the development of the total station remains a magnificent technological achievement.

High-Speed Plotter Technology *(page 336)*

The last article in this book, "Development of a High-Performance, Low-Mass, Low-Inertia Plotting Technology," is a delight to read, for it reflects accurately the character of the development team. No background will be given on the details of this, for the article is quite sufficient in this regard. Suffice it to say, the people on the team attacked the basic problem of a new plotter with imagination, enthusiasm, ingenuity, and a clear understanding of what they were trying to achieve.

This article also is another excellent example of serendipity. The team first started to develop a pocket-size plotter for a handheld calculator and wound up with a technique not only suitable for this application but suitable as well for many other fields such as electrocardiography, and for the development of a large plotter for engineering drafting.

The techniques and systems described represent a fresh approach to X-Y plotting, the full impact of which is still to be felt. Despite its simple elegance, the plotter is a significant contribution.

William R. Hewlett

May 1983

Contents

Volume 24, Number 2, October 1972

Volume 24, Number 5, January 1973

Volume 25, Number 2, October 1973

Volume 25, Number 9, May 1974

Volume 28, Number 3, November 1976

Volume 28, Number 9, May 1977

Volume 31, Number 9, September 1980

Volume 32, Number 10, October 1981

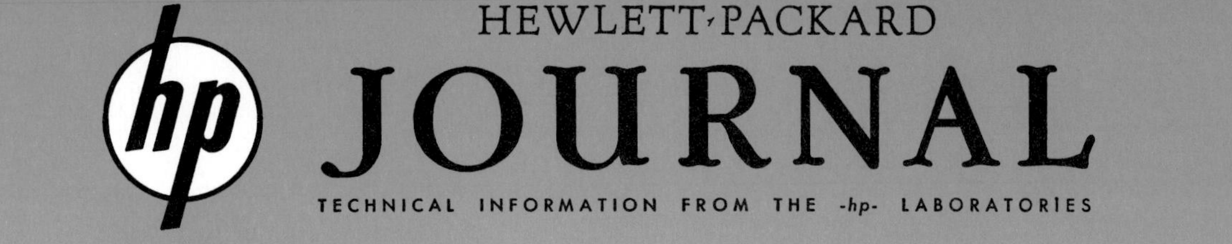

1

VOL. 1 No. 3

PUBLISHED BY THE HEWLETT-PACKARD COMPANY, 395 PAGE MILL ROAD, PALO ALTO, CALIFORNIA NOVEMBER, 1949

Design Notes on the Resistance-Capacity Oscillator Circuit

Part I

OVER a period of years the Hewlett-Packard Company has been requested to design and manufacture a large number of special-purpose resistance-capacity oscillators having such characteristics as compressed or expanded scales, unusually wide frequency range, high power output, high and low frequencies, special controls, etc. The following will describe some of the considerations involved in the design of the resistance-capacity circuit.

Briefly, the resistance-capacity oscillator circuit consists of a two-stage amplifier having both negative and positive feedback loops (Figure 2). The positive loop causes the circuit to oscillate and includes the resistance-capacity frequency-selective network which gives the circuit its name.

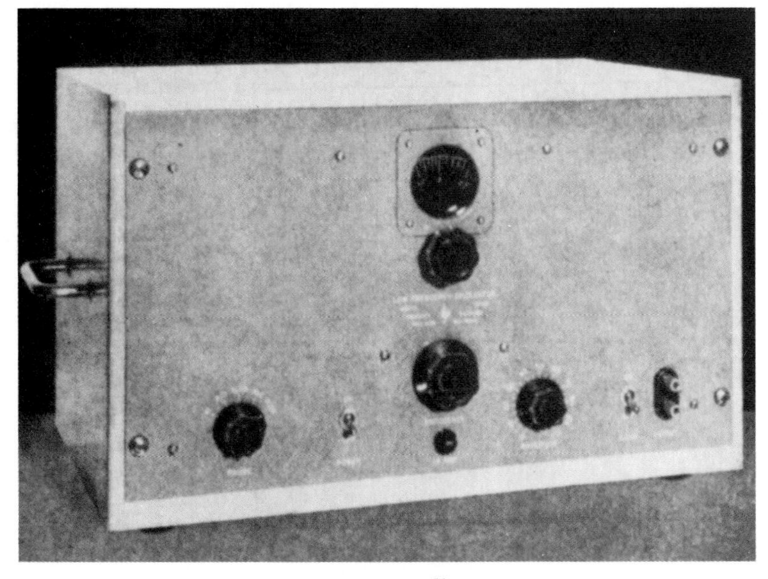

Figure 1. *Model 202B Low Frequency Oscillator Covering Frequency Range from ½ to 50,000 cps.*

THE "resonant" frequency of the RC network is given by the well-known expression $f_0=1/2\pi RC$. The negative feedback loop stabilizes the operation of the circuit by minimizing phase shift and confining operation to the linear portion of the characteristics of the tubes. One element of the negative feedback circuit is a non-linear ballast resistance which limits the amplitude of oscillation. The ballast element adjusts its resistance either higher or lower so as to compensate for any tendency of the oscillations to vary in amplitude. If the non-linear element is located in the cathode circuit, as in

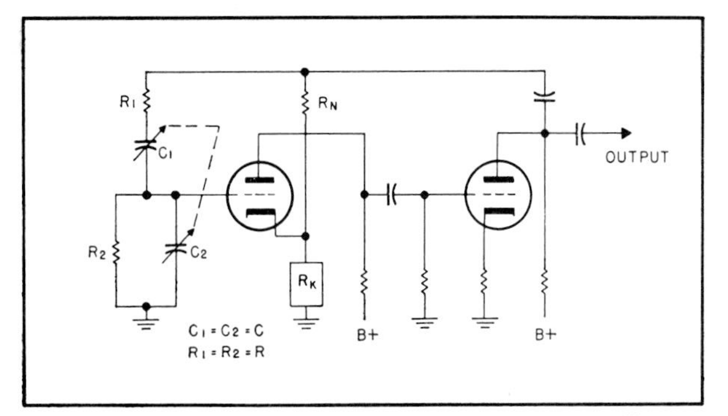

Figure 2. *Basic Circuit of Resistance Capacity Oscillator*

Figure 2, it must have a positive temperature coefficient. However, a negative temperature coefficient element also can be used, in which case the positions of R_k and R_n in Figure 2 would be reversed.

The resistance-capacity oscillator is commonly tuned with a tuning capacitor rather than a variable resistance because of the smooth frequency control and long life obtainable with such capacitors. Since the frequency of oscillation of the circuit is inversely proportional to the capacity C in the frequency-determining network (instead of to the square root of the capacity as in an LC oscillator), the RC oscillator can be made to cover a tuning span as wide as the capacity variation in a tuning capacitor. Thus, 10:1 frequency variations in a single sweep are easily obtained and it is customary to design the circuit to operate over 10:1 bands. A number of bands can be used with one circuit by changing the pairs of resistances in the frequency-determining network.

FREQUENCY RANGE

The resistance-capacity oscillator is inherently a wide-range oscillator and it is possible to design circuits to operate over a 100,000:1 frequency range in five 10:1 bands without undue loss of the desirable characteristics of the oscillator. At both the high and low ends of the range of such a circuit, however, some loss in performance does occur. In both cases this loss is the result of phase shift, a characteristic to which the circuit is very sensitive.

At the high frequency end of the range, the phase shift is introduced in the form of a lagging phase characteristic for the amplifier portion of the oscillator. The effect of this lag is that at a given setting of the tuning capacitor the frequency of oscillation is lower than that predicted by the "resonance" formula. This error occurs because the circuit must oscillate at a frequency for which the phase shift around the closed positive feedback loop is an integral multiple of 2π radians. Thus, if the amplifier shifts phase by $-\Delta\Phi$, the frequency of oscillation must shift to a lower frequency where the frequency-determining network will contribute a phase shift $+\Delta\Phi$ (Figure 3). In the usual multi-range oscillator the lag in phase results in an increasing error in calibration as the oscillator is tuned to the higher frequencies. This effect can be compensated to a degree, but the differences in individual instruments introduce capacity variations that limit the amount of practical compensation. It is interesting to note that a phase shift of only a fraction of a degree will cause calibration errors in the order of 1%, whereas in an amplifier the commonly used "3-db point" introduces a phase shift of 45 degrees.

At the low freqency end of the range the phase shift is introduced by the coupling capacitors, resulting in a leading phase characteristic for the amplifier. Consequently, at the lower frequency region of oscillation the actual frequency is higher than that predicted by the "resonance" formula.

In special cases it is possible to make use of these phase shift effects to obtain specific characteristics. For example, the amplifier can be intentionally given a lagging angle at low frequencies in order to increase the range of the circuit. Such an arrangement is shown in Figure 4 where a series RC network is connected across the amplifier output. By properly proportioning R and C, it is possible to increase the continuous sweep of the oscillator from 10:1 to more than 30:1. However, this range extension causes some deterioration of the performance of the circuit because of the load placed on the oscillator at the higher frequencies. This effect will be described in more detail later.

The above phase shift effects occur in the oscillator itself. Ordinarily, an isolating implifier is used following the oscillator and this amplifier introduces some further limitations, mostly in applications where it is necessary to use an output transformer for power or balanced-load reasons. The presence of an output transformer usually reduces the desirable frequency range of the oscillator to a range of about 1000:1.

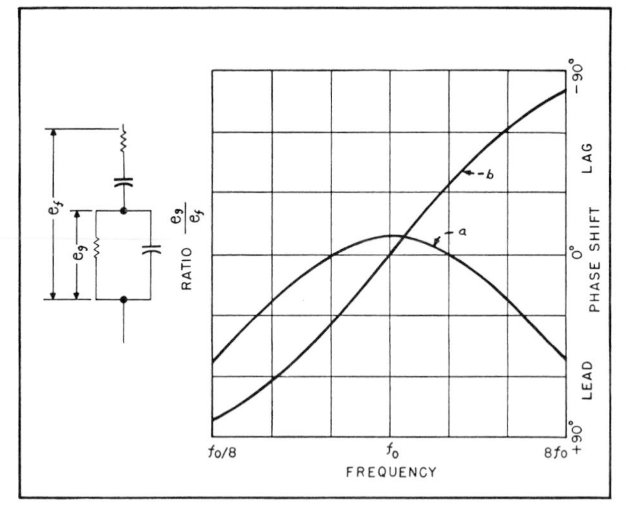

Figure 3. *Frequency (a) and Phase (b) Characteristics of Frequency Determining Network*

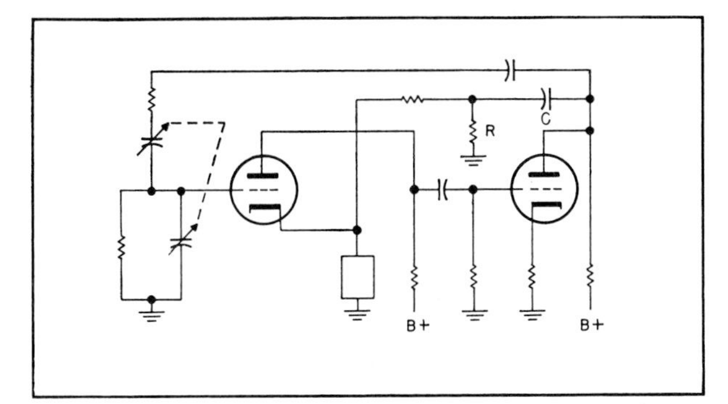

Figure 4. *Circuit for Increasing Frequency Range of Resistance Capacity Circuit*

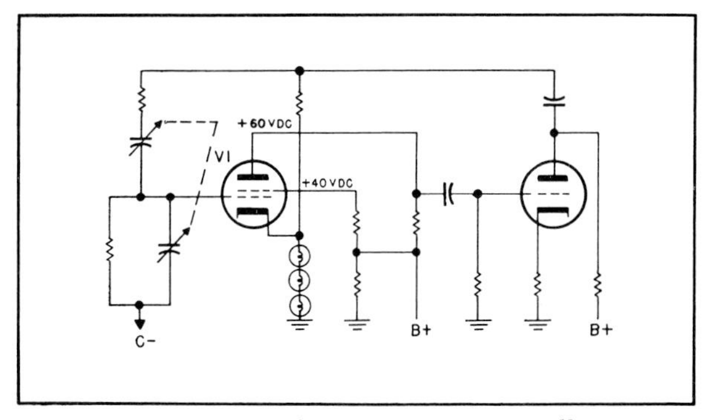

Figure 5. *Circuit for Low Frequency Oscillator*

LOW FREQUENCY OSCILLATORS

Circuits have been developed to permit the operation of the resistance-capacity oscillator at frequencies below one cycle per second and higher than one megacycle per second. However, it has not been practical to operate any one circuit over such a wide range.

In low frequency resistance-capacity oscillators one of the limiting factors is the operation of the ballast element in the negative feedback circuit. The most satisfactory ballast element from the standpoint of performance and cost is a small incandescent lamp. As used in higher frequency oscillators, the lamp will operate quite well down to frequencies corresponding to the low end of the audio spectrum. However, the thermal time constant of the lamp is such that at lower frequencies the lamp resistance tends to change in accordance with the variations in amplitude of the individual cycles of oscillation. This effect results in severe distortion of the output waveshape. Therefore, it is necessary to obtain a ballast element having relatively greater thermal inertia.

One solution is the use of certain types of negative temperature coefficient thermal resistors. However, tests on these devices indicate that their resistance varies widely with normal variations in ambient temperature within the oscillator cabinet, causing undesirable variations in performance.

A more satisfactory arrangement for obtaining the necessary thermal characteristics for low frequency operation is shown in Figure 5. The operating voltages on the tube electrodes are low so that a minimum of space current is drawn through the lamp in the cathode circuit. In addition, fixed bias is used on the control grid as aid in minimizing gas current. This arrangement allows the lamp to operate well down to frequencies as low as a fraction of a cycle per second because of the decrease in radiation from the lamp. Although the lamp is never operated where visible light is emitted, radiation nevertheless accounts largely for the cooling of the lamp. Since radiation is influenced by the fourth power of the lamp filament temperature, a lower operating level results in relatively higher thermal inertia. In actual practice three lamps in series are used to obtain the necessary value of resistance.

The operation of the resistance-capacity circuit at such low frequencies introduces an additional problem. Using tuning capacitors that are commercially available, the resistance values that are necessary to satisfy the "resonance" formula are in the order of 1 to 10 megohms for audio frequencies. However, at lower frequencies the resistance values become prohibitively high. For example, with a 1000-micromicrofarad capacitor the value of R for an oscillation frequency of 1 cps is approximately 160 megohms. Considering that the insulation resistance must be large compared to the frequency-determining resistance, it can be seen that both the input impedance of the tube and any leakage paths would affect the circuit.

In order to circumvent this stringent requirement, it is necessary to operate the circuit with several tuning capacitors in parallel. The impedance level of the circuit is inversely proportional to the number of capacitors used, making necessary the use of four or more standard four-gang capacitors to obtain the required capacitance in variable form. Through the use of this system it is possible to reduce the resistance values in the frequency-determining network to 40 megohms for one cycle per second operation and thereby to achieve lower insulation levels accordingly.

HIGH FREQUENCY OSCILLATORS

The higher frequency limit at which the resistance-capacity oscillator can be operated satisfactorily is determined primarily by the plate loading on the second tube of the oscillator. As can be seen from Figure 2, both the positive and negative feedback circuits load the plate of the second tube. The negative feedback loop constitutes a low-impedance load of about 4500 ohms on the tube. The impedance of the positive feedback loop decreases as the frequency is increased and is in the order of 3000 ohms at a phase angle of 45 degrees at one megacycle. These two loads are in parallel and the combination is in parallel with the plate feed resistor for the tube. This reactive and low-impedance loading reduces the gain of the circuit at high frequencies and introduces serious phase shift. As a result, the distortion increases and errors in calibration arise at the higher frequencies.

The above effects can be reduced by several methods. First, the gain of the second stage of the oscillator is kept high by the use of tubes with very high transconductance. Also, in some applications it is practical to reduce the capacity of the tuning capacitor and thereby achieve a higher impedance for the frequency-determining network. However, the limit to such reduction is determined by the stray capacity in the grid circuit of the first oscillator tube. This stray capacity is in parallel with part of the main tuning capacitor and therefore determines the minimum possible capacity in the frequency-determining network. Variations in this stray capacity will change the frequency of oscillation of the circuit. Therefore, it is desirable to minimize this effect by adding other fixed capacities in parallel with the stray capacity so that the strays constitute only a fraction of the minimum capacity. In order to have a tuning span of 10:1, it is necessary that the maximum capacity of the tuning capacitor be ten times as great as the sum of the capacities constituting the minimum capacity. These factors set a limit to which the impedance of the frequency-determining network can be increased for a given frequency of oscillation and tuning span.

At the higher frequencies the reduced gain of the oscillator and consequent reduction in negative feedback make the oscillator circuit more susceptible to drifts and variations caused by tube aging and supply voltage changes. As a result, it is common practice to operate the circuit from a regulated plate supply when the circuit is to be used at frequencies higher than approximately 150 kcs.

The above factors combine to limit the top practical frequency of the resistance-capacity oscillator to the order of one megacycle. However, the circuit will oscillate at frequencies of several megacycles.

DISTORTION

The resistance-capacity oscillator inherently is a generator of low distortion voltages owing to the negative feedback circuit and to the fact that the circuit is what might be termed a Class A oscillator.

The purity of the generated voltage is limited by the linearity of the transfer characteristics of the tubes. With proper operating voltages on the tube electrodes, the circuit will generate voltages having in the order of $\frac{1}{2}$ of 1% distortion, even with individual tubes that differ considerably from the average tube characteristic. By the selection of suitable individual tubes of the type used in the circuit, distortion can readily be held to less than $\frac{1}{4}$ of 1%. This figure is the order of the practical limit of distortion obtainable from the circuit over the sub-audio to middle ultrasonic frequency range.

The distortion obtained under these circumstances is mainly third harmonic. Second harmonic distortion is minimized by adjusting the dc voltages on the tube electrodes so that the second harmonic distortion generated by one tube of the oscillator is partially cancelled by the curvature of the other tube's transfer characteristic.

For applications requiring less than $\frac{1}{4}$ of 1% distortion, a frequency-selective amplifier can be used following the oscillator to eliminate some of the distortion generated by the oscillator. In these cases the tuning of the amplifier is variable and is tracked with the tuning of the oscillator in a manner similar to the system used between the oscillator and tuned rf amplifiers in a radio receiver. However, in the resistance-tuned application the selectivity of the amplifier is controlled by RC circuits which select only the harmonic voltages and apply them to the amplifier input as negative feedback. This system reduces the harmonics in the generated voltage without affecting the fundamental, resulting in an amplified voltage that has less distortion than the amplifier driving voltage. Practical reductions of 10 db in distortion over the major portion of the audio spectrum are possible with this arrangement.

In low frequency oscillators the distortion requirements usually are not as severe as in the audio and low ultrasonic regions. However, at frequencies as low as one or two cps it is possible to obtain distortions in the order of 1% with the high thermal inertia circuit (Figure 5) described above.

In high frequency oscillators of 50 kcs or more, the resistance-tuned circuit generates less than 1% distortion up to frequencies of several hundred kilocycles. At frequencies higher than this, the distortion increases rather rapidly owing to the heavy loading of the positive feedback network on the plate of the second oscillator tube and the consequent loss of negative feedback. As a result distortion is in the order of 4% at frequencies of one megacycle.

—Brunton Bauer.

The above is the first part of a two-part article by Mr. Bauer. The article will be concluded in an early issue of the *Journal.*

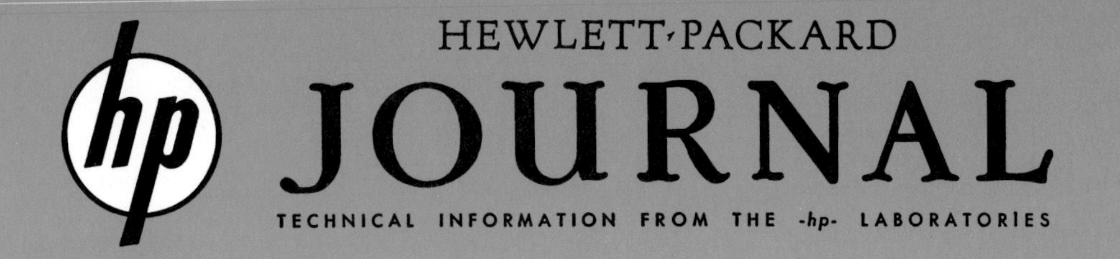

HEWLETT-PACKARD JOURNAL

TECHNICAL INFORMATION FROM THE -hp- LABORATORIES

VOL. 1 No. 4

PUBLISHED BY THE HEWLETT-PACKARD COMPANY, 395 PAGE MILL ROAD, PALO ALTO, CALIFORNIA DECEMBER, 1949

Design Notes on the Resistance-Capacity Oscillator Circuit

(The following is the concluding portion of Mr. Bauer's article that was begun in the November issue)

Part II

ACCURACY

OVERALL accuracy as applied to a variable-frequency oscillator is a general term that includes many factors such as the inherent stability of the circuit, mechanical stability and resetability of the tuning system, readability of the tuning dial, care with which the dial was calibrated, effects of aging on the various component parts, and the effects of power supply variations and ambient temperature changes. Some of these factors affect the short-time stability; others affect the long-time stability.

The accuracy specification of within 2% that is usually given for resistance-capacity oscillators is intended to include the majority of these factors. Consequently, the actual accuracy is different for different combinations of conditions and is generally better than this figure under normal operating conditions.

Figure 6. *Model 201B Audio Oscillator*

LONG-TIME stability—the stability over a period of several months or more—is a function of the quality of the circuit components and the mechanical stability of the tuning system. For best long-time stability it is desirable to use only wire-wound resistors in the frequency-determining network. However, at frequencies up to the mid-audio range the use of wire-wound resistors is impractical in most applications because of the high resistance values necessary. Composition resistors have been developed to the point where accuracies within 1% are practical and where their long-time stability is good. A very satis-

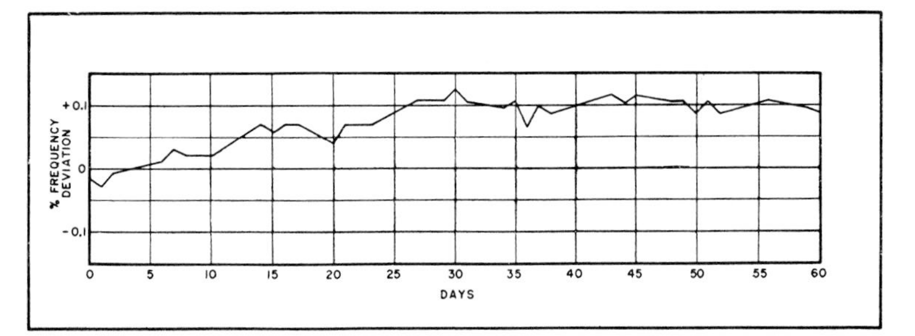

Figure 7. *Long Time Stability Curve of Circuit Using Wire-wound Resistors and Temperature Compensation*

factory type consists of inactive carbon deposited on a ceramic tube. Detailed descriptions* of these resistors have appeared in the literature. Through the use of such precision composition resistors, the long-time stability of the oscillator is increased to the point where accuracies within 1 to 2% are practical. Controls can be provided so that the calibration of the oscillator can be reset against a frequency standard. Such controls usually consist of resistance verniers located in the frequency-determining network. If the oscillator calibration is corrected from time to time with these controls, accuracies within 1% can be easily maintained.

At frequencies above a few kilocycles where wire-wound resistors can be used in the frequency-determining network, it is practical to increase the long-time accuracy to within approximately 0.5%. Ordinarily in these applications it is necessary to temperature-compensate the circuit to avoid the drifts that are associated with warm-up and ambient temperature changes. Figure 7 shows a long-time stability curve of a circuit in which wire-wound resistors and temperature-compensation are used. Where the maximum stability is required, it is desirable to reduce the span of the oscillator from 10:1 to 3:1 or 2:1. This reduction lessens the effects of stray circuit capacity by increasing the minimum value of the tuning capacity.

Some applications require oscillators having long-time accuracies within tolerances narrower than 0.5% and a number of circuits have been developed to meet this requirement. One of the most practical systems is the use of a precision fixed-frequency check oscillator in the same cabinet with the resistance-capacity oscillator. This arrangement allows convenient standardization of the calibration of the variable oscillator at intervals throughout its frequency range. The frequency comparison can be made by means of Lissajous figures on either a self-contained or external oscilloscope. At supersonic frequencies and above, a quartz plate is used to control the fixed-frequency oscillator while at lower frequencies a temperature-compensated tuning-fork is used. Standardizing of the variable oscillator at a number of check points can be accomplished with a simple capacitive vernier.

The use of a precision check oscillator allows an accuracy of 0.1% or better to be maintained throughout the life of the oscillator. However, because of the elaborate circuitry and the necessary use of a tuning drive commensurate in quality with the quality of the electrical systems, such oscillators tend to be large and involve considerable expense.

For low-frequency narrow-range oscillators of high long-time stability, it is often practical to incorporate an electron-ray or "tuning eye" in the circuit to permit checking the frequency of the generated voltage against the power line frequency. This arrangement will allow good accuracy over long periods of time. However, random fluctuations in the frequency of the power systems should be anticipated and may cause short-time errors approaching 1% in extreme cases.

Short-time stability — considered here to mean the stability over periods not exceeding one-half hour after sufficient warm-up—is a function of the effective Q of the circuit and of random effects such as regulation of the voltage supply, the effects of vibration, etc. In the resistance-capacity oscillator the frequency-determining network has an equivalent Q of 1/3. The action of the positive feedback loop increases the effective Q approximately 30

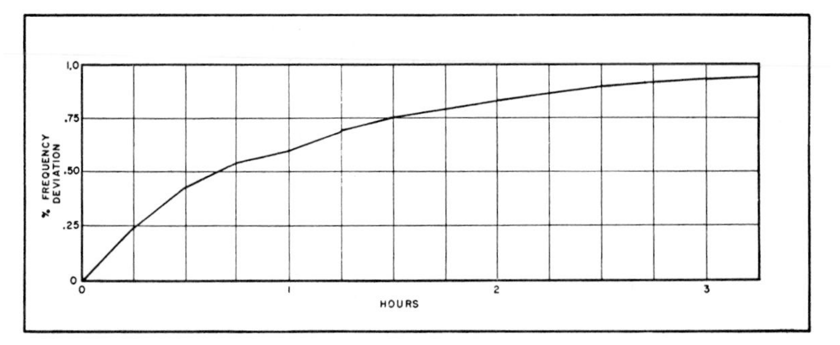

Figure 8. *Short Time Stability Curve of Resistance-Capacity Oscillator*

*P. R. Coursey, *Fixed Resistors For Use in Communication Equipment*, The Proceedings of the Institution of Electrical Engineers, Vol. 96, Part III, p. 169, May, 1949.

times, resulting in an effective operating Q of about 10. The short-time stability of the circuit is that obtainable with Q's of this order. This relatively low Q is contrasted with LC oscillators with which it is possible to obtain Q's of several hundred. A typical short-time stability curve is shown in Figure 8.

With further reference to short-time stability, the effect of power line voltage variations must often be considered. A number of studies of line-voltage conditions at the point of usage indicate that ±10-volt or more line variations on nominal 115-volt lines are the rule rather than the exception. Although the effects of line voltage variations are minimized by the negative feedback in the circuit, line voltage effects do increase at the higher frequencies where the gain of the circuit is less. Line voltage effects can be reduced by a factor of about three at these frequencies by the use of plate supply regulation. It is therefore customary to use such regulation at the higher frequencies when best stability is required.

The above discussion of accuracy concerns the oscillator circuit itself. It should be noted that in order to achieve the practical use of the accuracy of the circuit it is necessary to use an isolating amplifier between the oscillator circuit and the load.

SYNCHRONIZATION

Occasionally it is desirable to synchronize the resistance-capacity oscillator to obtain the accuracy and stability of a device such an an external frequency standard. Synchronization can be obtained on a 1:1 basis or as high as 12:1 by the arrangement shown in Figure 9. The synchronizing voltage is applied to the screen grid of the first oscillator tube through an isolating amplifier. With this arrangement the synchronizing voltage can be either sinusoidal or square. Synchronization on a 1:1 basis can also be obtained by coupling capacitively with a wire laid near the main tuning capacitor or by wrapping a few turns around the grid lead of the first tube with a wire connected to the synchronizing voltage.

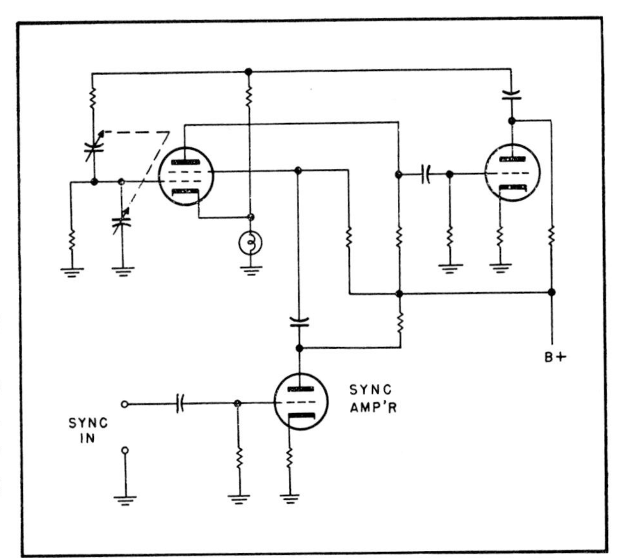

Figure 9. *Circuit for Synchronization of Resistance-Capacity Oscillator*

AMPLITUDE STABILITY

A form of stability not often considered but which is of importance in some bridge and magnetic circuits is that of constancy of oscillator output with time. The resistance-capacity oscillator inherently has good amplitude stability because of the negative feedback circuit. In applications where amplitude stability is of importance, it is desirable to remove all variable resistances from the circuit and to regulate the plate voltage supply. When these precautions have been taken, long-time variations in amplitude of less than 0.3% can be expected from the resistance-capacity circuit at medium frequencies with somewhat greater variations at the low and high frequencies.

SPECIAL CONTROLS

Although capacitive tuning has proved the more satisfactory method over a long period of time, tuning can also be accomplished by varying the resistance in the frequency-determining network, as indicated by the "resonance" formula. Applications wherein it is desired to remote-control an oscillator or wherein the frequency of oscillation must be varied in accordance with the rotation of a mechanical part indicate the use of a precision potentiometer or slide-wire for tuning purposes. Resistance tuning, while useful in some applications, is inferior to capacitive tuning owing to a combination of factors. Using composition controls, these factors include wear and poor resetability. Using wire-wound controls, these factors include wear and discrete breaks or steps in resistance as the control is rotated. These steps in resistance cause the incremental frequency changes to be much greater when only a small portion of the resistance of the control is in the circuit than when a large portion of the resistance is in the circuit. In addition, resistance tuning causes the impedance of the frequency-determining network to vary throughout each frequency band, resulting in undesirable variations in performance throughout the band. When variable resistors are used for remote tuning purposes, the oscillator should be calibrated with a specific length of a given cable connecting the resistance to the remainder of the circuit to insure proper calibration.

A type of oscillator frequently required is that with a vernier frequency control. Such controls fall into two classes, depending upon whether the control is calibrated or uncalibrated. In capacitively-tuned oscillators calibrated controls are provided as resistive verniers in series with the resistances of the frequency-determining network. The controls are calibrated in per cent deviation rather than directly in frequency, because a given incremental resistance change produces a different frequency increment at one setting of the main tuning capacitor than at any other setting. However, a given resistance change causes the same percentage variation in frequency at any setting of the tuning capacitor.

Capacitive frequency verniers are often used in capacitively-tuned circuits because of the smooth change in frequency obtainable. However, in such applications the range of control of such verniers varies widely, depending upon the setting of the main tuning capacitor. For this reason capacitive verniers are not calibrated in capacitively-tuned circuits. Capacitive verniers can be calibrated in per cent frequency deviation when used in circuits that are tuned by varying the resistance in the frequency-determining network. This arrangement is possible because the capacity in the circuit is constant.

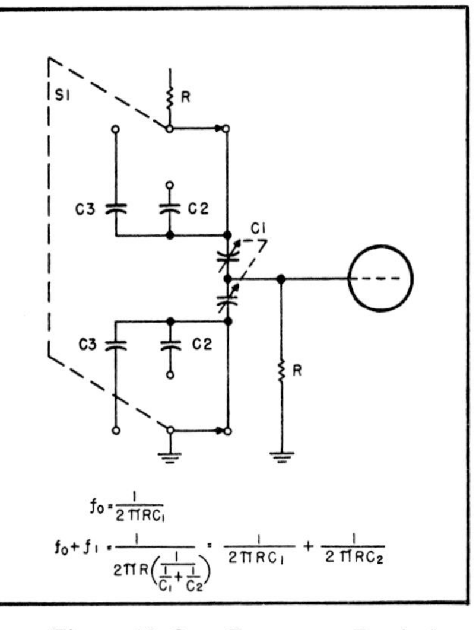

Figure 10. *Step-Frequency Deviation Control*

Figure 10 shows the elements of a deviation control that is used where deviations consisting of fixed steps are desired. A series of capacitors selectable by a switch are connected so that one of them will be in series with the main tuning capacitor on either side of the grid connection. By selecting the desired set of capacitors with a switch, a fixed-frequency step is obtained regardless of the setting of the main tuning capacitor. The action of the additional capacitors can be seen by referring to the expressions shown for the frequency of oscillation in Figure 10. There it is shown that the frequency of oscillation is increased by a factor corresponding to the frequency of oscillation of the resistance and vernier capacity considered separately.

SUMMARY

The characteristics discussed above are summarized for reference purposes in the following table.

—*Brunton Bauer.*

TYPICAL CHARACTERISTICS OF RESISTANCE-CAPACITY OSCILLATORS

CHARACTERISTICS	LOW FREQUENCY RANGE	MEDIUM FREQUENCY RANGE	HIGH FREQUENCY RANGE
Frequency Range	1/2 — 100 cps	100 cps — 100 kc	100 kc — 1 mc
Distortion	1% at 2 cps; increases below 2 cps	Less than 1%; less than 1/4% in special cases	Approx. 1% at 200 kc; approx. 3% at 1 mc
Long-time stability	Depends on circuit components; 2% typical; 1% in special cases	Depends on circuit components; 2% typical; 0.1% in special cases	Depends on circuit components; 2% typical; 0.5% in special cases
Short-time stability	0.3%	0.1%	Approx. 0.2% below 500 kc; poor above 500 kc
Amplitude stability	Within 0.5%	Within 0.3%	Within 1% below 500 kc; poor above 500 kc
Incremental control	Percentage calibrated resistive control; uncalibrated capacitive control	Percentage calibrated resistive control; uncalibrated capacitive control	Percentage calibrated resistive control; uncalibrated capacitive control

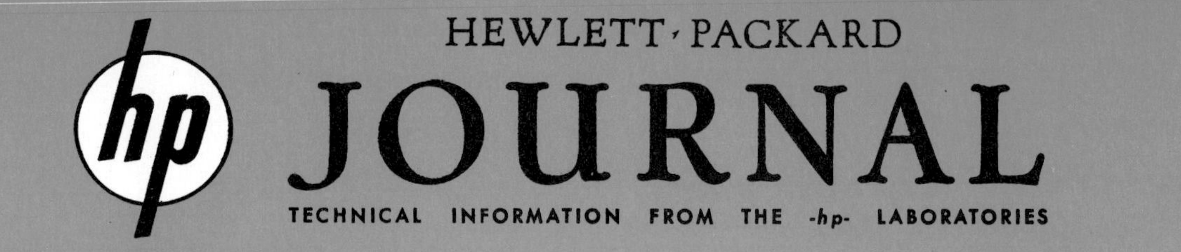

HEWLETT·PACKARD JOURNAL
TECHNICAL INFORMATION FROM THE -hp- LABORATORIES

VOL. 2 No. 5

PUBLISHED BY THE HEWLETT-PACKARD COMPANY, 395 PAGE MILL ROAD, PALO ALTO, CALIFORNIA JANUARY, 1951

The High-Speed Frequency Counter— A New Solution to Old Problems

ACCURATE frequency measurements have always been somewhat troublesome. To make such measurements, the state of the art has required the use of three separate equipments—a frequency standard, an interpolating system, and a detector. A thing to be hoped for was a new device that would automatically measure the frequency of an unknown with the same accuracy obtained with existing methods.

The -hp- Model 524A Frequency Counter is a new type of measuring instrument that performs this very function. The instrument directly measures and directly displays the frequency of an unknown. The unknown can lie anywhere within the range from 0.01 cps to 10 megacycles—a range that embraces the greater part of all frequency-measuring work.

The simplicity of operation of the frequency counter is worthy of special mention: when an unknown is connected to the input terminals, the measured frequency is displayed automatically by the arrangement shown in Figure 1. A decimal multiplier panel switch modifies the displayed value when the measurements have been made for less or more than one second.

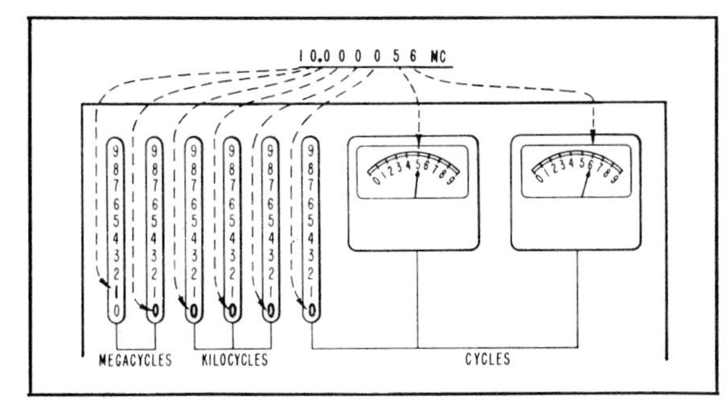

Figure 1. *Display arrangement of new -hp- Model 524A Frequency Counter. Columns of neon tubes behind panel slots illuminate the proper numerals which are reverse-printed on transparent plastic. Example shown represents a frequency of 10,000,056 cps counted for period of one second. Maximum rate for instrument is 10 mc; maximum counting period is 10 seconds.*

The Model 524A is based on pulse techniques and makes two types of measurements relating to frequency. First, the instrument counts unknown frequencies for a short accurately-known time interval of 10, 1, 0.1, 0.01, or 0.001 second as desired. The instrument first counts and then displays the count, repeating this process as long as desired. Second, the instrument measures period—the time consumed by one cycle of a voltage. Periods are counted in units of 1 microsecond up to a maximum of 100 seconds. Periods also are first measured and then displayed in a repeating process.

The accuracy of the -hp- Model 524A Frequency Counter is determined by a gating circuit that is accurate within 0.1 microsecond and by an internal crystal oscillator that is accurate within approximately 2 parts per million per week. Accuracy is thus determined primarily by the oscillator, which is equal in quality to a high-quality secondary frequency standard. For this reason, the -hp- Model 524A Frequency Counter is regarded as a new type of secondary frequency standard that has the advantages of displaying the measured frequency directly and of measuring frequencies as high as 10 megacycles. Where there is at hand a laboratory standard of better accuracy than that of the internal oscillator, the accuracy of the Model 524A can be increased by substituting the laboratory standard for the internal oscillator. A panel connector and switch are provided for this purpose.

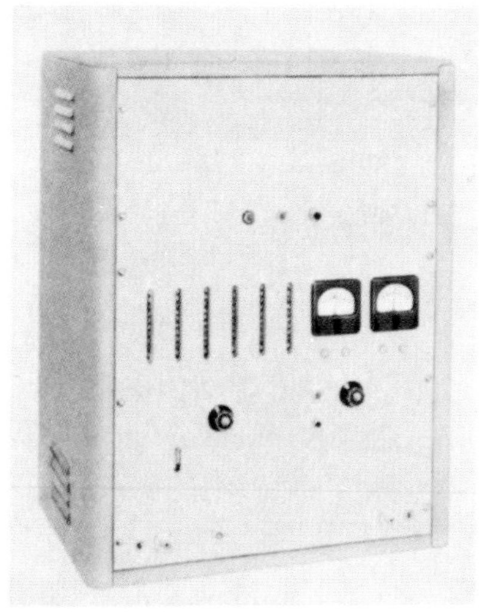

Figure 2. *New -hp- Model 524A Frequency Counter.*

A frequency counter having a range as wide as 10 megacycles is a powerful laboratory and production tool. Besides being a secondary frequency standard for the laboratory, the frequency counter can be used to calibrate oscillators, measure frequency drift, period, and for production applications involving calibrating and tuning. When used with a light source and phototube such as the -hp- Model 506A, the frequency counter can be used to measure precisely the speed of rotation of high-speed devices such as small motors and centrifuges.

CIRCUIT DESCRIPTION

For introductory purposes, the circuit of the Model 524A can be represented by the block diagram of Figure 3. Here, the unknown frequency is applied through a wide-band squaring amplifier to a fast gate controlled by a time base generator. When the fast gate is open, the unknown is applied to the counting circuits; when the fast gate closes, the counting circuits display the counted frequency until the time base generator triggers the resetting circuit. The counting circuits are then reset to zero and the gate is again opened so that the next sample of the unknown can be counted.

The heart and distinguishing feature of the Model 524A is its high-speed counting circuit, a development of the high-speed scaler described here in a recent issue[1]. For reliable 10-megacycle operation, the requirements of the circuit are that it must have a double-pulse resolving time of 0.1 microsecond and a triple-pulse resolving time of 0.2 microsecond, as well as being capable of continuous 10-megacycle operation. The filling of these requirements by the scaler described before has allowed the design of the Model 524A as a high-speed frequency counter.

The complete counting circuit consists of eight cascaded scalers with indicating systems. All scalers are decade types that generate one output pulse for every ten pulses received by the circuit. The input scaler is the high-speed 10-megacycle circuit that divides the incoming pulses by decades and indicates any remaining counts on a panel meter. The second circuit is similar to the first, except for slower speed operation. The remaining six scalers each indicate on a neon lamp the last digit of the quantity of pulses received from the preceding scaler.

TIME BASE GENERATOR

The circuit that determines the time of opening and closing of the fast gate is the time base generator. This circuit includes a 100 kc precision crystal-controlled oscillator and a series of precision frequency dividers that provide the following time bases or sampling times:

0.001 Second	1.000 Second
0.010 Second	10.00 Seconds
0.100 Second	

The longest sampling time of 10 seconds combined with the highest counting rate of 10 megacycles per second determine the maximum count that can be made by the instrument. The product of sampling time and maximum rate gives a total capacity for the instrument of one hundred million counts, which in practice is limited to 99,999,999 counts because of the eight-place indicating system.

The length of time that counts are displayed is designed to be equal in each instance to the sampling time. Display times that are less than 1 second are too short to permit the displayed data to be recorded in a notebook or data sheet, but these short display times have a practical value when performing operations such as tuning signal sources and oscillators. For such applications, the short sampling and display times cause the indicating system to give a continuous reading, thereby allowing immediate corrections to be made in the tuning operation.

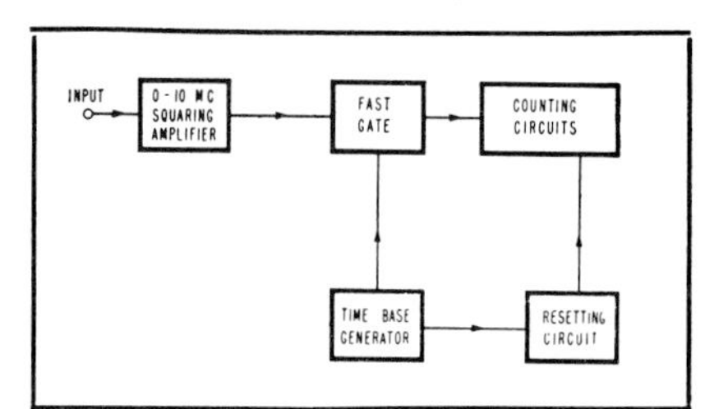

Figure 3. *Basic block diagram of -hp- Model 524A Frequency Counter.*

PERIOD MEASUREMENTS

Besides being arranged to measure frequency directly, the Model 524A is also arranged so that it can count the frequency of its internal 100 kc precision oscillator for a time interval equal to 10 periods of an unknown frequency. Thus, the instrument counts in units of 10 microseconds for a length of time equal to 10 cycles of the unknown, thereby displaying the period of 1 cycle of the unknown in units of 1 microsecond. For example, an unknown of 1 cps would cause the instrument to count the frequency of the internal 100 kc oscillator for a period of 10 cycles, giving the period of 1 cycle of the unknown as one million microseconds.

The value of period measurements becomes apparent when measuring low frequencies, where on direct frequency measurements the available

[1]A. S. Bagley, *A 10 MC Scaler for Nuclear Counting and Frequency Measurement,* Hewlett-Packard Journal, Vol. 2, No. 2, October, 1950.

sampling times would permit only a few cycles of the unknown to be measured. The ability to make period measurements extends the measurement range of the instrument to frequencies as low as 0.01 cps.

ACCURACY

Two factors influence accuracy: the reliability of the fast gate and the accuracy of the oscillator, whether the internal precision oscillator or an external frequency standard. The fast gate has been designed to open accurately within 0.1 microsecond. In counting terms, this reliability will give an accuracy of ±1 count. In frequency terms, the gate reliability will give an accuracy of within ±0.1 cycle if the sampling time is 10 seconds or within ±1 cycle if the sampling time is 1 second. On shorter sampling times, the accuracy will be correspondingly less.

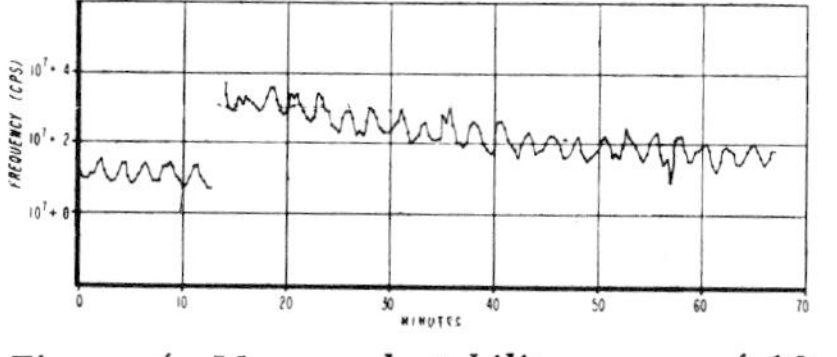

Figure 4. *Measured stability curve of 10 mc oscillator turned off for one minute after 13 minutes.*

Although the fast gate opens reliably within 0.1 microsecond, the signal to open the gate is given by the precision oscillator. The tolerance of the oscillator must therefore be added to the tolerance of the fast gate to obtain the overall accuracy for the instrument. Since the internal oscillator is accurate within approximately 2 parts per million per week, the Model 524A is basically accurate within 0.0005% ±1 count. When the instrument is used with an external frequency standard, the accuracy of the measurement becomes the accuracy of the standard diminished by 1 count.

A visual indication of the counting capabilities of the Model 524A is shown by the curve in Figure 4. The curve was plotted from data taken while using a high-stability laboratory frequency standard in place of the internal oscillator in the Model 524A and represents the type of measurement that can be made where high-precision standards are available. The source under measurement was a 10-megacycle oscillator whose crystal oven operation was investigated. The cyclic nature of the curve is caused by the cycling of the 10-megacycle crystal oven thermostat; the discontinuity resulted from turning off the oscillator for one minute.

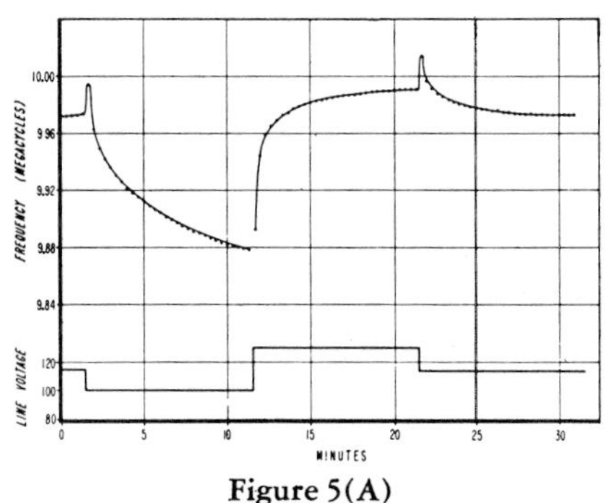

Figure 5(A)

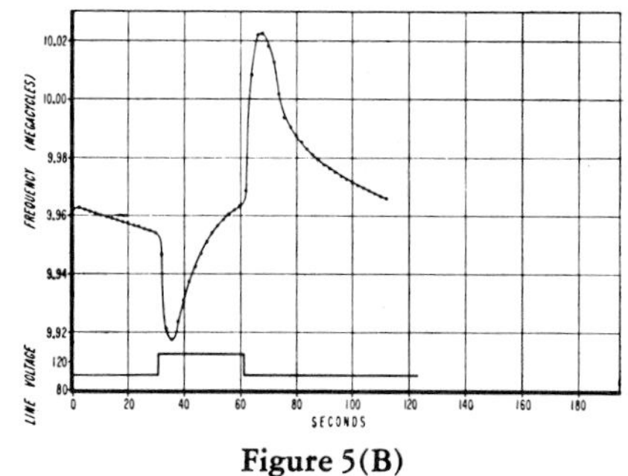

Figure 5(B)

Frequency shift in -hp- Model 650A Test Oscillator operating at 10 mc when subjected to ±15-volt line change. Measurements made at 20-second intervals. Discontinuous curve results from sudden change in frequency when line voltage is changed and is further investigated in Figure 5(b) where measurements are made at 2-second intervals to magnify discontinuous area. Such measurements are typical of those that can be made with Model 524A.

Note that the ordinate of Figure 4 is tremendously expanded, the whole vertical scale representing only 4 parts in ten million. From the curve it is apparent that the measurements were made within 1 part in one hundred million.

MEASUREMENTS

As lower and lower frequencies are measured, the tolerance of ±1 count becomes increasingly important on a percentage basis. Where a tolerance of 1 count in one hundred million is but 0.000001%, a tolerance of 1 count in one hundred thousand is 0.001%. Finally, a point is reached where measurements can be made with greater accuracy by reversing the frequency measurement process; that is, by determining frequency through a measurement of its period as described before. This transition point is illustrated in Figure 6, where curve A is a plot of theoretical accuracy obtainable on period measurements, and curve B a plot of accuracy obtainable on 10-second frequency counts. Curve A shows that highest accuracy is obtained on period measurements when the unknown frequency is low. Conversely, curve B shows that highest accuracy is obtained on direct frequency measurements when the unknown frequency is high. A transition point occurs at 316 cps where, to obtain best accuracy, the method of measurement must be changed from a direct measurement of frequency to a measurement of period.

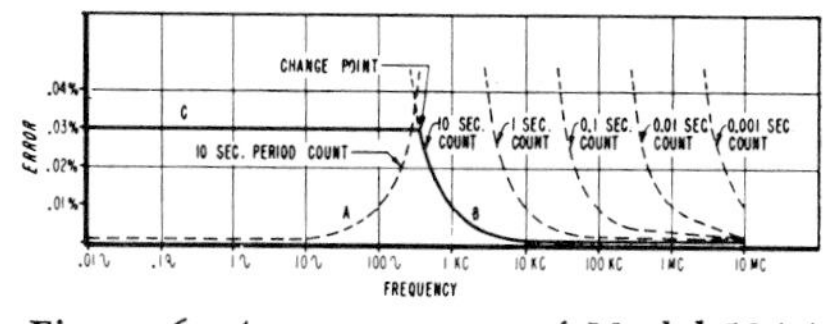

Figure 6. *Accuracy curve of Model 524A Frequency Counter for direct frequency and period measurements.*

Curve A in Figure 6 is a theoretical curve and its most accurate region can not be attained in actual practice because the rate-of-rise of low-frequency voltages is too slow to permit accurate triggering. However, for period measurements, it has been established that the accuracy of the Model 524A is within 0.03% on frequencies from 316 cps to as low as 0.01 cps.

COMPLETE CIRCUIT

The complete circuit of the Model 524A is represented by the block diagram of Figure 8. On direct frequency measurements, the unknown is applied through the input squaring

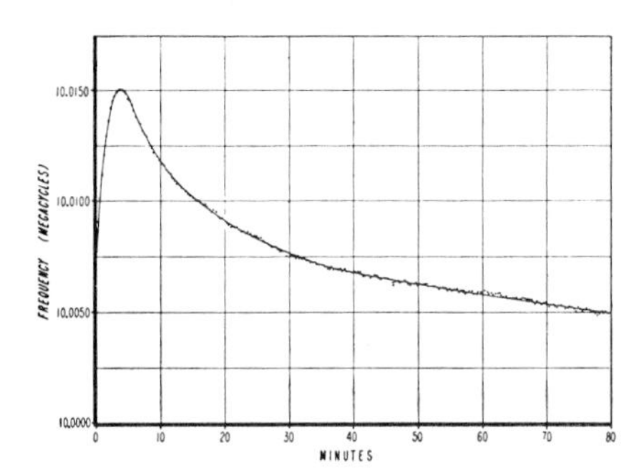

Figure 7. *Warm-up stability curve of -hp- Model 608A 10-500 mc signal generator with frequency dial set for 10-megacycle output. Measurements made with -hp- Model 524A at 20-second intervals.*

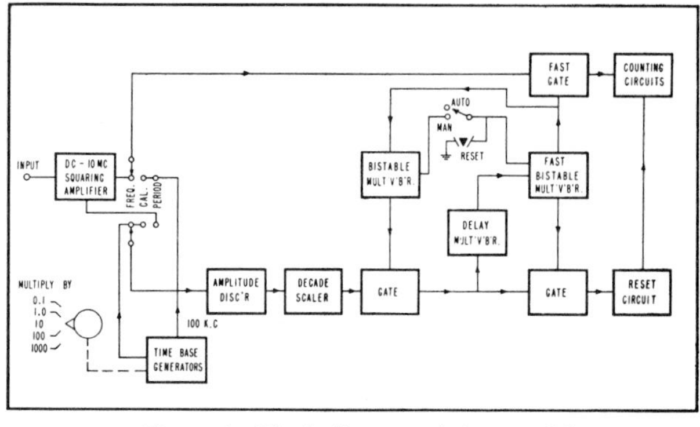

Figure 8. *Block diagram of -hp- Model 524A Frequency Counter.*

amplifier and fast gate to the counting circuits as described before. Depending upon the position of the multiplier switch in the time base circuits, a time base lying between 0.0001 second and 1 second is applied through the input switch to the amplitude discriminator. The amplitude discriminator is basically a regenerative amplifier designed for rapid rise of regenerative voltage once the driving voltage has reached a critical value. The circuit thus adds to the reliability of the instrument by insuring that triggering will always occur at the same point of the time base. The degree of discrimination of the circuit is sufficient to give a reliability of within 0.03% on time bases as long as 10 seconds.

Following the amplitude discriminator is a decade scaler that divides by a factor of ten the time bases generated by the time base circuits. The decade scaler operates into a gate that is always open on automatic measurements so that the output of the scaler is applied both to the gate for the resetting circuit and to the delay multivibrator that feeds into the fast bistable multivibrator.

The function of the fast bistable multivibrator is to open and close both the fast gate and the gate to the resetting circuit. The multivibrator is arranged so that when one gate is open the other is always closed. A high order of reliability is evident in the fact that opening or closing time combined with any jitter are such that the fast gate is reliable within 0.1 microsecond.

The operation of the gating circuitry is as follows: when a pulse comes from the decade scaler and passes through the first gate, it is applied both to the delay multivibrator and to the gate for the resetting circuit. If the gate is open, the pulse triggers the resetting circuits, clearing the counters and resetting them to zero.

Meantime, the pulse has triggered the delay multivibrator which delays the triggering of the fast bistable multivibrator by a fixed interval of 100 microseconds. The delay multivibrator also is carefully designed to achieve reliability. To insure an accurate delay interval, a 10 kc voltage from the time base circuits is superimposed upon the grid of the delay multivibrator in such a way that the delay multivibrator operates with the precision of the time base.

At the end of 100 microseconds, the delay multivibrator causes the fast bistable multivibrator to switch, opening the fast gate and closing the resetting gate. Since the counting circuits have been cleared and reset to zero, they are ready to count the unknown admitted through the fast gate.

The next pulse from the decade scaler finds the resetting gate closed, but triggers the delay multivibrator. After the fixed delay of 100 microseconds, the delay multivibrator operates through the fast bistable multivibrator to open the resetting gate and to close the fast gate. The counting circuits then display the counted value until the next pulse from the decade scaler causes the whole process to repeat.

The operation of the instrument on period measurements is similar to that on direct frequency measurements, except that the input switch is arranged so that the 100 kc frequency from the time base circuits is applied to the counters and 10 cycles of the unknown are used as the time base.

A third position of the input switch allows the operation of the overall circuit to be checked, using the 100 kc frequency from the time base circuits as the unknown.

—*A. S. Bagley*

SPECIFICATIONS FOR MODEL 524A FREQUENCY COUNTER

MAXIMUM COUNTING RATE: Ten million cycles per seond.

PRESENTATION: Total indication, 8 places. First six places on neon lamp banks. Last two places on two meters.

PERIOD OF COUNT: 0.001, 0.01, 0.1, 1, and 10 seconds, selected by panel control. Counting and display periods are equal and are automatically cycled. Panel push button allows counting for a single period with continuous display of count until push button again depressed.

LOW FREQUENCIES: Instrument can be switched so that low frequencies will operate as time bases. Period of low frequencies is displayed in microseconds.

ACCURACY: Direct frequency measurement: ±1 count ±tolerance of oscillator. Internal 100 kc crystal oscillator tolerance is 2 parts per million per week. External laboratory standards can be used to obtain higher oscillator accuracy.

PERIOD MEASUREMENT: Within 0.03% for frequencies up to 300 cps; within 1 microsecond for frequencies between 300 cps and 10 kc.

INPUT TO COUNTING CIRCUIT: Minimum required voltage is 1 volt peak.

INPUT IMPEDANCE: Approx. 100,000 ohms shunted by 30 mmf.

100 KC TIMING CIRCUIT: To use external 100 kc source requires 1 volt across 50,000 ohms shunted by 30 mmf.

POWER SOURCE: Operates from nominal 115-volt, 50/60 cycle supply. Requires approximately 400 watts.

CABLES SUPPLIED: 7′ 6″ power cord; permanently attached.

CONNECTORS: Signal and external standard connectors, BNC type jacks.

MOUNTING: Supplied in metal cabinet.

SIZE: Approx. 28″ high, 21¾″ wide, 14″ deep.

WEIGHT: Approx. 115 lbs. Shipping weight, approx. 175 lbs.

Price and delivery upon request.
Data subject to change without notice.

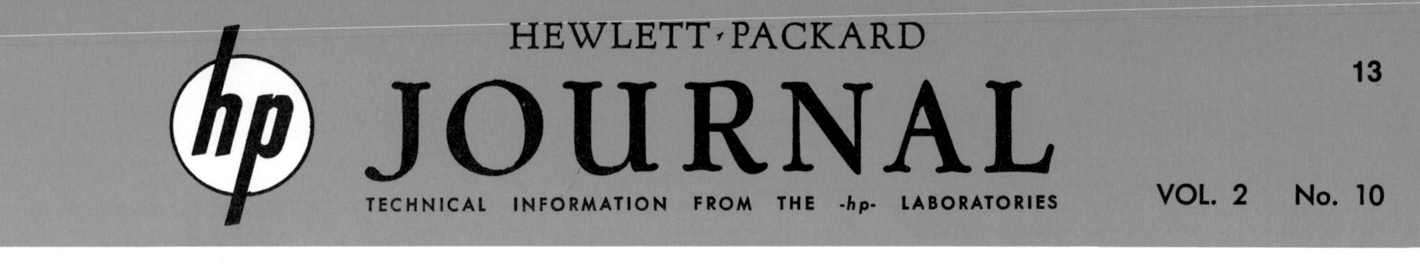
HEWLETT·PACKARD JOURNAL
TECHNICAL INFORMATION FROM THE -hp- LABORATORIES
VOL. 2 No. 10
PUBLISHED BY THE HEWLETT-PACKARD COMPANY, 395 PAGE MILL ROAD, PALO ALTO, CALIFORNIA JUNE, 1951

A New Generator of Frequencies Down to 0.01 CPS

IN contrast to ever-widening uses of higher and higher frequencies, much work goes on in that portion of the spectrum lying below the audio range. Medical, geological, servo, and vibration studies are all concerned with sub-audio frequencies. Servo applications in particular involve unusually low frequencies—often lying well below 1 cps.

The new *-hp-* Model 202A Low Frequency Function Generator has been designed to provide test voltages for such low-frequency work. The instrument generates frequencies from 1000 cps down to 0.01 cps. This range is covered in five decade bands, each of which has an effective range of approximately 13:1 to provide suitable overlap between bands.

The instrument generates three types of waveforms that are useful in low-frequency work: sinusoidal, square, and triangular. The particular waveform desired is selected by a panel switch. Also provided is a synchronizing pulse for triggering external equipment.

Output voltage from the instrument has been made ample for general-purpose work. For all three waveforms, a maximum of at least 30 volts peak-to-peak is provided. The synchronizing pulse is of 5 volts peak amplitude and occurs at a time corresponding to the crest of the sinusoidal output.

The output waveforms are provided from a system that can be operated either balanced or single-ended. The internal impedance of the output system has been made low—less than 100 ohms—although rated load for the system is 5000 ohms. The output amplifier is a stable d-c system having negligible drift; tests have shown the d-c component in the output to remain less than 0.1 volt over long intervals. However, an adjustment is available from the front panel to balance out this component if desired.

Figure 1. *New -hp- Model 202A Low Frequency Function Generator produces frequencies as low as 1 cycle per 100 seconds. Sine, square, and triangular wave outputs are provided.*

A number of interesting new ideas are represented in the circuitry of the Model 202A. To achieve a circuit capable of generating such low frequencies as 1 cycle per 100 seconds, a considerable departure has been made from established oscillator designs. However, the performance of the Model 202A is in all respects comparable to that of established laboratory-type audio oscillators. For example, transients of the type that frequently arise when chang-

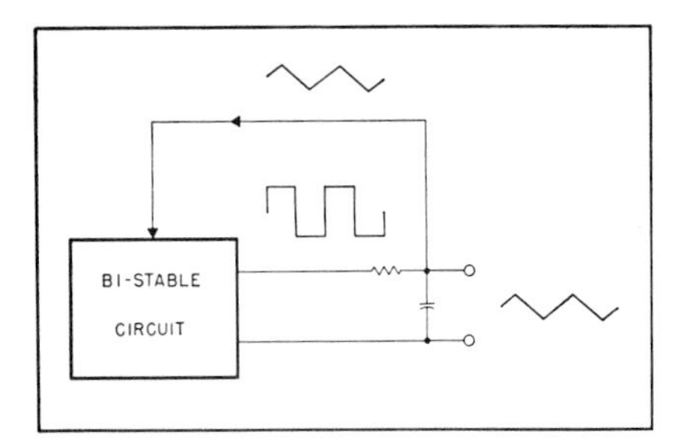

Figure 2. *Bi-stable circuit and integrator arranged to generate triangular waveform.*

ing frequency in l-f circuits are virtually non-existent. RMS distortion in the sinusoidal output is less than 1%, variation of output frequency with line voltage changes and warm-up is less than 1%, and dial calibration is accurate within 2%.

UNDERLYING CONSIDERATIONS

Formerly, low frequency generators have often made use of conventional multivibrator or phase shift techniques. While such techniques were useful to a point, they were not satisfactory for extremely low frequency applications. In some cases instability and drift were too large for many applications and very high values of circuit components were required. In other cases the need for amplitude control led to the use of avc circuits having long time-constants with resultant undesirable transients in the generated voltage. A real need existed for a circuit capable of stable operation in the range lying below about one cycle per second.

Before describing the circuitry of the Model 202A, it may be of interest to describe some of the fundamental considerations underlying the development of the generator circuit:

Among the possible methods of producing a sine wave is the method of synthesis. One possible synthesis[1] of a sine wave assumes the availability of a square wave. If a square wave is applied to an integrator, a triangular waveform is obtained. Now, a triangular waveform can be distorted into an approximation of a sine wave by use of non-linear elements. For example, if a generator of triangular waveforms is suitably loaded with a non-linear resistance whose value decreases as the instantaneous triangular voltage increases, an approximation of a sine wave can be obtained.

Such a synthesis is made practical in part by the use of the circuit of Figure 2. The generating system consists of a controlled bi-stable circuit and an integrating circuit. The bi-stable circuit drives the integrator with a square wave, while the triangular output of the integrator is coupled back to and triggers the bi-stable circuit. This arrangement gives a self-sustaining circuit that can be considered to be a very stable relaxation oscillator.

It is important that the triangular waveform generated at the integrator output have uniform characteristics at all fundamental frequencies generated by the instrument. This is accomplished by choosing the constants of the integrator so that an essentially pure triangular waveform is obtained. Specifically, the RC product of the integrator circuit is made more than 50 times the half-period of the lowest-frequency square wave generated. Under this condition, the sides of the triangular voltage are always linear within 1%, assuring an essentially pure triangular waveform at all times.

The bi-stable circuit that generates the square wave is a combination of a flip-flop and two amplitude-comparators, designed so that the flip-flop will be triggered accurately even for signal voltages that have low slope. This basic circuit is shown in Figure 3. The flip-flop configuration is apparent, while the grid of each tube is also coupled to its cathode through a diode and transformer. Normally, neither of these diodes is conducting; the voltage at which each diode does conduct is determined by pre-set bias or reference voltages.

To describe the operation of the circuit, assume that V2 is conducting because of suitable grid bias obtained from the voltage divider. When V2 is conducting, its plate current flows through the upper part of the voltage divider at the right, causing V3 to be cut off. Now, as a positive-going voltage of the triangular waveform from the integrator becomes equal to the bias (upper reference voltage) on diode

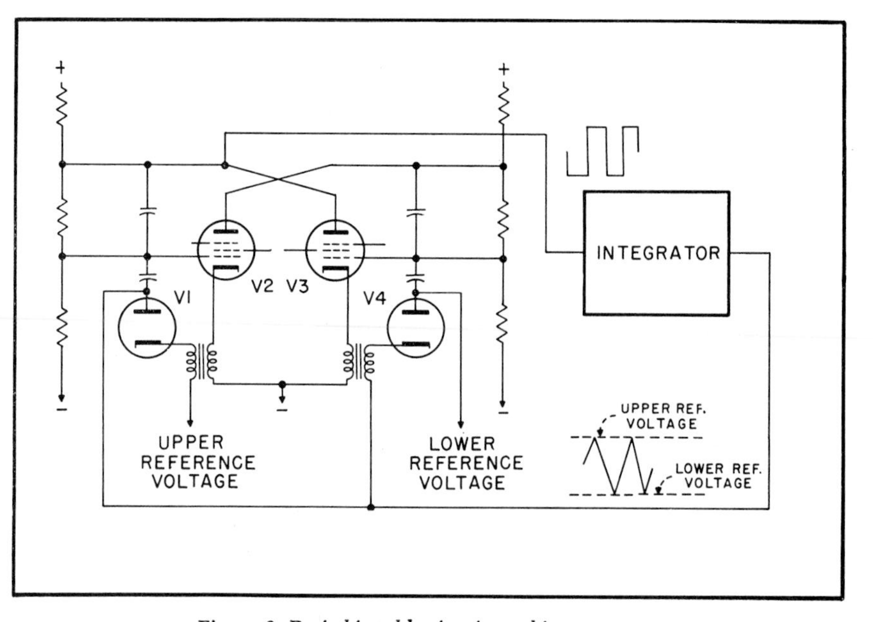

Figure 3. *Basic bi-stable circuit used in generator.*

[1] This basic generator circuitry is due to Dr. O. J. M. Smith of the University of California, Berkeley.

V1, V1 will conduct, thus coupling the grid of V2 to its cathode. When this occurs, oscillations quickly build up in V2, resulting in a large pulse voltage at the plate of V2. This pulse raises the grid of V3 above cut-off. V3 then conducts, producing a steep negative voltage front at the plate of V3. This negative voltage front is applied to the grid of V2, driving it below cut-off. The circuit is now stable until a negative-going voltage of the triangular waveform causes diode V4 to conduct. The reverse of the above action then occurs and the circuit reverts to its other stable condition.

The upper and lower reference voltages serve as amplitude controls that limit the excursions or peaks of the triangular waveform out of the integrator. Hence, the constants of the integrator can remain fixed for any one frequency band and the amplitude of the triangular waveform will always be constant. The amplitude control action is further aided by the use of the amplitude-comparators. In contrast to triggering methods that use a low slope trigger voltage directly, the amplitude-comparators generate a "whiplash" voltage that triggers the flip-flop.

INTEGRATOR

The integrator must give a linear voltage rise at its output over a period equal to one-half cycle of the square wave. Thus, the integrator time constant RC must be much larger than 50 seconds if a frequency of 1 cycle per 100 seconds is to be obtained. To obtain such a large time constant with highest practical values of R, C would normally be in the order of 100 microfarads—an impractical value for a precision capacitor.

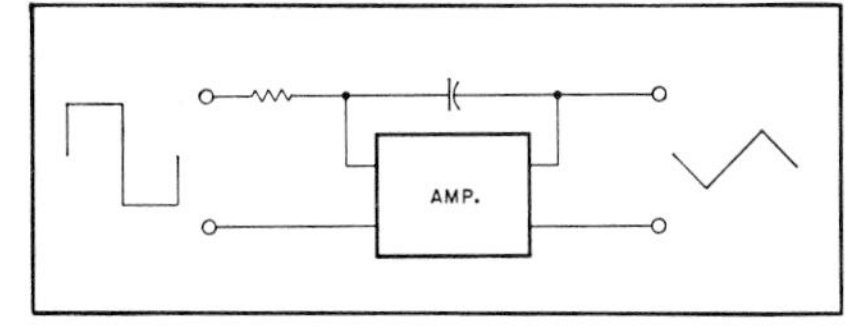

Figure 4. *Simple integrator converted into amplified time constant integrator.*

The electrical values of the components required in the integrator have been greatly reduced by converting the simple RC integrator into a Miller integrator where a high-gain amplifier is connected across the integrating capacitor (Figure 4). This amplifier reduces the value of the required time constant by a factor equal to the gain of the amplifier. In addition, the amplifier increases the voltage out of the integrator by the same factor. Thus, the slope of the voltage out of the integrator is still E/RC as with a simple integrator, but the linear range of the integrator is greatly increased.

FREQUENCY CONTROL

The slope E/RC of the voltage out of the integrator is obviously dependent upon the amplitude of the input square wave. This characteristic offers a means of controlling the frequency of the generator, because the time required for the triangular voltage-front to build up to a value that can trigger the bi-stable circuit is thus dependent upon the amplitude of the square wave applied to the integrator. Consequently, the frequency control for the instrument consists of a means of varying the amplitude of the integrator input. In production equipment, this control is a wire-wound potentiometer of high quality and high resolution. Band changes are accomplished by means of a switch that changes the capacity in the RC integrating network.

From the standpoint of accuracy, the frequency control system is good. Warm-up drift is less than 1%, while the dial calibration is accurate within 2%.

Since the output frequency is dependent upon the amplitude of the square wave applied to the integrator, the circuit would normally be susceptible to frequency drifts caused by tube aging and supply voltage changes. These drifts are avoided by clamping the square wave at its positive and negative extremes and by relating the reference voltages for the amplitude comparators to the same voltages used in the clamping. Thus, a change in clamping voltage will modify the amplitude of the square wave, but the reference voltages for the amplitude-comparators will also be modified by the same percentage. This arrangement reduces supply voltage and tube aging effects to the point where no discernable change in frequency occurs over reasonable limits.

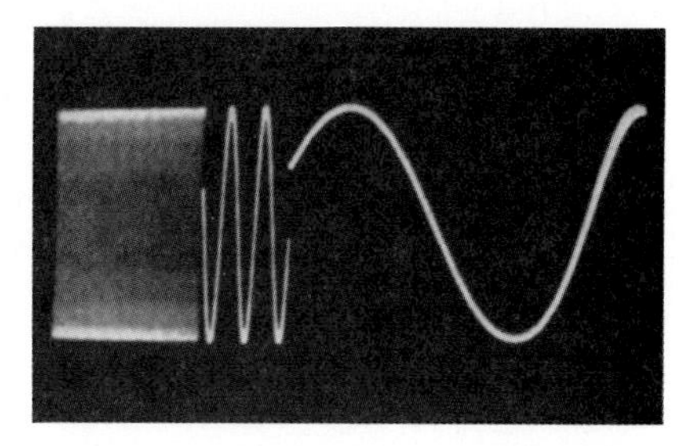

Figure 5. *Oscillogram of output waveform as output frequency changed twice. Freedom from transients is apparent.*

An interesting side light on the operation of the generator circuit as a whole is that it has practically no transient responses of the type that are common when changing frequency in l-f circuits. Usually, these transients arise from the use of a long time-constant avc system where a generator must operate for a number of cycles to provide the avc system with the proper correction voltages. These transients can be especially annoying at the low frequencies of the Model 202A. For example, if a generator must operate for 3 cycles to provide an avc system with a correction voltage, transients can exist for 3 cycles or up to 5 minutes each time frequency is changed. In the Model 202A, a change in frequency merely changes the amplitude of the square wave that drives the integrator. This change is at once reflected in a change in slope at the output of the integrator without introduction of transients. An oscillogram in Figure 5 shows the output sine wave as two changes in frequency were made, and the freedom from transients is apparent.

SHAPING THE WAVE

As stated earlier, a generator of triangular waveforms, when loaded with a non-linear resistance, can be made to produce a wave that is a close approximation of a sine wave. Such an arrangement is used in the Model 202A to shape the triangular waveform out of the integrator.

The basic shaping circuit is shown in Figure 6. If a triangular voltage is applied to the circuit of Figure 6 and if the bias voltages on the voltage divider at the left are suitably adjusted, the various diodes will conduct during different portions of a cycle. For example, as the triangular voltage-front rises above the axis, none of the diodes will conduct until the voltage at the output terminal overcomes the bias on V3, at which time V3 will conduct. The additional current flowing through R1 as a result of V3's conducting will alter the slope of the voltage appearing at the output terminals. As the input voltage increases more, it will cause diode V4 to conduct, again altering the slope of the output voltage.

On the negative-slope portion of the input voltage, diodes V4 and V3 will successively cease conducting. As the voltage passes below the axis, diodes V2 and V1 will conduct. All of these operations will alter the slope of the output voltage as indicated in the figure.

The practical shaping circuit used in the instrument has six duo-diodes that modify the triangular waveform. The resulting waveform, consisting of 26 linear segments, is a surprisingly close approximation of a sinusoid and has less than 1% distortion. In laboratory set-ups, the distortion has been adjusted to be less than 0.2%. Oscillograms of the triangular waveform and resulting sine wave are shown in Figure 7.

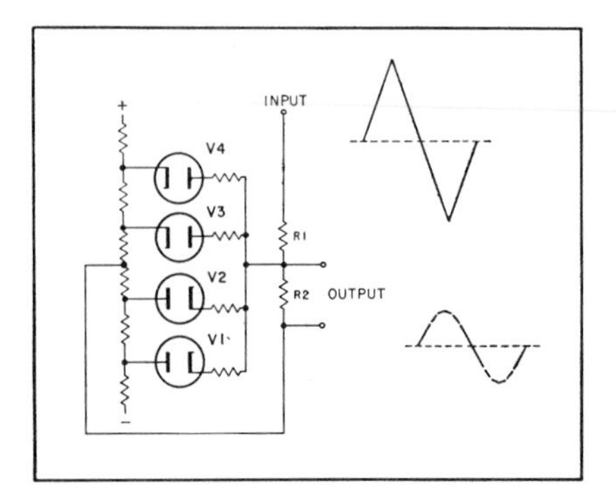

Figure 6. *Basic circuit for shaping triangular waveform into sine wave.*

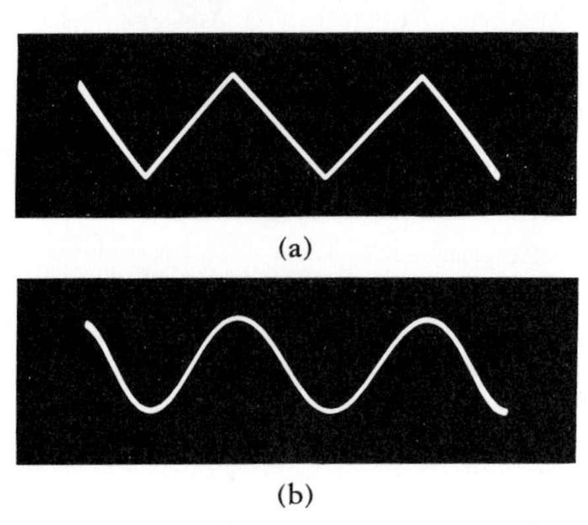

Figure 7. *Oscillograms of (a) triangular waveform applied to shaping circuit, and (b) resulting sine wave.*

OUTPUT AMPLIFIER

The output amplifier is a balanced type d-c amplifier stabilized with feedback. The output stage is a push-pull arrangement that can be operated either single-ended or balanced-to-ground. In order to accomplish this feature, the complete circuitry behind the output system is operated in a floating condition. The output system itself can be operated either floating or grounded, as desired.

The output amplifier is also provided with a suitable switching arrangement so that square and triangular waveforms, as well as the sinusoidal waveform, can be obtained at the output terminals. The sinusoidal waveform is obtained from the shaping circuits as described, while the square and triangular waveforms are obtained from the bi-stable and integrator circuits respectively.

GENERAL

The frequency calibration for the instrument is on a large 6-inch diameter dial, calibrated from approximately "0.8" to "10.4". Approximately 95 calibrated points are provided, all five ranges using the same calibrations with a suitable multiplier selected by the range switch. The dial drive system includes two knobs, one of which is a direct drive and the other a 6:1 vernier.

The main output terminals are conventional ¾-inch spaced binding posts. The Sync Out pulse, useful for triggering l-f sweeps, is provided at similar terminals.

The Model 202A is supplied in a rack-mounting style case. However, end frames are available to modify the rack-mounting style for bench use, although these frames are not essential for bench work. The end frames are provided with rubber mounting feet and with large trunk handles to facilitate carrying.

—R. H. Brunner

SPECIFICATIONS
MODEL 202A

Low Frequency Function Generator

FREQUENCY RANGE: 0.01 to 1,000 cps in five decade ranges with suitable overlap at each dial extreme.

DIAL ACCURACY: Within 2%.

FREQUENCY STABILITY: Within 1% including warm-up drift.

OUTPUT WAVEFORMS: Sinusoidal, square, and triangular. Selected by panel switch.

MAXIMUM OUTPUT VOLTAGE: At least 30 volts peak-to-peak across rated load for all three waveforms.

DISTORTION: Less than 1% rms distortion in sine wave output.

OUTPUT SYSTEM: Can be operated either balanced or single-ended. Output system is direct-coupled; d-c level of output voltage remains less than 0.1 volt over long periods of time. Adjustment available from front panel balances out any d-c.

FREQUENCY RESPONSE: Constant within 1 db.

HUM LEVEL: Less than 0.1% of maximum output.

SYNC PULSE: 5 volts peak, less than 10 microseconds duration. Sync pulse occurs at crest of sine wave output and at corresponding position with other waveshapes.

POWER: Operates from nominal 115-volt, 50/60 cycle source. Requires 175 watts.

CABLES SUPPLIED: 7½ foot power cable permanently attached.

DIMENSIONS: 10½" high, 19" wide, 13" deep.

MOUNTING: Relay rack style; end frames are available to convert to table mounting but are not essential.

WEIGHT: 35 lbs.; shipping weight, approx. 75 lbs.

PRICE: $450.00 f.o.b. Palo Alto, California. End frames $5.00 per pair f.o.b. Palo Alto, California (Specify No. 17).

Data subject to change without notice.

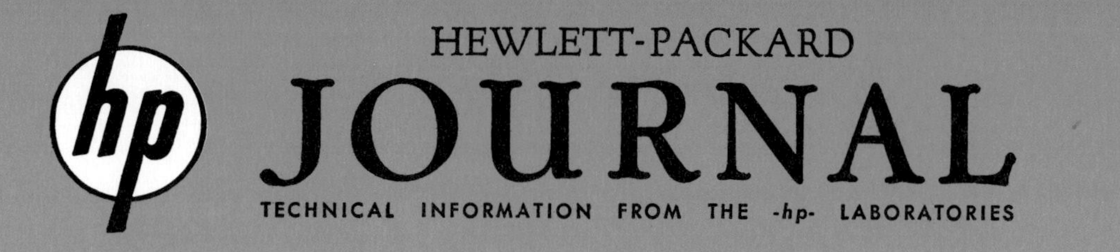

HEWLETT-PACKARD JOURNAL

TECHNICAL INFORMATION FROM THE -hp- LABORATORIES

Vol. 9 No. 10-11

PUBLISHED BY THE HEWLETT-PACKARD COMPANY, 275 PAGE MILL ROAD, PALO ALTO, CALIFORNIA JUNE-JULY, 1958

A Clip-On DC Milliammeter For Measuring Tube And Transistor Circuit Currents

IN typical electronics work direct currents are ordinarily measured either by measuring the current directly with a moving-coil type current meter or by measuring the voltage drop across a known resistance in the circuit. Although each of these methods has a long tradition, each also has substantial disadvantages. In the case of the indirect method with a voltmeter, the measurements are most often made across resistances having tolerances of 10% to 20% so that a minimum ambiguity of this same amount also exists in the current value obtained. Frequently, this ambiguity leads to selection of non-optimum component values with later difficulties. In the case of the direct method with a moving-coil instrument, the circuit must be opened and the current meter connected therein. This is an inconvenience that in practice often results in omission of the measurement with subsequent erroneous deduction and wasted effort. Additionally, in the continually-growing field of low-impedance circuits, where the current parameter assumes greater importance as in transistor work, the resistance of the moving-coil instrument is sometimes intolerable.

A new and much-easier-to-use current-measuring instrument that overcomes the foregoing objections has been developed in the form of a clip-on dc milliammeter which, as far as is known, is the first instrument of its type to be brought to production status. It makes the measurement of direct currents in low- as well as high-impedance circuits as easy as a measurement of voltage and at the same time covers the whole range of currents encountered in such work—from a fraction of a milliampere to 1 ampere in six ranges.

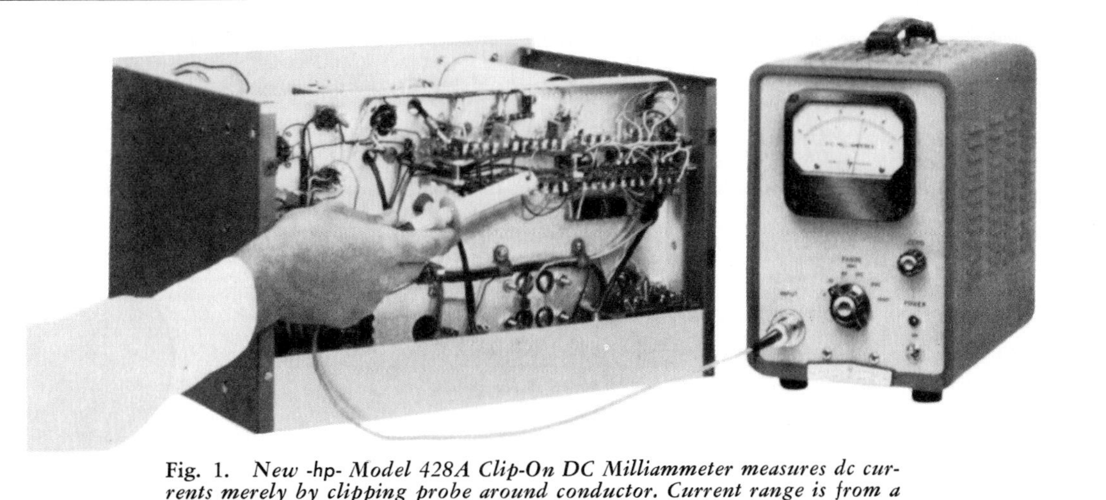

Fig. 1. *New -hp- Model 428A Clip-On DC Milliammeter measures dc currents merely by clipping probe around conductor. Current range is from a fraction of a milliampere to 1 ampere. Probe jaws are operated by flanges on probe body as shown in Fig. 3.*

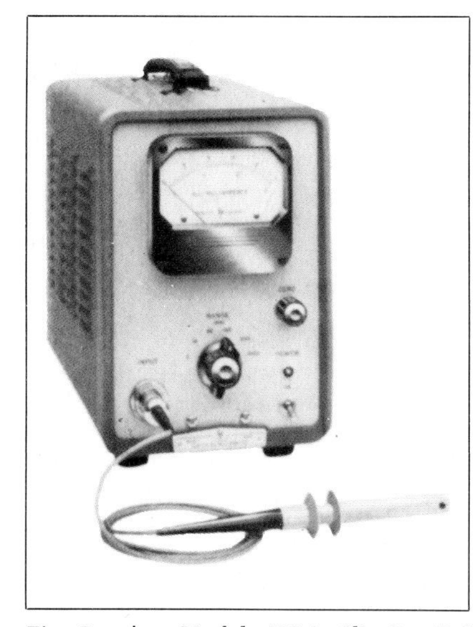

Fig. 2. -hp- *Model 428A Clip-On DC Milliammeter measures current in six 3:1 ranges from 3 ma full scale to 1 ampere full scale. Probe attaches to cabinet by 4' cable.*

Further, and especially valuable in the transistor field, it introduces no dc loading into the circuit being measured. Currents in low-impedance transistor circuits can thus be measured without disturbing static operating conditions. Also, the presence of alternating currents normally causes no error in the direct-current readings.

The instrument consists of a clip-on probe with associated electronic circuitry and indicating meter located in a small cabinet. The probe has been constructed in the -*hp*- penholder style in which flanges on the probe body open a set of split jaws for clipping around the conductor to be measured (Fig. 3). In typical practice the instrument can measure dc current in nearly any instance where a ½-inch length of conductor is available for the probe to clip around. Measurements can be made adjacent to a tube socket terminal, on transistor leads, on a conductor of a laced cable, and even on small composition resistors. At the same time remarkably few measuring precautions are required. Mainly, it is merely necessary to insure that the probe jaws are closed and are free of gap-causing particles, a condition which is easy to test for, since improper closure is indicated by a large earth's-field reading on the low range.

ELECTRICAL OPERATION

The new milliammeter operates by sensing the strength of the magnetic field produced by the current being measured. Since this sensing extracts no energy from the field, the milliammeter has the unusual property that it introduces no resistance (and very little inductance—less than 0.5 microhenry) into the circuit being measured. Freedom from loading is, of course, always desirable in a measuring instrument, but in transistor work where low-impedance circuits are common this property is especially valuable.

In some uses such as searching for shorts it is desirable to know the direction of current flow. The sensing elements in the probe are such that current direction can be determined, and the probe itself is marked with an arrow that shows current direction for an upscale reading. Connecting the probe on the conductor in the reverse orientation causes the meter pointer to move downscale but has no damaging effect on the instrument.

The probe contains a specially-developed magnetic amplifier which provides an ac output proportional to the magnetizing force produced by the measured dc current. The ac output is amplified and applied to a phase-sensitive detector whose output in turn feeds an indicating meter. The overall system including the magnetic amplifier is stabilized with 40 db of feedback for good long-time performance.

The ac signals produced by the magnetic amplifier in the probe occur at low levels so that the amplitude of the signal coupled into the measured circuit is slight. No specification for this amplitude has as yet been firmly established for production instruments but experience to date indicates it will be about 1/100 volt peak. The signal frequency has been selected to lie at a little-used region—40 kc—so that only rarely will the operation of the instrument be affected by the presence of this frequency and its harmonics in a circuit being measured. The presence of a frequency within about 20 cycles of 40 kc and its harmonics in the measured circuit will be indicated by a beat in the meter reading.

HIGHER SENSITIVITIES

On the instrument's most sensitive range (3 milliamperes full scale), readings can be made down to a minor division (0.1 ma) or so from the bottom of the scale, since the ultimate limitation of Barkhausen noise causes less than a half division reading at the bottom of the scale. In many cases, however, even higher sensitivities can be obtained by looping the conductor being measured through the probe two or more times to increase the effective magnetizing force of the current. Scale readings should then be divided by a factor equal to the number of turns made in the conductor.

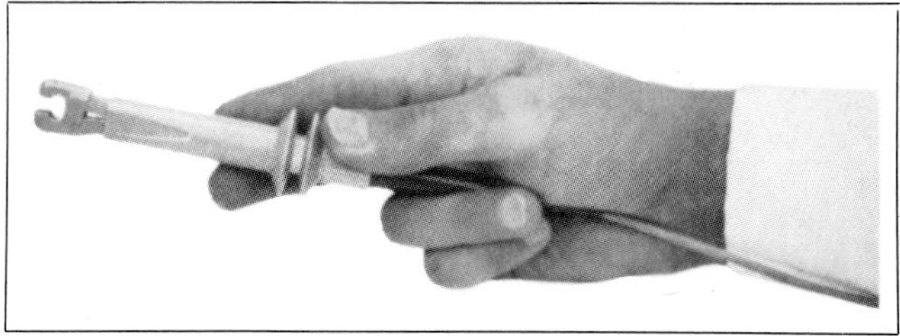

Fig. 3. *Probe jaws are opened by flanges on probe body. Spring return closes jaws when flanges are released.*

The sensitivity can be increased in this manner until reverse-coupled voltage becomes a limiting factor. Each additional turn will add the amount of reverse-coupled voltage described above for one turn. Even here, though, it should be noted that measurements can often be made at points in the circuit where a circuit capacitor will bypass the voltage, which is essentially 40 kc in composition.

HOMOGENEOUS AND NONHOMOGENEOUS EXTERNAL FIELDS

In order to make the instrument suitable for measuring small currents, considerable care has been taken to minimize the effects of external fields on the probe elements. The effect of external homogeneous fields is minimized by a balanced arrangement for the probe elements. In practice, the chief homogeneous field to be dealt with is the earth's field, but it is interesting to note that from the standpoint of small-current measurement the earth's field is relatively very strong. One way of illustrating this point is to note that the earth's field is roughly the same as the magnetizing force produced in the probe magnetic circuit by a conductor carrying 1 ampere.

The construction of the probe is such that no effects can be noticed from the earth's field other than on the most sensitive range. With the instrument set to this range, the most adverse changes in probe orientation will cause a shift in zero reading of less than ± 1 minor division, an effect which can then be compensated for any one orientation by adjustment of the zero control.

A high degree of shielding against nonhomogeneous fields is also provided so that measurements are essentially free from effects from currents in nearby conductors. Typically, a current of 1 ampere flowing in a conductor so close that it is touching the probe will produce an effect amounting to only about 3% of the most sensitive range, while if the conductor is 1 inch away the effect is scarcely perceptible.

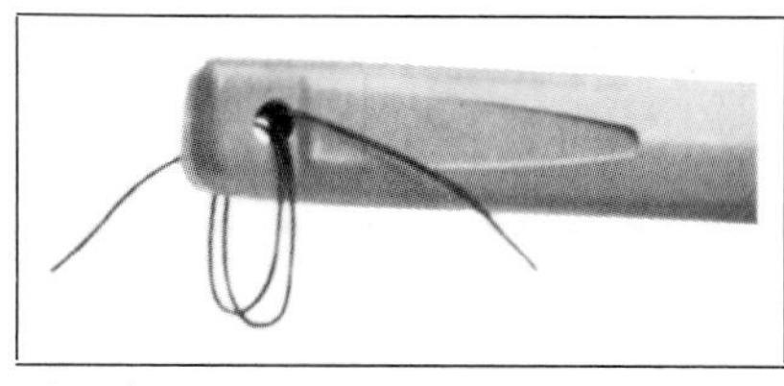

Fig. 4. *Measurement sensitivity can be increased by looping conductor through probe one or more times to increase effective magnetizing force of measured current. Probe accepts single conductors of up to $\frac{3}{16}$-inch overall diameter.*

ACCURACY

Rated accuracy for the instrument is $\pm 3\%$ of full scale ± 0.1 milliampere. This is an overall rating that includes the effect of operating the instrument from line voltages up to $\pm 10\%$ from a 115-volt center value, of meter movement tolerance, of temperature over normal room ranges, etc. The rating also applies when ac currents exist in the circuit being measured as long as their peak value does not exceed the full-scale value of the range used, and as long as they are not near 40 kc, as discussed earlier.

DEGAUSSING

The magnetic sensing elements in the probe are sufficiently free of remanence or residual induction effects that residual readings are not obtained in the normal course of measurements. If, however, large overloads of many amperes are applied such as from shorts in a measured circuit or rapid discharge of a capacitor, residual readings will result, and for this contingency a degausser is included in the instrument. The degausser is activated by a momentary-contact switch, which in use is manually held closed while the probe is withdrawn from the degausser aperture. On cabinet style instruments the degausser is located at the back of the cabinet and on rack instruments at the front.

GENERAL APPLICATIONS

A note is also in order concerning the general usefulness of the instrument since, although it was developed with low-impedance work principally in mind, it has turned out to be extremely useful in the general course of electronics work in higher-impedance circuits as well. It is useful for searching out shorts, checking tube electrode currents, checking power supply currents, and simply measuring currents for any number of purposes.

It has also been found very useful in production work where, with some advance programming of data for technicians, it becomes valuable for checking currents both in insuring proper operation of production equipment and in trouble-shooting.

—*Arndt Bergh, Charles O. Forge,* and *George S. Kan*

TENTATIVE SPECIFICATIONS
-hp-
MODEL 428A
CLIP-ON DC MILLIAMMETER

Current Range: Less than 0.3 milliampere to 1 ampere in 6 ranges with full-scale values of 3 ma, 10 ma, 30 ma, 100 ma, 300 ma, 1 ampere.

Accuracy: With line voltages of 115 volts $\pm 10\%$, overall accuracy is $\pm 3\%$ ± 0.1 ma. Errors include effects of earth's magnetic field, repeatability of probe closure, and noise and normal ambient room temperature ranges.

Inductance Introduced by Probe: Probe will cause less than 0.5 μh of inductance to be introduced into circuit being measured.

Effects of AC in Circuit being Measured: AC with a peak value less than 100% of full scale will have no effect on DC reading unless its frequency is within a few cycles of 40 kc and its harmonics.

Effects of Impedance of Circuit Being Measured: No errors will result if impedance of circuit being measured is 0.1 ohm or more at 40 kc.

Power Requirements: 115/230 volts, $\pm 10\%$, 100 watts.

Dimensions: Cabinet mount: 7½" wide, 11½" high, 14¼" deep; rack mount: 19" wide, 7" high, 12¼" deep.

Weight: Cabinet mount: 19 lbs. net, 24 lbs. shipping; rack mount: 24 lbs. net, 35 lbs. shipping.

Price: Model 428A cabinet mount: $475.00 f.o.b. Palo Alto, Calif. Model 428AR rack mount: $480.00 f.o.b. Palo Alto, Calif.

Data subject to change without notice

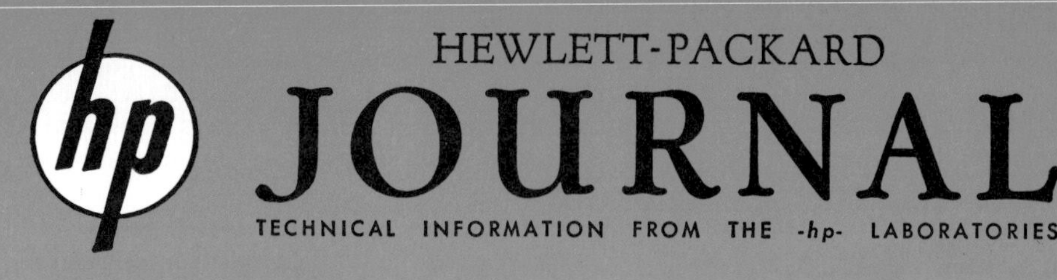

HEWLETT-PACKARD
JOURNAL
TECHNICAL INFORMATION FROM THE -hp- LABORATORIES
Vol. 11 No. 5-7

PUBLISHED BY THE HEWLETT-PACKARD COMPANY, 275 PAGE MILL ROAD, PALO ALTO, CALIFORNIA JAN.-MAR., 1960

A Versatile New DC - 500 MC Oscilloscope with High Sensitivity and Dual Channel Display

To -hp- Journal readers:

The accompanying article describes what we believe to be a fundamental breakthrough in the field of high frequency oscilloscopes. The instrument described in the article combines great bandwidth and high sensitivity with basic ease and simplicity of operation. It is in every sense of the word a general purpose instrument. From the user's standpoint it should be evaluated in these terms.

The techniques used to achieve these results are in a sense only incidental; nonetheless much of the article that follows is devoted to a detailed discussion of technique. This is consistent with the Journal policy of presenting a clear technical exposition of significant contributions of the Hewlett-Packard Research and Development groups.

Our experience with this oscilloscope in our own laboratories has been most promising. In one case, we discovered an important phenomenon which appears to hold great promise for use in the field of instrumentation. Without this high performance oscilloscope, the phenomenon would have gone unnoticed. Our own use for this instrument grows daily. We commend it to you for your consideration as a high-speed, high-sensitivity, simple-to-operate, general-purpose oscilloscope.

—Wm. R. Hewlett

THE illustration below shows a new oscilloscope which greatly extends for the engineer the frequency range over which he can obtain the information that only an oscilloscope can provide. This new oscilloscope is a general-purpose instrument with a frequency response extending up to 500 or more megacycles—i.e., to some four octaves higher than has previously been available in such an instrument. The increase in capabilities extends not only to frequency, however, but also to sensitivity since, if anything, the new oscilloscope has higher sensitivity over its entire range than previous "high frequency" type oscilloscopes have provided over frequency ranges that extended only into the lower megacycle region.

The value of the new oscilloscope as a measuring instrument for fast work can be judged from the fact that its 500-megacycle bandwidth gives it a rise time of 0.7 millimicrosecond. At the same time it has a maximum calibrated sensitivity of 10 millivolts/cm, which is increasable to about 3 millivolts/cm with an uncalibrated vernier. Moreover, these bandwidth and sensitivity characteristics are made available in each channel of a dual-channel plug-in type vertical amplifier. In the millimicrosecond region, the dual-channel provision is probably even more valuable than in the microsecond region because of the generally more difficult problem of time-relating phenomena on a millimicrosecond time scale. Fig. 2, for

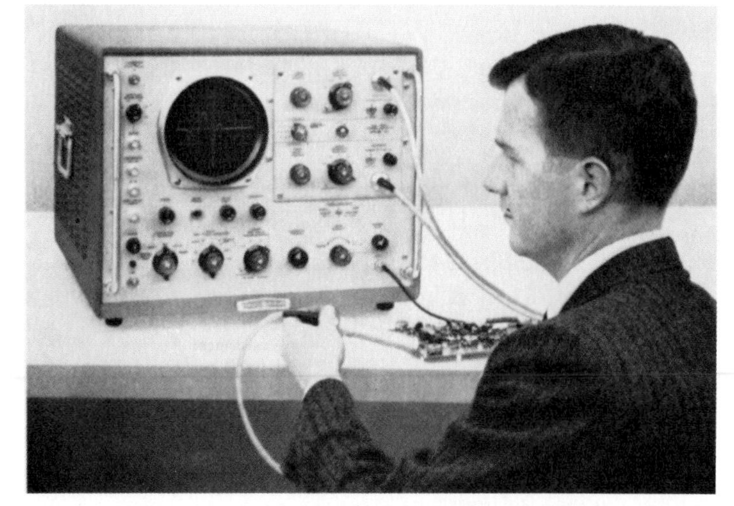

Fig. 1. *New -hp- Model 185A/187A Oscilloscope displays signals from 3 millivolts to 2 volts (x10 with adapter) over frequency range up to 500 megacycles. Besides very wide bandwidth and high sensitivity, instrument has several special advantages not found previously in oscilloscopes.*

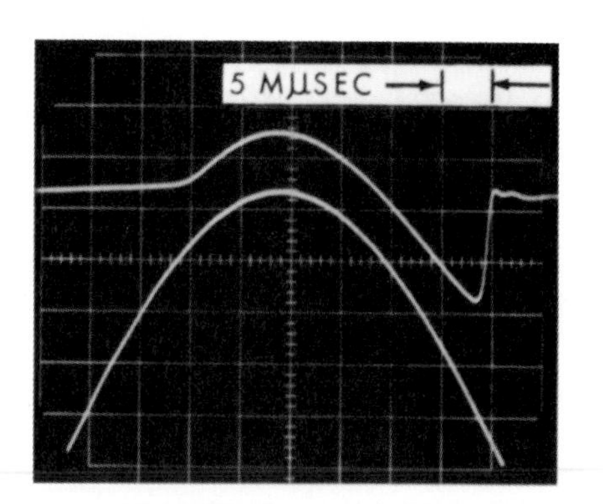

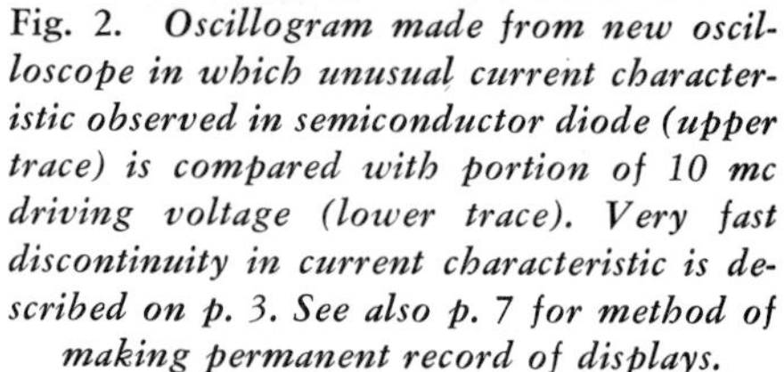

Fig. 2. *Oscillogram made from new oscilloscope in which unusual current characteristic observed in semiconductor diode (upper trace) is compared with portion of 10 mc driving voltage (lower trace). Very fast discontinuity in current characteristic is described on p. 3. See also p. 7 for method of making permanent record of displays.*

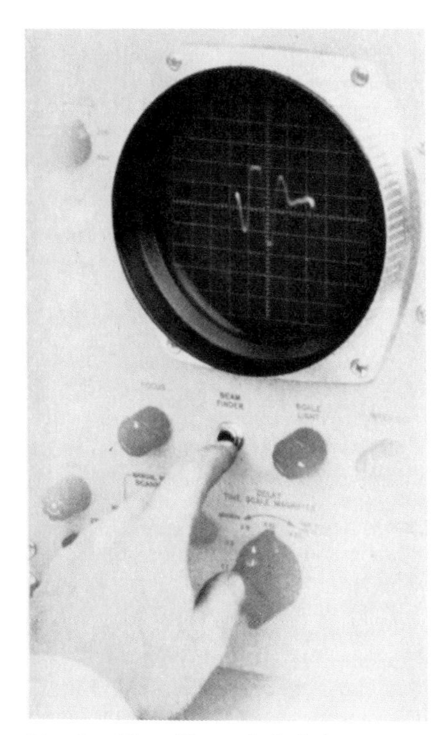

Fig. 3. *New "beam-finder" feature enables off-screen trace to be located immediately by pushing panel switch, thus overcoming long-time oscilloscope inconvenience. When switch is depressed, display indicates direction and amount beam is off-center, greatly simplifying adjustment of centering controls.*

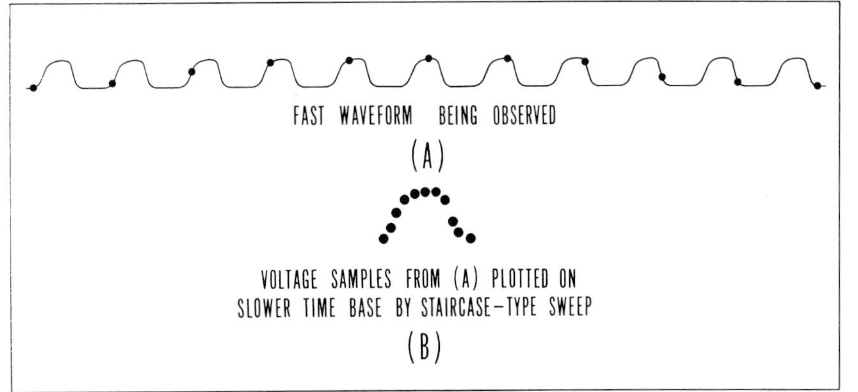

Fig. 4. *Drawing indicating technique new oscilloscope uses to "strobe" observed signals. Samples (indicated by circles in (A)) are taken of signal amplitude on successive signal cycles and replotted (B) on slower time-base by staircase sweep voltage.*

example, shows how the dual-channel display facilitates relating a fast phenomenon occurring in semiconductor diodes to a driving function.

Other characteristics that describe the degree to which the new oscilloscope has removed previous measurement limitations include the fact that it will trigger from signals with repetition rates higher than 50 megacycles. Calibrated sweeps are provided from 100 millimicroseconds/cm down to 0.1 millimicrosecond/cm, while a calibrated magnifier of up to x100 magnification is provided in combination with a delay control for delayed sweep applications. Both voltage and time calibrators are provided, the time calibrator providing frequencies of 500 and 50 megacycles for check purposes. The instrument has a large signal capacity, being capable of directly displaying signals up to 2 volts peak-to-peak, while accidental application of voltages up to 50 volts directly to the vertical inputs does not cause damage. An adapter increases both of these values by a factor of 10. In its display system, too, the instrument reverses previous trends in that it achieves about twice as much vertical deflection (10 cm) as that found in oscilloscopes an order of magnitude or more lower in frequency.

The instrument also incorporates a new arrangement for locating the crt beam when it is off-screen. In this arrangement a pushbutton on the panel limits the crt deflection voltages such that an off-screen trace is always brought on-screen and is done so in such a way as to greatly simplify adjusting the positioning controls.

SAMPLING TECHNIQUE

Two singular features of the new instrument are that it provides an output for operating an X-Y recorder and that the brightness of the trace is independent of the repetition rate of the observed signal, remaining as bright with low duty-cycle signals as with high. These characteristics, in addition to the primary advantages of very wide bandwidth and high sensitivity, have been achieved by designing the instrument to make use of a technique—the "sampling" technique — which is the electrical equivalent of the optical stroboscopic principle used in mechanical applications. As employed in the instrument, the sampling technique "observes" an external waveform by sampling the voltage amplitude of progressive points on successive cycles of the waveform while translating phase information concerning these points to a much slower time base equal to the number of waveform cycles needed to complete the sampling.

Fig. 4 demonstrates how the sampling technique is used to plot a waveform applied to the oscilloscope. During one recurrence of the waveform a sample is taken of the voltage amplitude at some point on the waveform. This sample is a very short pulse whose amplitude is proportional to the amplitude of the waveform at the moment of sampling. On ensuing recurrences, samples are taken at relatively later moments of time on the waveform.

These very short samples are then "stretched" in time into much longer samples which are amplified and applied to the crt as a Y-axis signal. The X-axis waveform, then, instead of being the customary sawtooth, is a staircase waveform which plots each of the Y-axis voltage samples as a discrete point, although the plotting density is high

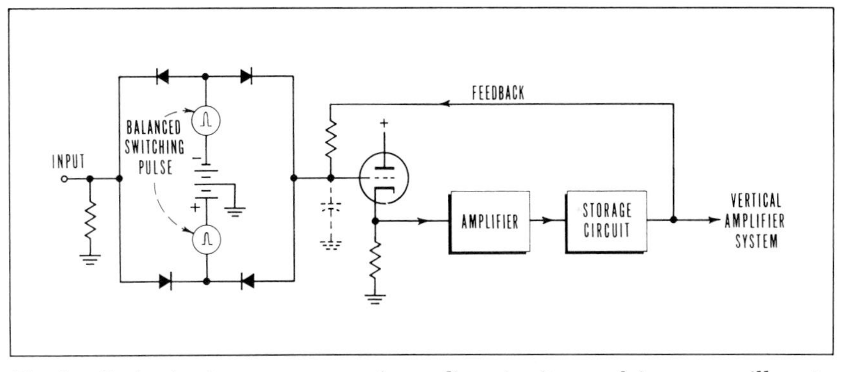

Fig. 5. *Basic circuit arrangement of sampling circuitry used in new oscilloscope. Servo-type circuit arrangement causes circuit to supply its own error signal if inaccurate sample should occur, thus making accuracy independent of sampling gate, switch pulse variations or gain variations.*

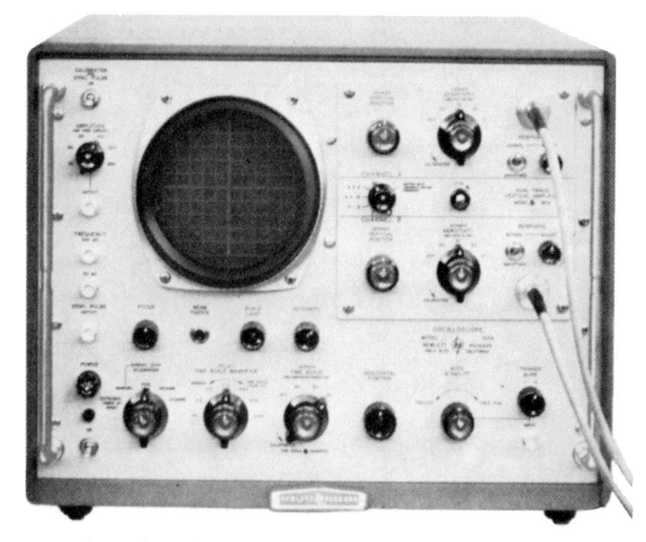

Fig. 6. *Panel view of -hp- Model 185A Oscilloscope with Model 187A Dual-Trace plug-in amplifier. Instrument uses new circuit technique, but controls are conventional and include presentation selector for separate or combined-channel displays.*

enough so that the overall plot appears continuous. Normally, 1,000 points are plotted per sweep (100 points per centimeter).

The sampling process, then, is a means of trading time for gain-bandwidth. It is one which requires fast circuits only for the sampling circuits, which are non-amplifying circuits located directly at the instrument input. The voltage samples themselves are translated to a longer time base for convenient amplification and display by conventional circuits. Since each sample is displayed on the crt until the next sample is taken, the brightness of the display becomes independent of the repetition rate or duty cycle of the waveform.

VOLTAGE ACCURACY INDEPENDENT OF SAMPLING

The particular design evolved for the sampling arrangement has a number of points of special interest both from an engineering and from a usage viewpoint. The sampling circuit itself (Fig. 5) consists of a very fast diode switch located between the input terminal and the input capacity of a cathode follower. Sampling is accomplished by momentarily closing the switch with a submillimicrosecond pulse from the fast pulse generator. This action stores a very short pulse of current proportional to the instantaneous signal amplitude in the cathode follower's input capacity, and the sample becomes the resulting change in voltage that occurs across the capacity during switching. The switch is composed of four

HIGH-SPEED EFFECT IN SOLID-STATE DIODES EXPLAINED WITH NEW OSCILLOSCOPE

One of the first applications of an early model of the new sampling oscilloscope described herein was to investigate a high-speed phenomenon observed in certain junction diodes. In working with diode harmonic generators, using the variation in capacitance of diodes with reverse bias as the mechanism of harmonic generation, A. F. Boff of the -*hp*- laboratories discovered that allowing certain diodes to conduct during a certain portion of the cycle increased the diode conversion efficiency at high harmonics by more than tenfold. Theories brought to bear did not explain the effect or why various diodes differed. Use of conventional oscilloscopes to observe the effect did not provide an answer.

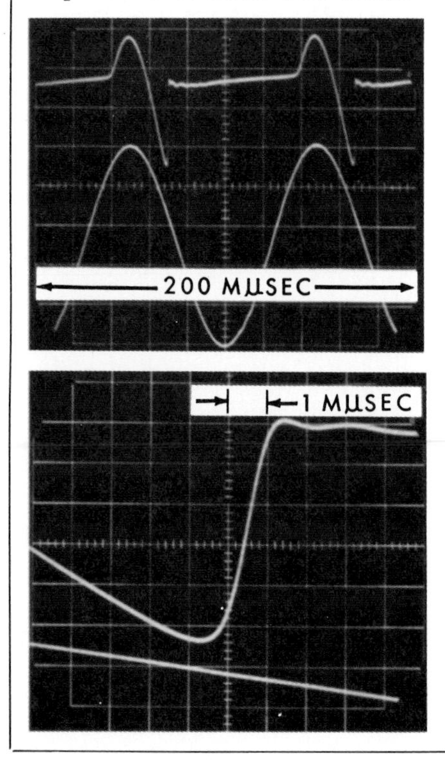

At about this time an early version of the sampling oscilloscope became available and was used to investigate the diode operating in the circuit shown. The oscilloscope revealed the diode current waveform to be as shown in the upper trace of the first oscillogram and provided the key to the mechanism. In the upper portion of the cycle the diode is conducting in the forward direction under the control of the applied voltage (lower trace). When the diode then becomes reverse-biased, the current reverses, being supplied by minority carriers stored during the forward portion of the cycle. The reverse current builds up to significant proportions, but when the supply of stored carriers is exhausted, the current drops very rapidly to zero. This mechanism would predict that the most effective diodes would have a long storage time and be of the graded junction type, and such has been found to be the case. The second oscillogram shows the phenomenon in greater detail using a sweep time of 1 millimicrosecond/cm. The rise time of the discontinuity is less than a millimicrosecond, showing the effect to be faster than the oscilloscope.

The effect was described in detail in the paper, "A New High-Speed Effect in Solid-State Diodes," given at the 1960 International Solid-State Circuits Conference by A. F. Boff of -*hp*-, Dr. John Moll of Stanford University and R. Shen of Harvard University.

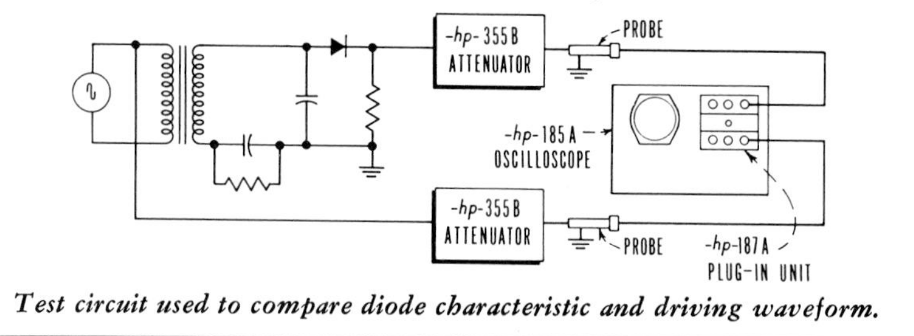

Test circuit used to compare diode characteristic and driving waveform.

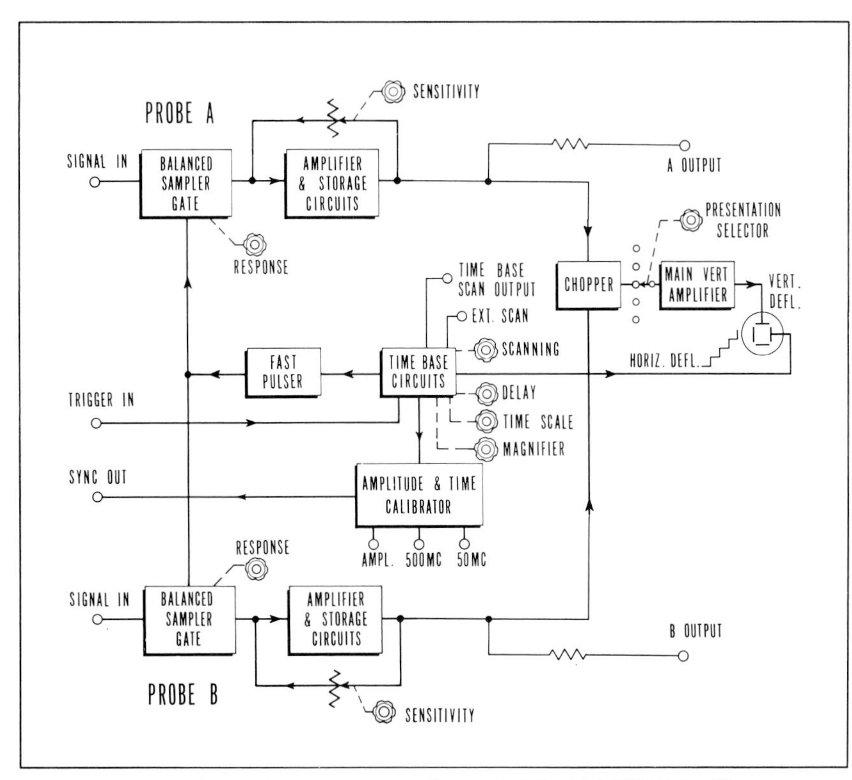

Fig. 7. *Basic circuit arrangement of -hp- Model 185A/187A Oscilloscope.*

matched diodes in a balanced arrangement whose bridge action reduces the effect of noise produced in the switch and gives the instrument a low noise level. The arrangement also minimizes the amount of switching pulse introduced into the measured circuit. Although this pulse would produce no error in the observed waveform, as a spurious pulse it would be undesirable in the measured system.

The sampling system has further been designed to give the instrument the very desirable property that its voltage accuracy is independent of the efficiency of the sampling process. More specifically, the voltage accuracy is independent of the characteristics of the sampling diodes, of the preamplifiers in the system, and, of particular importance, of the characteristics of the submillimicrosecond switching pulse, since in the present state of the art pulses of this speed do not have high stability with changes in temperature and time.

The independence from variations in the sampling switch characteristics has been obtained by enhancing the basic sampling circuitry with a feedback arrangement new to the sampling oscilloscope field. The net effect of this feedback can be determined from the basic circuit form indicated in Fig. 5. In this circuit the signal sample initially stored on the input capacity of the cathode follower is amplified and stored in a second storage circuit. The level in this second storage circuit is then compared with the signal itself by means of the feedback circuit and switch. If the stored level is not equal to the signal amplitude, an error signal will be applied to the system input during the moment of the next sample. The arrangement is such that the feedback signal and thus the signal in the second storage circuit will be brought into amplitude equality with the signal being measured through a nulling process. Because of this process, switching pulse, diode, and preamplification variations are removed as factors in the accuracy of the observation.

The feedback arrangement also gives a number of other advantages for the instrument including the wide dynamic range of the sampler. This occurs because only *changes* in the signal amplitude cause charge to be drawn through the sampling switch. The switch can thus sample large signals up to the value set by the reverse bias on the switch. This value is ± 2 volts, enabling large signals to be observed directly and giving the unit a dynamic range of about 1000:1, since the unit's noise level is 2 mv or less.

As further consequences of the fact that the feedback arrangement causes current to be drawn only on changes in the value of the observed signal, the sampler presents much less disturbance to the measured circuit than would a simple sampler and, further, a substantially smaller blocking capacitor than otherwise can be used at the sampler input when ac coupling is desired.

LOOP GAIN OPTIMIZATION

The foregoing indicates how the feedback in the sampling system corrects an inaccurate sample for the case where the sample-to-sample amplitude change in the measured signal is slight. This case is probably the general case, because the sampling density of 100 samples per centimeter is sufficient that the sample-to-sample change in the signal will normally be small for all but cases where a reduced sampling density may be used, as described later. Typical displays where the sample-to-sample change is small are those where the trace appears essentially continuous, as in Figs. 2, 8, and others.

For the case where a reduced sampling density may be used or for other possibilities where the signal is displayed by few samples such as the initial rise of the lower trace of Fig. 9, it is desirable that *each* sample be an entirely accurate sample, since otherwise the displayed rise of the signal will be erroneously lengthened.

To make the oscilloscope capable of making each sample an accurate sample regardless of scan density, the loop gain in the sampling system has been optimized to unity. For the loop consisting of the sampling and feedback circuits indicated in Fig. 5, it will be evident that unity loop gain causes each sample that the instrument makes to be an accurate sample, because any inefficiencies in the sampling portion of the loop are overcome by the amplification in the loop. Hence, optimum loop gain makes the display accuracy totally independent of sampling density. Further, the display will follow waveforms

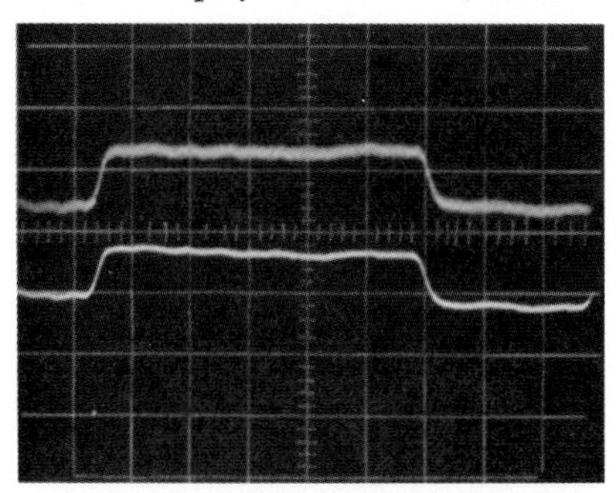

Fig. 8. *Double-exposure type oscillogram illustrating a 3 millivolt pulse with no smoothing (upper trace) and using smoothing feature (lower trace) to improve new oscilloscope's already small 2 millivolt rated noise value. External circuit random noise will be similarly improved by smoothing feature.*

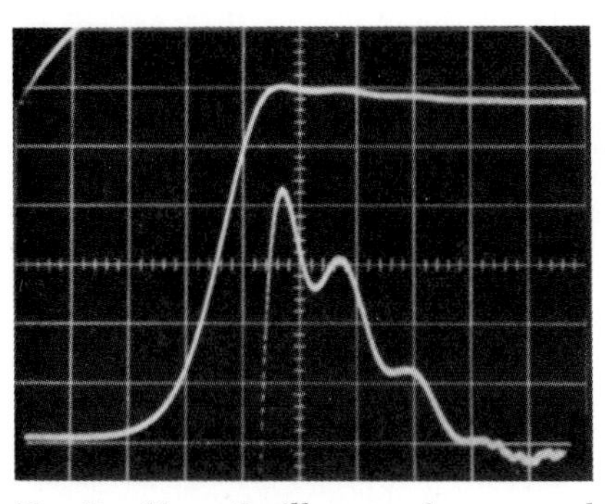

Fig. 9. *New Oscilloscope has unusual property of not overloading, as indicated by lower trace in this oscillogram which is same as upper trace except at 20 times as much sensitivity. See text for details.*

that are not uniform from cycle to cycle. Oscillation by the system is prevented by suitably synchronized gating in the loop.

EXTERNAL AND INTERNAL NOISE REDUCTION

Despite the fact that it is generally advantageous for the sampling system to have optimum loop gain, there are special situations where less than optimum loop gain can be employed to special advantage.

As described previously, a less than optimum value of loop gain will cause the sampling system to require more than one sample to build the displayed signal to proper amplitude. Now for cases in which the signal has *random* noise or jitter, reduced loop gain will act to reduce the apparent amount of such noise and jitter, since these can be considered to be signals that are not uniform from cycle to cycle. In other words reduced loop gain will actually cause the sampling system to discriminate against random noise and jitter in the display of the signal while having no effect on the signal itself, unless it is displayed by few samples. Similarly, the apparent noise level of the sampling circuit itself can be reduced by using less than optimum loop gain.

To permit a reduced loop gain to be used to advantage in measurements, a *Normal - Smoothed* switch is provided on the panel. In the *Normal* position the loop gain has its optimum value, while in the *Smoothed* position the loop gain is reduced such that random noise and jitter in the external signal will be reduced in the display by a factor of 3. The rated 2 millivolt noise level of the oscilloscope will also be reduced by this same factor to below 1 millivolt, and the rated 0.10 millimicrosecond jitter in the sampling will be reduced to below 0.05 millimicrosecond. The effect of smoothing on a sensitive display is shown in Fig. 8.

NON-OVERLOAD CHARACTERISTIC

An especially useful property of the sampling system used in the instrument is that it does not suffer the large-signal overload problem that is characteristic of conventional oscilloscopes. The reason for this is that the sampling system has the relatively long interval between samples to recover from overloads. As a result the oscilloscope can display small - amplitude phenomena existing along the top or bottom of large signals, as indicated in Fig. 9. Here, the upper trace shows a pulse displayed by one channel of the instrument using minimum sensitivity (200 mv/cm). The lower trace shows the detail at the top of the same pulse as displayed by the second channel using 20 times as much sensitivity (10 mv/cm). Ringing and overshoot on the large pulse can thus be examined in detail.

The non-overload property applies for signals up to the value that overcomes the reverse bias in the sampling system and thus disrupts the sampling process. This level is 2 volts.

SYNCHRONIZED DUAL CHANNELS

The dual channel capability in the new oscilloscope has been obtained by using two independent sampling channels which sample their respective inputs simultaneously. The input of the main vertical amplifier is then electronically switched between the two channels at a rate high enough to display the outputs of both channels on each sampling.

In fast-circuit work the dual presentation capability has proved very valuable, and considerable care has been taken to achieve close synchronization between the two channels. Time delay error between channels has been minimized by making both samplers operate at very nearly the same time. The pulses which close the diode samplers are generated in a fast pulse generator which is common to both channels, while physical layout and cables to the sampling probes have been made essentially identical. These measures have resulted in a time delay error between channels that is less than 0.10 mμsecond—i.e., the time required for light to travel approximately 1 inch. If desired, even this slight difference can be largely corrected by connecting both probes to the same fast signal and determining the proper correction factor from the resulting display.

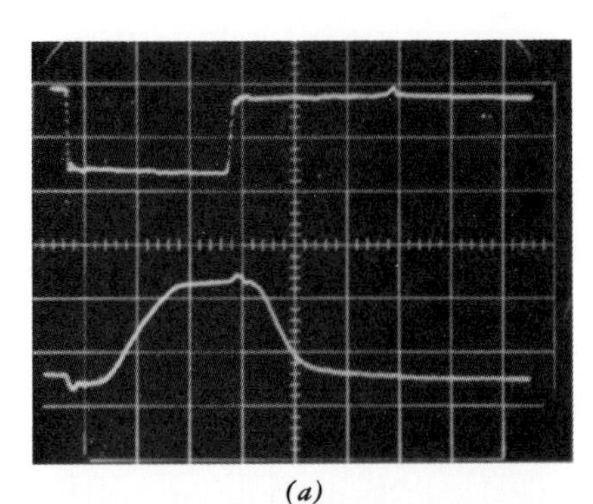

(a)

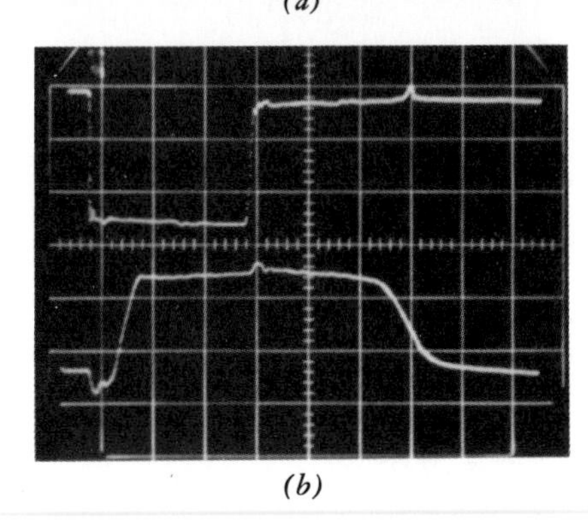

(b)

Fig. 10. *Oscillograms showing fast switching pulse (upper traces) with resulting mesa transistor output waveform for two degrees of saturation. Time scale is 5 millimicroseconds/cm.*

FAST-TRANSISTOR RESPONSE-TIME MEASUREMENTS

The close synchronization between channels coupled with the fast time scales of up to 0.1 mμsecond/cm enable fast-circuit time relationships and very small time differences in external signals to be measured. Fig. 10, for example, suggests how readily response-time measurements can be made on transistors. The waveforms shown are from a fast mesa transistor operating in the common-emitter test circuit of Fig. 11. Initially, the transistor is biased off and is turned on by a fast current pulse. The resulting transistor operation for two degrees of saturation is shown in Fig. 10.

Another measurement facilitated by the dual presentation feature is circuit delay. An interesting example of this type of measurement is demonstrated by the oscillograms of Fig. 12 which show the delay in the -*hp*- Model 355B DC - 500 megacycle coaxial attenuator at 0 and 20 db of attenuation. A test pulse derived from an avalanche transistor was passed through the attenuator while the input and output were monitored by the oscilloscope using a 1 mμsecond/cm sweep. Note that one signal in Fig. 12 (c) is 20 db below the other; i.e., delay measurements can be

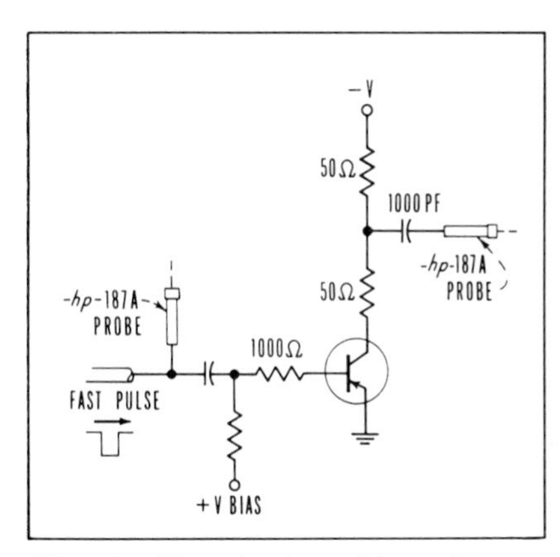

Fig. 11. *Test circuit used in making oscillograms in Fig. 10.*

made of two signals widely different in amplitude.

EXTERNAL TRIGGER ALTERNATIVES

In the design of conventional oscilloscopes it is customary to arrange for the signal being viewed to be delayed before being applied to the vertical deflection system. This delay permits the sweep circuits to become properly operative before displaying the signal. In the sampling oscilloscope, however, the "sweep" circuits serve a different-from-usual purpose in that they produce a pulse which is applied to the sampling switch directly at the vertical input terminal. Any inserted delay must therefore precede the sampling switch, but to insert delay at that point would give the oscilloscope the low impedance of a delay line as its input impedance rather than the high impedance that the balanced sampling switch has achieved. Consequently, other arrangements for achieving the required time-spacing between trigger and signal have been provided.

One arrangement is provided in the form of a special sync out pulse. This pulse is arranged to trail a trigger-in pulse by the 100 or so millimicroseconds needed for the sweep circuits to operate and produce the sampling pulse which samples the signal. This sync out pulse thus has the proper delay incorporated in it and can be used to trigger an external circuit under test. The sync out pulse has an amplitude of —1.5 volts and a rise time of 4 millimicroseconds. It is available both when the sweep circuit is free-running and when it is triggered. In the free-running mode the pulse has a 100 kc repetition rate, but in the triggered condition it occurs at the rate dictated by the external trigger. In each case this corresponds to the sampling rate of the oscilloscope.

A second alternative is to insert the necessary delay between an external trigger and the circuit it is triggering by means of a delay line. Such an external line is available as an accessory for the oscilloscope. It consists of an appropriate length of 50-ohm coaxial line housed in a convenient case, as indicated in Fig. 13. An adapter is also available (Fig. 18) to permit the trigger to be taken off ahead of the line to trigger the oscilloscope.

If the signal is in a coaxial circuit or if the signal must also serve as the trigger, the take-off adapter and delay line can also be used ahead of the signal input on the oscilloscope to delay the signal.

Other alternatives for achieving the needed lead-time in the trigger include selecting a trigger from an appropriately early stage in the external circuit. Excessive external trigger lead-time can be compensated for with instrument's delay control. When high-repetition rate signals of a megacycle or more are being studied, it is also often practical to use the oscilloscope's delay control to view the signal that follows the one that actually triggers the oscilloscope.

SCANNING OF LOW REPETITION-RATE SIGNALS

When a conventional oscilloscope is used to view a fast signal of low repetition rate, the trace becomes dim, frequently so much so that the presentation is no longer visible. The sampling oscilloscope, however, does not suffer this disadvantage. Its trace is as bright with a fast, low-rate signal as with any other, giving the instrument a tremendous advantage in fast-signal work. Low repetition rates do cause the beam to move more slowly than higher-rate signals because of the time needed to obtain the 1,000 samples that comprise the normal sweep, but this can be offset by reducing the number of samples taken per trace at some expense in resolution. The scan control on the oscilloscope is thus arranged to offer a choice of 50 or 200 (Coarse or Medium) samples per trace in addition to 1,000. This permits signals with repetition frequencies usually well below 50 pps to be observed. Lower rates can be observed by making a time-exposure photograph of the display or by making an external X-Y recording, as noted elsewhere herein. Fig. 14 demonstrates the change in resolution offered by the three sampling rates.

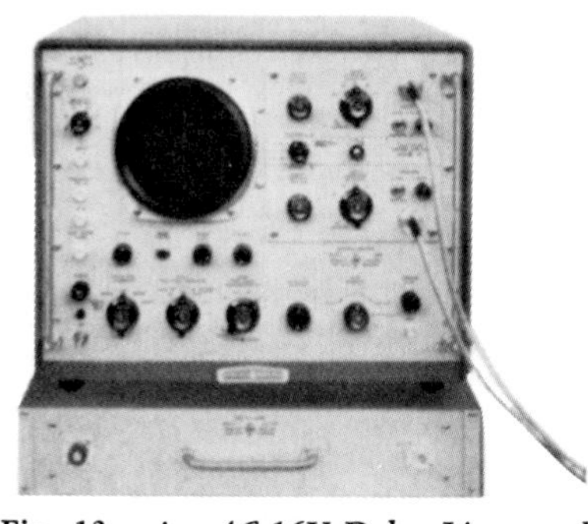

Fig. 13. -hp- *AC-16V Delay Line can be used to delay triggering of external circuit or to delay signal, is arranged to fit out of the way under oscilloscope.*

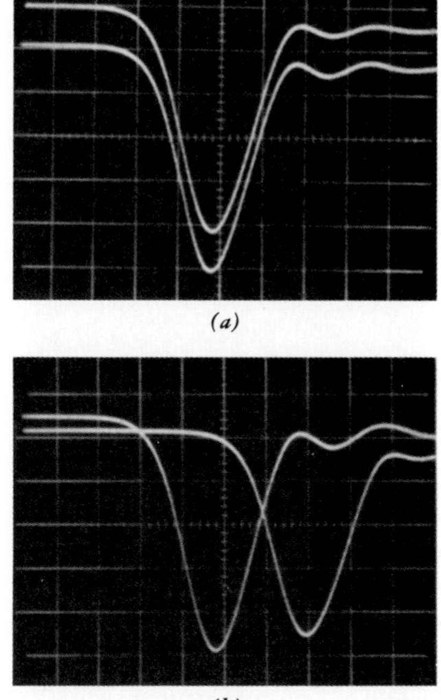

(a)

(b)

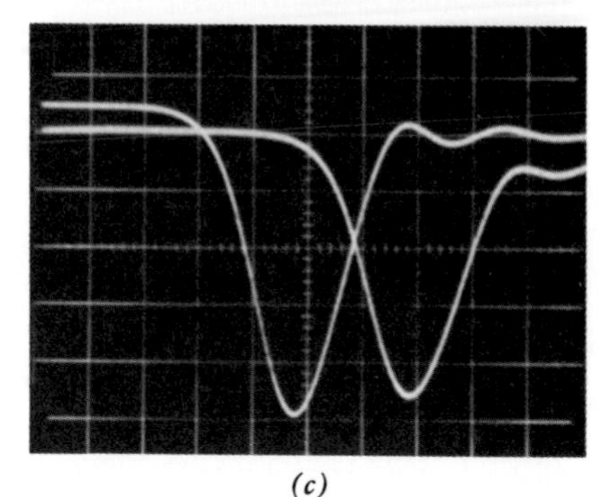

(c)

Fig. 12. *Oscillograms made from new oscilloscope demonstrating how readily time delay measurements can be made. Time scale is 1 mμsec/cm. (a) shows fast pulse at input to test device (-hp- 355B coaxial attenuator) as displayed by both oscilloscope channels; (b) compares pulse at attenuator input with same pulse at output for zero setting of attenuator; (c) same as (b) except attenuator set for 20 db and oscilloscope sensitivity increased correspondingly.*

TIME BASE AND MAGNIFICATION

Despite the unconventional nature of the sweep employed in the sampling

PERMANENT X-Y RECORDINGS OF DISPLAYED SIGNALS

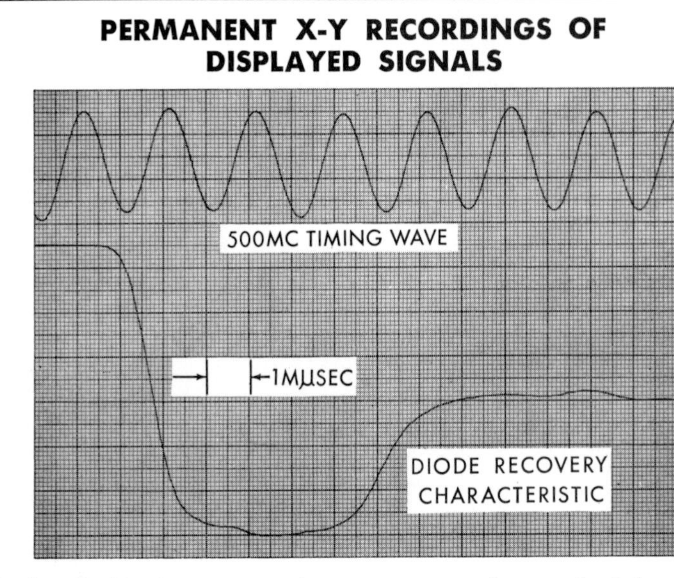

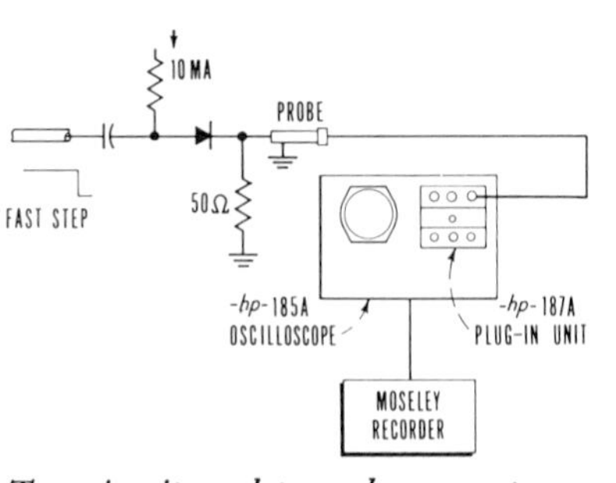

Test circuit used to make accompanying X-Y recording.

As described in the accompanying article, horizontal deflection in the new oscilloscope is accomplished by a staircase waveform whose steps are synchronized with the pulses that sample the signal being observed. In addition to this automatic scanning, however, provision is also made for scanning to be manually controlled by the operator. Under this arrangement, scanning occurs in proportion to the rotation of a panel control *(Scanning - Manual)*. Combined with this manual scanning provision, an output from each of the vertical signal channels in the form of dc voltages proportional to the vertical deflection of the trace is made available at terminals at the back of the instrument. A dc voltage proportional to the horizontal position of the trace is also available at a rear terminal.

The net result of this arrangement is that signal and horizontal deflection voltages in dc form are externally available and can be applied to an X-Y recorder to make permanent recordings of the signal or signals being displayed by the oscilloscope. Plotting is carried out by manually advancing the *Manual Scan* control. Even though the *Manual Scan* control is turned slowly enough to accommodate the recorder, the recording itself will have whatever time base that the controls on the oscilloscope are set to present. The curve reproduced here, for example, shows a very fast time function which was recorded using the outputs provided by the oscilloscope in this manner.

An additional terminal is provided at the back of the oscilloscope to permit scanning by the oscilloscope, and thus the external recording process as well, to be carried out under the control of an external dc voltage.

oscilloscope, the sweep controls are essentially identical to and provide a flexibility equal to that associated with well-equipped conventional oscilloscopes. The controls include a time base magnifier which provides up to x100 magnification. In this sampling oscilloscope magnification has the special characteristics that it causes no dimming of the trace and results in no loss of time base calibration accuracy. Neither does it cause a loss in resolution. The magnifier can be used with the basic time scale control to achieve sweep speeds of up to 0.1 millimicrosecond/cm or with the delay control in delayed-sweep applications. The delay control permits any point on the unmagnified sweep to be chosen for magnified display.

AMPLITUDE AND TIME-BASE CALIBRATORS

In order to provide for maximum convenience in checking the oscilloscope calibration, a two-frequency time-base calibrator is provided in addition to the customary amplitude calibrator. Two terminals are provided at the left side of the panel to make available a 500-megacycle and a 50-megacycle frequency for time-base comparison. These frequencies are accurate within 1% and are obtained by impulse-exciting two tuned circuits in the instrument.

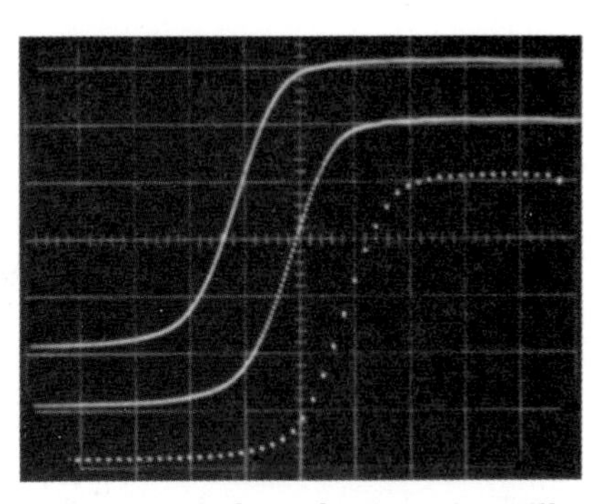

Fig. 14. *Sampling density of oscilloscope can be reduced to reduce scan time on low-repetition rate signals. Traces above compare resolution obtained with fine, medium and coarse scanning.*

The amplitude calibrator provides a square-wave calibrating waveform of amplitudes equal to the voltage ranges on the vertical channels at a voltage accuracy of within $\pm 3\%$.

PROBES AND ACCESSORIES

The physical arrangement of the signal probes provided on the vertical input channels is portrayed in Fig. 15. To make the probe adaptable to situations requiring minimum lead lengths such as in chassis wiring, the probe is equipped with a ground pin on a sliding type clamp. In cases where the clamp is not needed, such as when one of the adapters is being used, this clamp can be removed or placed at the back portion of the probe.

The adapters designed for use with the probe are shown in Fig. 17 and 18. Adapter (A) in Fig. 17 is the adapter shown being installed in Fig. 16 and changes the probe input to a type BNC male fitting. Adapter (B) changes the probe input to a type N female fitting. Adapter (C) is a divider that provides a 10:1 increase in the amplitude of signals that can be viewed and increases the oscilloscope input impedance to 1 megohm shunted by 3 mmf. Adapter (D) is a 1000-mmf 500-volt capacitor which is installed in place of the probe center pin to obtain ac coupling.

Adapter (E) is designed to permit

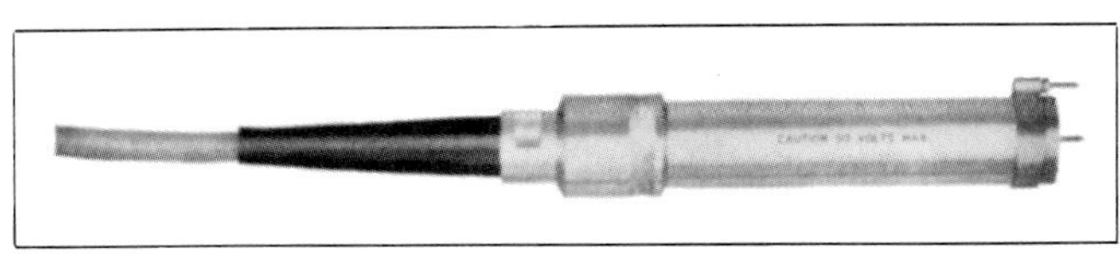

Fig. 15. *Physical form of signal probe. Each channel of plug-in unit is equipped with one such probe on five-foot cable.*

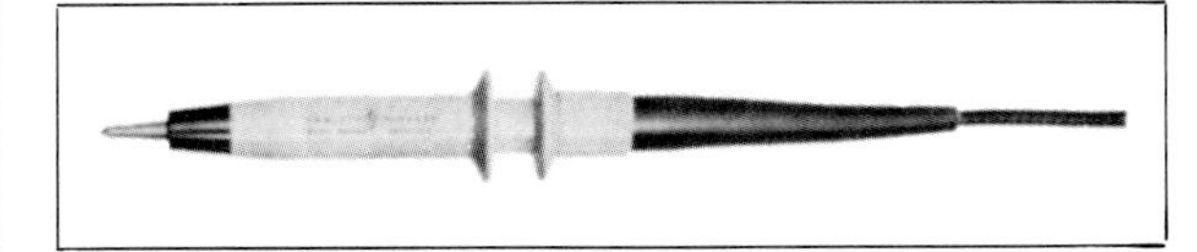

Fig. 19. *Sync probe provided on oscilloscope.*

the probe to monitor the voltage in a 50-ohm cable. The adapter consists of a 50-ohm T-section with type N connectors which can be connected into a 50-ohm line. The probe then connects to the special fitting at the side of the body of the adapter.

The final adapter (F) is the sync take-off adapter which is used in combination with the AC-16V Delay Line discussed earlier. The adapter gives a 6-db insertion loss for both signal and trigger outputs.

Fig. 19 shows the sync probe supplied with the instrument. The flanges on the probe body open the probe jaws in the manner that has proved popular on other *-hp-* oscilloscope probes.

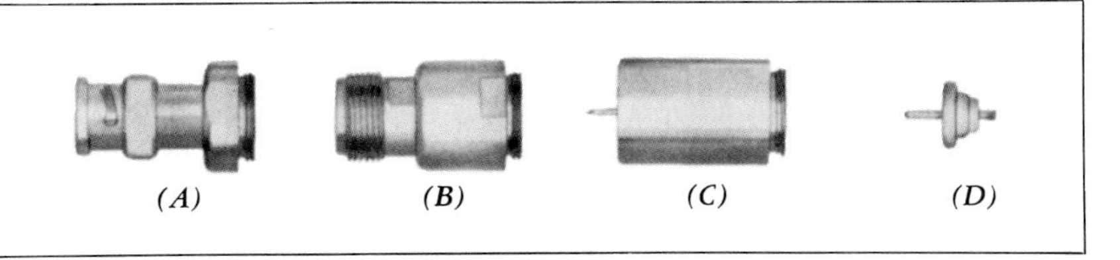

Fig. 17. *Adapters available for use with signal probe of Fig. 15.*

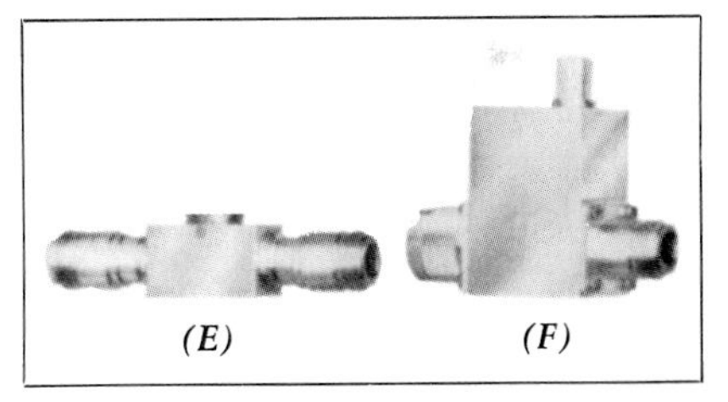

Fig. 18. *(right) Sync take-off adapter for use with AC-16V Delay Line of Fig. 13 permits trigger to be obtained for oscilloscope ahead of triggering of external circuit. (left) Adapter to permit oscilloscope signal probe to be connected across 50-ohm cable with type N fittings.*

OTHER PLUG-INS

Additional plug-in units for the new oscilloscope are in the planning stage. Correspondence concerning special measuring requirements is invited.

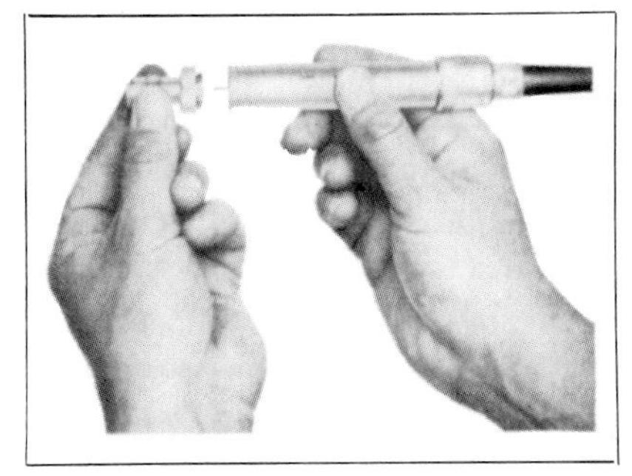

Fig. 16. *Adapter being installed to enable signal probe to be connected to BNC type fittings.*

ACKNOWLEDGMENT

The design and development of the 185A/187A Oscilloscope was a joint effort of several members of the Oscilloscope Division of the *-hp-* Research and Development Department. In addition to the undersigned, members of the group were Stewart Krakauer, Kay Magleby, Kenneth Miller, Richard Monnier, Victor Van Duzer, and Richard Woodbury. Valuable ideas and suggestions along the way were given by Bernard M. Oliver and Norman B. Schrock. The mechanical and production aspects of the instrument were ably handled by Charles Fitterer, Wallace Klingman, and Edna MacLean.

—Roderick Carlson

SPECIFICATIONS

-hp- MODEL 187A

DUAL TRACE AMPLIFIER

(When plugged into Model 185A Oscilloscope)

VERTICAL (Dual Channel)

Bandwidth: Greater than 500 mc at 3 db point. Less than 0.7 mμsec rise time.

Sensitivity: Calibrated ranges from 10 mv/cm to 200 mv/cm in a 1, 2, 5 sequence. Vernier control between steps increases sensitivity to 3 mv/cm.

Overshoot or Undershoot: Less than 5%.

Voltage Calibrator: 10 mv to 500 mv; accuracy ±3%.

Input: By means of input probe for each channel.

Noise: Less than 2 mv peak-to-peak; reduced by approximately 3:1 in smoothed (noise compensation) position of input switch.

Input Impedance: 100k shunted by 3μμf.

-hp- MODEL 185A OSCILLOSCOPE

HORIZONTAL

Sweep Speeds: 0.1 mμsec/cm to 100 mμsec/cm. Calibrated within ±5% using any combination of TIME SCALE and TIME SCALE MAGNIFIER settings with the exception of the first 50 mμsec of the 100 mμsec/cm TIME SCALE and first 20 mμsec of the 50 mμsec/cm TIME SCALE.

Time Scale: 4 ranges, 10, 20, 50, and 100 mμsec/cm. Vernier control between steps increases speed.

Time Scale Magnifier: x2, x5, x10, x20, x50, x100; may be used with any time scale setting.

Jitter: Less than 0.1 mμsec peak-to-peak; reduced by approximately 3:1 in smoothed (noise compensation) position of vertical input switch.

Sample Density: Fine (approx. 1000 samples/trace), medium (approx. 200 samples/trace), and coarse (approx. 50 samples/trace).

Manual Scan: Permits making X-Y pen-recordings.

Time Calibrator: 500 mc and 50 mc damped sine waves (frequency accuracy ±1%).

Minimum Delay: Less than 120 mμsec.

Variable Delay Range: Ten times the TIME SCALE setting less the display time. (Applies only when TIME SCALE is magnified.)

External Trigger: ±50 mv for 20 mμsec or longer, ±0.5 volt for 1 mμsec; approximately 120 mμsec in advance of signal to be observed.

"Sampling" Repetition Rate: 100 kc maximum.

Trigger Rate: 50 cps to at least 50 mc (holdoff circuit in operation above 100 kc).

Trigger Input Impedance: With Sync Probe, greater than 500 ohms; without probe, 50 ohms at panel. Capacitive coupling.

SYNC PULSE OUTPUT

Amplitude: Negative 1.5 volts into 50 ohms.

Rise Time: Approximately 4 mμsec.

Timing: Approximately 20 mμsec after the start of undelayed trace.

Repetition Rate: Approximately 100 kc, or the rate may be controlled by a fast-rise generator.

GENERAL

X-Y Recorder Output: Available in MANUAL SCAN for making pen-recording of waveforms: Horizontal Output: Zero at left to approximately 12 volts at right of CRT face; source impedance 2,000 ohms. Vertical Output: −1 volt at bottom to +1 volt at top of CRT face; source impedance 10,000 ohms.

Beam Finder: Facilitates location of beam that is off scale vertically or horizontally.

Cathode Ray Tube: 5 in. type 5AQP1.

Useful Deflection: 10 cm x 10 cm.

Power: 115/230 volts ±10%, 50 to 60 cps; approximately 250 watts.

Dimensions: Cabinet Mount: 14⅝ in. high, 19 in. wide, 22⅛ in. deep. Rack Mount: 12¼ in. high, 19 in. wide, 21 in. deep behind panel.

Weight: Net 75 lbs.

Accessories Furnished: 187A-76A BNC Adapter, 2 supplied. 185A-21A Sync Probe.

Accessories Available: 187A-76B Type N Adapter, 187A-76C 10:1 Divider, 187A-76D Blocking Capacitor, 187A-76E 50 ohm T Connector, 187A-76A Sync Take-off Adapter, AC-16V 120 mμsec Delay Line.

Price: -hp- Model 185A Oscilloscope, Cabinet Mount, $2,000.00. *-hp-* Model 187A Dual Trace Vertical Amplifier, $1,000.00.

Prices f. o. b. Palo Alto, California

Data subject to change without notice.

HEWLETT·PACKARD
JOURNAL

TECHNICAL INFORMATION FROM THE -hp- LABORATORIES
Vol. 15, No. 6
PUBLISHED BY THE HEWLETT-PACKARD COMPANY, 1501 PAGE MILL ROAD, PALO ALTO, CALIFORNIA FEBRUARY, 1964

Time Domain Reflectometry

THE MEASUREMENTS we make are limited by the tools we have at our disposal. Often measurements must be made by roundabout methods because no instruments exist with which to make them directly. Such difficulties are typical of pioneering work in any field. It is then the business of the instrument maker to devise tools to simplify frequent measuring needs.

Sometimes the cumbersome methods first required become so habitual, so ingrained, that the development of improved techniques is overlooked. This seems to be the case with respect to transmission and reflection measurement in the UHF and microwave range. Methods long in use here are now obsolete but are slow to give way to more modern direct methods.

The time-honored method of measuring reflections on a transmission line is to measure, as a function of frequency, the standing wave ratio (SWR) produced by the reflections. The resulting curves of magnitude and phase of the SWR can, in simple cases, be unscrambled to give the location and nature of the reflections, but the interpretation is difficult at best. With each alteration or attempted improvement, a completely new set of data must be taken and re-interpreted. The process is very time-consuming.

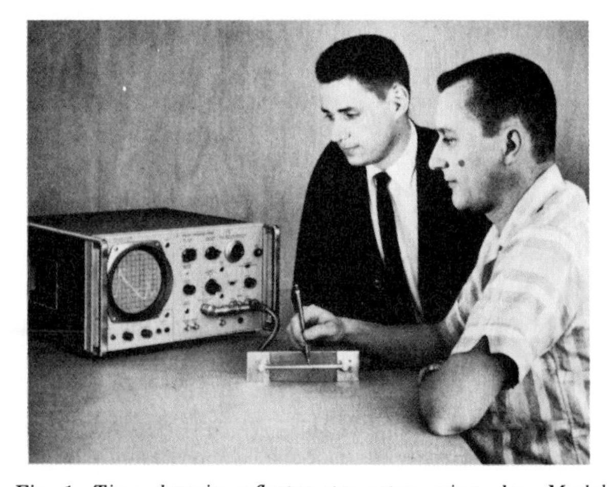

Fig. 1. Time domain reflectometer setup using –hp– Model 1415A plug-in unit with –hp– Model 140A Oscilloscope. Plug-in unit generates a fast step which is fed to external circuit. Energy reflected by discontinuities and impedance changes in external circuit returns to plug-in and is displayed on oscilloscope, permitting magnitude, sign and physical location of discontinuity to be read from scope face.

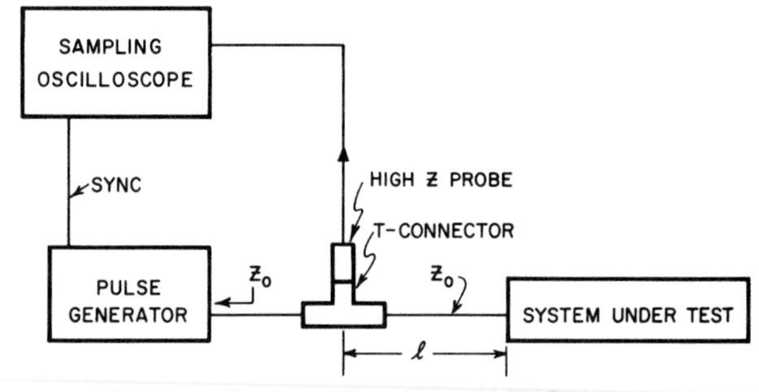

Fig. 2. *A Typical Time Domain Reflectometer.*

The swept-frequency reflectometer speeds the measurement considerably but does nothing to simplify the task of interpretation. Since no phase information is obtained, the results from even a single simple discontinuity are ambiguous.

The direct method of measuring reflections is to send out a pulse and listen for the echoes. This is what we do in radar, this is what the dolphin does, and the blind man who taps his cane. If the pulse is short enough each reflection produces a characteristic echo distinct from all others. The interpretation is extremely simple and the effect of changes can be seen instantly.

The pulse echo method has been used for many years for the location of faults in wide-band transmission systems such as coaxial cables. Here the time scale is such that microsecond pulses and megacycle bandwidths suffice. But in the laboratory setup, where reflections may be separated only an inch or less, nanosecond pulses and gigacycle bandwidths are needed. Thus pulse echo reflectometry as a laboratory tool has had to await the development of fast pulse generators and oscilloscopes. With these the era of time domain reflectometry has arrived and it is time for engineers to become familiar with this new measurement technique.

As shown in Figure 2, the time domain reflectometer consists merely of a fast rise-time pulse generator to

SEE ALSO:
TDR Plug-Ins, back page

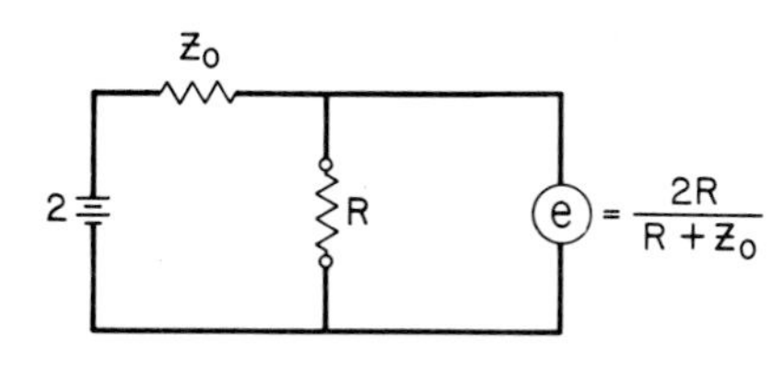

Fig. 3 *(at left). A simple ohmmeter using a two volt battery produces the same relation between terminal voltage and resistance R as a transmission line excited by a unit step.*

Fig. 5 *(at right). The reflections produced by simple resistive and reactive terminations.*

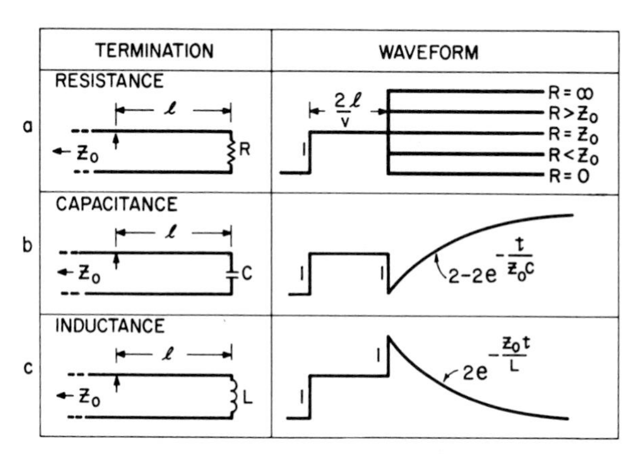

drive the system under test and a fast rise-time oscilloscope to display the reflections. No complicated directional couplers are needed to separate the incident and reflected waves; they are already separated by time, so a simple probe suffices to pick up the signals on the line. In order not to complicate the picture with re-reflections the probe should have a high impedance, and the generator should be a matched source. Reflections from the system under test are then viewed only once before being absorbed by the generator.

The time difference, t, between two successive echoes is 2s/v, where s is their separation along the line and v is the velocity of propagation. We will show later that if $t > \tau/2$, where τ is the system rise time, the echoes can be resolved. The minimum separation for accurate measurement is therefore $s = v\tau/4$. Since $v = 2$ to 3×10^{10} cm/sec and, at the present state of the art, $\tau < 10^{-10}$ seconds, we find s to be about 5 mm. Discontinuities separated by less than this distance produce seriously overlapping echoes and this complicates the interpretation. However, such close separation complicates the frequency domain measurement equally. To resolve such echoes, SWR measurements would have to be made over a frequency range of 3 to 10 Gc., i.e. over a band comparable to the spectrum of the pulse used in the time domain case.

Irregularities separated by more than the minimum distance produce distinct echoes. Each discontinuity then writes its own signature on the trace: a characteristic pulse shape that immediately tells what kind of discontinuity is present. Further, the position of the echo gives the location of the discontinuity. In exposed circuits one can touch the line to produce an added echo. Then, by running the point of contact along the line till this added echo coincides with the system echoes, one can literally put his finger on the troubles. In a coaxial cable one can produce a reflection by squeezing the cable.

Reflections occurring at intermediate points, unless quite large, may be ignored. Reflections from connectors, a great bugaboo of SWR measurements, are of no concern in time domain reflectometry. Nor do reflections, no matter how large, from later discontinuities affect the interpretation of a given echo. As a result, in cleaning up a system, one usually eliminates the first echo first, then the second, and so on to the end. Because the measurement is dynamic one can find the required cure for each echo very quickly experimentally. Often an entire system can be corrected in less time than would be required to make the first set of SWR measurements.

A modern sampling scope can accurately display signals of 100 μvolts or less in the presence of signals of 1 volt or more. It can therefore detect reflections that are down 80 db or more. This corresponds to $R \leq 10^{-4}$ and to SWR ≤ 1.0002, and is beyond the capability of the most elaborate frequency domain reflectometer or slotted line.

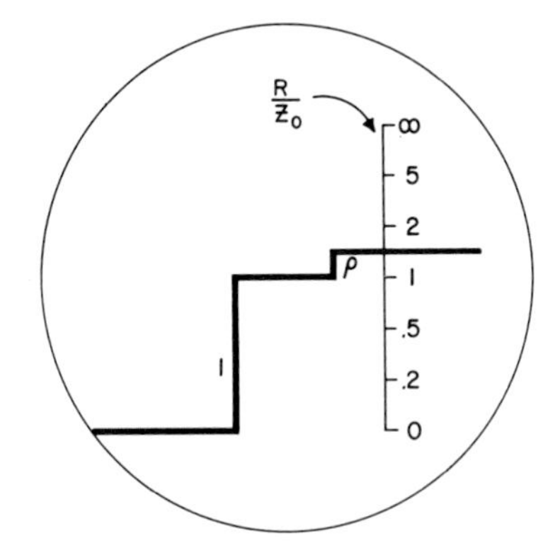

Fig. 4. *After the reflection from a resistive termination the trace has the height* $1 + \rho = 2R/(R + Z_0)$. *This is the scale law of the simple ohmmeter shown in Fig. 3. Such a scale on the scope face permits impedance to be read directly.*

ELEMENTARY ECHOES

The test signal may be either an impulse or a step function. If the system is linear the impulse response will be the derivative of the step response, so both waves contain essentially the same information. However, the step response is generally simpler in appearance, easier to interpret, and displays more clearly such things as gradual impedance variation with distance. Unless otherwise noted we will assume a step function test signal.

If the "system under test" in Figure 2 is simply a terminating impedance, Z(p), the reflection coefficient will be

$$\rho(p) = \frac{Z - Z_0}{Z + Z_0}.$$

The reflection of a step function will be a wave, f(t), whose spectrum $F(p) = \rho(p)/p$. At a time $2\ell/v$ after the incident unit step passed the probe the reflected wave returns. The total response h(t) for $t > 2\ell/v$ is the sum of the

FRONT PAGE PHOTO

Fig. 1 on the front page shows –hp– engineers Harley L. Halverson and George H. Blinn, Jr., using the –hp– reflectometer to determine the dimensions needed to achieve a desired impedance level in stripline. A stripline test section is constructed with the width of its center conductor varying linearly from one end to the other. By connecting this section to the reflectometer and then sliding a probe such as a pencil along the line, the point whose impedance is 50 ohms is determined from the reflectometer. The dimensions of the section at this point can then be measured. No measurements or assumptions about dielectric constant are required. The method was conceived by Wayne Grove of –hp– Associates.

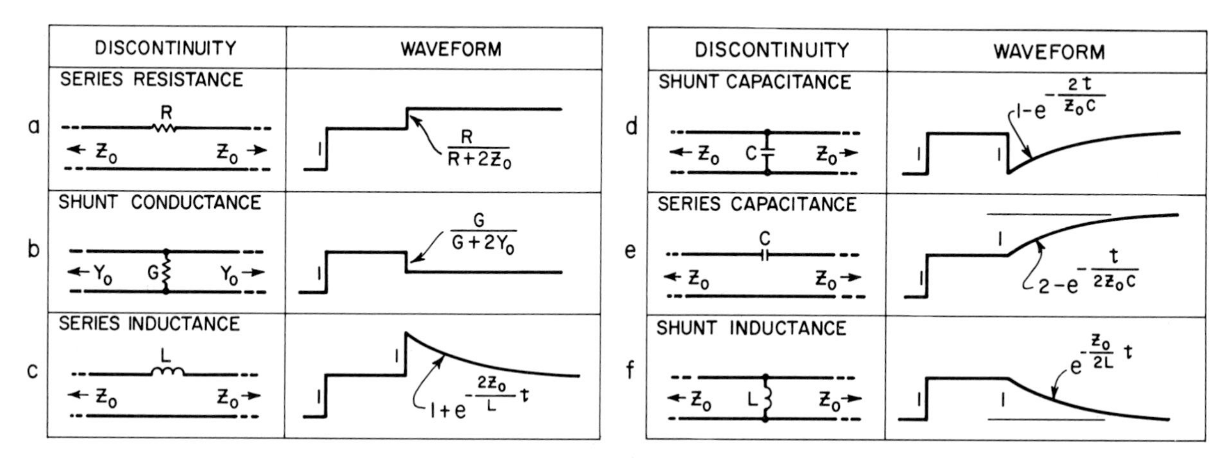

Fig. 6. *Typical reflections produced by simple discontinuities.*

incident unit step and the reflection, i.e.,

$$h(t) = 1 + f(t)\ ,\ t > 2\ell/v. \qquad (1)$$

Let us now analyze a few simple cases. We shall consider the coaxial cable to be lossless so that Z_o is constant and real.

If Z is a pure resistance then ρ is a constant independent of frequency. Thus f(t) is also a step function of magnitude $(R - Z_o)/(R + Z_o)$, and

$$h = 1 + \rho = \frac{2R}{R + Z_o} \qquad (2)$$

for $t > 0$. We may invert this relation to get $R/Z_o = h/(2 - h)$. (3)

Figure 3 shows a simple ohmmeter consisting of a 2-volt battery, an internal impedance Z_o, and a meter to read the terminal voltage, e, when an unknown, R, is connected. The expression for the terminal voltage (and therefore the ohmmeter scale law) is identical with (2). Thus we may use an ordinary ohmmeter scale normalized to Z_o at mid-scale, as shown in Figure 4, to read the terminating resistance. Alternatively we may compute the resistance from (3).

If the termination is a pure capacitance, $Z = \frac{1}{pC}$, $\rho = \frac{1 - pCZ_o}{1 + pCZ_o}$ and from $F(p) = \rho(p)/p$ we find

$$f(t) = 1 - 2\exp(-t/CZ_o).$$

Note that $|\rho| = 1$ at all frequencies but that its phase changes from 0 to π as p increases from zero to infinity. For high frequencies the capacitor looks like a short circuit, while for low frequencies it looks like an open circuit. Accordingly f(t) changes exponentially from -1 at $t = 0$ to $+1$ at $t = \infty$.

For an inductive termination,

$$Z = pL,\ \rho = \frac{pL - Z_o}{pL + Z_o},$$

and $f(t) = 2\exp(-Z_o t/L) - 1$. The reflection is the negative of that for a capacitance, changing from $+1$ at $t = 0$ to -1 at $t = \infty$. This is to be expected since an inductance looks like an open circuit for high frequencies and a short circuit for low frequencies. Again $|\rho| = 1$ at all frequencies.

The above cases for R, C, and L terminations are shown in Figure 5. The equations written by the curves are for $h(t) = 1 + f(t)$.

A simple discontinuity in a line adds an impedance, Z_d, if in series, or an admittance, Y_d, if in shunt. Thus the line carrying the incident wave faces an impedance $Z = Z_d + Z_o$ or an admittance $Y = Y_d + Y_o$. The reflection coefficient is therefore

$$\frac{Z_d}{Z_d + 2Z_o} \text{ or } -\frac{Y_d}{Y_d + 2Y_o}.$$

Figure 6 shows the various cases of pure discontinuities, and the corresponding waveforms produced. In all cases one may draw the waveform immediately by determining whether the reflection coefficient at infinite frequency and at zero frequency is $+1$, 0, or -1. The result gives the value of f(t) at zero and infinite time respectively. The transition is an exponential whose time constant is determined by the value of the reactive element and the impedance it faces.

The first two cases of Figure 6 can occur because of poor contacts, bad insulation, or circuit loading. When $R << Z_o$ or $G << Y_o$, the reflection has the magnitude $R/2Z_o$ or $G/2Y_o$. If we take $\rho = 10^{-4}$ as a reasonable lower limit and $Z_o = 50$ ohms we find we can detect a series resistance of one hundredth of an ohm, or a shunt resistor of one megohm!

The middle two cases of Figure 6 frequently occur as a result of cable defects, connector tolerances, or dimensional errors in design. Often the added reactance or susceptance is so small that the time constant of the exponential is much less than the pulse rise time. This important case will be discussed in the next section.

The last two cases in Figure 6 generally occur only from accidental open or short circuits, in which case the time constant will be short, or from design elements in a high pass structure, in which case the time constant may be quite long.

EFFECTS OF FINITE RISE TIME

The waveforms derived in the previous section are ideal in that they assume a zero rise-time step. The actual observed waveform is the same as would be obtained by passing the ideal waveform through a filter whose step response is that of the generator-oscilloscope combination. The impulse

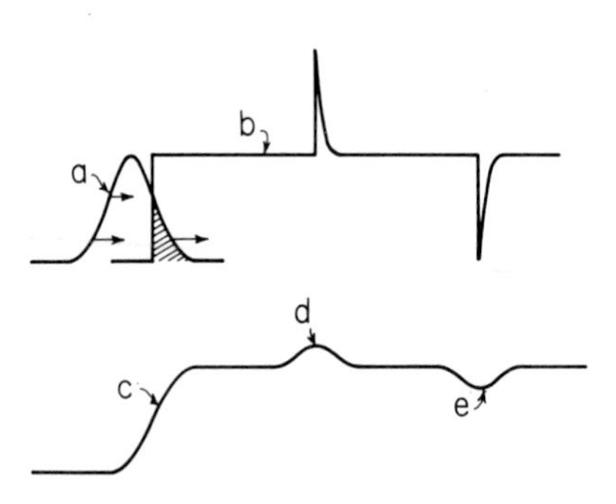

Fig. 7. *As the reversed system impulse response (a) scans the ideal curve (b) the integral of the product (shaded area) produces the actual response shown in the lower curve.*

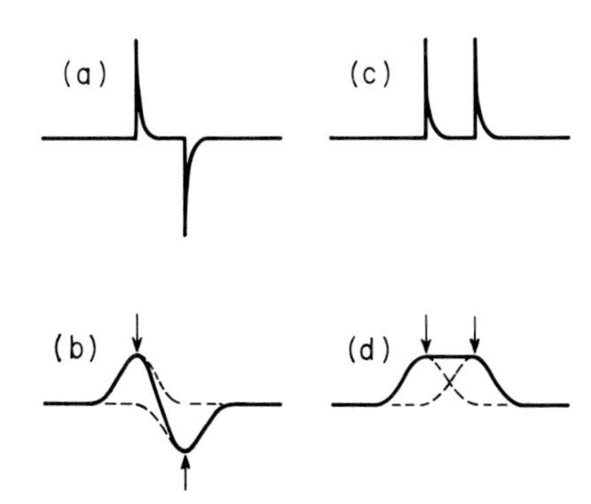

Fig. 8 *(at left). Pairs of closely spaced reflections as they would be seen with a zero rise time system, (a) and (c), and (b) and (d) as seen on a system having the impulse response shown by the dotted curves. Height measurements indicated by the arrows are still accurate, but further overlap would cause error.*

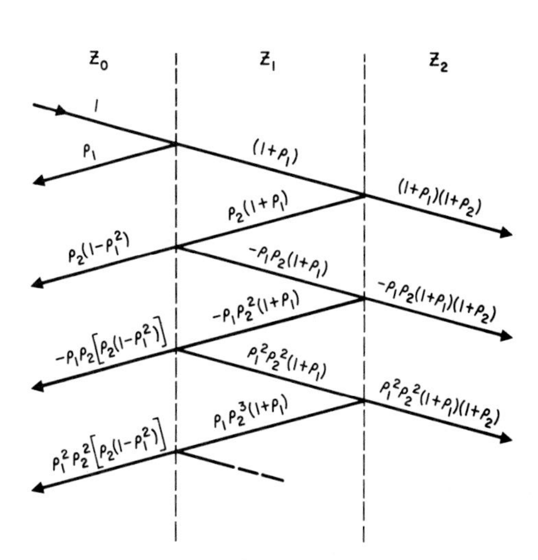

Fig. 9 *(at right). The successive reflections and transmissions produced by a pair of discontinuities in impedance. The first, from Z_0 to Z_1 produces a reflection coefficient $\rho_1 = (Z_1 - Z_0)/(Z_1 + Z_0)$ while the second from Z_1 to Z_2 produces $\rho_2 = (Z_2 - Z_1)/(Z_2 + Z_1)$. The sum of all the reflections is*

$$\rho = \frac{\rho_1 \rho_2}{1 + \rho_1\rho_2} = \frac{Z_2 - Z_0}{Z_2 + Z_0}.$$

The sum of all the transmissions is $1 + \rho$.

response of this equivalent filter is therefore the derivative of the observed step response.

As is well known, the output of a linear filter is the convolution of the input with the filter impulse response[1,2] (that is, the integral of the product of the input wave and the impulse response reversed in time, as the two functions scan past each other). Each step in the input integrates the impulse response to produce the step response; each input impulse produces an output impulse response. Figure 7 shows a typical impulse response (a) reversed in time and scanning an ideal step and pair of reflections (b). The step integrates the impulse response to produce the actual observed step response (c). The two reflections scan the impulse response to produce a bump (d) and a dip (e) each of which is a replica of the impulse response and has the same area as the reflection producing it.

The reflections in Figure 7 are typical of a small series inductance and shunt capacitance and have the areas $L/2Z_0$ and $CZ_0/2$ respectively. Rather than trying to estimate the area of (d) or (e) in order to find L or C, it is easier to estimate the peak amplitude, a, of the observed reflection and the maximum slope, m, of the step response. Since the latter is also the peak value of the response to a unit area impulse, it follows that *a* is *m* times the area, $L/2Z_0$ or $CZ_0/2$, of the actual impulse. Therefore,

$$L = 2Z_0\frac{a}{m} \qquad (4)$$

$$C = \frac{2\ a}{Z_0 m} \qquad (5)$$

[1] B. M. Oliver, "Square Wave and Pulse Testing of Linear Systems," Hewlett-Packard Journal, Vol. 7, No. 3, November, 1955.

[2] "Table of Important Transforms," Hewlett-Packard Journal, Vol. 5, No. 3-4, Nov-Dec., 1953. (Also reprinted in expanded form.)

If $Z_0 = 50\Omega$, $m = 10^{10}$/sec, and the smallest observable value of a is 10^{-4}, we find the minimum detectable $L = 10^{-12}$ henry, and $C = 4 \times 10^{-16}$ farad!

Reflections closer together than the width of the impulse response (the rise time of the step response) will overlap. Figure 8 shows two reflections (a) of opposite sign and (c) of the same sign. The corresponding responses are shown in (b) and (d). Note that the peak of the impulse response from one reflection occurs just before or after the response from the other. Thus height measurements taken at the arrows will give correct measures of L or C. Closer separation will not permit accurate measurement. We can take the practical limit of resolution to be one half the width of the impulse response, i.e., a little more than half the nominal rise time. Obviously any period of overshoot or ringing will further lengthen the minimum separation required for accurate measurement.

MULTIPLE REFLECTIONS

Whenever more than one discontinuity is present multiple reflections occur, which, if ignored, can cause errors of interpretation. To see at what point these become serious let us consider the case of three transmission lines of different impedances, Z_0, Z_1 and Z_2, connected in tandem. Two discontinuities having reflection coefficients $\rho_1 = (Z_1 - Z_0)/(Z_1 + Z_0)$ and $\rho_2 = (Z_2 - Z_1)/(Z_2 + Z_1)$ will be present and the series of reflections these cause is shown as the set of arrows emerging to the left in Figure 9.

Ignoring multiple reflections we would expect only two reflections of amplitudes ρ_1 and ρ_2. The first actually occurs but the second is diminished by the factor $(1 - \rho_1^2)$. There follows a geometric series of reflections each weaker than the last by the factor $-\rho_1\rho_2$. From this we can conclude that, since each disturbing multiple reflection involves at least two more reflections than the primary echoes, we can neglect multiple reflections to the extent that $\rho_i\rho_j << 1$ for all *i, j*.

The successive reflections (and transmissions) in Figure 9 are of course delayed by multiples of τ, the propagation time of the middle line. Before adding the coefficients in the frequency domain, each should be multiplied by $e^{-n\tau p}$ where n is the number of crossings of the middle line that has occurred. In the time domain each coefficient produces a distinct impulse or added step. Figure 10 shows the manner in which the successive reflections accumulate for several impedance combinations. The convergence to the final value is in an alternating geometric series if ρ_1 and ρ_2 have the same sign, and in a monotone series if they are of opposite sign.

From curves (a) and (f) we notice that, even with a matched generator, reflections can more than double or cancel the voltage or current on a line. In the cases shown the overshoot is 25% and is the greatest amount obtainable using a different impedance uniform line as a termination.

IMPEDANCE PROFILES

If we read the height of the traces of Figure 10 on the "ohmmeter" scale of

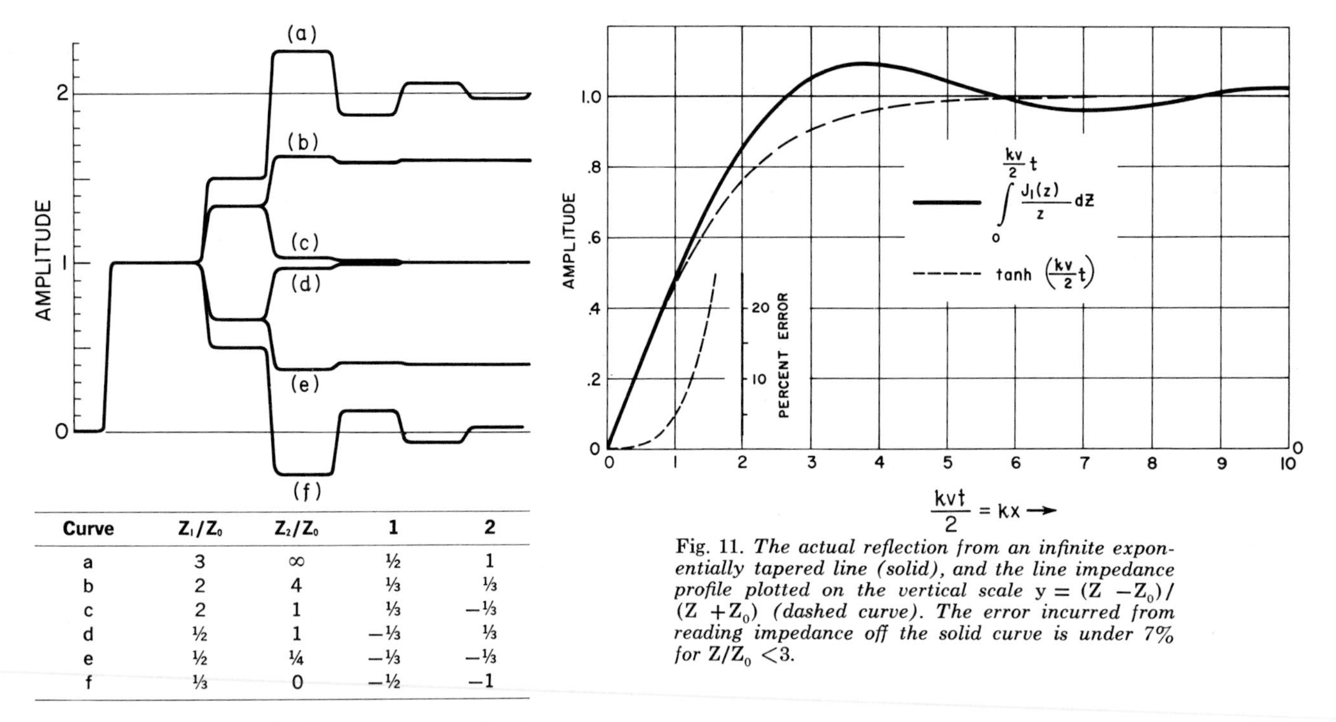

Curve	Z_1/Z_0	Z_2/Z_0	1	2
a	3	∞	½	1
b	2	4	⅓	⅓
c	2	1	⅓	−⅓
d	½	1	−⅓	⅓
e	½	¼	−⅓	−⅓
f	⅓	0	−½	−1

Fig. 10. *Incident wave and reflection pattern produced by multiple reflections shown in Figure 9.*

Fig. 11. *The actual reflection from an infinite exponentially tapered line (solid), and the line impedance profile plotted on the vertical scale* $y = (Z - Z_0)/(Z + Z_0)$ *(dashed curve). The error incurred from reading impedance off the solid curve is under 7% for* $Z/Z_0 < 3$.

Figure 4, then before the first reflection all traces coincide at the value Z_0. After the first reflection all traces have the value Z_1, and all assume the final value Z_2. If we take distance along the x-axis to represent distance along the line, the scale factor being the ratio of sweep speed to v/2, then cases (b) through (e) may be interpreted as a reasonably accurate profile of impedance along the line. Under certain conditions this interpretation is quite accurate and useful.

In order for the impedance profile to be accurate, multiple reflections must be small. In the foregoing example the greatest error occurs immediately after the second reflection.

Using the ohmmeter scale to read traces (c) and (e) immediately after the second reflection yields the result $Z_2 = 4.4\ Z_0$ and $Z_0/4.4$ instead of $4\ Z_0$ and $Z_0/4$: an error of only 10%. For smaller total impedance change the error is less.

For the impedance profile to be valid, all enduring reflections must be produced by impedance changes in a physical structure. Discontinuities such as series resistance and capacitance, or shunt conductance and inductance produce permanent reflections that do not represent changes in the characteristic impedance of the line. Small shunt capacitive or series inductive discontinuities on the other hand produce short spike reflections before and after which the impedance profile may still be valid.

Often a time domain reflectometer is used to inspect cable quality or the reflections from some other supposedly smooth line. Here the reflections, though possibly numerous, are normally all small and the impedance profile concept may be used with impunity. For small variations about Z_0 we find from (3) that

$$\Delta Z = 2aZ_0 \qquad (6)$$

where a is the amplitude of the departure from the incident unit step height.

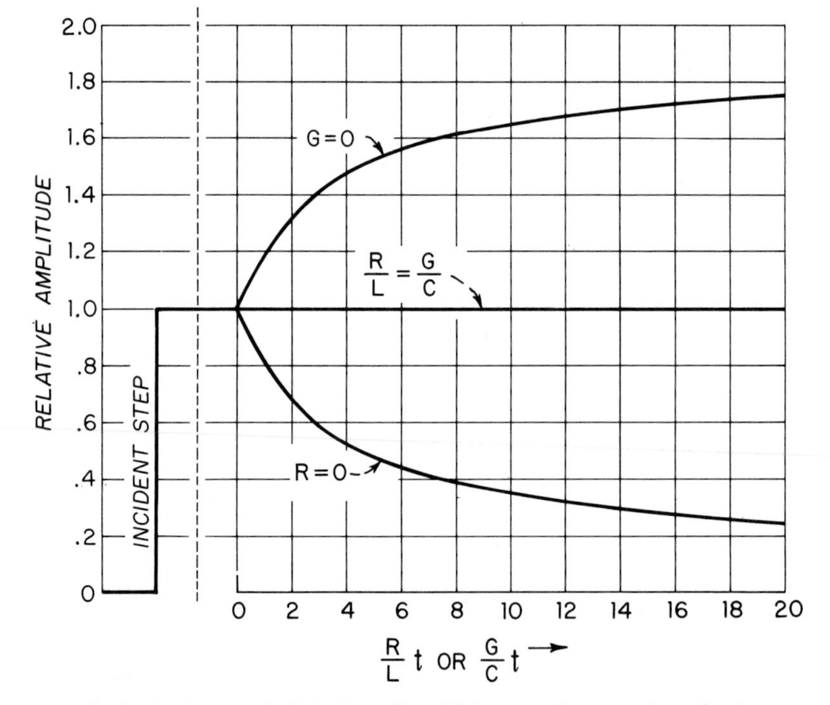

Fig. 12. *Reflection from an infinite lossy line. With no conductance the reflection eventually corresponds to an open circuit; with no resistance, a short circuit. The distortionless line with* $R/L = G/C$ *produces no reflection.*

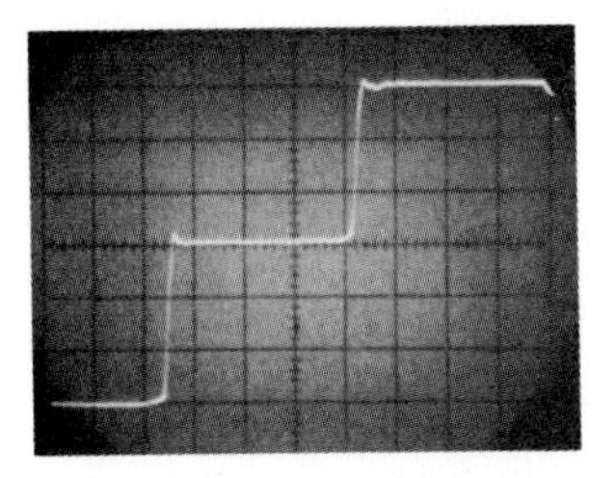

Fig. 13. *Three feet of RG-9/U open circuited. Sweep speed 2.5 ns/division. The round trip delay is 3.8 divisions or 9.5 nanoseconds. The propagation velocity is thus 193 meters/microsecond or 64% of the velocity of light. The dielectric constant is therefore $(1/.64)^2$ or 2.4.*

Fig. 14. *Double exposure showing the incident step (below) and the reflection from an open circuit at the end of 15 feet of RG-9/U. The slower rise and rounding of the top caused by cable loss are clearly evident. 2.5 ns/cm sweep speed.*

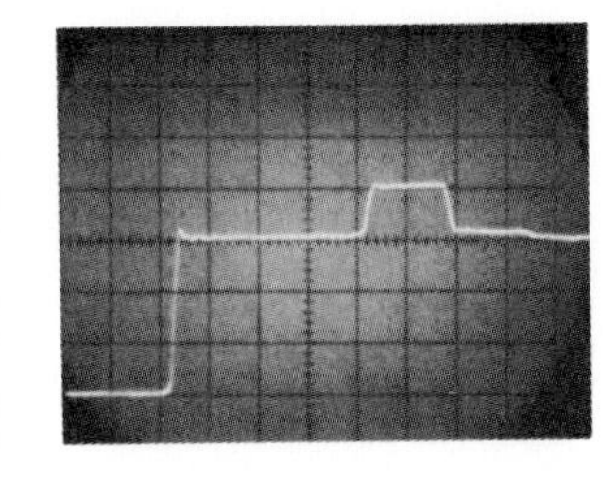

Fig. 15. *Fifteen inches of 100-ohm cable spliced into a 50-ohm cable to give the case depicted in Fig. 10c. The slight multiple reflection can be seen following the main bump. Aside from this, the trace is an accurate impedance profile of the line.*

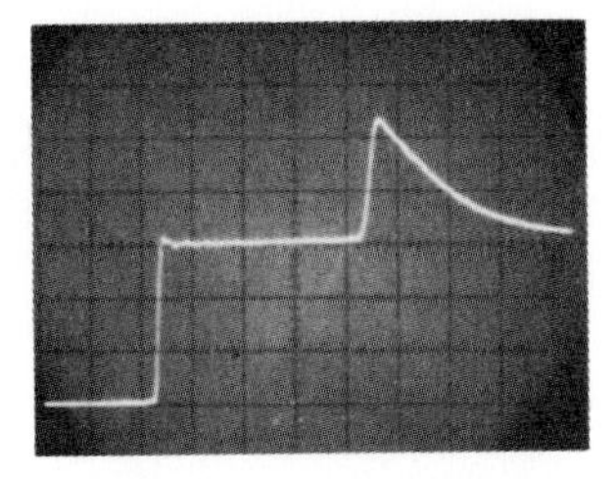

Fig. 16. *Reflection from a lumped inductance of 330 nanohenries in series with a 50-ohm line. Sweep speed is 2.5 ns/cm. The time constant is $L/2Z_0$ or 3.3 nanoseconds. The finite rise time of the step prevents the voltage from quite doubling.*

TAPERED SECTIONS

When large impedance changes are necessary, and narrow band operation is permissible, quarter-wave matching sections are often used. (Figure 10 (b) and (e) show the profile that would be obtained for a matching section between impedances having a 4 to 1 ratio.) When wide-band operation is desired a section of line having a tapered impedance is commonly used. Such sections can be designed to produce very little reflection above some minimum frequency and thus be good high-pass impedance transformers.

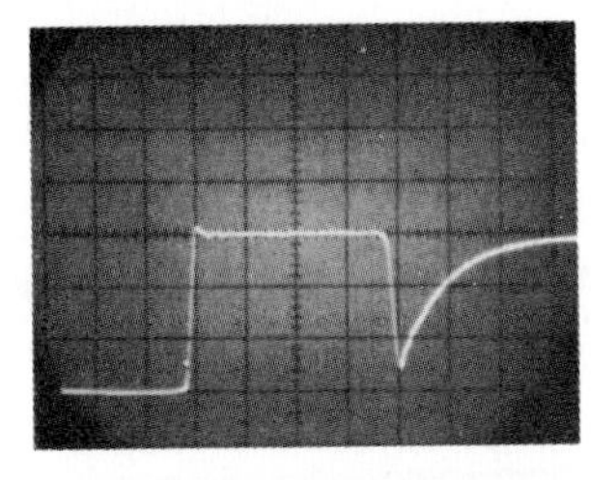

Fig. 17. *Reflection from 100 picofarads shunting a 50-ohm line. Sweep speed 2.5 ns/cm. Time constant is $CZ_0/2$ or 2.5 nanoseconds. The finite rise time prevents the voltage from dropping quite to zero.*

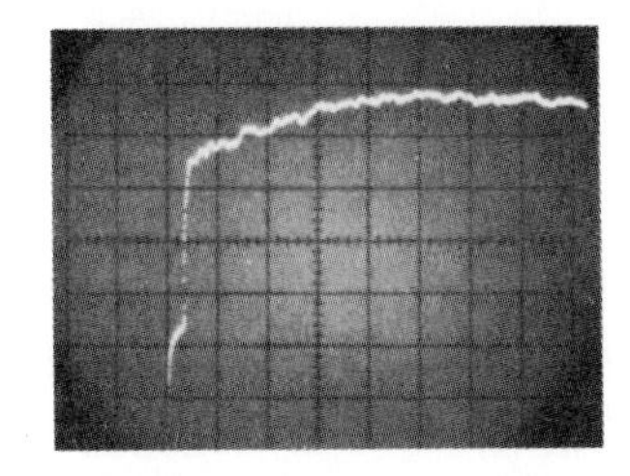

Fig. 18. *A long section of lossy 60-ohm line. The jump from 50 ohms to 60 ohms is evident at the start, followed by an irregular rise due to loss and cable defects. On the vertical scale shown, the incident step would be 35 cm high. Much more vertical magnification could be used without distortion.*

If the impedance of an infinite tapered section varies exponentially with distance, x, the impedance at any point will be

$$Z = Z_0 e^{kx} \left\{ \sqrt{1 + \left(\frac{kv}{2p}\right)^2} \pm \left(\frac{kv}{2p}\right) \right\} \quad (7)$$

where the upper sign applies for waves travelling in the positive direction of x, the lower for waves travelling in the negative direction. The bracket has unity magnitude and assumes complex conjugate values for the two cases. Thus when joined at $x = 0$ to a uniform line of impedance Z_0, there is a mismatch in angle of impedance and

$$\rho = \frac{\frac{kv}{2}}{\sqrt{p^2 + \left(\frac{kv}{2}\right)^2} + p} \quad (8)$$

The inverse transform of this expression is the reflection produced by a unit impulse and is

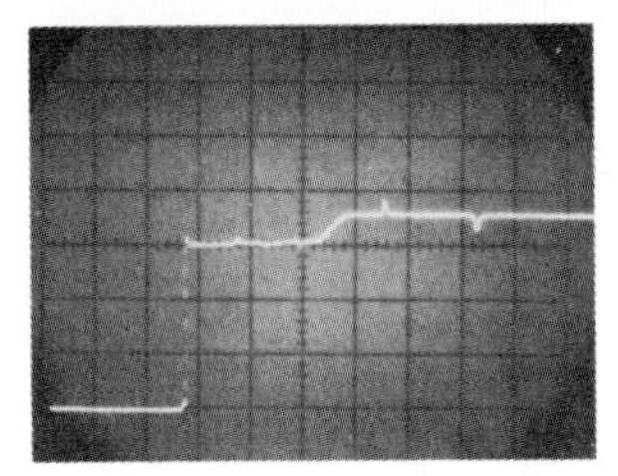

Fig. 19. *A 50-ohm (unbalanced) to 200-ohm (balanced) balun followed by a section of twin lead tapering from 200 to 300 ohms. The balun produces only the minor reflections before the slope, but changes the reference impedance level to 200 ohms. The slope is produced by the tapered section. Small inductive and capacitive discontinuities produce the two pips in the 300-ohm portion of the trace. The former was produced by slitting the twin lead and spreading the conductors, the latter by the top of a metal partition over which the lead was draped.*

$$f'(t) = \frac{1}{t} J_1 \left(\frac{kv}{2} t\right) \quad (9)$$

A step function therefore produces the integral of (9) or:

$$f(t) = \int_0^{kvt/2} \frac{J_1(z)}{z} dz \quad (10)$$

Figure 11 shows the actual reflection (solid curve) as well as the shape it would have to have (dashed curve) to be a true impedance profile. The error made by interpreting the reflection as impedance on the ohmmeter scale is shown as the dotted curve. For $kx < 1.4$, i.e., for impedance ratios less than $e^{1.4} \approx 4$, the error is less than 15%.

The step function response, Si(x), of an ideal low pass filter naturally contains no frequencies above the cutoff. If the reflection from a tapered section had the shape of the sine-integral (or the step response of any completely low pass filter) this would imply no reflection (and hence complete transmission) above a critical frequency. Using a time domain reflectometer and a razor blade one could presumably empirically shape the impedance Z(x) of a strip line, say, to produce the desired reflection.

LOSSY LINES

Distributed series resistance and shunt conductance, unless properly balanced, will cause the characteristic impedance to be complex and cause reflection at the junction with a distortionless line. For the lossy line we have

$$Z = \sqrt{\frac{R + pL}{G + pC}} = \sqrt{\frac{L}{C}} \sqrt{\frac{p + R/L}{p + G/C}}$$

$$= Z_0 \frac{p + \eta}{p + \sigma}.$$

When connected to a line of impedance Z_0, the reflection coefficient

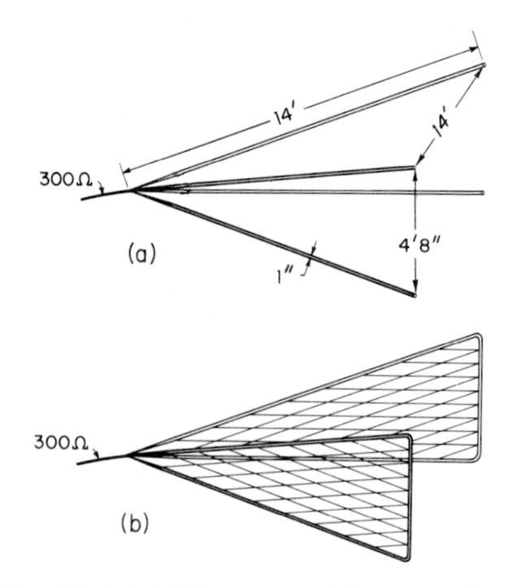

Fig. 20. *(a) A V-antenna with two conductors per side tapered to give a constant 300-ohm impedance for the first two feet after which the impedance rises. (b) The same antenna with end conductors and mesh added to make side surfaces rather than rods. The sides form a conical 300-ohm transmission line.*

$$\rho = \frac{\sqrt{p+\eta}-\sqrt{p+\sigma}}{\sqrt{p+\eta}+\sqrt{p+\sigma}}$$

results. Taking the Fourier transform we find for the reflection of a unit impulse

$$f'(t) = \frac{1}{t}e^{-\frac{\eta+\sigma}{2}t}\,I_1\left(\frac{\eta-\sigma}{2}t\right) \qquad (11)$$

Again, the step response is the integral of (11). For $\sigma = 0$ we find

$$f(t) = 1 - e^{-\frac{\eta t}{2}}\left\{I_o\left(\frac{\eta t}{2}\right) + I_1\left(\frac{\eta t}{2}\right)\right\} \qquad (12)$$

while for $\eta = 0$

$$f(t) = e^{-\frac{\sigma t}{2}}\left\{I_o\left(\frac{\sigma t}{2}\right) + I_1\left(\frac{\sigma t}{2}\right)\right\} - 1 \qquad (13)$$

These two functions are shown in Figure 12.

As time goes on, the line with only series resistance ($\sigma = 0$) looks more and more like an open circuit while the line with only shunt conductance ($\eta = 0$) looks more and more like a short circuit.

Normally one encounters lengths and losses such that only the initial part of these curves is seen. The initial slope is found from (11)

$$f'(0) = \frac{\eta-\sigma}{4} = \frac{1}{4}\left(\frac{R}{L} - \frac{G}{C}\right) \qquad (14)$$

After a time t, this slope will produce a departure a $= tf'(0)$ from the height of the incident unit step. If $G = 0$ we can interpret this departure in terms of accumulated series resistance. Thus from (6)

$$\Delta Z = 2\,a\,Z_o = \frac{1}{2}\frac{R}{\sqrt{LC}}\,t = R\,\frac{vt}{2} \qquad (15)$$

R is the resistance per unit length and vt/2 is the distance the incident wave had travelled for the time displayed. A similar result can be derived for G if $R = 0$, but no such simple picture exists if neither R nor G is negligible.

How well an actual physical system can display the theoretical responses discussed above may be seen from the photographs reproduced as Figures 13 to 21.

SOME EXPERIMENTAL TECHNIQUES

Using a time domain reflectometer one soon develops techniques for improving the speed and accuracy of measurements. Reflections are quickly located by touching or squeezing along the line as mentioned earlier. If no discontinuity is evident at or near the spot located, the observed echo may be a multiple reflection. This can often be checked by altering a few line lengths and noting if the echo in question moves with respect to the others.

When looking at reflections far down a system or after many earlier reflections, one can always check the effective incident step size and rise time by shorting or opening the line at the point in question to produce unity reflection. The resulting step should be used as the reference rather than the incident step first seen by the scope.

Highly accurate evaluation of shunt irregularities may be made by duplicating the observed reflection at some nearby point using known components. Series irregularities may be evaluated by cancelling the reflections they produce using known components in shunt at the same point. Thus a small shunt capacitance will cancel the reflection of a small series inductance; a small shunt conductance will cancel a small series resistance, etc.

Such techniques enable one largely to dispense with quantitative measurements of the trace and deal directly in terms of the test circuit elements and locations. The oscilloscope becomes, in a sense, a qualitative indicator telling where to look for trouble and what kind it is. The remedy is then determined by a quick succession of trials, guided by the display.

In narrow-band resonant structures TDR is of little use because the reflections are apt to be overlapping oscillatory transients of long duration. Here the time domain picture is complex and the frequency domain picture is usually simpler. But in the wide-band structures which comprise the majority of cases today, TDR is incomparably faster and simpler than conventional SWR or reflectometer methods.

ACKNOWLEDGMENT

The author wishes to acknowledge the contributions of Harley L. Halverson of the *–hp–* laboratories who was quick to appreciate the potentialities of time domain reflectometry and who has done much with an earlier paper[3] to popularize the technique.

—B. M. Oliver

[3] Harley Halverson, "Time Domain Reflectometry," Cleveland Electronics Conference, April 16, 1963.

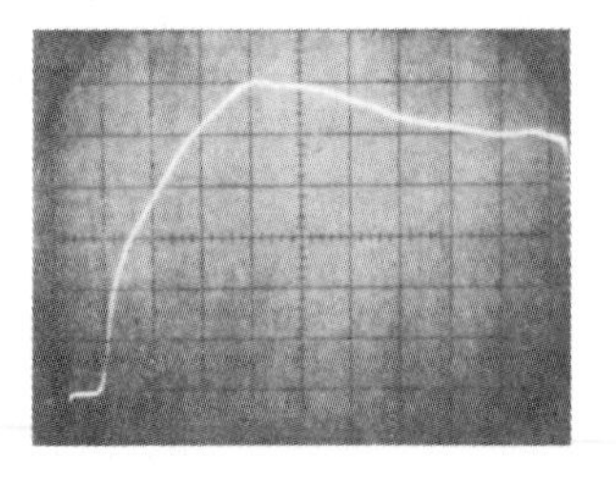

Fig. 21. *(a) The reflection from the antenna of Fig. 20(a). The tapering impedance causes a 20% overshoot (as in Figs. 10a and 11) and there is a rapid initial rise caused by too little fringing capacity at the open end.*

(b) The reflection from the modified antenna of Fig. 20 (b). No reflection occurs at the throat; in fact none occurs until the wave reaches the open end. The increased end capacity due to the end rods and the constant characteristic impedance produce a smooth reflection with very little overshoot. The reflection may be approximated by an exponential rise of 16-nanosecond time constant. Thus the return loss is 3 db at 10 Mc and falls at 6 db/octave above this frequency to become 14 db at 50 Mc, 20 db at 100 Mc, and 26 db at 200 Mc. Sweep speed is 5 ns/cm; incident step height is 5 cm.

TIME DOMAIN REFLECTOMETRY WITH A PLUG-IN FOR THE 140A OSCILLOSCOPE

TIME DOMAIN reflectometry measurements are easily made with the –*hp*– Model 140A general-purpose Oscilloscope and a plug-in designed specifically for such measurements. To make a time domain reflectometry measurement with this plug-in, –*hp*– Model 1415A, it is only necessary to connect the system to be tested to a front panel coaxial connector and to set the controls for the desired time base and amplitude range.

As shown in the block diagram, the voltage step from the internal generator in the plug-in passes through the sampling channel to the external system under test. Reflections from any impedance mismatches in the external system travel back through the sampling gate and are displayed on the cathode-ray tube in the oscilloscope main frame. The reflections subsequently are absorbed by the matched source impedance of the step generator.

Sweep magnifier controls enable any portion of the train of reflections to be

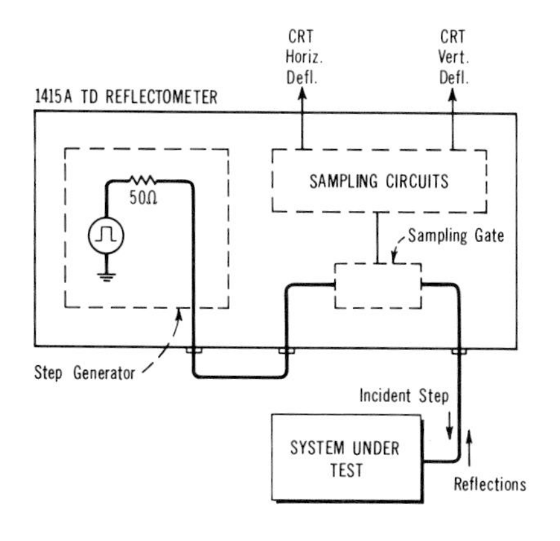

Voltage step generator in –hp– 1415A Plug-in connects directly to feed-through sampling channel. Plug-in also has rate generator and necessary synchronizing functions.

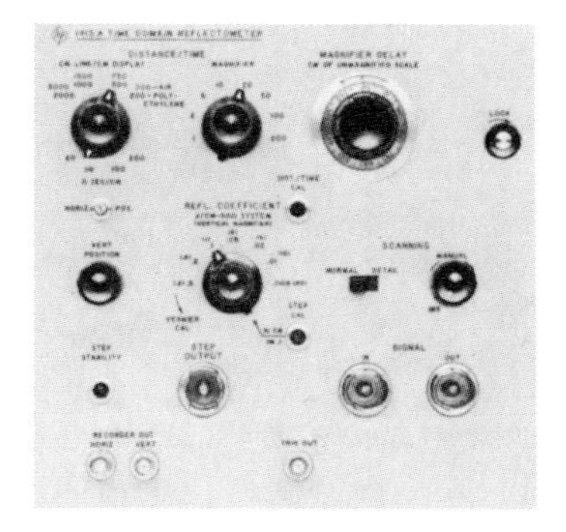

Self-contained –hp– 1415A Time Domain Reflectometer Plug-in for –hp– 140A Oscilloscope makes detailed measurements on transmission systems up to 1000 feet in length (incident step top is flat within 1% for 2000 nanosecs).

expanded for detailed examination. The controls are calibrated to indicate the distance to the section of line under examination, simplifying identification of the physical location of observed discontinuities. The wide dynamic range of the sampling channel allows expansion in the vertical dimension so that reflections that are even as small as 0.1% of the incident step may be examined in detail.

Overall system rise time is less than 150 picoseconds, enabling resolution of discontinuities that are spaced about 1 cm apart in coaxial systems.

The plug-in also provides outputs that permit replicas of the displayed waveform to be recorded in permanent form on an external X-Y recorder. For this application, the horizontal sweep is generated manually with a front panel control. Manual control of the sweep also enables the amplitude of any part of the observed waveform to be measured precisely with a digital voltmeter connected to the Y-axis recorder output.

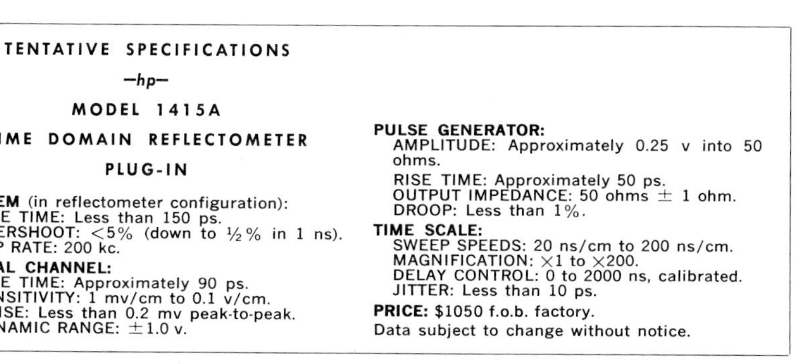

TENTATIVE SPECIFICATIONS
–hp–
MODEL 1415A
TIME DOMAIN REFLECTOMETER
PLUG-IN

SYSTEM (in reflectometer configuration):
RISE TIME: Less than 150 ps.
OVERSHOOT: <5% (down to ½% in 1 ns).
REP RATE: 200 kc.

SIGNAL CHANNEL:
RISE TIME: Approximately 90 ps.
SENSITIVITY: 1 mv/cm to 0.1 v/cm.
NOISE: Less than 0.2 mv peak-to-peak.
DYNAMIC RANGE: ±1.0 v.

PULSE GENERATOR:
AMPLITUDE: Approximately 0.25 v into 50 ohms.
RISE TIME: Approximately 50 ps.
OUTPUT IMPEDANCE: 50 ohms ± 1 ohm.
DROOP: Less than 1%.

TIME SCALE:
SWEEP SPEEDS: 20 ns/cm to 200 ns/cm.
MAGNIFICATION: ×1 to ×200.
DELAY CONTROL: 0 to 2000 ns, calibrated.
JITTER: Less than 10 ps.

PRICE: $1050 f.o.b. factory.
Data subject to change without notice.

TDR WITH -hp- SAMPLING SCOPES

The –*hp*– Model 188A Plug-in for the 185A/B Sampling Oscilloscopes realizes the highest resolution yet attained in a TDR system. When the incident step is supplied by the –*hp*– Model 213B tunnel-diode Pulse Generator, overall system rise times approaching 120 picoseconds are achieved with the 188A.

Well-defined time domain reflectometry measurements also are made with the 187A/B plug-ins and the 185A/B scopes. The high impedance probes of these plug-ins are well-fitted for bridging 50-ohm lines (see Fig. 2, page 1). System rise time with the 185B/187B is less than ½ nsec when the 213B Pulse Generator supplies the incident step, or 2 nsec if the 185B sync pulse serves by itself as the step.

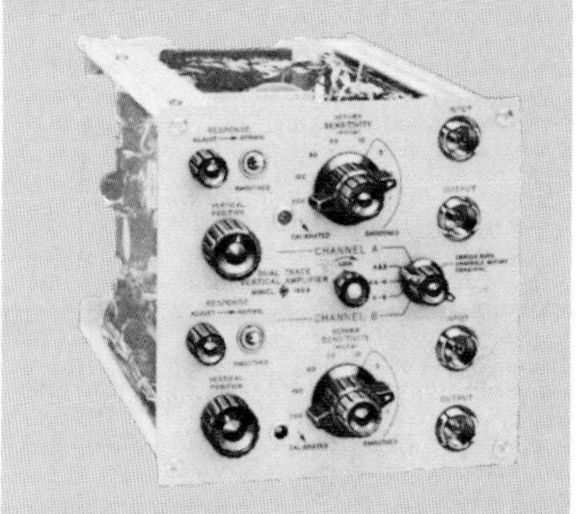

Fast-rise feed-through sampling channels of –hp– 188A dual-channel plug-in for–hp– 185A/B Sampling Oscilloscopes bridge 50-ohm lines without attenuation.

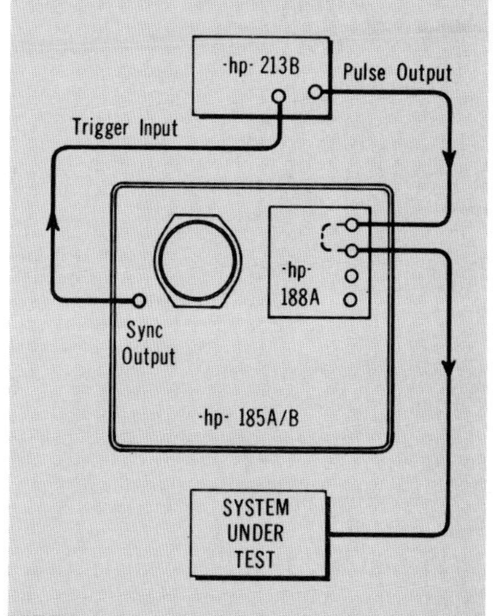

TDR measurements are easily made with –hp– 188A plug-in since plug-in allows incident step and returning reflections to pass through in opposite directions.

HEWLETT-PACKARD
JOURNAL
TECHNICAL INFORMATION FROM THE -hp- LABORATORIES
Vol. 15, No. 9
PUBLISHED BY THE HEWLETT-PACKARD COMPANY, 1501 PAGE MILL ROAD, PALO ALTO, CALIFORNIA
MAY, 1964

A 0-50 Mc Frequency Synthesizer with Excellent Stability, Fast Switching, and Fine Resolution

DIGITAL FREQUENCY SYNTHESIS

About twelve years ago the digital frequency counter made its commercial debut. It revolutionized the art of frequency measurement; and a tedious, complex process of multiple heterodyning was reduced to a simple act of plugging in the signal and reading the answer. The tremendous time saving more than justified the cost of the counter and these instruments have become ubiquitous. They are found today in applications undreamed of a decade ago.

With the advent of the digital frequency synthesizer I believe we face an analogous situation. Here we have a single instrument that can generate any frequency over a range far greater than most oscillators or signal generators — a frequency known to quartz crystal accuracy of a part in 10^{10} or better. This one instrument can do the work of a whole battery of general- and special-purpose generators and do it better and more conveniently. The initial model has been engineered to the most exacting specifications and construction to meet the need for a high-performance instrument. Later lower-priced models will provide high performance over a reduced frequency range.

It will be very interesting to watch the applications develop for frequency synthesizers. Almost certainly they will become as common as counters. Almost certainly there are applications in your own work where a synthesizer would save time and money and do a better job.

Bernard M. Oliver
Vice President, Research and Development
Hewlett-Packard Company

SEE ALSO:

1 and 10 Mc Synthesizers, p. 3
Synthesizer Switching Speed, p. 5.
Notes on Synthesizer Applications, p. 7.
Synthesizer Design Leaders, p. 8

AS THE ELECTRONIC art has advanced, so has the need for variable-frequency signal sources of high stabilities — stabilities comparable to those obtained from high-quality frequency standards. Such sources are valuable in sophisticated communications work, radio sounding, radar, doppler systems, automatic and manual testing of frequency-sensitive devices, numerous timing situations, spectrum analysis, stability studies, and many other areas.

The sources that provide highest frequency-stability are single-frequency sources — "frequency standards." Today, some of these are refined to the extent that they use atomic resonance for maximum long-term stability with a highly-developed quartz oscillator as a "flywheel" for maximum short-term stability. Having these high-quality standards in hand, it is natural to look for some method to translate their stability to other desired frequencies. This translation, when the operation is something more than a

Fig. 1. *New –hp– Model 5100A/5110A Frequency Synthesizer generates high-stability signals in 0.01 cps increments from dc to 50 megacycles. Generated frequency can be controlled by panel keyboard or by remote control. Programmed control tapes can be used to specify frequency patterns for automatic testing, fast receiver tuning, NMR work, etc.*

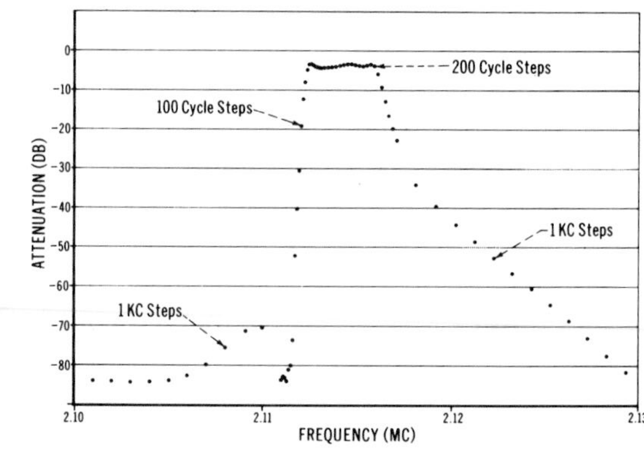

Fig. 2. *Amplitude response of crystal filter made using synthesizer in setup of Fig. 4, which permits automatic plotting if desired. Phase plots of many devices can be similarly made to check device stability.*

Fig. 3. *-hp- Model 5100A Synthesizer (upper unit) with Model 5110A Synthesizer Driver (lower unit). One Driver unit can operate up to four Synthesizer units.*

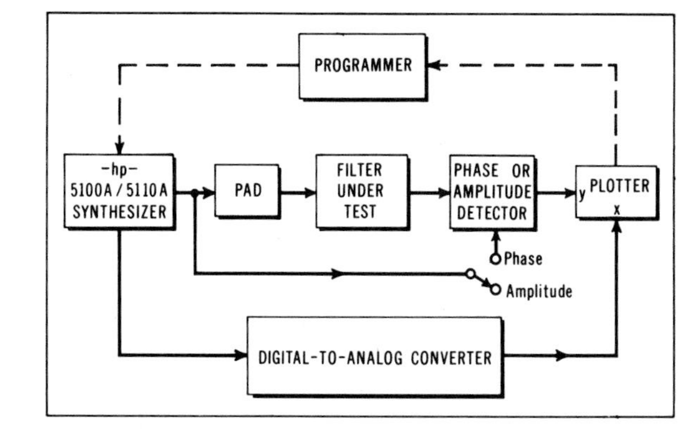

Fig. 4. *Equipment setup for making response record of Fig. 2. Phase plots can be made using phase detector. System can make measurements automatically using suitable programmer.*

single arithmetic operation, is commonly known as frequency synthesis. Hence, a variable-frequency synthesizer is an instrument that translates the frequency stability of a single frequency, usually one from a frequency standard, to any one of many other possible frequencies, usually over a broad spectrum. Such an instrument may provide any one of thousands, even billions, of frequencies. In everyday usage the word "variable" is usually omitted from the name, and the instruments are called merely "frequency synthesizers."

The two basic approaches to frequency synthesis are known as "direct" (or "true") and "indirect." Direct synthesis simply performs a series of arithmetic operations on the signal from the frequency standard to achieve the desired output frequency. The indirect method involves the use of tunable oscillators which are phase-locked to harmonics of signals derived from the standard.

The direct-synthesis approach has the pronounced advantages of permitting fine resolution and fast switching in the same instrument, as well as fail-safe operation and an extremely clean output signal. For these reasons it has been selected as the design approach for a sophisticated new synthesizer developed by the *-hp-* Frequency and Time Laboratories.

The new synthesizer, for example, provides frequencies from 0.01 cps to 50 megacycles in digital increments as fine as 0.01 cps! — a total of 5 billion discrete frequencies. At any frequency the output is a spectrally-pure signal — any non-harmonic spurious signal is more than 90 db below the desired signal.

The output frequency can be selected by front-panel pushbuttons or by remote electronic control for work involving automatic testing or fast frequency tracking. Under electronic control, the transition from one frequency to any other can be accomplished in much less than 1 millisecond, as described later.

The stability of the frequencies is derived from a self-contained frequency standard which has excellent short-term stability and a maximum drift rate of 3 parts in 10^9 per day. The high-frequency short-term stability, which includes noise as well as spurious signals, is quite comparable to that of a quality frequency standard. An external standard can also be used if desired.

PHYSICAL ARRANGEMENT

The Synthesizer is divided into two units, (Fig. 3), both of which are completely solid-state. The upper unit is the Synthesizer proper and the lower the Driver. Manual selection of the output frequency is made by the keyboard frequency control on the Synthesizer; remote electronic control is permitted by a switch at the lower left of the unit.

The pushbutton keyboard allows single-button frequency steps as great as 10 megacycles in the left column and as fine as 0.01 cps in the right. An extra row of "S" (Search) buttons across the top of the keyboard permits continuous variation of the output frequency either manually or remotely by external voltages. This gives continuous frequency coverage and facilitates frequency-search work. Manual searching is provided for in the form of a panel control which is calibrated from 0 to 10, corresponding to the full-scale frequency rating of the column being searched.

SYSTEM OPERATION

A simplified block diagram of the overall instrument is shown in Fig. 5. The Driver (Model 5110A) contains a frequency standard, a spectrum generator, and appropriate selection networks to provide a series of fixed frequencies between 3 and 39 Mc to the Synthesizer unit. The Synthesizer unit (Model 5100A) contains harmonic generators and suitable mixers, dividers, and amplifiers to derive the desired output frequency as a function of the fixed frequencies.

The fine resolution portion of the instrument is particularly interesting and also serves to illustrate the method of synthesis used. As shown in the right-hand portion of Fig. 5, there are seven identical mixer-divider units, each of which corresponds to a place or position in the final output frequency number. In each of these units, and in the eighth unit as well, a frequency of 24 Mc is used as a carrier input, as shown.

In the right-hand unit, which produces what ultimately becomes the highest resolution digit (10^{-2} cps), the 24 Mc carrier is added to a 3.0 Mc frequency in frequency-adder "A" to produce 27.0 Mc. In "B" the 27.0 Mc frequency is added to a frequency of from 3.0 to 3.9 Mc, depending on the setting of the panel pushbutton or remote control-circuit. Selection of a "2" in this particular digit position, for example, electronically selects a signal of 3.2 Mc from the Driver.

The output of "B" is a frequency of 30.0 to 30.9 Mc, which is divided in "C" to produce 3.00 to 3.09 Mc. This frequency is applied to the second unit, where it adds with the 24 Mc carrier as before and the process repeats. If the process is followed through, it will be seen that the frequencies noted in the block diagram are obtained at the outputs of the various adders and dividers. In essence, each mixer-divider unit, through a frequency division process, moves a given digit one place to the right in the final frequency and at the same time inserts a new number in the displaced position. The final result is that the output of the eighth unit is a frequency of between 30,000,000.00 and 30,999,999.99 cps, depending on the output frequency selected[1].

In the following two operations the signal is added to a frequency of 330 Mc, and the resultant again added to an appropriate frequency between 30 and 39 Mc to yield a frequency of between 390 and 400 Mc. One of the five frequencies from 350 to 390 Mc is then subtracted from this to yield the desired 0.01 cps to 50 Mc output frequency.

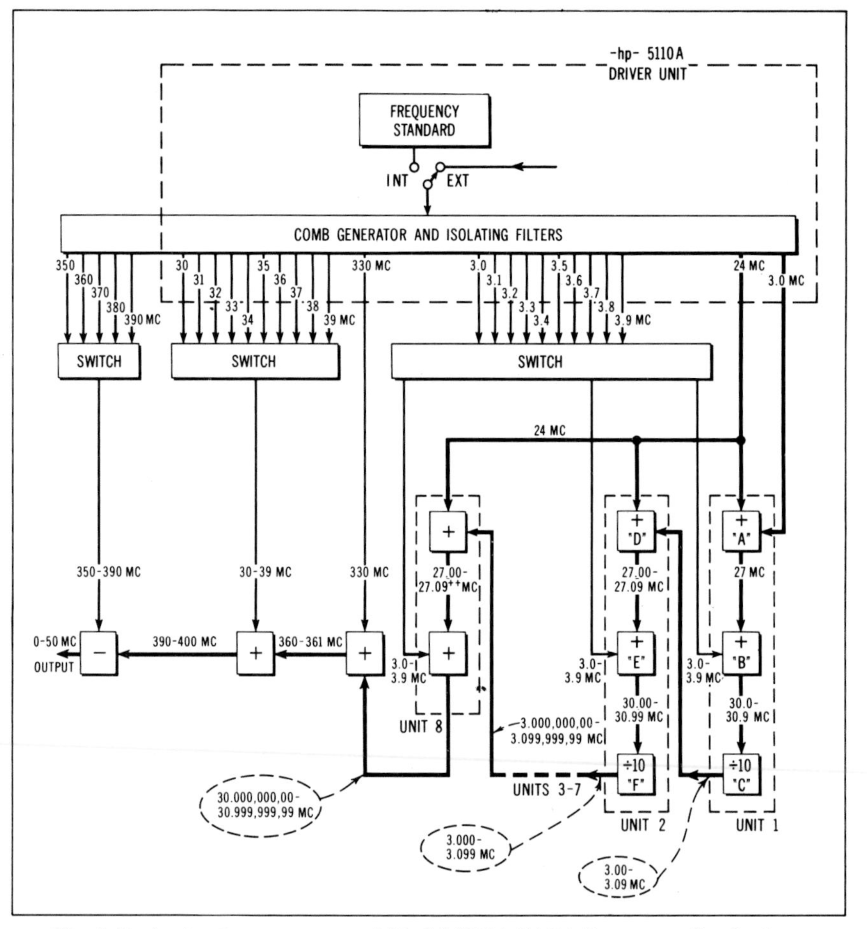

Fig. 5. *Basic circuit arrangement of Model 5100A/5110A Frequency Synthesizer. Direct-type synthesis design approach is used to achieve fast switching with fine resolution and fail-safe type circuit.*

SEARCH OSCILLATOR

The search feature has proved useful in several ways. Besides facilitating searching for an unknown frequency, it permits smooth frequency modulation of the output, phase-locking the synthesizer into another system, sweep operation with a sweep width smaller than 0.1 cps, and the capability of placing the sweep accurately anywhere within the instrument range. The search oscillator permits the output frequency to be continuously varied over the frequency range of any one column except the left-hand two (megacycles and tens of megacycles columns). Searching can be done either manually by a panel control or electronically by an external voltage of —1 to —11 volts.

In the circuit, the search oscillator is a 3.0 to 4.0 Mc variable oscillator that is substituted for the 3.0 to 3.9 Mc fixed frequencies available to the filter-divider units (1-8 in Fig. 5). Its manual control is calibrated to 3% accuracy and its voltage control capability has a 5% linearity specification. The RMS Δf contributions of about 1 cycle for one-second averaging when this oscillator is searching in the 100 kc step position limits the synthesizer's short-term stability to that extent, but in search work this is presumably of little consequence. At any rate, the instability effects are reduced as the digital position of insertion is made less significant because of the frequency dividers involved. Insertion in the 10-cycle step position does not result in a significant reduction of the synthesizer's short-term stability.

[1] Synthesis methods similar to this have been considered by several people including some at Hewlett-Packard. Especially significant work in this area was carried on for many years by H. Hastings and R. Stone at the U. S. Naval Research Laboratory.

SIGNAL PURITY

One of the important design objectives for this system was the elimination of non-harmonically-related spurious signals to a —90 db specification. A parallel design objective was the reduction of noise to as low a level as possible, since noise appears as a small random phase modulation which, in critical high-stability work, adversely affects the frequency stability of the signal. Some sources of noise are standard oscillator instabilities, filter instabilities, thermal and current noise introduction at low level points, 1/f reactance noise in tuned circuits, electrical contact problems, and semiconductor breakdown.

Fig. 6 shows the low value of the phase noise distribution typical of the new synthesizer (the AM noise is some 20 db further down within 100 kc). The synthesizer's contribution to a "perfect" external standard would be somewhat less than indicated for the noise closer than 100 cycles. This close-in noise is

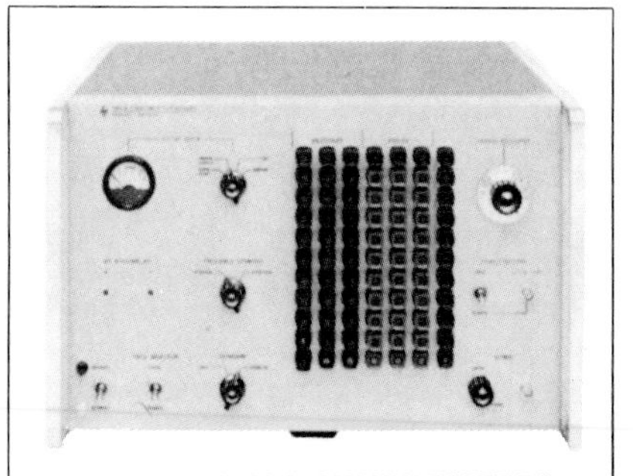

1 AND 10 MC SYNTHESIZERS

Two additional synthesizers, based on the same design approach used in the 50-Mc synthesizer described in this issue, will soon be placed in production. The chief difference in these units is that their upper frequency limits are 1 and 10 Mc, respectively. Their general performance, while not precisely the same as the first, has been kept at a high level. Frequency selection is also made either by panel pushbuttons or by remote control.

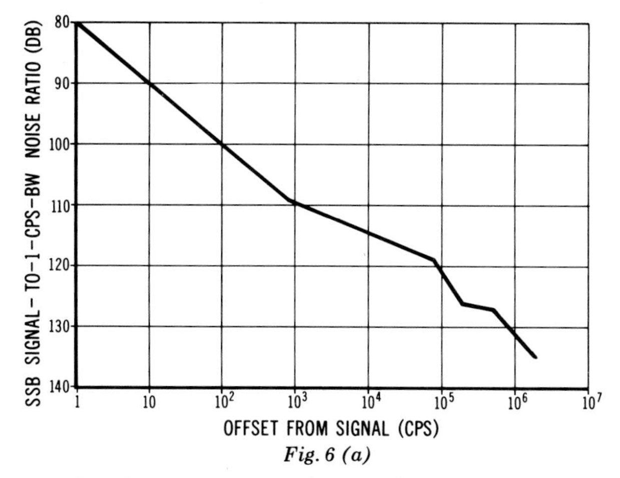

Fig. 6 (a)

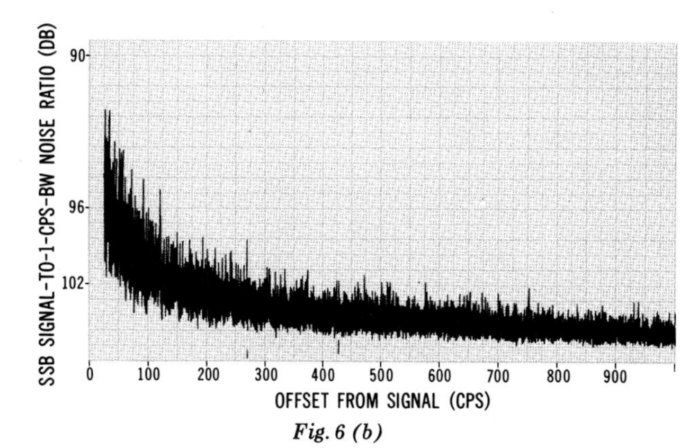

Fig. 6 (b)

Fig. 6. *Typical low value of phase noise in output of new synthesizer. (b) shows "close-in" noise in greater detail. AM noise is much below these levels. Both records are of single side-band made with tunable selective system.*

also less for lower output frequencies. Multiplication to X-band and analysis with 1-cycle-bandwidth analyzer shows the spectral width to be much less than 1 cycle (Fig. 7).

In an attempt to give a rough indication of the synthesizer's noise performance, we have specified that the phase noise in an arbitrarily-selected 30 kc noise bandwidth (excluding 1 cycle at the center) centered on the signal will be more than 54 db down (see specifications). The contribution within 100 cycles is less at lower frequencies, so the specification is most conservative there. In a like bandwidth the AM noise will be more than 74 db down for output frequencies above 100 kc.

In order to characterize the synthesizer's performance for timing applications, we have specified the RMS fractional frequency deviations in terms of this same 30-kc bandwidth. One advantage of this characterization is that the same number holds after processing the signal through a perfect frequency multiplier or divider. The specifications given are most conservative at the lower output frequencies for the longer averaging times. The specifications can be viewed as applying to the synthesizer including the internal standard or as RMS contributions to a very high quality external standard. For averaging times of a second or longer, the synthesizer can convert the stability of the best single-frequency standards to the 50 Mc region without significant deterioration.

LEVEL STABILITY

The level stability of the equipment is of considerable importance in some applications such as in frequency multiplication where level instability can be converted to phase instability. The specifications include both the effects of frequency change and environmental conditions. Fig. 8 shows a sample instrument output level as a function of frequency. Improved response from dc to 100 kc is provided by a low-level, high-impedance output. Level stabilities as a function of time and of line voltage are shown in Fig. 9.

SWITCHING SPEED

The provision for fast electronic frequency selection makes the Synthesizer's resolution and stability available for such tasks as automatic digital frequency tracking, automatic testing of frequency-sensitive devices, automatic special communications systems, as well as simple remote control or readout of frequency.

When the LOCAL-REMOTE switch is thrown to REMOTE, the switching power is transferred from the front panel control to three remote-control connectors on the back panel of the synthesizer. Frequency control is then accomplished by connecting the switching-power line to the lines corresponding to the desired digits. With electronic control, extremely fast frequency selection can be accomplished and with virtually no dead time, as shown in the accompanying group of oscillograms.

INTERNAL STANDARD

The quality of the internal 1 Mc standard is indicated by its aging rate of less than 3 parts in 10^9 per day, which is only about one order of magnitude below that of the finest crystal standards. A standby feature keeps the internal crystal oven turned on as long as ac line-power is available. The standard is well protected from line voltage variations, the worst effect being a momentary frequency shift on the order of a few parts in 10^{11} with a fast change from low to high line (+20%). There is less than 2 parts in 10^{10} frequency shift per °C change in the ambient temperature.

The internal standard can be adjusted by an external voltage to permit locking the synthesizer into some other system. A range of ±5 parts in 10^8 frequency control can be exercised with a ±5-volt external source. The short-term stability of the standard is adequate to provide the short-term stability specified for the overall instrument.

When an external standard is used in place of the internal standard, a crystal filter and other circuits in the synthesizer improve to a greater or lesser extent the noise and spurious signal modulation present on that standard. A measure of this improvement is indicated in

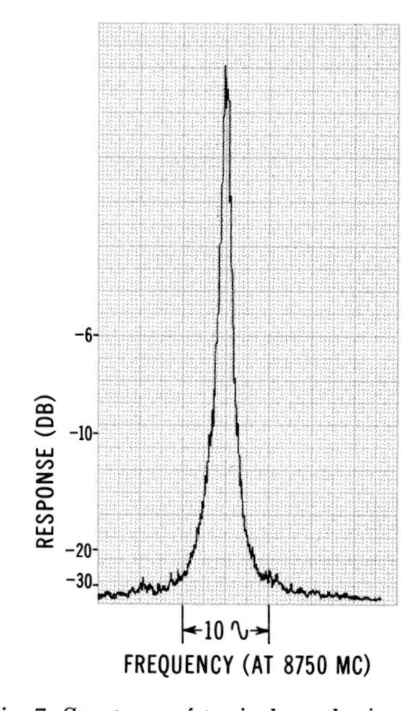

Fig. 7. *Spectrum of typical synthesizer signal when multiplied to X-band. Narrow width here of <1 cps indicates high quality of original synthesizer signal.*

(a) (Upper trace) Synthesizer frequency switched from 20.003 Mc to 20.006 Mc and back with 20 Mc subtracted to display switching in greater detail. This switching detail is typical of any of the six right-hand places. Sweep speed is 200 μsec/cm.

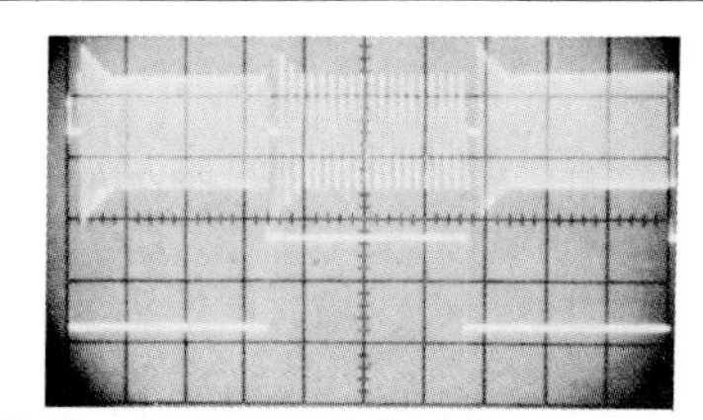

(b) Synthesizer switched between 20.03 Mc and 20.06 Mc with 20 Mc subtracted. Sweep is 200 μsec/cm. Lower trace in all cases is switching waveform applied to synthesizer.

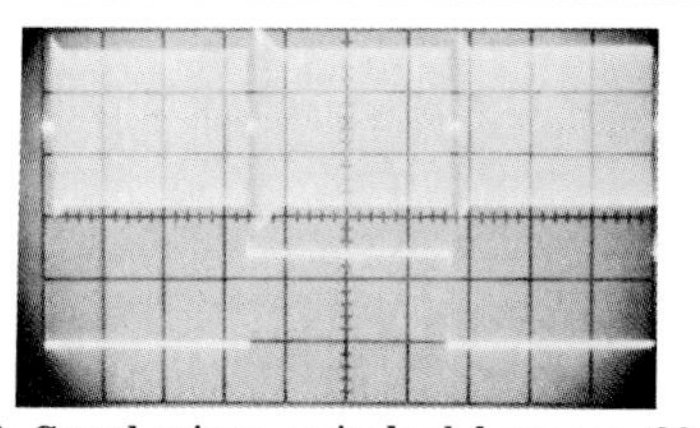

(c) Synthesizer switched between 20.3 Mc and 20.6 Mc. No subtraction. Sweep is 200 μsec/cm.

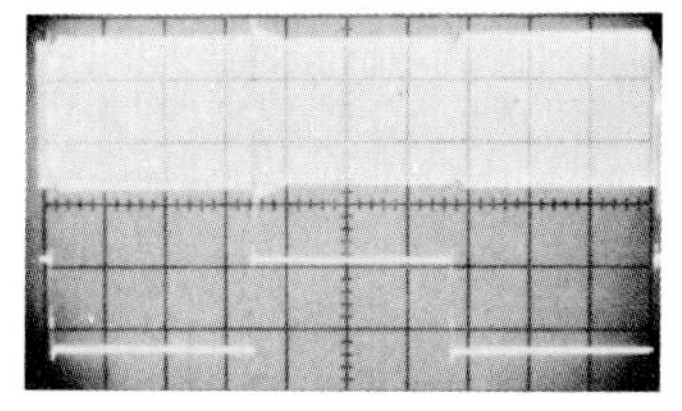

(d) Synthesizer switched between 23 Mc and 26 Mc. Sweep is 100 μsec/cm.

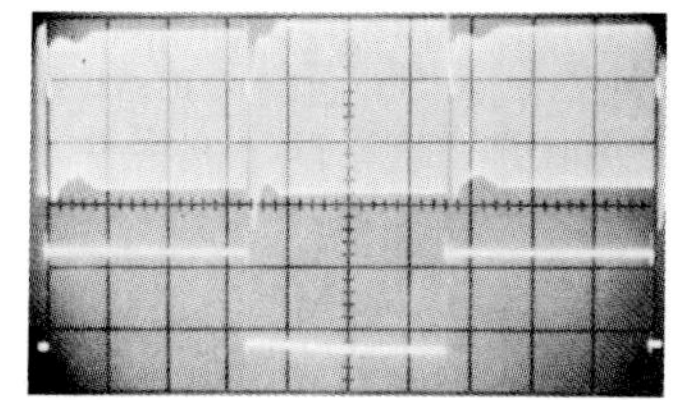

(e) Synthesizer switched between 10 Mc and 20 Mc. Sweep is 100 μsec/cm.

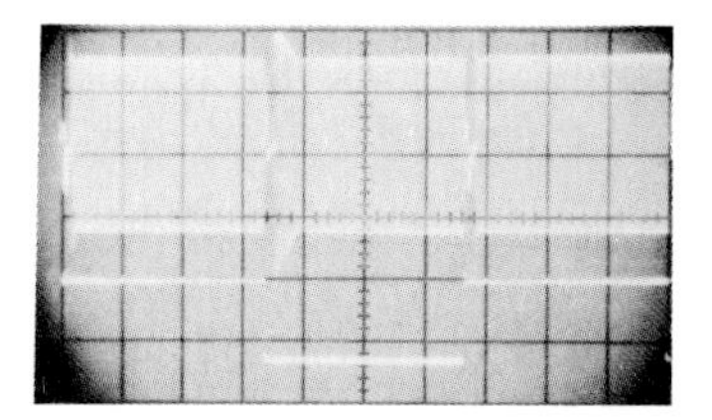

(f) Synthesizer switched between 19.99 Mc and 20 Mc. Sweep is 200 μsec/cm.

SYNTHESIZER SWITCHING SPEED

The accompanying oscillograms show the tremendous speed with which the synthesizer can change its output frequency under electronic control. The upper waveform in each case is the synthesizer output as it is changed from one frequency to another and back again. The lower waveform is the external switching voltage applied to the synthesizer. The oscillograms have been selected to show switching both within a single place (digit column) and between various places of the output frequency.

In all cases a fast sweep speed of 200 microseconds/cm or faster has been used in order to adequately portray the speed of switching and the virtual non-existence of switching transients. Any frequency instabilities at the start of the new frequencies are not revealed by such oscillograms, but these have been determined by other means to be very slight and brief in duration.

The switching waveform was derived by applying an external positive voltage of +2 volts in series with a contact closure as the "off" signal. A simple contact opening without an external voltage can also be used as the turn-off signal at a moderate reduction in switching speed and turn-on clarity.

Fig. 10 which shows the conversion of modulation on the driving standard to modulation on the synthesizer output at both a high and a low frequency.

For some applications it is desirable to bypass the crystal filter in the synthesizer so that the external standard may be shifted or frequency-modulated. Therefore a pair of curves is also shown in Fig. 10 for operation without the filter.

The improvement resulting when using the filter may cause a measurement ambiguity if the synthesizer output is checked for short-term stability against an external driving standard. Such a comparison would usually be expected to show the synthesizer contribution, but in this case, where the signal from the synthesizer may be an improved version of that from the driving standard, the measurement may really be a measurement of the noise on the standard. In other words, *evaluating the synthesizer contributions requires an extremely high-quality standard.*

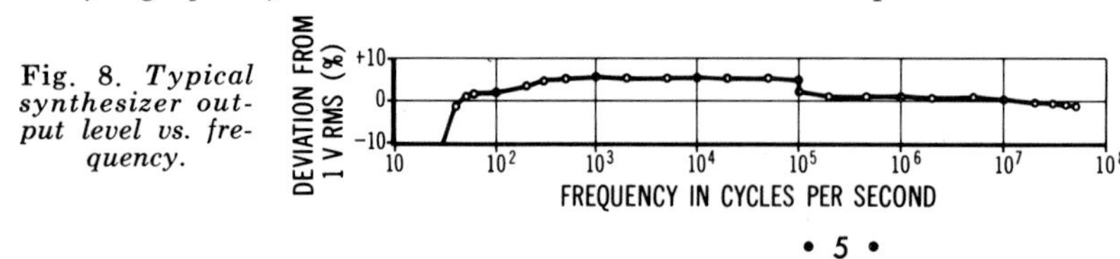

Fig. 8. *Typical synthesizer output level vs. frequency.*

ENVIRONMENTAL PERFORMANCE

The synthesizer is specified to operate over an ambient temperature range that is quite wide — 0 to 55°C. Consequently, the effects due to ambient temperature changes of usual amounts are small. For example, exclusive of the frequency standard, the phase shift per °C is typically 6° + .25°/Mc. Converting this to fractional frequency error, we have

$$\frac{\Delta f}{f} = N\left(\frac{5}{F} + 0.2\right) \times 10^{-12},$$

where N is the rate of temperature change in degrees per hour and F is the output frequency in megacycles. At 50 megacycles with a 1°C/hour rate of ambient temperature change, this would amount to but 3×10^{-13} frequency error.

The internal frequency standard has a typical frequency shift of ±1 part in 10^{10} per °C (normal for a high quality standard), so that the *frequency* stability of the system above 100 kc will normally be limited by the frequency standard. The best available quartz standards have a temperature dependence of about 20×10^{-13} per °C. The Synthe-

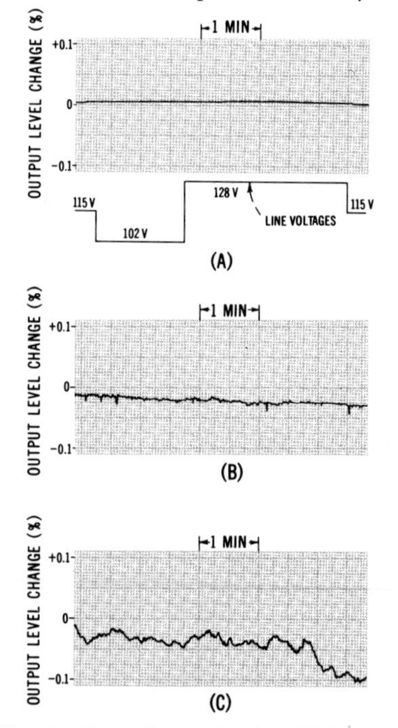

Fig. 9. *Recordings showing high output voltage stability of new synthesizer. (a) is typical of outputs above 100 kc, (b) below 100 kc. Line voltage in (a) and (b) was varied widely as shown in (a). For comparison, (c) shows stability of a quality LC oscillator operating on a constant-voltage regulated line.*

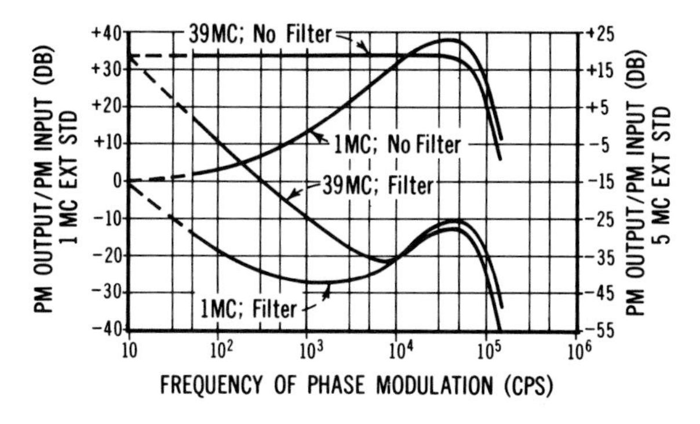

Fig. 10. *Noise "cleanup" produced by synthesizer on a frequency synthesized from external signal compared with the external signal. Curves are given both for case where synthesizer's internal filter is used and when not used.*

sizer and Driver performance are presently being checked at 55°C before shipment as a quality assurance measure. No damage will result from non-operating exposure to —40° to +75°C.

Tests on sample instruments show that these instruments can be expected to meet the specifications under a combination of both 95% relative humidity and 40°C temperature even after a few 24-hour cycles of low to high humidity.

The specifications on short-term stability are given for a vibration-free environment. The typical slight degradation of short-term stability for the system exclusive of the frequency standard is indicated in Fig. 11. No damage was experienced on a sample instrument-pair tested from 10 to 55 cycles with .010-inch peak-to-peak excursions in the three principle directions.

The radiated and conducted interference caused by the system falls within the limits allowed by MIL-I-16910A, and the equipment has been designed and sample-tested to meet the susceptibility conditions of MIL-I-6181D and

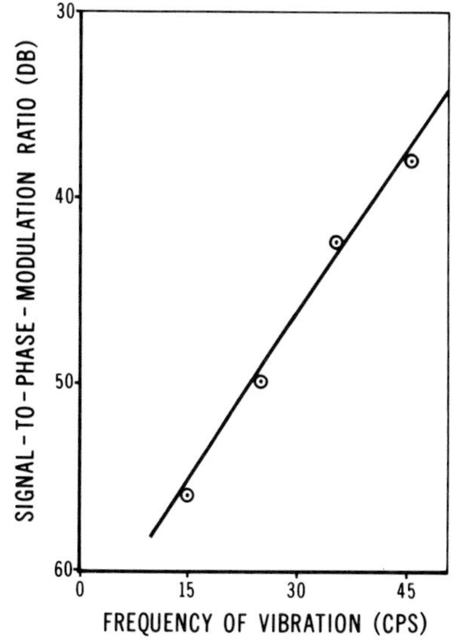

Fig. 11. *Synthesizer exhibits only slight vibration effects. Curve shown is for .010" p-p excursion. Synthesizer operates under temperature and humidity extremes.*

MIL-I-26600. This means that there should not be any RF interference problems when operating the synthesizer near other reasonably well-designed equipment.

Another important consideration is the spurious signal production due to external sources of low frequency (power line) magnetic fields. The synthesizer system has been carefully designed so that a field of at least 0.3 gauss is required to cause a —90 db spurious *modulation* of the output signal. Some electronic instruments produce considerably more than this, however. Spurious signals *at* the frequency of the magnetic field (and its second harmonic) will be considerably worse.

SOLID-STATE MODULES

Plated-through printed circuit board construction is used throughout. Modular construction has been used which should be a great help in any maintenance, since it is relatively easy to trace a trouble to an offending module. Modules are interchangeable with others of like kind.

ACKNOWLEDGMENT

This instrument development has been one of the largest yet undertaken by Hewlett-Packard. To bring it to production status has required more than 40 engineering man-years of development with a highly concentrated effort extending over a period of almost three years. The bulk of the electrical design work was done by David E. Baker, John N. Dukes, John E. Hasen, Albert P. Malvino, Walter R. Rasmussen (a five-year sustained effort), Hans H. Junker, Alexander Tykulsky, and the undersigned. The mechanical and production aspects were ably handled by Lawrence A. Lim, William Powell, and Theodore G. Pichel. There are, of course, many others who made valuable contributions and their efforts are greatly appreciated.

—*Victor E. Van Duzer*

SPECIFICATIONS

— hp —
MODEL 5100A
FREQUENCY SYNTHESIZER

OUTPUT FREQUENCY: DC to 50 Mc.

DIGITAL FREQUENCY SELECTION: 0.01 cps through 10 Mc per step. Selection by front panel pushbutton or by remote switch closure. Any change in frequency may be accomplished in less than 1 millisecond.

OUTPUT VOLTAGE: 1 volt rms ±1 db from 100 kc to 50 Mc. 1 volt rms +2 db, —4 db from 50 cps to 100 kc, into a 50-ohm resistive load. Nominal source impedance is 50 ohms. 15 millivolts rms minimum open circuit from 100 kc down to DC at separate rear output connector; source impedance of 10K ohm with shunt capacitance approximately 70 pf.

SIGNAL-TO-PHASE NOISE RATIO: Greater than 54 db in a 30 kc band centered on the signal (excluding a 1-cps band centered on the signal).*

SIGNAL-TO-AM-NOISE RATIO: (Above 100 kc): Greater than 74 db in a 30-kc band.*

RMS Fractional Frequency Deviation (With a 30-kc noise bandwidth):*

AVERAGING TIME	OUTPUT FREQUENCY			
	1 Mc	5 Mc	10 Mc	50 Mc
10 milliseconds	$3x10^{-8}$	$6x10^{-9}$	$3x10^{-9}$	$6x10^{-10}$
1 second	$3x10^{-10}$	$6x10^{-11}$	$3x10^{-11}$	$1x10^{-11}$

SPURIOUS SIGNALS: Non-harmonically related signals are at least 90 db below the selected frequency.

HARMONIC SIGNALS: 30 db below the selected frequency (when terminated in 50 ohms).

SEARCH OSCILLATOR: Provides continuously variable frequency selection with an incremental range of 0.1 cps through 1 Mc. Manual or external voltage (—1 to —11 volts) control with linearity of ±5%.

WEIGHT: Net 75 lbs (34 kg); shipping 127 lbs (58 kg).

EQUIPMENT FURNISHED: 05100-6180 Decade Test cable, 05100–6066 Output cable, 05100–6157/8 cable assembly connects 5100A Synthesizer to 5110A Driver. Permits rack mounting of up to two 5100A's immediately above and/or below the 5110A Driver. Special-length cable assembly available for other mounting arrangements.

*Performance data stated are based on internal frequency standard or highest quality external standard.

— hp —
MODEL 5110A
SYNTHESIZER DRIVER

INTERNAL FREQUENCY STANDARD: 1 Mc Quartz Oscillator.

AGING RATE: < ±3 parts in 10^9 per day.

STABILITY: As a function of ambient temperature: $\pm 2 \times 10^{-10}$ per °C from 0°C to +55°C.

As a function of line voltage: $\pm 5 \times 10^{-11}$ for a ±10% change in line voltage (rated at 115 or 230 volts rms line voltage).

Short term: Adequate to provide the 5100A performance noted above (1×10^{-11} rms for 1 second averaging on direct output for 30 kc noise bandwidth).

PHASE LOCKING CAPABILITY: A voltage control feature allows 5 parts in 10^8 frequency control for locking to an external source. –5 to +5 volts required from phase detector (not supplied).

EXTERNAL FREQUENCY STANDARD:
INPUT REQUIREMENTS: 1 or 5 Mc, 0.2 v rms minimum, 5 v maximum across 500 ohms. Stability and spectral purity of 5100A Frequency Synthesizer will be partially determined by the characteristics of the external standard if used.

WEIGHT: Net 54 lbs (25 kg); shipping 60 lbs (27 kg).

GENERAL

OPERATING TEMP. RANGE: 0 to +55°C.

INTERFERENCE: Complies with MIL–I–16910A.

POWER: 115 or 230 volts ±10%, 50 to 400 cycles, 35 watts each unit (independent supplies).

PRICE: Model 5100A Frequency Synthesizer, $10,250. Model 5110A Synthesizer Driver, $5000.

Prices f.o.b. factory.

Data subject to change without notice.

NOTES ON THE APPLICATION OF FREQUENCY SYNTHESIZERS

In the digital frequency synthesizer we have a frequency standard whose output frequency can be selected by either manual or electronic command to very high resolution in less than a millisecond. Such an instrument constitutes a most powerful tool. In communications work, for example, the synthesizer's excellent spurious-frequency performance makes it well suited to use as the master oscillator in a transmitter and as the local oscillator in a receiver. If the transmitter and the RF section of the receiver are untuned, an extremely fast switching system can be used to change the local oscillator (synthesizer) frequency to achieve communications systems of high integrity.

Again, the synthesizer can greatly facilitate surveillance work if it is used as the local oscillator in a receiver* designed to accurately determine the frequency of remote transmitters. The ease and speed with which the synthesizer frequency can be changed will allow monitoring of a multiplicity of channels with a single receiver by sequencing the local oscillator (synthesizer) through the desired channels.

Sequencing the synthesizer output through a group of desired frequencies can also permit a single instrument to operate as an automatic calibrator for a multiple-frequency setup such as a multiple transmitter installation. The arrangement can provide for phase-locking the transmitter frequencies to the synthesizer by a circuit with a time constant long enough to maintain the transmitter frequency for the duration of the sequencing cycle.

In HF communications work, dependable long-distance communications requires the use of a frequency near the maximum usable frequency, which is determined by ionospheric conditions. Since these conditions can change rapidly, test transmissions over the HF spectrum are used at frequent intervals to insure loop operation. The fast switching and electronic control of the synthesizer make it a natural part of such a "radio sounding" system.

*In receiver work a double-balanced mixer is recommended since it will discriminate against spurious responses and will further improve the effective noise level.

MICROWAVE/SPACE COMMUNICATIONS

The effective use of the microwave spectrum for communications requires frequency sources having extremely good fractional frequency-stability so that the receiver bandwidth can be minimized. With a 3 kc information bandwidth at 10 kMc, a frequency stability of 3 parts in 10^8 for the duration of a message is desirable for double-sideband work. For single-sideband work the requirement is about 1 part in 10^9 for the same conditions. Obviously, a synthesizer must be used in such a communications system to make it practical. The high spectral purity of the synthesizer permits multiplication of its output even to X-band with a signal-to-noise ratio ample for such applications.

Determining the velocity of far-out space vehicles through Doppler frequency measurements involves operation at X-band with receivers having IF bandwidths of but a few cycles to minimize noise levels. As the vehicle velocity changes, the receiver's local oscillator frequency must be changed to keep the received signal in the center of the IF bandwidth. Here again, the synthesizer is ideal because of its stability and because its frequency can be changed in known and selectable increments.

AUTOMATIC TESTING

Fig. 2 in the accompanying article shows the amplitude response of a narrow-band crystal filter plotted automatically in permanent-record form. Fig. 4 of that article shows the system used to make the record. The system provides for either manual change of frequency or automatic change of frequency under the control of a simple tape-operated programmer. Under automatic operation, a point is plotted automatically as soon as the X-Y plotter's positioning servos null. The programmer is then commanded to advance to the next test point. Where the response is a slow function of frequency, the programmer dictates relatively large steps in frequency. Where the response changes rapidly, small frequency steps are made.

This same type of system could be used for plotting the in-band phase response by substituting a leveler and phase comparator for the amplifier indicated in the diagram.

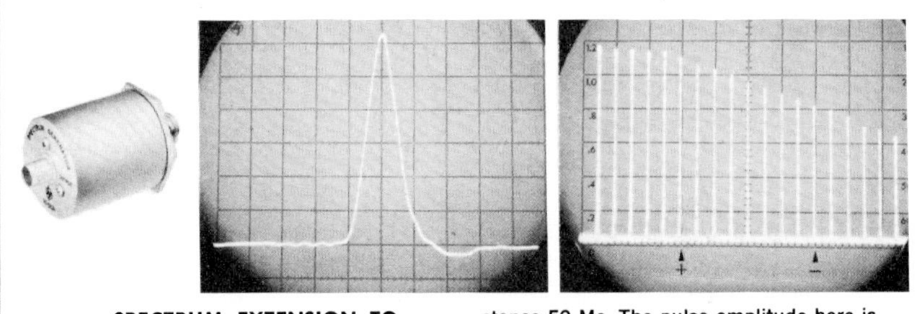

SPECTRUM EXTENSION TO ABOVE 500 MC

Many readers will find in the accompanying oscillograms an unusual and excellent example of a time-frequency transform pair. The oscillograms are the more unusual, however, in that they are actually of times and frequencies in the nanosecond-kilomegacycle region. They were made possible by the use of the -hp- 185B Sampling Oscilloscope and a new -hp- Spectrum Analyzer soon to be described.

In addition to their technical interest, though, these oscillograms demonstrate how the frequency output of the new synthesizer can be extended to at least 500 Mc through the use of a simple spectrum generator. The generator is a passive device which can be driven directly from the synthesizer.

The time-plot oscillogram shows the output pulse typically produced by the generator when driven from the synthesizer. The pulse has the same repetition frequency as the synthesizer frequency, in this instance 50 Mc. The pulse amplitude here is about a volt and the width about one nanosecond. Width is essentially independent of driving frequency, which can range down to 10 Mc.

The generator's corresponding frequency spectrum is shown from 50 Mc to 1000 Mc in the second oscillogram. The amplitude of the components varies from about 75 millivolts at the 50 Mc component to about 2.5 millivolts at the 1000 Mc component, measured across a 50-ohm load. The spectrum is not rated above 500 Mc, although a 1000 Mc spectrum is shown here and usable outputs to 2000 Mc are usually available.

Being harmonics of the synthesizer's frequency, the generator's output frequency components have the full precision and stability of the synthesizer. Further, their resolution is reduced only by the number of the harmonic used. Even at 500 Mc, then, harmonics can be adjusted in 0.1 cps steps by changing the synthesizer's 50-Mc output in 0.01 cps steps. Commercial bandpass filters are available to select a desired harmonic range from the generator.

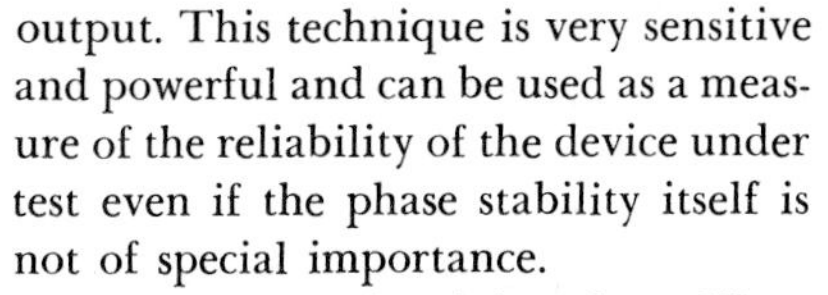

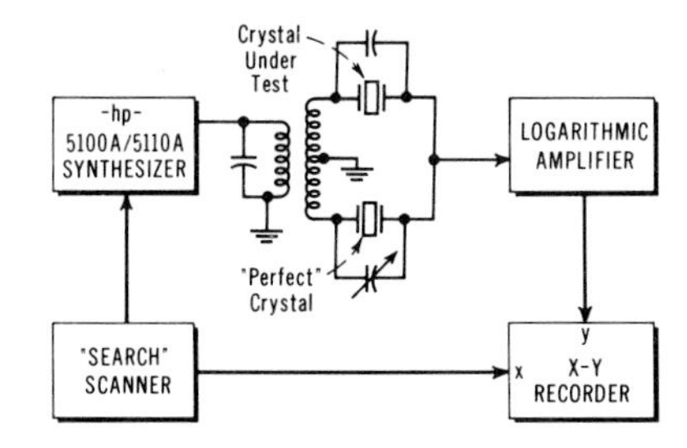

Arrangement for plotting spurious modes in crystals.

A similar automatic arrangement could be used for testing and sorting frequency-sensitive components with great precision and speed.

SPECTRUM ANALYSIS/NMR

The ability of the synthesizer to provide a signal of extremely high frequency stability, when coupled with the fineness of resolution provided by the instrument, greatly facilitates analysis of spurious modes in resonant devices. A setup for plotting spurious modes in quartz plates is shown in the accompanying illustration. The arrangement shown allows observation of modes which are very close to the main response mode and which are 60 to 70 db down.

Nuclear magnetic resonance methods are increasingly used to determine, among other things, the qualitative and quantitative make-up of materials. In this method the strength of an applied dc magnetic field and the frequency (usually 2 to 100 Mc) of a simultaneously-applied RF field needed to produce a nuclear resonance in the material must be known. The dc magnetic field can be controlled at a defined reference value with great stability by previously-developed means. The synthesizer now provides the RF excitation frequency at the high precision needed to greatly enhance the precision of NMR measurements. The ease of frequency control in the synthesizer allows for automatically testing for the presence and quantity of several elements in a sample and does so with such speed as to make NMR in-process control a real possibility.

Lastly, a most interesting case is the use of the synthesizer in the examination of the atomic spectral lines of a cesium beam tube.*

FREQUENCY AND LEVEL STABILITY MEASUREMENTS

The excellent frequency stability of the synthesizer can be used as a standard in measuring the frequency stability of other signal sources. In frequency-stability measurements, the synthesizer signal can be subtracted from the signal under test, thereby translating instabilities to a lower frequency where they can be measured by a frequency counter, low-frequency analyzer, or other means.

It is also interesting to note that the synthesizer can be used in measuring the phase stability of such devices as IF amplifiers, frequency multipliers, frequency dividers, trigger circuits, and resonant devices. In such work a phase comparator is used to synchronously mix the input signal (supplied by the synthesizer for stability) and the output signal of the unit under test. By adjusting the phase of the signals to a quadrature relation, any phase perturbations introduced by the unit under test will be readily observable at the comparator output. This technique is very sensitive and powerful and can be used as a measure of the reliability of the device under test even if the phase stability itself is not of special importance.

Measurements involving the calibration of voltmeters, power meters, and attenuators must depend on a signal source with high stability of output level. The level stability (typically 0.01% over a few-minute period) of the synthesizer is about an order of magnitude better than that available from high-quality generators operating on a regulated power line.

For versatile coverage from 50 to 500 Mc the *-hp-* 10511A Spectrum Generator accessory shown elsewhere herein can be used to provide more than 25 mv at any desired frequency over this range. Fixed or tunable filters are commercially available to eliminate the undesired harmonics of the output spectrum.

-hp-'s Dymec division has a system available for tunable coverage over the range from 0.8 to 12.4 Gc using microwave oscillators phase locked to harmonics of the synthesizer's output frequency.

The synthesizer phase noise and spurious signals will be deteriorated by at least the harmonic multiplication factor in any frequency multiplication scheme. Parametric harmonic generation in particular must be carefully evaluated.

GENERAL

The performance and versatility of the new synthesizer are such that it is reasonable to expect that many new and significant applications will come to light as scientists and engineers contemplate the potentialities of this system.

—Victor E. Van Duzer

*Leonard S. Cutler, "Examination of the Atomic Spectral Lines of a Cesium Beam Tube with the -hp- Frequency Synthesizer," **Hewlett-Packard Journal**, Vol. 15, No. 4, December, 1963.

SYNTHESIZER DESIGN LEADERS

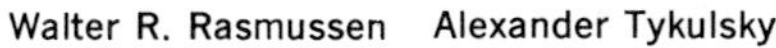

Victor E. Van Duzer — Hans H. Junker — Theodore G. Pichel — Walter R. Rasmussen — Alexander Tykulsky

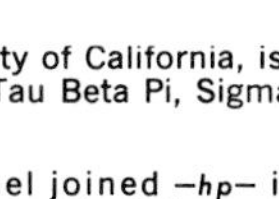

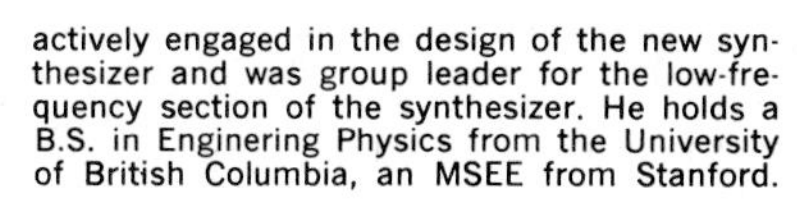

Vic Van Duzer joined -hp- in 1958 as a development engineer. He worked on the -hp- sampling oscilloscope, later became group leader for the initial work on the -hp- 215A Pulse Generator. For the past three years he has been the section manager for the frequency synthesizer. He holds a BSEE from the University of Illinois, an MSEE from Stanford, and has received patents with others pending.

Hans Junker joined -hp- in 1960, working on the development of the 104AR and other frequency standards. He was group leader for the driver section of the frequency synthesizer and is presently group leader of the Atomic Standards group. He holds a BSEE and MSEE from the University of California, is a member of Eta Kappa Nu, Tau Beta Pi, Sigma Xi, and Phi Beta Kappa.

Ted Pichel joined -hp- in 1956 and has worked as a product design engineer on -hp- digital recorders, the -hp- delay generator, frequency standards and on the new synthesizers. He received a B.E. in mechanical engineering from Yale University and has done engineering in the aircraft field on jet and missile equipment. He holds patents and has others pending.

Wally Rasmussen joined -hp- in 1958 and was assigned to an investigation project on frequency synthesis. Since 1961 he has been actively engaged in the design of the new synthesizer and was group leader for the low-frequency section of the synthesizer. He holds a B.S. in Enginering Physics from the University of British Columbia, an MSEE from Stanford.

Al Tykulsky joined -hp- in early 1962 as group leader for the UHF section of the new synthesizer. He has had experience in the design of multichannel FM and radio-relay equipment, in UHF circuitry for TV, and prior work in the design of VHF and UHF synthesizers. He has a BSEE from CCNY and an MSEE from Rutgers, and has patents relating to SSB generation, duplex communications, a pattern generator, and a sweeping oscillator.

HEWLETT-PACKARD JOURNAL
TECHNICAL INFORMATION FROM THE -hp- LABORATORIES
Vol. 15, No. 11
PUBLISHED BY THE HEWLETT-PACKARD COMPANY, 1501 PAGE MILL ROAD, PALO ALTO, CALIFORNIA
JULY, 1964

A New Performance of the "Flying Clock" Experiment

A new experiment has been made to compare clocks in the U.S. and Europe to higher precision. The results will be used by the various official agencies of the two continents in improving their time synchronization.

Fig. 1. *Data being logged in Switzerland at the Observatoire de Neuchatel by Alan S. Bagley of the -hp- Frequency and Time Laboratories. Atomic clock in left foreground carried time to Switzerland aboard transatlantic passenger flight.*

THE PROBLEM of correlating the time of day at widely separated locations with great accuracy is one that has absorbed chronologists, navigators, astronomers and cartographers for hundreds of years. As vehicles grow faster and the range of exploration reaches far into outer space, more accurate time determinations become increasingly necessary. Precise timing and coordination of events far apart may determine the success of precise mapping operations, satellite orbital placement, astronomical observations, or missile landings.

In recent years, intercontinental time of day comparisons gradually reached an accuracy of about one millisecond through use of h-f radio signals, the limit being imposed by propagation-time uncertainties of high frequency waves. Considerably higher accuracy is, however, possible. For example, two recent means have emerged to achieve intercontinental time of day correlations with accuracy of the order of a few microseconds.

One method has been to fly an accurate clock, set precisely to a given time standard, from point to point, effectively bringing the time standard to each. U. S. government experiments have used quartz oscillator-driven clocks as well as "atomic" clocks in this manner. In 1960, for example, Reder, Brown, Winkler, and Bickart, of the U. S. Army Signal Research and Development Labora-

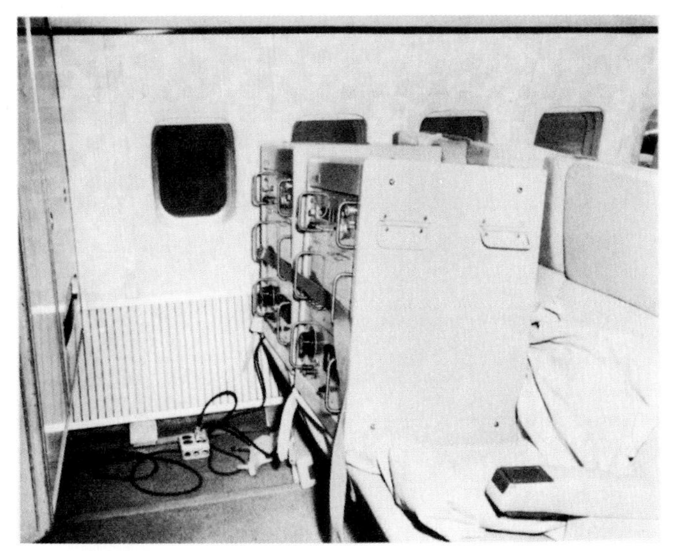

Fig. 2. *The two atomic clocks being fastened in passenger seats aboard airliner. Operating power in flight was supplied by plane.*

tory employed the flying clock method[1] to achieve synchronization of within about 5 microseconds among Pacific Ocean Area stations. In this instance, a KC-135 jet tanker, carrying an atomic clock, flew the time from station to station.

[1] F. Reder, P. Brown, G. Winkler, and C. Bickart, "Final Results of a World-Wide Clock Synchronization Experiment," Proc. of the 15th Annual Symposium on Frequency Control, 1961.

In a second method, Telstar I was used as a relay to synchronize clocks[2] between the U. S. and Great Britain. An accuracy of about one microsecond was achieved between these points.

[2] J. McA. Steele, Wm. Markowitz, and C. A. Lidback, "Telstar Time Synchronization" **Conference on Precision Electromagnetic Measurements,** June 23-25, 1964. To be published in **IEEE Transactions on Instrumentation and Measurement.**

See also: Wm. Markowitz, "International Frequency and Clock Synchronization," **Frequency,** Vol. 2, No. 4, July-August, 1964.

Recently, engineers from the Frequency and Time Laboratories of the Hewlett-Packard Company performed the flying clock experiment anew, this time between the U. S. and continental Europe (Switzerland) using a newly-developed atomic frequency standard and clock. The equipment successfully compared time of day standards of the U.S. with those of Switzerland to an accuracy of about one microsecond. The equipment was sufficiently light and small that it was flown as a "passenger" on regularly scheduled passenger airlines.

The -*hp*- atomic clock became available as the result of the development in the -*hp*- Frequency and Time Laboratories of a new cesium beam frequency standard of very high performance. The new standard was to be described in a paper at the International Conference on Chronometry, a conference held every five years at Lausanne, Switzerland. In this way a convenient opportunity arose on the same jour-

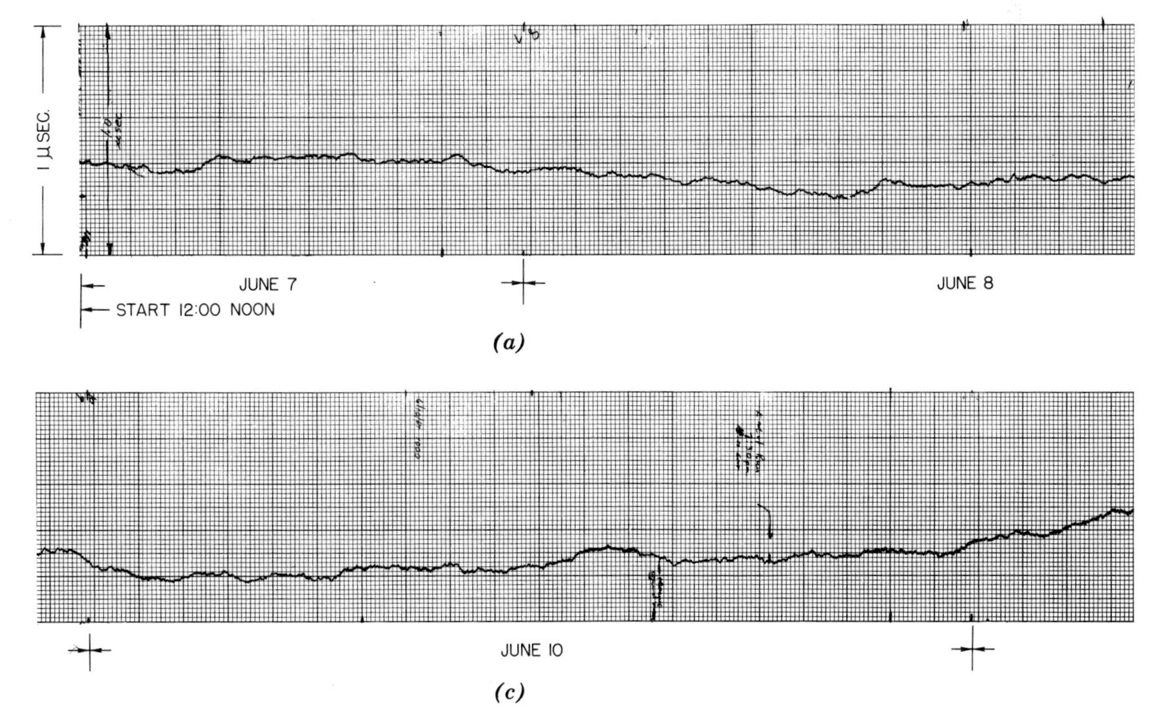

Fig. 3. *Strip recording made at exhibit associated with International Conference on Chronometry in Lausanne, Switzerland. Record is of the time difference between the two atomic clocks used in the experiment and is thus a*

ney to use the new standard to compare time of day standards in the U.S. and Switzerland and also to compare the performance of the new standard with the instruments used by both countries for their primary frequency standard determinations.

Two of the new *-hp-* cesium-beam frequency standards were used. Each was combined with an *-hp-* Model 115BR electronic clock which, among other functions, provides accurate electrical "ticks" from the standard frequency for comparison purposes. A special standby battery power supply was included, resulting in a complete atomic clock. Since the equipment would have to be powered from whatever source might be at hand, the supply was designed to accept power from a variety of frequencies and voltages, including dc voltages. Even an auto cigar-lighter plug was provided.

As preparation for the experiment, two newly-manufactured standards were carefully checked and their C-fields adjusted to the prescribed value. The standards were then turned on, the first time they had ever operated. Measurements were begun to compare their frequencies by means of VLF phase comparisons with the frequencies of National Bureau of Standards stations WWVL and WWVB at Boulder, Colorado. No adjustments of any sort were made on the standards. A 12-day comparison with these stations showed the frequency of the standards to be within 5 parts in 10^{12} of the United States Frequency Standard.

The plan for the experiment was first to transport the standards to Washington to compare their time reading against the official U. S. time standards at the U. S. Naval

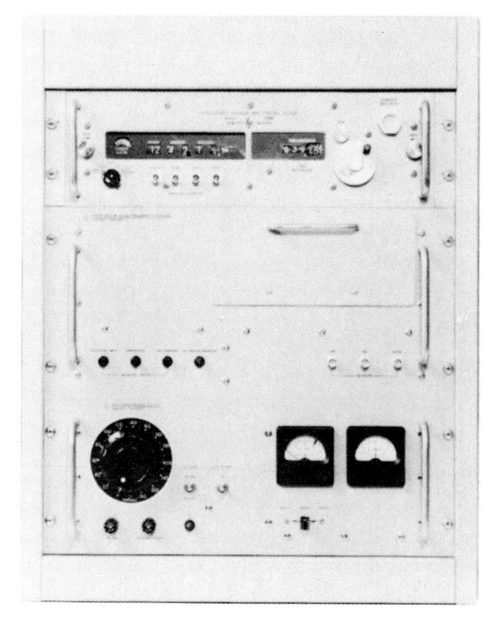

Fig. 4. *Panel view of one of "atomic" clocks used in experiment. Cesium standard in middle drives electronic clock (top) which integrates standard frequency and also produces very accurate electrical "ticks" based on driving frequency. Power supply unit at bottom provides for operation from a range of ac and dc voltages.*

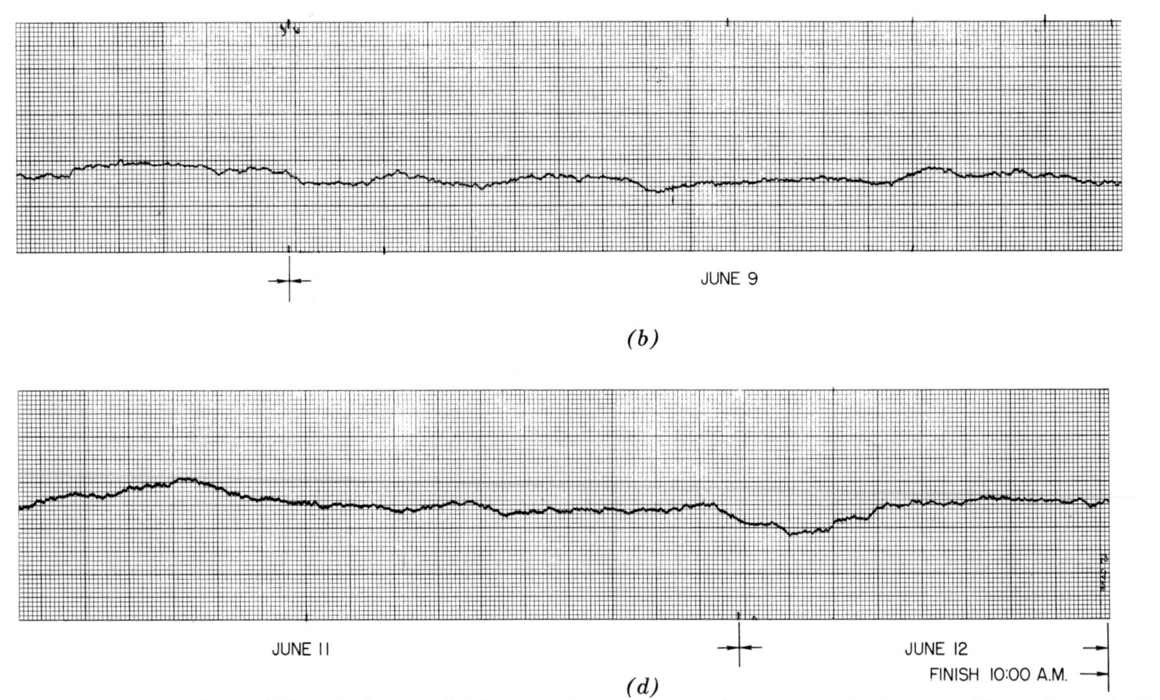

(b)

(d)

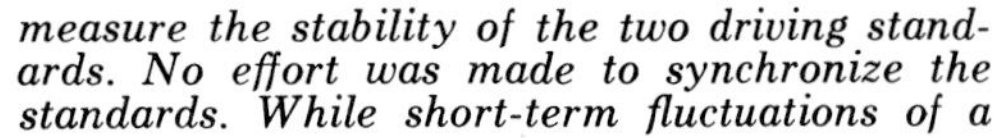

measure the stability of the two driving standards. No effort was made to synchronize the standards. While short-term fluctuations of a larger magnitude are evident, the record shows the standards were within 2 parts in 10^{13} of one another as averaged over the full 5-day period.

Fig. 5. *Frequency comparison being made between one standard and the "long" cesium beam standard (foreground) at the Laboratoire Suisse de Recherches Horlogeres, Neuchatel, Switzerland. This long standard is considered one of the world's best cesium standards.*

Observatory. The next part was to transport the standards to Switzerland for a comparison against the Swiss time standards and to make measurements of the time of arrival there of WWV time ticks. The clocks would then be returned to NBS station WWV for time comparisons that would permit calculating propagation time. These would include the time of occurrence of the WWV ticks and a comparison against WWV clocks. The standards would next be taken to the Naval Observatory for a second check there against the U. S. master clock to determine the tolerance on the time measurements.

Lastly, a stop would be made at NBS, Boulder, Colo., to compare the standards against the long cesium-beam standard maintained there and to make further time checks.

On Friday, June 5, 1964, the experiment was begun by taking the standards to San Francisco International Airport by station wagon. At the airport they were placed in passenger seats on a regularly-scheduled DC-8 flight bound for Washington, D. C. Power was provided during the flight by the DC-8's electrical system.

In Washington on June 5, the units were transported to the U. S. Naval Observatory for a comparison against the official U. S. time standards. Records were kept of the comparison for future use.

SWITZERLAND

That evening, the clocks were taken by rental station wagon to New York. During this and other ground-transport times throughout the trip, the standards operated from external storage batteries or from internal batteries while being hand-carried. At New York they were placed aboard a regular transatlantic flight, again operating from the plane's power. The next morning, Saturday, June 6, the standards arrived in Switzerland, and were transferred to the Observatoire de Neuchatel. Time of day comparisons were then made at this Swiss observatory, which maintains the Swiss national standard. In addition, one standard was driven to the Laboratoire Suisse de Recherches Horlogeres Neuchatel for a frequency comparison with the long cesium-beam standard maintained there.

At this time in Neuchatel, data were recorded in raw form because of tight schedules. Some time later, however, calculations revealed the time difference between Swiss and U. S. time of day standards. This had already been known, from HF radio comparisons, within about a millisecond. The comparison made with the Hewlett-Packard flying clocks established the value within about one microsecond.

Late Saturday afternoon the standards were taken from Neuchatel and, after an overnight stop, arrived at the exhibit of the International Conference on Chronometry in Lausanne, Switzerland, on the morning of Sunday, June 7.

At noon on Sunday a chart recording was begun of the difference between the time "ticks" produced by the two standards. This record was made continuously during the exhibit in public view until the morning of Friday, June 12. The record, reproduced here, showed that the two *-hp-* standards were within 2 parts in 10^{13} of each other as averaged over the full five days. Again, no effort was made to standardize or synchronize the standards before making the record or at any time before or during the whole experiment.

An opportunity also arose during the exhibit to make a high-precision comparison of the frequency of the cesium-beam standards with that of a hydrogen maser; this is described in an accompanying article.

On Saturday, June 13, the standards were returned to the Observatory in Neuchatel for further measurements against the "long" cesium-beam standard.

The trip, in fact, permitted the *-hp-* standards to be compared against two of the best of the world's long cesium-beam standards: the one in Neuchatel and, later, that of the U. S. Bureau of Standards in Boulder, Colorado, which is used with other standards to maintain the U. S. Frequency Standard.

Since the long-beam standard is actually located at the Laboratoire Suisse de Recherches Horlogeres Neuchatel, a telephone line was used in the comparison. The –*hp*– standards were measured as being $(+2 \pm 6) \times 10^{-12}$ and $(+6 \pm 5) \times 10^{-12}$, respectively, with respect to the long-beam standard.

Fig. 6. *Leonard S. Cutler (left), –hp– Frequency and Time Laboratories, and Dr. Jacques Bonanomi, director of Observatoire de Neuchatel, Switzerland.*

UNITED STATES

On Monday morning, June 15, the standards were flown to New York, again as "passengers" on a regular flight. From New York they were taken to station WWV at Beltsville, Maryland, arriving early on Tuesday, June 16. Upon arrival, measurements of the time of occurrence of WWV time ticks were made to establish the propagation time between Neuchatel and WWV to enable the other time comparisons to be made. This measurement yielded a propagation time of 23,709 ±200 microseconds, the tolerance being mainly the uncertainty in establishing the time of arrival of ticks in Neuchatel. Prior to the experiment the propagation time had been estimated at 22,800 microseconds.

Later Tuesday, the standards were taken to the Naval Observatory for further time checks. On the basis of these measurements the tolerance of approximately 1 microsecond in the correlation between the time of day in the United States and Switzerland was established.

In addition, however, measurements showed that the time kept by the –*hp*– standard was still within 1.5 microseconds of the time kept by the Naval Observatory. In terms of frequency, this measurement meant that the –*hp*– standard was within 2×10^{-12} of the weighted mean of the standards utilized by the Naval Observatory for the complete period from June 5 to June 16.

The time difference on June 16 between station WWV ticks and the Naval Observatory master clock was also measured, the value being 2397 microseconds.

The next morning, Wednesday, June 17, the standards while operating on airplane power were flown from Washington D. C., to Denver, Colorado, and then driven by car to the Bureau of Standards in Boulder. Checks were there made against the long cesium-beam standard referred to earlier and known as U. S. Frequency Standard No. 3. These checks showed that the frequency of the two –*hp*– standards were $(-3.9 \pm 3) \times 10^{-12}$ and $(-2.7 \pm 3.3) \times 10^{-12}$ with respect to the NBS standard. The measurements were based on 44 and 45 200-second samples, respectively. Measurements were also made of propagation time between Washington and Boulder.

On Thursday the standards were returned to Denver, placed aboard a regular airline and operated on airplane power en route to San Francisco, and finally taken to the –*hp*– F and T laboratories in Palo Alto.

TABLE OF MAJOR CHARACTERISTICS
-hp-
Cesium Beam Frequency Standard

OUTPUT FREQUENCIES:	5 MHz, 1 MHz, and 100 kHz, in A-1 or UT_2 time scale as specified.
ACCURACY:	±2 parts in 10^{11}.
LONG-TERM STABILITY:	Within ±1 part in 10^{11}.
SHORT-TERM STABILITY:	RMS Fractional Frequency Deviation $\frac{(\Delta f_{rms})}{f}$ for 1-second averaging is less than 1 part in 10^{10}.
NOISE-TO-SIGNAL RATIO:	At least 87 db below rated 5 Mc output; filter bandwidth is approximately 125 cps.
CESIUM BEAM TUBE LIFE:	10,000 hours operating guaranteed.

GENERAL

The experiment is considered to have been highly successful with the following information established with an accuracy not previously attained:

(a) The time of day was correlated in Switzerland with that of the United States to an accuracy of about one microsecond.

(b) The propagation time between station WWV and Neuchatel was measured to a tolerance of about 200 microseconds.

(c) The two –*hp*– standards were compared with two well-known long

(continued on p. 8)

FLYING CLOCKS *(from p. 5)*

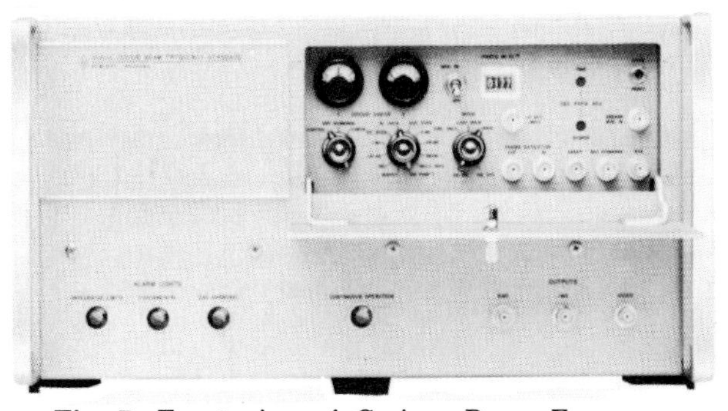

Fig. 7. Front view of Cesium Beam Frequency Standard. Cesium beam standards utilize a quantum-mechanical effect to produce ultra-stable frequencies. Two standards were combined with electronic clocks and power supplies to form an "atomic" clock with stabilities of a few parts in 10^{12}.

cesium-beam standards and found to be in agreement with them within a few parts in 10^{12}, the second series of measurements being made after a trip of ten thousand miles and an elapsed time of some 13 days.

(d) The two –hp– standards agreed with one another within a few parts in 10^{12} after the full trip.

(e) A measurement to high precision was made to compare the frequencies of a hydrogen maser and a cesium beam standard.

(f) Lastly, it is understood that the results of the time measurements will be used to obtain better synchronization by the various agencies in the United States and Europe.

The standards operated without flaw for the full trip, although one associated circuit suffered an electronics failure and momentary loss of power. This invalidated the clock reading of that unit but did not affect the accuracy of the cesium-beam tube, as indicated by the final frequency readings made at NBS, Boulder.

Many people contributed generously of their efforts to the experiment. These include Dr. George E. Hudson, David Allen, and James Barnes of the Bureau of Standards, Boulder; Dr. Wm. Markowitz and C. A. Lidback of the U. S. Naval Observatory, Washington; Dr. Jacques Bonanomi, Director of the Observatoire de Neuchatel; and Dr. Peter Kartaschoff of the Laboratoire Suisse de Recherches Horlogeres Neuchatel. LaThare N. Bodily of the –hp– Frequency and Time Division made many measurements on the Washington-to-Palo Alto portion of the experiment. Appreciation is expressed to –hp–'s Frequency Standard Group in preparing the equipment needed for the experiment.

Special thanks are also given to the people of United Air Lines and Swissair who made the needed arrangements.

— Alan S. Bagley and Leonard S. Cutler

Copies of the paper, "A Modern Solid State Portable Cesium Beam Frequency Standard," presented to the International Conference on Chronometry, are available from the editor, **Hewlett-Packard Journal.**

AUTHORS

Alan S. Bagley

Al Bagley became affiliated with –hp– while earning his MS degree at Stanford University in 1949. His first project was the development of a high-speed scaler, and he subsequently became project leader on a program applying the scaler circuitry to a frequency counter, leading to the industry's first high-speed counter, the –hp– Model 524A. He has been project leader on many of –hp–'s first generation of counters and also on the well-known –hp– 560 series digital recorders. In 1958, he was appointed engineering manager of the electronic counter group in the –hp– R and D Department and later became manager of the –hp– Frequency and Time Division when that activity assumed divisional status.

Leonard S. Cutler

Len Cutler joined Hewlett-Packard in 1957, following several years of experience in development supervision of frequency-measuring instruments. At –hp– he has been section leader of the frequency standards group and was responsible for development of the –hp– high frequency counter time bases. He was also responsible for the development of the –hp– Models 103, 104, 106, and 107 Frequency Standards and the –hp– 5060A Cesium Beam Standard. At present, he is director of quantum electronics in –hp–'s Physics Research and Development Group. He holds a BS and MS in physics from Stanford University and is presently completing work on his doctorate in physics.

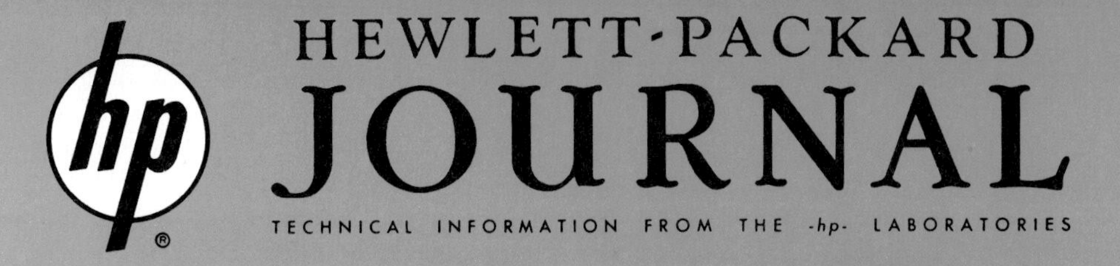

Vol. 15, No. 12

PUBLISHED BY THE HEWLETT-PACKARD COMPANY, 1501 PAGE MILL ROAD, PALO ALTO, CALIFORNIA AUGUST, 1964

A New Microwave Spectrum Analyzer

A new spectrum analyzer displays up to 2,000 Megacycles of the spectrum at a time for RFI checking, spectrum surveillance, and other wide-band investigations.

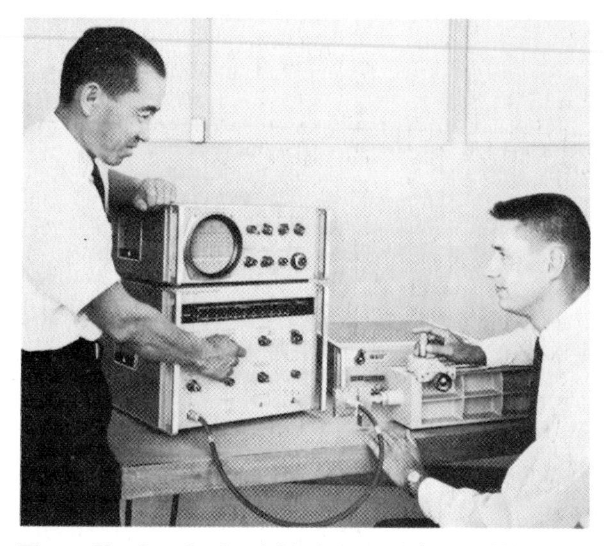

Fig. 1. *Hewlett-Packard Model 8551A/851A Microwave Spectrum Analyzer enables evaluation of signals widely separated both in amplitude and frequency. Analyzer here displays harmonic and spurious outputs of frequency doubler throughout range from 10 Mc to above 10 Gc. Doubler generates 3800 Mc output in response to 1900 Mc input and Analyzer clearly shows how mistuning can cause certain diodes to produce non-harmonically related parametric oscillations.*

MICROWAVE spectrum analyzers are swept-frequency receivers with visual display. The basic function of a spectrum analyzer is to plot input signal amplitude as a function of frequency while the analyzer automatically tunes through a selected range of frequencies.

Spectrum analyzers are potentially useful in the laboratory as well as in the field because they provide quantitative information about several signals simultaneously. For instance, they facilitate adjustment of devices that operate with several different frequencies at the same time, such as parametric amplifiers. A spectrum analyzer connected to an antenna can display a plot of all detectable electromagnetic radiations over a selected frequency band, indicate the relative strength of each signal, and show individual frequency components of any one of the signals to provide useful information on sideband structure.

It has not always been possible, however, to make measurements with spectrum analyzers in a simple, straightforward manner. Broad-band analyzers often include image responses and a

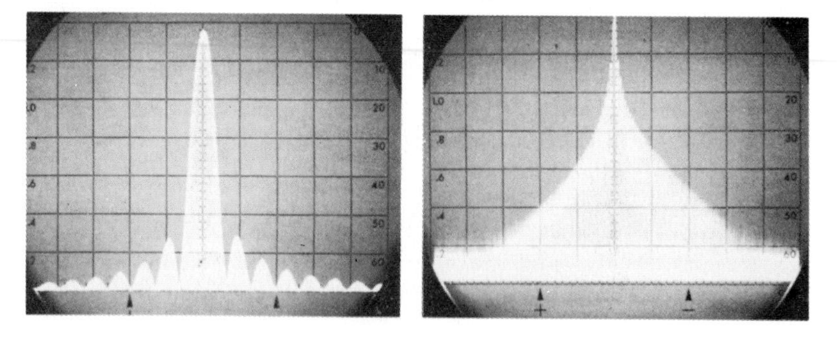

Fig. 2. *Wide measurement range of new -hp- Spectrum Analyzer is indicated by this pair of photos. First photo shows spectrum of RF pulse train as displayed with linear vertical deflection and 30 Mc spectrum width centered at 1500 Mc. Amplitudes of components beyond 12 Mc from center are not easily evaluated. Second photo shows same spectrum with logarithmic vertical deflection characteristics and 300 Mc spectrum width. Components out to 75 Mc from center are easily evaluated though they are more than 50 db down from center frequency (vertical calibration is 10 db/cm).*

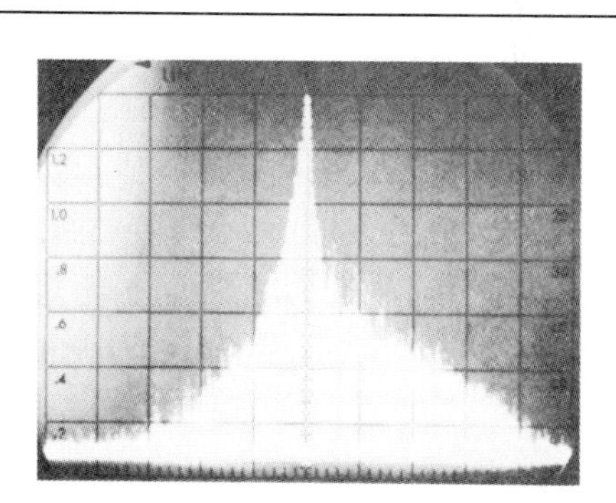

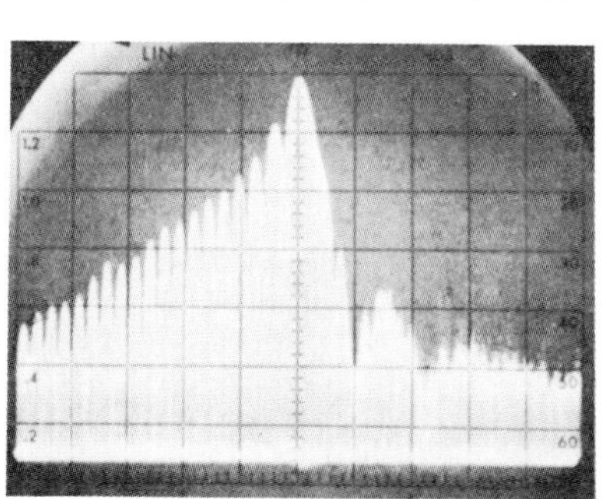

SPECTRUM SIGNATURES

Oscillograms of output frequency components of L-band radar illustrate usefulness of large dynamic display range of –hp– Spectrum Analyzer for spectrum signature measurements. Left photo, with 10 Mc/cm spectrum width, shows that radar generates measurable power in sidebands over more than 70 Mc range. Photo also shows transients that would escape notice on slower acting graphic recorders. Right photo, made with spectrum width of 3 Mc/cm, indicates how calibrated spectrum widths enable measurement of spacing between groups of frequency components; spacing of just under 1 Mc between envelope maxima indicates pulse width of slightly more than 1 μsec.

variety of spurious responses resulting from harmonic and intermodulation products. Marker generators have been necessary to provide quantitative information in an indirect way.

It was for the purpose of eliminating operational difficulties and of making spectrum analysis a useful, flexible laboratory tool that Hewlett-Packard embarked upon the development of a spectrum analyzer and as a result, a spectrum analyzer is now available that not only reduces spurious responses significantly, but which is easier to use, in which all basic functions are completely calibrated, and which has a wider range of capabilities.

THE –HP– SPECTRUM ANALYZER

The new Hewlett-Packard 8551A/851A Spectrum Analyzer, in brief, is a multi-band instrument that responds to signals throughout a 10 Mc to 40 Gc frequency range and that is capable of sweeping over frequency ranges as wide as 2000 Mc. The instrument can sweep over narrow ranges as well, the necessary stability being provided by an unusual phase-lock system that locks the local oscillator to a stable reference oscillator even while tuning. Exceptionally flat frequency response is achieved, variations being less than ±5 db over the full 2000 Mc sweeping range and ±1 db over any 200 Mc range using fundamental mixing (±2 db when working on local oscillator second harmonic).

The display circuits enable simultaneous display of signals having a wide 60 db amplitude range. The input mixer has a dynamic range greater than 95 db and a new RF attenuator increases the input range up to +30 dbm (1 watt).

The instrument establishes a new high with respect to freedom from spurious and residual responses. Operating features include a "Signal Identifier," that quickly identifies the frequency range of all displayed responses. The new –hp– Spectrum Analyzer requires far fewer controls than other wide-band analyzers, and the controls are calibrated so that frequency and relative amplitude of signals may be read directly without use of markers. The display dynamic range and wide frequency response of this instrument enable direct measurements on signals widely separated both in frequency and in amplitude.

WIDE FREQUENCY RANGE

The Hewlett-Packard approach to the design of a microwave spectrum analyzer is to use a backward-wave oscillator as the first local oscillator in a swept-tuned, broad coverage, triple-conversion receiver. The backward-wave oscillator, which has been found to have better spectral purity than other voltage-tuned microwave tubes, can be tuned manually from 2 to 4 Gc and sweep-tuned electrically around the selected center frequency.

As shown in the simplified block diagram of Fig. 3, the BWO signal heterodynes with the input signals in the front-end mixer. Wide coverage from 10 to 40,000 Mc is obtained since there are no frequency sensitive circuits ahead of the first mixer. Any signals that generate a 2 Gc mixing product with the BWO signal then pass through the first IF amplifier (2 Gc) and subsequently appear as a vertical deflection on the cathode-ray tube.

The cathode-ray beam sweeps horizontally in synchronism with the BWO frequency sweep. A particular signal frequency component therefore appears as a vertical deflection at some point along the horizontal scale of the CRT graticule, as shown on page 8. The frequency of the signal component can be read from its position on the CRT, with reference to the settings of the calibrated *Tune* and *Spectrum Width* controls.

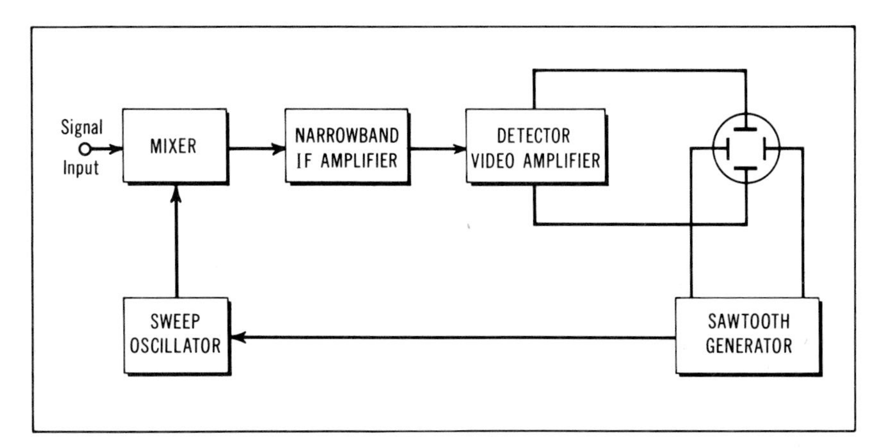

Fig. 3. *Simplified block diagram of wide-band spectrum analyzer.*

Fig. 4. *Hewlett-Packard Model 8551A/851A Microwave Spectrum Analyzer is packaged in two modules that may be used on bench or mounted in standard relay racks with brackets supplied. Standard -hp- 5-inch CRT bezel mounts oscilloscope cameras directly. Pointer on frequency dial shows frequency of BWO first local oscillator on topmost scale. On lower scales, pointer shows frequency of response appearing at center of CRT display.* Frequency *switch selects appropriate scales to correspond to input frequency range; number in window at left shows which harmonic of BWO generates responses in that frequency range. If frequency range of any response is not known,* Signal Identifier *(see text) quickly indicates proper scale to be used for measuring frequency of response.*

The first IF amplifier, a cavity-tuned triode for large dynamic range, has a center frequency of 2000 Mc. The high 1st IF frequency provides wide image separation (4000 Mc), an important feature in a class of instruments where image frequencies have been troublesome. Wide image separation also makes it easier to use RF preselection filters for selecting or eliminating certain bands of frequencies.*

Spurious responses resulting from mixer operation have been a major cause of operating difficulties in previous spectrum analyzers. Considerable effort was expended at Hewlett-Packard in the design of a wideband mixer that would generate a minimum amount of intermodulation. The new mixer responds to signals up to 0 dbm without signal compression, achieving a wide dynamic range. Particular attention also was paid to achieving flat frequency response in the mixer over the broad sweep widths.

A newly-designed 0-to-60 db attenuator precedes the mixer to reduce strong signals below the point where they would overdrive the mixer and create spurious responses. Although waveguide-beyond-cut-off attenuators have been used previously, this attenuator is significant because of its low insertion loss, which is 0 db at the low frequency end, and no more than 2 db at 10 Gc. No cable patching is required to reduce the attenuation to zero when examining weak signals.

CALIBRATED CONTROLS

A major design effort was applied toward achieving calibrated controls for the new analyzer. Calibrated controls make it possible to read signal frequency and relative amplitude from the cathode ray tube graticule, and also assure measurement repeatability.

The tuning dial pointer indicates the frequency of a signal appearing at the center of the CRT. The *Spectrum Width* control, calibrated in kc or Mc per centimeter of CRT horizontal deflection, selects the range of the frequency sweep which is centered at the frequency selected by the *Tune* control.

The CRT graticule is calibrated vertically with three scales that correspond to the IF amplifier characteristics selected by the *Vertical Display* switch. The amplifier response can be linear, or proportional to the square of input signal voltage and thus indicative of power, or logarithmic. The 7 cm CRT graticule Log Scale is calibrated in db at 10 db per cm so that with the switch in the *Log* position, it is possible to view simultaneously signals that have more than a 60 db (1,000,000 to 1) range.

The input RF attenuator, the IF attenuator, the IF Bandwidth, and the Sweep Time switches also are calibrated.

RESOLUTION

The resolution of an analyzer, i.e., its ability to separate closely spaced signals on the frequency scale, is an important characteristic that is determined by the IF amplifier passband. Two adjacent signals merge into one response if the passband is broad enough to pass both signals at the same time, but will be plotted as two separate signals if the passband is narrow compared to the signal separation.

Although a narrow passband is desirable for high resolution, it also slows the rise time of the IF amplifier. If the passband is too narrow, the analyzer may sweep across a signal too fast for the IF voltage to build up to a steady-state level.

* Interdigital bandpass filters, —hp— Series 8430A, are available for use with the Spectrum Analyzer.

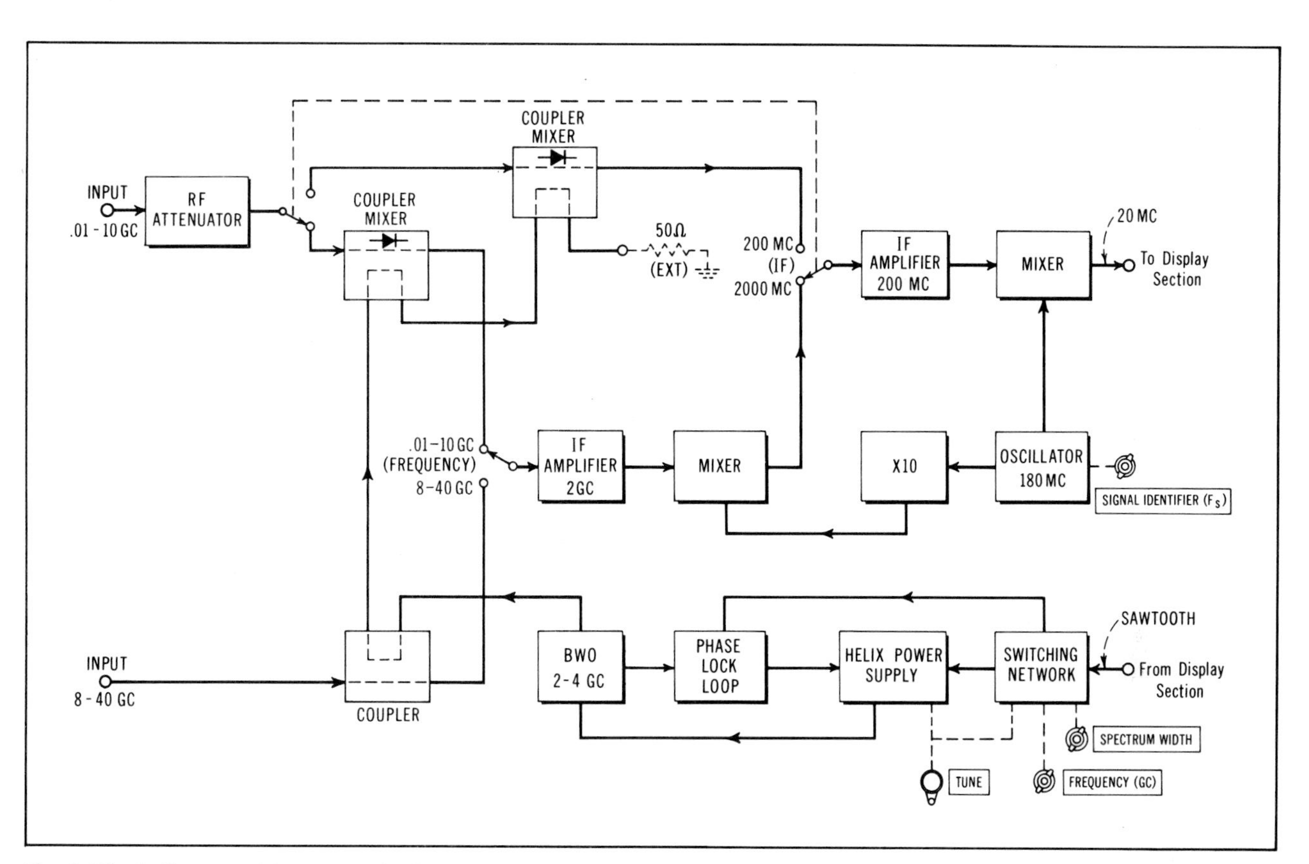

Fig. 5. *Block diagram of Spectrum Analyzer RF Section. Only two inputs are needed to cover entire 0.01 to 40 Gc frequency range. Newly-developed RF attenuator does not require separate input connector (see text). Input for 8–40 Gc signals sends BWO signal to external waveguide mixer and receives 2–Gc IF signal over same connecting cable.*

The narrowest practical bandwidth therefore depends on the rate at which the instrument changes frequency, which is a function of both sweep speed and sweep width. For this reason, the bandwidth of the *–hp–* analyzer is not fixed but can be switched to be 1, 3, 10, 100, or 1000 kc. Ordinarily, a trial and error procedure is used to find the optimum bandwidth but this can be done automatically by a new operating feature on the *–hp–* analyzer. An *Auto Select* position on the *I.F. Bandwidth* switch programs the instrument to select the optimum bandwidth for each combination of spectrum width and sweep rate.

Examination of the spectra of short pulses, on the other hand, requires a broad passband. If the pulse width is shorter than the risetime of the IF amplifiers, the pulse will be attenuated because of the slow risetime. The *–hp–* analyzer provides a 1 Mc bandwidth with fast risetime to reduce the attenuation of short pulses. Although the broad bandwidth may allow individual frequency components to overlap on the display, the spectrum envelope is preserved and this is of prime interest here.

SWEEP-TUNED PHASE LOCK

Short-term stability of the electrically-tuned BWO is 1 part in 10^5, which means short term instabilities of typically 10 to 30 kc, undetectable on wide frequency sweeps. This instability becomes significant on very narrow sweeps, however.

To provide the stability necessary for narrow sweeps ($<$1 Mc/cm), a tunable phase-lock system locks the backward wave oscillator to a harmonic of a stable 10 Mc reference oscillator. The reference oscillator sweeps in synchronism with the BWO and the phase lock system impresses the stability of the 10 Mc oscillator on the BWO. Short term stability of the BWO is better than 1 kc p-p when locked.

A further refinement allows the analyzer to be tuned while the BWO is phase-locked. The 10 Mc oscillator itself cannot be tuned over a wide range, but as the instrument is tuned, the reference oscillator tunes through its range over and over again while the phase detector automatically passes from harmonic to harmonic of the oscillator to retain phase lock. The BWO thus can be tuned through its entire range while phase-locked.

SIGNAL IDENTIFIER

Wide-band spectrum analyzers that use a swept first local oscillator without front-end pre-selection respond to signals that are anywhere within the bandwidth of the front end (e.g., 10 Mc to above 40 Gc). All signal frequencies within this range that develop suitable mixing products with any local oscillator harmonic will appear on the display. Identification of these responses has been one of the major problems with previous wide-band analyzers.

To quickly identify the frequency of any displayed response, the *–hp–* Spec-

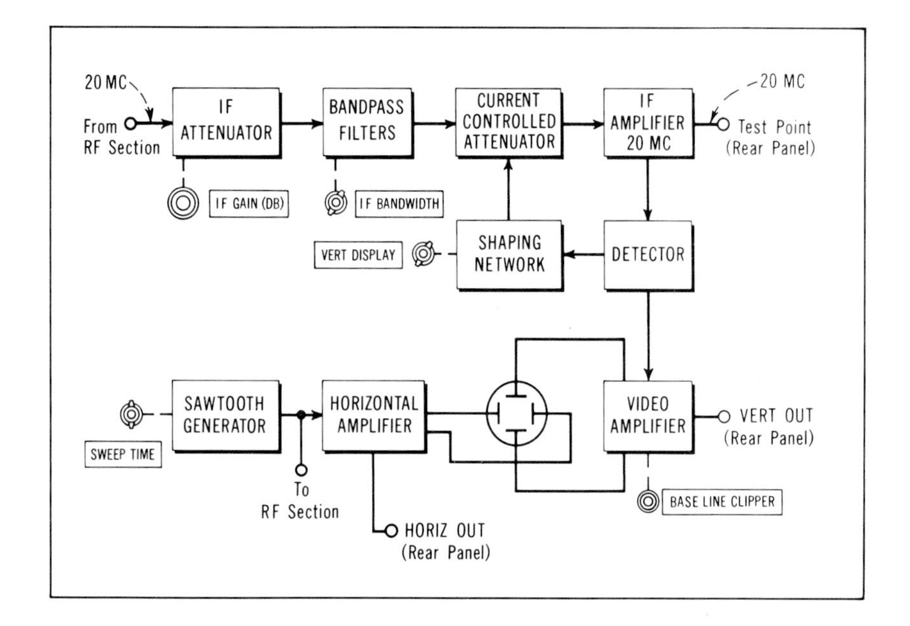

Fig. 6. *Block diagram of Spectrum Analyzer Display Section.*

trum Analyzer has a signal identifier. The identifier works by shifting the frequency of the second and third local oscillators. When the shift is made, a particular input signal must beat with a slightly different first local oscillator frequency in order to pass through the narrow-band 20 Mc third IF. This means, then, that a signal develops the proper IF frequency at a different point in the frequency sweep and as a result, the signal appears at a different point along the CRT horizontal scale. The amount of horizontal displacement depends not only on the frequency shift in the second and third LO but also on the harmonic number of the BWO harmonic that is beating with the signal.

The *Signal Identifier* is a multiposition switch that steps the frequency shift in the 2nd and 3rd LO signals to progressively greater degrees until the response under examination moves exactly 2 cm along the horizontal scale of the CRT. The switch is labeled to show which harmonic number then is producing the response and thus which frequency scale should be used (or apply the formula $F_s = nF_{LO} \pm 2$ Gc). Furthermore, whether the response shifts to the right or to the left on the horizontal scale depends on whether the signal lies above or below the particular harmonic. The direction of shift indicates which one of each pair of frequency scales is to be used.

OTHER FEATURES

A block diagram of the –*hp*– 8551A Spectrum Analyzer RF Section is shown in Fig. 5. Only two input connectors are required to cover the wide range of input frequencies from 10 to 40,000 Mc. One input is coaxial and responds to all frequencies from 10 Mc to 10 Gc. Since higher frequencies are most often encountered in waveguides, external waveguide mixers are available for frequencies from 8 to 40 Gc. The external mixers, connected to the other analyzer input through a single coaxial cable that carries both the local oscillator (LO) and IF signals, can be attached to the RF system at any convenient point. External mixers make it unnecessary to plumb waveguide directly into the analyzer.

The sawtooth voltage that sweeps the CRT is applied to the BWO helix power supply to frequency sweep the BWO. The amplitude of the sawtooth is controlled by the *Spectrum Width* control, which selects the width of the frequency sweep, and also by the *Frequency* (range) switch so that the frequency sweep of the indicated BWO harmonic is that shown by the *Spectrum Width* control. Because of the frequency vs. voltage characteristics of the backward wave oscillator, the sawtooth is reshaped to have a logarithmic rather than a linear rise. The center frequency of the backward wave oscillator is selected by changing the dc level of the helix voltage. A vernier *Spectrum Width* control enables continuous reduction in spectrum width all the way down to zero. At zero spectrum width, of course, the instrument functions as a fixed-tuned receiver.

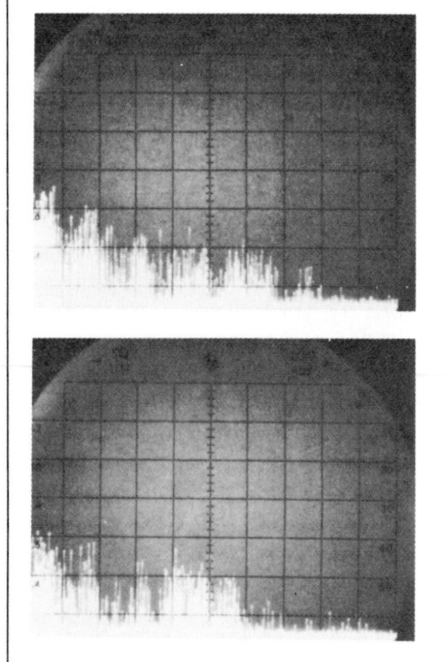

EMC/RFI

Electromagnetic compatibility (radio frequency interference) measurements are readily made with –hp– Spectrum Analyzer connected to one of standard EMC/RFI antennas. Photos here show radiations from automobile ignition system with hood up (top photo) and slight improvement obtained with hood down (bottom photo). Photos demonstrate how shielding effort can be evaluated on the spot with a spectrum analyzer.

The front-end mixer shown at the top of the block diagram of Fig. 5 is for use with input signals that are close to 2 Gc in frequency. This mixer feeds the 200 Mc second IF directly, bypassing the 2 Gc IF, so that there can be no feedthrough of 2 Gc signals. This arrangement is selected when the instrument controls are set for the 1.8-4.2 Mc range. Image separation here, however, is 400 Mc but sensitivity is about 10 db higher than with the wideband mixer.

After conversion to a 20 Mc intermediate frequency the signal travels to the *-hp-* Model 851A Spectrum Analyzer Display Section, shown in the block diagram of Fig. 6. In the all solid-state Display Section, the 20 Mc signal passes through the precision IF attenuator, through the filter sections, and through the current-controlled attenuator that shapes the amplifier gain vs. amplitude characteristics.

The detected IF signal is supplied through a video amplifier to the vertical deflection plates and to a rear panel connector for recording of CW signals, or for use as an AM-detected or slope-demodulated FM output

SPECIFICATIONS

-hp-

MODEL 8551A/851A SPECTRUM ANALYZER

MODEL 8551A RF SECTION

FREQUENCY CHARACTERISTICS

FREQUENCY RANGE:
COAXIAL INPUT: 10 Mc to 10 Gc
WAVEGUIDE INPUT: 8.2 to 40 Gc (accessory mixers and taper sections required).

SPECTRUM WIDTH: 10 calibrated spectrum widths from 100 kc to 2 Gc in a 1, 3, 10 sequence to 1 Gc. Vernier allows continuous adjustment between calibrated ranges and can be used to reduce width to zero.

SWEPT-FREQUENCY LINEARITY*:
±5% when LO is stabilized and swept 10 Mc or less;
±10% (typically ±5%) when LO is swept more than 10 Mc.

IMAGE SEPARATION: 4 Gc

PHASE LOCKING: Internal phase lock provided for stabilizing LO. LO sweep tracks sweep of voltage-tuned 10-Mc reference oscillator.

PHASE LOCK RANGE: Unit can be phase locked for spectrum widths up to N × 10 Mc, where N is harmonic of LO.

PHASE LOCK TUNING: Reference oscillator automatically tracks with TUNE control over 2-Gc LO range.

TUNING: Selectable continuous coarse, fine, and stabliized (phase locked) tuning determines center frequency about which LO is swept. Tuning is by single front-panel TUNE control.

FINE TUNING. Frequency change of LO fundamental is 10 Mc ±2 Mc per revolution of TUNE control up to ±20 Mc maximum accumulative error across band. Setability of signal on CRT with TUNE control: ±50 kc (fundamental mixing).

STABILIZED TUNING: Frequency change of LO fundamental is 10 Mc ±1 Mc per revolution of TUNE control; maximum accumulative error across the band: ±2 Mc. Setability of signal on CRT with TUNE control: ±5 kc (fundamental mixing).

TUNING ACCURACY: ±1% of LO fundamental or harmonic.

SENSITIVITY (10 kc IF bandwidth):

10 Mc to 2 Gc,	−95 dbm
1.8 to 4.2 Gc,	−100 dbm
2 to 4 Gc,	−85 dbm
4 to 6 Gc,	−95 dbm
6 to 10 Gc,	−80 dbm
8.2 to 18 Gc,	−80 dbm
18 to 26 Gc,	−75 dbm
26 to 40 Gc,	−65 dbm

With source stability better than 1 kc, greater sensitivity can be achieved by use of narrower IF bandwidth.

SIGNAL IDENTIFIER: 14-position switch shifts display in proportion to harmonic of LO used in mixing; direction of shift depends on whether signal frequency is higher or lower than LO harmonic.

*Correlation between BWO frequency and sweep position on Model 851A CRT as a percentage of selected spectrum width.

COAXIAL-INPUT ATTENUATOR:
RANGE: 0 to 60 db in 10–db steps.
INSERTION LOSS: 0 at 10 Mc; less than 2 db at 10 Gc.

MAXIMUM INPUT POWER (for 1–db compression):

COAXIAL INPUT:

INPUT ATTENUATOR SETTING	MAX INPUT 2 Gc IF	TYPICAL MAX INPUT 200 Mc IF
0 db	0 dbm	− 5 dbm
10 db	+ 10 dbm	+ 5 dbm
20 db	+ 20 dbm	+ 15 dbm
30 db	+ 30 dbm	+ 25 dbm
40 thru 60 db	+ 30 dbm	+ 30 dbm

WAVEGUIDE INPUT:
11521A EXTERNAL MIXER (8.2-12.4 Gc): typically −10 dbm
11517A EXTERNAL MIXER (12.4-40 Gc): typically −15 dbm

FREQUENCY RESPONSE:
COAXIAL INPUT: ±1 db over any 200–Mc range using fundamental mixing, ±2 db over any 200–Mc range using 2nd-harmonic mixing, ±5 db over any 2–Gc range (typically ±3 db) except when signal or LO is within 60 Mc of 2 Gc; includes mixer and RF attenuator response with attenuator setting ≥10 db.

IF OUTPUT FREQUENCY: 20 Mc

RESIDUAL LO FM: 1 kc peak-to-peak or less when phase locked; otherwise, approximately 30 kc peak-to-peak.

RESIDUAL RESPONSES (no input signal): Less than −90 dbm (−85 dbm when LO is within 60 Mc of 2 or 4 Gc) referred to signal input on fundamental mixing.

LO NOISE SIDEBANDS: Greater than 60 db below CW signal level 50 kc or more away from signal.

LO OUTPUT: Approximately 1 mw available. Output connector, female type N on rear panel.

ACCESSORIES AVAILABLE:
11517A P-K-R Band Waveguide Mixer, $160.00.
11518A P-Band Taper Section, $65.00.
11519A K-Band Taper Section, $65.00.
11520A R-Band Taper Section, $65.00.
11521A X-Band Waveguide Mixer, $75.00.

PRICE: -hp- MODEL 8551A, $7,100.00.

-hp-

MODEL 851A DISPLAY SECTION

DISPLAY CHARACTERISTICS

VERTICAL DISPLAY:
AMPLITUDE CHARACTERISTICS: Linear, square (power), or logarithmic.
DYNAMIC RANGE: Linear, 70:1, square, 70:1; log, 60 db.
ACCURACY: Linear, ±3% of full scale; square, ±5% of full scale*; log, ±2 db*.

SWEEP RATE: Six calibrated rates from 3 msec/cm to 1 sec/cm in 1, 3, 10 sequence. Vernier provides continuous adjustment between calibrated rates and extends slowest rate to at least 3 sec/cm.

*Except pulse spectrums on 1-Mc IF bandwidth.

SWEEP RATE ACCURACY: ±3%.

SWEEP SYNCHRONIZATION:
INTERNAL: Sweep runs free.
LINE: Sweep synchronized with power-line frequency.
EXTERNAL: Sweep synchronized with externally applied signal +3 to +15 volts peak.
SINGLE SWEEP: Sweep actuated by front panel pushbutton.

IF BANDWIDTH:
MANUAL: Bandwidths of 1, 3, 10, 100 kc, and 1 Mc can be selected.
AUTO SELECT: One of above bandwidths automatically selected for best resolution of a CW signal with each combination of spectrum width and sweep rate.
BANDWIDTH ACCURACY: Individual bandwidths are calibrated within ±20%; bandwidth repeatability and stability typically better than ±3%.

IF INPUT:
CENTER FREQUENCY: 20 Mc.
INPUT IMPEDANCE: 50 ohms, nominal.
INPUT REQUIRED FOR 6–cm VERTICAL DISPLAY:
1–Mc bandwidth, −56 ±3 dbm.
100–kc bandwidth, −66 ±6 dbm.
10–kc bandwidth, −86 ±6 dbm.
3–kc bandwidth, −86 ±6 dbm.
1–kc bandwidth, −77 ±6 dbm.

MAXIMUM CW INPUT SIGNAL: −14 dbm.

IF GAIN SET: Two-section attenuator provides 0 to 80 db attenuation in 1–db steps. One section provides 0 to 70 db attenuation in 10–db steps; the other, 0 to 10 db in 1–db steps. Vernier provides continuous adjustment between 1–db steps.

IF GAIN SET ACCURACY: 70–db section, ±0.5 db; 10–db section, ±0.1 db.

GENERAL

OUTPUT SIGNALS: Vertical and horizontal signals applied to CRT available for external monitoring.
VERTICAL: 0 to −4 volts; output impedance, 4700Ω.
HORIZONTAL: 10 v p-p ±0.3 v; sweep approximately symmetrical about zero. Output impedance 4700Ω.

CATHODE-RAY TUBE: 7.5-kv post-accelerator tube with P2 medium-persistence phosphor (others optional) and internal graticule. Light blue filter supplied.

INTERNAL GRATICULE: Parallax-free 7 x 10 cm, marked in cm squares with 2-mm subdivisions on major vertical and horizontal axes.

RFI: Conducted and radiated leakage limits are below those specified in MIL-I-16910.

PRICE: -hp- MODEL 851A, $2,400.00.

OPTIONS:
07. P7 phosphor in lieu of P2 (amber filter supplied), no additional charge.
31. P31 phosphor in lieu of P2 (green filter supplied), no additional charge.

Prices f.o.b. factory.
Data subject to change without notice.

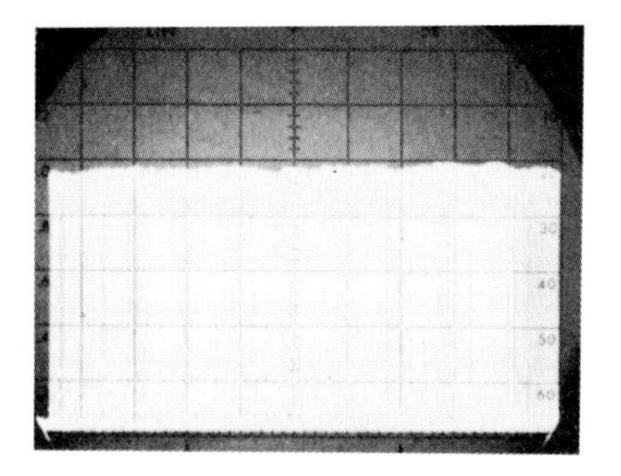

Fig. 7. *Frequency response of –hp– Spectrum Analyzer throughout 1000 Mc range is shown in this time exposure photograph. Exposure was made while leveled signal generator was tuned from 850 Mc to 1850 Mc. Analyzer display was in "Log" mode, showing that frequency response variations are well within ±2 db.*

when the analyzer is in the fixed frequency (zero sweep width) mode. The 20 Mc IF signal, sampled ahead of the detector, also is available at a rear panel connector.

Horizontal deflection is controlled by a linear sawtooth voltage from the sweep generator. The sweep rate can be fast enough for flickerless viewing on the CRT or slow enough for X-Y recording (the sweep voltage is available at a rear panel connector for external use). A single sweep capability is provided for CRT photography.

Other design features include a baseline clipper which blanks the cathode ray tube at the zero voltage level to prevent an overly bright baseline. The cathode ray tube has an internal graticule, similar to the well known internal graticule of the –hp– oscilloscopes, which eliminates parallax as a reading error.

The instrument is particularly well-shielded against radio frequency interference, both from the effect that external RFI sources may have on instrument performance and on the amount of leakage from the instrument itself.

SUMMARY

The –hp– Spectrum Analyzer is finding new applications in situations requiring the display of several different frequencies simultaneously, and is thus becoming a useful laboratory tool for tuning up and studying the performance of solid-state, microwave devices. The clean performance of the new analyzer also makes it useful for frequency surveillance and radio frequency spectrum control by enabling a wide range of RF signals at a particular location to be displayed simultaneously.

The broad frequency range and calibrated performance of the –hp– 8551A/851A Microwave Spectrum Analyzer promise to make frequency domain analysis become a general-purpose laboratory tool, much as time domain analysis has been enabled by modern calibrated oscilloscopes. As recent competitive activity shows, the new analyzer already is setting a new performance standard for the industry.

Fig. 8. *External mixer connects Spectrum Analyzer to waveguide systems. Two mixers are available, one for X band and one for P, Q and R bands. Taper sections match various waveguide sizes to ridged waveguide of mixer.*

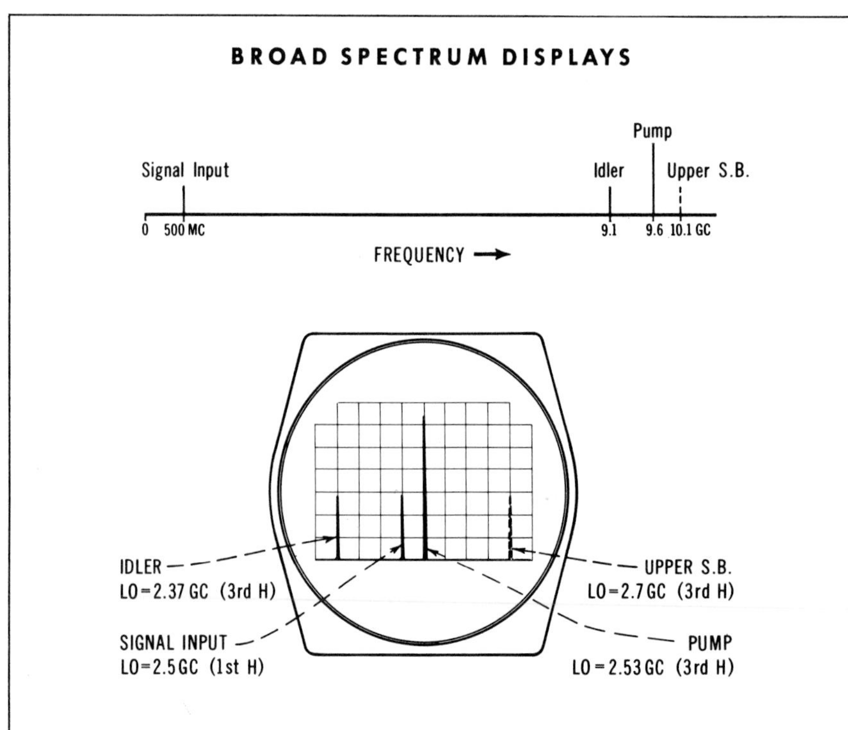

Drawing indicates how Spectrum Analyzer can serve in laboratory to display several signals widely separated in frequency at same time. Upper scale shows relative positions of parametric amplifier signals in frequency spectrum.

Spectrum Analyzer displays all signals at once by responding to input signal on fundamental range, and showing higher frequency signals as responses to local oscillator harmonics.

ACKNOWLEDGMENTS

The –hp– Model 8551A/851A Microwave Spectrum Analyzer was developed under the direction of Arthur Fong in the R and D Laboratories of the Hewlett-Packard Microwave Division. George C. Jung was project supervisor for the Display unit and the undersigned for the RF unit. Others in the electrical design team included R. W. Anderson, R. H. Bauhaus, J. D. Cardon, G. J. Eiler, and R. F. Rauskolb. The mechanical design team included E. H. Phillips, J. L. Boortz, and D. R. Veteran. We give particular credit to our product designers R. C. Given, W. R. Hanisch, J. M. Hedquest, and E. C. Hurd and to A. E. Inhelder for the industrial design. Our appreciation is also extended to Dr. H. C. Poulter, Manager of the Hewlett-Packard Microwave Division R and D Labs, for his advice and encouragement.

—Harley L. Halverson

SPECTRUM SURVEILLANCE

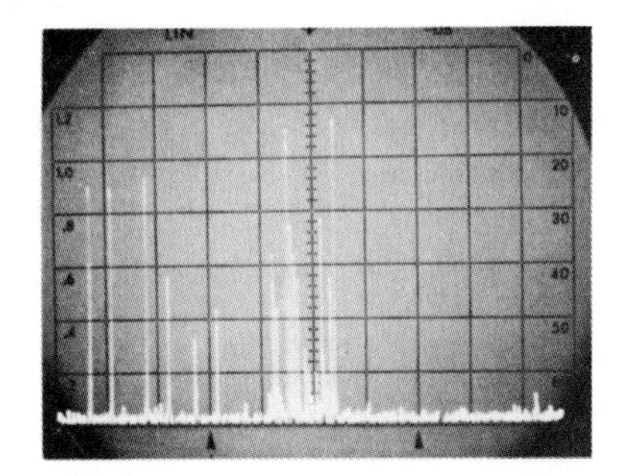

A

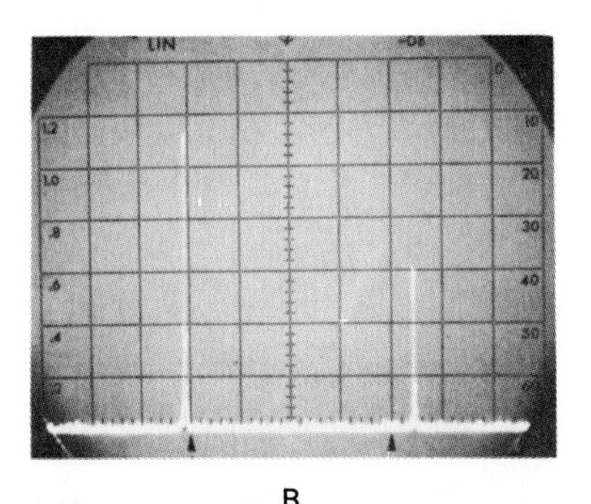

B

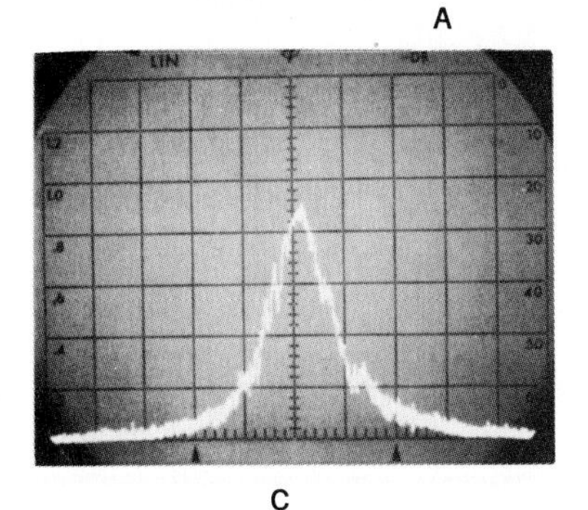

C

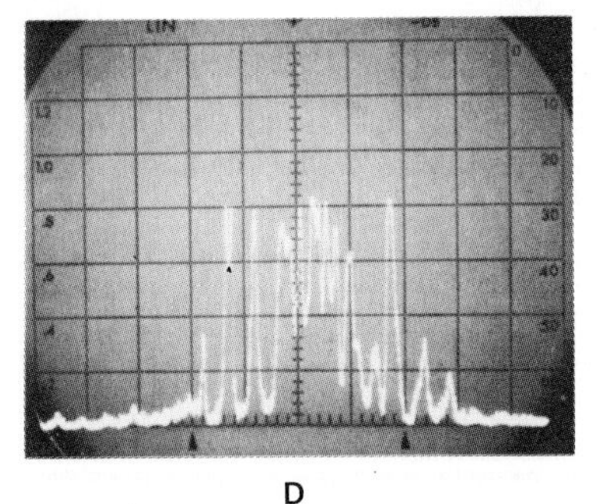

D

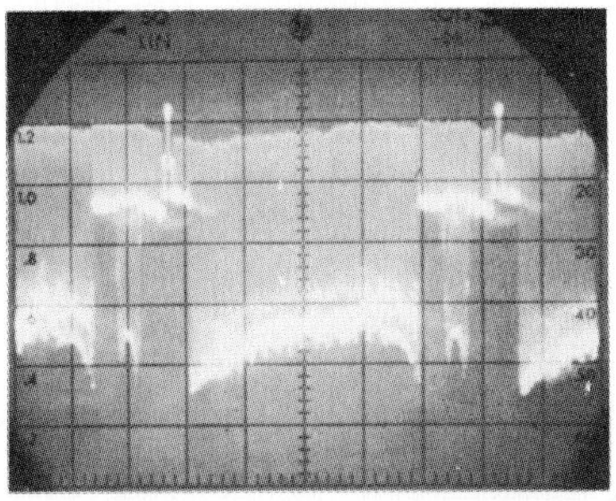

E

This series of photos is indicative of Spectrum Analyzer usefulness in spectrum surveillance. Photo A displays RF spectrum from 50 Mc to 150 Mc at Palo Alto, California, as detected by –hp– Spectrum Analyzer connected to simple dipole antenna. FM stations are in center of display and pairs of responses at left represent TV stations. Photo B shows expanded view of one TV station with Analyzer set to sweep through 10 Mc range centered at 78 Mc. Calibrated sweep of 1 Mc/cm shows 4.5 Mc separation between video and sound carriers. Spectrum analyzers also provide information on sideband structure besides showing frequencies and relative power levels of all radiations within selected frequency ranges. Photos C and D show sound channel at two instants in time, illustrating characteristic behavior of FM signal when viewed in frequency domain; spectrum width here is 100 kc. Photo E shows amplitude modulation of video carrier when SPECTRUM WIDTH VERNIER is turned to zero while Analyzer is tuned to viedo carrier; Spectrum Analyzer functions here as fixed-tuned receiver.

DESIGN LEADERS

Arthur Fong

Art Fong first worked on the development of spectrum analyzers along with other microwave test equipment at the MIT Radiation Laboratory during World War II. A graduate of the University of California, he came to –hp– in 1946 where he was initially engaged in the development of the –hp– Model 803A VHF Impedance Bridge and other high-frequency measuring devices, including the widely-used –hp– Model 650A Test Oscillator. He later headed up the Signal Generator Group and was responsible for the well-known –hp– Models 606A, 614A, 620A Signal Generators among others. He is presently section manager in charge of Spectrum Analyzer development.

Harley L. Halverson

Harley Halverson joined Hewlett-Packard in 1957 while finishing up his M.S. degree at Stanford University. Between Stanford and graduation from South Dakota State College in 1954, Harley spent two years in the Air Force where he was in charge of an electronic instrument laboratory. At Hewlett-Packard, he worked on the –hp– 355-Series VHF attenuators and the –hp– Model 606A HF Signal Generator before becoming project supervisor of the –hp– Model 8551A Spectrum Analyzer RF Unit. One of the early proponents of nanosecond time delay reflectometry at Hewlett-Packard, Harley adopted the technique for wide-band production testing of RF attenuators.

George C. Jung

George Jung joined the –hp– engineering laboratories of Hewlett-Packard in 1959 after doing research work in Canada. His initial contribution to Hewlett-Packard test equipment was the double-bridge technique used in the –hp– 431 series Microwave Power Meters. He also worked on the –hp– 344A Transistorized Noise Figure Meter before assuming project leadership of the –hp– Model 851A Spectrum Analyzer Display unit. He is a graduate of the Middlebaar Technische of Holland and served two years in the Dutch Air Force, both as a pilot and as a supervisor of a landing-control radar installation. George also spent some years in the radar department of the Netherlands' equivalent of the FAA.

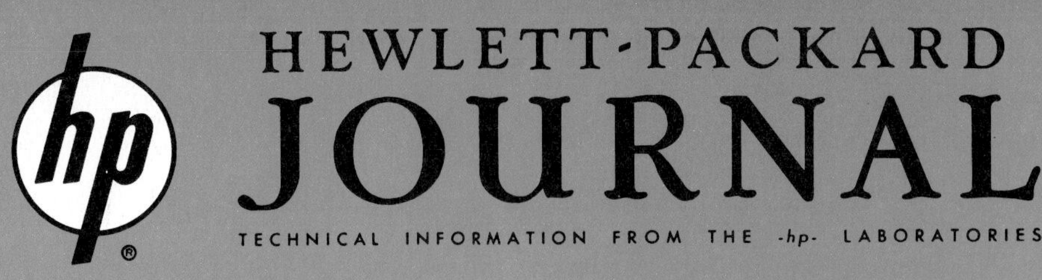

Vol. 16, No. 4

PUBLISHED BY THE HEWLETT-PACKARD COMPANY, 1501 PAGE MILL ROAD, PALO ALTO, CALIFORNIA DECEMBER, 1964

Microwave Harmonic Generation and Nanosecond Pulse Generation with the Step-Recovery Diode

–hp–'s affiliate, **hp associates,** was established four years ago to perform research, development and manufacturing in the semiconductor field. After beginning with the development of specialized silicon, germanium, and gallium arsenide diodes, **hpa** has gone on to achieve industry leadership in metal-on-semiconductor ("hot carrier") technology and is contributing to the advance of the art in optoelectronics and solid-state microwave devices.

One of **hpa**'s developments having general interest to design engineers is the Step Recovery Diode. This device has made possible advances in fast pulse work and is unique as an efficient generator of high-order harmonics. These capabilities are described in the following article.

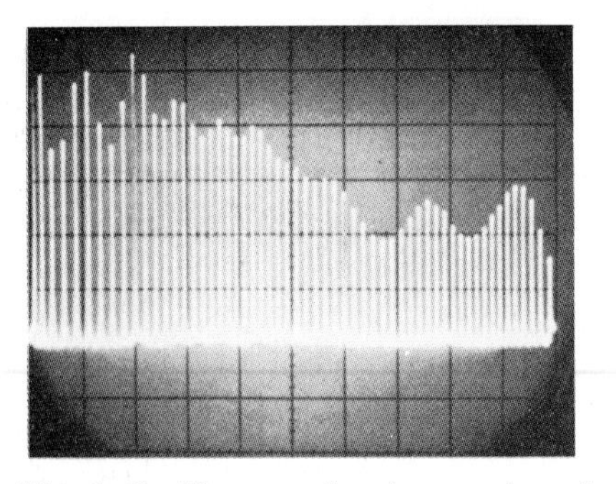

Fig. 1. *Oscillogram showing portion of a harmonic spectrum available from a typical Step Recovery Diode. Harmonics generated by 50 Mc driving signal (Fig. 5) and singly detected by square-law detector.*

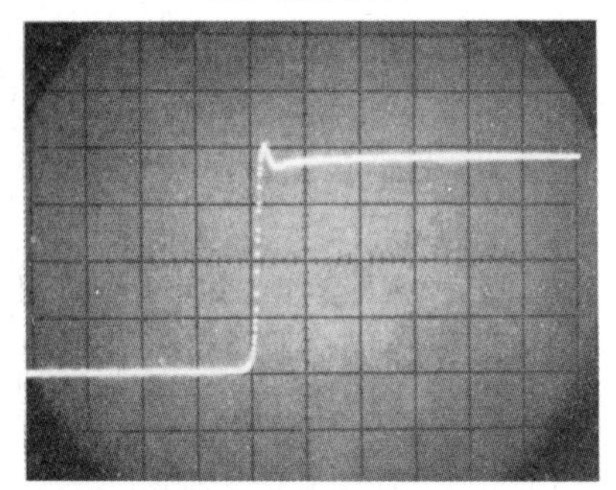

Fig. 2. *Oscillogram of very fast step (300 picoseconds) formed using Step Recovery Diode to steepen pulse-front as described in text. Pulse amplitude here is 4 volts.*

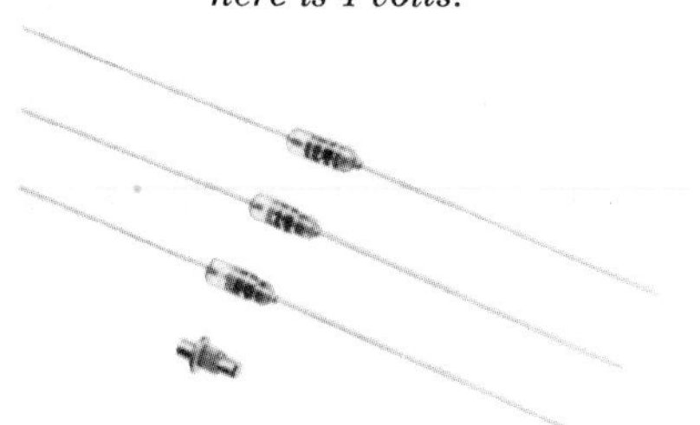

Fig. 3. *Step Recovery Diodes in glass and ceramic packages. Ceramic package is designed to be especially useful with coaxial structures but can also be used in other ways as in Fig. 8.*

ANY *p-n* junction semiconductor diode can be made to conduct heavily in the reverse direction for a brief time immediately following forward conduction. This momentarily augmented reverse conductivity results from the presence of stored minority carriers which had been injected and stored during forward conduction. In the past such reverse storage-conduction has been considered as detrimental in many applications, and so-called "fast-recovery diodes" were developed to reduce it.

Recently, *hp associates* developed a very different class of semiconductor diodes. These were designed to enhance storage and to achieve an abrupt transition from reverse storage-conduction to cutoff. The abruptness of this transition is such that it can be used to switch tens of volts or hundreds of milliamperes in less than a nanosecond. Such a combination of switching speed and power-handling range is not possible with any other existing device. It enables the diodes to generate milliwatts of power at X-band or to provide fast pulses at amplitudes of tens of volts across 50 ohms. As power generators, the diodes will generate high-order harmonics in the microwave region with greater efficiency and simplicity than is possible by any other means. In pulse work, the diodes will generate fractional-nanosecond pulses in which amplitude and timing can be freely controlled, as described in the latter

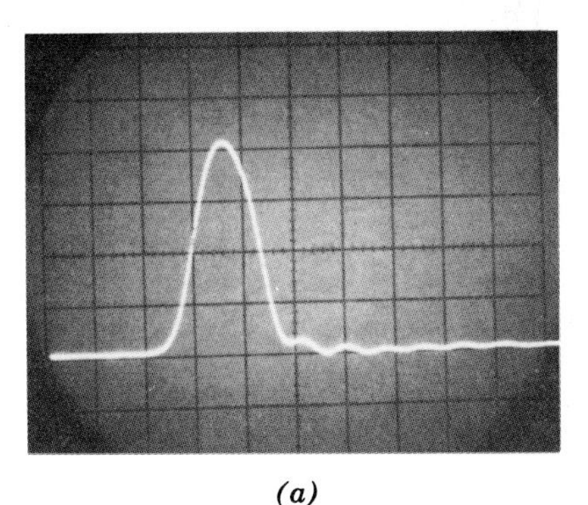

(a)

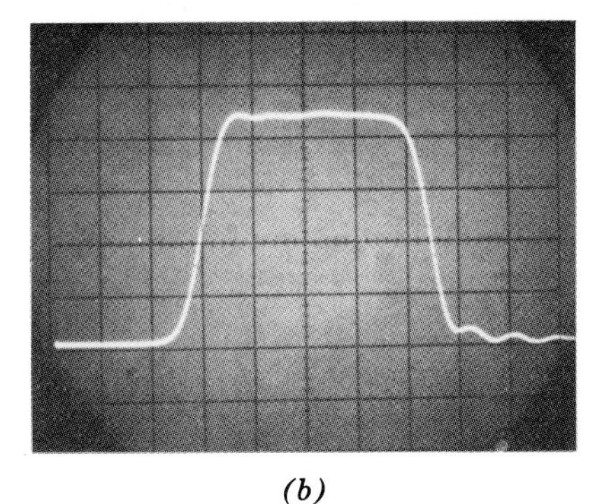

(b)

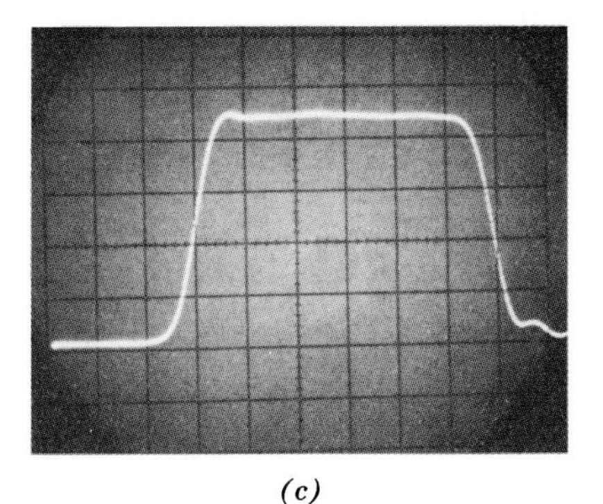

(c)

Fig. 4. *Examples of fast pulses of appreciable amplitude generated with Step Recovery Diodes. Pulse height is about 8 volts (2 v/cm). Sweep time is 1 nsec/cm.*

part of this article. Figs. 2 to 4 indicate the high multiplication factors and fast-pulse application of the diodes.

Because their conductance during reverse storage conduction follows a step function, these diodes have been termed "Step Recovery Diodes" by *hpa*. They are also known as "snap-off" or "charge-storage" diodes. Their step-recovery characteristic is illustrated in the oscillogram of Fig. 5 which shows the diode current while the diode was excited by a 50-megacycle sinusoid. In the forward direction the diode conducts in a conventional way. As the excitation becomes negative, however, the diode current also reverses until the supply of stored minority carriers becomes exhausted. At this point cessation of reverse current occurs very rapidly — in less than a nanosecond — resulting in a fast current step that is rich in harmonics. In fact, harmonics to above the 100th can be obtained at power levels approaching a milliwatt if not higher.

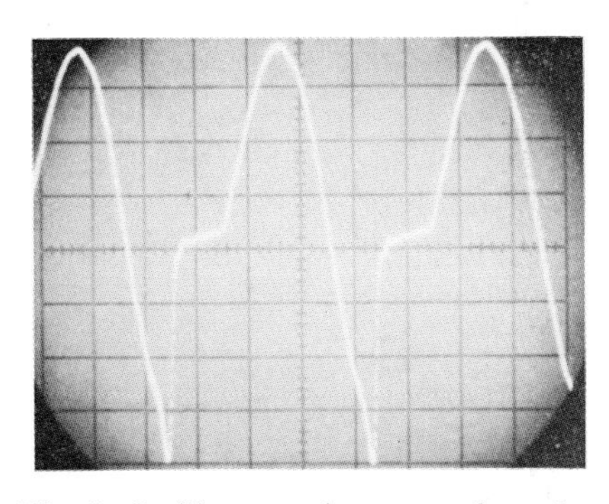

Fig. 5. *Oscillogram of current through a Step Recovery Diode when driven by a sine wave (50 Mc). Abrupt transition on negative half-cycle occurs when stored charge becomes depleted and is an efficient mechanism for generating harmonics (Fig. 1).*

In practical work the diodes can be used in a single-stage multiplier to produce harmonics in the microwave region with a simplicity and freedom from noise unmatched by other arrangements or by conventional harmonic-generating diodes such as varactors. The presently-known state of the art in high-order multiplication with the Step Recovery Diode is shown in Table I. The variation in performance from band to band presumably results from various degrees of design effort.

Even the performance shown in Table I is expected soon to become obsolete as a result of work in progress at *hp associates* and elsewhere. Step Recovery Diodes should soon become available which will be capable of approaching 1 watt in the 1 to 2 Gc range and 50 to 100 milliwatts in the 8 to 12 Gc range.

In harmonic-generating work the high efficiency of the Step Recovery Diode compared to conventional diodes occurs because of the basically-different generating mechanism involved in the two cases. Conventional diodes, such as varactors, are operated within their reverse saturation region so as to avoid both forward conduction and avalanche breakdown. Under these conditions the diode acts as a voltage-variable capacitor having some dissipation. The resulting capacitance-voltage characteristic is smooth and thus does not generate appreciable high-order harmonic power. For such diodes the conversion efficiency in this mode is usually found to decrease approximately at the rate of $1/n^2$, where n is the harmonic number. For this reason efficient harmonic generation with varactors usually requires a cascade of doublers and triplers with attendant idlers and complications.

Variable-resistance (point-contact) diodes also suffer from poor efficiency when used for other than small multiplication factors. Tube multipliers, of course, are noisy and unstable among other disadvantages.

On the other hand, the Step Recovery Diode has a reverse-bias capacitance that varies only slightly with bias voltage. The contribution of this mechanism to harmonic generation is thus negligible, and the step-recovery mechanism is the important one. Fur-

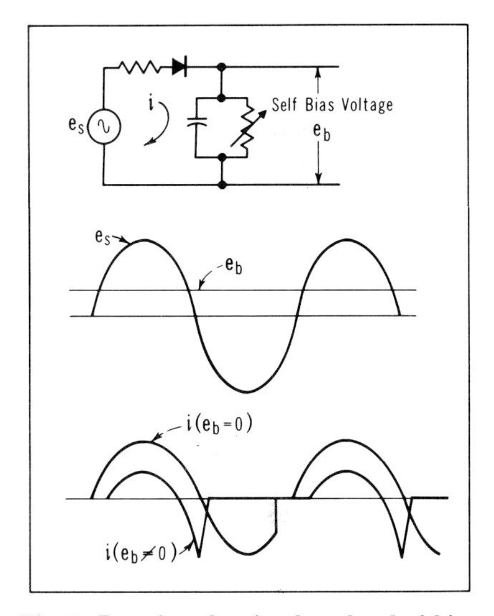

Fig. 6. *Drawing showing how level of bias affects the amplitude of the diode current discontinuity and thus the efficiency of harmonic generation.*

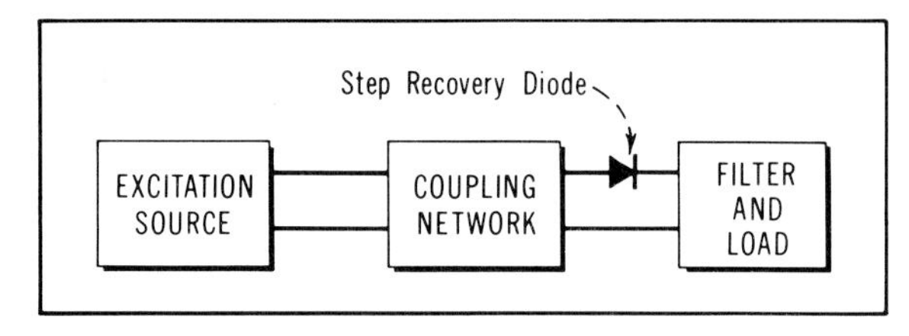

Fig. 7. *Functional block diagram of harmonic multiplier.*

TABLE I. PRESENT STATE OF THE ART IN HIGH ORDER MULTIPLICATION WITH STEP RECOVERY DIODES

F_{out} (Gc)	Multiplication order (n)	Maximum P_{out} (mw)	Conversion Loss (db/n)
1– 2	10–30	200	0.5
2– 4	10–30	100	0.75
4– 8	10–40	50	1.0
8–12	9–84	20	1.5

ther, the output power of the Step Recovery Diode can be shown by Fourier analysis to decrease only as 1/n when n is larger than 5. As a result, the Step Recovery Diode is the most practical means available for generating moderate-power harmonics and is probably the only practical means for generating high-order harmonics.

DIODE BIAS

Bias is required for these diodes in harmonic generators because they would otherwise conduct during the entire drive-frequency cycle and no electrical discontinuity would be produced. Under bias, however, and with lifetime much longer than the excitation period, the time integral of forward current will be slightly larger than that of the reverse current. A large discontinuity may then be produced when conduction ceases, as shown in Fig. 6. In a circuit that is optimized for efficient harmonic generation, the current waveform may not be as simple as is indicated in this figure since strong oscillatory components will be supported, and the conduction angle will be strongly influenced by the reactance elements that are present.

HARMONIC MULTIPLIER DESIGN

The functional block diagram of a single-stage multiplier is shown in Fig. 7.

The design of the Step Recovery Diode harmonic generator circuit should:

1. Resonate the diode at the output frequency.
2. Provide a broadband reactive termination for unwanted harmonics.
3. Match input power down to the diode impedance (1–10Ω).
4. Tune out the circuit susceptance at the input frequency.
5. Provide high Q energy storage at the output frequency.
6. Provide appropriate bias for the diode, as described above.

These criteria are met by either of the two designs shown in Figs. 9 and 10. In Fig. 9, a broadband input circuit is utilized to permit making changes in frequency without retuning the input. Fig. 10 shows a resonant input circuit which is simpler for fixed frequency applications and still decouples harmonics efficiently from the source. In either circuit self-bias of the diode is satisfactory, but in some applications it may be desirable to use fixed bias, and to control the bias resistance.

The diode in its associated package and the output circuit should be resonant at the output frequency, a condition normally difficult to calculate since it depends upon geometric details of diode mounting, the diode parameters, the output circuit and the driving reactance. The diode should not be resonant at other harmonic frequencies. In practice one usually provides some reactive tuning device at the diode to adjust for best operation. One such device is shown in Fig. 11.

Energy storage at the output frequency is needed to develop voltage on the diode for maximum efficiency and to eliminate undesired adjacent harmonics. In the examples of Figs. 9 and 10, a double-tuned, quarter-wave-stub, iris-coupled cavity is indicated. Any of many other configurations of waveguide coaxial cavity could be used.

Fig. 8. *Broadband stripline circuit used in bench work with Step Recovery Diode. Either glass or ceramic style diodes can be accommodated as shown.*

The cavities should have high Q to reduce insertion loss at the desired output frequency while simultaneously eliminating adjacent harmonics. An alternative description of the requirement is to say that enough energy should be stored so that it is not depleted appreciably by the load in the interval between impulses from the drive frequency. This means suppressing the AM modulation of the output at the drive frequency. The degree of suppression desired will depend upon the application and will determine the amount of output filtering required. To reduce neighboring sidebands in a X20 multiplier by 20 db, for example, a loaded Q of about 300 is needed.

Adjustment of multipliers using the Step Recovery Diode is a simple procedure. The output resonator should be pre-aligned to resonate at the proper frequency, and the input frequency and power level set. The diode self-bias resistance, the output circuit resonance with the packaged diode, and the input matching circuit are then adjusted for maximum power output and efficiency.

Additional application information is given in the *hpa* Application Note "Harmonic Generation with Step Recovery Diodes."

A TYPICAL SINGLE-STAGE X20 MULTIPLIER

A single-stage X20 Step Recovery Diode multiplier was designed using the principles previously cited. An efficiency in excess of 10%, or 2/n where n is the harmonic number, was achieved by this technique using various *hpa* Step Recovery Diodes. The circuit of the multiplier and associated filter are shown schematically in Fig. 12.

The Step Recovery Diode is placed in series with the input resonator which is of the shorted type. The only adjustment of the diode reactance and microwave circuit impedance required is by use of the sliding short shown in the figure. The 2000 Mc output filter is a six-resonator interdigital structure with a bandwidth of 20 Mc and a 2 db insertion loss. The estimated total circuit losses of the circuit described are approximately 2.5 to 3 db.

The operation of the circuit is simple and replacing a diode requires a minimum of retuning time to achieve a stable maximum output power. The tuning requires only that the variable resistor and capacitor be adjusted for best input match and the sliding short be positioned for maximum output power.

Various *hpa* Step Recovery Diodes were substituted in this X20 multiplier and Table II shows a summary of the average experimental results obtained with the different types of diodes. In addition, single-stage X16 multipliers have been built which have conversion efficiencies greater than 20% and outputs greater than 100 mw at 1.5 Gc. At 10 Gc output levels in excess of 20 mw have been obtained when driven by 1.25 Gc.

DIODE SELECTION

With respect to Step Recovery Diode requirements for efficient harmonic

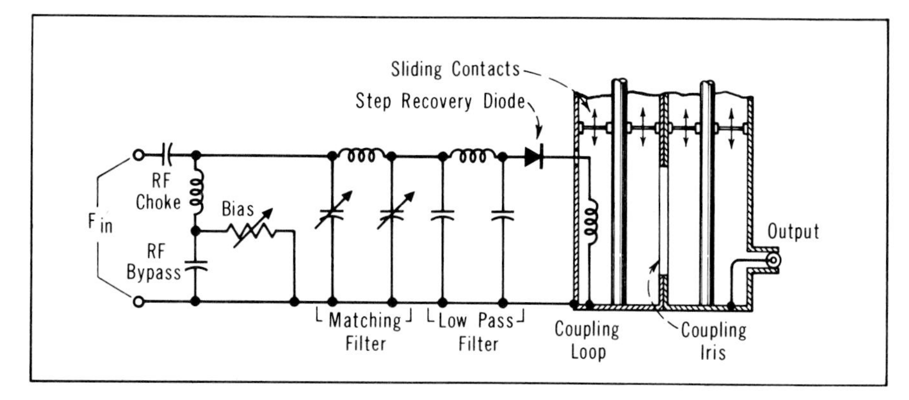

Fig. 9. *Basic circuit of a broadband type harmonic generator.*

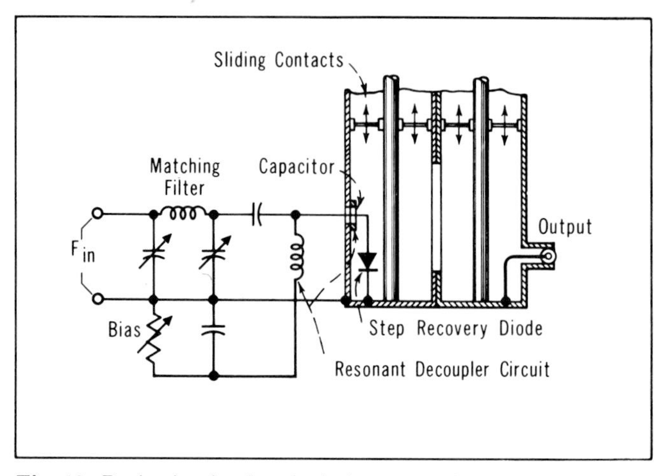

Fig. 10. *Basic circuit of a single-frequency harmonic generator.*

generation, it is usually desirable that the minority carrier lifetime τ be at least three times the period of the excitation frequency, and that the transition time t_t be shorter than ½ the period of the output frequency. A long value of τ allows for a large amount of stored charge for conversion to harmonic power. It also simplifies establishing appropriate bias conditions for the diode.

The conversion efficiency depends on the speed and amplitude of the transition time t_t. Step Recovery Diodes with transition times less than one-half of the period of the desired output frequency achieve efficiencies in excess of 1/n. The efficiency increases as the transition time becomes short compared to the period of the output frequency until it is one-tenth or less. Although theoretical predictions other than the 100% efficiency permitted by the Manley-Rowe relations are lacking, diode efficiencies up to 30% have been achieved experimentally in the previously cited X20 multiplier.

TABLE. II. AVERAGE OPERATING RESULTS ×20 MULTIPLIER

Input Frequency 100 Mc — Output Frequency 2000 Mc		
hpa Diode Type	**P_{out} (mw)**	**Diode Conv. Eff. (%)**
0112	20	15
0151, 0251	5	25
0152, 0252	5	20
0153, 0253	15	30
0154, 0254	15	15
0132	80	10

The probable practical limit of efficiency for high-order multiplication obtained by extrapolating data to the limit of zero transition time is approximately 40%.

Fig. 13 shows the efficiency obtained using the *hpa* X20 multiplier from a 100 Mc input to a 2000 Mc output. The efficiency at the diode is plotted against transition time t_t which is measured in units of the period T of the output frequency. The region $t_t/T < 0.2$ is extrapolated. Approximately 40 different diodes were used to obtain the data for this graph. Data on many more diodes in C-band multipliers confirm the general relation shown here.

For a given output frequency, it is also desirable to select a diode with the

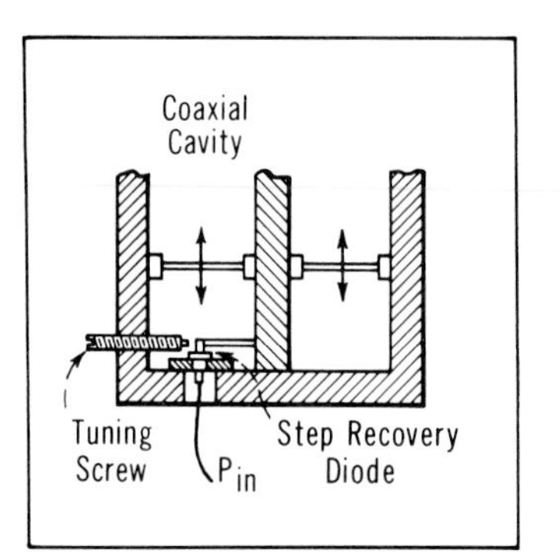

Fig. 11. *Drawing showing use of tuning screw to resonate Step Recovery Diode at desired output frequency.*

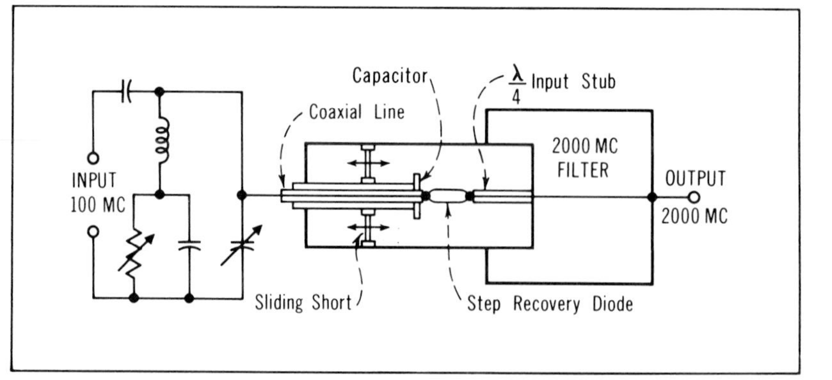

Fig. 12. *Basic circuit of a ×20 harmonic multiplier using a Step Recovery Diode providing a 2000-Mc output at the efficiencies shown in Table II.*

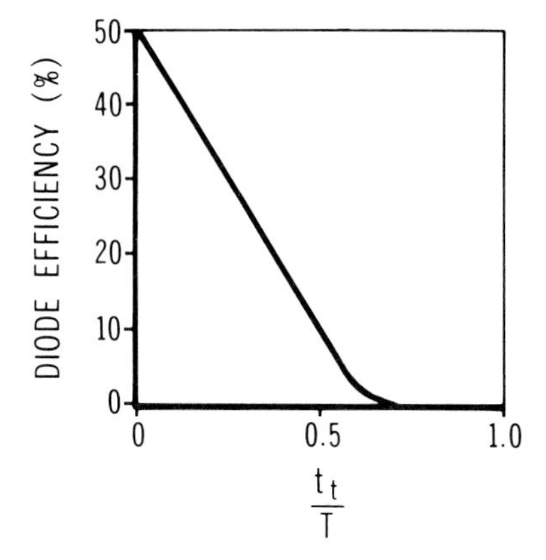

Fig. 13. *Efficiency of multiplier in Fig. 12 as a function of diode transition time.*

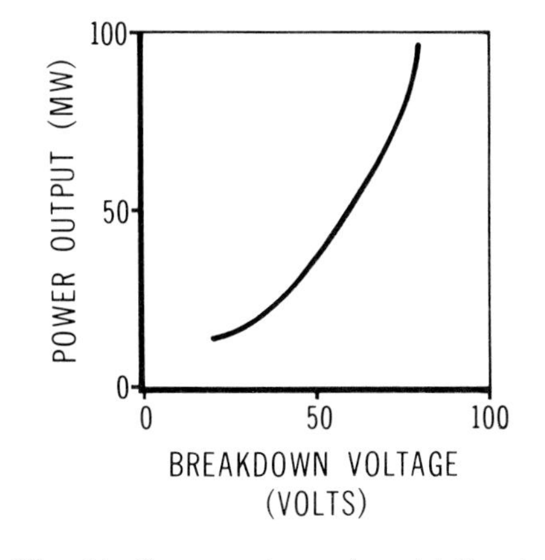

Fig. 14. *Power output of multiplier in Fig. 12 as a function of diode breakdown voltage.*

lowest values of diode series resistance R_s and junction capacitance C_j to achieve good circuit efficiency.

The effect on power output of diode breakdown voltage and of input drive level in a particular harmonic multiplier are shown in Figs. 14 and 15.

PULSE SHAPING AND GENERATION

In fast pulse work its unique characteristics make the Step Recovery Diode a very important device. It can be used to square the rise and fall of fast pulses or to produce short, fixed delays. Impulses, in turn, can be generated from the fast pulses.

The value of the diodes in fast pulse work arises from a combination of their fast transition times, which are presently less than 100 picoseconds, and the large voltage which they can switch—more than 10 volts at 50 ohms. No other device presently offers this speed, amplitude variation, and convenience.

PULSE-FRONT SQUARING

Fig. 17 shows several basic circuits using the Step Recovery Diode for pulse-front squaring, for generating fixed delays, and for trailing-edge squaring. Fig. 17 (a) shows a basic circuit for pulse-front squaring in which the diode is placed in shunt with the load. It can be seen from this circuit that, following the application of the driving pulse which acts as a reversing current, the diode will reverse-conduct for a period of time. It then abruptly exhausts its stored minority carriers so that the abrupt transition to cutoff occurs, thereby generating a step in current and voltage. This step can now be used as the sharpened front of the output pulse. Very fast output pulse rise times of a fraction of a nanosecond can be generated by this means in one or more stages, depending upon the speed of the driving pulse. Sufficient

DESIGN LEADERS

Robert D. Hall

Bob Hall joined **hpa** in 1963 to work on an advanced program concerned with integrated microwave semiconductor components. While at **hpa,** he has applied network synthesis techniques to obtain improved performance from diode switches, frequency multipliers, mixers and other components and has developed novel methods for constructing the necessary circuits. A member of the American Physical Society, Bob obtained a BA degree in Physics at Reed College and an MS degree in Mathematics at Stanford. Prior to joining **hpa,** he spent five years as a research engineer concerned with solid-state microwave components during which time he made important contributions to the theory of tunnel diode oscillators and amplifiers. He holds several patents.

Stewart M. Krakauer

Stew Krakauer joined the **-hp-** Oscilloscope Division in 1958 where he participated in the development of the **-hp-** Model 185A/187A Sampling Oscilloscope. He subsequently became head of applications engineering in the **-hp-** Semiconductor Laboratories and contributed to the development of other fast-switching and high-frequency circuits that use semiconductor devices. He transferred to **-hp- associates** in 1962 where he now heads the applications group. Stew earned a BSEE degree at Cooper Union and an MEE from Polytechnic Institute of Brooklyn. Prior to joining **-hp-**, he worked on electronic instrumentation as applied to inertial navigation and to physiological and radiological research.

bias current will generally be required to delay the transition until the applied pulse comes to a steady-state value.

Figs. 17 (b) and (c) show the use of fast-recovery diodes to avoid small pedestal voltages and dc offsets. In these circuits the fast-recovery diode isolates the load from the Step Recovery Diode until the voltage across the Step Recovery Diode has begun its transition. Fig. 17 (c) also shows the use of two stages to increase the speed of pulse-front squaring. Since the pulse rise time that is applied to the second stage is faster than the excitation pulse, less storage is required in the second-stage diode. Hence, its transition speed can be faster. The actual rise times will depend upon the impedance of the associated circuitry. However, tens of volts or hundreds of milliamperes can be switched by this arrangement with fractional nanosecond risetimes.

The voltage and current levels are established by the external circuit and are limited only by the reverse breakdown voltage of the diode and diode dissipation. Thus, the diode can switch many *watts* without exceeding its rated dissipation.

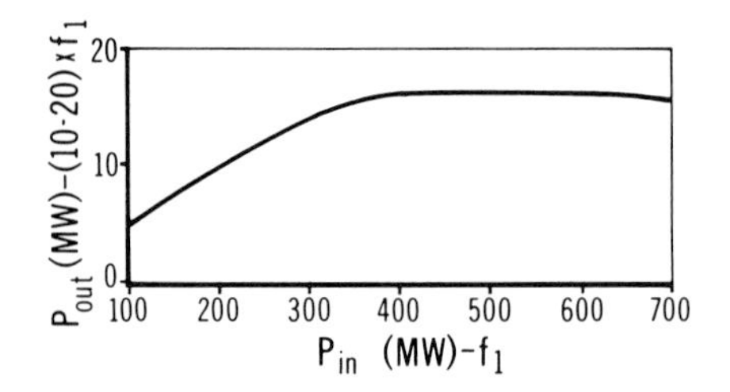

Fig. 15. *Typical power output at 6000 Mc as a function of input power for a diode with 28-volt breakdown voltage. Multiplication of input frequency is from 10 to 20 times.*

In addition to pulse-front squaring, the circuits can be used to generate impulses by differentiating the output pulse-front.

PULSE DELAY

In Fig. 17 (a) it can be seen that the time interval between the application of the drive pulse and the occurrence of the sharpened pulse-front will depend on the stored charge and hence on the forward bias current. Accordingly, this time interval can be varied from about 1 to 1,000 nanoseconds by varying the bias current. Thus, a convenient method is available for generating delay. The method is simple and has a sharply-defined end point. The effect of temperature changes on the delay can be compensated by making the bias current suitably temperature-dependent.

TRAILING-EDGE SQUARING

To square the trailing edge of pulses, the Step Recovery Diode can be placed in series with the load, as shown in Fig. 17 (d). Very fast trailing edges are thus possible. Combinations of leading- and trailing-edge squaring can be used to obtain square, fast pulses, as shown in Fig. 4.

GENERAL

The Step Recovery Diode is unique in its efficient conversion of power up to high harmonic order in a single stage. It exceeds the performance of the conventional varactor diode because its functional non-linearity is

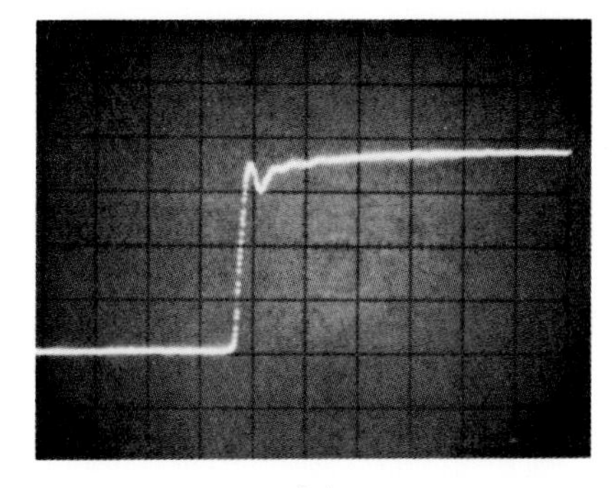

(a)

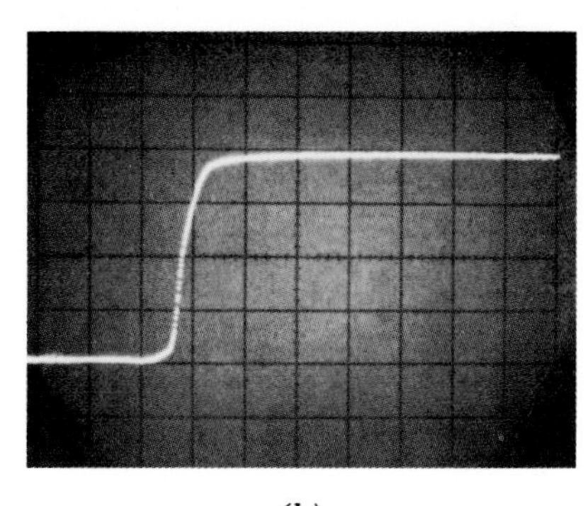

(b)

Fig. 16. (a). *Very fast step (200 picoseconds) obtained with Step Recovery Diode; (b) driving step. Sweep time 2 nsec/cm.*

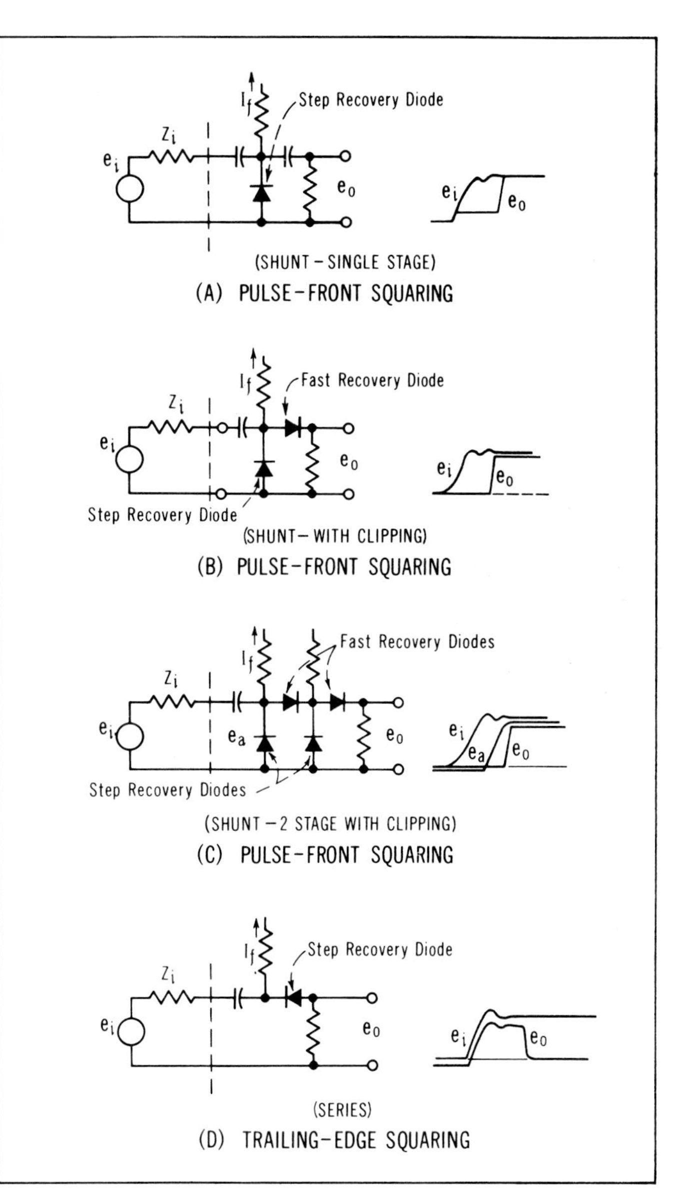

Fig. 17. *Circuit arrangements using Step Recovery Diodes to sharpen leading and trailing edges of pulses and to obtain pulse delay.*

greater. It also has the distinct advantage over all other multiplying techniques that it is attractively simple. It can accomplish in a single stage what requires a series of stages of other multipliers and does so without the complexity and complication of idlers and with very little parametric up-conversion of noise.

In pulse applications the diode can be used to achieve pulses with a combination of speed and amplitude not obtainable with other devices.

ACKNOWLEDGMENT

Several people have contributed to the design of the **hpa** Step Recovery Diodes, but the undersigned wish to cite particularly the contributions of M. M. Atalla and Mason A. Clark.

—Robert D. Hall and
Stewart M. Krakauer

REFERENCES

[1] S. Krakauer, "Harmonic Generation, Rectification, and Lifetime Evaluation with the Step Recovery Diode," Proc. IRE, Vol. 50, July, 1962.

[2] R. Hall, "Harmonic Generation with Step Recovery Diodes," **hpa** Application Note No. 2.

[3] Moll, Krakauer, and Shen, "P-N Junction Charge Storage Diodes," Proc. IRE, Vol. 50, pp 43-53; January, 1962.

HEWLETT-PACKARD JOURNAL

TECHNICAL INFORMATION FROM THE -hp- LABORATORIES

CORPORATE OFFICES • 1501 PAGE MILL ROAD • PALO ALTO, CALIFORNIA 94304 • VOL. 16 NO. 7, MARCH 1965

The Linear Quartz Thermometer — a New Tool for Measuring Absolute and Difference Temperatures

A linear-temperature-coefficient quartz resonator has been developed, leading to a fast, wide-range thermometer with a resolution of .0001° C.

Fig. 1. New linear Quartz Thermometer (foreground) uses quartz resonator as sensor to measure temperatures from −40°C to +230°C at resolutions up to .0001°C. Thermometer can measure temperature at many-meter distances, and digitally-presented data can be recorded and processed by existing hardware. One calibration point of Thermometer is established by freezing-point of tin (+231.88°C).

IT HAS LONG been recognized that the temperature dependence of quartz crystal resonators was a potential basis for the accurate measurement of temperature. In practice, however, it has not previously been satisfactory to make wide-range temperature-measuring systems based on quartz resonators because of the large non-linearity in the temperature coefficient of frequency of available quartz wafers. Recently, however, an orientation in quartz was predicted and verified by Hammond[1] in the -hp- laboratories which resulted in a crystal wafer having a linear temperature coefficient over a wide temperature range. This new orientation, the "LC" (linear coefficient) cut, has permitted development of a "quartz thermometer" that measures temperatures automatically, quickly, and with very high resolutions on a direct digital display. Temperatures can be measured over a range from −40°C to +230°C to a resolution of .0001°C in 10 seconds — or faster with proportionately less

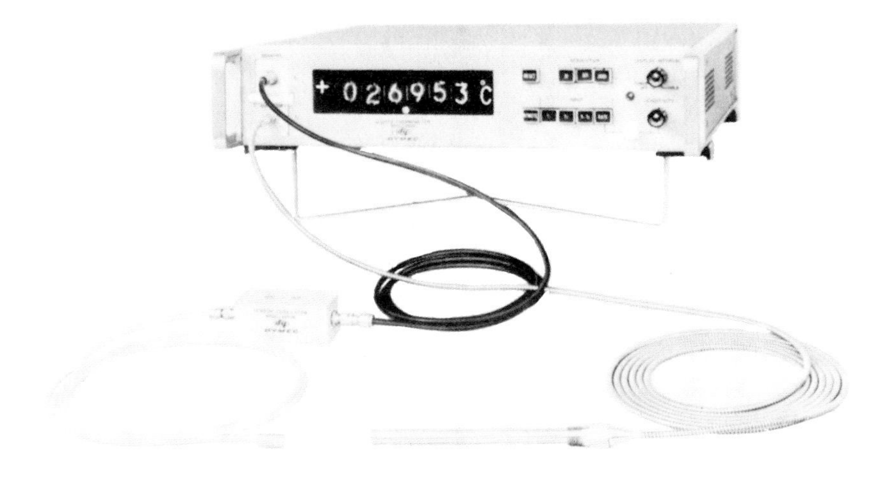

Fig. 2. *Two-channel version of Quartz Thermometer can measure temperature sensed by either probe or difference between probes. Sensor oscillator is normally located within cabinet but is self-contained and can be located externally for remote measurements.*

resolution or in repetitive measurements.

Two versions of the thermometer have been designed, one of which has two inputs to enable differential measurements of temperature. The two versions of the instrument have each been designed to read directly in Fahrenheit (—40°F to +450°F) or in Centigrade (Celsius) but not both.

It is apparent on its face that a temperature-measuring instrument with the above capabilities has great value in many fields, but the instrument also has a number of additional characteristics that are of much interest. It is possible, for example, for the instrument to make measurements through connecting wires at distances of up to 10,000 feet, with no adverse effect on measurement accuracy caused by lead length. Other unusual and interesting characteristics are discussed later.

The thermometer's temperature-sensing quartz resonator is located in a small sensor probe which connects through a length of cable to its oscillator. The oscillator is located in the main cabinet but can be physically removed as a unit to permit measuring temperatures at a distance from the cabinet. The cabinet otherwise contains what is essentially a special frequency counter which displays the measured temperature directly in numerical form on a digital readout. The temperature measurements are made automatically, either repetitively or initiated singly with a panel push button in the manner of a frequency counter. Repetitive readings can be made from 4 per second to 1 per 15 seconds. Three styles of sensor probes have been designed to accommodate various measurement situations including measurements in high-pressure environments. The time constant of each of the probes is only 1 second.

RESONATOR TEMPERATURE COEFFICIENT

Quartz wafers are widely used in oscillator circuits to hold the operating frequency constant. In this application, temperature has been the principal factor influencing the stability of quartz resonators. It was discovered some time ago, however, that the temperature coefficient of frequency is both a function of the angle at which the resonator is sliced from the parent crystal and of the temperature itself. In 1962, Bechmann, Ballato and Lukaszek of the U. S. Signal Corps Lab, at Ft. Monmouth, N. J. reported an analysis of the first, second, and third order temperature coefficients of frequency of a number of quartz resonator designs.[2] This analysis made it possible to calculate the first three coefficients of a third-order expansion of the temperature-dependent frequency for a quartz crystal plate of generalized orientation:

$$f(T) = f(0)(1 + \alpha T + \beta T^2 + \gamma T^3)$$

Bechmann used this approach to study resonator orientations with a zero first-order temperature coefficient ($\alpha = 0$).

Recently, Hammond, Adams, and Schmidt[1] of the Hewlett-Packard Company used this same approach to determine that an orientation existed in which the second and third order terms went to zero ($\beta = \gamma = 0$) while the first order term remained finite. This orientation occurred for a thick-

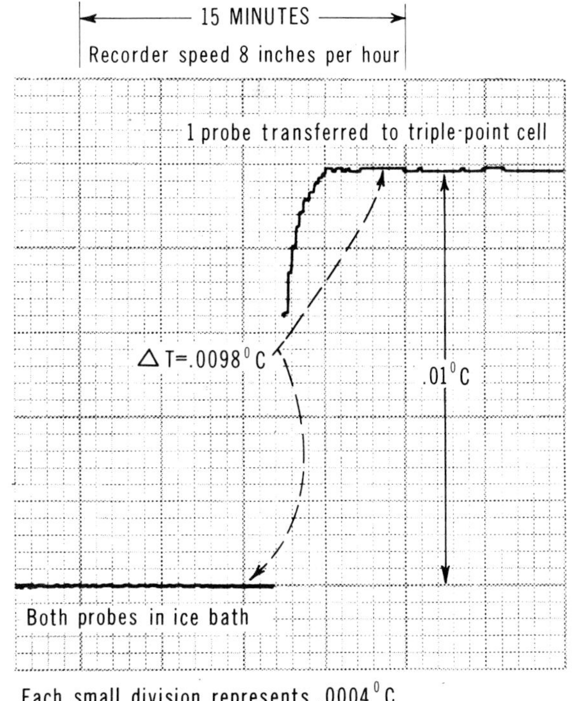

Fig. 3. *Recording of measurement made with two-channel Quartz Thermometer of difference between melting point of ice and triple-point of water. By definition, the latter is +.01° Celsius (Centigrade) and is considered to be .01° above ice melting point.*

ness-shear mode of operation designated the "LC" cut, for Linear Coefficient of frequency with temperature. This cut is the basis for the sensor design used in the new Quartz Thermometer.

Resonators sliced in the LC orientation from high-quality synthetic single-crystal quartz exhibit a temperature coefficient of 35.4 ppm/°C. For use in the new thermometer, the resonators are ground to the precise thickness and orientation required to achieve the linear mode while exhibiting a frequency slope of 1000 cps/°C. This slope is achieved at a third overtone resonance near 28 Mc/s.

Fig. 4. *Two-channel Quartz Thermometer measuring a small temperature difference. Six-digit display permits such difference measurements to be measured with a resolution of .0001°C.*

QUARTZ PROPERTIES

Quartz-crystal wafers have certain desirable properties which make them valuable as resonators in frequency standards and time-keeping systems and which are equally important for temperature-sensing systems. Chief among these are quartz's high purity and chemical stability. Further, quartz is a hard material that cannot be deformed beyond its elastic limit without fracture. It has almost perfect elasticity and its elastic hysteresis is extremely small.

THE LINEAR COEFFICIENT QUARTZ RESONATOR

For digital thermometry, the ideal quartz crystal resonator should have a linear frequency-temperature characteristic. This characteristic can be well-represented over rather wide temperature ranges for nearly all types of quartz resonators by a third-order polynomial in temperature. The two degrees of orientational freedom which are significant in quartz resonators are just sufficient to adjust the second- and third-order terms to zero if a region of solution exists.

An analysis was made of the frequency temperature behavior of the three possible thickness modes for all possible orientations in quartz using Bechmann's constants.[1] A single orientation of wave propagation was found for which the second- and third-order temperature coefficients are simultaneously zero. The accompanying diagram shows the analytically-determined loci of zero second-order and third-order temperature coefficients for the lowest frequency shear mode, the C mode, in a primitive orientational zone in quartz. These loci cross at $\phi = 8.44°$ and $\theta = 13.0°$. Experimental studies indicate the actual orientation of zero second- and third-order terms to be $\phi = 11.17°$ and $\theta = 9.39°$. The discrepancy between observed and predicted orientations is consistent with the accuracy of the elastic and expansion constants used in the analysis. A resonator cut at this orientation and operated on the C mode, has been designated the **LC** cut to indicate **L**inear **Co**efficient of frequency with respect to temperature.

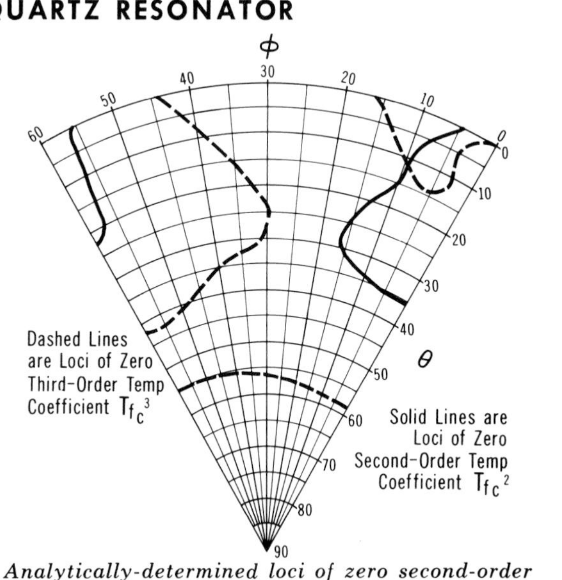

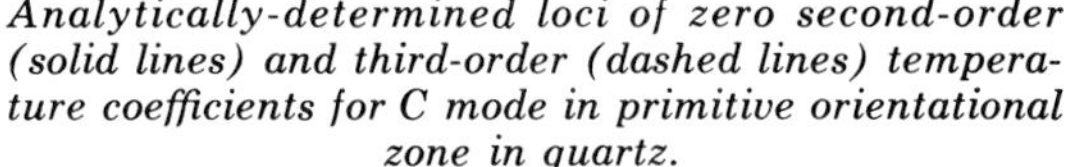

Analytically-determined loci of zero second-order (solid lines) and third-order (dashed lines) temperature coefficients for C mode in primitive orientational zone in quartz.

At this orientation, the nonlinearity is restricted to the fourth- and higher-order terms which have been experimentally shown to be less than a few millidegrees over the temperature range from 0 to 200°C.

It is interesting to note that the frequency-temperature relationship was linearized with respect to the International temperature scale. If a new temperature scale is adopted in the future, minor adjustments can be made in the two orientational parameters of the LC cut to linearize relative to the new temperature scale. —*Donald L. Hammond*

[1] R. Bechmann, A. D. Ballato, and T. J. Lukaszek, "Higher Order Temperature Coefficients of the Elastic Stiffness and Compliances of Alpha Quartz," **Proc. IRE,** Vol. 50, No. 8, August, 1962.

Fig. 5. *Close-up view of quartz sensor mounted on header. Sensor is later sealed in helium atmosphere.*

Quartz, unlike the platinum or nickel used in resistance thermometers, can be found in a natural state that has a high degree of purity. Alpha quartz, the crystalline formation that exhibits a piezo-electric effect, is generally found in Brazil, but American-made synthetic quartz is now available. Both exhibit impurity levels of less than 10 ppm (an almost negligible amount). Its ordered crystalline structure resists the plastic deformation that causes drift and retrace errors in resistance materials and permits the great frequency-stability found in quartz crystal resonators. The short-term variations of indicated temperature in the quartz thermometer, for example, are much less than .0001°C.

Quartz's asymmetrical structure also provides control of its temperature characteristic through angular orientation that is unavailable in the amorphous resistance materials. Platinum, for example, has a fixed deviation from the best straight line of .55% over the same range as the quartz thermometer, which is currently held to less than .05%.

SENSOR CONSTRUCTION

After the deposition of gold electrodes on the surface of the quarter-inch diameter quartz wafer, each wafer is brazed to three small ribbons which support it inside a TO-5 size transistor case (Fig. 5). Thus mounted, the quartz is remarkably immune from both drift and breakage. Drop tests have shown that an acceleration of more than 10,000 g's is required to fracture the crystal, and that no discernible shift in calibration occurs short of the point of fracture. Vibration levels of 1000 g's from 10 c/s to 9000 c/s have had no measurable effect.

The wafer case is hermetically sealed in a helium atmosphere which provides both a good heat conduction path and a passive atmosphere for long term resonator stability. The wafer itself dissipates only 10 μW internally, an amount of heat that contributes less than 0.01°C error when the sensor is in water flowing at 2 ft./sec.

Since slope and linearity are controlled closely during manufacture by the orientation and thickness of the crystal and its gold electrodes, quality control is more precise than is possible with resistance thermometer materials. No detectable change in slope or linearity after manufacture has occurred in crystals tested to date. There is a characteristic "aging" effect that causes the frequency (measured at the ice point) to change about 0.01°C per month but, because no slope changes occur, recalibration at the ice point alone is usually sufficient.

FREQUENCY TO DIGITAL CONVERSION

The block diagram of the new thermometer in Fig. 6 shows how electronic counter techniques are used with the quartz resonator/oscillator to obtain a digital display of temperature.[3, 4] The sensor oscillator output is compared to a reference frequency of 28.208 Mc/s. By design, this frequency is also the sensor frequency at zero degrees. As mentioned previously, the frequency of the quartz sensor was chosen so that a slope of 1000 cps/°C is obtained, i.e., a temperature of 200°C produces a sensor frequency that differs from the reference by 200 kc/s. The difference frequency is detected in the mixer circuit, converted into a pulse series and passed to the electronic display decades. Here, it is counted for a fixed length of time and the resulting count is displayed on "Nixie" tubes to provide a numerical readout.

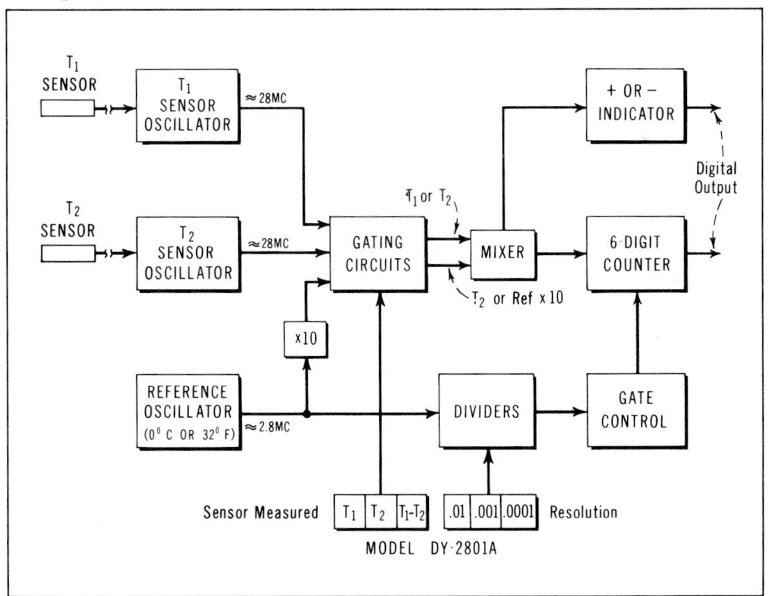

Fig. 6. *Block diagram of circuit arrangement of two-channel Thermometer. Single-channel Thermometer circuit is similar in principle but does not include second sensor channel, variable gate times, or difference-temperature capability.*

The duration of the count is controlled by the reference oscillator, which drives the gate control-circuit through a frequency divider chain. The count accumulates at a rate of 1000 c/s per second per °C difference from 0°C. The counting or gating interval then determines the resolution.

The optional Fahrenheit scale is obtained by increasing the gate-time by a factor of 9/5 and using a reference crystal 1.778 kc/s lower in frequency than the Centigrade unit. This lower frequency reduces the zero-reading point by 17.78°C, equivalent to 32°F.

The reference crystal is mounted in a temperature-controlled oven to achieve a long-term drift of only a few parts in 10^7 per month, or less than 0.005°C change in zero-setting per month. Short-term stability is such that changes from reading to reading are less than 0.2 millidegrees, unnoticeable on the 4-digit instrument and noticeable only when using the 0.0001° resolution scale of the dual-channel instrument.

Fig. 7. Single-channel Thermometer. Sensor oscillator can also be used externally for remote measurements.

DIFFERENTIAL TEMPERATURE MEASUREMENTS

The higher resolution of the two-channel unit over the single-channel unit has been achieved by using a 6-digit readout and longer gating (sampling) intervals than the 0.01 second of the single-channel instrument. In the two-channel unit, one of three gating intervals can be selected by front panel push buttons. The 0.1-second gate provides temperature measurements in least increments of one-hundredths of a degree, the 1-second gate millidegree increments, and the 10-second gate tenths of millidegrees.

In this instrument, the preset number in the time base dividers is adjustable with decade switches on the rear of the instrument. These provide a means of compensating for variations in the frequency/temperature slope of individual probes during the calibration of the instrument. Being a digital technique, the adjustment is not a potential source of drift error. The two probes normally are supplied with slopes matched to better than 0.05%. The zero temperature frequencies are matched closely enough so that trimmers on the individual sensor oscillators can provide an exact zero indication.

The two-channel instrument, which can be programmed externally, may be switched to indicate the absolute temperature sensed by either probe. Differential temperature measurements are obtained by switching the instrument to measure the beat frequency between the two sensor oscillators.

To identify on which side of zero the measured temperature lies, the two-channel instrument has a polarity indicator circuit that compares the sensor oscillator output to the reference frequency. The comparison determines which signal is higher in frequency and turns on a "+" or "—" indicator accordingly. In the differential mode, the polarity indicates whether probe T_1 is higher than or lower than probe T_2. The polarity indicator can also be incorporated in the single-channel unit.

SENSOR PROBES

The outer shell of the three sensor probes is fabricated from type 304 stainless steel for chemical stability. In these probes the quartz wafer is situated parallel to and about .01 inch away from the flat circular end of the probe and is sealed in a helium atmosphere, as mentioned previously. The 12-foot connecting cable has a dielectric and outer sheath of type TFE Teflon which can withstand temperatures up to 250°C. The stainless steel shell on the probes is thin and of small diameter, resulting in a low thermal mass of less than 10^{-3}BTU/°F (equivalent in heat capacity to less than 0.5 gm of water) for the short probe.

The low thermal conductivity of the thin-wall shell also results in a low value of stem conduction error. This error is, for example, less than 0.01% of the temperature difference between the tip and the threaded end of the longer probe as measured in essentially stationary water.

Much work has also been done in the design of the probes to achieve a low thermal time constant, which is less than 1 second as measured in warm water moving at 3 fps.

The two longer probes are equipped at the cable end with a fitting that adapts the probes for insertion into pipes and tanks. The fittings have a standard ¼" National Pipe Thread. All probes may be used in pressures up to 3,000 psi.

SENSOR OSCILLATOR

The sensor oscillator is in a small die-cast aluminum case, as shown in Fig. 2, with a waterproof coaxial connector at each end. DC power for the oscillator is supplied over the same cable that carries the oscillator frequency back to the instrument. The solid-state oscillator may be operated within a temperature range of —20°C to +70°C and shifts the indicated temperature by less than 0.001° per degree of ambient change. A trimmer is provided to shift the frequency of the oscillator slightly (±50 cps) so that a pair of probes can be exactly matched.

The oscillator-amplifier combination provides virtually complete isolation between the sensor and any variations due to cable length and load. When some distance is required between the point of measurement and the instrument, a standard 70-ohm coaxial cable can be used for an extension; losses as high as 20 dB can be tolerated which permits a cable length, using RG-59/U, as great as 1000 ft. For greater extensions, lower loss cables or booster amplifiers can be used, and it appears likely that lengths beyond 10,000 feet are entirely feasible.

The cable between the sensor and oscillator is necessarily fixed at ½ wavelength of 28 Mc/s to reflect the crystal impedance directly to the oscillator.

DIGITAL OUTPUT

An additional advantage of Quartz Thermometers is that the measurement information is provided in a digital form that is readily recorded or

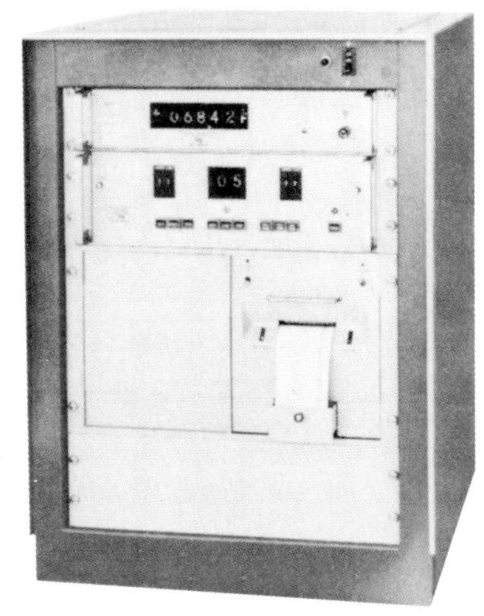

Fig. 8. *Quartz Thermometer (upper unit) used with Scanner and Printer to measure and record up to 100 temperatures. Measurements can be recorded on paper tape as shown, or on punched tape, cards or magnetic tape.*

transmitted for processing. BCD outputs for direct coupling into digital recorders, such as the *-hp-* Model 562A, are included.

Analog records such as that shown in Fig. 3 may be obtained by using a strip-chart recorder in conjunction with the analog output from the *-hp-* Model 562A Printer or the Model 580A Digital-Analog Converter.

REMOTE SENSING

Since the sensor oscillator output is a frequency-modulated radio-frequency signal, it opens the possibility for telemetric transmission by direct radiation from the sensor oscillator alone at the 28.2 Mc/s frequency. Transmission in the range of existing telemetry receivers may be had by multiplying the frequency four times to 112.8 Mc/s. Telemetry receivers can be made to provide a frequency output proportional to temperature by using the quartz thermometer reference oscillator as a beat frequency oscillator (BFO) and coupling the output signal to an electronic counter. In this way one can realize the full precision of the Thermometer.

The ability of the Quartz Thermometer to transmit data in FM form over long cable runs also opens the possibility of improved precision in measuring temperatures at great depths and at many points in the ocean. The measurement information could be transmitted either by the extended cable connection mentioned above or by radio telemetry from floating buoys.

RATE INPUT

Since the Quartz Thermometer has all the elements of an electronic frequency counter, it was convenient to arrange the two-channel instrument for use as a frequency counter. A separate input terminal and a sensitivity control for input signals are on the front panel, and a front panel switch converts the instrument to a frequency counter having a maximum counting rate of 300 kc/s.

CALIBRATION TECHNIQUES

The temperature-sensing quartz resonators are calibrated after being sealed in their cases but prior to mounting in probe shells. Calibration is accomplished by mounting each res-

BRIEF SPECIFICATIONS

-dy- MODELS 2800A and 2801A

QUARTZ THERMOMETERS

TEMPERATURE RANGE: —40 to +230°C (—40 to +450° F with Option M1).

RESOLUTION:

MODEL NO.	READING① RESOLUTION	FULL SCALE DISPLAY		SAMPLE PERIOD (Sec.)	
	(°C or °F)	°C Readout	°F Readout	°C	°F
2800A	.1	230.0	450.0	.01	.018
2801A	.01 .001 .0001	0230.00 230.000 *(2)30.0000	0450.00 450.000 *(4)50.0000	.1 1.0 10.0	.18 1.8 18.0

① ±1 digit ambiguity in least significant displayed digit for all sample periods.

* 'Hundreds' digit not displayed when using .0001° resolution.

ACCURACY: Best accuracy determined by 'Linearity' and 'Short-Term Stability,' listed below.

LINEARITY (ABSOLUTE): Better than ±.02°C (.04°F) from 0 to +100°C, referred to straight line through 0 and 100°C.
Better than ±.15°C (.27°F) from —40 to +230°C, referred to straight line through 0 and 200°C.

MATCHED PROBES: With DY-2801A, probes are matched to better than ±.1°C at 200°C, or .05% of operating temperature (when zero set at icepoint).

STABILITY

SHORT-TERM: Maximum variation from reading-to-reading at constant probe temperature less than ±.0002°C (.0004°F) for absolute measurements. No reading-to-reading variation for differential measurements (excluding normal display ambiguity of ±1 count in least significant digit).

LONG-TERM: Zero drift less than ±.01°C (.018°F) at constant probe temperature for 30 days.

TEMPERATURE CYCLING: When used repeatedly over range from —40 to +230°C, reading at 0°C will not change by more than ±.05°C (.09°F). Error decreases with reduced operating range.

AMBIENT TEMPERATURE: Reading changes by less than .001°C per °C change in ambient temperature for instrument or sensor oscillator.

RESPONSE TIME: Response to step function of temperature, measured by inserting probe into water at dissimilar temperature flowing at 2 fps:

63.2% of final value in <1.0 sec.
99.0% of final value in <4.5 sec.
99.9% of final value in <6.9 sec.

THERMAL MASS:

DY-2850A (SHORT) PROBE: Equivalent to less than 0.5 gm of water.

SELF-HEATING: Less than 10μw; contributes less than .01°C error with probe immersed in water flowing at 2 fps.

INTEGRATION TIME: Defined by sample period selected; see 'Resolution.'

SAMPLE RATE: Readings are taken in response to pushbutton or external signal, or automatically at self-regulated intervals. Time interval between readings can be adjusted at front panel from approximately .2 to 5 sec.

POWER REQUIRED: 115/230 v ±10%, 50 to 60 cps { DY-2800A, 65 w. / DY-2801A, 85 w.

PROBE ENVIRONMENT:

MEASURAND: Gases and liquids non-reactive with 304 stainless steel. Probes can be supplied in other materials such as 316 stainless steel on special order.

TEMPERATURE: —200 to +250°C (—330 to +480°F). Instrument zero may require readjustment on returning to proper operating range from temperatures below about —100°C.

PRESSURE: 3000 psi maximum.

SHOCK: To 10,000 g, without change in calibration. Equivalent to 10-inch drop onto steel surface.

VIBRATION: To 1000 g at 1000 cps.

NUCLEAR RADIATION:

X-Ray or Gamma Radiation: Levels to 10^8 roentgens are permissible. Radiation at this level may cause a frequency offset in the order of 1 ppm, which can be removed by heating probe to +250°C.

Neutron Radiation: Low energy (e.g. .025 ev) thermal neutrons will not affect crystal. Higher energy neutron radiation can be expected to cause permanent damage to crystal.

SENSOR OSCILLATOR ENVIRONMENT: Ambient temperatures from —20 to +70°C (—4 to +158°F). May be totally immersed; constituent materials are aluminum, nickel, neoprene, Teflon, epoxy paint.

INSTRUMENT ENVIRONMENT: Ambient temperatures from 0 to +55°C (+32 to +130°F), at relative humidity to 95% at 40°C.

PRICE:

MODEL DY-2800A QUARTZ THERMOMETER: With one Temperature Sensor, Model DY-2850C (unless otherwise specified), $2,250.

MODEL DY-2801A QUARTZ THERMOMETER: With two Temperature Sensors, Model DY-2850C (unless otherwise specified), $3,250.

Prices f.o.b. Palo Alto, California

Data subject to change without notice

Fig. 9. *Sensor probes presently designed for Quartz Thermometer. Probes can be use in pressures up to 3,000 psi. Fittings on longer probes facilitate measurements in tanks, pipes, etc.*

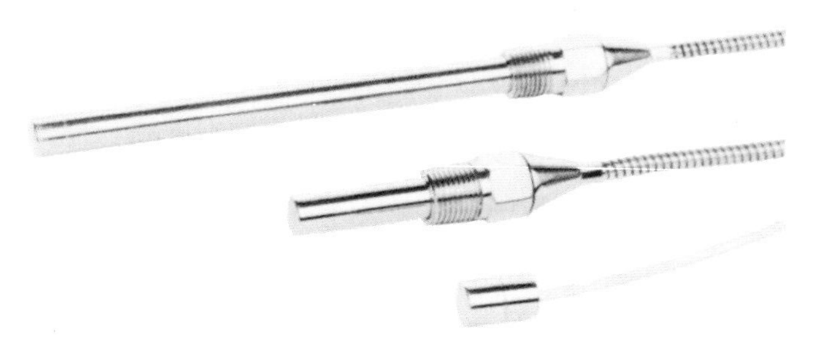

onator case in a test probe and operating these in groups of 50 in a temperature-controlled oil bath.

Each probe is connected to a separate sensor oscillator, and the oscillators are scanned by a scanner similar to the one in Fig. 8. The frequencies of the resonators are printed sequentially on an *-hp-* 562A Recorder at each calibration temperature.

Eight calibration temperatures are used ranging from —40°C to +240°C in 40° steps. The calibration baths are monitored by a transfer standard calibrated against an NBS-certified platinum resistance thermometer and Mueller bridge. The transfer standard is regularly checked for drift in a triple-point cell, a certified tin-freezing-point standard, and against the freezing point of triple-distilled mercury. If excessive drift is noted, the transfer standard can be recalibrated against the certified thermometer. These checks, plus the excellent short-term stability of the temperature-controlled baths, achieve a calibration accuracy of 0.01°C at all eight points.

The crystals are individually classified into groups on the basis of the zero degree frequency and the average slope. The grouping permits the matching of probes for use in the dual-channel instrument.

The data for each sensor are applied to a computer program which provides preset numbers for the best average slopes to use over both the 0°C to 100°C range and the full —40°C to +230°C range. The computer also prints a table of deviations for each range in 10° steps so that corrections for the residual non-linearity can be applied in critical applications to the displayed readings. The maximum deviations permitted are 0.02°C for the 0°–100° range and 0.15°C for the full range of the instrument.

ACKNOWLEDGMENT

The design of the Quartz Thermometer is based on the development of the linear coefficient quartz resonator by Donald L. Hammond in the *-hp-* Physics R and D Laboratory. Herschel C. Stansch as project manager was responsible for the instrument development. The members of the design team included Cleaborn Riggins, John M. Hoyte, Kenneth G. Wright, F. Glenn Odell, Willis C. Shanks, and Malcolm W. Neill. We also wish to acknowledge the services of Donald E. Norgaard and Robert J. Moffat.

—Albert Benjaminson

REFERENCES

[1] D. L. Hammond, C. A. Adams, P. Schmidt, Hewlett-Packard Co., "A Linear Quartz Crystal Temperature Sensing Element," presented at the 19th Annual ISA Conference, Oct. 12-15, 1964, New York. **Preprint No. 11, 2-3-64.**

[2] See footnote in separate article, p. 3.

[3] H. C. Stansch, "A Linear Temperature Transducer with Digital Readout" presented at the 19th Annual ISA Conference, Oct. 12-15, 1964, New York. **Preprint No. 11, 2-2-64.**

[4] Albert Benjaminson, "The Quartz Crystal Resonator as a Linear Digital Thermometer," presented at 4th Annual Conference of the Temperature Measurements Society, Feb., 1965.

DESIGN LEADERS

Donald L. Hammond

Albert Benjaminson

Don Hammond joined *-hp-* in 1959 as manager of the Quartz Crystal Department where he established the *-hp-* facility for precision quartz-crystal resonator production. Don participated in the research and development which led to the high-precision resonators in the *-hp-* 106A and 107A Quartz Oscillators and the crystal "flywheel" used in the *-hp-* 5060A Cesium Beam Frequency Standard, and he also developed the linear quartz temperature sensor. In 1964, Don was transferred to the position of general manager of the *-hp-* Physics R & D group where he directs research and development on quantum-electronics, electro-acoustics, and high-vacuum devices.

Don received BS and MS degrees in Physics from Colorado State University and has done further graduate study at Columbia University. He has also directed research in industry and at government laboratories in the areas of frequency-control devices, synthesis of crystalline quartz, solid-state properties of quartz, and the theory of vibrations applied to piezo-electric resonators.

Al Benjaminson joined the Hewlett-Packard Dymec Division in 1959 as a development engineer. He was made Engineering Manager of RF System Development in 1960 and in this position was responsible for the Dymec 5796, 2650, and 2654 Oscillator Synchronizers, for the 2590 Microwave Frequency Converter, and for the 2365 Tunable VLF Receiver. In 1964 he was made Engineering Manager for transducer development in which capacity he is responsible for the Quartz Thermometer development.

During World War II, Al was stationed at the Naval Research Lab, Radio Materiel School in Washington as an electronics instructor. He attended the Polytechnic Institute of Brooklyn on a part-time basis and then obtained his BEE degree at the University of Adelaide, South Australia, during the post-war years. He has worked as a radio design engineer on commercial products and was also an engineering section head for military airborne electronic systems before joining Dymec.

THE RF VECTOR VOLTMETER—AN IMPORTANT NEW INSTRUMENT FOR AMPLITUDE AND PHASE MEASUREMENTS FROM 1 MHz TO 1000 MHz

A broadband two-channel millivoltmeter and phasemeter simplifies many measurements heretofore often neglected. Included are device gain and loss, impedance and admittance, length inequalities in transmission paths, and precision frequency comparisons.

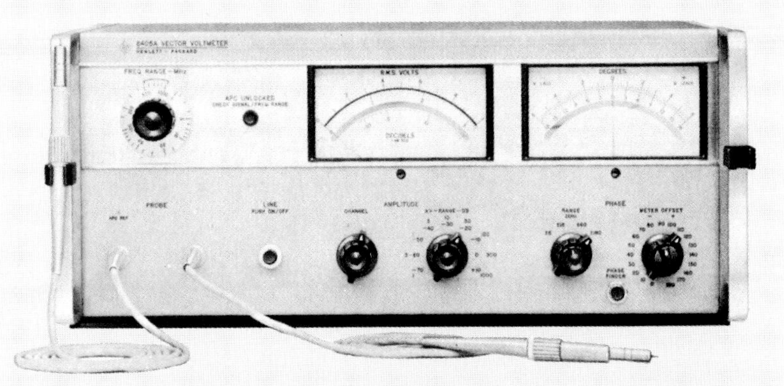

Fig. 1. *-hp- Model 8405A Vector Voltmeter measures amplitudes and phase simultaneously. Instrument has frequency range of 1 MHz to 1 GHz, sensitivity of 100 μV full-scale, dynamic range of 95 dB, phase resolution of 0.1°, and is simple to operate. Thus it makes feasible many measurements which formerly were difficult or impossible.*

AN IMPORTANT NEW INSTRUMENT, which seems certain to become one of the major electronic measuring instruments, has recently been developed by the -hp- Microwave Division. The RF Vector Voltmeter (Fig. 1) is a two-channel millivoltmeter and phasemeter: it measures the voltage in channel A, and simultaneously measures the phase angle between the fundamental components of the signals in channels A and B; it may then be switched to measure the voltage in channel B so that gain or loss may be determined. It makes these measurements over a broad frequency range (1 to 1000 MHz) in a part of the spectrum where information is often peculiarly difficult to obtain.

Voltage and phase are so fundamental in electrical engineering that the new Vector Voltmeter has an extraordinary number of applications. It can, for example, measure complex or vector parameters such as impedance or admittance, amplifier gain and phase shift, complex insertion loss or gain, complex reflection coefficient, two-port network parameters, and filter transfer functions. It can also be used as a selective receiver and as a design tool: possible applications are detecting RF leakage, measuring antenna characteristics, detecting Miller effects in tuned RF amplifiers, tuning feedback amplifiers, measuring the electrical length of cables, measuring group delay, and many others.

Although adequate voltmeters for measuring amplitudes over a wide frequency range have been available for some time, there has been no equally convenient means for measuring phase. Consequently, simultaneous measurements of voltage and phase have not always been easy to make. Most systems which are able to measure phase angles require several control adjustments for each measurement, and many of them

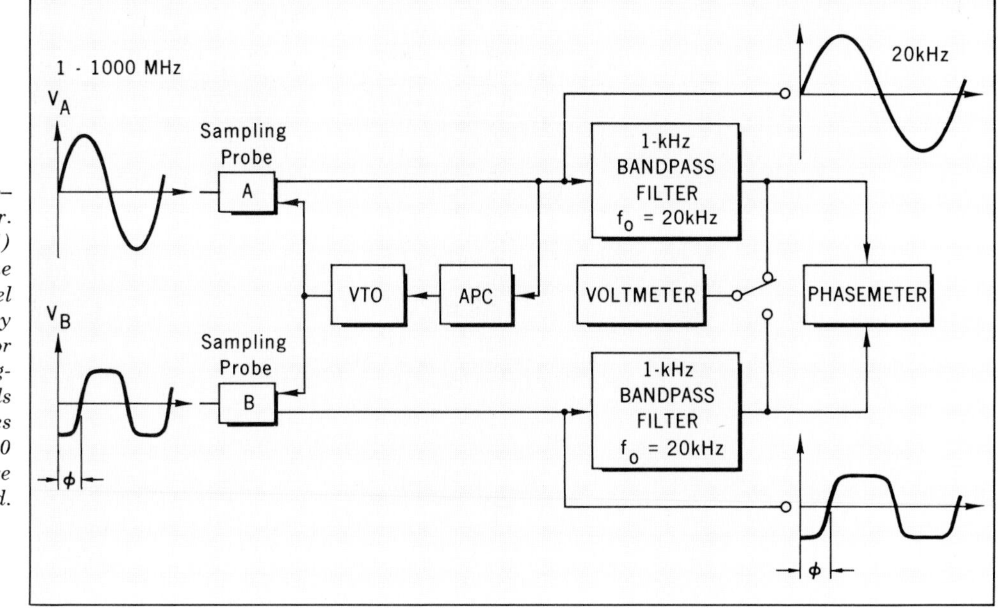

Fig. 2. Block diagram of –hp– Model 8405A Vector Voltmeter. Automatic phase control (APC) uses phase-locked loop to tune and phase-lock meter to channel A signal. APC adjusts frequency of voltage-tuned local oscillator (VTO) which triggers sampling-type mixers in probes. RF signals are reconstructed from samples at intermediate frequency of 20 kHz, where voltage and phase are measured.

are limited in frequency range, sensitivity, and dynamic range.

The new Vector Voltmeter (VVM), on the other hand, operates over a frequency range of 1 MHz to 1 GHz. It has high sensitivity and wide dynamic range. Its phase resolution is 0.1° at any phase angle at all frequencies, and it operates with the simplicity of a voltmeter: the operator merely selects appropriate meter ranges, touches two probes to the points of interest, and reads voltage and phase on two meters.

As a voltmeter, the VVM has nine voltage ranges, which have full-scale sensitivities of 100 μV to 1 V rms. Its dynamic range is 95 dB, which means that it can measure gains or losses of up to 95 dB. The 10:1 voltage dividers supplied with the instrument enable it to measure voltages up to 10 V.

As a phasemeter, the VVM will measure phase angles between +180° and −180°. It has four ranges: ±180°, ±60°, ±18°, and ±6°. The phase meter can be offset up to ±180° in 10° steps so that any phase angle may be read on the ±6° range, which has 0.1° resolution. For example, a phase angle of +145° can be measured with 0.1° resolution by selecting a phasemeter offset of +140° or +150° and using the ±6° range. Phase readings are independent of the voltage levels in the two channels.

The reference signal for the phase measurement is channel A. An automatic phase control circuit (APC) tunes and phase-locks the instrument to the channel A signal. The frequency range of the APC is selected by means of a front-panel control; there are 21 overlapping ranges, each more than an octave in width. In making a measurement, the operator selects a frequency range which includes the frequency of the signal which is driving the circuit under test. The APC then tunes the instrument automatically and essentially instantaneously (10 milliseconds), and keeps it tuned even if the input frequency drifts or sweeps at moderate rates (up to 15 MHz/second).

In the input probes of the VVM are sampling-type mixers which convert the RF signals to a 20-kHz intermediate frequency, where the voltage and phase measurements are made. Feedback stabilization of the mixers keeps the voltage conversion loss at 0 dB despite environmental influences, and common local-oscillator drive for both samplers keeps the phase difference between the IF signals equal to the phase difference between the RF signals.

The RF waveforms are reconstructed at the intermediate frequency: the fundamental components of the RF waveforms are converted to 20 kHz, the second harmonics to 40 kHz, the third harmonics to 60 kHz, and so on, up to the highest harmonic of the input signal which falls within the 1-GHz bandwidth of the samplers. Outputs are provided directly from the sampling mixers in both channels. Since the input waveforms are preserved in the IF signals, the VVM can be used to convert many low-frequency oscilloscopes, wave analyzers, and distortion analyzers to high-frequency sampling instruments for signals of moderate harmonic content. A similar sampling principle was originally employed by -hp- in sampling-type oscilloscopes.[1]

For the voltage and phase measurements, the IF signals from the sampling mixers are filtered so that only their 20-kHz fundamentals remain, and the amplitudes of these fundamentals and the phase angle between them are measured and displayed on the two front-panel meters (see block diagram, Fig. 2). Since only the fundamentals are measured, the amplitude and phase readings are not affected by the harmonic content of the input signals. The narrow-bandwidth IF filters (1 kHz) also reduce thermal noise at the meter inputs. The dc meter signals for both voltage and phase are available at the rear panel and can be used to drive recorders.

[1] Roderick Carlson, 'The Kilomegacycle Sampling Oscilloscope,' **'Hewlett-Packard Journal,'** Vol. 13, No. 7, March, 1962.

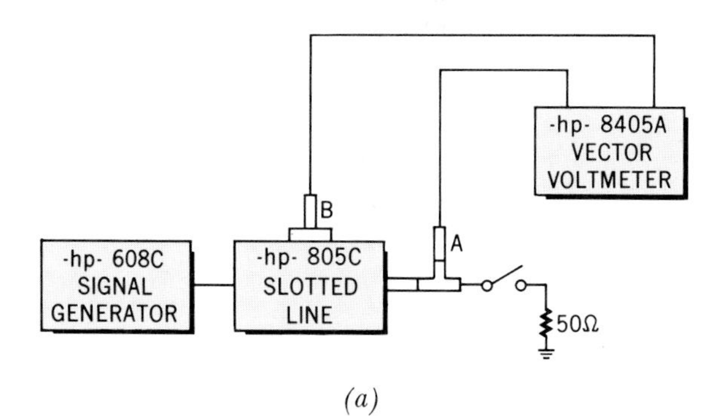

(a)

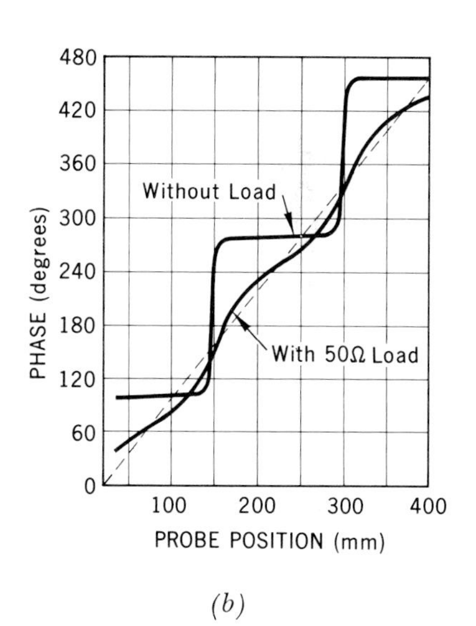

(b)

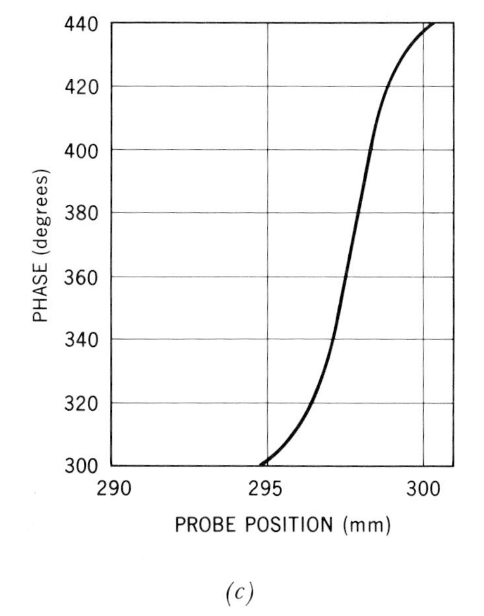

(c)

Fig. 3. *Demonstration of phase-measuring capabilities of Vector Voltmeter. Phasemeter readings vs. slotted-line probe positions for setup of (a) are plotted in (b). Steepest portion of (b) curves is shown expanded in (c). Maximum rate of change of phase of 50°/mm, or 2° per human hair diameter, is easily measured. With slotted line terminated in 50 ohms, maximum deviation of phase from linear, theoretically 22.8° for measured VSWR of 2.26, is measured as 22°. Frequency is 1.003 GHz.*

PHASE-MEASURING CAPABILITIES

Figs. 3(b) and 3(c) demonstrate the phase-measuring capabilities of the Vector Voltmeter. They show, first of all, how the high phase resolution of the VVM makes possible very precise measurements of length. Fig. 3(b) also includes an example of the phase-measurement accuracy of the instrument.

To obtain the data for Figs. 3(b) and 3(c), a 1-GHz signal was applied first to an unloaded slotted line and then to the same slotted line with a 50-ohm load [see block diagram, Fig. 3(a)]. Probe A of the VVM was placed at the output of the slotted line, and probe B was attached to the movable slotted-line probe. Fig. 3(b) is a plot of the phasemeter readings versus the position of the slotted-line probe. The measured curve closely follows the theoretical curve for an open-circuited lossless line.

Without the 50-ohm load, the standing-wave ratio on the line was 50.5. This was determined by measuring the maximum and minimum voltages on the line with the voltmeter of the VVM switched to channel B. The phase-vs-position curve is the step-like curve of Fig. 3(b), and Fig. 3(c) shows one of the steep portions of this curve with an expanded horizontal scale. The maximum rate of change of phase can be determined from Fig. 3(c) to be 50° per millimeter, or 0.05° per micrometer. Thus, a change equal to the diameter of a human hair in the position of the slotted-line probe was accompanied by about a 2° phase change, and was easily resolved by the high-resolution (0.1°) phasemeter.

With the 50-ohm load, the VSWR was 2.26. Had the VSWR been 1.0, the phase-vs-position curve would have been linear, as shown by the dashed line in Fig. 3(b). The theoretical maximum deviation from linear of the phase curve for a VSWR of 2.26 is

$$\Delta\phi = \text{arc sin}\frac{2.26 - 1}{2.26 + 1} = 22.8°.$$

The measured maximum deviation shown in Fig. 3(b) is about 22°.

AMPLIFIER MEASUREMENTS

Fig. 4(b) shows curves of gain, phase, and group delay versus frequency for a transistor amplifier stage operating in the 10-to-12-MHz range. The curves were measured with the Vector Voltmeter in the setup of Fig. 4(a). Compared with previously-available methods, the time and effort required to take the data were minimal.

Two sets of curves are shown in Fig. 4(b). With the switch shown in Fig. 4(a) in the closed position, the gain of the second amplifier stage was reduced to zero. The solid curves of Fig. 4(b) were obtained with the switch open (second stage gain >1) and the dashed curves were obtained with the switch closed. The difference between the curves shows that the impedance seen by the first stage has been changed by the Miller effect of the collector-to-base capacitance of the second transistor and the gain of the second stage.

Besides amplitude and phase curves, Fig. 4(b) shows group-delay curves, in which delay distortion produced by the Miller effect is apparent. A group delay curve can be obtained either by plotting the slope of the phase curve, or directly from the phasemeter. By changing the input frequency in increments of 2.78 kHz, or 27.8 kHz, or 278 kHz, etc., the group delay can be read directly from the corresponding changes in the phasemeter readings. The scale factors will in this case be 1 μs, 100 ns, or 10 ns, etc., per degree, since 1 $\mu s = 1$ degree at 2.78 kHz, and so on. Group delay information is very useful in cable testing, where constant time delay for all frequencies is desirable.

MEASUREMENTS OF TRANSISTOR AND NETWORK PARAMETERS

Another important application for the new VVM is measuring transistor gain and other transistor parameters. The wide frequency range of the VVM, and its ability to measure very small signals, make it well-suited for transistor measurements.

Fig. 5(a) shows a test setup which is being used at -hp- to measure transistor scattering parameters, or *s*-parameters. The *s*-parameters contain the same in-

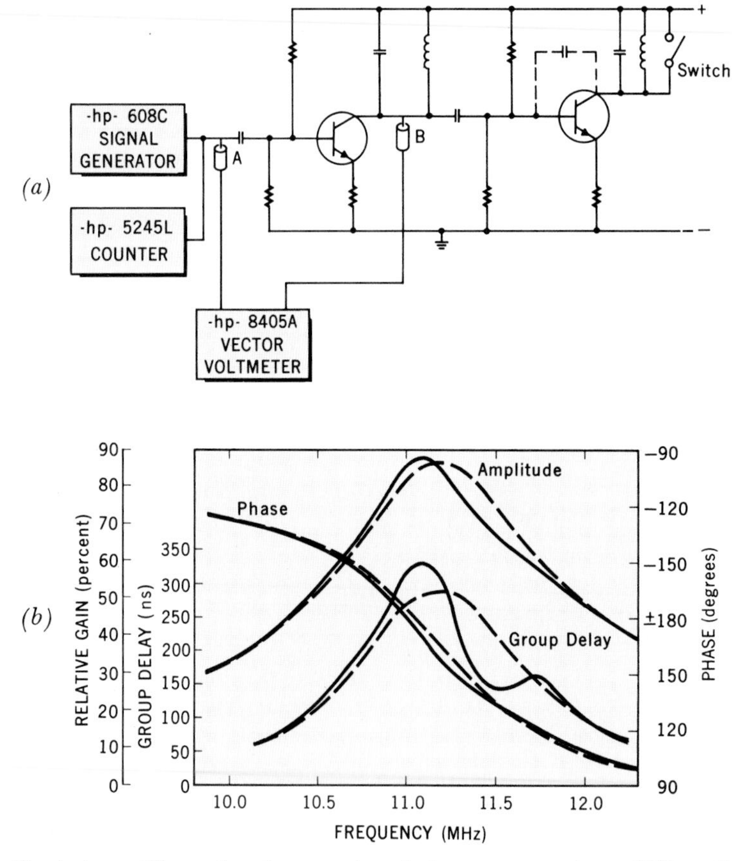

Fig. 4. *Typical amplifier gain, phase, and group delay measurements, made with Vector Voltmeter in setup shown in (a). Solid curves of (b) were taken with switch on second amplifier stage open, so that second-stage gain was greater than one. Dashed curves of (b) were measured with switch closed, second-stage gain = 0. Difference between curves shows Miller effect of second-stage collector-base capacitance. Group delay curves can be obtained by differentiating phase curve, or by changing input frequency in increments of 2.78 × 10^nkHz and determining delay from corresponding phasemeter changes with scale factor of 10^{-n} μs per degree, where n = 0, ±1, ±2, . . .*

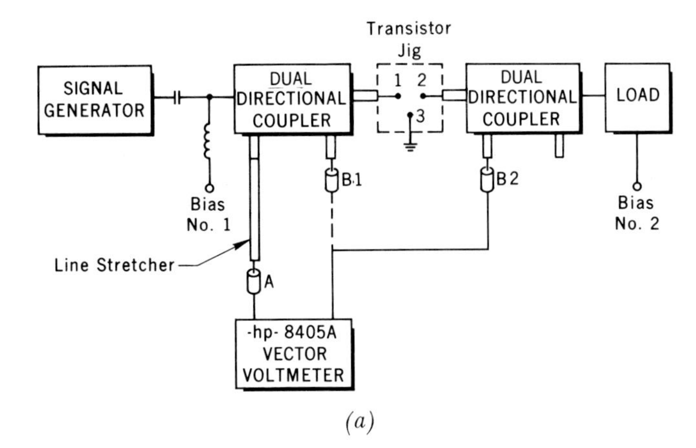

(a)

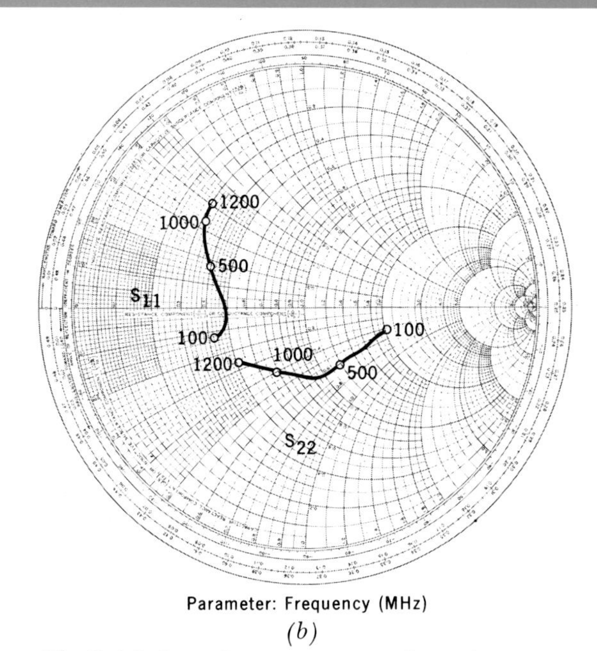

Parameter: Frequency (MHz)

(b)

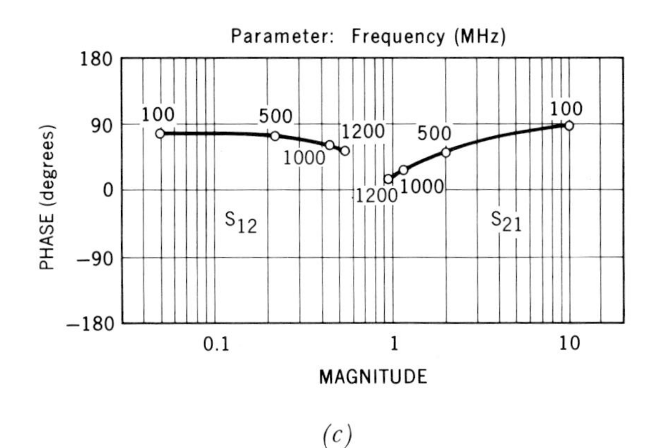

(c)

Fig. 5. (a) *Setup for measurement of transistor scattering parameters, or s-parameters. Input reflection coefficient s_{11} is measured with probe B in position B_1. Forward gain s_{21} is measured with probe B in position B_2. Output reflection coefficient s_{22} and reverse gain s_{12} are measured by turning transistor around in special jig and making same measurements as for s_{11} and s_{21}, respectively. (b) Amplitude and phase of s_{11} and s_{22} measured with Vector Voltmeter are plotted on Smith chart using polar coordinates. Normalized input and output impedances can then be read on impedance scales. (c) s_{12} and s_{21} for same transistor. Transistor was in grounded-emitter configuration with 50-ohm source and load impedances.*

formation as other common types of two-port network parameters, such as *y*-, *z*-, *h*-, or *a*-parameters, but are much easier to measure and to work with at high frequencies because, unlike the other parameters, *s*-parameters are not defined in terms of short circuits or open circuits, which are difficult to obtain at high frequencies. Now that transistor gain-bandwidth products greater than 1000 MHz are becoming common, new methods for specifying transistor high-frequency performance are coming into use. The *s*-parameters will probably be employed for this purpose more often in the future.[2]

The parameter s_{11} is the complex reflection coefficient at the input, or port 1, of a two-port network, with the network terminated in equal source and load impedances, usually 50 ohms. The reflection coefficient at port 2 is s_{22}.

The parameter s_{21} is the complex transducer gain or loss from input to output, or port 1 to port 2, of a two-port network, again with equal source and load impedances. The reverse gain is s_{12}.

Fig. 5(b) shows a Smith-chart plot of input and output reflection coefficients s_{11} and s_{22} as a function of frequency for a high-frequency transistor. The measurements were made over a wide measurement range from 100 to 1200 MHz with the new Vector Voltmeter, using the setup of Fig. 5(a). The Smith chart is useful for plotting s_{11} and s_{22} because the amplitude and phase of these reflection coefficients can be plotted using the polar coordinates of the chart, and then the normalized input reactance and resistance of the network can be read directly from the reactance and resistance scales.

Fig. 5(c) shows plots of reverse and forward gain s_{12} and s_{21} obtained with the same transistor as Fig. 5(b), in the circuit of Fig. 5(a).

All of the measurements discussed here, as well as many others, some of which are described briefly on pages 7, 10 and 11 can be made quickly and easily with the new Vector Voltmeter. In the past, these measurements were difficult to make, and often were not made at all, because of the difficulty of obtaining phase information.

[2] George E. Bodway, 'Two-Port Power Flow Analysis of Linear Active Circuits Using the Generalized Scattering Parameters'; to be published.

SAMPLING MIXERS

Fig. 7 is a block diagram of the sampling-type harmonic mixers, which are located in the probes. These mixers are similar to those used in -hp- sampling oscilloscopes. They operate on a stroboscopic principle, sampling a high-

Fig. 6. *Design team for –hp– Model 8405A Vector Voltmeter included (l. to r.) Fritz K. Weinert, final-phase project leader, William R. Hanisch, Allen Baghdasarian, Siegfried H. Linkwitz, Jeffrey L. Thomas, and Roderick Carlson, initial-phase project leader.*

frequency periodic input signal at a slightly different phase at each sampling instant and reconstructing a low-frequency image of the signal from the samples. The time between sampling pulses is determined by the frequency of the voltage-tunable local oscillator (VTO), which is controlled by the phase-locked loop.

In operation, the sampler gate is opened for about 300 picoseconds. The input voltage at this time is stored in a 'zero-order hold' circuit until the next sample. The output waveform is a faithful replica of the input, constructed in small steps.[3] Negative feedback is employed to stabilize the voltage conversion loss at 0 dB (output amplitude is same as input amplitude) and to give a high input impedance.

The two probes are ac-coupled and permanently attached to the instrument with 5-foot cables. Loading of the system under test is minimized by the high input impedance of the probes (0.1 megohm shunted by 2.5 pF; with divider, 1 megohm shunted by 2 pF).

AUTOMATIC PHASE CONTROL

The phase-locked loop, shown in Fig. 8, tunes the instrument to the signal frequency. The loop is preceded by a high-gain amplifier-limiter which

[3] This process is essentially distortionless: distortion introduced by the sample-and-hold system does not appear in the IF signals until the 50th harmonic, which is usually much higher than the highest significant harmonic in the input signal.

THE VECTOR VOLTMETER AS A PRECISION FREQUENCY COMPARATOR

Adjusting a precision oscillator so that its frequency is the same as that of a standard calls for a precise frequency comparison between two highly stable signal sources. Such frequency comparisons are also needed in studies of aging effects, or long-term stability, in precision oscillators. In these comparisons, frequency differences of a few parts in 10^{12} are significant and must be detected.

Most methods for comparing the frequencies of two stable oscillators require long time periods to achieve the required precision. For example, it takes about one day to compare two 5-MHz frequency standards to a precision of one part in 10^{13}, by the best of these methods.

By using the Vector Voltmeter to detect the phase difference between the two oscillators, the time required to achieve a precision of one part in 10^{13} can be reduced to a few minutes, at typical standard frequencies of 1 MHz or more. The block diagram shows the measurement arrangement.

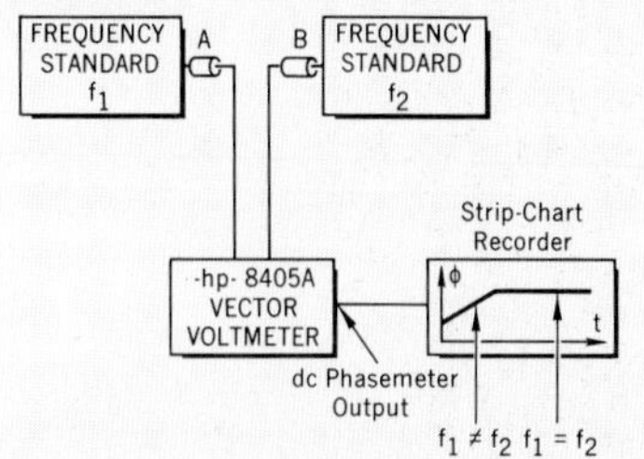

If the frequencies of the two oscillators are the same their phase difference will be constant. If the frequencies differ, the phasemeter reading will change at a rate given by

$$\frac{\Delta\phi}{\Delta t} = 360\ \Delta f$$

where $\Delta\phi$ = phase change in degrees, indicated by VVM

Δt = time in seconds required for phase change $\Delta\phi$

Δf = frequency difference in Hz between input signals.

The direction of the phase change tells which frequency is higher: clockwise rotation of the phasemeter pointer indicates that the frequency in channel B is higher than that in channel A.

The phase change and direction of change can be recorded on a strip-chart recorder by connecting the recorder to the dc phasemeter output jack on the rear panel of the VVM. The record shown is a typical recorder trace for two 1-MHz oscillators with a frequency offset of 2.3×10^{-5} Hz, or 2.3 parts in 10^{11}. The time

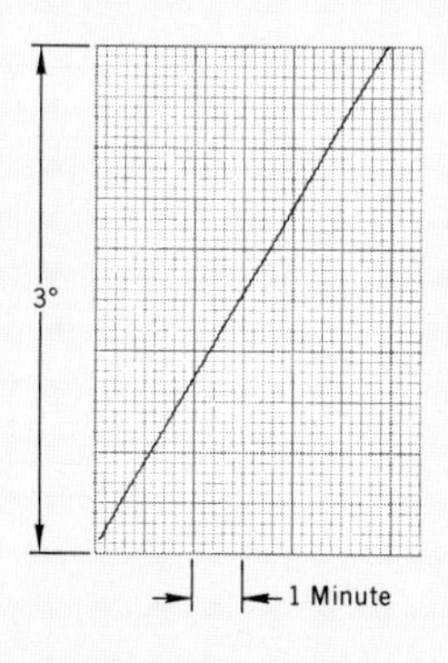

scale is 12 seconds per division, and the full-scale phase difference is 3°. The slope of the trace can be determined within less than one minute, whereas older methods would have required much longer to achieve this precision.

When the Vector Voltmeter is used as a precision frequency comparator, the two oscillators must have low noise, the oscillator frequencies must fall within the range of the VVM (1 MHz to 1 GHz), and the oscillator frequencies must differ by less than a few hertz. Oscillators whose frequencies differ by more than a few hertz should first be tuned coarsely using a counter or an oscilloscope.

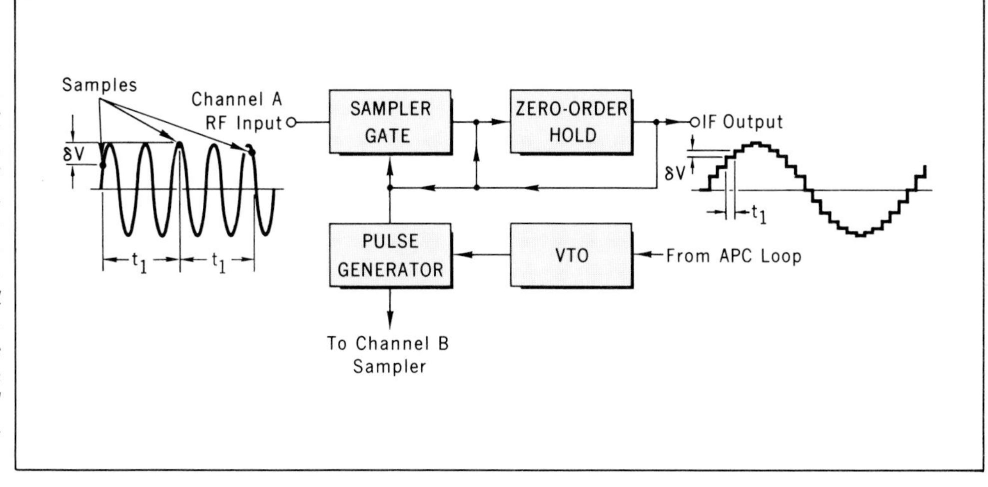

Fig. 7. *Block diagram of sampling-type harmonic mixers used in new VVM. Mixers operate on stroboscopic principle, sampling RF signal at different points in cycle at successive sampling instants. RF waveforms are reconstructed in small steps at intermediate frequency: fundamental component of RF signal is transposed to 20 kHz, second harmonic to 40 kHz, and so on. Feedback keeps IF voltage equal to RF voltage.*

delivers a constant output regardless of the input voltage.

When an RF signal is applied to channel A and the instrument is not tuned properly, the IF is not 20 kHz, and the search generator produces a ramp voltage which adjusts the frequency of the VTO. This changes the time between samples and, consequently, changes the intermediate frequency. When the IF reaches 20 kHz, the loop locks and controls the VTO so as to correct for changes in VTO frequency, signal frequency, or phase modulation.

When the loop is locked, the difference between the signal frequency and a harmonic of the VTO frequency is exactly the 20-kHz reference oscillator frequency:

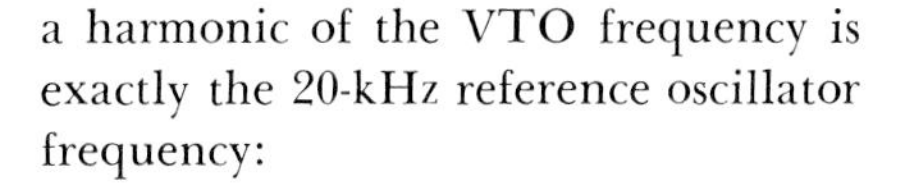

$$f_{sig} - nf_{VTO} = \pm 20 \text{ kHz}.$$

The 20-kHz IF can be either the 'inverted' mode or the 'noninverted' mode, depending upon whether the signal is 20 kHz below or 20 kHz above a VTO harmonic. The IF phase difference is identical to the RF phase difference only for the noninverted mode. For the inverted mode, the phase angle is correct, but is lagging when the RF phase angle is leading. A sideband decision circuit detects the sideband mode and starts the search generator again if the IF mode is inverted. The time required to complete the tuning operation is about 10 milliseconds.

Overall gain of the phase-locked loop is a linear function of the harmonic number to which the signal is locked. A variable attenuator adjusts the loop gain to an optimal value for any signal frequency so that the gain will be sufficient to ensure phase lock but not so high that the loop oscillates. The attenuator control knob is labeled FREQUENCY RANGE, and has 21 overlapping octave-wide bands.

METER CIRCUITS

The voltmeter and phasemeter cir-

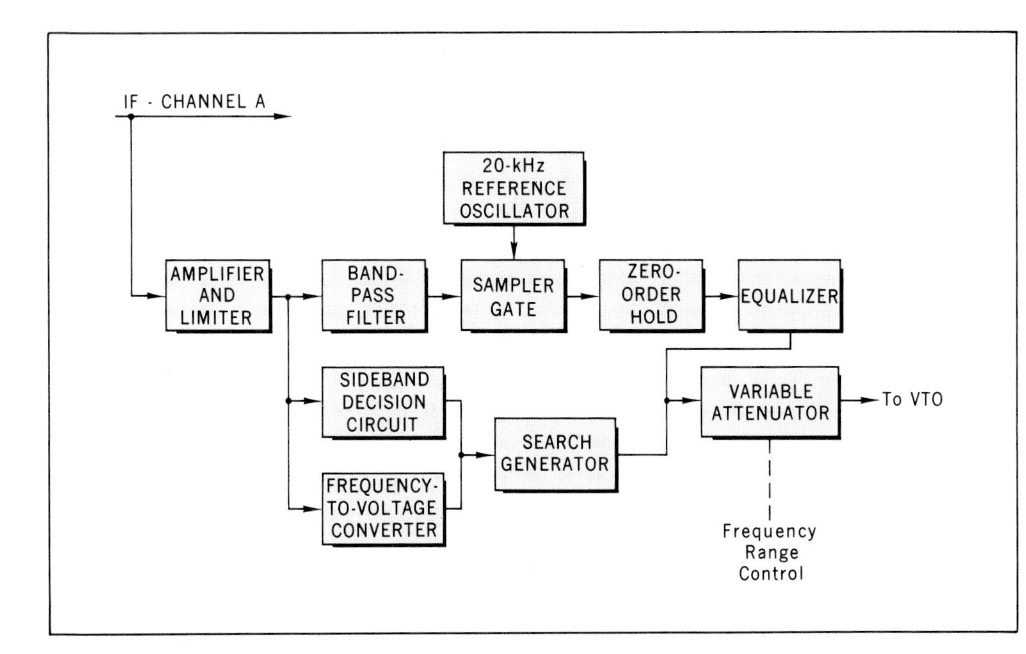

Fig. 8. *Block diagram of automatic phase control (APC) circuit, which tunes and phase-locks Vector Voltmeter to channel A signal. APC loop adjusts frequency of voltage-tuned local oscillator (VTO) which generates sampling pulses for mixers, thus keeps IF at 20 kHz. APC requires only 10 ms to tune meter, and remains locked even if input frequency changes at rates up to 15 MHz/s. Sideband decision circuit ensures that $f_{sig} - nf_{VTO}$ is always +20 kHz, never −20 kHz.*

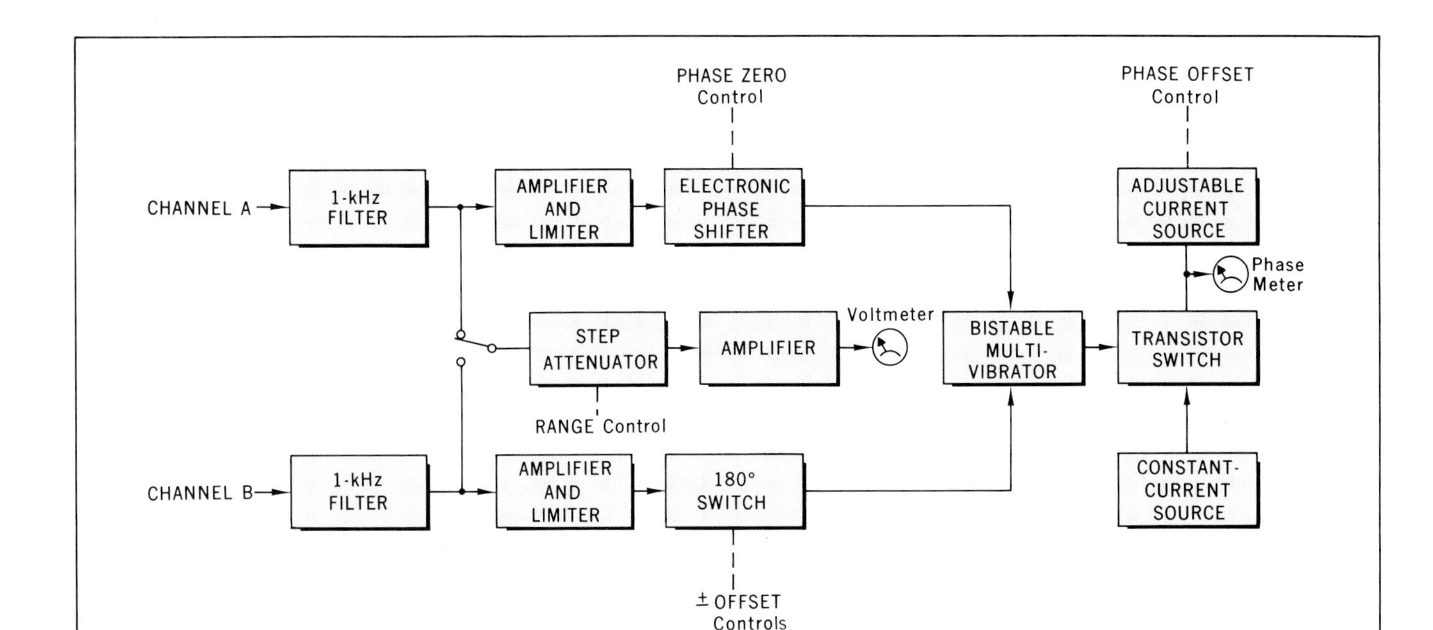

Fig. 9. *Block diagram of voltmeter and phasemeter circuits. Amplifier-limiters make phase readings independent of signal levels.*

cuits are shown in Fig. 9. The 20-kHz phasemeter has identical amplifiers and limiters in both channels so that the meter reading is independent of the input signal levels.

The phase detector is a bistable multivibrator which is triggered to one of its stable states by channel A and to the other by channel B. The multivibrator operates a transistor switch, which turns the meter current on and off. Another meter input current provides the phase offset, which is adjustable in 10° steps. This kind of phase detector has a very linear characteristic and gives precise phase offset steps in spite of extreme environmental conditions or intermediate-frequency shifts.

ACKNOWLEDGMENTS

The Vector Voltmeter design was initiated by a study[4] made by Chu-Sun Yen, Kay B. Magleby, and Gerald J. Alonzo of the -*hp*- Advanced Research and Development Laboratories. The design group for the Vector Voltmeter has included Roderick Carlson, Allen Baghdasarian, William R. Hanisch, Siegfried H. Linkwitz, Jeffrey L. Thomas, Giacomo J. Vargiu and the undersigned.

—*Fritz K. Weinert*

[4] Chu-Sun Yen, 'Phase-Locked Sampling Instruments,' IEEE Transactions on Instrumentation and Measurement, Vol. IM-14, Nos. 1 and 2, March-June, 1965.

DESIGN LEADERS

RODERICK CARLSON

Rod Carlson joined –hp– in 1958 as a development engineer. He participated in the design of the –hp– 160A Oscilloscope and was project leader for the development of the –hp– 185A Sampling Oscilloscope. Later he became section manager for sampling oscilloscope development. He transferred to the –hp– Microwave Division in 1964, and was the project leader during the initial development of the –hp– 8405A Vector Voltmeter. He then became manager of the signal analysis section of the Microwave Laboratory, the section concerned with wave and spectrum analyzers, broadband detectors, and power measurement.

Rod holds a BSEE degree from Cornell University. He is a member of IEEE, Tau Beta Pi, Eta Kappa Nu, and Phi Kappa Phi. Before joining –hp–, Rod spent five years as an instrumentation engineer for Cornell Aeronautical Laboratory, dealing with aircraft stability and control under flight conditions.

FRITZ K. WEINERT

Fritz Weinert graduated Magna Cum Laude from Ingenieurschule Gauss, Berlin, Germany, with a degree in electrical communications engineering and precision mechanics. Beginning in 1947, he was associated with several German firms as a development engineer for carrier telephone systems, and as project engineer for a variety of projects dealing with electronic test instruments, antennas, and fields. After coming to the United States in 1960, Fritz spent four years as a project engineer in the development of RFI instrumentation.

Fritz joined –hp– in 1964, becoming project leader for the final development of the 8405A Vector Voltmeter. He is now a project leader in the network analysis section of the –hp– Microwave Laboratory. Fritz holds patents and has published papers dealing with pulse circuits, tapered-line transformers, digital tuned circuits, and shielding systems. He has taught undergraduate electronics and mathematics.

SELECTED VECTOR VOLT-

- **COMPLEX INSERTION LOSS OR GAIN**
- **COMPLEX REFLECTION COEFFICIENT**
- **COMPLEX IMPEDANCE OR ADMITTANCE**
- **TWO-PORT NETWORK PARAMETERS**
- **ANTENNA IMPEDANCE AND PHASE CHARACTERISTICS**

The arrangement shown in (a) is limited to the frequency range of the directional couplers (usually >200 MHz). The cable or line-stretcher may be needed at the higher frequencies to compensate for phase shift in the directional couplers and other circuitry. A simpler arrangement, useful at lower frequencies, is shown in (b).

Reflection coefficients, input parameters, and impedances are measured with probe B in position B1. Transmission parameters, loss, and gain are measured with probe B in position B2. Input impedance can be determined by plotting magnitude and phase of reflection coefficient on Smith chart and reading normalized impedance on resistance and reactance scales.

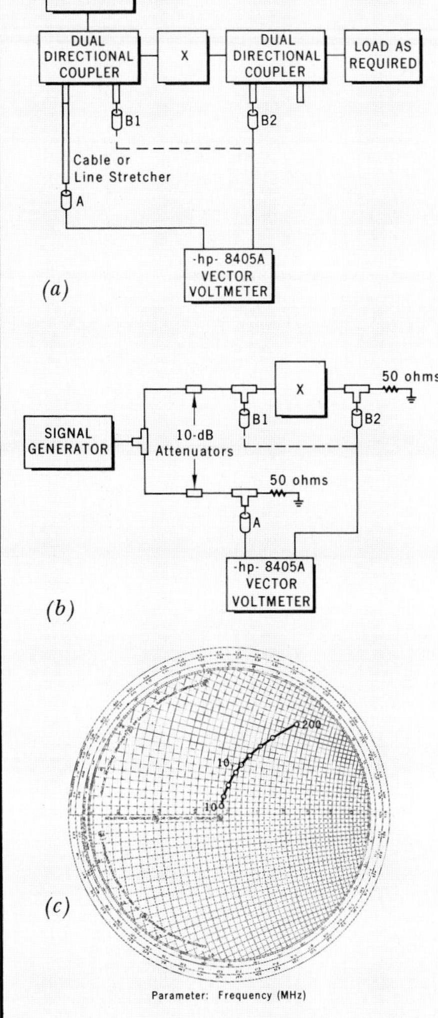

Expanded Smith chart plot of reflection coefficient of 50-ohm ±1% metal film resistor attached to BNC connector. Circuit was (a) with probe B in position B1, resistor as device X, and only one directional coupler.

- **GAIN AND PHASE OF ONE OR MORE AMPLIFIER STAGES**
- **GROUP DELAY AND DISTORTION**
- **COMPLEX TRANSMISSION COEFFICIENTS**
- **FILTER TRANSFER FUNCTIONS**
- **ATTENUATION**

Measurements of gain, phase shift, and group delay of any device can be made by placing one probe (A or B) of the Vector Voltmeter at the input of the device and the other probe at the output. The difference between channel A and channel B voltmeter readings in dB is the gain or loss. Phasemeter reading is the phase shift. Group delay is the slope of the phase-vs-frequency curve.

If signal frequency must be measured more accurately than is possible with the signal-generator dial, a counter may be used to measure frequency, or a frequency synthesizer may be used as a signal generator.

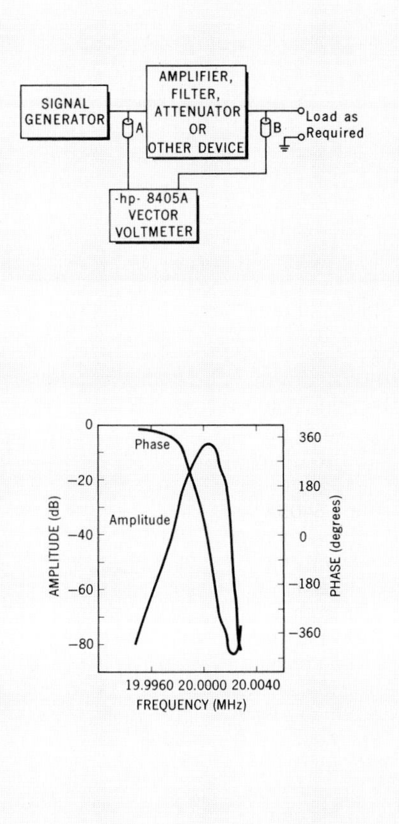

Gain and phase shift of *-hp-* Model 8442A 20-MHz Crystal Filter as measured with new RF Vector Voltmeter.

- **OPEN-LOOP GAIN OF FEEDBACK AMPLIFIERS**
- **GAIN AND PHASE MARGINS**

Closed-loop gain of a feedback amplifier is $\frac{A}{1 - A\beta}$ (a complex number).

$A\beta$ is open-loop gain. If $-A\beta = -1$, the feedback is positive and oscillations occur.

Important quantities in feedback amplifier design are gain margin and phase margin, which are measures of the degree of stability of an amplifier. Gain margin is the magnitude of $-A\beta$, in dB, at the frequency for which the phase of $-A\beta$ is $-180°$. Phase margin is the difference between $-180°$ and the phase of $-A\beta$ at the frequency for which the magnitude of $-A\beta$ is 0 dB. Typical gain margins are −10 dB to −40 dB, typical phase margins greater than 30°.

The Vector Voltmeter greatly simplifies the design of feedback amplifiers and oscillators by giving both amplitude and phase of open-loop gain simultaneously and quickly.

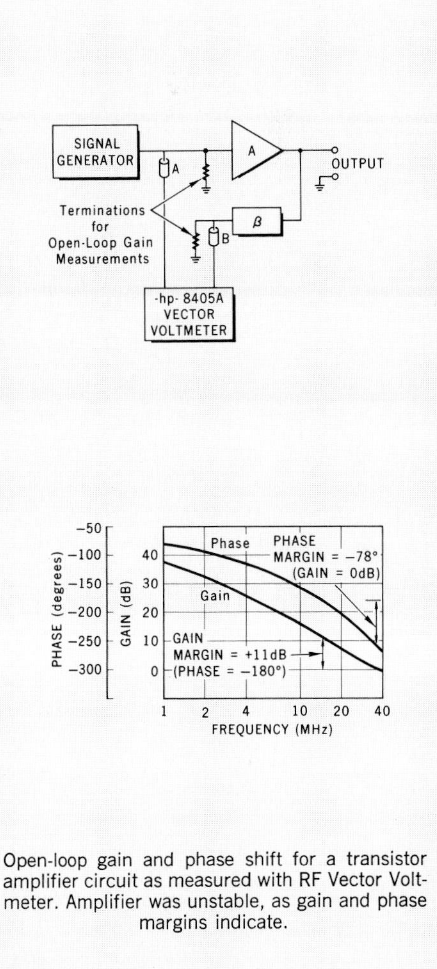

Open-loop gain and phase shift for a transistor amplifier circuit as measured with RF Vector Voltmeter. Amplifier was unstable, as gain and phase margins indicate.

METER MEASUREMENTS

- **AMPLITUDE MODULATION INDEX**
- **RF DISTORTION**
- **CONVERSION OF LOW-FREQUENCY INSTRUMENTS TO SAMPLING INSTRUMENTS FOR OBSERVATION AND MEASUREMENT OF HIGH-FREQUENCY SIGNALS**

Device X is any signal source, 1 MHz to 1 GHz. The Vector Voltmeter converts the fundamental of the RF signal to a 20-kHz IF, the second harmonic to 40 kHz, and so on, so RF waveforms are preserved in the IF signals. The IF output can be used as the input to a low-frequency oscilloscope, distortion analyzer, wave analyzer, or other instrument.

For amplitude-modulated signals, the voltmeter is synchronized to the carrier frequency f_c and the sidebands $f_c \pm \Delta f$ are reproduced at the IF as 20 kHz $\pm\Delta f$. Modulation index can be measured using an oscilloscope.

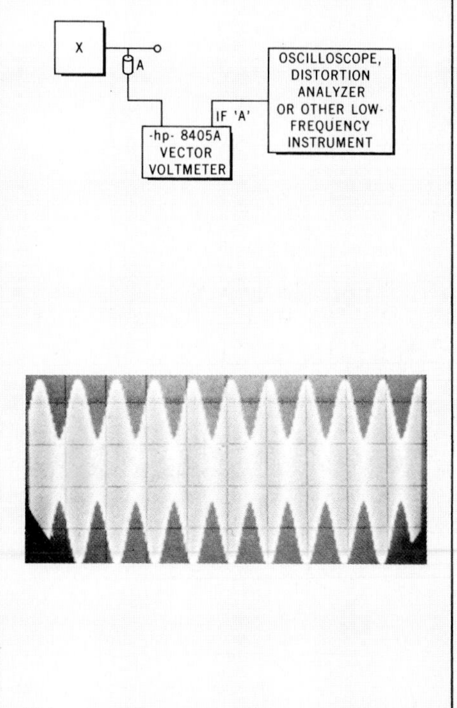

20-kHz IF output of Vector Voltmeter observed on *-hp-* 120B Oscilloscope (bandwidth = 450 kHz). Input to VVM was 300-MHz carrier, amplitude modulated by 1-kHz signal. Oscilloscope was synchronized to modulating signal only.

- **ELECTRICAL LENGTH OF CABLES**
- **PHASE TRACKING BETWEEN SIGNAL PATHS**

The electrical length of a cable can be adjusted precisely using the phase resolution of the Vector Voltmeter. One arrangement for doing this is shown in the block diagram.

To cut a cable to an electrical length of one-quarter wavelength at frequency f, the signal generator is first tuned precisely to frequency f. Next, with a short circuit at the output of the directional coupler, the system is calibrated by adjusting the PHASE ZERO control of the VVM until the phasemeter reads 180°. Then the short circuit is replaced by the cable and the cable length is adjusted until the phasemeter again reads 180°. The electrical length of the cable is then one-quarter wavelength.

A cable can be adjusted to the same electrical length as another cable by 1) connecting the first cable to the directional coupler and noting the phasemeter reading, and 2) connecting the second cable and cutting it until the phasemeter reading is the same as for the first cable.

Another method for adjusting two cables to the same length is simply to drive both cables with the same signal source and measure the phase difference between the cable output signals with the VVM. Zero degrees phase difference indicates equal electrical lengths. Phase tracking between any two signal paths can be measured in the same way, that is, by driving both paths with the same source and measuring the phase difference at the path outputs with the VVM.

If the cable or cables must be longer than one-quarter wavelength at the frequencies within the range of the VVM, cable length must first be determined to within one-quarter wavelength by other means (e.g., time domain reflectometry).

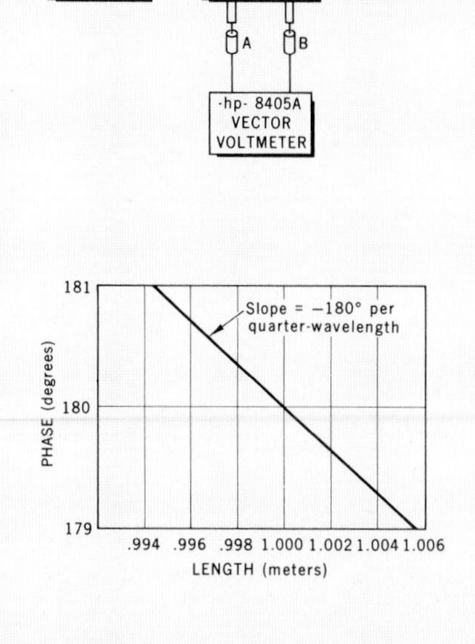

Phase versus length of cable for system shown in block diagram, calibrated at 75 MHz (quarter-wavelength = 1 m). Phase resolution of VVM is 0.1°, allowing length to be determined within 0.6 mm at 75 MHz, or more accurately at higher frequencies.

- **SELECTIVE RECEIVER**
- **NEAR-FIELD ANTENNA CHARACTERISTICS**
- **RF LEAKAGE**

The Vector Voltmeter can be used as a selective receiver by synchronizing channel A to the desired frequency or signal, and equipping the channel B probe with an antenna. Meter bandwidth of the VVM is 1 kHz. RF leakage from any device can be detected by this technique. Antenna characteristics can be measured also.

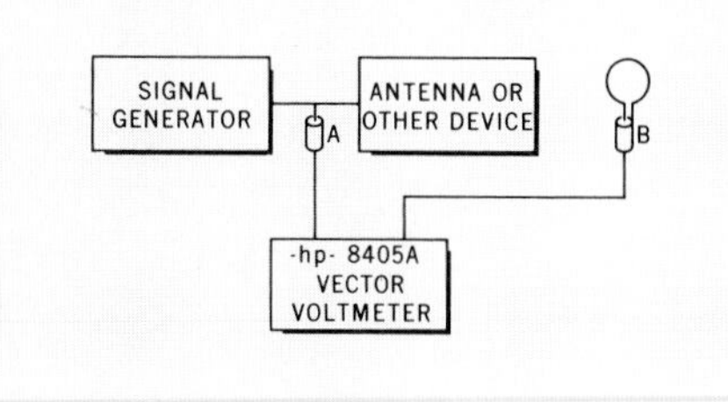

- **COMPLEX IMPEDANCE AND ADMITTANCE (AT FREQUENCIES BELOW 100 MHz)**

Two simple techniques for measuring impedances at lower frequencies are shown in the accompanying diagrams. These methods are useful if the probe and circuit impedances are negligible in comparison with the unknown and if the reactance of the current transformer or resistor is small.

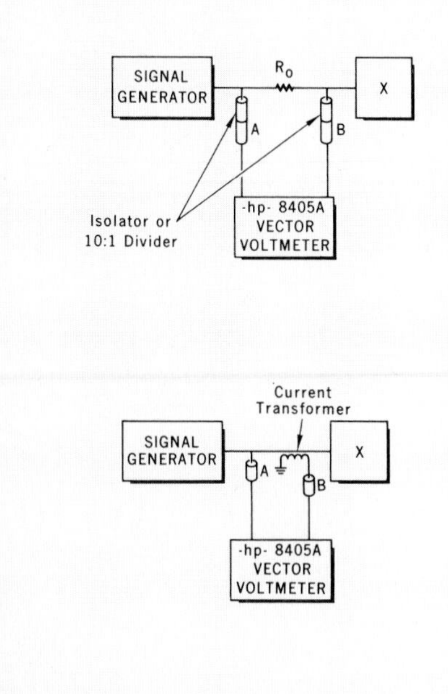

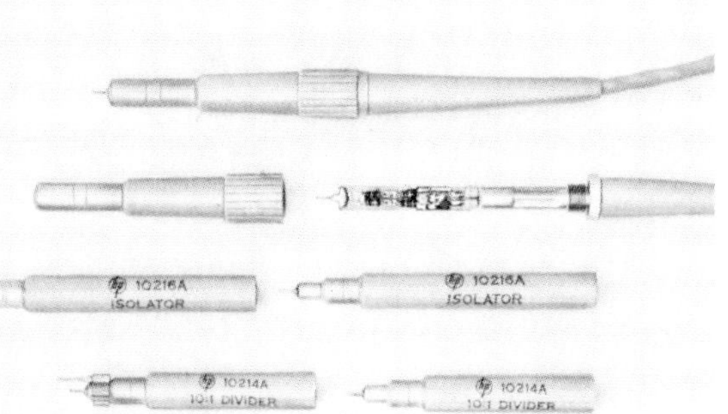

Fig. 10. *Probes and accessories for –hp– Model 8405A Vector Voltmeter. Top to bottom: assembled probe; exploded view of probe, showing sampler circuit; isolators, 10:1 dividers, and ground clips supplied with probes. Isolators (or dividers) and ground clips are used when probing by hand in arbitrary circuits. (Tees and adapters are available for probing in 50-ohm coaxial systems.) Dividers extend upper limit of voltmeter range to 10 V. See Specifications for details.*

SPECIFICATIONS
–hp–
MODEL 8405A
VECTOR VOLTMETER

Instrument Type: Two-channel sampling RF millivoltmeter-phasemeter which measures voltage level in two channels and simultaneously displays the phase angle between the two signals.

Frequency Range: 1 MHz to 1 GHz in 21 overlapping octave bands (lowest band covers two octaves).

Meter Bandwidth: 1 kHz.

Tuning: Automatic within each band. Automatic phase control (APC) circuit responds to the channel A input signal. Search and lock time, approximately 10 millisec; maximum sweep speed, 15 MHz/sec.

Voltage Range:
Channel A: 1.5 mV to 1 V rms from 1 to 5 MHz; 300 μV to 1 V rms, 5 to 500 MHz; 500 μV to 1 V rms, 500 MHz to 1 GHz; can be extended by a factor of 10 with 10214A 10:1 Divider.
Channel B: 100 μV to 1 V rms full scale (input to channel A required); can be extended by a factor of 10 with 10214A 10:1 Divider.
Meter Ranges: 100 μV to 1 V rms full scale in 10-dB steps.

Full-Scale Voltage Accuracy: Within ±2% 1 to 100 MHz, within ±6% to 400 MHz, within ±12% to 1 GHz, not including response to test-point impedance*.

Voltage Response to Test Point Impedance:* +0, −2% from 25 to 1000 ohms. Effects of test-point impedance are eliminated when 10214A 10:1 Divider or 10216A Isolator is used.

Residual Noise: Less than 10 μV as indicated on the meter.

Phase Range: 360°, indicated on zero-center meter with end-scale ranges of ±180, ±60, ±18, and ±6°. Meter indicates phase difference between the fundamental components of the input signals.

Resolution: 0.1° at any phase angle.

Meter Offset: ±180° in 10° steps.

Phase Accuracy: Within ±1°, not including phase response vs. frequency, amplitude, and test-point impedance.*

Phase Response vs. Frequency: Less than ±0.2° 1 MHz to 100 MHz, less than ±3° 100 MHz to 1 GHz.

Phase Response vs. Signal Amplitude: Less than ±2° for an amplitude change from 100 μV to 1 V rms.

Phase Response vs. Test Point Impedance*: Less than ±2° 0 to 50 ohms, less than −9° 25 to 1000 ohms. Effects of test-point impedance are eliminated when 10214A 10:1 Divider or 10216A Isolator is used.

Isolation Between Channels: Greater than 100 dB 1 to 400 MHz, greater than 75 dB 400 MHz to 1 GHz.

Input Impedance (nominal): 0.1 megohm shunted by approximately 2.5 pF; 1 megohm shunted by approximately 2 pF when 10214A 10:1 Divider is used; 0.1 megohm shunted by approximately 5 pF when 10216A Isolator is used. ac coupled.

Maximum ac Input (for proper operation): 3 V p-p (30 V p-p when 10214A 10:1 Divider is used).

Maximum dc Input: ±150 V.

20 kHz IF Output (each channel): Reconstructed signals, with 20 kHz fundamental components, having the same amplitude, waveform, and phase relationship as the input signals. Output impedance, 1000 ohms in series with 2000 pF, BNC female connectors.

Recorder Output:
Amplitude: 0 to +1 Vdc ±6% open circuit, proportional to voltmeter reading. Output impedance, 1000 ohms; BNC female connector.
Phase: 0 to ±0.5 Vdc ±6%, proportional to phase meter reading; less than 1% effect on Recorder Output and meter reading when external load is ≥10,000 ohms; BNC female connector.

RFI: Conducted and radiated leakage limits are below those specified in MIL-I-6181D and MIL-I-16910C except for pulses emitted from probes. Spectral intensity of these pulses is approximately 60 μV/MHz; spectrum extends to approximately 2 GHz. Pulse rate varies from 1 to 2 MHz.

Power: 115 or 230 V ±10%, 50 to 400 Hz, 35 watts.

Weight: Net 30 lbs. (13,5 kg). Shipping 35 lbs. (15,8 kg).

Dimensions:
16¾ in. wide, 7⁹⁄₃₂ in. high, 18⅜ in. deep (425 x 185,2 x 467 mm) overall. Hardware furnished for conversion to rack mount 19 in. wide, 6³¹⁄₃₂ in. high, 16⅜ in. deep behind panel (483 x 177,2 x 416 mm).

Accessories Furnished:
10214A 10:1 Divider (two furnished) for extending voltmeter range. Voltage error introduced is less than ±6% 1 MHz to 700 MHz, less than ±12% to 1 GHz; if used on one channel only, phase error introduced is less than ±(1 + 0.015f/MHz)°.
10216A Isolator (two furnished) for eliminating effect of test-point impedance on sampler*. Voltage error introduced is less than ±6% 1 to 200 MHz, response is 3 dB down at 500 MHz; if used on one channel only, phase error introduced is less than ±(3 + 0.185f/MHz)°.
10213-62102 Ground Clips (six furnished) for 10214A and 10216A.
5020-0457 Probe Tips (six furnished).

Accessories Available:
10218A BNC Adapter, converts probe tip to male BNC connector, $6.00.
10220A Adapter, for connection of Microdot screw-on coaxial connectors to the probe, $3.50.
10221A 50-ohm Tee, with GR874 RF fittings, for monitoring signals in 50-ohm transmission line without terminating the line, $40.00.
11529A Accessory Case, for convenient storage of accessories, includes two compartmented shelves and snap-shut lid, $8.50.
1250-0778 Adapter, both connectors type N male (UG-57B/U).
1250-0780 Adapter, type N male and BNC female (UG-201A/U).
1250-0846 Adapter, Tee, all connectors type N female (UG-28A/U).
General Radio type 874-W50 50-ohm Load (also available from –hp– under part no. 0950-0090).
General Radio type 874-QNP Adapter, GR 874 and type N male (also available from –hp– under part no. 1250-0847).
General Radio type 874-QNJA Adapter, GR 874 and type N female (also available from –hp– under part no. 1250-0240).
General Radio type 874-QBPA Adapter, GR 874 and BNC male (also available from –hp– under part no. 1250-0849).
General Radio type 874-QBJA Adapter, GR 874 and BNC female (also available from –hp– under part no. 1250-0850).

Complementary Equipment:
774D Dual Directional Coupler, 215 to 450 MHz, $200.00.
775D Dual Directional Coupler, 450 to 950 MHz, $200.00.
8491A (Option 10) 10-dB Coaxial Attenuator, $50.00.
8491A (Option 20) 20-dB Coaxial Attenuator, $50.00.

Price: Model 8405A, $2500.00.

Prices f.o.b. factory.

Data subject to change without notice.

* Variations in the high-frequency impedance of test points as the probe is shifted from point to point influence the samplers and can cause the indicated amplitude and phase errors. These errors are different from the effects of any test-point loading due to the input impedance of the probes.

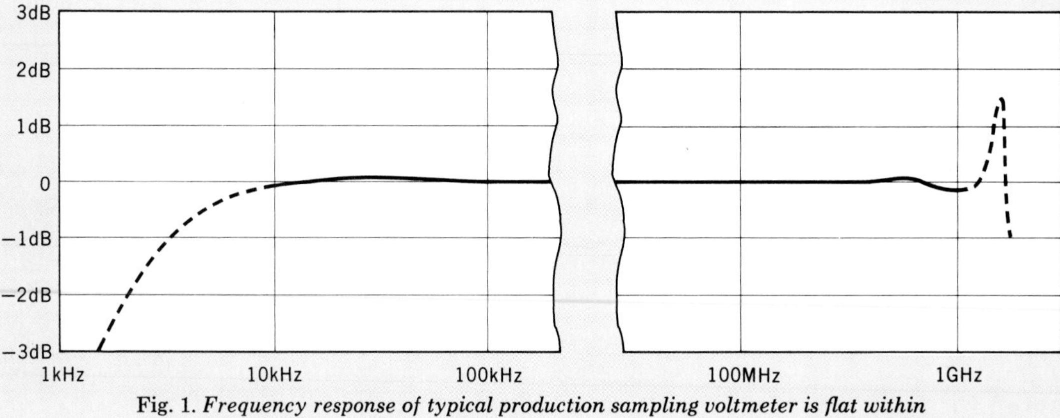

Fig. 1. *Frequency response of typical production sampling voltmeter is flat within ±1% from 10 kHz to 1 GHz. Useful sensitivity extends from 1 kHz to over 2 GHz.*

A SENSITIVE NEW 1-GHz SAMPLING VOLTMETER WITH UNUSUAL CAPABILITIES

A voltmeter operating on the principle of incoherent sampling measures over wide frequency and voltage ranges while providing an output usable for signal analysis.

SAMPLING HIGH-FREQUENCY WAVES in order to construct low-frequency equivalents of them is a powerful technique for observing and measuring broadband signals. The sampling oscilloscope,[1,2] introduced about seven years ago, can display repetitive waveforms which contain frequency components up to several gigahertz. A more recent development, the RF vector voltmeter,[3] can measure amplitudes and phase angles simultaneously and automatically at frequencies up to one gigahertz. Other sampling instruments are being investigated at -*hp*- for frequency ranges as high as X band (12.4 GHz).

A sampling technique has been used by the -*hp*- Loveland Division to achieve exceptional sensitivity, frequency response, and accuracy in a new broadband voltmeter (Fig. 2). In addition to its basic voltage-measuring function, the sampling operation and flat frequency response of the new voltmeter give it many capabilities not found in more conventional RF millivoltmeters. Peak voltages, amplitude modulation envelopes, true rms values, pulse height information, and probability density functions of broadband signals can be determined by observing the output of the sampling circuit. Much of this information has never before been accessible for broadband signals. Other uses for the instrument include broadband power measurements and leveling of the outputs of broadband signal generators.

[1] Roderick Carlson, 'A Versatile New DC-500 MC Oscilloscope with High Sensitivity and Dual Channel Display,' **Hewlett-Packard Journal,** Vol. 11, No. 5-7, Jan.-Mar., 1960.

[2] Wayne M. Grove, 'A New DC-4000 MC Sampling 'Scope Plug-in with Signal Feed-Through Capability,' **Hewlett-Packard Journal,** Vol. 15, No. 8, April, 1964.

[3] Fritz K. Weinert, 'The RF Vector Voltmeter — An Important New Instrument for Amplitude and Phase Measurements from 1 MHz to 1000 MHz,' **Hewlett-Packard Journal,** Vol. 17, No. 9, May, 1966.

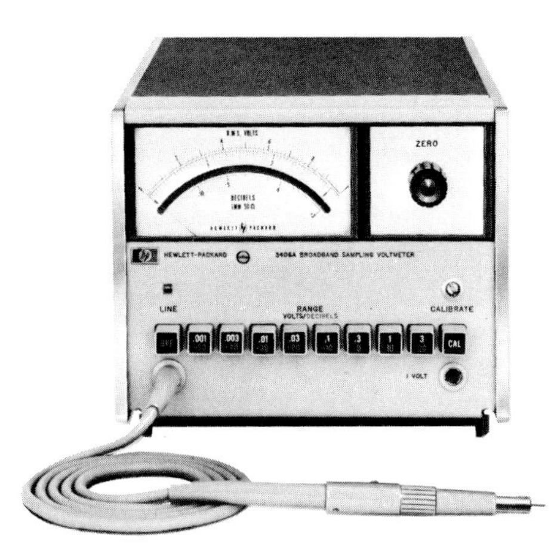

Fig. 2. *Broadband sampling voltmeter, –hp– Model 3406A, has 50-μV sensitivity, 20-μV resolution, flat frequency response from 10 kHz to 1 GHz. Accuracy is ±3% to 100 MHz, ±5% to 700 MHz, ±8% to 1 GHz. Sampling circuit output is available on rear panel, giving instrument unusual capabilities for analysis of broadband signals.*

The specified frequency range of the voltmeter is 10 kHz to 1 GHz, and the frequency response of a typical production instrument is flat within one percent over this range. Useful sensitivity extends from 1 kHz to 2 GHz or more. Voltage measurements are accurate within ±3% of full scale from 100 kHz to 100 MHz, ±5% from 10 kHz to 700 MHz, and ±8% to 1 GHz.

The sampling voltmeter responds to the absolute average values of unknown voltages, and is calibrated to read both the rms value of a sine wave and dBm in 50-ohm systems. It has eight voltage ranges from 1 mV full scale to 3 V full scale, and its sensitivity is high enough to measure voltages as small as 50 μV. Voltage scales are linear, and resolution is 20 μV on the 1 mV range.

Unlike some RF millivoltmeters, which are rms-responding on the lower ranges and gradually change to peak-detecting on the higher ranges, the new voltmeter is average-responding on all ranges. This means that its measurements of non-sinusoidal voltages are more accurate because its detector law does not change with the amplitude of the input signal. The absolute average value of any input signal can be determined simply by multiplying the meter reading by $\sqrt{8}/\pi$ (the ratio of absolute average to rms values of a sine wave).

Instead of the coherent, waveform-preserving sampling method used in most sampling instruments, the new voltmeter uses an incoherent technique which does not preserve the input waveform. In this type of sampling, which was developed in the -hp- Loveland Laboratory, the input voltage is sampled at irregular intervals which have no relationship to any of the frequency components of the input signal. Enough samples are taken, however, so that the average, peak, and rms values of the samples closely approximate the average, peak, and rms values of the input voltage. Thus the information that is relevant to the voltage-measuring function is preserved, while waveform, which is irrelevant, is not preserved. Details of both coherent and incoherent sampling methods can be found on pages 4 and 5.

Incoherent sampling is especially advantageous in a voltmeter, because it gives the meter the sensitivity, accuracy, and broad frequency range of a sampling instrument, yet it is less costly than coherent techniques and, unlike coherent sampling, it does not require that the input signal be periodic. The sampling voltmeter operates equally well with sinusoidal, pulsed, random, or frequency-modulated signals.

The sampling circuit of the voltmeter is located in its probe, which is ac-coupled and permanently attached to the instrument with a 3-foot cable. Also located on the probe is a pushbutton which, when pressed, causes the voltmeter to retain its reading until the button is released. This memory system

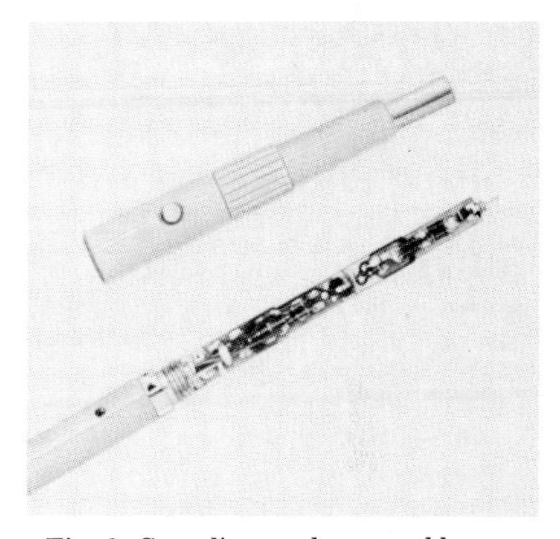

Fig. 3. *Sampling probe assembly contains sampling-pulse generator and four-diode sampling bridge. Photo also shows pushbutton which, when depressed, causes meter to retain reading until button is released. This memory device eliminates need to hold probe in circuit and read meter at same time.*

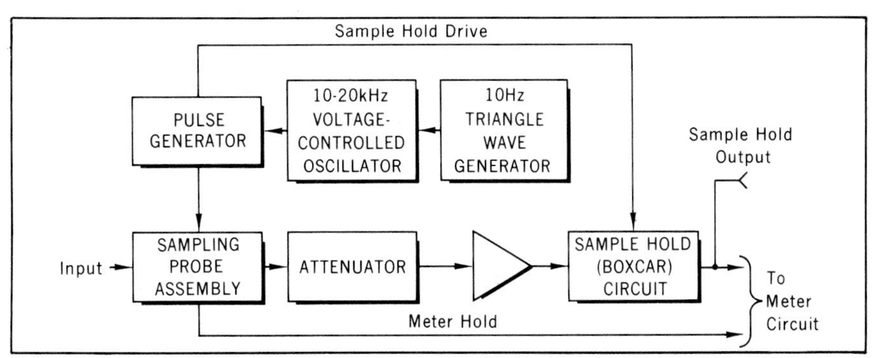

Fig. 4. *Block diagram of sampling circuits of sampling voltmeter. Incoherent sampling is used, i.e., sampling intervals are not correlated with input signal. Incoherent intervals are generated by 'smearing' sampling rate from 10 kHz to 20 kHz at 10-Hz rate. Sample hold circuit retains constant voltage proportional to sample until next sampling instant.*

simplifies measurements in awkward positions where it is difficult to place the probe and at the same time read the meter. Fig. 3 is a photograph of a disassembled probe, showing the sampling circuit and the memory pushbutton.

Other conveniences of the new voltmeter, besides the memory pushbutton already mentioned, include pushbutton range selection, rapid recovery from overloads, and a front-panel calibrator and zero receptacle. The meter recovers within five seconds from an overload of 30 V peak-to-peak (about 10,000:1 on its most sensitive, 1 mV range). The front-panel receptacle allows the instrument to be zeroed in the presence of an RF field, or to be calibrated using its internal, 1 V ±0.75% calibrator.

SAMPLER OPERATION

Fig. 4 is a simplified block diagram of the sampling circuits. The incoherent sampling intervals are generated by 'smearing' the sampling rate. The basic sampling frequency is varied from 10 kHz to 20 kHz by a 10-Hz triangle wave. This sampling rate is uncorrelated with practically all input signals. (It is not uncorrelated with identical, phase locked waveforms, so that voltages within the voltmeter cannot be measured.)

The 10-Hz triangular voltage varies the frequency of a voltage-controlled oscillator. The output of this oscillator drives a pulse generator which in turn triggers a sampling-pulse generator located in the probe. The sampling pulses, which are balanced pulses of approximately 250 picoseconds duration, turn on the diodes in a sampling bridge located in the probe, thereby allowing a sampling capacitor to charge to a voltage proportional to the input signal. The sampler output is a train of pulses whose amplitudes are

COHERENT AND INCOHERENT SAMPLING

Most sampling instruments, including the sampling oscilloscope and the vector voltmeter,* sample *coherently*. On the other hand, the broadband voltmeter described in the accompanying article samples *incoherently*. The reason for the difference is that most sampling instruments must preserve the waveform of an input signal, whereas the voltmeter needs only a measure of magnitude, such as the rms or the average value of the signal.

Coherent sampling is analogous to the familiar stroboscopic technique, by which an oscillating or repetitive motion is apparently 'slowed down' by observing it only at discrete times, instead of continuously. The observations, or samples, may be taken by flashing a light, by observing the oscillating object through a slit in a rotating disc, or by some other means.

Consider a stroboscopic observation of a tuning fork in motion. The apparent motion of the tuning fork can be made arbitrarily slow by adjusting the sampling rate, which in this case is either the rate at which the light flashes or the speed of rotation of the disc. The tuning fork may vibrate back and forth many times between glimpses, but so long as its position on each glimpse is only slightly advanced from its position on the preceding one, it seems to be moving much more slowly than it really is. If the slow motion were recorded on movie film it would, of course, be possible to determine the peak, the average, and the rms values of the tuning fork's excursions from its center position. This could be done simply by measuring the excursion on each frame of film and computing the peak, average, and rms values of the resulting collection of samples by standard techniques.

Now, if the film were cut apart and then spliced back together randomly, all time-sequence information about the movement of the tuning fork would be lost. However, certain information would be retained. The peak excursion would not change, and a little reflection will show that the average and rms values of the excursions would also be the same. In fact, even information about the probability of the fork's being at a given excursion would be retained. The same information could have been obtained by randomly flashing the light or by randomly opening a shutter. So long as a sufficient number of pictures were taken, the peak, average, and rms excursions could still be found. This kind of sampling, in which statistics are preserved but time-sequence information is not preserved, is *incoherent* sampling.

The difference between coherent and incoherent sampling for a high-frequency

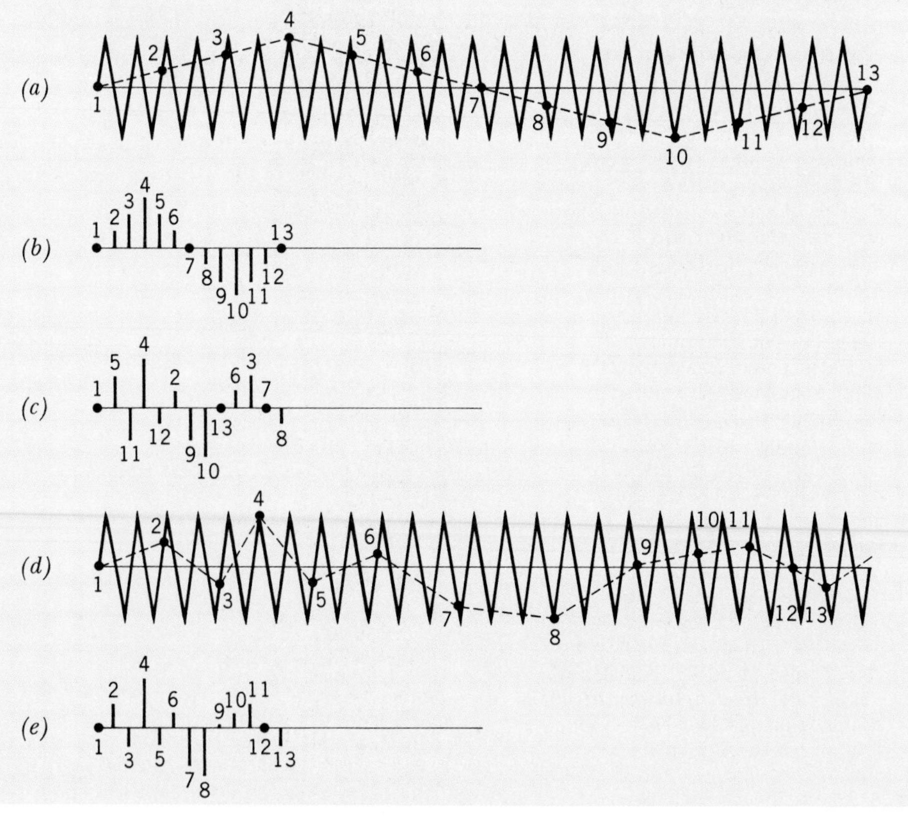

*See footnotes, p. 2.

proportional to the input voltage at the sampling instants.

The output of the sampler is fed through attenuators and amplifiers to the 'boxcar' circuit, which is a zero-order hold with clamp (modified pulse-stretcher). The bandwidth of the cable and amplifiers is narrow compared to the bandwidth of the sampling pulses, so by the time the samples reach the boxcar circuit, they have become pulses of about 5 microseconds duration, similar to the pulses illustrated in Fig. 5.

The boxcar circuit output is clamped to ground for 2 microseconds following the sampling pulse and then becomes a steady voltage that is proportional to the height of each sample taken. The output of this circuit is available from a connector at the rear of the instrument and is labeled 'Sample Hold Output.'

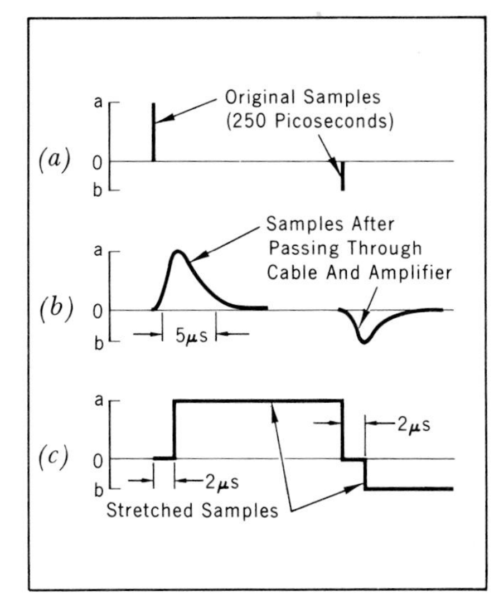

Fig. 5. *Sampling pulses which open sampling gate in probe are approximately 250 picoseconds wide, as shown in (a). When samples reach sample hold circuit after passing through cable and amplifier, they are about 5 μs wide, as shown in (b). Sample hold output is clamped to zero volts for 2 μs following each sampling instant, then becomes constant voltage proportional to sample.*

FREQUENCY RESPONSE

The exceptionally flat frequency response of the sampling voltmeter is shown in Fig. 1. This response was measured in a 50-ohm system, with the sampling probe inserted in a 50-ohm tee. Broadband or high-frequency measurements would normally be made in this configuration. At lower frequencies, probing would probably be done by hand, and the probe would be equipped with a divider or an isolator (see Specifications).

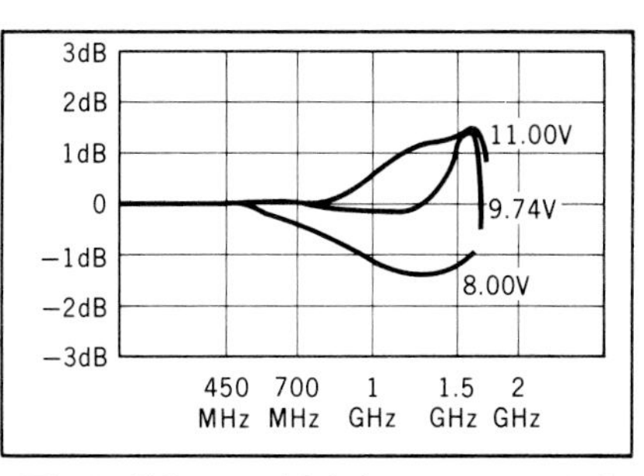

Fig. 7. *Effect on high-frequency rolloff of changing bias on sampling gate in probe (see Fig. 6).*

In production instruments, the high-frequency response is adjusted for optimum flatness by changing the bias on the sampling bridge in the probe. Since the sampling pulse does not have vertical leading and trailing edges, reducing the bridge bias makes the pulse wider, and vice versa (see Fig. 6). Wider pulses result in lower high-frequency response. The probe by itself tends to peak at the high frequencies, so very close cancellation can be obtained by making the sampling pulse longer. Fig. 7 shows the frequency response of a typical production instrument as a function of sampling bridge bias.

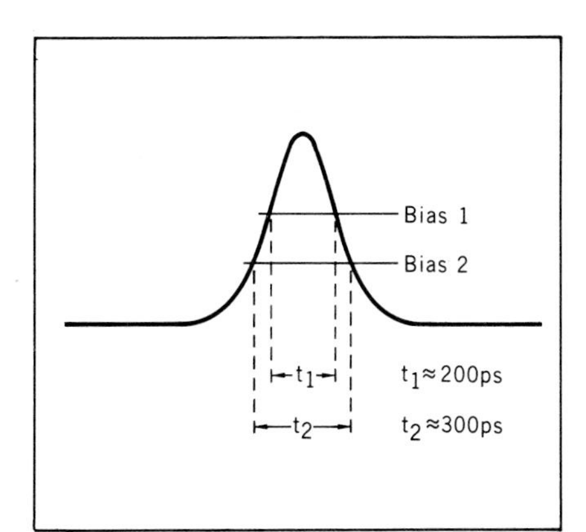

Fig. 6. *Width of sampling pulse is adjusted by changing bias on sampling gate in probe. High-frequency response of voltmeter can be adjusted in this way, since wider pulses mean lower cut-off frequency, and vice versa.*

Temperature variations produce very little change in the flatness of the frequency response. Fig. 8 shows the environmental performance of the response of a typical production unit.

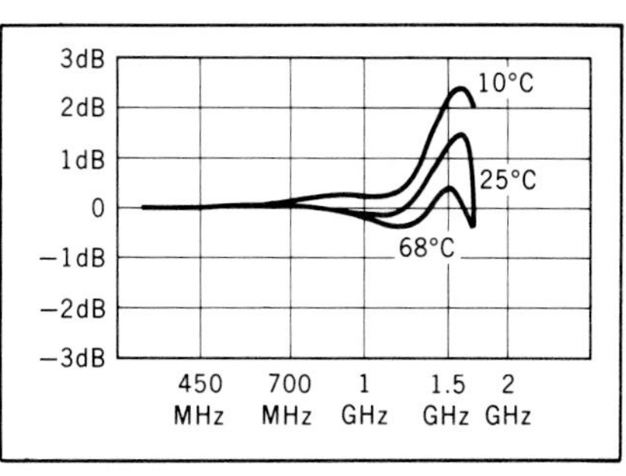

Fig. 8. *High-frequency response of sampling voltmeter is relatively insensitive to temperature changes. Temperature variation from 10°C to 68°C produces only ±2% changes in 1-GHz response.*

wave is shown in illustrations (a) through (e). In (a), samples are taken at regular intervals, and at such a rate that a lower-frequency equivalent of the original signal can be reconstructed from the samples. In (b) the samples are shown with their correct amplitudes, polarities, and relative phases (order). In (c) the same samples are shown scrambled, so that only their amplitudes and polarities are preserved. The average, peak, and rms values of the (c) group are the same as the average, peak, and rms values of the (b) group.

In (d) the original high-frequency wave is shown sampled incoherently. The interval between samples is not constant, and the waveform cannot be reconstructed from the samples, which are shown at (e). However, the group of samples in (e) is statistically equivalent to the groups of samples in (b) and (c). So long as all three groups contain a large enough number of samples, they have the same peak, average, and rms values.

In order for the technique of incoherent sampling to work in all situations it is necessary that there be no correlation between the sampling times and the motion or signal under observation. If the sampling frequency were a subharmonic of the frequency of the motion or signal being measured the motion would be completely stopped; thus, all of the samples would have exactly the same height and it would be impossible to determine the peak, average, rms, and so on.

In the new broadband sampling voltmeter, the basic sampling signal is frequency-modulated by a 10-Hz triangular wave, so that the sampling frequency varies between 10 kHz and 20 kHz, at a 10-Hz rate. This produces non-uniform sampling intervals which are, for all practical purposes, uncorrelated with all input signals.

SAMPLE HOLD OUTPUT INFORMATION

The sample hold output voltage is a low-frequency pulse train which, despite the lower frequency and difference in waveshape, has the same average, peak, and rms values as the input signal. This output makes it possible, therefore, to obtain information about broadband signals by using only low-frequency instruments. Fig. 9 shows oscillograms of typical sample hold outputs for sinusoidal and random input signals having frequency components up to 1 GHz.

Amplitude modulation envelopes can be observed at the sample hold output if the modulation frequency is sufficiently low compared to the sampling frequency, which is 10-20 kHz. Modulation envelopes can be observed with any low-frequency oscilloscope (e.g., -*hp*- Model 130C) for carrier frequencies up to 2 GHz or more and modulating frequencies up to 1 or 2 kHz. Oscillograms of typical displays are shown in Fig. 10.

Peak measurements and pulse-height analyses may also be made by observing the sample hold output with a low-frequency oscilloscope. The crest factor of the input signal can be as high as 10 (4.5 on 1 V range, 1.4 on 3 V range) without affecting the calibration of the sample hold output.[4]

If a true-rms-reading voltmeter is connected to the sample hold output, the true rms value of the input signal can be measured. Previously, high-frequency true-rms measurements could only be made with a power meter, which is much less sensitive than a voltmeter. Conventional RF millivoltmeters can also measure true rms values for small signals, but these voltmeters gradually change to peak detectors as the amplitude of the input signal increases, whereas the detector law of the sampling voltmeter is the same on all ranges. The sampling voltmeter is also more sensitive than a conventional instrument.

The statistics of the unclamped portion of the sample hold output closely approximate the statistics of the input signal. Consequently, it is possible to determine the statistical characteristics of broadband signals by applying appropriate low-frequency techniques to the sample hold output. For example, the probability density and probability distribution of, say, random noise with 1-GHz bandwidth can be determined by analyzing the sample hold output. Probability information, of course, is helpful in dealing with any signal, but it is especially necessary when the signal is random. Yet, up to now it has often been neglected or assumed, because of the impossibility of measuring it for broadband signals.

[4] Crest factor of an ac waveform is the ratio of its peak voltage to its rms voltage; e.g., a crest factor of 10 for an ac pulse waveform corresponds to a duty cycle of 0.01. See 'The Significance of Crest Factor,' **Hewlett-Packard Journal**, Vol. 15, No. 5, Jan. 1964.

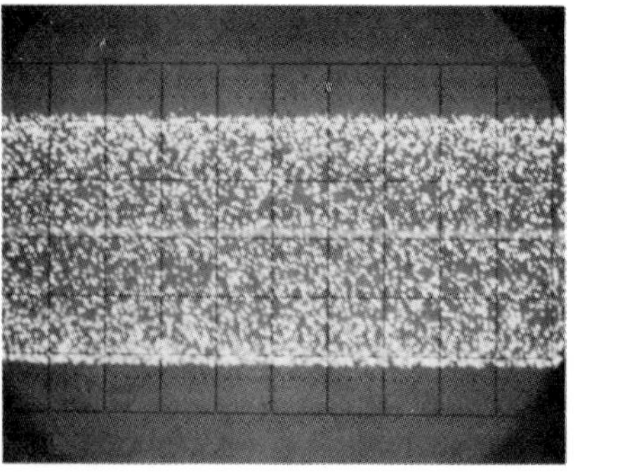

(*a*)

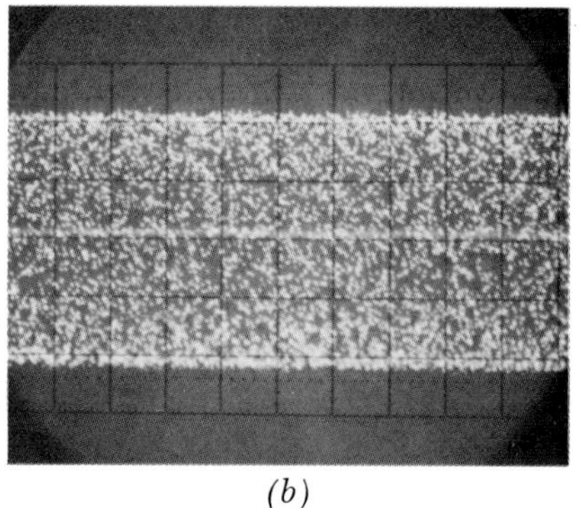

(*b*)

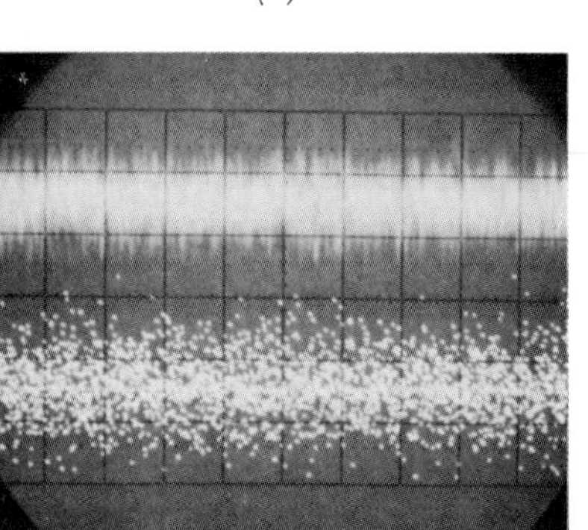

(*c*)

Fig. 9. *Time-exposure oscillograms of sample hold output of sampling voltmeter. Sample hold signals are statistically equivalent to input signals, but can be observed and measured with low-frequency instruments. Input signals were (a) 1-MHz sine wave, (b) 1-GHz sine wave, (c) random noise with upper frequency limit of 150 MHz. Upper trace in (c) shows noise input to voltmeter corresponding to sample hold output shown in lower trace. Noise source was two cascaded amplifiers.*

DC OUTPUT AND POWER MEASUREMENTS

With its probe inserted in a 50-ohm tee, the voltmeter can monitor the voltage across a 50-ohm load, and power readings can be taken directly from the dBm scale. Power levels as small as one nanowatt can be measured in this way. Hence the sampling voltmeter is a much more sensitive power monitor than the more conventional power meter and directional coupler.

In addition to the sample hold output, the sampling voltmeter also has a dc output at which a dc voltage proportional to the meter reading is available. This output is primarily for driving a recorder but, because of the very flat frequency response of the voltmeter, it also has other uses.

Using the voltmeter to monitor the voltage across a load, leveled voltage output over the 10-kHz-to-1-GHz range can be obtained from any signal generator which operates in this range and has a dc modulation input. The dc output of the voltmeter is fed back through appropriate shaping networks to the dc modulation input of the signal generator, causing the generator output to be as constant as the frequency response of the voltmeter.

METER CIRCUITS

Fig. 11 is a block diagram of the meter circuits. The output of the boxcar circuit is detected and filtered to produce a dc voltage which is a measure of the absolute average value of the input signal. The output signal-to-noise ratio of an average-reading detector is a nonlinear function of the input signal-to-noise ratio,[5] so the gain of the detector for the average value of

[5] This nonlinearity is different from that of a typical diode detector, which is a square-law device for small signals and a linear device for large signals. The detector in the sampling voltmeter is linear, and the nonlinearity is caused by the presence of the noise. See W. R. Bennett, 'Response of a Linear Rectifier to Signal and Noise,' Bell System Technical Journal, Vol. 23, No. 1, Jan. 1944. See also B. M. Oliver, 'Some Effects of Waveform on VTVM Readings,' **Hewlett-Packard Journal**, Vol. 6, No. 10, June, 1955.

(a)

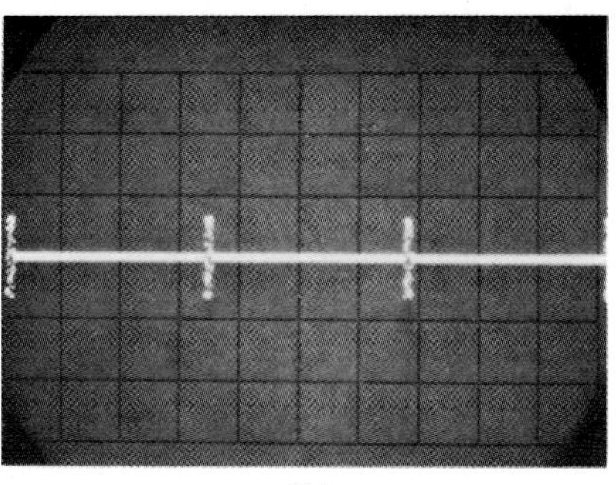

(b)

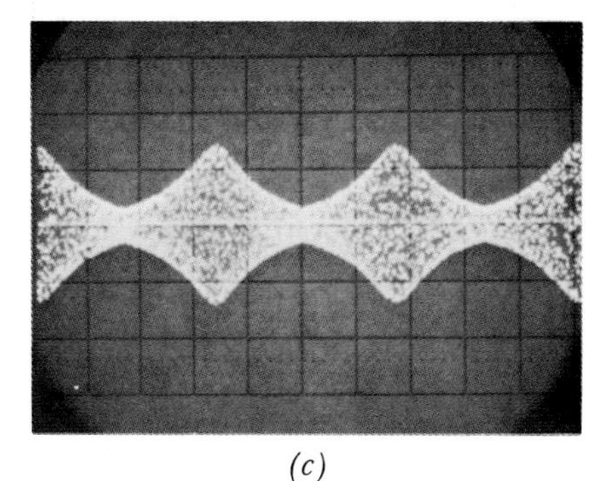

(c)

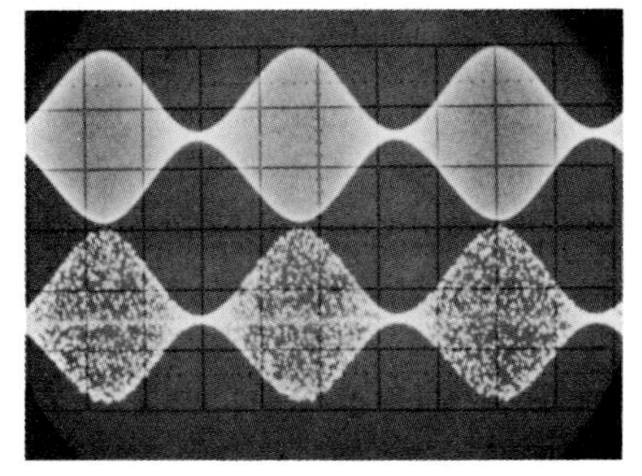

(d)

Fig. 10. *Amplitude modulation envelopes can be observed at sample hold output for carriers up to more than 1 GHz and modulation frequencies up to about 1 kHz. Oscillograms shown are time exposures. Carriers were all sinusoidal. (a) carrier: 65 MHz; modulation: 300-Hz triangle wave. (b) carrier: 2 GHz; modulation: 30-Hz pulse train. (c) carrier: 2 GHz; modulation: 30-Hz triangle wave distorted by PIN diode modulator. (d) carrier: 1 MHz; modulation: 30-Hz sine wave. Upper trace in (d) shows input to voltmeter corresponding to sample hold output in lower trace. Oscilloscope was synchronized to modulating signal only.*

the signal is nonlinear. To make the meter's voltage scales linear, a nonlinear circuit is placed between the detector and the meter. The resulting gain is essentially constant from 50 μV to full scale.

Noise in voltmeters often causes considerable meter jitter and loss of sensitivity and linearity on the lower ranges. These effects have been greatly reduced in the sampling voltmeter. On the one millivolt range, the inherent noise of the system plus thermal noise amounts to about 150 to 200 microvolts. This noise is not dependent upon the source impedance of the signal being measured. It has an essentially constant mean value, so its effects on meter readings can be corrected easily. The mean value of the noise is subtracted from the output in a noise suppression circuit, thereby giving the voltmeter much greater sensitivity.

Since the noise is random and the

DESIGN LEADERS

JOHN T. BOATWRIGHT

RONALD K. TUTTLE

FRED W. WENNINGER, JR.

ROGER L. WILLIAMS

John Boatwright received his BS degree in electrical engineering from Massachusetts Institute of Technology in 1960. He continued his studies while working with a firm in Cambridge, and then joined –hp– in 1961 as a development engineer. He is now an engineering section manager in the development laboratory of the –hp– Loveland Division. John holds several patents in the field of space communications, and has a patent pending on random sampling.

Ron Tuttle received his BS degree in chemical engineering and his MS degree in electrical engineering from the University of California at Berkeley in 1959 and 1961, respectively. After joining the –hp– Frequency and Time Division in 1961, he worked as a circuit designer on the 5260A Frequency Divider project. In 1964 he transferred to the –hp– Loveland Division, where he has contributed to the design and testing of the 3406A Sampling Voltmeter.

Fred Wenninger received his BS degree in physics, and his MS and PhD degrees in engineering from Oklahoma State University in 1959, 1962, and 1963. He joined the –hp– Loveland Division in 1963, and since 1965 he has been an engineering group leader at Loveland. He has a patent pending on random sampling. Prior to his joining –hp–, Fred's professional activities included micrometeorite research, satellite instrumentation design, and development of an electronic brain stimulator. He also served for several years as a part-time instructor in mathematics and physics at Oklahoma State.

Roger Williams joined the –hp– Loveland Division in 1963 as a summer student, after receiving his BS degree in engineering from Harvey Mudd College. He subsequently worked with –hp– in Palo Alto while continuing his studies at San Jose State College. In 1964, he returned to the Loveland Division, and did the product design work for the 3406A Sampling Voltmeter. Roger is continuing his studies towards the MS degree at Colorado State University.

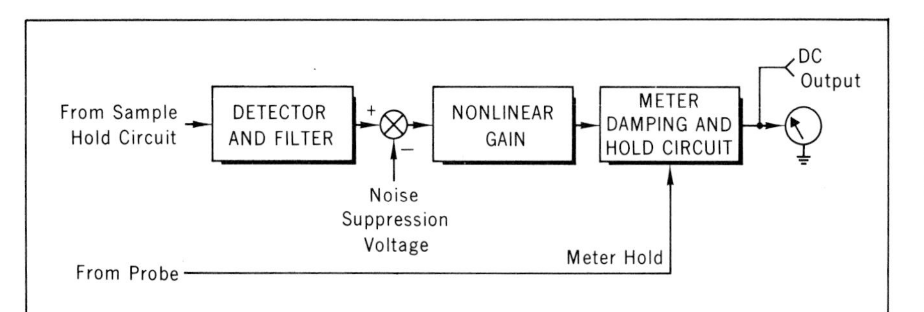

Fig. 11. *Block diagram of meter circuits of sampling voltmeter. Instrument is average-responding on all voltage ranges. Damping and noise-cancellation circuits reduce effects of noise. Nonlinear gain circuit gives meter linear voltage scales.*

gain of the circuit is highest when the signal is smallest, the meter would be very jittery if damping were not introduced. However, damping sufficient to reduce the jitter to a usable level would cause the response to be very sluggish. For this reason, a nonlinear damping circuit is employed. The nonlinear damping circuit provides heavy damping for small variations in the input and drastically reduces the meter jitter due to noise. The damping is decreased for large variations in the input signal, so that the overall response of the instrument is fairly rapid.

Because a finite number of samples are taken in any given time interval, there is a certain variance in the sample hold output, and the nonlinear damping serves to reduce meter jitter due to this effect as well as that due to noise. The pushbutton memory circuit mentioned earlier is also incorporated in the nonlinear damping circuit.

ACKNOWLEDGMENTS

The sampling voltmeter is the direct result of a late-hour discussion between John T. Boatwright and the author. Mr. Boatwright, who is now a section manager in the Loveland Laboratory, provided substantial contributions in the initial stages of the project. Ronald K. Tuttle designed the accessories and contributed in the circuit design area. Roger L. Williams was responsible for the product design.

Special credit is also due Marco R. Negrete, Loveland Laboratory manager, Ronald W. Culver, engineering aid, and B. M. Lovelace, production engineer.

—Fred W. Wenninger, Jr.

SPECIFICATIONS
-hp-
MODEL 3406A BROADBAND SAMPLING VOLTMETER

VOLTAGE RANGE: 1 mV to 3 V full scale in eight ranges; decibels from −50 to +20 dBm (0 dBm = 1 mW into 50 ohms); absolute average-reading instrument calibrated to rms value of sine wave.

FREQUENCY RANGE: 10 kHz to 1 GHz; useful sensitivity from 1 kHz to beyond 2 GHz.

FULL-SCALE ACCURACY WITH CALIBRATOR:
±3%, 100 kHz–100 MHz
±5%, 10 kHz–700 MHz
±8%, 5 kHz–1 GHz
±1 dB, 4 kHz–1.2 GHz
±4 dB, 2 kHz–1.5 GHz

INPUT IMPEDANCE: 100,000 ohms at 100 kHz. Capacity approximately 2 pF. Input capacity and resistance will depend upon accessory tip used. (approximately 8 pF with 11072A isolator tip supplied.)

SAMPLE HOLD OUTPUT: Provides ac signal whose unclamped portion has statistics that are narrowly distributed about the statistics of the input, inverted in sign (operating into >200 kΩ load with <1000 pF).
Noise:
Typically 175 μV rms.
Accuracy with Calibrator:
0.01 V Range and Above: Same as full-scale accuracy of instrument.
0.001 V to 0.003 V Range: Value of input signal can be computed by taking into account the residual noise of the instrument (see references in footnote 5, p. 6).
Jitter:
Typically ±2% peak of reading (with -hp- Model 3400A true-rms voltmeter connected to Sample Hold Output).

Crest Factor:
0.001 V to 0.3 V: 20 dB full scale (inversely proportional to meter indication); 1 V: 13 dB; 3 V: 3 dB.

DC RECORDER OUTPUT: Adjustable from zero to 1.2 mA into 1000 ohms at full scale, proportional to meter deflection.

METER:
Meter scales: Linear voltage, 0 to 1 and 0 to 3; decibel, −12 to +3. Individually calibrated taut-band meter.
Response Time: Indicates within specified accuracy in <3 s.
Jitter: ±1% peak (of reading).

GENERAL:
Calibrator Accuracy: ±0.75%.
Overload Recovery Time: Meter indicates within specified accuracy in <5 s. (30 V p-p max.)
Maximum Input: ±100 Vdc, 30 V p-p.
RFI: Conducted and radiated leakage limits are below those specified in MIL-I-6181D and MIL-I-16910C except for pulses emitted from probes. Spectral intensity of these pulses is approximately 50 nV/$\sqrt{Hz}$; spectrum extends to approx. 2 GHz.
Temperature Range:
Instrument 0°C to +55°C.
Probe +10°C to +40°C.
Power: 115 or 230 volts ±10%, 50 Hz to 1000 Hz, approximately 17 watts.
Dimensions: Standard ½ module 6½ in. high, 8⅞ in. wide, 11½ in. deep (165 x 225 x 292 mm).

WEIGHT: Net, 12 lbs. (5,4 kg); Shipping, 15 lbs. 6,8 kg).

PRICE: $650.00.

ACCESSORIES

ACCESSORIES FURNISHED: 11072A isolator tip.
8710-0084 nut driver for tip replacement.
5020-0457 replacement tips.
10213-62102 ground clips and leads.

ACCESSORIES AVAILABLE:
11064A Basic Probe Kit, $100.00 consists of the following:
11063A 50-ohm 'T'.
11061A 10:1 divider tip.
10218A BNC adapter.
0950-0090 50-ohm termination.
11071A Probe Kit $185.00 consists of all the above plus:
11073A Pen-type probe,
10219A Type 874A adapter.
10220A Microdot adapter.
11035A Probe-tip kit.
11061A: 10:1 Divider
As well as dividing the input voltage by a factor of ten this accessory eliminates the effects of source impedance variations.
Accuracy (divider alone):
±5% 1 kHz to 400 MHz.
±12% 400 MHz to 1 GHz.
Max. Input: 150 V p-p ac; 600 V dc.
11063A 'TEE': Should be used whenever measurements are made in 50Ω systems.
VSWR: ≤1.15 at 1 GHz (bare probe in tee). Useful to about 1.5 GHz.
Insertion Power Loss: <4% up to 1 GHz.
11072A: Isolator
Essentially eliminates effects of source impedance variations.
Increases probe capacitance by approximately 6.5 pF. Recommended frequency range is 10 kHz to 250 MHz.
11073A: Pen-type Isolator
Recommended frequency range is 10 kHz to 50 MHz. Various accessories adapt the 11073A to alligator jaws and other tips which facilitate point-to-point measurements. Increases probe capacity by approximately 7 pF.
10218A: Probe-to-Male-BNC Adapter
Recommended frequency range is 10 kHz to 250 MHz.

Prices f.o.b. factory.
Data subject to change without notice.

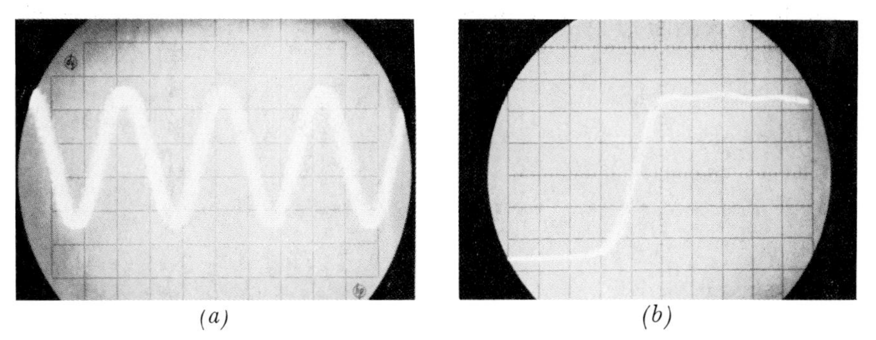

(a) (b)

Fig. 1. *Oscillograms of signals at frequencies never before viewed with oscilloscopes. (a). 18-gigahertz (18 x 10^9 cps) sine wave. (b). Step with 20-picosecond (20 x 10^{-12}s) rise time. Rise time of step response is about 30 ps, indicating that oscilloscope bandwidth is considerably greater than 12.4 GHz. Oscillograms were made using new sampling plug-ins discussed in text. (a) Vertical: 100 mV/cm, Horizontal: 20 ps/cm. (b) Vertical: 50 mV/cm, Horizontal: 20 ps/cm.*

AN ULTRA-WIDEBAND OSCILLOSCOPE BASED ON AN ADVANCED SAMPLING DEVICE

The state of the oscilloscope art has taken a significant forward step with the development of a new oscilloscope that operates from DC to 12.4 GHz and displays signals as small as 1 millivolt.

SAMPLING oscilloscopes have a combination of wide bandwidth and high sensitivity that has never been matched by any real-time oscilloscope. This is because, up to now at least, it has been much easier to acquire, amplify, and display a narrow sample of a high-frequency repetitive waveform than to amplify and display the entire waveform.[1] This situation shows no signs of being reversed; in fact, the opposite is true. Four new sampling plug-ins for *-hp-* general-purpose and variable-persistence oscilloscopes have been developed, and among them is a vertical amplifier with a sensitivity of 1 millivolt per centimeter and a bandwidth greater than 12.4 GHz, more than three times the widest bandwidth previously attained.[2] The ultra-wideband vertical amplifier is based upon a remarkable new sampling device developed by ***hp associates***. This device and the 'integrated' design approach which produced it are described in the article beginning on p. 12.

Short, sharp pulses in high-speed computers and high-frequency pulsed radars, both present and future, are well within the range of the new oscilloscope, as are many microwave signals which have never before been observable. The two oscillograms of Fig. 1 could not have been made without the new sampling plug-ins; they are displays of an 18-GHz sine wave and a voltage step having a rise time of approximately 20 picoseconds (20 x 10^{-12} second). Overall rise time of the step display is about 30 ps, indicating that the plug-in's rise time is less than 28 ps, equivalent to a bandwidth of more than 12.4 GHz. By comparison, rise times of pulses in the fastest computers are now about one nanosecond and are getting faster.

Fig. 3 shows how the new sampling plug-ins are related to each other, to the oscilloscope main frames, and to some auxiliary equipment, which is

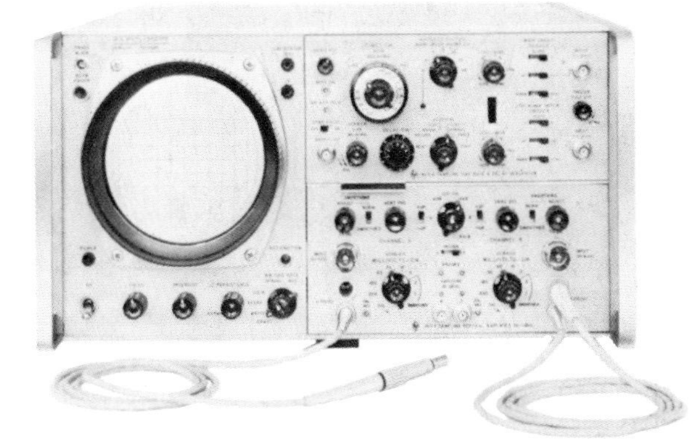

Fig. 2. *Model 1425A Sampling Time Base and Delay Generator and Model 1410A 1-GHz Sampling Vertical Amplifier installed in Model 141A Variable Persistence Oscilloscope. Sampling gates in 1-GHz vertical amplifier are in probes. GR874 50Ω inputs lead to internal delay lines and trigger amplifiers for internal triggering.*

[1] Principles of sampling oscilloscope operation have been presented several times in these pages. See, for example, R. Carlson, 'The Kilomegacycle Sampling Oscilloscope,' **Hewlett-Packard Journal**, Vol. 13, No. 7, Mar., 1962, and 'Coherent and Incoherent Sampling,' **Hewlett-Packard Journal**, Vol. 17, No. 11, July, 1966. Instead of displaying every cycle of a repetitive waveform, the sampling oscilloscope forms an apparently continuous trace from a series of dots, or samples, taken usually many cycles apart, each representing the input voltage at a slightly different point in its cycle.

[2] The four new sampling plug-ins fit the -hp- Model 140A General-purpose Oscilloscope and the -hp- Model 141A Variable-persistence Oscilloscope.

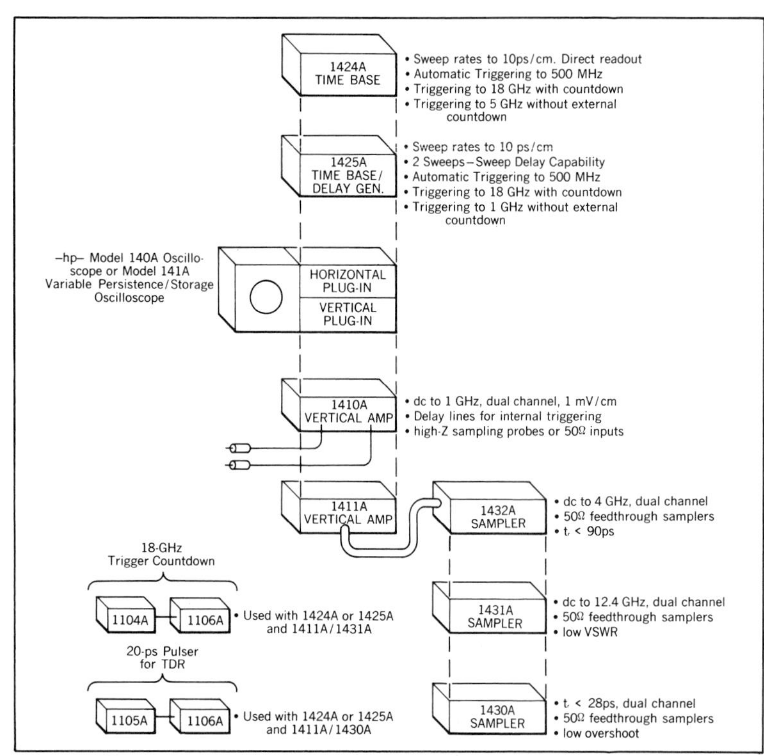

Fig. 3. *New sampling plug-ins, remote samplers, pulse generator and trigger countdown unit for –hp– Model 140A and 141A Oscilloscopes are all solid-state. Units can be combined in various ways to form sampling oscilloscopes with bandwidths as high as 12.4 GHz, more than three times widest bandwidth previously attained.*

plug-ins and auxiliary equipment. Besides wider bandwidth and faster rise time, other capabilities of one or more combinations of instruments are automatic triggering, sweep delay, triggering on CW signals up to 5 GHz without an external countdown, and a choice of high-impedance sampling probes or 50Ω inputs with internal delay lines in a single vertical amplifier plug-in. All of these capabilities will be discussed at greater length later in this article.

Of the two vertical amplifier plug-ins, one is a dc-to-1-GHz unit, and the other is a general-purpose unit which operates in combination with one of three wideband remote samplers. Both vertical amplifiers are dual-channel units and both have maximum calibrated sensitivities of 1 mV/cm, much better than any wideband real-time oscilloscope.

Of the two horizontal plug-ins, one is a single-time-base unit with sweep speeds from 10 ps/cm to 500 μs/cm. The other horizontal plug-in is a dual-sweep time base and delay generator with a similar range of sweep rates.

also new. All of the plug-ins and other new instruments contain only solid-state active devices.

Two of the four plug-ins are time bases and two are vertical amplifiers, and either vertical amplifier may be used with either time base. The main frames are compact laboratory oscilloscopes (9 inches high), one having variable persistence and storage, and the other designed for general-purpose service where variable persistence is not required.

The auxiliary instruments, described in detail on p. 9, make it possible to take full advantage of the wide bandwidth and fast rise time of the new sampling oscilloscopes. These instruments are a tunnel-diode pulse generator for time domain reflectometry and other uses requiring fast pulses, and an 18-GHz trigger countdown for triggering on high-frequency CW signals. Rise time of the pulser is approximately 20 ps, making it one of the fastest now in existence.

Table I lists the major capabilities of each of the possible combinations of

TABLE I. CAPABILITIES OF NEW SAMPLING PLUG-INS FOR –hp– MODEL 140A AND 141A OSCILLOSCOPES

Instrument Combination			Vertical: Bandwidth Rise Time Sensitivity	Horizontal: Triggering	Horizontal: Sweeps**
Vertical Plug-in/ Sampler	Horizontal Plug-in	Auxiliary Instruments			
1410A/ 50Ω Inputs* or High-Z Probes	**1424A**		**dc to 1 GHz 350 ps 1 mV/cm**	**Automatic to 500 MHz, Internal* or External Level Select to 1 GHz, Internal* or External (CW to 5 GHz, External)**	**One**
1410A/ 50Ω Inputs* or High-Z Probes	**1425A**		**dc to 1 GHz 350 ps 1 mV/cm**	**Automatic to 500 MHz, Internal* or External Level Select to 1 GHz, Internal* or External**	**Main, Delaying, Main Delayed**
1411A/ 1432A	**1424A**		**dc to 4 GHz 90 ps 1 mV/cm**	**Automatic to 500 MHz Level Select to 1 GHz (CW to 5 GHz)**	**One**
1411A/ 1432A	**1425A**	**18-GHz Trigger Countdown**	**dc to 4 GHz 90 ps 1 mV/cm**	**Automatic to 500 MHz Level Select to 1 GHz (CW to 18 GHz with countdown)**	**Main, Delaying, Main Delayed**
1411A/ 1431A	**1424A**	**18-GHz Trigger Countdown**	**dc to 12.4 GHz ~28 ps 1 mV/cm**	**Automatic to 500 MHz Level Select to 1 GHz (CW to 5 GHz, or 18 GHz with countdown)**	**One**
1411A/ 1431A	**1425A**	**18-GHz Trigger Countdown**	**dc to 12.4 GHz ~28 ps 1 mV/cm**	**Automatic to 500 MHz Level Select to 1 GHz (CW to 18 GHz with countdown)**	**Main, Delaying, Main Delayed**
1411A/ 1430A	**1424A**		**dc to ~12.4 GHz 28 ps 1 mV/cm**	**Automatic to 500 MHz Level Select to 1 GHz (CW to 5 GHz)**	**One**
1411A/ 1430A	**1425A**		**dc to ~12.4 GHz 28 ps 1 mV/cm**	**Automatic to 500 MHz Level Select to 1 GHz**	**Main, Delaying, Main Delayed**
1411A/ 1430A	**1424A or 1425A**	**20-ps Pulse Generator**	**TDR system with <40 ps rise time. Resolves discontinuities spaced <¼ inch apart.**		

* 50Ω Inputs lead to built-in delay lines for internal triggering.
** Sweep rates: 10 ps/cm to 500 μs/cm.

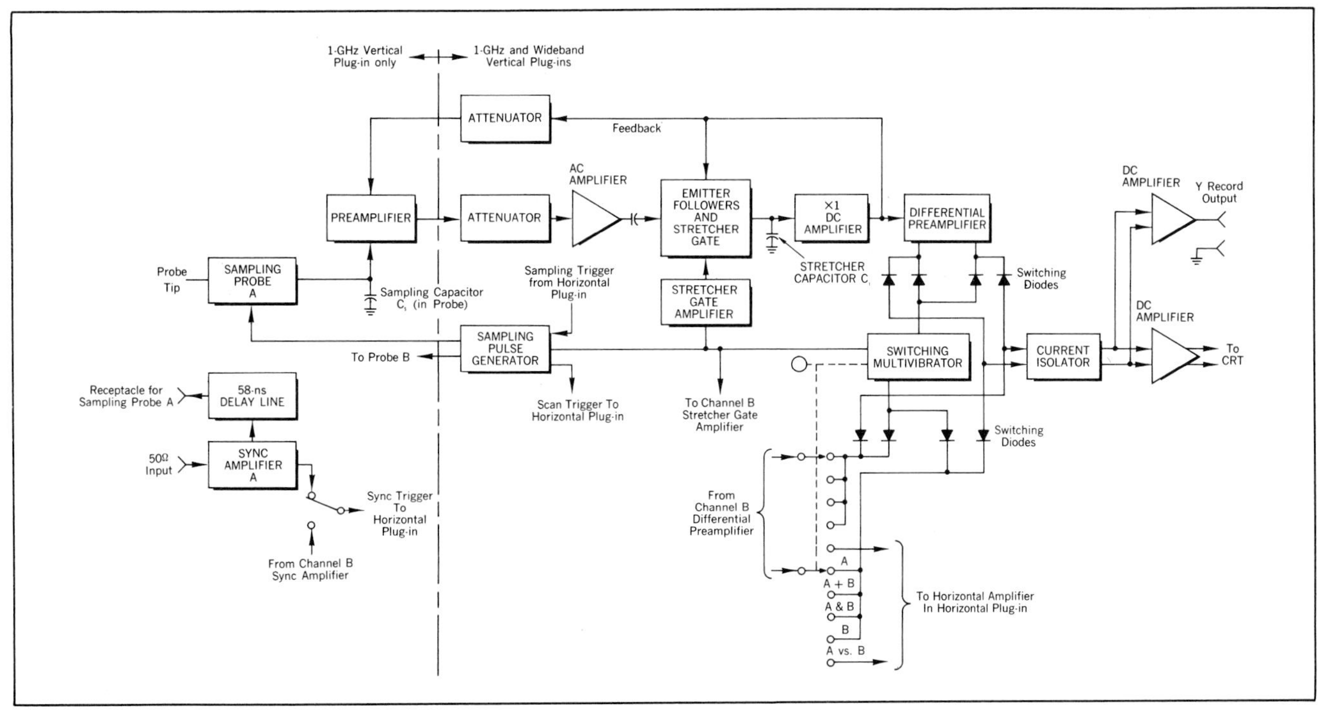

Fig. 4. *Block diagram of Model 1410A 1-GHz Sampling Vertical Amplifier. Right side of diagram is identical in 1-GHz plug-in and in Model 1411A Sampling Vertical Amplifier.*

Both plug-ins have trigger circuits which permit automatic triggering on a wide range of pulsed, CW, or other signals having frequencies between 50 Hz and more than 500 MHz. 'Automatic' triggering means that a baseline is displayed when the trigger signal is absent; then when a trigger signal is applied to the time base, the sweep is automatically synchronized to it. Reliable automatic triggering on a wide range of signals eliminates many trigger adjustments that would otherwise have to be made in the process of setting up a trace. These two plug-ins mark the first time that a sampling oscilloscope has had this capability, and therefore the first time that triggering a sampling oscilloscope has been as uncomplicated as triggering a real-time instrument.

A separate UHF countdown trigger circuit in the single-sweep plug-in permits this plug-in to trigger on CW signals having frequencies up to more than 5 GHz. The dual-sweep plug-in has no countdown, but will trigger reliably up to 1 GHz or more.

1-GHz VERTICAL AMPLIFIER

Fig. 2 is a photograph of the 1-GHz dual-channel vertical amplifier installed in the variable-persistence main frame along with the sampling time base and delay generator. Input signals to this plug-in are sampled by hot-carrier-diode sampling gates located in two high-impedance (100 kΩ, 2pF) probes. Delay lines and trigger amplifiers built into the plug-in permit the oscilloscope to be triggered by either of its input signals (internal triggering) and still display the leading edge of the triggering signal. When the delay lines are used, the input signals are fed into two front-panel 50Ω inputs and the sampling probes are plugged into receptacles on the front panel. Fig. 4 is a block diagram of the 1-GHz plug-in, showing a delay line and trigger amplifier at the lower left.

Although the plug-in's response extends to dc, there are no high-gain dc amplifiers in the signal path; the feedback loop shown in Fig. 4 makes them unnecessary. This means that the stability of the instrument can be and is high, because it is determined by passive components and a low-gain dc amplifier. Observed drift is less than 3 mV/hr.

A critical factor in making the plug-in all solid-state was the availability of low-leakage field-effect transistors for the stretcher gate and the dc amplifier shown in Fig. 4. Leakage in these transistors is so low that the sampling rate can be as low as one sample per second without resulting in excessive droop in the voltage on the stretcher capacitor. A field-effect transistor was also chosen

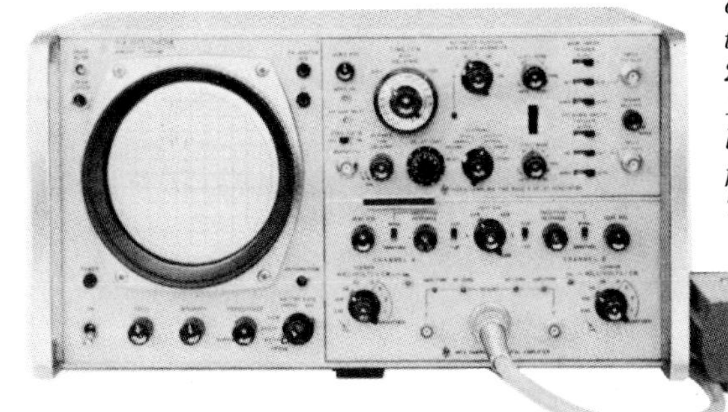

Fig. 5. *Model 1425A Sampling Time Base and Delay Generator and Model 1411A Sampling Vertical Amplifier with Model 1430A 28-ps Sampler installed in Model 141A Variable Persistence Oscilloscope. Remote samplers are feedthrough units, useful for TDR and for observing signals without terminating them.*

for the input stage of the preamplifier (Fig. 4) because of its low-noise characteristics.

Five display modes are possible for the vertical amplifier: channel A only, channel B only, channel A and channel B (alternate samples), channel A and channel B added algebraically and, for phase measurements or X-Y plots, channel A versus channel B. When both channel A and channel B are displayed, a switching multivibrator (Fig. 4) controls two groups of diodes which switch between channels in synchronism with the sampling process. During one sampling interval the latest sample in channel A is displayed and during the next interval the latest sample in channel B is displayed. With this arrangement there is no chopper noise like that often found in real-time dual-channel displays.

Recorder outputs on the front panel of the 1-GHz vertical amplifier supply approximately 0.1 V/cm from a 500Ω source for driving strip-chart or X-Y recorders. The gain and dc level of the recorder output can be adjusted independently.

WIDEBAND VERTICAL AMPLIFIER AND SAMPLERS

Everything to the right of the colored line in the block diagram of Fig. 4 is identical in both the 1-GHz vertical amplifier and the wideband vertical amplifier. Sensitivities, display modes, recorder outputs, and internal operation are the same for both. The wideband unit, however, has no built-in delay lines and, instead of probes, uses one of the three dual-channel feedthrough samplers. Fig. 5 is a photograph of the wideband unit installed in the main frame, and Fig. 6 is a block diagram of the samplers.[3]

Depending upon which of the three wideband samplers is used, the bandwidth of the wideband vertical amplifier plug-in can be either 12.4 GHz or 4 GHz. Two of the samplers are ultra-wideband units. One has a rise time of less than 28 ps and optimum pulse response (overshoot <5%) but a VSWR that increases with frequency (3 at 12.4 GHz); the other has a bandwidth of 12.4 GHz and a VSWR typically less than 1.8 at 12.4 GHz, but has 5% to 10% more overshoot in the pulse response. The third sampler is a 4-GHz, 90-ps unit for applications where the widest bandwidths are not needed and lower cost is attractive. A five-foot cable (10-foot cable optional) connects the plug-in and the sampler so that measurements can be made at remote locations. Input signals are not terminated by the feedthrough samplers, so time domain reflectometry and signal monitoring are straightforward.

[3] See article, p. 12, for a description of the wideband sampling devices used in these samplers.

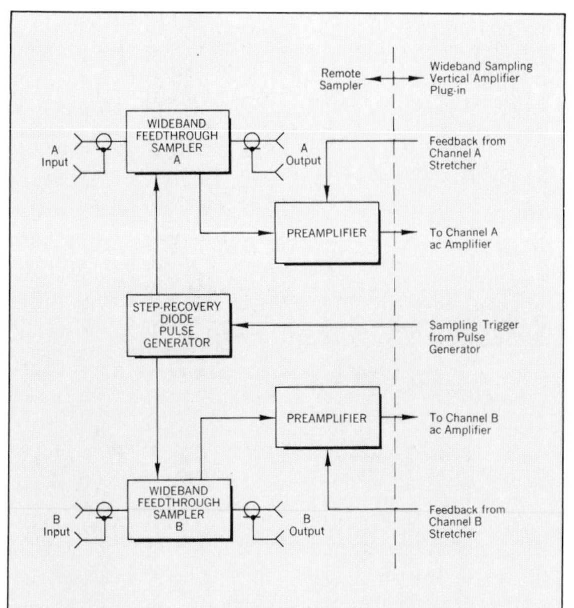

Fig. 6. *Block diagram of dual-channel remote samplers used with Model 1411A Sampling Vertical Amplifier. Ultra-wideband feedthrough samplers are described beginning on p. 12.*

Oscillograms of 8-GHz and 18-GHz sine waves, taken using the low-VSWR (CW-optimized) sampler, are shown in Figs. 7 and 1(a) respectively. Note that time jitter is less than 10 ps, even at the highest frequencies.

Step response of the pulse-optimized sampler is shown in Figs. 1(b) and 8 for two different time scales. The flat top and absence of excessive overshoot are evident in Fig. 8. Fig. 9 shows the reflection from the pulse-optimized sampler for an incident step having a rise time of 20 ps. The vertical scale is calibrated to read reflection coefficient with a scale factor of 0.1/cm, and the sampler reflection is only 6%.

TRIGGERING

Triggering of both the single-sweep time base[4] and the dual-sweep time base and delay generator[5] can be either automatic or manually adjustable and, when used with the 1-GHz vertical plug-in, either internal or external. Except for an extra UHF countdown in the single-sweep unit, all trigger circuits are identical circuits based on a newly designed tunnel-diode thresh-

[4] See Fig. 10. [5] See Figs. 2 and 5.

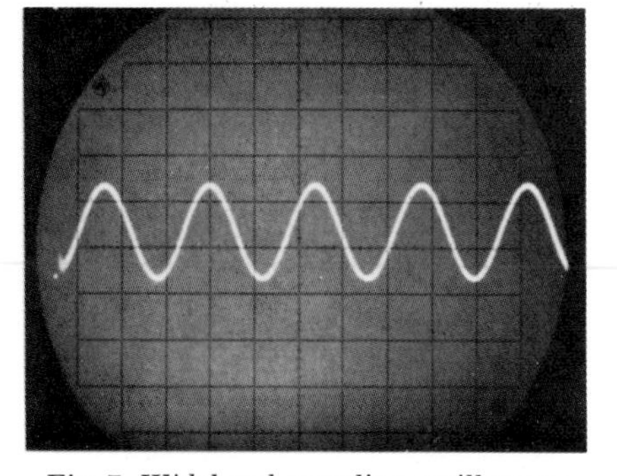

Fig. 7. *Wideband sampling oscilloscope display of 8-GHz sine wave. See Fig. 1(a) for display of 18-GHz sine wave. Vertical: 20 mV/cm. Horizontal: 50 ps/cm.*

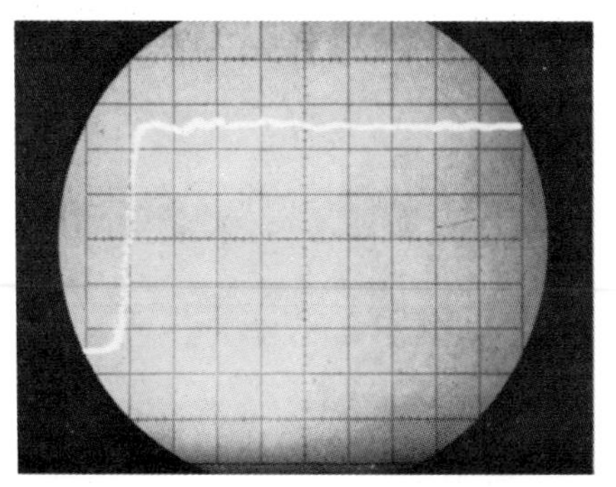

Fig. 8. *Response of pulse-optimized Model 1430A Sampler to step with 20-ps rise time has 30-ps rise time, small overshoot, flat top. Vertical: 50 mV/cm; Horizontal: 100 ps/cm. See Fig. 1 (b).*

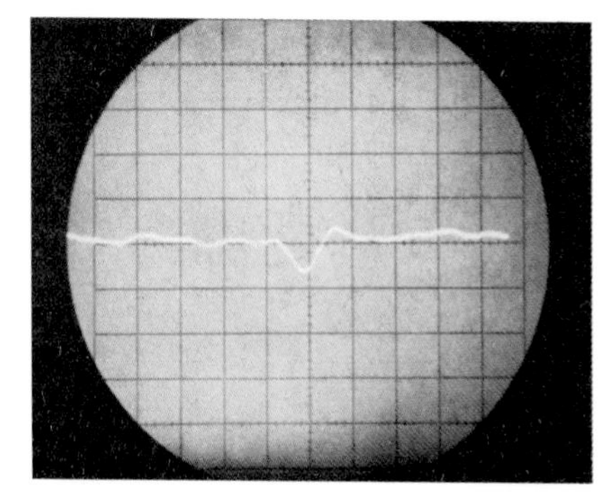

Fig. 9. *Reflection from wideband sampler in 40-ps TDR system is only 6%. Vertical: reflection coefficient = 0.1/cm; Horizontal: 100 ps/cm.*

old detector. The detector produces an output to start the sampling process when the incoming trigger signal crosses the level set by the LEVEL control. A SLOPE switch determines whether triggering will occur on the positive or negative slope of the trigger signal.

The threshold detector operates in one of three modes, depending upon the setting of the MODE control which varies the supply current to the detector. Turning the control clockwise increases the supply current. For low supply currents the detector is bistable, that is, the incoming signal must both trigger and reset the detector. As the current is increased the trigger circuit becomes monostable; that is, it is triggered by the incoming signal, but it resets itself. For still higher currents the circuit becomes astable and oscillates.

The bistable mode is used to trigger on the trailing edges of pulses or on pulses that are so long that the trigger circuit would normally re-arm and trigger again before the end of the pulse. The monostable mode is used to trigger on short pulses and on sine waves up to about 100 MHz. This mode is more sensitive than the bistable mode, especially on pulses shorter than about 30 ns. In the astable mode the detector oscillates at 10 to 40 MHz, depending upon the MODE control setting, and any incoming sine wave alters this frequency so that it is a sub-harmonic of the incoming frequency. This type of circuit is called a 'countdown', because for triggering to occur the incoming frequency must be greater than the oscillation frequency. At about 100 MHz the astable mode is as sensitive as the monostable mode, but at 1000 MHz the astable mode is about twenty times more sensitive.

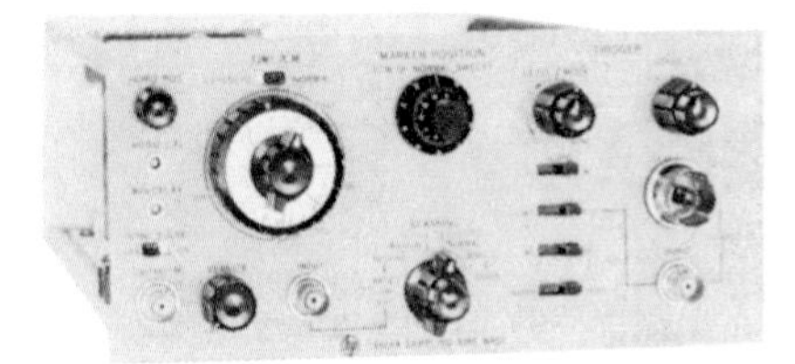

Fig. 10. *Single-sweep Model 1424A Sampling Time Base Plug-in has calibrated marker position control and direct readout of both magnified and unmagnified sweep rates. Another new horizontal plug-in is dual-sweep time base and delay generator, shown in Figs. 2 and 5.*

To prevent double triggering on complex waveforms in which the desired trigger level and slope appear more than once each cycle, both horizontal plug-ins have a variable hold-off control which can be used to increase the minimum time between samples.

SWEEP DELAY AND SWEEP EXPANSION

Like automatic triggering, sweep delay is a capability which has never before been possible for a sampling oscilloscope, but which is now available in one of the new horizontal plug-ins. The value of sweep delay for examining complex waveforms has long been recognized, of course, and real-time oscilloscopes have had delay generators for some time.

The two sweeps of the new sampling time base and delay generator operate in three sweep display modes: the displayed sweep can be a main sweep, a delaying sweep, or the main sweep delayed by an interval determined by the settings of the delay controls. Sweep delay is normally used to select any portion of a complex waveform for display on an expanded, faster time base. Therefore the main sweep rate is normally faster than the delaying sweep rate, although this need not be true.

Without sweep delay, sweep magnification was the only means for expanding details of a sampling-oscilloscope display, and both of the new horizontal plug-ins still have magnifiers. However, sweep magnifiers are usually limited to expansions of 100:1 or less, whereas expansions of 10,000:1 or more are possible with sweep delay.

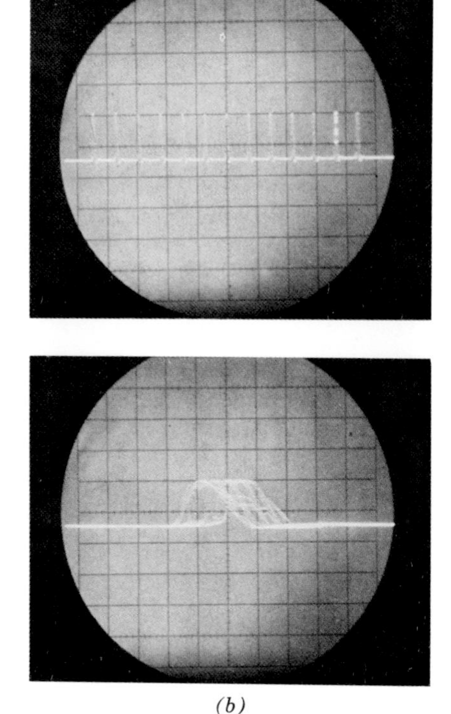

(b)

Fig. 11 (a). *Pulse train displayed with main sweep of sampling time base and delay generator. Bright spot on 12th pulse indicates center of area to be magnified. (b) 12th pulse magnified × 20. Note rate jitter, which is also magnified (see Fig. 12). Vertical: 100 mV/cm; Horizontal: (a) 100 ns/cm, (b) 5 ns/cm.*

Another limitation of magnification alone becomes evident if the portion of the waveform to be magnified does not always occur at precisely the same time after the beginning of the sweep, as would be the case, for example, in a pulse train in which the period between pulses varies randomly. When the waveform is magnified, this 'rate jitter' will also be magnified and the signal may be difficult to observe. Fig. 11(a) is an oscillogram of a pulse-train display using the main sweep of the new delay-generator plug-in as the time base. The bright dot on the 12th pulse indicates the point about which magnification will occur when the MAGNIFIER control is turned to one of its six magnification settings (×2 to ×100). Rate jitter in the 12th pulse of Fig. 11(a) is evident in the magnified display of Fig. 11(b).

With the delay-generator plug-in, rate jitter can be eliminated from the display. Fig. 12(a) shows the pulse train of Fig. 11 displayed using the delaying (slow) sweep as the time base. Here the bright dot on the 12th pulse indicates the time at which the fast main sweep will start when the SWEEP switch is turned to MAIN DELAYED. The amount of calibrated delay is continuously variable from 50 ns to 5 ms. Fig. 12(b) shows the same pulse train using the delayed main sweep triggered normally (i.e., not automatically) after the delay interval. When triggered normally, the main sweep is merely armed at the end of the delay interval and is not triggered until the selected pulse occurs. The resulting display of Fig. 12(b) is entirely free of jitter. Had the main sweep been triggered automatically at the end of the delay interval, the display would have been identical to Fig. 11(b), with the rate jitter still present.

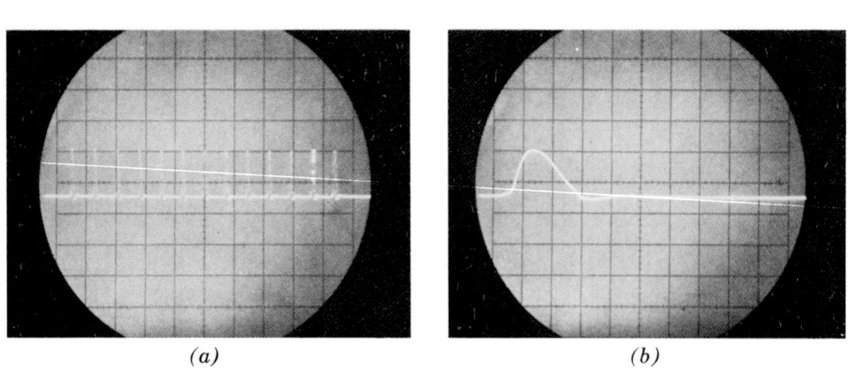

(a) (b)

Fig. 12(a). *Pulse train displayed with delaying sweep of sampling time base and delay generator. Bright spot on 12th pulse indicates start of delayed main sweep. (b) 12th pulse displayed with main sweep armed at end of delay interval and triggered on 12th pulse. Note absence of rate jitter. Vertical: 100 mV/cm; Horizontal: (a) 100 ns/cm, (b) 5 ns/cm, (c) 5 ns/cm.*

SYNC PULSE

Sync pulse outputs in both horizontal plug-ins provide pulses of about 1.5 V amplitude and 1 ns rise time. These pulses are useful as pretriggers to drive a pulse generator or as driving pulses for a circuit being tested. They are exceptionally clean and flat-topped, and make excellent test pulses for time domain reflectometry.

In the single-sweep plug-in, the sync pulses are synchronized with the sweep. In the delay generator plug-in, they are synchronized with the main sweep so that in addition to the above uses, they can be used as a calibrated-delay pulse source simply by setting the main sweep delay controls to the desired delay interval.

—Allan I. Best,
Darwin L. Howard and
James M. Umphrey

DESIGN LEADERS

ALLAN I. BEST

DARWIN L. HOWARD

JAMES M. UMPHREY

After spending three years in the U. S. Army, Al Best attended the University of California, graduating in 1960 with a BSEE degree. He then joined –hp– as a development engineer and contributed to the design of the 185B Sampling Oscilloscope, later assuming responsibility for the 185B after it went into production. After doing further work in sampling oscilloscope design, he transferred to Colorado Springs in 1964 and became design leader for the new sampling plug-ins for the 140A Oscilloscope. Al has one patent pending on delayed-sweep sampling time bases and another on a fast-ramp linearizer. Currently he is an engineering group leader in the –hp– Colorado Springs Laboratory, responsible for TDR and certain aspects of sampling.

Dar Howard joined –hp– in 1958 after graduating from the University of Utah with a BSEE degree. As a development engineer in the –hp– Microwave Laboratory, he contributed to the design of the 344A Noise Figure Meter and the 415C SWR Meter, and did further work on a phase-locked RF signal generator. In 1962, he received his MS degree in electrical engineering from Stanford University through the –hp– Honors Cooperative Program.

Dar transferred to the –hp– Oscilloscope Division (now at Colorado Springs) in 1963. As an engineering group leader, he organized the design group for the new sampling plug-ins for the 140A and 141A Oscilloscopes. In 1965 he assumed his present position of engineering manager, Colorado Springs Division.

Before deciding to become an electrical engineer, Dar attended Brigham Young University as an accounting major for one year, and then spent four years in the U. S. Navy as an electronics technician.

Jim Umphrey received his BSEE degree in 1961 from Stanford University, then joined –hp– as a development engineer, working on the 187B and 187C Sampling Vertical Amplifiers and the 213B Pulse Generator. He transferred to Colorado Springs in 1964 and eventually assumed responsibility for the design of the 1104A/1106A Trigger Countdown, the 1105A/1106A Pulse Generator, and the wideband sampling vertical amplifier plug-ins for the 140A Oscilloscope. He is now an engineering group leader in the –hp– Colorado Springs Laboratory, with responsibility for a number of sampling and pulse-generator projects.

Jim received his MSEE degree from Stanford in 1961 on the –hp– Honors Cooperative Program. He is a member of Tau Beta Pi and IEEE.

CONDENSED SPECIFICATIONS

SAMPLING PLUG-INS FOR —hp— MODEL 140A OSCILLOSCOPE AND —hp— MODEL 141A VARIABLE-PERSISTENCE OSCILLOSCOPE

MODEL 1410A SAMPLING VERTICAL AMPLIFIER

MODE OF OPERATION: Channel A only; B only; A and B; A and B added algebraically; A vs. B.

POLARITY: Either channel may be displayed either positive or negative up in any mode.

RISE TIME: Less than 350 ps.

BANDWIDTH: dc to 1 GHz.

OVERSHOOT: Less than 5%.

SENSITIVITY: Calibrated ranges from 1 mV/cm to 200 mV/cm.

ISOLATION BETWEEN CHANNELS: Greater than 40 dB to 1 GHz.

INPUT IMPEDANCE:
Probes: 100 kΩ shunted by 2 pF, nominal.
GR Type 874 Inputs: 50Ω ±2% with 58-ns internal delay lines for viewing leading edge of fast rise signals.

DYNAMIC RANGE: ±2 V.

DRIFT: Less than 3 mV/hr after warmup.

TRIGGERING: Internal or external when using 50Ω inputs. Internal triggering selectable from Channel A or B. External triggering necessary when using probes.

TIME DIFFERENCE BETWEEN CHANNELS: <100 ps.

RECORDER OUTPUTS: Front panel outputs provide 0.1 V/cm from a 500Ω source. Gain adjustable from approximately 0.05 V/cm to 0.2 V/cm. dc level adjustable from approximately −1.5 V to +0.5 V.

PRICE: $1600.

MODEL 1411A SAMPLING VERTICAL AMPLIFIER

(Used with 1430A, 1431A, or 1432A Sampler)

MODE OF OPERATION, POLARITY, SENSITIVITY, RECORDER OUTPUTS: Same as 1410A.

ISOLATION BETWEEN CHANNELS: 40 dB over bandwidth of sampler.

PRICE: $700.

MODEL 1430A SAMPLER

(used with 1411A)

RISE TIME: Approximately 28 ps. (Less than 35 ps observed with 1105A/1106A pulse generator and 909A 50Ω load.)

BANDWIDTH: dc to approximately 12.4 GHz.

OVERSHOOT: Less than ±5%.

DYNAMIC RANGE: ±1 V.

INPUT CHARACTERISTICS:
Mechanical: Amphenol GPC-7 precision 7 mm connectors on input and output.
Electrical: 50Ω feedthrough, dc coupled. Reflection from sampler is approximately 10%, using a 40-ps TDR system. VSWR <3:1 at 12.4 GHz.

TIME DIFFERENCE BETWEEN CHANNELS: Less than 5 ps.

CONNECTING CABLE LENGTH: 5 ft. (10 ft. optional).

PRICE: $3000 ($3035 with 10-ft. cable).

MODEL 1431A SAMPLER

(used with 1411A)

BANDWIDTH: dc to greater than 12.4 GHz (less than 3 dB down from a 10-cm dc reference).

RISE TIME: Approximately 28 ps.

VSWR: dc to 8 GHz <1.4:1
8 to 10 GHz <1.6:1
10 to 12.4 GHz <2.0:1

DYNAMIC RANGE: ±1 V.

INPUT CHARACTERISTICS:
Mechanical: Same as 1430A.
Electrical: Same as 1430A except reflection from sampler is approximately 5%, using a 40-ps TDR system.

PHASE SHIFT BETWEEN CHANNELS: Less than 10° at 5 GHz, typically less than 2° at 1 GHz.

CONNECTING CABLE LENGTH: 5 ft. (10 ft. optional).

PRICE: $3000 ($3035 with 10-ft. cable).

MODEL 1432A SAMPLER

(used with 1411A)

RISE TIME: Less than 90 ps.

BANDWIDTH: dc to 4 GHz.

OVERSHOOT: Less than ±5%.

DYNAMIC RANGE: ±1 V.

INPUT CHARACTERISTICS:
Mechanical: GR Type 874 connectors used on input and output.
Electrical: 50Ω feedthrough, dc coupled. Reflection from sampler is approximately 15% using a 90-ps TDR system.

TIME DIFFERENCE BETWEEN CHANNELS: Less than 25 ps.

CONNECTING CABLE LENGTH: 5 ft. (10 ft. optional).

PRICE: $1000 ($1035 with 10-ft. cable).

MODEL 1424A SAMPLING TIME BASE

SWEEP RANGE: 24 ranges, 10 ps/cm to 500 μs/cm. Sweeps from 1 ns/cm to 500 μs/cm may be expanded up to 100 times and read out directly.

MARKER POSITION: Intensified marker indicates point about which sweep is expanded; 10-turn, calibrated control.

TRIGGERING: (Less than 1 GHz.)
Internal (with 1410A):
Automatic:
Pulses: 75 mV amplitude for pulses 2 ns or wider for jitter less than 20 ps.
CW: 25 mV amplitude from 60 Hz to 500 MHz for jitter less than 10% of input signal period. (Usable to 1 GHz with increased jitter.)
Level Select:
Pulses: 35 mV amplitude for pulses 2 ns or wider for jitter less than 20 ps.
CW: 25 mV amplitude from dc to 300 MHz (increasing to 200 mV at 1 GHz) for jitter less than 1% of input signal period + 10 ps.
External:
Automatic:
Pulses: At least 100 mV amplitude for pulses 2 ns or wider for jitter less than 20 ps.
CW: 50 mV from 60 Hz to 500 MHz for jitter less than 10% of input signal period. (Usable to 1 GHz with increased jitter.)
Level Select:
Pulses: At least 50 mV amplitude required of pulses 2 ns or wider for jitter less than 20 ps.
CW: 50 mV from dc to 1 GHz for jitter less than 1% of input signal period + 10 ps. Jitter is less than 30 ps for signals of 10 mV at 1 GHz.
Slope: Positive or negative.
External Trigger Input:
50Ω, ac or dc coupled.
Jitter: Less than 10 ps on 1 ns/cm range, and less than 20 ps (or 0.005% of unexpanded sweep speed, whichever is larger) at 2 ns/cm and slower, with signals having rise times of 1 ns or faster.

TRIGGERING: (Greater than 1 GHz.)
Jitter less than 20 ps for 25 mV input, 500 MHz to 5 GHz.

SCANNING:
Internal: X axis driven from internal source. Scan density continuously variable.
Manual: X axis driven by manual scan control knob.
Record: X axis driven by internal slow ramp; approximately 60 seconds for one scan.
External: 0 to +15 V required for scan; input impedance, 10 kΩ.
Single Scan: One scan per actuation; scan density continuously variable.

SYNC PULSE OUTPUT:
Amplitude: Greater than 1.5 V into 50Ω.
Rise Time: Approximately 1 ns.
Overshoot: Less than 5%.
Width: Approximately 1 μs.
Relative Jitter: Less than 10 ps.
Repetition Rate: One pulse per sample.

PRICE: $1200.

MODEL 1425A SAMPLING TIME BASE AND DELAY GENERATOR

MAIN SWEEP:
Range: 13 ranges, 1 ns/cm to 10 μs/cm.
Magnifier: 7 calibrated expansion ranges, X1 to X100. Increases fastest calibrated sweep speed to 10 ps/cm. Pushbutton returns magnifier to X1.
Magnified Position: 10-turn control with intensified marker that indicates sweep expansion point.

TRIGGERING: (For both Main and Delaying Sweep.)
Internal:
Automatic:
Pulses: 150 mV amplitude for pulses 2 ns or wider for jitter less than 20 ps.
CW: 50 mV amplitude from 200 Hz to 500 MHz for jitter less than 10% of input signal period. (Usable to 1 GHz with increased jitter.)

Level Select:
Pulses: 70 mV amplitude for pulses 2 ns or wider for jitter less than 20 ps.
CW: 50 mV amplitude from 200 Hz to 300 MHz (increasing to 400 mV at 1 GHz) for jitter less than 1% of input signal period + 10 ps.
External: Same as 1424A, except low end of CW triggering range is 200 Hz in Automatic and Level.
External Trigger Input:
50Ω ac coupled (2.2 μF).
Slope: Same as 1424A.
Jitter: Same as 1424A.

DELAYING SWEEP:
Range: 15 ranges, 10 ns/cm to 500 μs/cm.
Delay Time: Continuously variable from 50 ns to 5 ms.

SWEEP FUNCTIONS: Main sweep, delaying sweep, main sweep delayed.

SCANNING: Same as 1424A except no external scan input.

SYNC PULSE OUTPUT: Same as 1424A. Pulse always synchronized to main sweep trigger circuit; pulse delay and rate are variable.

PRICE: $1600.

18-GHz TRIGGER COUNTDOWN
MODEL 1104A COUNTDOWN SUPPLY
MODEL 1106A TUNNEL DIODE MOUNT

INPUT:
Frequency Range: 1 GHz to 18 GHz.
Sensitivity: Signals 100 mV or larger, and up to 12.4 GHz, produce less than 20 ps of jitter (200 mV required to 18 GHz).
Input Impedance (1106A): 50Ω Amphenol GPC-7 input connector. Reflection from input connector is less than 10% using a 40-ps TDR system.

OUTPUT:
Center Frequency: Approximately 100 MHz.
Amplitude: Typically 150 mV.

PRICE: 1104A, $200; 1106A, $550.

20-ps PULSE GENERATOR
MODEL 1105A PULSE GENERATOR SUPPLY
MODEL 1106A TUNNEL DIODE MOUNT

OUTPUT:
Rise Time: Approximately 20 ps. Less than 35 ps observed with –hp– Model 1411A/1430A 28-ps Sampler and –hp– Model 909A 50Ω termination.
Overshoot: Less than ±5% as observed on 1411A/1430A with 909A.
Droop: Less than 3% in first 100 ns.
Width: Approximately 3 μs.
Amplitude: Greater than +200 mV into 50Ω.
Output Characteristics (1106A):
Mechanical: Amphenol GPC-7 precision 7 mm connector.
Electrical: dc resistance — 50Ω ±2%. Source reflection — less than 10%, using a 40-ps TDR system. dc offset voltage — approximately 0.1 V.

TRIGGERING:
Amplitude: At least ±0.5 V peak required.
Rise Time: Less than 20 ns required. Jitter less than 15 ps when triggered by 1 ns rise time sync pulse from 1424A or 1425A Sampling Time Base. Jitter increases with slower trigger rise times.
Width: Greater than 2 ns.
Input Impedance: 200Ω, ac coupled through 20 pF.
Repetition Rate: 0 to 100 kHz; free runs at 100 kHz.

PRICE: 1105A, $200; 1106A, $550.

Prices f.o.b. factory
Data subject to change without notice

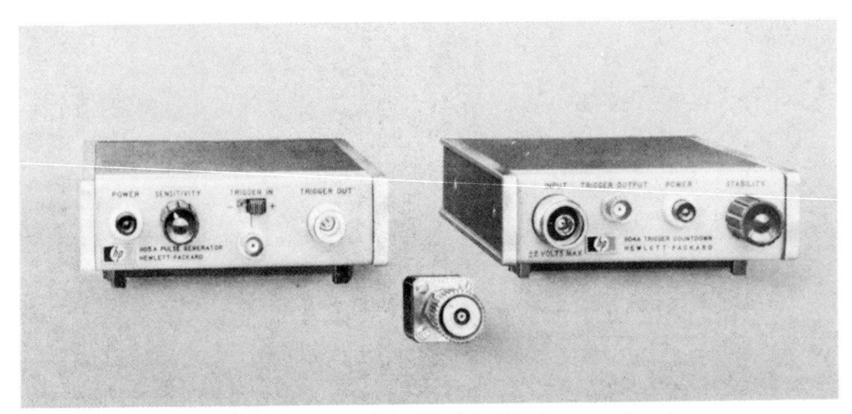

Fig. 1. Tunnel diode mount, –hp– Model 1106A, c., and pulse generator supply, –hp– Model 1105A, l., form pulse generator with rise time of approximately 20 ps, one of fastest in existence, useful for high-resolution TDR work. Same tunnel diode mount is used with countdown supply, –hp– Model 1104A, r., to allow new sampling plug-ins to display sine waves up to 18 GHz.

ULTRA-FAST TRIGGERING AND ULTRA-RESOLUTION TDR

To take full advantage of the 12.4-GHz bandwidths of the wideband oscilloscopes described in the preceding article, it is naturally necessary to synchronize the oscilloscope time bases to signals having frequencies of 12.4 GHz and above. Similarly, to take advantage of the 28-ps rise times of the samplers for high-resolution time-domain reflectometry (TDR), it is necessary to have a step generator which has an ultra-fast rise time less than 28 ps. A single tunnel diode mount has been developed to meet both of these needs. Used with a countdown supply, the tunnel diode permits either sampling time base to trigger on signals up to 18 GHz. Used with a pulse generator supply, the same tunnel diode produces a 3-μs pulse with a rise time of 20 ps, fast enough to resolve cable discontinuities spaced as closely as a few millimeters. Fig. 1 is a photograph of the tunnel diode mount and the two supplies.

The pulse generator has what is probably the fastest rise time available today—so fast, in fact, that there are no instruments to measure it exactly. Twenty picoseconds is a conservative estimate. In spite of its fast rise time, however, its overshoot is less than 5%, a remarkably small value for such a fast pulse. With the 28-ps sampling oscilloscope, the pulse generator forms a time domain reflectometer with a rise time of less than 40 ps, which is about the time it takes for light to travel 1/2 in. in air. Fig. 2 is a TDR oscillogram showing reflections from a special section of air line in which the diameter of the center conductor has been reduced in three sections for demonstration purposes. Centers of the three sections are 5/8 in. apart and their lengths are 1/8 in., 1/4 in., and 1/2 in. All three sections show up clearly and are readily resolvable.

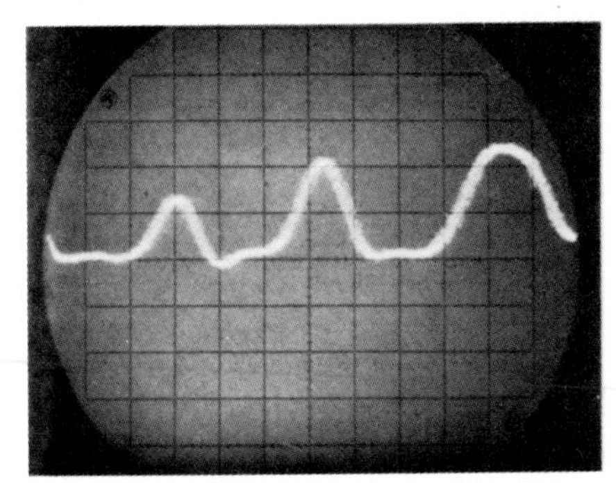

Fig. 2. Oscillogram taken with 40-ps TDR system showing reflections from short length of line with three discontinuities. TDR system can resolve discontinuities spaced as closely as 1/4 in. apart.

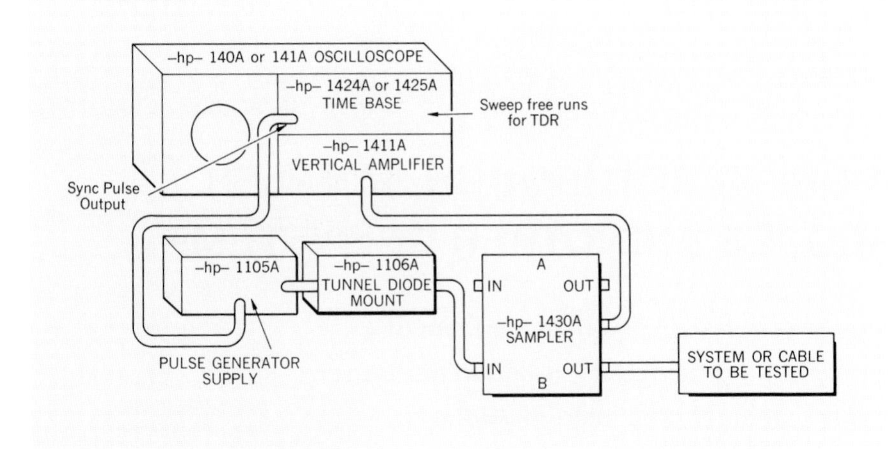

Fig. 3. *40-ps TDR system, using -hp- Model 1105A/1106A 20-ps Pulse Generator. Capabilities of system are shown by Fig. 2.*

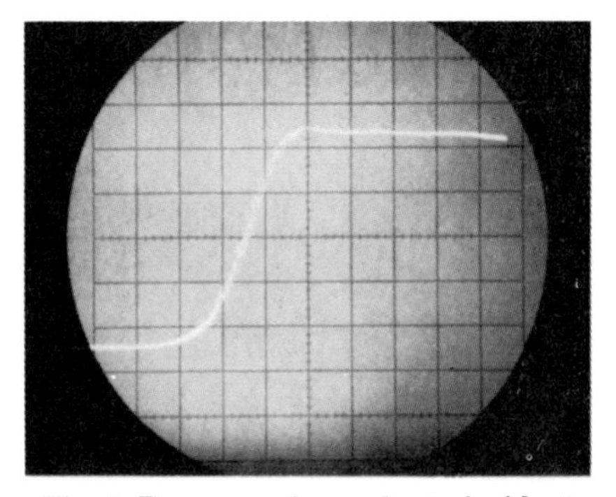

Fig. 4. *Response of sampler to incident step in TDR System of Fig. 3. Typical rise time is less than 35 ps. Vertical: reflection coefficient = 0.2/cm; Horizontal: 20ps/cm.*

Fig. 3 is a block diagram of the 40-ps TDR system. Notice that the time base of the oscilloscope is allowed to free run, and the pulse generator is triggered by the sync pulse output of the time base. Fig. 4 shows a typical step response for this TDR system. Rise time is about 35 ps.

A typical tunnel-diode characteristic is shown in Fig. 5. In the 20-ps pulse generator, the diode is originally biased at point A on this curve. The trigger causes the diode current to rise above point B, moving the diode into its negative resistance region B-D. The operating point then jumps very rapidly to point C, producing the 20-ps pulse. Fig. 6 shows the output waveform.

The 18-GHz countdown unit delivers a subharmonic of the input signal to the time base. Center frequency of the countdown output is 100 MHz.

For the 18 GHz countdown unit, the tunnel diode is biased at a point slightly above point B in Fig. 5, where it is astable. Its free-run rate varies, becoming a subharmonic of any high-frequency input signal.

SAMPLING UNITS DESIGN TEAM

Contributors to development of new sampling vertical amplifier and time base plug-ins and new pulse generator and countdown are, front row, l. to r., Norman L. O'Neal, Allan I. Best, James N. Painter, Edward J. Prijatel, W. A. Farnbach, James M. Umphrey; back row, l. to r., Robert B. Montoya, Gordon A. Greenley, Jeffrey H. Smith, Howard L. Layher, Jay A. Cedarleaf, all of the -hp- Colorado Springs Division. Not shown: Donald L. Gardner.

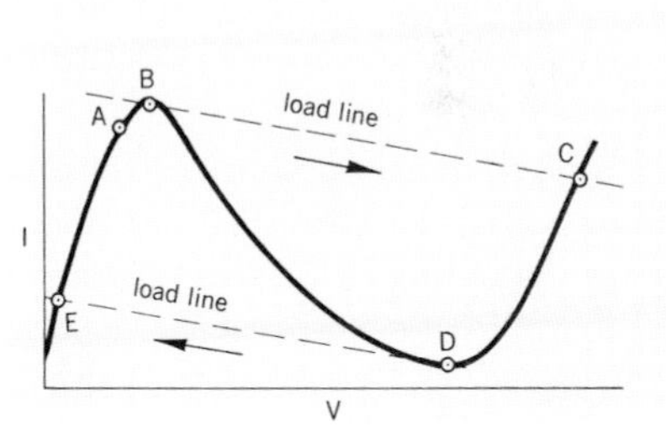

Fig. 5. *Typical tunnel diode characteristic, showing negative-resistance region BD. To generate pulse with 20-ps rise time, diode is biased at A. Trigger input raises diode current to unstable point B. Operating point then shifts rapidly to C, giving fast pulse as shown in Fig. 6.*

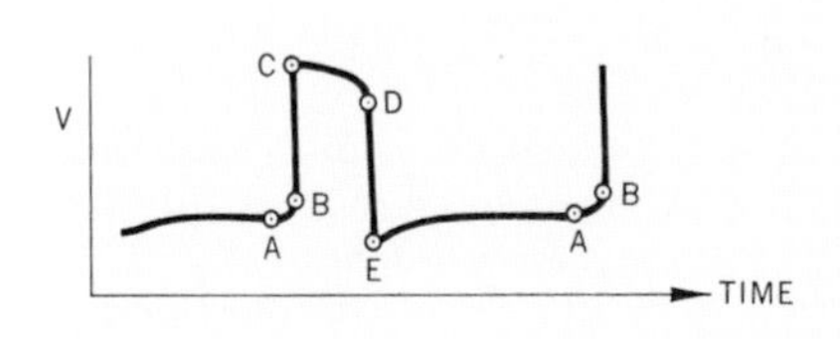

Fig. 6. *Waveform corresponding to operating point sequence of Fig. 5.*

A DC TO 12.4 GHz FEEDTHROUGH SAMPLER FOR OSCILLOSCOPES AND OTHER RF SYSTEMS

An important circuit development in the form of an ultra-wideband sampling device is leading to major new capabilities in electronic instrumentation.

DESPITE its relatively recent appearance, the diminutive object on this month's cover has already had considerable impact on high-frequency electronic instrumentation, and is likely to have an increasing effect in the future. The object is an ultra-wideband sampling device developed by *hp associates* and now used in such diverse broadband systems as sampling oscilloscopes (see p. 2), phase-locked automatic transfer oscillators, and vector measurement systems.[1] Bandwidth of the device is specified as greater than 12.4 GHz, more than three times the bandwidth of any previous sampling device.

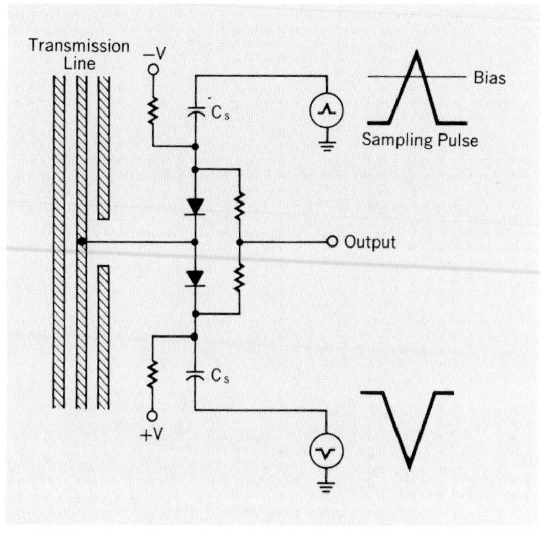

Fig. 2. *In two-diode feedthrough sampling circuit used in wideband sampling device, normally back-biased diodes are gated on by sampling pulses for short periods, allowing sampling capacitors C_s to acquire voltages proportional to signal appearing on transmission line.*

High-frequency sampling techniques make possible many broadband systems, including those just mentioned, that would not be feasible otherwise.

[1] 'The Impact of Ultra-Wideband Sampling and Associated Developments on Electronic Instrumentation,' Session 23, Western Electronic Show and Convention, August 26, 1966.

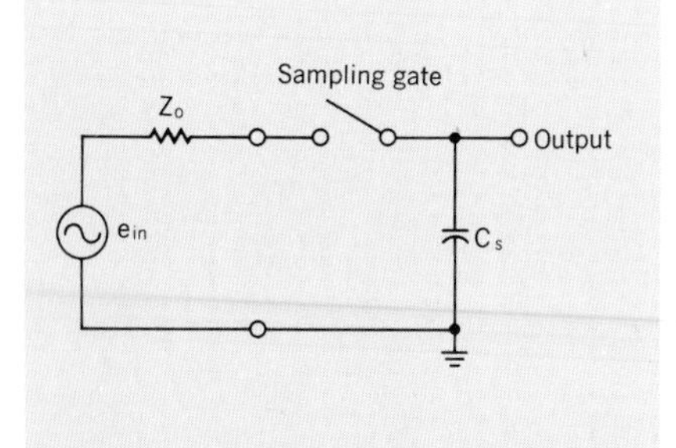

Fig. 1. *Idealized sampling circuit. System to be sampled is represented by voltage e_{in} and impedance Z_0. Sample is taken by closing gate for short period, allowing sampling capacitor C_s to charge to fraction of e_{in}. Sample is stored on C_s when gate is opened.*

The systems which use sampling techniques are quite varied, but it is significant that the requirements for broadband sampling devices are nearly the same regardless of the application. Consequently, the use of the new sampling device in RF instrumentation is not limited to any specialized class of instruments, and will in all likelihood spread to systems which are not now thought of as sampling systems and to other RF systems which have yet to be conceived.

Development of the wideband sampling device required a large investment in time, funds, and engineering ingenuity. So that nothing would be overlooked which might aid—or frustrate—the development effort, a design approach was taken which can be best described as 'functionally integrated'.[2]

[2] The words 'functionally integrated' should not be confused with monolithic integrated circuit technology. There are no integrated circuits in the sampling device.

This means that the effects of every element of the sampling device on every other element were considered in the design, including the effects of parasitic elements and the practical considerations of cost, manufacturability, and repairability. From this approach has come a device which embodies sophisticated microwave system design and advanced diode design and packaging.

SAMPLER OPERATION

The basic elements of a sampling circuit are shown in the idealized circuit of Fig. 1. In this diagram, the system to be sampled is represented by a voltage generator e_{in} and an impedance Z_o, and the sampler consists of a sampling gate and a sampling capacitor C_s. When a sample is to be taken, the switch or gate is closed for a short period, allowing the sampling capacitor C_s to charge

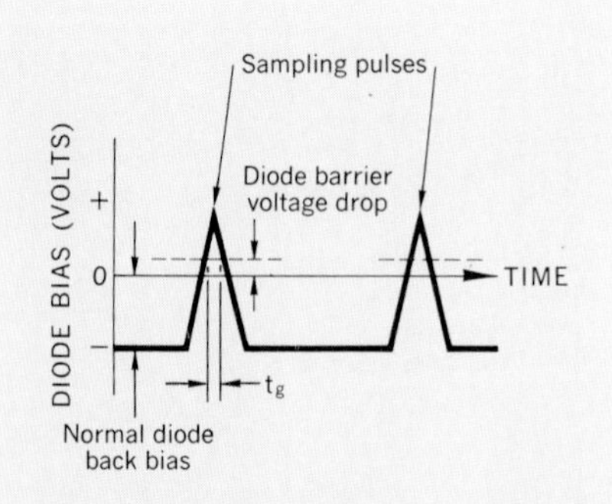

Fig. 3. *Gate time t_g is time for which impedance of diodes is low in circuit of Fig. 2. Gate time is inversely proportional to bandwidth of sampling circuit. Note that gate time can be less than rise time of sampling pulse. In wideband sampling device, t_g is less than 28 ps, equivalent to bandwidth of more than 12.4 GHz.*

to some fraction of the input voltage e_{in}. The switch is then opened, leaving the sample of the input stored on C_s.

If the voltage on C_s is assumed to be reset to zero before each new sample, a useful measure of the efficiency of the circuit is the sampling efficiency η, defined as the ratio of the voltage on C_s after each sample, e_{sample}, to the input voltage, e_{in}:

$$\eta = \frac{e_{sample}}{e_{in}}$$

Bandwidth of the sampler is defined as the frequency at which η is $1/\sqrt{2}$ times its dc or low-frequency value.

The new wideband sampling device is a realization of the basic two-diode feedthrough sampling circuit of Fig. 2. Signals to be sampled appear between the center and outer conductors of the feedthrough transmission line.

The diodes, which act as the sampling gates, are normally back biased. When a sample is to be taken, the diodes are gated on by balanced sampling pulses, thereby providing a low impedance path through the diodes and the sampling capacitors C_s to ground. Bandwidth of this circuit is inversely proportional to the time for which the diode impedance is low. This time interval, called the gate width t_g, is related to the bandwidth BW approximately by the formula

$$BW(GHz) \cong \frac{350}{t_g(ps)}$$

A bandwidth of 12.4 GHz corresponds to a t_g of approximately 28 picoseconds. When the sampler is used in a sampling oscilloscope like the one described in the article begining on p. 2, the rise time of the oscilloscope is also approximately equal to the gate width. Fig. 3 shows the relationships of the gate width to the sampling pulse, the diode bias, and the diode voltage drop.

WIDEBAND SAMPLING DEVICE

As a result of the 'functionally integrated' design approach which produced it, no part of the wideband sampling device has only a single function; all parts work together in such a way that the device must be considered as a whole in order to be understood. All elements, including parasitics, have been accounted for and, where possible, made integral parts of the sampling circuit. A number of normally separate parts have been combined to form unified parts, and two of the signals present — the input signal and the sampling pulses — occupy the same transmission line without interfering with each other.

Fig. 4 is a drawing of the wideband feedthrough sampling device. The device consists of a two-diode sampler located at the center of a biconical cavity which forms a part of the RF transmission line. To make room for the diodes, the two cone-shaped faces of the cavity are truncated. The RF line is perpendicular to the axis of the cones, and the sides of the cavity act as the RF ground conductor. The RF center conductor passes through the center of the cavity, and the two diodes are placed in contact with it. The characteristic impedance of the RF line is maintained at 50Ω throughout the sampling structure, the only discontinuity being a portion of the diode capacitance at the sampling node.

The diodes are specially designed low-storage hot-carrier diodes in low-

Fig. 4. *Cutaway drawing of wideband sampling device. (a) dielectric-filled biconical cavity, (b) cone-shaped face of cavity, (c) top of truncated cone, (d) RF center conductor, (e) hot-carrier diode and sampling capacitor in three-terminal package, (f) sampling pulse input line, (g) plug (non-functional).*

capacitance, low-inductance, three-terminal packages. The diodes and their packages embody a number of noteworthy features.

1. *An advanced diode fabrication technique using an extremely small semiconductor chip.* One benefit is a chip with a remarkably small capacitance of less than 0.08 pF.
2. *A diode package with a series inductance of less than 200 pH and an unusually low package capacitance of approximately 0.08 pF.* The insulating portion of the diode package is made of a material which has a low dielectric constant; this lowers the package capacitance and helps prevent this capacitance from reducing bandwidth, as described next.
3. *A three terminal diode package with a built-in sampling capacitor.* Having the sampling capacitor an integral part of the diode package not only reduces lead inductances, but also relocates the diode package capacity so that it appears across the RF line instead of in parallel with the diode chip. This means that the package capacity can be partially masked (i.e., made an integral part of the RF line) by filling the cavity with a dielectric which has approximately the same dielectric constant as the package material.

Design of the cavity was complicated by the requirement of minimum inductance between the ground of the RF line at the sampling axis and the ground points of the sampling capacitors. This inductance was eliminated by splitting the ground conductor of the RF line and making the RF and sampling-capacitor ground points physically the same point (AA′ in Fig. 5). This technique is basic to the operation of the device and plays a key role in the reduction of sampling circuit inductance which would otherwise limit the bandwidth.

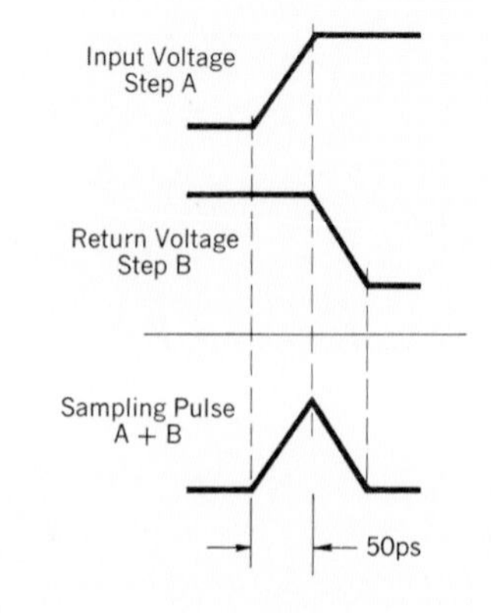

Fig. 6. *Sampling signal input is voltage step which turns diodes on, then travels outwards from center of biconical cavity to short circuit at outer edges of cavity. Step is reflected at outer edges and equal and opposite step travels back to center of cavity and out sampling signal line, turning diodes off. Round trip time is 50 ps.*

The split ground configuration also provides an ideal entry point for the sampling signal, since it is possible to develop a voltage (for a short time) across the impedance between the two sides of the RF ground conductor. The sampling signal enters the cavity at a point next to one of the diodes (upper diode in Fig. 4) through a section of 50-ohm microcoax transmission line. The center conductor of the sampling line continues across the cavity, contacting the opposite truncated cone face near its center. Hence, when a sample is to be taken, the sampling voltage is introduced between the centers (approximately) of the two cone faces. This means that the sampling signal appears as a potential difference between two points on the RF ground conductor, as shown schematically in Fig. 5.

A biconical cavity driven from its center has a constant characteristic impedance. To the sampling signal, the cavity appears to be a shorted section of 50-ohm line, driven from the center of the cavity and shorted at the outer edges. This shorted 'stub' is part of the network which forms the sampling pulse.

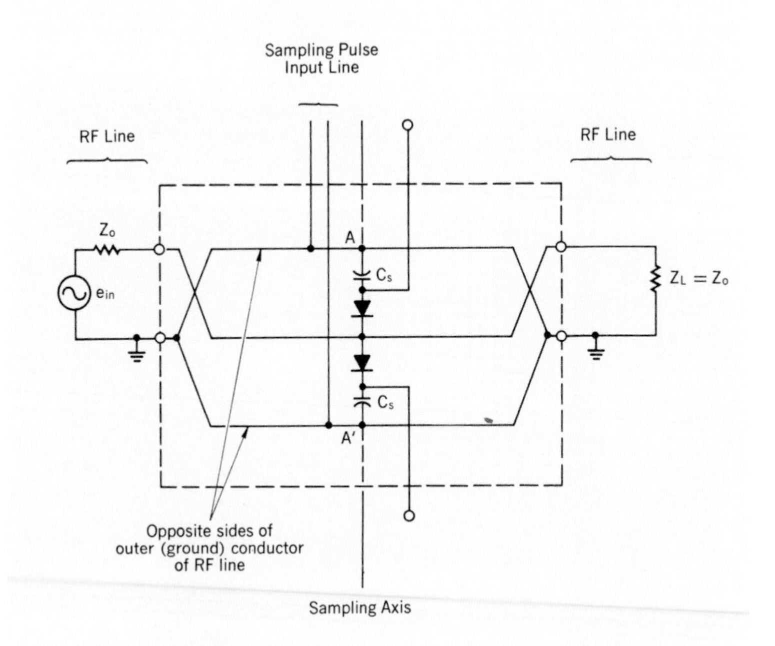

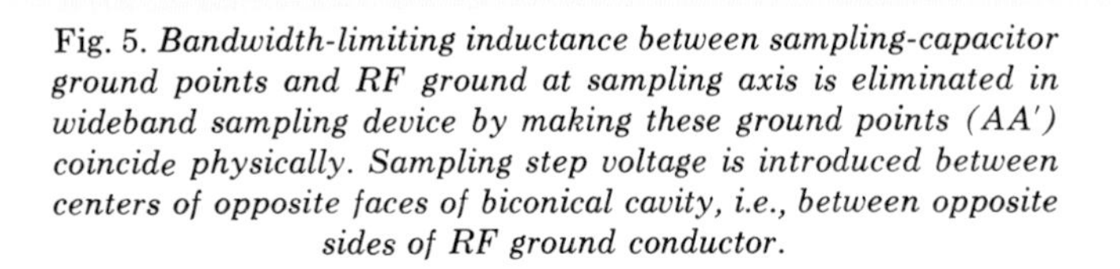
Fig. 5. *Bandwidth-limiting inductance between sampling-capacitor ground points and RF ground at sampling axis is eliminated in wideband sampling device by making these ground points (AA′) coincide physically. Sampling step voltage is introduced between centers of opposite faces of biconical cavity, i.e., between opposite sides of RF ground conductor.*

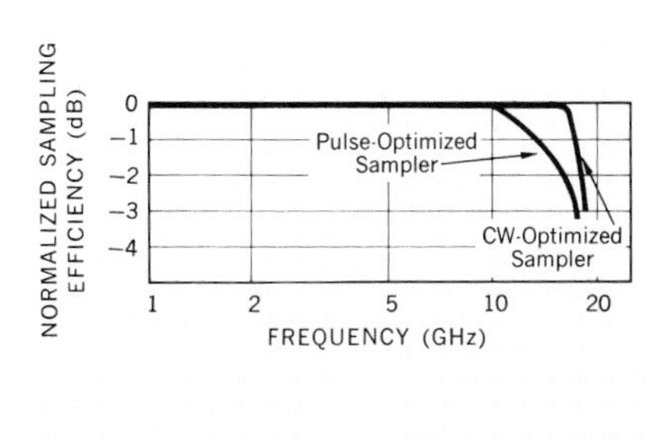

Fig. 7. *Frequency response of the two types of wideband sampling devices. Bandwidths of 16 GHz are typical.*

To begin the sampling process, a voltage step is introduced into the biconical cavity through the sampling pulse line. Coupling through the sampling capacitors and turning the diodes on as it enters the cavity, the step travels out towards the short circuit at the outer edges. A reflection occurs when the step reaches the short, and an equal and opposite step travels back to the center of the cavity and out the sampling line, turning the diodes off as it passes their location (see Fig. 6). The time taken by the step to travel from the center of the cavity to the outer edge and back to the center is about 50 picoseconds. However, as shown in Fig. 3, the gate width t_g can be less than 50 picoseconds because the sampling pulse does not have zero rise and fall times. In the 12.4-GHz sampler, t_g is about 28 ps.

Care has been taken to introduce the sampling signal into the cavity in such a way that the RF center conductor always lies on an equipotential plane with respect to the sampling signal. This means that the voltage on the RF line is not affected by the presence of the sampling signal, and that the sampling signal is not affected by the presence of the RF center conductor. Although there are two modes of transmission on the RF line — the input signal and the sampling signal — these two modes are electrically isolated from each other. Typical isolation is greater than 40 dB.

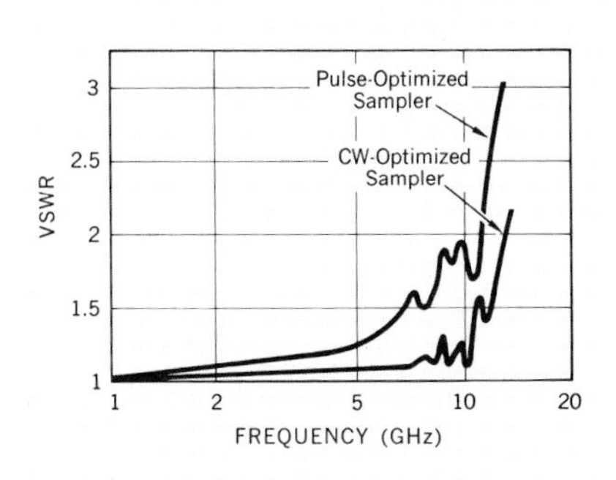

Fig. 8. *VSWR of the two types of wideband sampling devices.*

PULSE AND CW SAMPLERS

While the partial masking of the diode package capacitance was a key factor in achieving the wide bandwidth of the sampling device, it was not necessary to mask totally both the package capacitance and the diode chip capacitance in order to meet the bandwidth specification. In the basic sampler, the unmasked diode capacitance at the sampling node causes the voltage standing wave ratio at the RF input to increase with frequency, approaching 3:1 at 12.4 GHz. For CW applications, which call for minimum VSWR rather than good transient response, a compensated, or CW-optimized, version of the sampler has been designed. Compensation is accomplished by incorporating the unmasked diode capacitance in a low-pass T-filter network. The filter takes the form of a modified RF center conductor in the truncated section of the biconical cavity. Measured frequency response, VSWR, and step response for both the uncompensated (pulse-optimized) and the compensated (CW-optimized) versions of the sampler are shown in Figs. 7, 8, and 9. Compensation increases the bandwidth and reduces the VSWR of the sampler to about 1.8:1 at 12.4 GHz, but causes 5 to 10% more overshoot in the step response. Notice in Fig. 7 that the 3-dB bandwidths of both samplers are considerably greater than 12.4 GHz. Typical bandwidths are now 14–16 GHz, and there is ample reason to believe that they will eventually be extended to over 18 GHz.

ACKNOWLEDGMENTS

The author wishes to acknowledge the helpful suggestions of Dr. Bernard M. Oliver, *-hp-* vice president for research and development, and the work of the many individuals at *hp associates* who contributed to the development of the wideband sampler.

—Wayne M. Grove

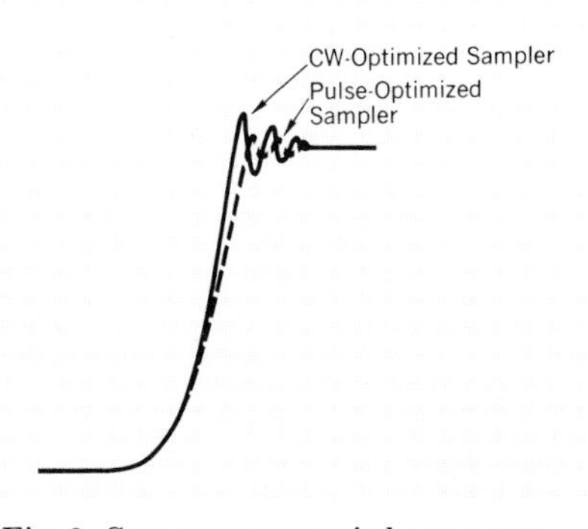

Fig. 9. *Step responses of the two types of wideband sampling devices. CW-optimized sampler has wider bandwidth and lower VSWR, but has greater overshoot in step response.*

WAYNE M. GROVE

Wayne Grove's first project after joining the -hp- Oscilloscope Division in 1961 was the design of the transistorized high-voltage power supply for the 140A Oscilloscope. Moving next into the study of high-frequency sampling techniques, he was responsible for the design and development of the 188A 4-GHz Sampling-Oscilloscope Vertical Amplifier. In 1963 Wayne joined hp associates, and was involved in the development of the hpa broadband microwave switch. He next became the design leader for the 12.4-GHz sampling device and the 1106A Tunnel Diode Mount, and contributed to the development of other microwave components including mixers and detectors. In mid-1966 he assumed his present position of research and development manager of the photoconductor department of hp associates.

Wayne received his BS degree in electrical engineering from Iowa State University in 1961 and his MS degree, also in electrical engineering, from Stanford University in 1963. He is a member of Tau Beta Pi, Eta Kappa Nu, Phi Kappa Phi, and IEEE. He holds several patents on wideband sampling and tunnel-diode pulsers.

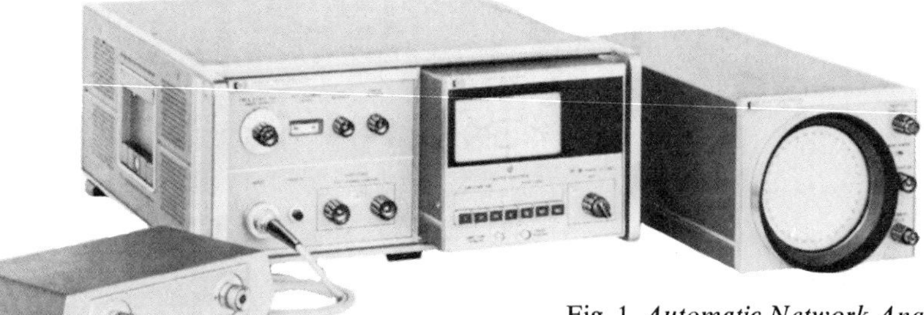

Fig. 1. *Automatic Network Analyzer for measuring complex impedances, gain, loss, and phase shift from 0.11 to 12.4 GHz consists of (l. to r.), Model 8411A Harmonic Frequency Converter, Model 8410A Main Frame, and a plug-in display module (either Model 8413A Phase-Gain Indicator or Model 8414A Polar Display). Network Analyzer makes swept or single-frequency measurements. See Fig. 8 for other system components.*

An Advanced New Network Analyzer for Sweep-Measuring Amplitude and Phase from 0.1 to 12.4 GHz

The information obtainable with a new network analyzer greatly improves microwave design practices, especially where phase information is important.

A NEW MICROWAVE NETWORK ANALYZER developed in the -hp- microwave laboratory promises to be of major importance in many electronic fields, especially those concerned with the phase properties of microwave systems and components. The new instrument sweep-measures the magnitude and phase of reflection and transmission coefficients over the range from 110 MHz to 12.4 GHz. This makes it possible for the analyzer to completely characterize active and passive devices, since nearly every parameter of interest for high-frequency devices can be measured including gain, attenuation, phase, impedance, admittance and others.

The new analyzer represents a major step in the continuing trend to automation in microwave measurements, a trend recognized in several articles in this publication and elsewhere*. Systems that are especially aided by the kinds of automated measurements the analyzer makes are the modern systems that emphasize phase properties, such as electronically-scanned radar and monopulse and doppler radar. Similarly, optimum use of the new high-frequency solid-state devices that make systems such as phased-array radars economically practical is dependent on sophisticated measurements. The reason for this de-

* See references on page 9.

-hp- Journal readers:

We believe you who work with frequencies above 100 MHz will be especially interested in this issue because it discusses an important new system that measures gain, phase, impedance, admittance and attenuation on a swept basis *from 110 MHz to 12.4 GHz. In other words the system will measure all network parameters not only of passive networks and devices but also of transistors and even of negative real impedances. Readout is on a meter or on a scope which presents measured performance over a whole frequency band at a glance.*

The new system leads to wider use of the familiar quantities we usually call reflection and transmission coefficients. These coefficients are also known as 'scattering parameters', and using them in combination with the new system leads to more sophisticated design techniques including computerized design. An informative article about scattering parameters begins on p. 13.

Obviously, the new system is a powerful tool for the engineer. In addition, it has important implications for the whole microwave engineering field in the future. This is also discussed in this issue (p. 11) by Paul Ely, engineering manager of our microwave laboratory.

I invite your attention to what we believe is unusually important microwave information.

Sincerely, Editor

PRINTED IN U.S.A.

pendence is that these solid-state devices can best be utilized in new functions if they can be completely characterized and understood.

The analyzer characterizes networks by measuring their complex small-signal parameters. The particular types of parameters measured are called the scattering or "s" parameters. These parameters have proved a valuable tool for the design engineer because of their inherent ease of measurement, their design advantages and the intuitive insight they provide. A separate article in this issue deals with their theory and describes new design practices developed with them at Hewlett-Packard.

Network Analyzer Concept

The concept of the network analyzer follows naturally from network-parameter theory. Measuring s-parameters is a matter of measuring (a) the ratio of the magnitudes and (b) the relative phase angles of response and excitation signals at the ports of a network with the other ports terminated in a specified 'characteristic' or reference impedance. It is not difficult to define the basic elements of a network analyzer system to perform these measurements (Fig. 3). First, a source of excitation is required. Then a transducer instrument is needed to convert the excitation signal and the response signals produced by the unknown to a set of output signals containing the network information (a dual-directional coupler for measuring the complex reflection coefficient s_{11} is illustrated). Next, an instrument capable of measuring magnitude ratio and phase difference is used to extract the pertinent information from the test signals. A readout mechanism to present the data completes the basic network analyzer.

The above is the concept that has been followed in designing the new network analyzer. A further refinement of the concept is the use of a plug-in readout. Although the network parameter data are the same for each application (i.e., magnitude and phase), the form in which the data are most useful depends upon the application.

Table I System Components

MODEL	FUNCTION	RANGE
8410A Network Analyzer Main Frame	Mainframe for readout modules, includes tuning circuits, IF amplifiers, and precision IF attenuator.	0.11 to 12.4 GHz when used with Model 8411A.
8411A Harmonic Frequency Converter	Converts 2 RF input signals 0.11 to 12.4 GHz into 20-MHz IF signals.	0.11 to 12.4 GHz when used with the 8410A. Impedance 50 ohms.
8413A Phase-Gain Indicator	Plug-in module for 8410A Mainframe provides meter display of relative amplitude and phase between input signals, auxiliary outputs for scope or X-Y recorder.	Full scale ±3, 10, 30 dB and ±6, 18, 60, 180 degrees. Auxiliary outputs 50 mV/dB and 10 mV/degree.
8414A Polar Display Unit	Plug-in module for 8410A Mainframe. CRT polar display of amplitude and phase. X-Y outputs for high resolution polar and Smith Chart impedance plots.	Internal graticule CRT for nonparallax viewing. Amplitude calibration in five linear steps. Phase in 10° intervals through 360°. Smith Chart overlays for direct impedance readout (normalized to 50 ohms).
8740A Transmission Test Unit	Simplifies RF input and test device connection for attenuation or gain test. Accepts RF input signal from source and splits into reference and test channels for connection to 8411A and the unknown device. Calibrated line stretcher balances out linear phase shift when test device is inserted.	0.11 to 12.4 GHz. Impedance 50 ohms.
8741A Reflection Test Unit	Wide-band reflectometer, phase balanced for swept or spot frequency impedance tests below 2 GHz. Accepts RF input and provides connections for unknown test device and 8411A. Movable reference plane.	0.11 to 2.0 GHz.
8742A Reflection Test Unit	Ultra-wide band reflectometer, phase balanced for impedance tests above 2.0 GHz. Movable reference plane.	2.0 to 12.4 GHz

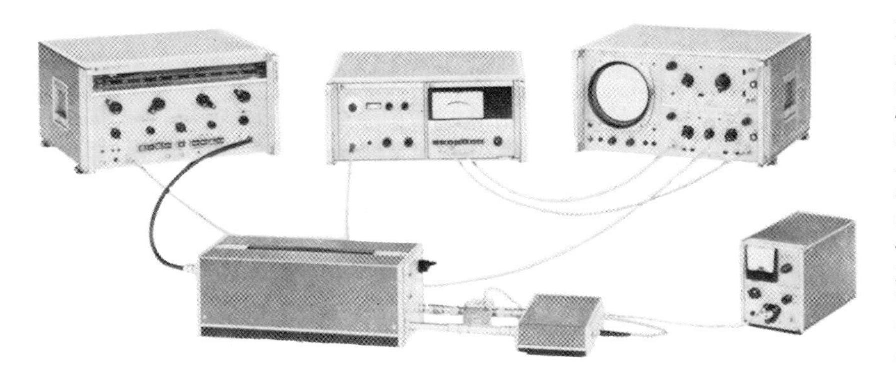

Fig. 2. *Typical test setup using new Network Analyzer (top center) to sweep-measure transmission of microwave filter. Magnitude and phase are measured on Analyzer meter and presented as a function of frequency on oscilloscope. Magnitude and phase can also be presented in polar form on a Polar Display scope which plugs in, in place of Phase Gain Indicator and will feed external recorder.*

Other pieces of auxiliary equipment will, in general, be added to complete a specific measurement. Examples would be bias supplies for active devices and matched loads for termination purposes.

Here then is a very flexible system that defines completely the complex parameters of an active or passive network. It provides this information, much of which was previously very difficult or prohibitively expensive to obtain, over a huge frequency range with an ease and rapidity that consistently intrigues those who see it the first time. Specific features of the network analyzer are the following:

1. One system measures both magnitude and phase of all network parameters from 110 MHz to 12.4 GHz. The measurements can be made at a single frequency or on a swept frequency basis over octave bandwidths.
2. The analyzer combines wide dynamic range with high measurement resolution. Direct dynamic display range is 60 dB in magnitude and 360° of phase. Precise internal attenuators and a calibrated phase offset allow expanded measurements with better than 0.1 dB resolution in magnitude and 0.1° in phase.
3. It is accurate. Precision components are used throughout to assure basic accuracy. The two-channel comparative technique removes error terms caused by the source and variations common to both channels.
4. A choice of display allows the data to be presented in the most useful form for the specific measurement. The measured data are also provided in analog form for external oscilloscope, recorder, or digital display.

Frequency Translation by Sampling

Figs. 1 and 8 show the elements of the analyzer system and Table I lists the elements, their functions, and their frequency ranges. The basic analyzer (Fig. 1) consists of three units: a main frame, either of two plug-in display modules, and a harmonic frequency converter. The transducer instruments for reflection and transmission (Fig. 8) complete the system.

The key technique that allows the new microwave network analyzer to measure complex ratio is the technique of frequency translation by sampling. The block diagram of the basic analyzer shown in Fig. 4 is helpful to understand this technique. Sampling as used in this system is a special case of heterodyning, which translates the input signals to a lower, fixed IF frequency where normal circuitry can be used to measure amplitude and phase relationships. The principle is to exchange the local oscillator of a conventional heterodyne system with a pulse generator which generates a train of very narrow pulses. If each pulse within the train is narrow compared to a period of the applied RF signal, the sampler becomes a harmonic mixer with equal efficiency for each harmonic. Thus sampling-type mixing has the advantage that a single system can operate over an extremely wide input frequency range. In the case of the network analyzer this range is 110 MHz to 12.4 GHz.

In order to make the system capable of swept frequency operation, an internal phase-lock loop keeps one channel of the two-channel network analyzer tuned to the incoming signal. Tuning of the phase-lock loop is entirely automatic. When the loop is unlocked, it automatically tunes back and forth across a portion of whatever octave-wide frequency band has been selected by the user. When any harmonic of the tracking-oscillator frequency falls 20 MHz below the input frequency, i.e., when $f_{in} - nf_{osc} = 20$ MHz, the loop stops searching and locks. Search and lock-on are normally completed in

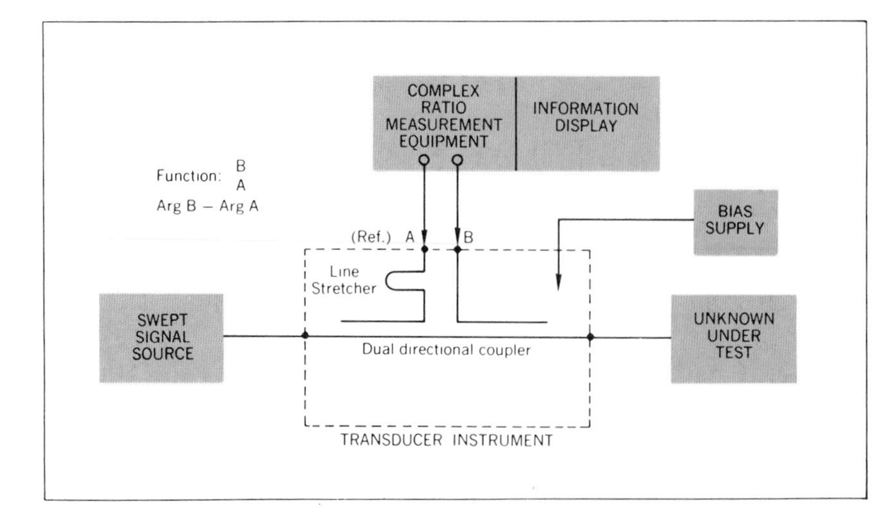

Fig. 3. *Network Analyzer concept follows from network theory, as explained in text.*

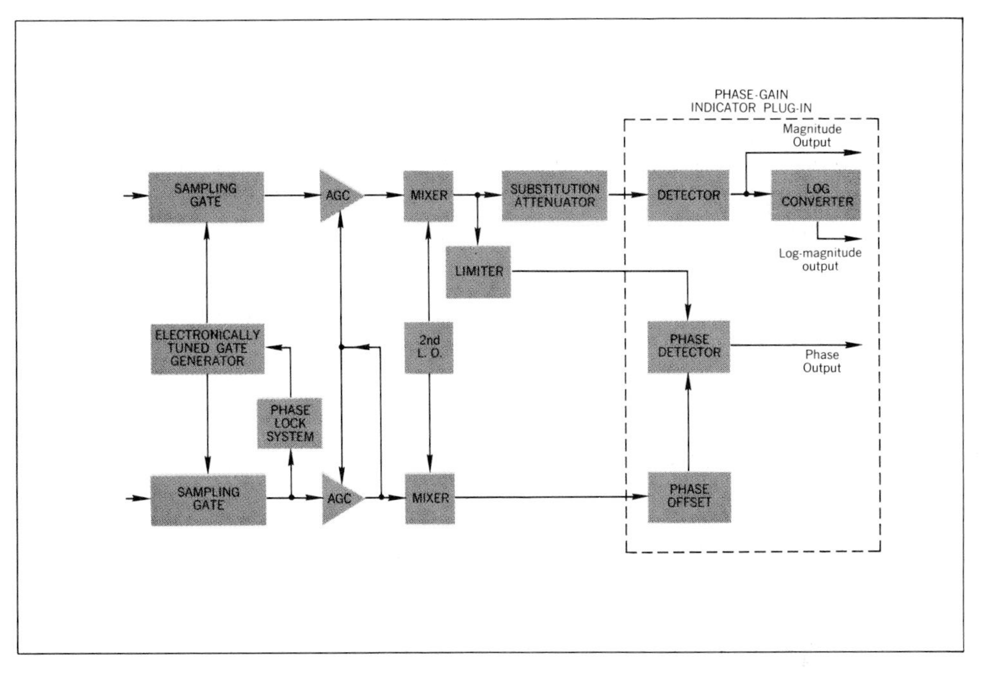

Fig. 4. *Basic system used in Analyzer to achieve frequency translation by a sampling technique.*

about 20 μs. The loop will remain locked for sweep rates as high as 220 GHz/sec (a rate corresponding to about 30 sweeps per second over the highest frequency band, 8 to 12.4 GHz).

The IF signals reconstructed from the sampler outputs are both 20-MHz signals, but since frequency conversion is a linear process, these signals have the same relative amplitudes and phases as the microwave reference and test signals. Thus gain and phase information are preserved, and all signal processing and measurements take place at a constant frequency.

Referring again to Fig. 4, the IF signals are first applied to a pair of matched AGC (automatic gain control) amplifiers. The AGC amplifiers perform two functions: they keep the signal level in the reference channel constant, and they vary the gain in the test channel so that the test signal level does not change when variations common to both channels occur. This action is equivalent to taking a ratio and removes the effects of power variations in the signal source, of frequency response characteristics common to both channels, and of similar common-mode variations.

Before the signals are sent to the display unit, a second frequency conversion from 20 MHz to 278 kHz is performed. To obtain the desired dB and degree quantities, the phase-gain indicator plug-in display unit (Fig. 4) contains a linear phase detector and an analog logarithmic converter which is accurate over a 60 dB range of test signal amplitudes. Ratio (in dB) and relative phase can be read on the meter of the display unit if desired, but the plug-in also provides calibrated dc-coupled voltages proportional to gain (as a linear ratio or in dB) and phase for display on the vertical channels of an oscilloscope or X-Y recorder. If the horizontal input to the oscilloscope or recorder is a voltage proportional to frequency, the complete amplitude and phase response of the test device can be displayed.

Polar Display Unit

The Polar Display Unit (Fig. 5) converts polar quantities of magnitude and phase into a form suitable for display on a CRT. This is accomplished by using two balanced-modulator phase detectors. The phase of the test channel is shifted 90° with respect to the reference channel before being applied to the balanced modulator. The output of one modulator is proportional to A sin θ. This signal is amplified and fed to the vertical plates of

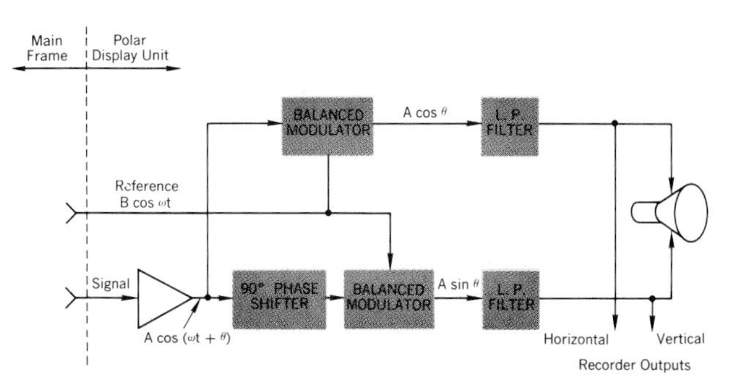

Fig. 5. *Block diagram of basic Polar Display Unit which converts polar magnitude and phase information to be presented on its self-contained CRT.*

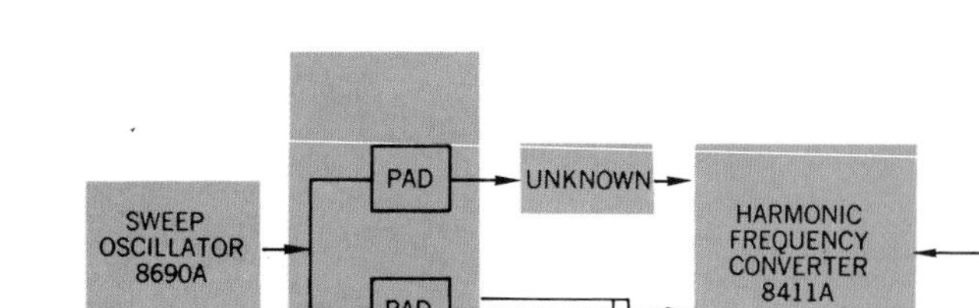

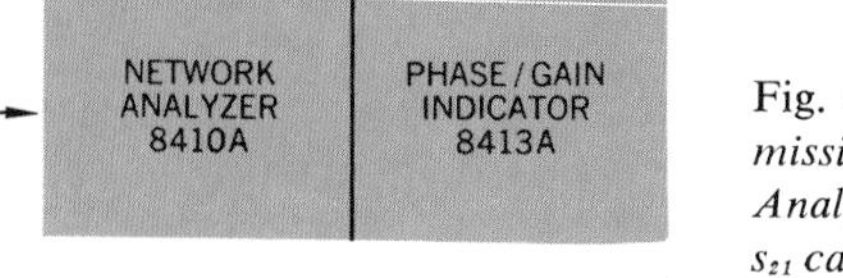

Fig. 6. *Block diagram of transmission test with new Network Analyzer. S-parameters s_{12} and s_{21} can be measured thus.*

the CRT. The output of the other modulator is proportional to A cos θ and this signal is applied to the horizontal plates of the CRT. Thus, the polar vector can be displayed in rectangular coordinates of an oscilloscope or an X-Y recorder.

Transmission Measurements

Fig. 6 illustrates the measurement of the transmission coefficients s_{21} and s_{12} with the network analyzer. As explained on p. 13, these parameters are the forward and reverse transmission gain of the network when the output and input ports, respectively, are terminated in the reference or characteristic impedances. Transmission measurements are used to determine bandwidth, gain, insertion loss, resonances, group delay, phase shift and distortion, etc. For these measurements a swept-frequency source provides an input to the transmission test unit, which consists of a power divider, a line stretcher and two fixed attenuators. The transmission test unit has two outputs, a reference channel and a test channel, which track each other closely in amplitude and phase from dc to 12.4 GHz. The device to be measured is inserted in the test channel, as shown in Fig. 6. Variations in the physical length of test devices can be compensated for by a mechanical extension of the reference channel of the test unit. Thus the magnitude and phase of the transmission coefficient is measured with respect to a length of precision air-line. Of course gain- and phase-difference measurements between similar devices can also be made by inserting a device in each channel. Excess electrical length in the test device can be compensated for by the line stretcher which acts as an extension to the electrical length of the reference channel.

Since the impedance levels in both reference and test channels are 50 ohms, the ratio of the voltage magnitudes applied to the test and reference channels of the harmonic frequency converter is proportional to the insertion gain (or loss), s_{12} or s_{21}, of the device with respect to the reference impedance 50 ohms. The phase between these voltages is likewise the insertion phase shift. When insertion parameters are being measured, the quantities of greatest interest are a logarithmic measure of gain (dB) and transfer phase shift. To obtain these quantities, the network analyzer is used with the phase-gain indicator plug-in.

Reflection Measurements

Complex reflection coefficient, admittance, and impedance measurements are made using the set-up shown in Fig. 7. In this case the signal from the swept-frequency source drives a reflection test unit consisting of a dual directional coupler and a line stretcher. Only two reflection test units are needed to cover the analyzer's entire frequency range — one for frequencies between 0.11 and 2.0 GHz, and one for frequencies from 2.0 to 12.4 GHz.

For reflection measurements, the polar display plug-in with its built-in internal-graticule (parallax-free) CRT is most convenient. A Smith chart overlay for this display converts reflection coefficients directly to impedance or

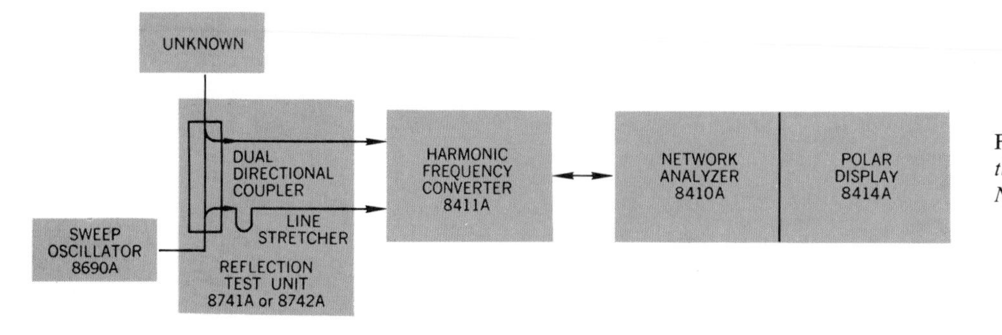

Fig. 7. *Block diagram of reflection (impedance) test with new Network Analyzer.*

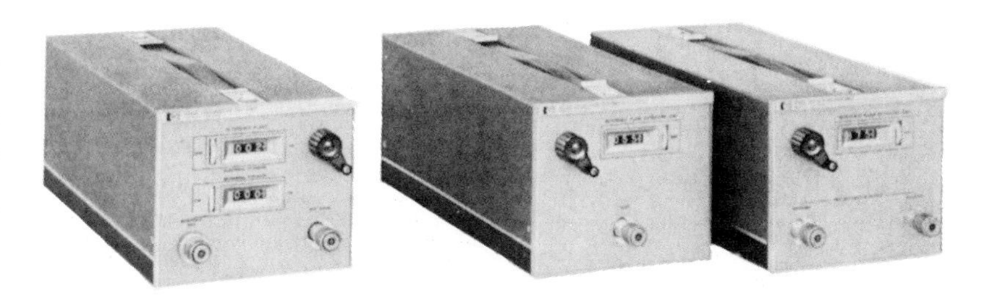

Fig. 8. *Model 8740A Transmission Test Unit or Models 8741A and 8742A Reflection Test Units contain the calibrated line stretchers, attenuators, and directional couplers needed for network analysis.*

admittance. The line stretchers within the test units allow *the plane at which the measurement is made to be extended past the connector to the unknown device.* Thus the Smith Chart display can reveal the impedance or admittance within the test device as frequency is varied without the necessity of graphical manipulations of data plotted on a Smith chart. Seeing the impedance locus of a device over an octave-wide frequency range plotted on this display and watching it change as a tuning adjustment or some other condition is varied is truly an impressive experience for anyone who has ever had to use older methods.

Design considerations

In designing the new analyzer and in achieving some of its performance characteristics, several interesting circuit innovations were devised. Space limitations preclude a detailed treatment, but a summary of some of the salient innovations is given below.

a. A wide-band phase-lock loop was designed to enable the system to sweep rapidly. Maximum sweep rate, which is determined by the loop bandwidth, is about 220 GHz per second.

b. A voltage-controlled oscillator was devised to permit the harmonic frequency converter to tune over more than an octave in frequency (Fig. 9). With the varactors in Fig. 9 connected to the emitters, the voltage swings are small, permitting a low dc bias voltage to be used to get a large value of capacitance. Since the oscillator period is proportional to the varactor capacitance, a large tuning range results.

c. The fast voltage-step needed to obtain fast sampling in the harmonic frequency converter is initiated in a step-recovery diode that operates in a 25-ohm line. To obtain a step of adequate voltage to accommodate the external sampled signal, it is necessary to drive this diode with substantial current. The current is provided by the basic power amplifier shown in Fig. 10. The amplifier follows the local oscillator and consists of emitter followers in a binary tree configuration. Each of the four output transistors supplies nearly 200 mA peak-to-peak over the range from 60 to 150 MHz.

d. In the IF circuits of the signal and reference channels of the main part of the analyzer, AGC action is required but with small relative amplitude and phase change between channels. To achieve this, AGC amplifiers were devised which remove up to 20 dB of power variation while giving less than 1 dB of differential amplitude change and less than about 2° of differential phase change. AGC action is obtained from the current-dependent incremental impedance characteristic of a silicon diode.

e. Amplitude and phase change in the phase/gain in-

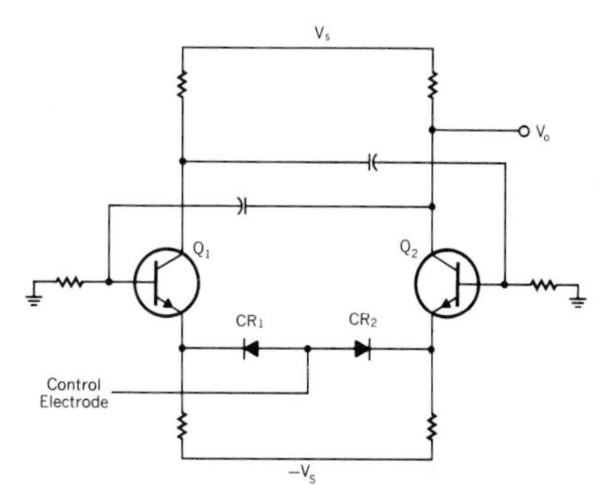

Fig. 9. *Emitter-coupled multivibrator is used for voltage-controlled local oscillator. Tuning range is 60–150 MHz.*

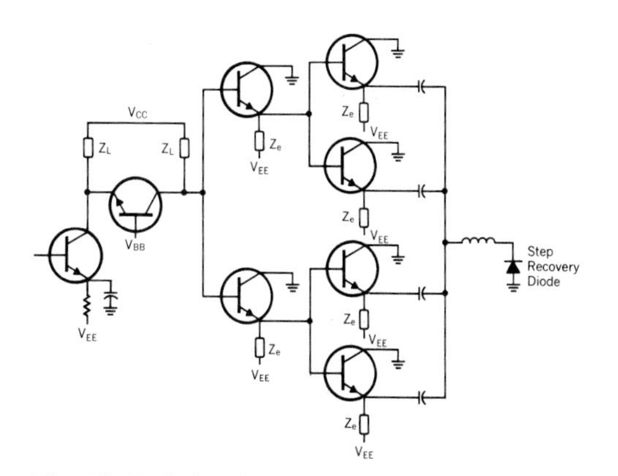

Fig. 10. *Wide-band power amplifier provides at least 0.75 amp p-p over frequency range of 60–150 MHz.*

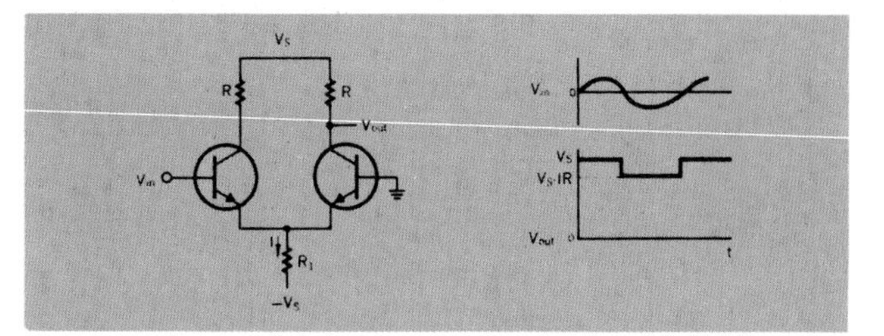

Fig. 11. *Limiting amplifier with two transistors switching total current I. Output voltage is dependent only on V_s and R.*

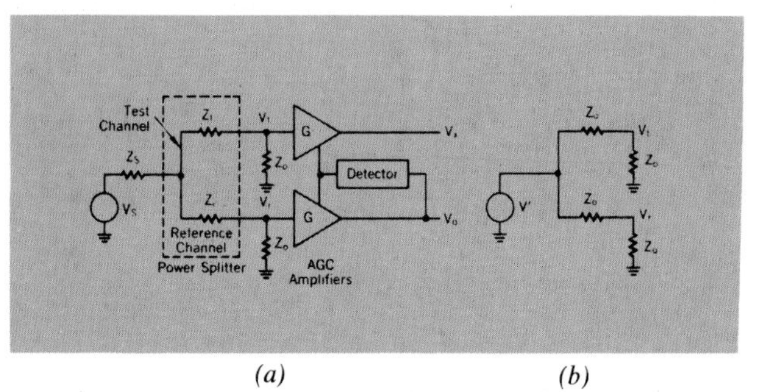

(a) *(b)*

Fig. 12(a). *Equivalent configuration of power divider and AGC amplifiers for calculating ratio.* 12(b). *Simplified equivalent with resultant zero-impedance source V'.*

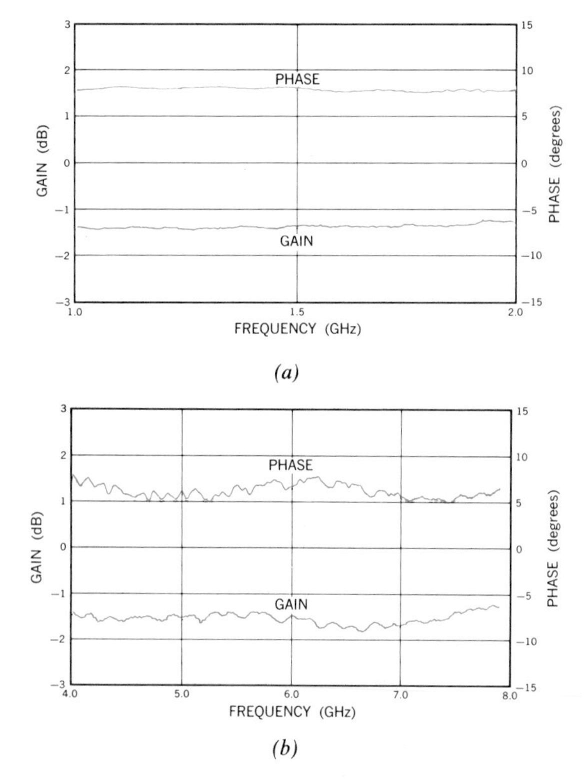

Fig. 13(a). *Phase and gain responses typical of Network Analyzer between 1 GHz and 2 GHz: Analyzer is accurate within ±0.1 dB and 1.0° in swept measurements. Accuracy in single-frequency measurements is better.* (b). *Phase and gain responses typical of Network Analyzer between 4 GHz and 8 GHz.*

dicator unit were reduced by using a series of limiters of the type shown in Fig. 11. To prevent added delay when the amplifier starts to limit, the transistors are cut off but not allowed to saturate. A single limiter exhibits less than 1° of phase shift when passing from linear operation to limiting. Output voltage is dependent only on V_s and R.

f. A major engineering contribution occurred in the form of two wide-band directional couplers used in the reflection test units. The couplers have 30 to 40 dB of directivity over their frequency ranges of 0.1 to 2 GHz and 2 to 12.4 GHz. This represents a combination of performance characteristics heretofore unattainable.

g. Normally, a power divider operates with its three ports matched. In the transmission test unit a precision power divider was devised which operates with the source port matched but with the output ports mismatched. The ratio calculation performed by the AGC amplifiers (Fig. 12a) has the effect of making V′ a low-impedance source, so that the two channels do not interact with each other. If Z_t and Z_r in Fig. 12(b) are made equal to Z_o, standing waves are not present.

Performance

Typical measurement accuracies for the 1-to-2-GHz frequency range are shown in Fig. 13(a) which is a plot of the network analyzer's amplitude and phase responses over this range. Gain and phase measurements accurate within ±0.1 dB and ±1° appear reasonable for swept measurements. For single-frequency measurements, the accuracy is much better — comparable to that of standards-laboratory instruments.

Fig. 13(b) shows the amplitude and phase responses of the analyzer from 4 GHz to 8 GHz. The slightly-reduced calibration accuracy apparent in Fig. 13(b) can be attributed principally to the increased reflection coefficient of the harmonic frequency converter (wideband sampler) at higher frequencies.

Phase errors caused by changes in the amplitude of the signal in the test channel are shown in Fig. 14. Greatest accuracy in phase measurements is obtained for signal levels within ±20 dB of mid-range. In this range, phase ambiguities are less than ±1°.

Fig. 15 shows the gain and phase stability of the network analyzer. Over a period of six hours, total drift did not exceed 0.05 dB and 0.2° under normal room-temperature variations.

Gain and phase accuracies at low signal levels are limited by the signal-to-noise ratio at the output of the harmonic frequency converter. Noise in the test channel is below −80 dBm, which means that accurate measurements can be made for test-channel amplitudes down to −70 dBm or less.

More typical measured data are presented in the s-parameter article (p. 13).

Acknowledgments

It is a pleasure to acknowledge the contributions of the following members of the Microwave Division.

Network Analyzer Main Frame:
Kenneth S. Conroy, George M. Courreges, Wayne A. Fleming, Robert W. Pace.

Harmonic Frequency Converter:
William J. Benham, Richard T. Lee.

Phase/Gain Indicator Unit:
Donald G. Ferney, David R. Gildea, Alan L. Seely.

Polar Display Unit:
Larry L. Ritchie, William A. Rytand.

Transmission and Reflection Test Units:
Jean Pierre Castric, Wilmot B. Hunter, George R. Kirkpatrick, Richard A. Lyons, Auber G. Ryals.

Industrial Design:
Ned R. Kuypers.

We also wish to thank Microwave Division engineering manager Paul C. Ely, Jr. for his encouragement and helpful suggestions.

—Richard W. Anderson and Orthell T. Dennison

REFERENCES

1. J. K. Hunton, and N. L. Pappas, 'The -hp- Microwave Reflectometers,' **'Hewlett-Packard Journal,'** Vol. 6, No. 1-2, Sept.-Oct. 1954.
2. P. D. Lacy, and D. E. Wheeler, 'A New 8-12 kMc Voltage-Tuned Sweep Oscillator for Faster Microwave Evaluations,' **'Hewlett-Packard Journal,'** Vol. 8, No. 6, Feb., 1957.
3. J. K. Hunton, and E. Lorence, 'Improved Sweep Frequency Techniques for Broad Band Microwave Testing,' **'Hewlett-Packard Journal,'** Vol. 12, No. 4, Dec., 1960.
4. S. B. Cohn, 'Microwave Automation,' **'the microwave journal,'** March, 1962, p. 13.
5. J. A. Young, 'Formula for Success,' **'the microwave journal,'** Feb., 1966, p. 58.

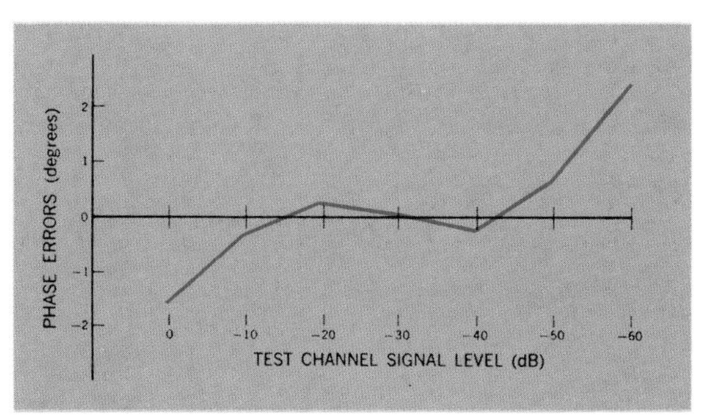

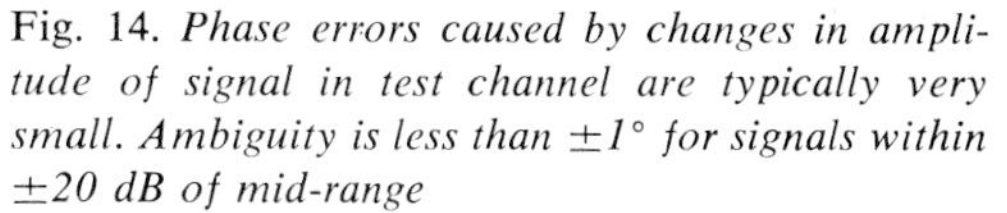
Fig. 14. *Phase errors caused by changes in amplitude of signal in test channel are typically very small. Ambiguity is less than ±1° for signals within ±20 dB of mid-range*

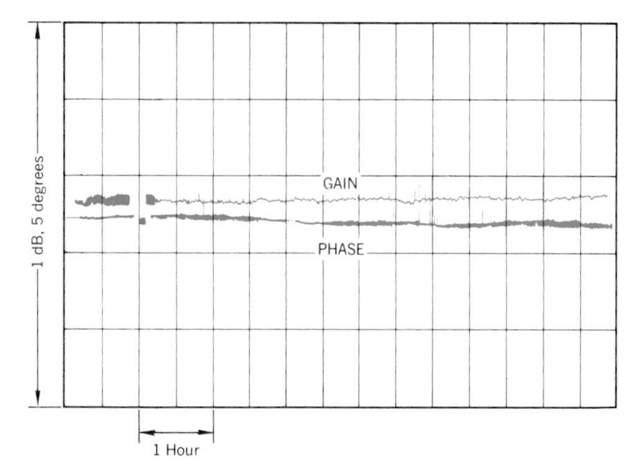

Fig. 15. *Typical gain and phase stability of Network Analyzer. Total drift under normal room-temperature variations over six-hour period was <0.05 dB and <0.2°.*

Richard W. Anderson

Dick Anderson joined the *-hp-* Microwave Division in 1959 after receiving his BSEE degree from Utah State University. He has contributed to the development of a variety of microwave instruments and devices, and he is now manager of the network analyzers section of the *-hp-* Microwave Laboratory. In 1963 he received his MS degree in electrical engineering from Stanford University on the *-hp-* Honors Cooperative Program.

Dick is active in the IEEE Group on Microwave Theory and Techniques. He holds several patents and has published a number of technical papers, his most recent being 'Sampler Based Instruments for Complex Signal and Network Analysis,' presented at WESCON 1966.

Orthell T. Dennison

Ted Dennison joined the *-hp-* Microwave Laboratory in 1960. He contributed to the design of the 415C and 415D SWR Meters, and directed the development of the 416B Ratio Meter and the later stages of the development of the 690-series Sweep Oscillators. Since 1963 he has been project leader of the 8410A Network Analyzer program.

Ted received his BSEE degree in 1960 from Utah State University and his MSEE degree in 1964 from the University of Santa Clara. At Santa Clara, he specialized in control systems and solid-state design. Ted is a member of IEEE, Sigma Tau, and Phi Kappa Phi.

Condensed SPECIFICATIONS

-hp-

8410A Network Analyzer (Operating with 8411A)

INSTRUMENT TYPE: Measures relative amplitude and phase of two RF input signals; choice of two plug-in display modules for meter readout (8413A), or CRT polar display (8414A).

FREQUENCY RANGE: 0.11 to 12.4 GHz.

TUNING: Automatic over octave band selected by front panel switch.

SWEPT OPERATION: Sweeps in octave bands; apply sweep reference voltage for fast sweep operation (compatible with sweep reference out of Model 8690A Sweep Oscillators).

INPUT IMPEDANCE: 50Ω, SWR < 1.4 to 8 GHz, < 2.0 to 12.4 GHz; connectors precision 7mm coax.

CHANNEL ISOLATION: >70 dB, 0.11 to 6.0 GHz; >60 dB, to 12.4 GHz.

AMPLITUDE

INPUT POWER RANGE:

REFERENCE CHANNEL: −20 to −40 dBm (±4 dBm).

TEST CHANNEL: −10 (±2) dBm maximum. Not to exceed reference channel power by more than 10 dB.

DYNAMIC RANGE:

REFERENCE CHANNEL: 20 dB or more.

TEST CHANNEL: At least 60 dB.

TEST CHANNEL NOISE: Less than −78 dBm equivalent input noise (measured on 8413A Meter).

AMPLITUDE CONTROL: Adjusts gain of test channel relative to reference channel.

RANGE: 69 dB total in 10 and 1 dB steps; vernier provides continuous adjustment over at least 2 dB.

ACCURACY: ±0.1 dB per 10 dB step, not to exceed ±0.2 dB cumulative. ±0.05 dB per 1 dB step, not to exceed ±0.1 dB cumulative.

GENERAL

PHASE CONTROL: Vernier provides continuous phase reference adjustment over at least 90°.

OUTPUTS: Two rear panel auxiliary outputs provide 278 kHz IF signals; outputs may be used for signal analysis, special applications, and convenient test points; modulation bandwidth nominally 10 kHz.

-hp-

8413A PHASE-GAIN INDICATOR (Installed in 8410A)

INSTRUMENT TYPE: Plug-in Meter display unit for 8410A. Displays relative amplitude in dB between reference and test channel inputs or relative phase in degrees. Pushbutton selection of meter function and range.

AMPLITUDE

RANGE: ±30, 10 and 3 dB full scale.

ACCURACY: ±3% of end scale.

LOG OUTPUT: 50 millivolts per dB up to 60 dB total; bandwidth 10 kHz nominal depending on signal level; source impedance 1 kΩ; accuracy, ±3% of reading.

LINEAR OUTPUT (rear panel): 0 to 1 V maximum; 10 kHz bandwidth; 200Ω source impedance.

MAXIMUM DRIFT:

LOG: <±0.1 dB per degree C.

LINEAR: <±5 mV per degree C.

PHASE

RANGE: ±180, 60, 18, 6 degrees full scale.

ACCURACY: ±2% of end scale.

OUTPUT: 10 millivolts per degree; 10 kHz bandwidth; 1 kΩ source impedance. Accuracy ±2% of reading.

MAXIMUM DRIFT: ±0.2 degree per degree C.

PHASE OFFSET: ±180 degrees in 10 degree steps.

ACCURACY: ±0.3 degree per 10 degree step, not to exceed ±1.5 degrees cumulative.

PHASE RESPONSE VERSUS SIGNAL AMPLITUDE: 4 degrees maximum phase change for 60 dB amplitude change in test channel.

-hp-

8414A POLAR DISPLAY (Installed in 8410A)

INSTRUMENT TYPE: Plug-in CRT display unit for 8410A. Displays amplitude and phase data in polar coordinates on 5" cathode ray tube.

RANGE: Normalized polar coordinate display; magnitude calibration 20% of full scale per division. Scale factor is a function of GAIN setting on 8410A. Maximum scale factor is 3.16 decreasing to at least 0.0316; phase calibrated in 10° increments over 360° range.

ACCURACY: Error circle on CRT <3 mm radius.

OUTPUTS: Two dc outputs provide horizontal and vertical components of polar quantity. Maximum output ±10 volts, <100Ω source impedance, bandwidth (3 dB) 10 kHz.

DRIFT: Beam center drift, <0.2 mm/°C. Measurement drift, amplitude less than 2% of reading/°C, phase less than 0.2°/°C. Auxiliary outputs, <±10 mV/degree C.

BEAM CENTER: Pressing BEAM CENTER simulates zero-signal input to test channel. Allows convenient beam position adjustment for reference.

GENERAL

CRT: 5 inch, 5 kV post accelerator tube with P-2 phosphor; internal polar graticule.

MARKER INPUT (rear panel): Accepts frequency marker output pulse from *-hp-* 8690-Series and 690-Series Sweep Oscillators, −5 volts peak. Markers displayed as intensified dot on CRT display.

BLANKING INPUT (rear panel): Accepts −5 volt blanking pulse from *-hp-* 8690-Series and 690-Series Sweep Oscillators to blank retrace during swept operation.

BACKGROUND ILLUMINATION: Controls intensity of CRT background illumination for photography. Eliminates need for ultraviolet light source in oscilloscope camera when photographing internal graticule.

-hp-

8740A TRANSMISSION TEST UNIT

INSTRUMENT TYPE: RF power splitter and calibrated line stretcher for convenient transmission tests with 8410A. Provides reference and test channel RF outputs for connection to unknown device and the 8411A Converter.

FREQUENCY RANGE: dc to 12.4 GHz.

FREQUENCY RESPONSE: Reference and test channel outputs track a mean value within ±0.5 dB amplitude and ±3 degrees phase to 8.0 GHz; ±0.75 dB and ±5 degrees to 12.4 GHz (includes frequency response of 8411A Converter).

OUTPUT IMPEDANCE: 50 ohms, reflection coefficient 0.07 (1.15 SWR, 23 dB return loss) dc to 8 GHz; 0.11 (1.25 SWR, 19 dB return loss) 8.0 to 12.4 GHz.

REFERENCE PLANE EXTENSION:

Electrical; 0 to 30 centimeters.

Mechanical; 0 to 10 centimeters.

Both extensions calibrated by digital indicators.

-hp-

8741A AND 8742A REFLECTION TEST UNITS

INSTRUMENT TYPE: Wideband reflectometer, phase-balanced for swept or spot frequency impedance tests with 8410A. Calibrated variable reference plane.

FREQUENCY RANGE: 0.11 to 2.0 GHz (8741A); 2.0 to 12.4 GHz (8742A).

FREQUENCY RESPONSE: Incident and reflected outputs from reflectometer track a mean value within ±0.5 dB amplitude, ±3 degrees phase (8741A); ±0.75 dB, ±5 degrees over any octave (8742A) (includes frequency response of 8411A Converter).

IMPEDANCE: 50 ohms.

RESIDUAL REFLECTION COEFFICIENT: <0.01, 0.11 to 1.0 GHz; <0.02, 1.0 to 2.0 GHz (8741A); <0.03, 2.0 to 12.4 GHz (8742A).

REFERENCE PLANE EXTENSION: 0 to 15 cm (8741A); 0 to 17 cm (8742A); calibrated by digital dial indicator.

PRICES: 8410A $1700; 8411A $2200; 8413A $825; 8414A $825; 8740A $1100; 8741A and 8742A less than $1500 each.

MANUFACTURING DIVISION:

-hp- Microwave Division
1501 Page Mill Road
Palo Alto, California 94304

Prices f.o.b. factory
Data subject to change without notice

The Engineer, Automated Network Analysis and the Computer – Signs of Things to Come

THREE NEW DEVELOPMENTS hold exciting possibilities for microwave measurement and design. The new developments referred to are the Network Analyzer (p. 2), the new design procedures using scattering parameters (p. 13), and an unusual new instrumentation computer recently developed by the –hp– Dymec division. The Analyzer and the s-parameter design procedures are new tools that can logically and analytically solve problems that until now had to be attacked empirically. The instrumentation computer can magnify the power and potential of these tools manyfold. Combined as tomorrow's instrument system, they offer the microwave engineer capabilities he has never had before.

To show the implications for other future instrumentation systems, we will trace the history of these three developments in our laboratory and look at the prototype system evolving therefrom.

The Network Analyzer project started several years ago with some basic ideas fortified by intuition but with little information about the increasing importance of its applications. During the course of the Analyzer project, the development of a related instrument, the RF Vector Voltmeter,[1] was completed, and several of our engineers found that with it their efforts to advance high-frequency solid-state circuit performance were helped by using the scattering-parameter characterization of the active devices. The scattering-parameter design procedures[2] evolved from these first measurements.

After a time, when the first prototypes of the new Network Analyzer were completed, they were pressed into service measuring active devices in chip form and thin-film components for our hybrid microcircuits. These measurements were coupled with the newly-evolved design procedures and resulted in several important advances in high-frequency integrated circuits.

As the s-parameter design techniques were applied to increasingly difficult problems, it became apparent that the design calculations and transformations could be carried out to advantage by computer. Programs were developed on the B-5500 computer at the nearby Stanford Computational Center used by our engineers. The results were quite rewarding. Later, when a computer time-share terminal was installed in the lab, several shorter programs were written and we found that the direct link between the computer and the engineer offered many advantages.

Computerized HF Transistor Measurement

The next step was obvious: as we began to think about using the newly developed instrumentation computer, we became more and more enthusiastic about the possibilities of combining the three developments. These concepts have grown until we are now evaluating the prototype system shown in the diagram. The prototype system consists of a 0.1 to 2 Gc programmable source, the 8410 Network Analyzer system, a prototype transducer instrument for transistor parameter measurement, programmable transistor bias supplies and the 2116 Instrumentation Computer. The computer controls all the conditions of measurements including test frequency, transistor bias, and the parameter to be measured. The transistor itself is inserted in a removable fixture which can be adapted to any transistor package configuration. The 8410 Network Analyzer reads into the computer through a digital voltmeter.

A not insignificant advantage of the use of a computer in this system is the capability to correct measurement errors. The system is first calibrated with unity and zero s-parameter standards. The computer calculates and stores the error vector for each frequency and as the data is taken it subtracts out these values. The accuracy and dynamic range of measurement are greatly improved by this procedure.

In operation the engineer can select one or a number of frequencies and bias conditions and can ask the computer to do a variety of operations such as:

1. Measure and correct the s-parameters.
2. Convert to h, y, or z-parameters.

[1] Fritz K. Weinert, 'The RF Vector Voltmeter — an Important New Instrument for Amplitude and Phase Measurements from 1 MHz to 1000 MHz,' **'Hewlett-Packard Journal,'** Vol. 17, No. 9, May, 1966.

[2] George E. Bodway, 'Two-Port Power Flow Analysis of Linear Active Circuits Using the Generalized Scattering Parameters,' Hewlett-Packard Company Internal Report.

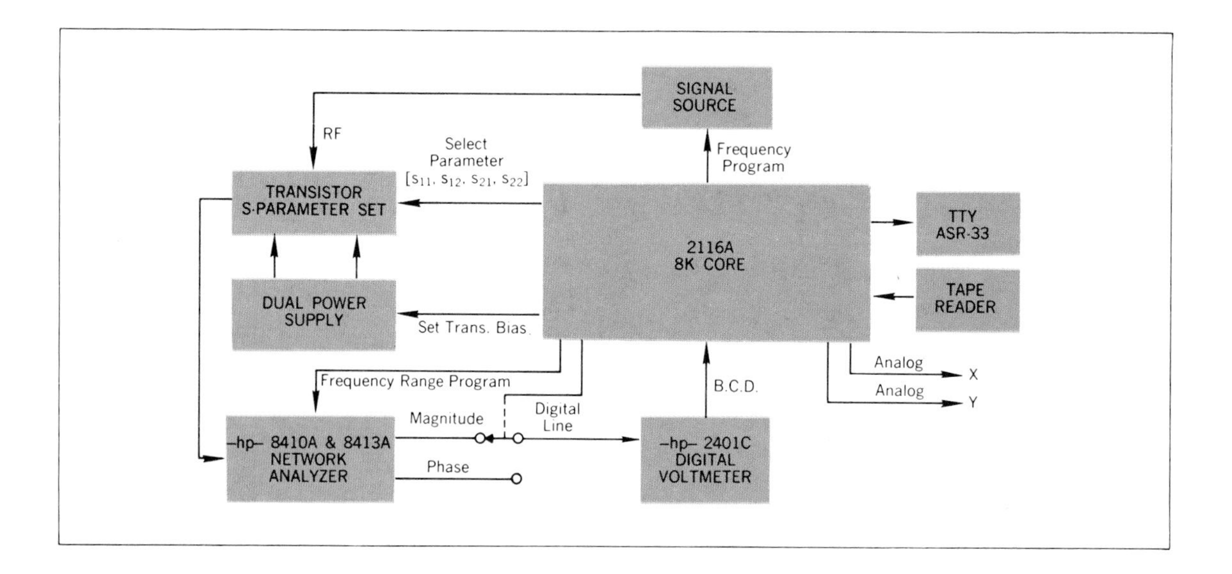

3. Calculate maximum available power gain $G_{A\ max}$.
4. Calculate the terminal conditions to achieve $G_{A\ max}$.
5. Find gain-bandwidth product f_t.
6. Find maximum frequency of oscillation f_{max}.
7. Find the biases for maximum f_t and for f_{max}.

The engineer communicates with the system through a typewriter and the computer control panel.

The programs are being improved to incorporate additional design analysis calculations and eventually some network synthesis operations. Several other organizations are working on computer-aided circuit design for high frequency applications. Bell Telephone Laboratories and Texas Instruments, for example, have particularly significant efforts in these areas. We can look for many advances in the design and synthesis techniques as this work proceeds. However, the concept as it has been developing at HP is somewhat unusual in that the experimental measurements, and the design computations are combined in one on-line system controlled directly by the engineer. The man-machine communication is in the normal language the engineer uses to solve these problems.

Future Measurement Systems

The possibilities for this instrument system to add significantly to the power of our engineers are impressive. But beyond the value of this particular system are the implications for the whole field of high frequency instrumentation. It is easy to imagine a number of measurement systems using the on-line computer to provide new and valuable capabilities. The computer and the rest of the system combine to form a new instrument of greatly magnified capabilities, particularly if the man-machine interface is also designed for the specific function of this new instrument. We can look for several levels of microwave instrumentation systems: some will just control the measurement and perhaps correct the data; others will go on to analyze the data; the most sophisticated will carry on from the analysis into synthesis operations.

These beguiling concepts offer the opportunities for many new contributions in instrumentation developments at Hewlett-Packard and our engineers are approaching these opportunities with enthusiasm.

— Paul C. Ely, Jr.

Paul C. Ely, Jr.

Paul Ely has been active in the microwave and radar instrumentation field for fifteen years. At -hp-, Paul has been associated with the development of the -hp- 690 series sweep oscillators as well as with microwave spectroscopy work, and was section manager of the group that did the early development work on the 8410 Network Analyzer discussed in this issue. At present he is engineering manager for the -hp- Microwave Division laboratory.

Paul holds a BS in Engineering Physics from Lehigh University and received a MS in EE from Stanford University under the -hp- Honors Co-op Plan. He is active in IEEE work, having served on national and sectional committees. Presently he is secretary-treasurer of the IEEE PG on Instrumentation and Measurement and is a director of the San Francisco Bay Area Science Fair.

A Computer for Instrumentation Systems

Problems of interconnection, programming, and environment arise in the design of systems containing both computers and instruments. They are solved in advance by this new integrated-circuit computer.

DIGITAL COMPUTERS AND LABORATORY INSTRUMENTS ordinarily make strange bedfellows. They not only have difficulty making contact with each other and working together, but most computers don't even feel comfortable in the unsympathetic everyday world in which instruments live.

In recent years, however, a growing need has arisen for a computer that can work efficiently in instrumentation systems. These systems are becoming more numerous and more complex. Many of them would not be feasible without the data processing and flexible control that a computer can provide.*

Because the need for an instrumentation computer was not being filled by others, *-hp-* decided to go ahead and build one, feeling that a computer to work with instruments could probably be built best by an instrumentation manufacturer. Our new computer is an integrated-circuit machine which has the computing power and special capabilities needed to work well in instrumen-

* See p. 11 for descriptions of some of these systems.

tation systems. It does away with three discouraging problems that have plagued systems designers in the past:

—the interface problem, or how to connect the instruments with the computer so that efficient data communication can take place

—the software problem, or how to program the computer efficiently, using programming systems that were not designed with instruments in mind

—the environmental problem, or how to keep the computer working in the unfriendly environments in which instruments operate.

A closer look at these problems and at how the instrumentation computer eliminates them will be the best introduction to the new machine and what it can do for the systems designer.

Convenient Computer/Instrument Interfaces . . .

Laboratory instruments are unfamiliar input/output devices for most computers. Seldom is there any ready-made, convenient means for connecting an instrument to a computer; it usually takes an expensive, custom-designed interface. This is true even of some common computer input/output devices. For example, if a computer is not specifically designed to work with a magnetic tape unit, and many are not, it may take several thousand dollars worth of engineering talent and equipment, plus several months, to produce an interface that will allow the computer and a tape unit to work together. If a number of instruments have to be interfaced, or if the system configuration has to be changed, the user's problems are multiplied many times, of course.

. . . Provided by Flexible Input/Output System

The new instrumentation computer is different; it has an extremely flexible input/output system which greatly simplifies interfacing to a large number of instruments. Within the main frame of the basic computer is a box which accepts up to 16 plug-in interface cards (Fig. 1). Most instruments and input/output devices can be interfaced to the computer with one card each; a few require two cards. Accessory modules raise the maximum number of interface cards to 48.

Eventually, there will be plug-in interface cards for all *-hp-* instruments which either provide a digital output or can be programmed.* At present, cards have been designed for more than 20 instruments, including counters, nuclear scalers, electronic thermometers, digital voltmeters, ac/ohms converters, data amplifiers, and input scanners. Interface cards have also been designed for most kinds of input/output devices, such as magnetic tape recorders, teletypewriters, paper tape readers and punches, and dataphones. Interfaces for printers and card readers and punches are under development.

Efficient Programming . . .

Programming systems, or 'software' in computer jargon, which have not been designed to work with instruments can be made to do so only reluctantly and inefficiently. The inability of previously available software to cope with instruments is never more obvious than when the computer and its instruments have to exchange data or control signals.

* Instruments which provide analog outputs can be interfaced with the computer via a digital voltmeter or an analog-to-digital converter. Instruments which require analog programming can be interfaced via a digital-to-analog converter.

Fig. 1. *Operator of new -hp- 2116A Computer changes one of the plug-in interface cards which make it easy to connect the computer with laboratory instruments and other input/output devices. This and other features of the new machine greatly simplify the task of augmenting an instrumentation system with a general-purpose computer.*

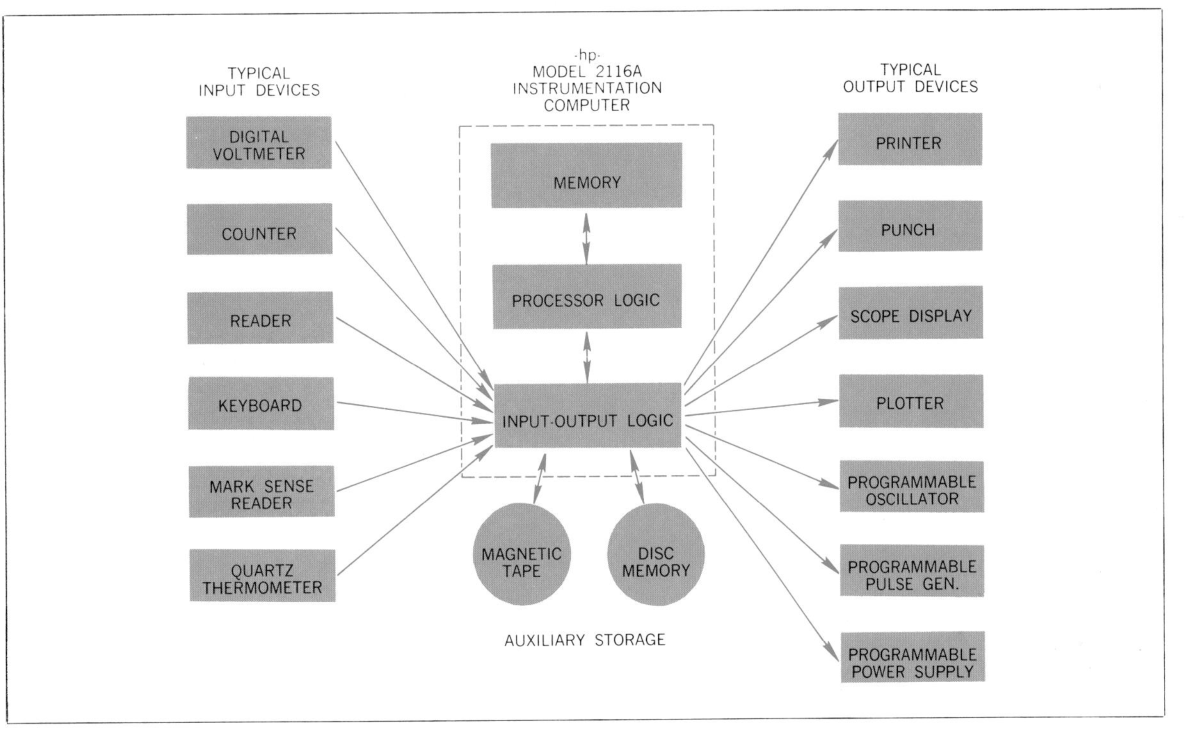

Fig. 2. *Some of the input/output devices that can be connected to –hp– 2116A Computer by means of plug-in cards. More are listed in text.*

Typically, computers in instrumentation systems process data in real time — as soon as it is acquired — rather than at some later time. The computer is 'on line' continually, processing data according to the program prepared by the user. When an instrument is ready to transmit data to the computer, the computer must interrupt its regular program, determine which instrument is requesting service, and load the new data into the proper place in the memory. Although this seems simple enough, it is not. Unless the computer has been designed to work with instruments, it takes a surprisingly long time for the computer to scan all of the instruments and find the one which has the data to transmit. This time can usually be put to better use. What's more, someone — a programmer — has to write the programs which tell the computer how to talk to each instrument. Data formats and codes often differ from instrument to instrument, so programs for different instruments may have to be quite different. In fact, the software problem may outweigh the interface problem when a system is being assembled or changed.

... Provided by Multi-Channel Interrupt System ...

To allow all of its instruments to be serviced rapidly and in real time, the instrumentation computer has a multi-channel priority interrupt system. 'Multi-channel' is the key word here. The priority of each instrument is determined by the slot occupied by its interface card, and each slot, or channel, is assigned a different location in the computer's memory. When one or more instruments signal the computer that they are ready to transmit data, the computer does not have to scan the instruments in order of their priorities to find out which one has produced the interrupt. While other machines may require a millisecond or more just to determine which instrument is requesting service, the instrumentation computer locates the highest-priority interrupting instrument and its service subroutine in a few microseconds, transfers the data, and goes on to the next-lower-priority instrument. When all instruments requesting service have been serviced, the computer returns to its regular program, having spent a minimum of time away from it. Alternatively, interrupt signals from one or more instruments can be inhibited by the computer, or the computer can signal an instrument to make a measurement.

For those special occasions when even a few microseconds can't be spared, the new computer will soon have direct memory data channels. One of these direct channels can be assigned under program control to any input/output channel, permitting data to be transferred directly to memory without going through the normal channels (i.e., through one of the two accumulator registers). The maximum data transfer rate for a direct memory channel is 600,000 16-bit words per second, whereas the maximum rate for normal channels is about 60,000 words per second.

. . . and Compatible Software

Software for the instrumentation computer has been designed to make full use of the flexibility of the input/output (I/O) hardware. A modular control system allows programs to be written without concern for the specific operating requirements of individual I/O devices, and a 'software configurator' is furnished which allows the user to modify his control system easily to fit different I/O hardware configurations. Systems can be upgraded (say by switching from a low-speed to a high-speed tape punch) without changing the program. In other words, programming of the instrumentation computer is very nearly independent of the I/O devices used.

Programs for the new computer can be written in either of two programming languages, FORTRAN or assembly language. The computer comes equipped with a FORTRAN compiler which operates in the basic memory of 4096 16-bit words. The compiler is a program which converts any program written in ASA Basic FORTRAN — a universally accepted programming language — to the binary machine language that the computer understands. Actually, the new computer's FORTRAN is an augmented version of ASA Basic FORTRAN, allowing more flexibility in programming.

Assembly language, the other programming system, is a symbolic language which is closely related to the computer's hardware. The assembler, a program which converts assembly language programs to binary machine language, also operates in the basic 4096-word core memory.

The example below illustrates the two programming languages.

An interesting fact not shown in the example is that the instrumentation computer's FORTRAN compiler can produce an assembly-language listing of a FORTRAN program at the same time that it translates the program to machine language. This unusual capability makes it easy for a programmer to write part of a program in FORTRAN and part in assembly language, and then combine the two parts. It is often most efficient to write part of a program for the specially-designed instrumentation computer in assembly language, which is more closely related to the design of the machine than is FORTRAN, a machine-independent language.

The new computer's modular basic control system includes the following software modules:

—a relocating loader, which loads, combines, and initiates the execution of programs or parts of programs prepared by the FORTRAN compiler and the assembler.

—an I/O control, a general I/O device control program

—I/O drivers, which control specific I/O devices.

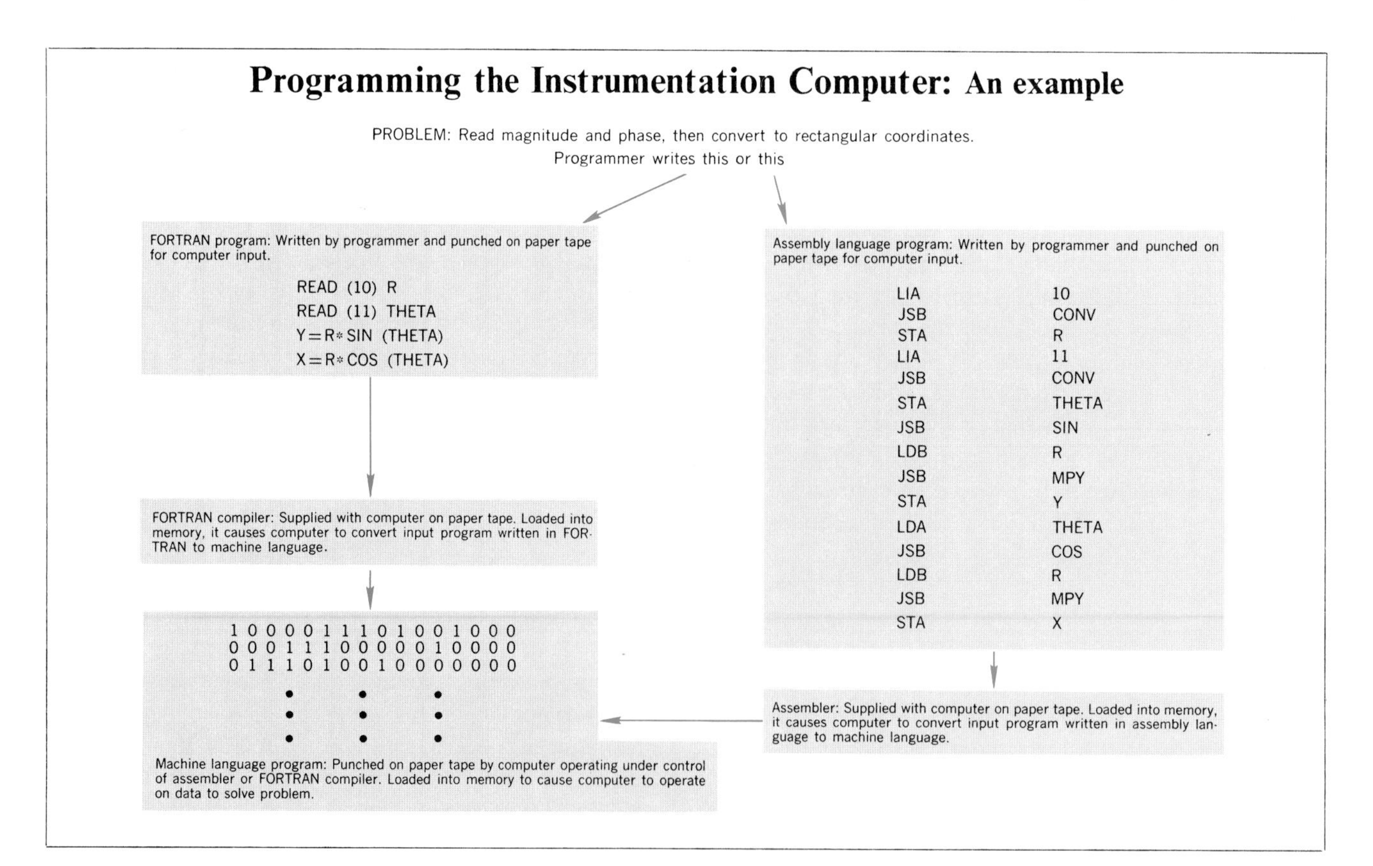

Two other software packages provided with the basic control system are:

—a 'prepare control system' program, used to combine or modify the elements of the basic control system.

—debugging routines.

Other software elements which have been designed for the instrumentation computer include hardware diagnostic programs for troubleshooting, a symbolic editor program which makes it easy to edit or change any program, and a library of subroutines for various mathematical operations. All of this software is provided on punched paper tape. All of it works in the computer's basic memory of 4096 16-bit words, although the memory can be expanded to 16,384 words if desired and the software will take full advantage of the additional memory.

Environment Not a Problem

Unfriendly environments in which instruments must commonly work are no problem for the instrumentation computer. Unlike most computers, it will not balk at temperatures ranging from 0°C to 55°C, line voltage varying ±10%, line frequency varying between 50 and 70 Hz, and humidity up to 95%. The computer needs no air conditioning in order to operate in such environments. It can also function normally under conditions of electromagnetic interference and vibration that would seriously hamper most computers.

What It Can Do

Some general categories of problems which can be solved easily by the new computer and an appropriate combination of standard *-hp-* instruments are:

1. **Data reduction problems,** in which a large number of data points are taken, but only a few answers are required. A computer can perform mathematical operations, such as integration and convolution, thereby greatly reducing the amount of data requiring human analysis later on. Time savings are possible too, because the computer can be on line, processing data while measurements are being made.

2. **Data transformation and report preparation.** Frequently the data produced by an instrument is not in the best form for the user. The computer can perform operations such as curve-fitting and linearization, and then present the data graphically, or in other required forms.

3. **Problems which require fast results** for feedback during a test. When feedback is available, the amount of data taken can often be reduced substantially by not taking redundant data. The computer can decide on the basis of previous data what new measurements should be made, and then direct the proper instruments to make the measurements.

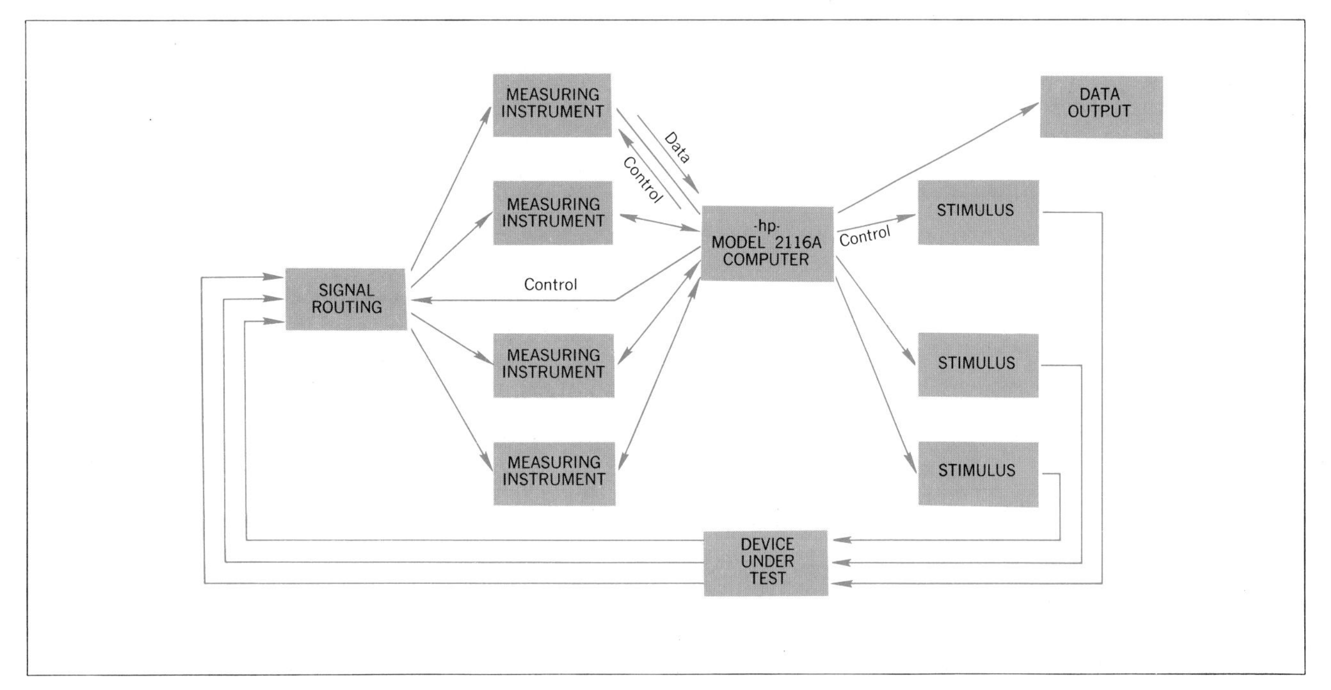

Fig. 3. *Typical application for -hp- 2116A Computer is controlling detailed tests of transistors, printed circuit cards, or other devices, then presenting processed data in any required form.*

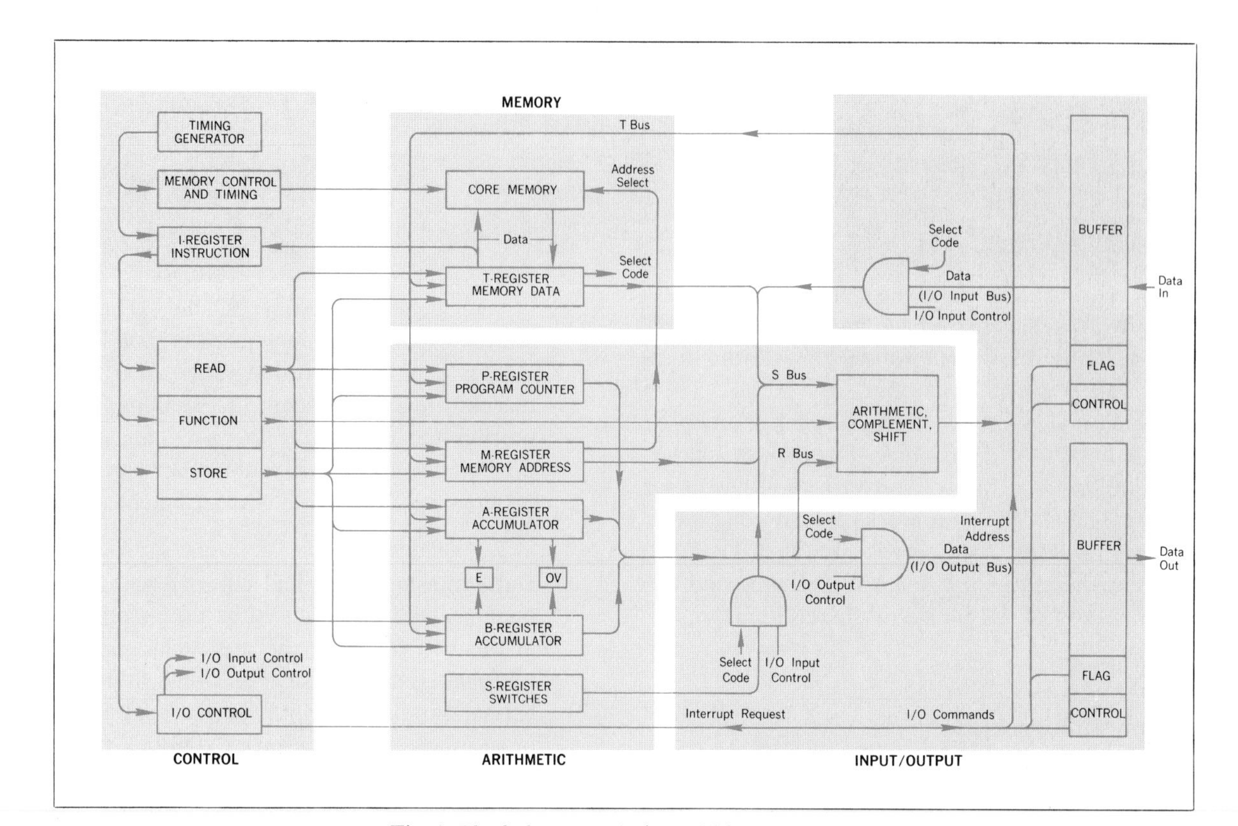

Fig. 4. *Block diagram of –hp– 2116A Computer.*

4. **System and process control.** The computer can receive data from instruments which are monitoring a system or process, operate on this data, and make adjustments to the process in order to optimize its performance. Computer control is automatic and continuous, and the computer program can be changed easily when necessary.

5. **Automatic testing.** The computer can cause a device under test to be stimulated, monitor its response, and reduce the results to any desired form (Fig. 3). It can perform complex tests quickly and automatically, and can even keep records of test results for statistical analysis.

Instrumentation computers have been integrated into several different systems to perform these and other kinds of tasks. Descriptions of four of these systems can be found on page 11.

What's Inside

So far, I've tried to describe the new computer from the point of view of a designer of instrumentation systems, emphasizing how this computer differs from others, why it differs, and what it will do for the systems designer. Now, I'd like to take you inside the machine for a closer look at certain aspects of its design.

Fig. 4 is a block diagram of the instrumentation computer, showing its four major sections, which are:

1. an arithmetic section
2. a memory
3. a control unit
4. an input/output (I/O) system.

Arithmetic operations and temporary storage of data and instructions are accomplished in the nine internal registers indicated in Fig. 4. Eight of these are flip-flop (integrated circuit) registers. The ninth is a row of toggle switches for manual data entry. The contents of all but one of the flip-flop registers are available to the programmer, and are displayed on the front panel.

The A and B registers, called accumulators, execute and hold the results of the arithmetic and logical operations called for by programmed instructions. These registers operate independently, giving the programmer considerable freedom in program design. (Many small com-

Fig. 5. *All elements of –hp– 2116A Computer are accessible from front of cabinet. Memory modules (the two cube-shaped objects protruding from top of backplane) can be changed in five minutes. Computer is fully enclosed for bench use, weighs only 230 lbs. (10.4 kg) and stands 31½ inches high (782.4 mm). It can be rack mounted.*

puters have only one accumulator; others have two, but they are not independent.) These two registers can be addressed by any memory reference instruction as memory locations 00000 and 00001, thus permitting inter-register operations, such as 'add (B) to (A), 'compare (B) with (A),' etc., using a single-word instruction.

In addition to the nine internal registers, there is an input/output buffer register for each instrument or device connected to the computer. These I/O buffers are located on the plug-in interface cards. Each interface card also contains a flag flip-flop, which is set by the external device to indicate that it is ready to transmit data, and a control flip-flop, which is set by the computer to inhibit or request data transfer, or to cause the external device to do whatever it is supposed to do.

Memory

The instrumentation computer has a coincident-current core memory system capable of storing 4096 16-bit words. A second 4096-word memory module can be installed within the computer main frame (Fig. 5), and external modules can be added to expand the memory capacity to 16,384 words. Another memory option is a seventeenth bit in each word, to be used for parity checking (error detection). Cycle time of the memory, the time required to read and write one word, is 1.6 microseconds.

Cores used in the memory have low temperature coefficients, and are also temperature compensated in order to meet the wide environmental specifications necessary for operation in instrumentation systems. Lithium core material is used because of its insensitivity to temperature variations.

The memory is organized into 'pages' of 1024 words each. One page is designated the 'base' page. There is only one base page, regardless of how many 4096-word memory modules there are. A one-word program instruction stored on one page can order the computer to do something with any other word stored on either the same page or the base page, but not with words stored on the other pages. Words stored on the other pages have to be called out, or 'addressed,' indirectly, using more than one word. Thus it is always possible to address 2048 words (two pages) directly. This is an unusually large direct-addressing capability for this size of computer, and it makes for more efficient, faster-running programs.

Instruction Repertoire

Choosing a set of instructions is the most critical decision in the design of any computer. The computer will be wired to respond to whatever instructions are chosen, and it will respond only to these instructions. If a machine is easy to program, if it carries out programs efficiently, it is because the instructions have been well designed. A look at the instruction repertoire, therefore, will tell a knowledgeable person more about a computer than anything else.

There are 68 basic instructions in the instrumentation computer's repertoire. Fourteen are memory reference instructions, which are used when information is to be obtained from the memory or stored there, forty-one are register reference instructions, used to alter or test the contents of the registers, and the last thirteen are input/output instructions.

Diagrams showing the 68 instructions and how they are coded into 16-bit words are presented on page 9. There isn't space in this article to discuss all of the instructions in detail, but the diagrams shouldn't be difficult to understand, especially if you know something about computers. Notice that each instruction has a three-letter symbol (mnemonic) which is used in writing assembly-language programs.

Of major significance in the design of the instructions is the manner in which several register reference instructions can be combined into a single 16-bit word. The in-

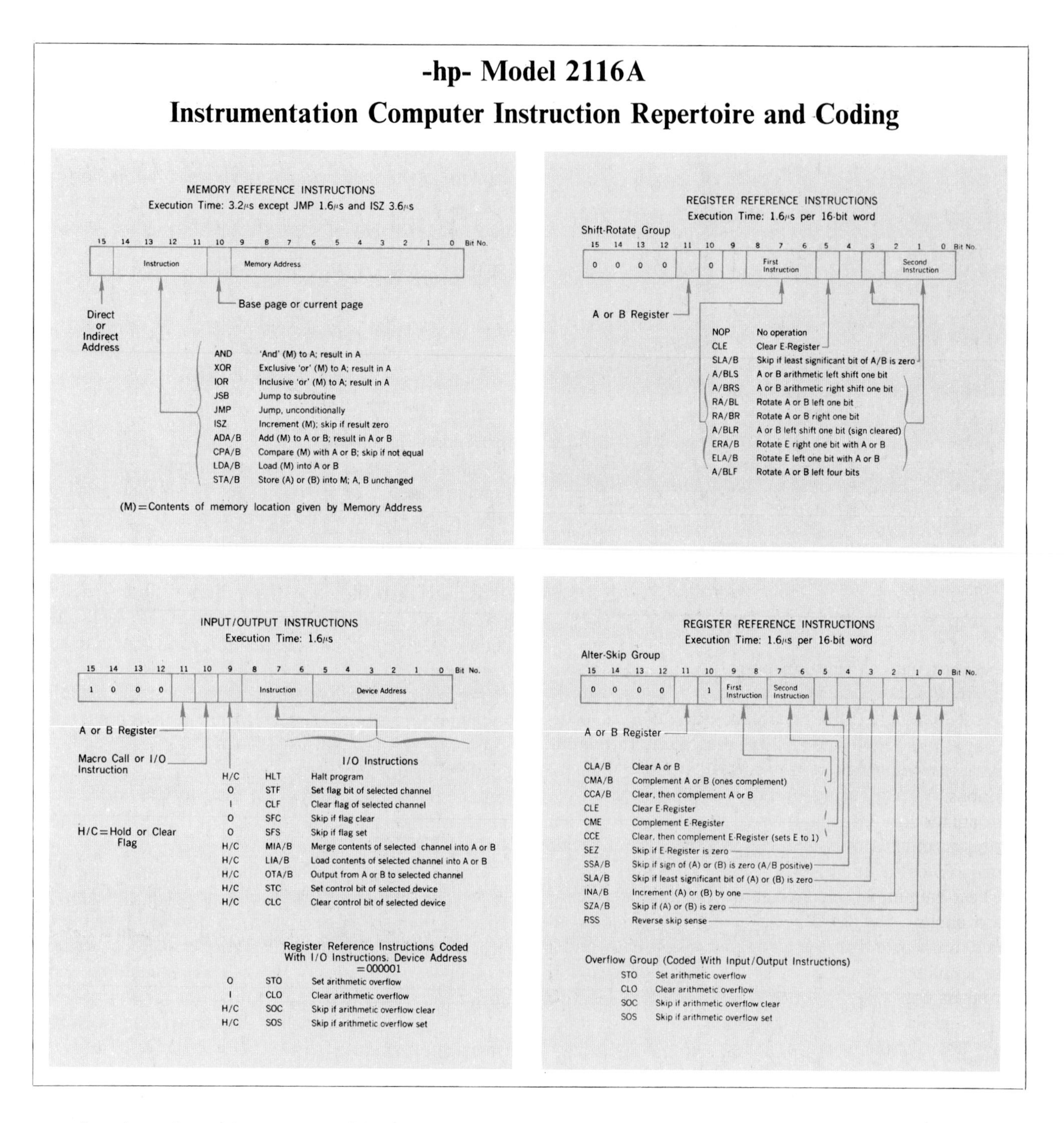

structions in each word are executed in time sequence, reading from left to right in the diagrams above, and it takes only one memory cycle, 1.6 microseconds, to execute all of the instructions in a word. This powerful capability of combining instructions makes the instrumentation computer's repertoire equivalent to over 1000 useful one-word, one-cycle instructions. It lets the programmer write extremely compact, efficient programs.

Several of the individual instructions are tailor-made for instrumentation work. For example, one of the 'shift-rotate' group of register reference instructions is ALF (or BLF), which causes the word in the A (or B) register to be shifted four places to the left in one memory cycle. This operation is needed often when translating data from the binary-coded-decimal format favored by instrument designers to the straight binary format used by the computer.

If this instruction (ALF) is written twice in a row, it will fit twice into one 16-bit word and will cause the computer to exchange the most significant 8 bits of a word with the least significant 8 bits in only one memory cycle. This operation is needed in changing from 16-bit words to the 8-bit characters used by teletype or punched-tape units, and vice versa. To be able to do it in one memory cycle instead of eight represents quite a lot of time saved.

Acknowledgments

So many people have contributed to the design of the instrumentation computer that it would be impossible to list them all individually. I am most grateful for the dedication and the creative efforts of all my colleagues in the following groups:

Group	Leaders
Logic Design	*Edward R. Holland and Eugene R. Stinson*
Memory	*Robert L. Gray and Joseph Olkowski, Jr.*
Input/Output	*Richard C. Reyna*
Mechanical Design	*Tor Larsen*
Programming	*Roy L. Clay*
Applications	*John Koudela, Jr.*

I also wish to acknowledge the valuable consulting services of Samuel N. Irwin, and the support of Henry A. Doust, Jr.'s engineering services group, Joseph B. Dixon's prototype shop, and Norman A. Day, Jr.'s layout group.

—*Kay B. Magleby*

Kay B. Magleby

Kay Magleby received his BS degree from the University of Utah in 1957, and his MS and PhD degrees from Stanford University in 1960 and 1964, respectively. He came to *-hp-* in 1958 and became the first *-hp-* engineer to work on sampling oscilloscopes. He subsequently served as project leader for the development of several sampling-oscilloscope plug-ins. Then, after moving from the oscilloscope division to the advanced research and development group, he contributed to the 8405A Vector Voltmeter project and to research projects in digital techniques, eventually starting the 2116A Computer project. In 1965, he transferred to the *-hp-* Dymec Division as head of the computer development group.

Kay holds several patents in sampling and digital instrumentation. He is active in the IEEE Instrumentation and Measurements Group and is a member of Phi Kappa Phi, Tau Beta Pi, Sigma Xi, and Eta Kappa Nu. He has a wife and three children, and enjoys skiing and woodworking.

SPECIFICATIONS
-hp- Model 2116A COMPUTER

TYPE

General-purpose digital computer, with input/output system and modular software organized for flexible application in on-line instrumentation systems.

MEMORY

TYPE. Magnetic core.

SIZE: 4096 16-bit words. Expandable to 8192 words (in main frame) with plug-in 4096-word module and associated cards, Option M4. Maximum memory size 16,384 words. (Parity bit included in standard stack for use with Option M2, Memory Parity Check.)

ADDRESSING: Memory is organized in 1024-word pages, 2048 words directly addressable.

SPEED: 1.6 microsecond cycle time.

LOADER PROTECTION: Last 64 locations of memory reserved for Basic Binary Loader. Front panel switch, in 'Protect' position, prevents alteration of contents of these locations.

MEMORY PARITY CHECK (Option M2): Permits parity checking within memory. Consists of one plug-in card for each 4K of memory.

MEMORY TEST (Option M3): Enables memory to be tested independently of program control. Consists of one plug-in card.

ARITHMETIC Parallel, binary, fixed point, two's complement.

SPEED

Add	3.2 μs
Subtract	4.8 μs
Multiply	130 μs
Divide	200 μs
Floating point add	375 μs
Floating point subtract	375 μs
Floating point multiply	750 μs
Floating point divide	1.1 ms

(Above are subroutine operations except for Add. Times shown are approximate.)

REGISTERS

Eight internal hardware (flip-flop) registers and Switch register. Contents of all registers except Instruction and Switch register displayed by front panel lamps.

A-REGISTER. Accumulator, input/output. (16 bits.)

B-REGISTER: Accumulator, input/output. (16 bits.)

E-REGISTER: Extend register, links A and B register; indicates carry from A or B register. (1 bit.)

OV-REGISTER: Overflow register, indicates overflow from A or B register. (1 bit.)

T-REGISTER: Transfer register, temporarily holds data transferred in or out of memory. (16 bits.)

P-REGISTER: Program counter. (15 bits.)

M-REGISTER: Memory address register, holds address of next memory location to be accessed. (15 bits.)

I-REGISTER: Instruction register, decodes Memory Reference instructions, holds indicators for zero/current page and direct/indirect addressing. (6 bits, 10-15.)

S-REGISTER: Toggle switches on front panel for manual data entry. Contents of register indicated by switch positions. (16 bits.)

INSTRUCTIONS

68 basic, one-word instructions, in three types:

Memory Reference	(2-cycle)	14
Register Reference	(1-cycle)	41
Input/Output	(1-cycle)	13

Register Reference instructions are micro-operations, can be combined to form over 1000 one-word, single-cycle instructions.

INPUT/OUTPUT

NUMBER OF CHANNELS: 48, 16-bit parallel interrupting channels, with priority control, utilized through plug-in I/O interface cards (1 per channel). 44 channels available for I/O devices; 4 channels reserved for processor options.

MAIN FRAME CAPACITY: 16 channels for I/O devices. Power for interface cards provided from internal supply. (Peripherals draw power directly from 115/230v line.)

INTERRUPT RESPONSE: Servicing of interrupt request (execution of first useful instruction) begins within 3 μs with one I/O channel in use, or within 7 μs for highest priority channel in multiple-channel system.

DATA FORM

PUNCHED TAPE: ASCII. Parity not used, 8th level always punched. (1-inch tape.)

MAGNETIC TAPE: IBM-compatible, 7-channel NRZI. (½-inch tape).

SOFTWARE

Software (punched tape) available consists of:
Compiler, ASA Basic FORTRAN (Extended)
Assembler
Symbolic Editor
Basic Control System
I/O Device Handling Routines
Cross-reference Symbol Table Generator
Hardware Diagnostics

Basic Control System is modular, includes configurator (Prepare Control System) to permit adaptation by user to different I/O arrangements. Also includes Debugging Routines.)

PRICES: 2116A Computer (4096-word memory, no I/O options) $22,000
Memory parity check. Option M2, $1000
Memory test, Option M3, $420
8192-word memory (basic 4K plus 4K additional), Option M4, $8000
I/O Options, $1000 to $15,000

MANUFACTURING DIVISION: -hp-Dymec Division
395 Page Mill Road
Palo Alto, California 94306

Successful Instrument-Computer Marriages

Instrumentation computers are designed to be easy to incorporate into any system which contains electronic, chemical, or medical instruments. Here are four remarkably varied examples of how these computers are being used.

Fig. 1. *Computing data acquisition system can be used in a variety of scientific and industrial applications, including jet engine testing (see Fig. 2).*

Computing Data Acquisition System

Data acquisition systems are the simplest type of instrumentation system. An elementary data acquisition system converts data from a number of inputs to a form suitable for printing by an output recorder. In more complex systems, some processing of the raw data is done, and the operation of the system may be controlled to some extent by the data.

Because it can easily carry out complex programs, and because its programs can be changed easily, an instrumentation computer in a data acquisition system makes control of the system extremely flexible. It also provides rapid, local data processing, thereby eliminating the loss of time inherent in remote data processing.

A typical computing data acquisition system is shown in Fig. 1. Such systems are used, for example, in testing jet engines: the analog inputs are physical parameters such as pressure, temperature, fuel flow, and engine speed, and the computer outputs are operating parameters such as efficiency and power. The computer not only provides immediate results to help the operator set up the test, but also controls some portions of the test, thereby making the checkout more automatic.

Fig. 2 is a block diagram of the system of Fig. 1, illustrating its use in jet engine testing. Surprisingly, this computing system costs little more than a less flexible

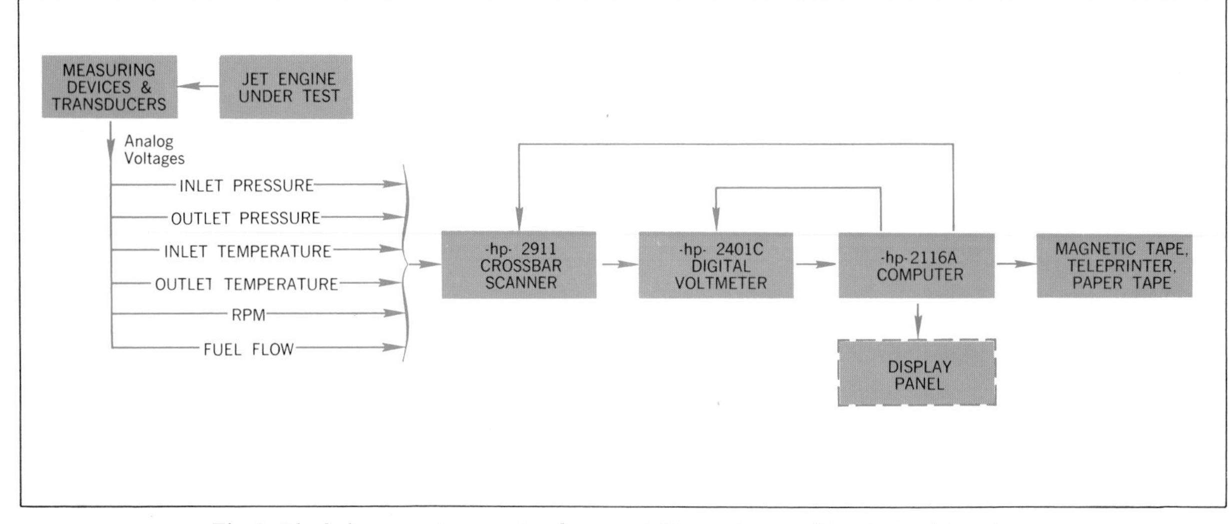

Fig. 2. *Block diagram of computing data acquisition system used for testing jet engines. Computer calculates engine efficiency, power, and other performance data.*

Fig. 3. *–hp– 2116A Computer speeds gas chromatography by analyzing outputs of several chromatographs (one is visible at right), freeing chemist to do less tedious tasks.*

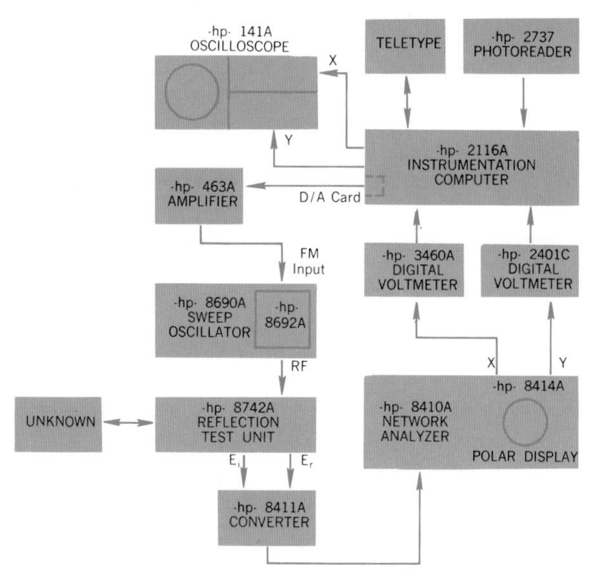

Fig. 4. *Block diagram of microwave impedance-measuring system including instrumentation computer. Computer controls test instruments, refines measurements, and presents results in several forms.*

Fig. 5. *–hp– 2116A Computer pretests logic cards for other –hp– 2116A Computers. Tests take only minutes per card, would take months without computer. Computer also keeps statistics for quality control.*

noncomputing system capable of performing some, but not all, of the same tasks.

Gas Chromatograph System

A gas chromatograph is a versatile chemical instrument which provides an analytical chemist with information about the composition of an unknown sample of material. However, it takes a considerable amount of interpretation and analysis to extract this information from the output of the chromatograph, and analytical chemists who do this type of work spend a large portion of their time on data reduction. An instrumentation computer in a chromatograph system can not only free the chemist from time-consuming data reduction, but because it can analyze the outputs of many instruments, it can also greatly increase the number of samples that a chemist can test in a day. Fig. 3 shows part of a developmental system in which an instrumentation computer will be used to analyze the outputs of up to 36 gas chromatographs.

Microwave Impedance-Measuring System

A block diagram of an impedance-measuring system including an instrumentation computer and a network analyzer (Hewlett-Packard Journal, Feb., 1967) is shown in Fig. 4. This system measures the reflection coefficient of an unknown device as a function of frequency, then calculates impedance, admittance, standing wave ratio, return loss, and mismatch loss. Residual errors in the system are measured with a calibrating short in place of the unknown, and the computer automatically subtracts these errors from the measurements. The refined results are stored in the computer memory and displayed on the oscilloscope. The network analyzer provides a Smith Chart display of the raw data.

Logic Module Test System

Final testing of a complex system (e.g., a computer) can be greatly simplified by pre-testing the modules or cards that make up the system. However, manual testing of logic modules can be extremely expensive and time consuming, because there are so many inputs and outputs to be checked. A computer makes pre-testing practical, because it can automatically stimulate the modules and monitor their responses. A complete test of a module with 16 inputs would require 2^{16}, or 64,000, different tests. At one minute per test, it would take a technician six months to test one module. An instrumentation computer can do it in less than one minute. The computer can also keep statistics on the tests for quality control.

Logic modules for –hp– instrumentation computers are tested by the computer system illustrated in Fig. 5. The response of the module under test is compared with that of a reference circuit and the operator is alerted if the test device's response is not within specified limits. ■

Fifty years ago this Marchant calculator was touted as 'The Last Word in Calculators.' Indeed it was! It was the culmination of a history in which many agonies are recorded. For instance, in 1842 Babbage found his funds cut off during the development of a sophisticated 'Analytical Engine.' To raise more, he and his friend Lady Lovelace played the horses. They lost and the machine was never completed.

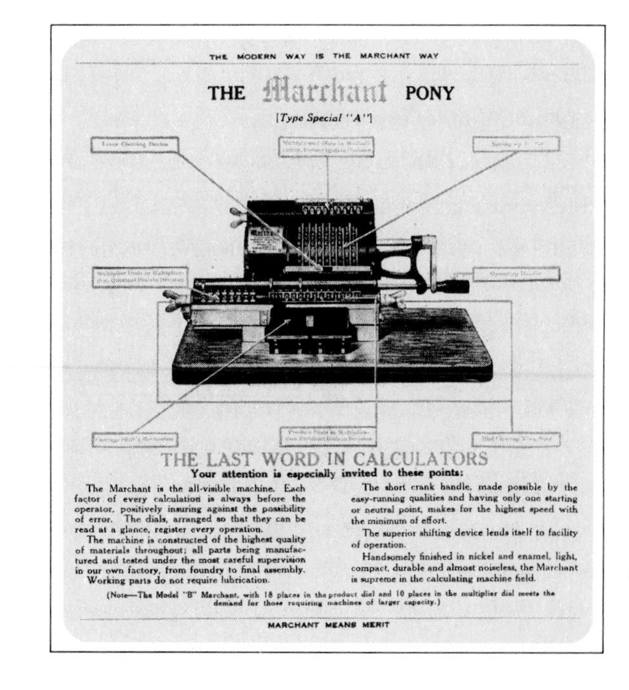

Then came Thomas Hill in 1857 who forgot that 'bodies set in motion tend to remain in motion.' His 'Arithmometer' was accurate to within a few spins of the digit wheels.

Some machines were too slow. An experienced accountant could add figures in his head faster than some of the machines proposed.

A former bank clerk, Wm. Seward Burroughs, made the first practical calculator in 1872. D. E. Felt, in 1879 began selling a machine he called the 'Comptometer.' A 'Multiplier' developed by E. D. Barbour in 1872 was based on the famous 'Napier's bones' principle.

Automatic, motor driven models began to appear in the 1920's when, in the U.S., Burroughs, Monroe, Marchant and Friden became recognized standards. In addition to the four basic operations of add, subtract, multiply and divide, some of these machines include square root.

Electronic desktop calculators first made their appearance about 5 years ago. There are various sizes, capabilities and modes of operation. Although more expensive than the mechanical calculators, they are quiet, fast, versatile and virtually maintenance free.

Now we have a new, more powerful desktop calculating device. Our Model 9100A is more a computer than a calculator. It may very well be the 'First Word' in a new breed of calculating machines.

Laurence D. Shergalis

A New Electronic Calculator with Computerlike Capabilities

By Richard E. Monnier

MANY OF THE DAY-TO-DAY COMPUTING PROBLEMS faced by scientists and engineers require complex calculations but involve only a moderate amount of data. Therefore, a machine that is more than a calculator in capability but less than a computer in cost has a great deal to offer. At the same time it must be easy to operate and program so that a minimum amount of effort is required in the solution of typical problems. Reasonable speed is necessary so that the response to individual operations seems nearly instantaneous.

The HP Model 9100A Calculator, Fig. 1, was developed to fill this gap between desk calculators and computers. Easy interaction between the machine and user was one of the most important design considerations during its development and was the prime guide in making many design decisions.

CRT Display

One of the first and most basic problems to be resolved concerned the type of output to be used. Most people want a printed record, but printers are generally slow and noisy. Whatever method is used, if only one register is displayed, it is difficult to follow what is happening during a sequence of calculations where numbers are moved from one register to another. It was therefore decided that a cathode-ray-tube displaying the contents of three registers would provide the greatest flexibility and would allow the user to follow problem solutions easily. The ideal situation is to have both a CRT showing more than one register, and a printer which can be attached as an accessory.

Fig. 2 is a typical display showing three numbers. The X register displays numbers as they are entered from the keyboard one digit at a time and is called the keyboard register. The Y register is called the accumulator since the results of arithmetic operations on two numbers, one in X and one in Y, appear in the Y register. The Z register is a particularly convenient register to use for temporary storage.

Numbers

One of the most important features of the Model 9100A is the tremendous range of numbers it can handle without special attention by the operator. It is not necessary to worry about where to place the decimal point to obtain the desired accuracy or to avoid register overflow. This flexibility is obtained because all numbers are stored in 'floating point' and all operations performed using 'floating point arithmetic'. A floating point number is expressed with the decimal point following the first digit and an exponent representing the number of places the decimal point should be moved—to the right if the

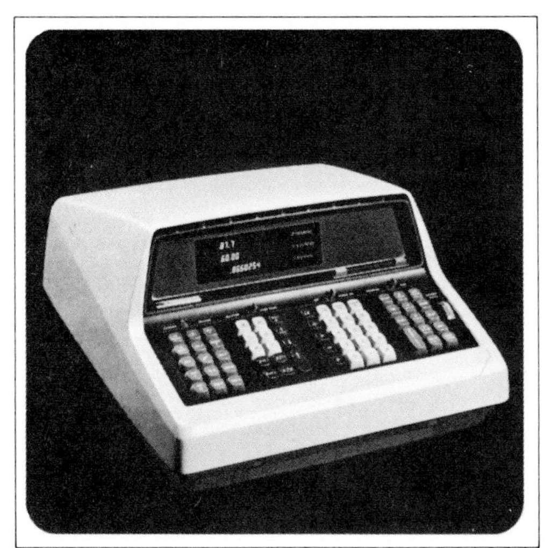

Fig. 1. *(left) This new HP Model 9100A Calculator is self-contained and is capable of performing functions previously possible only with larger computers.* Fig. 2. *(right) Display in fixed point with the decimal wheel set at 5. The Y register has reverted to floating point because the number is too large to be properly displayed unless the digits called for by the DECIMAL DIGITS setting is reduced.*

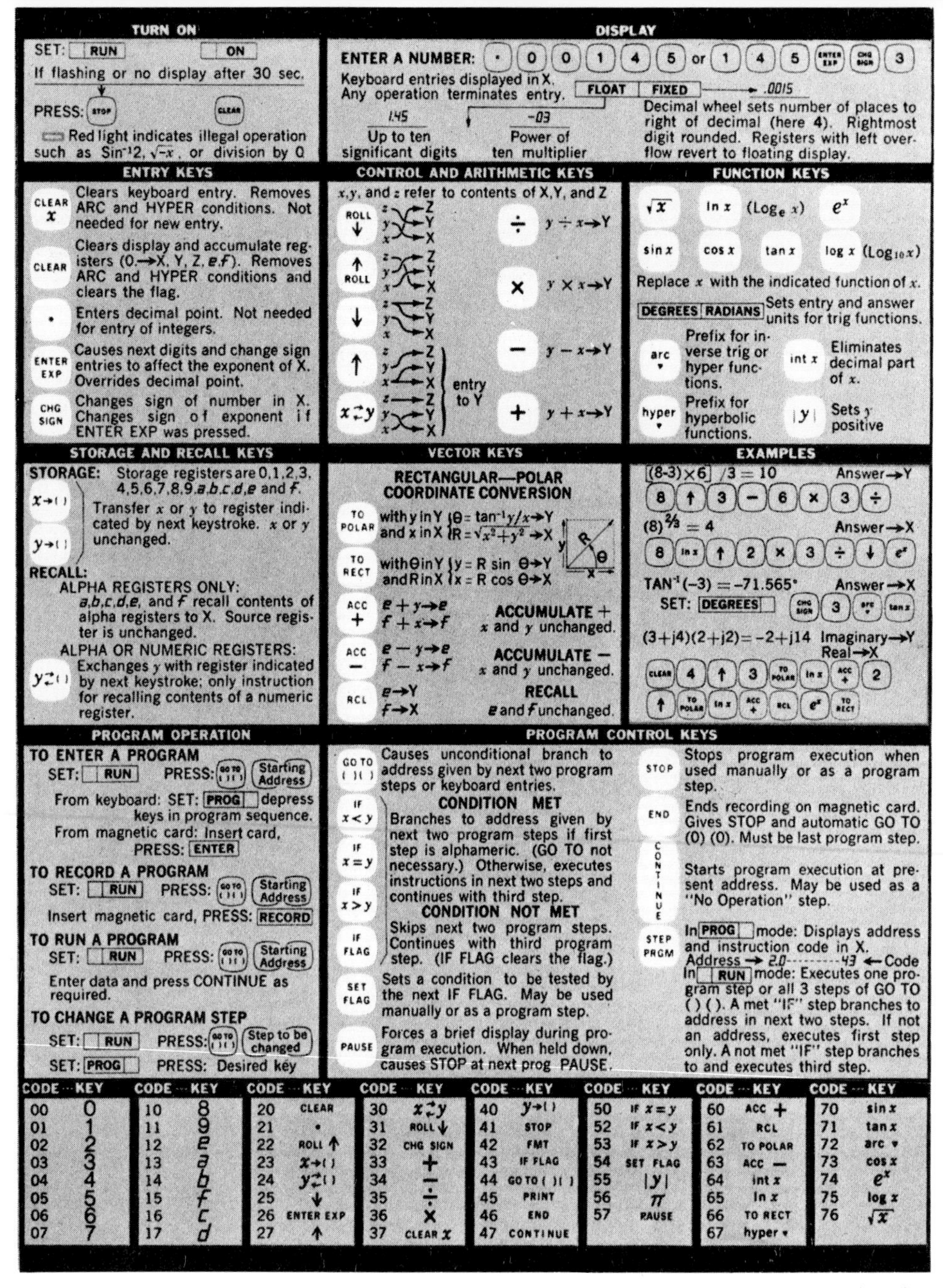

CODE	KEY	CODE	KEY	CODE	KEY	CODE	KEY	CODE	KEY	CODE	KEY	CODE	KEY	CODE	KEY
00	0	10	8	20	CLEAR	30	x⇄y	40	y→( )	50	IF x=y	60	ACC +	70	sin x
01	1	11	9	21	.	31	ROLL ↓	41	STOP	52	IF x<y	61	RCL	71	tan x
02	2	12	e	22	ROLL ↑	32	CHG SIGN	42	FMT	53	IF x>y	62	TO POLAR	72	arc
03	3	13	a	23	x→( )	33	+	43	IF FLAG	54	SET FLAG	63	ACC −	73	cos x
04	4	14	b	24	y⇄( )	34	−	44	GO TO ( ) ( )	55	|y|	64	int x	74	eˣ
05	5	15	f	25	↓	35	÷	45	PRINT	56	π	65	ln x	75	log x
06	6	16	c	26	ENTER EXP	36	×	46	END	57	PAUSE	66	TO RECT	76	√x
07	7	17	d	27	↑	37	CLEAR x	47	CONTINUE			67	hyper		

Fig. 3. *Pull-out instruction card is permanently attached to the calculator and contains key codes and operating instructions.*

exponent is positive, or to the left if the exponent is negative.

$$4.398\,364\,291 \times 10^{-3} = .004\,398\,364\,291.$$

The operator may choose to display numbers in FLOATING POINT or in FIXED POINT. The FLOATING POINT mode allows numbers, either positive or negative, from 1×10^{-99} to $9.999\,999\,999 \times 10^{99}$ to be displayed just as they are stored in the machine.

The FIXED POINT mode displays numbers in the way they are most commonly written. The DECIMAL DIGITS wheel allows setting the number of digits displayed to the right of the decimal point anywhere from 0 to 9. Fig. 2 shows a display of three numbers with the DECIMAL DIGITS wheel set at 5. The number in the Y register, $5.336\,845\,815 \times 10^5 = 533\,684.5815$, is too big to be displayed in FIXED POINT without reducing the DECIMAL DIGITS setting to 4 or less. If the number is too big for the DECIMAL DIGITS setting, the register involved reverts automatically to floating point to avoid an apparent overflow. In FIXED POINT display, the number displayed is rounded, but full significance is retained in storage for calculations.

To improve readability, 0's before the displayed number and un-entered 0's following the number are blanked. In FLOATING POINT, digits to the right of the decimal are grouped in threes.

Pull-Out Instruction Card

A pull-out instruction card, Fig. 3, is located at the front of the calculator under the keyboard. The operation of each key is briefly explained and key codes are listed. Some simple examples are provided to assist those using the machine for the first time or to refresh the memory of an infrequent user. Most questions regarding the operation of the Model 9100A are answered on the card.

Data Entry

The calculator keyboard is shown in Fig. 4. Numbers can be entered into the X register using the digit keys, the π key or the ENTER EXP key. The ENTER EXP key allows powers of 10 to be entered directly which is useful for very large or very small numbers. 6.02×10^{23} is entered 6 0 2 ENTER EXP 2 3. If the ENTER EXP key is the first key of a number entry, a 1 is automatically entered into the mantissa. Thus only two keystrokes ENTER EXP 6 suffice to enter 1,000,000. The CHG SIGN key changes the sign of either the mantissa or the exponent depending upon which one is presently being addressed. Numbers are entered in the same way, regardless of whether the machine is in FIXED POINT or

FLOATING POINT. Any key, other than a digit key, decimal point, CHG SIGN or ENTER EXP, terminates an entry; it is not necessary to clear before entering a new number. CLEAR X sets the X register to 0 and can be used when a mistake has been made in a number entry.

Control and Arithmetic Keys

ADD, SUBTRACT, MULTIPLY, DIVIDE involve two numbers, so the first number must be moved from X to Y before the second is entered into X. After the two numbers have been entered, the appropriate operation can be performed. In the case of a DIVIDE, the dividend is entered into Y and the divisor into X. Then the ÷ key is pressed causing the quotient to appear in Y, leaving the divisor in X.

One way to transfer a number from the X register to the Y register is to use the double sized key, ↑, at the left of the digit keys. This repeats the number in X into Y, leaving X unchanged; the number in Y goes to Z, and the number in Z is lost. Thus, when squaring or cubing a number, it is only necessary to follow ↑ with × or × ×. The ↓ key repeats a number in Z to Y leaving Z unchanged, the number in Y goes to X, and the number in X is lost. The ↑ ROLL key rotates the number in the X and Y registers up and the number in Z down into X. ROLL ↓ rotates the numbers in Z and Y down and the number in X up into Z. x⇄y interchanges the numbers in X and Y. Using the two ROLL keys and x⇄y, numbers can be placed in any order in the three registers.

Functions Available from the Keybord

The group of keys at the far left of the keyboard, Fig. 4, gives a good indication of the power of the Model 9100A. Most of the common mathematical functions are available directly from the keyboard. Except for |y| the function keys operate on the number in X replacing it with the function of that argument. The numbers in Y and Z are left unchanged. √x is located with another group of keys for convenience but operates the same way.

The circular functions operate with angles expressed in RADIANS or DEGREES as set by the switch above the keyboard. The sine, cosine, or tangent of an angle is taken with a single keystroke. There are no restrictions on direction, quadrant or number of revolutions of the angle. The inverse functions are obtained by using the arc key as a prefix. For instance, two key depressions are necessary to obtain the arc sin x: arc sin x. The angle obtained will be the standard principal value. In radians:

$$-\frac{\pi}{2} \leq \text{Sin}^{-1}\, x \leq \frac{\pi}{2}$$

$$0 \leq \text{Cos}^{-1}\, x \leq \pi$$

$$-\frac{\pi}{2} < \text{Tan}^{-1}\, x < \frac{\pi}{2}$$

The hyperbolic sine, cosine, or tangent is obtained using the hyper key as a prefix. The inverse hyperbolic functions are obtained with three key depressions. $\text{Tanh}^{-1}\, x$ is obtained by arc hyper tan x. The arc and hyper keys prefix keys below them in their column.

Log x and ln x obtain the log to the base 10 and the log to the base e respectively. The inverse of the natural log is obtained with the e^x key. These keys are useful when raising numbers to odd powers as shown in one of the examples on the pull-out card, Fig. 3.

Two keys in this group are very useful in programs. int x takes the integer part of the number in the X register

Fig. 4. *Keys are in four groups on the keyboard, according to their function.*

which deletes the part of the number to the right of the decimal point. For example int(—3.1416) = —3. |y| forces the number in the Y register positive.

Storage Registers

Sixteen registers, in addition to X, Y, and Z, are available for storage. Fourteen of them, 0, 1, 2, 3, 4, 5, 6, 7, 8, 9, a, b, c, d, can be used to store either one constant or 14 program steps per register. The last registers, e and f, are normally used only for constant storage since the program counter will not cycle into them. Special keys located in a block to the left of the digit keys are used to identify the lettered registers.

To store a number from the X register the key x→( ) is used. The parenthesis indicates that another key depression, representing the storage register, is necessary to complete the transfer. For example, storing a number from the X register into register 8 requires two key depressions: x→( ) 8 . The X register remains unchanged. To store a number from Y register the key y→( ) is used.

The contents of the alpha registers are recalled to X simply by pressing the keys a, b, c, d, e, and f. Recalling a number from a numbered register requires the use of the y⇄( ) key to distinguish the recall procedure from digit entry. This key interchanges the number in the Y register with the number in the register indicated by the following keystroke, alpha or numeric, and is also useful in programs since neither number involved in the transfer is lost.

The CLEAR key sets the X, Y, and Z display registers and the f and e registers to zero. The remaining registers are not affected. The f and e registers are set to zero to initialize them for use with the ACC + and ACC − keys as will be explained. In addition the CLEAR key clears the FLAG and the ARC and HYPER conditions, which often makes it a very useful first step in a program.

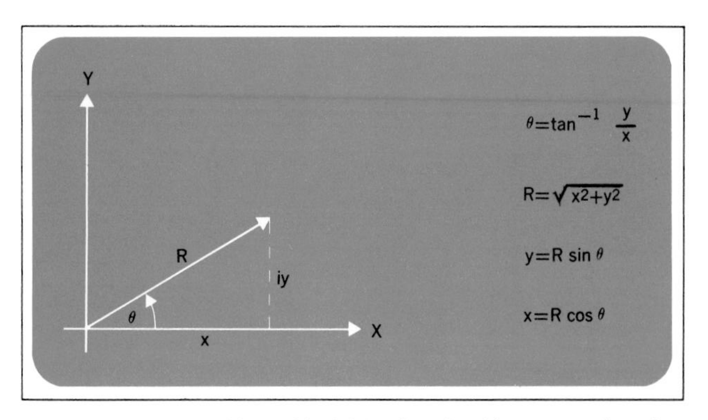

Fig. 5. Variables involved in conversions between rectangular and polar coordinates.

Coordinate Transformation and Complex Numbers

Vectors and complex numbers are easily handled using the keys in the column on the far left of the keyboard. Fig. 5 defines the variables involved. Angles can be either in degrees or radians. To convert from rectangular to polar coordinates, with y in Y and x in X, press TO POLAR. Then the display shows θ in Y and R in X. In converting from polar to rectangular coordinates, θ is placed in Y, and R in X, TO RECT is pressed and the display shows y in Y and x in X.

ACC+ and ACC— allow addition or subtraction of vector components in the f and e storage registers. ACC+ adds the contents of the X and Y register to the numbers already stored in f and e respectively; ACC— subtracts them. The RCL key recalls the numbers in the f and e registers to X and Y.

Illegal Operations

A light to the left of the CRT indicates that an illegal operation has been performed. This can happen either from the keyboard or when running a program. Pressing any key on the keyboard will reset the light. When running a program, execution will continue but the light will remain on as the program is completed. The illegal operations are:

- Division by zero.
- $\sqrt{x}$ where $x < 0$
- ln x where $x \leq 0$; log x where $x \leq 0$
- $\sin^{-1} x$ where $|x| > 1$; $\cos^{-1} x$ where $|x| > 1$
- $\cosh^{-1} x$ where $x < 1$; $\tanh^{-1} x$ where $|x| > 1$

Accuracy

The Model 9100A does all calculations using floating point arithmetic with a twelve digit mantissa and a two digit exponent. The two least significant digits are not displayed and are called guard digits.

The algorithms used to perform the operations and generate the functions were chosen to minimize error and to provide an extended range of the argument. Usually any inaccuracy will be contained within the two guard digits. In certain cases some inaccuracy will appear in the displayed number. One example is where the functions change rapidly for small changes in the argument, as in tan x where x is near 90°. A glaring but insignificant inaccuracy occurs when an answer is known to be a whole number, but the least significant guard digit is one count low: 2.000 000 000 $\simeq$ 1.999 999 999.

Accuracy is discussed further in the 'Internal Programming' article in this issue. But a simple summary is: the answer resulting from any operation or function will lie within the range of true values produced by a variation of ± 1 count in the tenth digit of the argument.

Programming

Problems that require many keyboard operations are more easily solved with a program. This is particularly true when the same operations must be performed repeatedly or an iterative technique must be used. A program library supplied with the Model 9100A provides a set of representative programs from many different fields. If a program cannot be found in the library to solve a particular problem, a new program can easily be written since no special experience or prior knowledge of a programming language is necessary.

Any key on the keyboard can be remembered by the calculator as a program step except STEP PRGM. This key is used to 'debug' a program rather than as an operation in a program. Many individual program steps, such as 'sin x' or 'to polar' are comparatively powerful, and avoid the need of sub-routines for these functions and the programming space such sub-routines require. Registers 0, 1, 2, 3, 4, 5, 6, 7, 8, 9, a, b, c, d can store 14 program steps each. Steps within the registers are numbered 0 through d just as the registers themselves are numbered. Programs can start at any of the 196 possible addresses. However 0-0 is usually used for the first step. Address d-d is then the last available, after which the program counter cycles back to 0-0.

Registers f and e are normally used for storage of constants only, one constant in each register. As more constant storage is required, it is recommended that registers d, then c, then b, etc., are used starting from the bottom of the list. Lettered registers are used first, for the frequently recalled constants, because constants stored in them are more easily recalled. A register can be used to store one constant or 14 program steps, but not both.

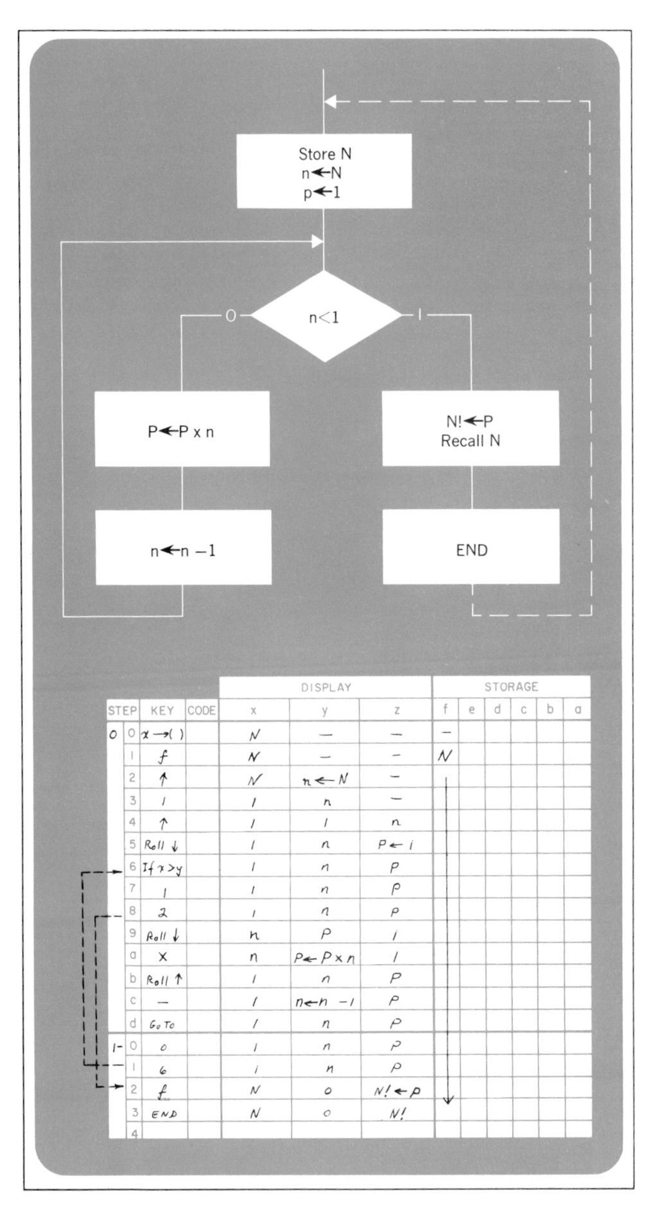

STEP		KEY	CODE	DISPLAY x	y	z	STORAGE f	e	d	c	b	a
0	0	x→( )		N	—	—	—					
	1	f		N	—	—	N					
	2	↑		N	n←N	—						
	3	1		1	n	—						
	4	↑		1	1	n						
	5	Roll ↓		1	n	P←1						
	6	If x>y		1	n	P						
	7	1		1	n	P						
	8	2		1	n	P						
	9	Roll ↓		n	P	1						
	a	×		n	P←P×n	1						
	b	Roll ↑		1	n	P						
	c	—		1	n←n −1	P						
	d	Go To		1	n	P						
1-	0	0		1	n	P						
	1	6		1	n	P						
	2	f		N	0	N!←P						
	3	END		N	0	N!						
	4											

Fig. 6. *Flow chart of a program to compute N! (top). Each step is shown (bottom) and the display for each register. A new value for N can be entered at the end of the program, since END automatically sets the program counter back to 0-0.*

Branching

The bank on the far right of the keyboard, Fig. 4, contains program oriented keys. [GO TO ( ) ( )] is used to set the program counter. The two sets of parentheses indicate that this key should be followed by two more key depressions indicating the address of the program step desired. As a program step, 'GO TO' is an unconditional branch instruction, which causes the program to branch to the address given by the next two program steps. The 'IF' keys in this group are conditional branch instructions. With [IF x<y] [IF x=y], and [IF x>y] the numbers contained in the X and Y registers are compared. The indicated condition is tested and, if met, the next two program steps are executed. If the first is alphameric, the second must be also, and the two steps are interpreted as a branching address. When the condition is not met, the next two steps are skipped and the program continues. [IF FLAG] is also a very useful conditional branching instruction which tests a 'yes' or 'no' condition internally stored in the calculator. This condition is set to 'yes' with the SET FLAG from the keyboard when the calculator is in the display mode or from a program as a program step. The flag is set to a 'no' condition by either asking IF FLAG in a program

```
5.336845815  05     z temporary
0.           00     y accumulator
2.d---------36      x keyboard
```

Fig. 7. *Program step address and code are displayed in the X register as steps are entered. After a program has been entered, each step can be checked using the STEP PRGM key. In this display, step 2-d is 36, the code for multiply.*

Fig. 8. *Programs can be entered into the calculator by means of the magnetic program card. The card is inserted into the slot and the ENTER button pressed.*

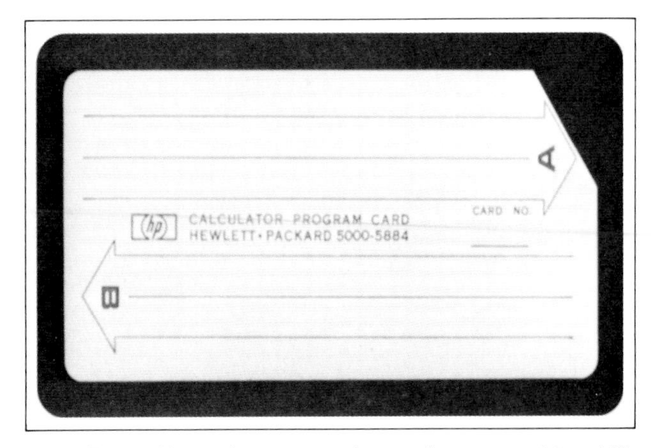

Fig. 9. *Magnetic programming card can record two 196-step programs. To prevent accidental recording of a new program over one to be saved, the corner of the card is cut as shown.*

or by a CLEAR instruction from the keyboard or from a program.

Data Input and Output

Data can be entered for use in a program when the machine is in the display mode. (The screen is blank while a program is running.) A program can be stopped in several ways. The STOP key will halt the machine at any time. The operation being performed will be completed before returning to the display mode. As a program step, STOP stops the program so that answers can be displayed or new data entered. END must be the last step in a program listing to signal the magnetic card reader; when encountered as a program step it stops the machine and also sets the program counter to 0-0.

As a program step, PAUSE causes a brief display during program execution. Nine cycles of the power line frequency are counted—the duration of the pause will be about 150 ms for a 60 Hz power line or 180 ms for a 50 Hz power line. More pauses can be used in sequence if a longer display is desired. While a program is running the PAUSE key can be held down to stop the machine when it comes to the next PAUSE in the program. PAUSE provides a particularly useful way for the user and the machine to interact. It might, for instance, be used in a program so that the convergence to a desired result can be observed.

Other means of input and output involve peripheral devices such as an X-Y Plotter or a Printer. The PRINT key activates the printer, causing it to print information from the display register. As a program step, PRINT will interrupt the program long enough for the data to be accepted by the printer and then the program will continue. If no printer is attached, PRINT as a program step will act as a STOP. The FMT key, followed by any other keystroke, provides up to 62 unique commands to peripheral equipment. This flexibility allows the Model 9100A to be used as a controller in small systems.

Sample Program — N!

A simple program to calculate N! demonstrates how the Model 9100A is programmed. Fig. 6 (top) shows a flow chart to compute N! and Fig. 6 (bottom) shows the program steps. With this program, 60! takes less than ½ second to compute.

Program Entry and Execution

After a program is written it can be entered into the Model 9100A from the keyboard. The program counter is set to the address of the first program step by using the

GO TO ( ) ( ) key. The RUN-PROGRAM switch is then switched from RUN to PROGRAM and the program steps entered in sequence by pushing the proper keys. As each step is entered the X register displays the address and key code, as shown in Fig. 7. The keys and their codes are listed at the bottom of the pull-out card, Fig. 3. Once a program has been entered, the steps can be checked using the STEP PRGM key in the PROGRAM mode as explained in Fig. 7. If an error is made in a step, it can be corrected by using the GO TO ( ) ( ) key without having to re-enter the rest of the program.

To run a program, the program counter must be set to the address of the first step. If the program starts at 0-0 the keys GO TO ( ) ( ) 0 0 are depressed, or simply just END since this key automatically sets the program counter to 0-0. CONTINUE will start program execution.

Magnetic Card Reader-Recorder

One of the most convenient features of the Model 9100A is the magnetic card reader-recorder, Fig. 8. A program stored in the Model 9100A can be recorded on a magnetic card, Fig. 9, about the size of a credit card. Later when the program is needed again, it can be quickly re-entered using the previously recorded card. Cards are easily duplicated so that programs of common interest can be distributed.

As mentioned earlier, the END statement is a signal to the reader to stop reading recorded information from the card into the calculator. For this reason END should not be used in the middle of a program. Since most programs start at location 0-0 the reader automatically initializes the program counter to 0-0 after a card is read.

The magnetic card reader makes it possible to handle most programs too long to be held in memory at one time. The first entry of steps can calculate intermediate results which are stored in preparation for the next part of the program. Since the reader stops reading at the END statement these stored intermediate results are not disturbed when the next set of program steps is entered. The stored results are then retrieved and the program continued. Linking of programs is made more convenient if each part can execute an END when it finishes to set the program counter to 0-0. It is then only necessary to press CONTINUE after each entry of program steps.

Richard E. Monnier

Dick Monnier has a BSEE from the University of California at Berkeley, 1958, and an MSEE from Stanford University, 1961.

Dick has been with HP since 1958 when he joined the group working on sampling oscilloscopes. He has been project leader on the Models 120B, 132A and 140A oscilloscopes and the Model 191A television waveform monitor prior to the transfer of the project to HP's Colorado Springs Division. Since early 1966 he has been Section Leader responsible for the 9100A Calculator project in Palo Alto.

Dick holds several patents and is a Member of IEEE.

SPECIFICATIONS

HP Model 9100A

The HP Model 9100A is a programmable, electronic calculator which performs operations commonly encountered in scientific and engineering problems. Its log, trig and mathematical functions are each performed with a single key stroke, providing fast, convenient solutions to intricate equations. Computer-like memory enables the calculator to store instructions and constants for repetitive or iterative solutions. The easily-readable cathode ray tube instantly displays entries, answers and intermediate results.

OPERATIONS

DIRECT KEYBOARD OPERATIONS INCLUDE:

ARITHMETIC: addition, subtraction, multiplication, division and square-root.

LOGARITHMIC: log x, ln x and e^x.

TRIGONOMETRIC: sin x, cos x, tan x, $\sin^{-1}x$, $\cos^{-1}x$ and $\tan^{-1}x$ (x in degrees or radians).

HYPERBOLIC: sinh x, cosh x, tanh x, $\sinh^{-1}x$, $\cosh^{-1}x$, and $\tanh^{-1}x$.

COORDINATE TRANSFORMATION: polar-to-rectangular, rectangular-to-polar, cumulative addition and subtraction of vectors.

MISCELLANEOUS: other single-key operations include—taking the absolute value of a number, extracting the integer part of a number, and entering the value of π. Keys are also available for positioning and storage operations.

PROGRAMMING

The program mode allows entry of program instructions, via the keyboard, into program memory. Programming consists of pressing keys in the proper sequence, and any key on the keyboard is available as a program step. Program capacity is 196 steps. No language or code-conversions are required.

A self-contained magnetic card reader/recorder records programs from program memory onto wallet-size magnetic cards for storage. It also reads programs from cards into program memory for repetitive use. Two programs of 196 steps each may be recorded on each reusable card. Cards may be cascaded for longer programs.

SPEED

Average times for total performance of typical operations, including decimal-point placement:

add, subtract: 2 milliseconds
multiply: 12 milliseconds
divide: 18 milliseconds
square-root: 19 milliseconds
sin, cos, tan: 280 milliseconds
ln x: 50 milliseconds
e^x: 110 milliseconds

These times include core access of 1.6 microseconds.

GENERAL

WEIGHT: Net 40 lbs. (18,1 kg.); shipping 65 lbs. (29,5 kg.).
POWER: 115 or 230 V ±10%, 50 to 60 Hz, 400 Hz, 70 watts.
DIMENSIONS: 8¼" high, 16" wide, 19" deep.

ACCESSORIES FURNISHED AT NO CHARGE:

09100-90001 Operating and Programming manual, $5.00.
09100-90002 Program library binder containing sample programs, $30.00.
5060-5919 Box of 10 magnetic program cards, $10.00.
09100-90003 Pad of 100 program sheets, $2.50.
09100-90004 Magnetic card with pre-recorded diagnostic program, $2.50.
9320-1157 Pull-out instruction card mounted in calculator, $5.00.
4040-0350 Plastic dust cover, $2.50.

ADDITIONAL ACCESSORIES AVAILABLE:

5000-5884 Single magnetic card, $2.00
09100-90000 Box of 5 program pads, $10.00.

PRICE: HP Model 9100A, $4900.00.

PERIPHERALS: Printer, X-Y plotter and input/output interface will be available soon.

MANUFACTURING DIVISION: LOVELAND DIVISION
P.O. Box 301
815 Fourteenth Street N.W.
Loveland, Colorado 80537

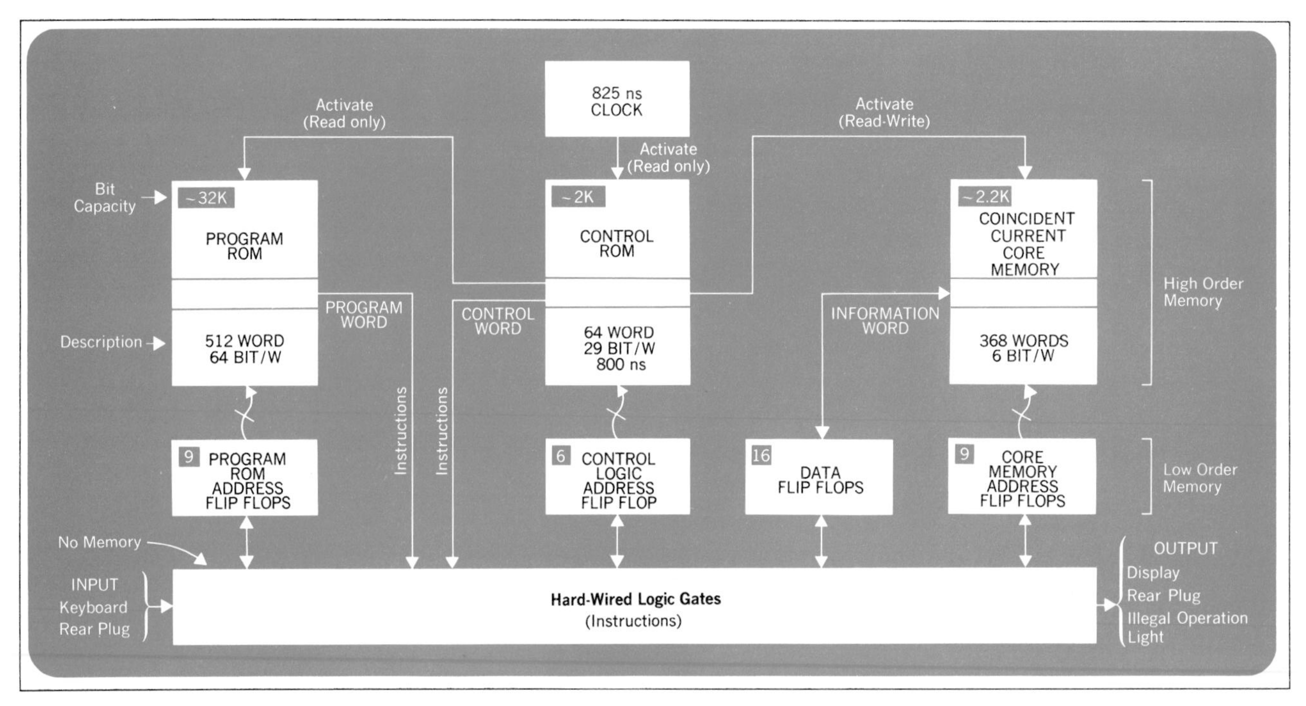

Fig. 1. *Arithmetic processing unit block diagram. This system is a marriage of conventional, reliable diode-resistor logic to a 32,000-bit read-only memory and a coincident current core memory.*

Hardware Design of the Model 9100A Calculator

By Thomas E. Osborne

ALL KEYBOARD FUNCTIONS IN THE MODEL 9100A are implemented by the arithmetic processing unit, Figs. 1 and 2. The arithmetic unit operates in discrete time periods called clock cycles. All operations are synchronized by the clock shown at the top center of Fig. 1.

The clock is connected to the control read only memory (ROM) which coordinates the operation of the program read only memory and the coincident current core read/write memory. The former contains information for implementing all of the keyboard operations while the latter stores user data and user programs.

All internal operations are performed in a digit by digit serial basis using binary coded decimal digits. An addition, for example, requires that the least significant digits of the addend and augend be extracted from core, then added and their sum replaced in core. This process is repeated one BCD digit at a time until the most significant digits have been processed. There is also a substantial amount of 'housekeeping' to be performed such as aligning decimal points, assigning the proper algebraic sign, and floating point normalization. Although the implementation of a keyboard function may involve thousands of clock cycles, the total elapsed time is in the millisecond region because each clock cycle is only 825 ns long.

The program ROM contains 512 64-bit words. When the program ROM is activated, signals (micro-instructions) corresponding to the bit pattern in the word are sent to the hard wired logic gates shown at the bottom of Fig. 1. The logic gates define the changes to occur in the flip flops at the end of a clock cycle. Some of the micro-instructions act upon the data flip flops while others change the address registers associated with the program ROM, control ROM and coincident current core memory. During the next clock cycle the control ROM may ask for a new set of micro-instructions from the program ROM or ask to be read from or written into the coincident current core memory. The control ROM also has the ability to modify its own address register and to issue micro-instructions to the hard wired logic gates. This flexibility allows the control logic ROM to execute special programs such as the subroutine for unpacking the stored constants required by the keyboard transcendental functions.

Control Logic

The control logic uses a wire braid toroidal core read only memory containing 64 29-bit words. Magnetic logic of this type is extremely reliable and pleasingly compact.

The crystal controlled clock source initiates a current pulse having a trapezoidal waveform which is directed through one of 64 word lines. Bit patterns are generated by passing or threading selected toroids with the word lines. Each toroid that is threaded acts as a transformer to turn on a transistor connected to the output winding of the toroid. The signals from these transistors operate the program ROM, coincident current core, and selected micro-instructions.

Coincident Current Core Read/Write Memory

The 2208 (6 x 16 x 23) bit coincident current memory uses wide temperature range lithium cores. In addition, the X, Y, and inhibit drivers have temperature compensated current drive sources to make the core memory insensitive to temperature and power supply variations.

The arithmetic processing unit includes special circuitry to guarantee that information is not lost from the core memory when power is turned off and on.

Power Supplies

The arithmetic processing unit operates from a single —15 volt supply. Even though the power supply is highly regulated, all circuits are designed to operate over a voltage range of —13.5 to —16.5 volts.

Display

The display is generated on an HP electrostatic cathode ray tube only 11 inches long. The flat rectangular face place measures 3¼″ x 4¹³⁄₁₆ inches. The tube was specifically designed to generate a bright image. High contrast is obtained by using a low transmissivity filter in front of the CRT. Ambient light that usually tends to 'wash out' an image is attenuated twice by the filter, while the screen image is only attenuated once.

Fig. 3. *Displayed characters are generated by modulating these figures. The digit 1 is shifted to the center of the pattern.*

All the displayed characters are 'pieces of eight.' Sixteen different symbols are obtained by intensity modulating a figure 8 pattern as shown in Fig. 3. Floating point numbers are partitioned into groups of three digits and the numeral 1 is shifted to improve readability. Zeros to the left of the most significant digit and insignificant zeros to the right of the decimal point are blanked to avoid a confusing display. Fixed point numbers are automatically rounded up according to the decimal wheel setting. A fixed point display will automatically revert to floating point notation if the number is too large to be displayed on the CRT in fixed point.

Multilayer Instruction Logic Board

All of the hard wired logic gates are synthesized on

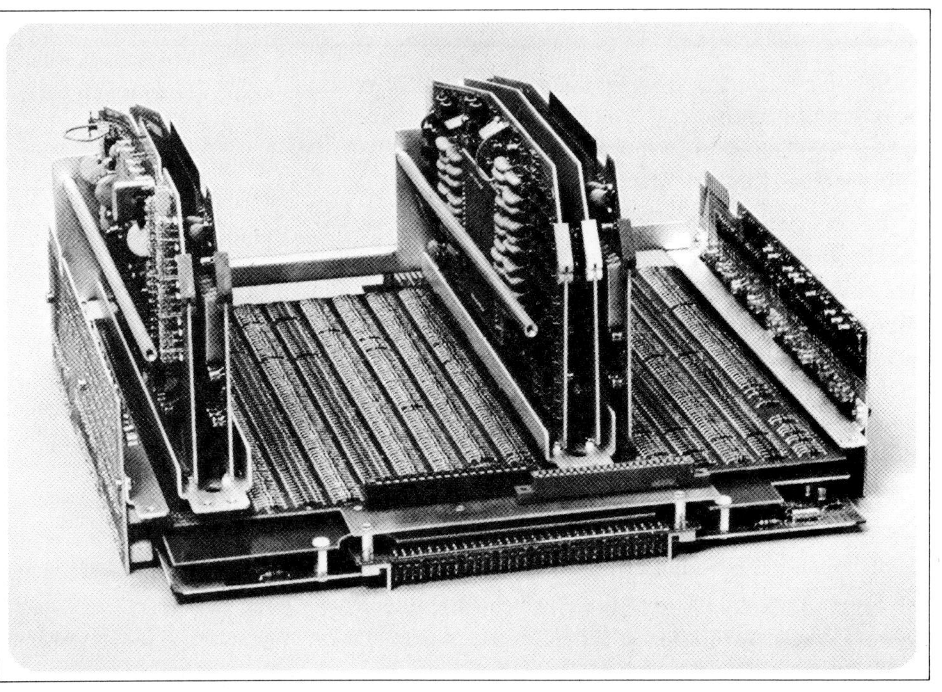

Fig. 2. *Arithmetic unit assembly removed from the calculator.*

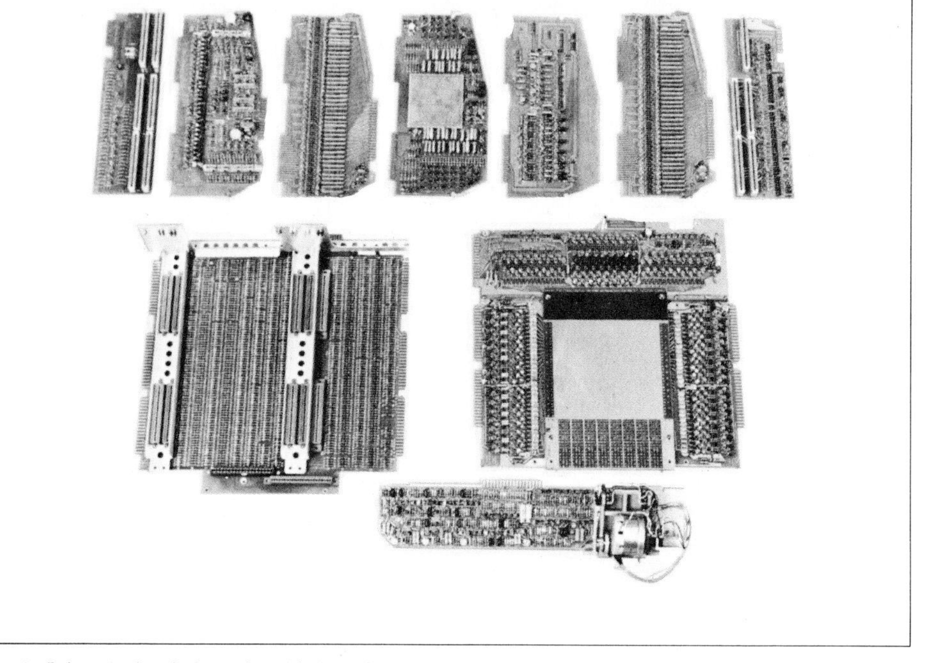

Fig. 4. *Printed circuit boards which make up the arithmetic unit are, left to right at top, side board, control logic, flip-flop, core and drivers, core sense amplifiers and inhibit, flip-flop, and side board. Large board at the lower left is the multilayer instruction board and the program ROM is at the right. The magnetic card reader and its associated circuitry is at the bottom.*

the instruction logic board using time-proven diode-resistor logic. The diodes and resistors are located in separate rows, Fig. 4. All diodes are oriented in the same direction and all resistors are the same value. The maze of interconnections normally associated with the back plane wiring of a computer are located on the six internal layers of the multilayer instruction logic board. Solder bridges and accidental shorts caused by test probes shorting to leads beneath components are all but eliminated by not having interconnections on the two outside surfaces of this multilayer board. The instruction logic board also serves as a motherboard for the control logic board, the two coincident core boards and the two flip flop boards, the magnetic card reader, and the keyboard. It also contains a connector, available at the rear of the calculator, for connecting peripherals.

Flip Flops

The Model 9100A contains 40 identical J-K flip flops, each having a threshold noise immunity of 2.5 volts. Worst case design techniques guarantee that the flip flops will operate at 3 MHz even though 1.2 MHz is the maximum operating rate.

Program Read Only Memory

The 32,768 bit read only program memory consists of 512 64-bit words. These words contain all of the operating subroutines, stored constants, character encoders, and CRT modulating patterns. The 512 words are contained in a 16 layer printed-circuit board having drive and sense lines orthogonally located. A drive line consists of a reference line and a data line. Drive pulses are inductively coupled from both the reference line and data line into the sense lines. Signals from the data line either aid or cancel signals from the reference line producing either a 1 or 0 on the output sense lines. The drive and sense lines are arranged to achieve a bit density in the ROM data board of 1000 bits per square inch.

The program ROM decoder/driver circuits are located directly above the ROM data board. Thirty-two combination sense amplifier, gated-latch circuits are located on each side of the ROM data board. The outputs of these circuits control the hard wired logic gates on the instruction logic board.

Side Boards

The program ROM printed circuit board and the instruction logic board are interconnected by the side boards, where preliminary signal processing occurs.

The Keyboard

The keyboard contains 63 molded plastic keys. Their markings will not wear off because the lettering is imbedded into the key body using a double shot injection molding process. The key and switch assembly was specifically designed to obtain a pleasing feel and the proper amount of tactile and aural feedback. Each key operates a single switch having gold alloy contacts. A contact closure activates a matrix which encodes signals on six data lines and generates an initiating signal. This signal is delayed to avoid the effects of contact bounce. An electrical interlock prevents errors caused by pressing more than one key at a time.

Magnetic Card Reader

Two complete 196 step programs can be recorded on the credit card size magnetic program card. The recording process erases any previous information so that a card may be used over and over again. A program may be protected against accidental erasure by clipping off the corner of the card, Fig. 9, page 8. The missing cor-

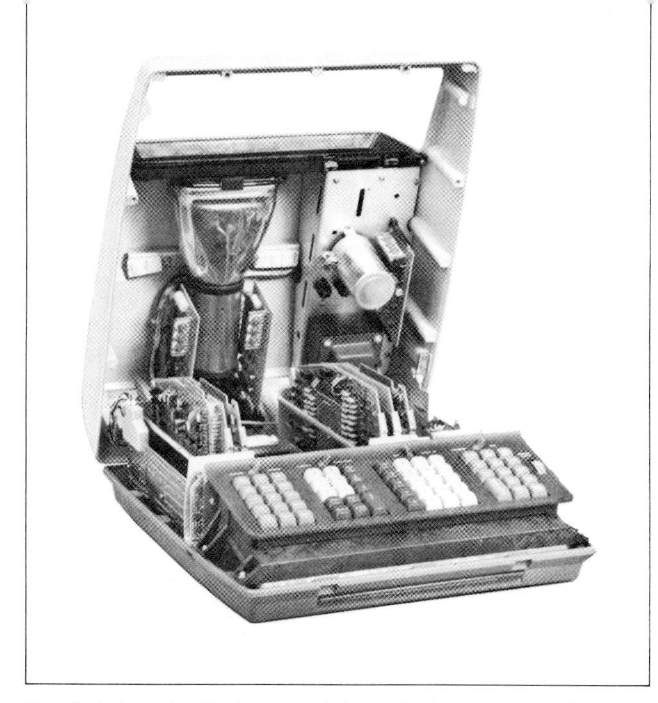

Fig. 5. *Internal adjustments of the calculator are easily accessible by removing a few screws and lifting the top.*

ner deactivates the recording circuitry in the magnetic card reader. Program cards are compatible among machines.

Information is recorded in four tracks with a bit density of 200 bits per inch. Each six-bit program step is split into two time-multiplexed, three-bit codes and recorded on three of the four tracks. The fourth track provides the timing strobe.

Information is read from the card and recombined into six bit codes for entry into the core memory. The magnetic card reading circuitry recognizes the 'END' program code as a signal to end the reading process. This feature makes it possible to enter subroutines within the body of a main program or to enter numeric constants via the program card. The END code also sets the program counter to location 0-0, the most probable starting location. The latter feature makes the Model 9100A ideally suited to 'linking' programs that require more than 196 steps.

Packaging and Servicing

The packaging of the Model 9100A began by giving the HP industrial design group a volume estimate of the electronics package, the CRT display size and the number of keys on the keyboard. Several sketches were drawn and the best one was selected. The electronics sections were then specifically designed to fit in this case. Much time and effort were spent on the packaging of the arithmetic processing unit. The photographs, Figs. 2 and 5 attest to the fact that it was time well spent.

The case covers are die cast aluminum which offers durability, effective RFI shielding, excellent heat transfer characteristics, and convenient mechanical mounts. Removing four screws allows the case to be opened and locked into position, Fig. 5. This procedure exposes all important diagnostic test points and adjustments. The keyboard and arithmetic processing unit may be freed by removing four and seven screws respectively.

Any component failures can be isolated by using a diagnostic routine or a special tester. The faulty assembly is then replaced and is sent to a service center for computer assisted diagnosis and repair.

Reliability

Extensive precautions have been taken to insure maximum reliability. Initially, wide electrical operating margins were obtained by using 'worst case' design techniques. In production all transistors are aged at 80% of rated power for 96 hours and tested before being used in the Model 9100A. Subassemblies are computer tested and actual operating margins are monitored to detect trends that could lead to failures. These data are analyzed and corrective action is initiated to reverse the trend. In addition, each calculator is operated in an environmental chamber at 55°C for 5 days prior to shipment to the customer. Precautions such as these allow Hewlett-Packard to offer a one year warranty in a field where 90 days is an accepted standard.

Thomas E. Osborne

Tom Osborne joined HP as a consultant in late 1965 with the responsibility for developing the architecture of the Model 9100A. Previous to joining HP, he had designed data processing equipment, then formed Logic Design Co., where he developed a floating point calculator upon which the Model 9100A is based.

Tom graduated from the University of Wyoming in 1957 with a BSEE, and was named 'Outstanding Electrical Engineer' of his class. He received his MSEE from the University of California at Berkeley.

Tom enjoys flying as a pastime, he is an ardent theater-goer and a connoisseur of fine wines. He is a member of Sigma Tau and Phi Kappa Phi honorary fraternities, and a member of IEEE.

Internal Programming of the 9100A Calculator

By David S. Cochran

EXTENSIVE INTERNAL PROGRAMMING has been designed into the HP Model 9100A Calculator to enable the operator to enter data and to perform most arithmetic operations necessary for engineering and scientific calculation with a single key stroke or single program step. Each of the following operations is a hardware subroutine called by a key press or program step:

- Basic arithmetic operations
  - Addition
  - Subtraction
  - Multiplication
  - Division
- Extended arithmetic operations
  - Square root
  - Exponential — e^x
  - Logarithmic — ln x, log x
  - Vector addition and subtraction
- Trigonometric operations
  - Sin x, cos x, tan x
  - Arcsin x, arccos x, arctan x
  - Sinh x, cosh x, tanh x
  - Arcsinh x, arccosh x, arctanh x
  - Polar to rectangular and rectangular to polar coordinate transformation
- Miscellaneous
  - Enter π
  - Absolute value of y
  - Integer value of x

In the evolution of internal programming of the Model 9100A Calculator, the first step was the development of flow charts of each function. Digit entry, Fig. 1, seemingly a trivial function, is as complex as most of the mathematical functions. From this functional description, a detailed program can be written which uses the microprograms and incremental instructions of the calculator. Also, each program must be married to all of the other programs which make up the hard-wired software of the Model 9100A. Mathematical functions are similarly programmed defining a step-by-step procedure or algorithm for solving the desired mathematical problem.

The calculator is designed so that lower-order subroutines may be nested to a level of five in higher-order functions. For instance, the 'Polar to Rectangular' function uses the sin routine which uses multiply which uses add, etc.

Addition and Subtraction

The most elementary mathematical operation is algebraic addition. But even this is relatively complex — it requires comparing signs and complementing if signs are unlike. Because all numbers in the Model 9100A are processed as true floating point numbers, exponents must be subtracted to determine proper decimal alignment. If one of the numbers is zero, it is represented in the calculator by an all-zero mantissa with zero exponent. The difference between the two exponents determines the offset, and rather than shifting the smaller number to the right, a displaced digit-by-digit addition is performed. It must also be determined if the offset is greater than 12, which is the resolution limit.

Although the display shows 10 significant digits, all calculations are performed to 12 significant digits with the two last significant digits (guard digits) absorbing truncation and round-off errors. All registers are in core memory, eliminating the need for a large number of flip-flop registers. Even with the display in 'Fixed Point' mode, every computed result is in storage in 12 digits.

Multiplication

Multiplication is successive addition of the multiplicand as determined by each multiplier digit. Offset in the digit position flip-flops is increased by one after completion of the additions by each multiplier digit. Exponents are added after completion of the product. Then the product is normalized to justify a carry digit which might have occurred.

Division

Division involves repeated subtraction of the divisor

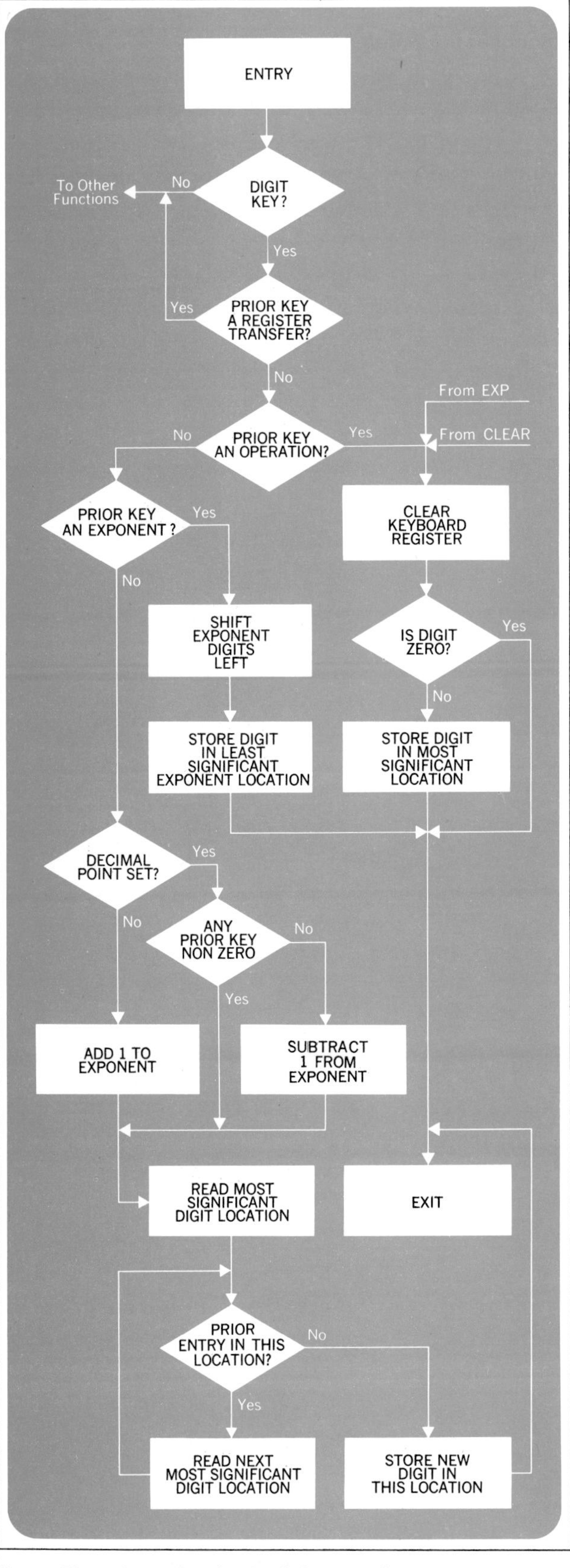

Fig. 1. *Flow chart of a simple digit entry. Some of these flow paths are used by other calculator operations for greater hardware efficiency.*

from the dividend until an overdraft occurs. At each subtraction without overdraft, the quotient digit is incremented by one at the digit position of iteration. When an overdraft occurs, the dividend is restored by adding the divisor. The division digit position is then incremented and the process continued. Exponents are subtracted after the quotient is formed, and the quotient normalized.

Square Root

Square root, in the Model 9100A, is considered a basic operation and is done by pseudo division. The method used is an extension of the integer relationship.

$$\sum_{i=1}^{n} 2i - 1 = n^2$$

In square root, the divisor digit is incremented at each iteration, and shifted when an overdraft and restore occurs. This is a very fast algorithm for square root and is equal in speed to division.

Circular Routines

The circular routines (sin, cos, tan), the inverse circular routines (arcsin, arccos, arctan) and the polar to rectangular and rectangular to polar conversions are all accomplished by iterating through a transformation which rotates the axes. Any angle may be represented as an angle between 0 and 1 radian plus additional information such as the number of times $\pi/2$ has been added or subtracted, and its sign. The basic algorithm for the forward circular function operates on an angle whose absolute value is less than 1 radian, but prescaling is necessary to indicate quadrant.

To obtain the scaling constants, the argument is divided by 2π, the integer part discarded and the remaining fraction of the circle multiplied by 2π. Then $\pi/2$ is subtracted from the absolute value until the angle is less than 1 radian. The number of times $\pi/2$ is subtracted, the original sign of the argument, and the sign upon completion of the last subtraction make up the scaling constants. To preserve the quadrant information the scaling constants are stored in the core memory.

The algorithm produces $\tan \theta$. Therefore, in the Model 9100A, $\cos \theta$ is generated as $\frac{1}{\sqrt{1 + \tan^2\theta}}$ and $\sin \theta$ as $\frac{\tan \theta}{\sqrt{1 + \tan^2\theta}}$.

Sin θ could be obtained from the relationship $\sin \theta = \sqrt{1 - \cos^2\theta}$, for example, but the use of the tangent relationship preserves the 12 digit accuracy for very small angles, even in the range of $\theta < 10^{-12}$. The proper signs of the functions are assigned from the scaling constants.

For the polar to rectangular functions, $\cos \theta$ and $\sin \theta$ are computed and multiplied by the radius vector to obtain the X and Y coordinates. In performing the rectangular to polar function, the signs of both the X and Y vectors are retained to place the resulting angle in the right quadrant.

Prescaling must also precede the inverse circular functions, since this routine operates on arguments less than or equal to 1. The inverse circular algorithm yields arctangent functions, making it necessary to use the trigonometric identity.

$$\sin^{-1}(x) = \tan^{-1}\frac{x}{\sqrt{1 - x^2}}.$$

If a $\cos^{-1}(x)$ is desired, the arcsin relationship is used and a scaling constant adds $\pi/2$ after completion of the function. For arguments greater than 1, the arccotangent of the negative reciprocal is found which yields the arctangent when $\pi/2$ is added.

David S. Cochran

Dave Cochran has been with HP since 1956. He received his BSEE from Stanford in 1958 and his MS from the same school in 1960. He has been responsible for the development of a broad range of instruments including the Model 3440A Digital Voltmeter and the 204B Oscillator. He has been working on development of electrostatic printing. Dave developed the internal programming for the Model 9100A Calculator.

He holds six patents in the instrumentation field and is the author of several published papers. Dave is a registered professional engineer, a member of IEEE and Tau Beta Pi.

Exponential and Logarithms

The exponential routine uses a compound iteration algorithm which has an argument range of 0 to the natural log of 10 (ln 10). Therefore, to be able to handle any argument within the dynamic range of the calculator, it is necessary to prescale the absolute value of the argument by dividing it by ln 10 and saving the integer part to be used as the exponent of the final answer. The fractional part is multiplied by ln 10 and the exponential found. This number is the mantissa, and with the previously saved integer part as a power of 10 exponent, becomes the final answer.

The exponential answer is reciprocated in case the original argument was negative, and for use in the hyperbolic functions. For these hyperbolic functions, the following identities are used:

$$\sinh x = \frac{e^x - e^{-x}}{2}, \cosh x = \frac{e^x + e^{-x}}{2}, \tanh x = \frac{e^x - e^{-x}}{e^x + e^{-x}}.$$

Natural Logarithms

The exponential routine in reverse is used as the routine for natural logs, with only the mantissa operated upon. Then the exponent is multipled by ln 10 and added to the answer. This routine also yields these $\log_{10}$ and arc hyperbolic functions:

$$\text{Log}_{10}\, x = \frac{\ln x}{\ln 10}; \sinh^{-1}(x) = \ln(x + \sqrt{x^2 + 1});$$

$$\cosh^{-1}(x) = \ln(x + \sqrt{x^2 - 1}); \tanh^{-1}(x) = \ln\sqrt{\frac{1+x}{1-x}}.$$

The $\sinh^{-1}(x)$ relationship above yields reduced accuracy for negative values of x. Therefore, in the Model 9100A, the absolute value of the argument is operated upon and the correct sign affixed after completion.

Accuracy

It can be seen from the discussion of the algorithms that extreme care has been taken to use routines that have accuracy commensurate with the dynamic range of the calculator. For example; the square root has a maximum possible relative error of 1 part in 10^{10} over the full range of the machine.

There are many algorithms for determining the sine of an angle; most of these have points of high error. The sine routine in the Model 9100A has consistent low error regardless of quadrant. Marrying a full floating decimal calculator with unique mathematical algorithms results in accuracy of better than 10 displayed digits.

Computer-Testing the HP Model 9100A Calculator

By Charles W. Near

WITH SEVERAL THOUSAND DIODES AND TRANSISTORS plus three different memory arrays making up the 9100A Calculator, the probability is near zero that an untested calculator will operate perfectly at initial turn-on. Aging and pretesting of the transistors add greatly to reliability, but the total test time per calculator remains high unless rigorous testing is performed on the logic subassemblies. More than a million test measurements per calculator are required to assure acceptable performance of the 40 flip-flops, the braid memory control logic, the core memory and the main read only memory.

Use of a digital computer, Fig. 1, to control this testing offers a number of important advantages. Many tests may be performed in a short time and the test data stored in computer memory. The computer can analyze the test data and provide instantaneous print-out of fault diagnostics. By using programmable power supplies, the actual operating margins may be measured and printed, thereby simulating adverse environments or detecting processing changes which may influence the performance range of the assembly. Test automation also assures that the test is repeated identically on a day to day run or run to run basis for greatest accuracy.

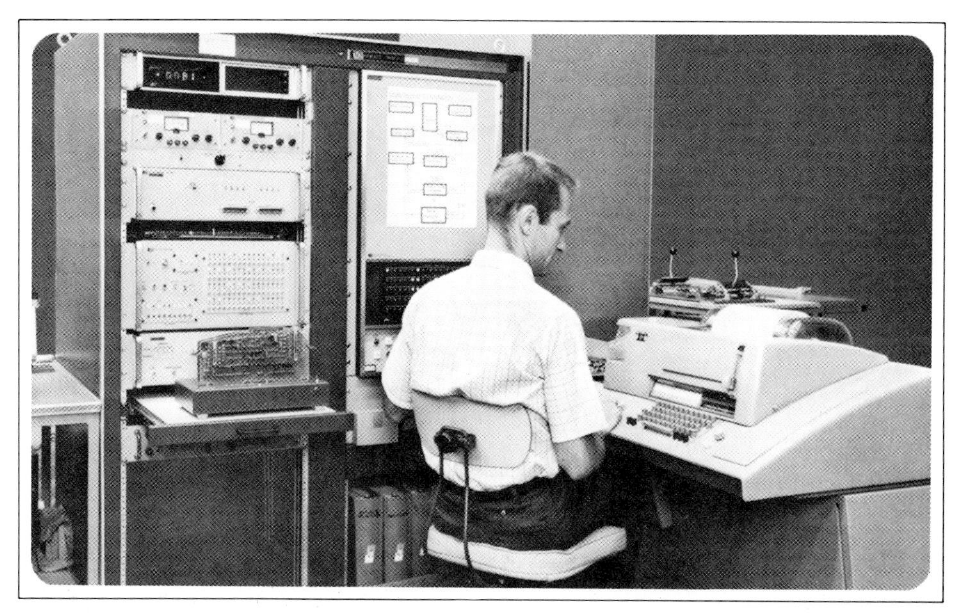

Fig. 1. *Computer test station for the Model 9100A Calculator uses an HP Model 2116B computer, a teleprinter and test interface units for each logic assembly to be tested. Braid memory control logic is under test.*

Test System

An HP Model 2116B Computer with 8192 words of core memory has been coupled to two scanners, a digital voltmeter and three programmable power supplies for production testing of the calculator, Fig. 2. The scanners have 100 inputs each with reed relay contacts capable of 200 scan points per second. A Model ASR-35 teleprinter provides a printed output of test results and permits communication with the system during test.

A test interface unit has been built for each logic subassembly to be tested. Each interface unit contains circuits to produce dynamic stimuli compatible with calculator operation as well as to measure timing, voltage levels and logic states under computer control. A relay

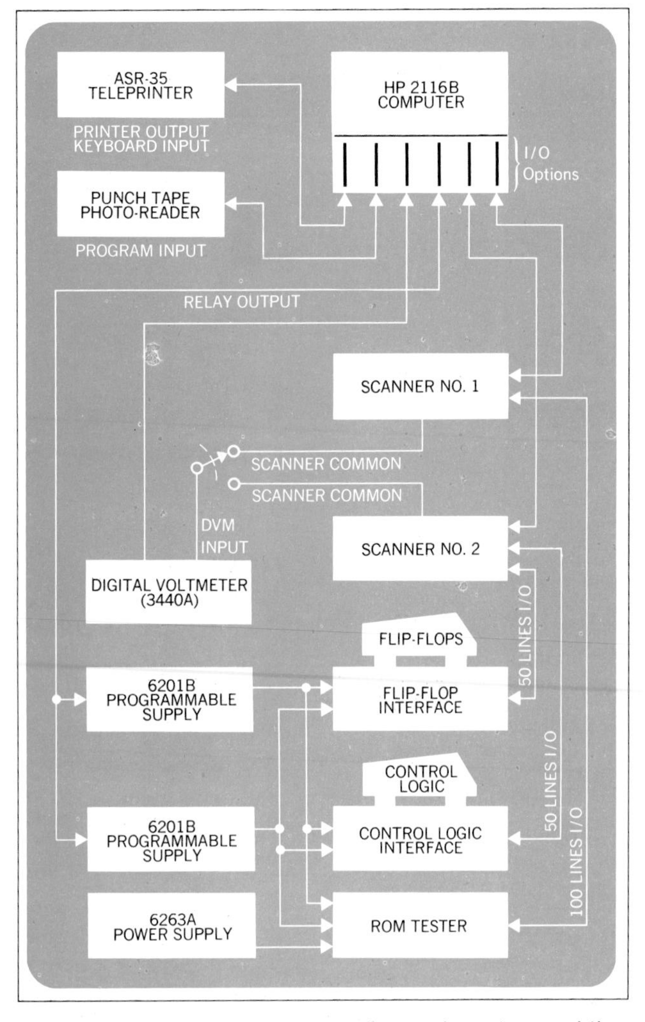

Fig. 2. *System for pretesting logic subassemblies. Programmable power supplies check operating margins.*

output register in the computer directly programs the power supplies, and the two general purpose registers provide input/output for the scanners and interface units.

Dynamic Testing

Response times of the calculator circuits are fully as important in testing as are the logic levels. For example, in testing the JK flip-flop a minimum width clock pulse is generated at a minimum time after shifting the input levels, thereby simulating a 'worst-case' operation. Similarly, a fixed duration forcing pulse applied to the output collector of each flip-flop detects any transistor that exceeds its stored base charge specification. The identifying numbers of the defective flip-flop and transistor are printed out together with the voltage and load at which failure occurred. If a collector voltage is outside of the programmed limits, the voltage will be printed out and again identified. The message from the teleprinter is placed with the tested board for repair or stockpiling and the next board is inserted. For the 2250 measurements made on a flip-flop board, test time is 30 to 40 seconds.

Woven Wire Memory

Testing of the Control Logic board requires verification of all bits stored in a 64 word by 29 bit toroid braid memory. Since the crystal-controlled clock is on this board, this clock is used in the tester interface, Fig. 2, to test the braid memory dynamically. As the word address is changed in a given clock time, the output to a test flip-flop is enabled on the next cycle, and the flip-flop state is read into computer memory. By switching the

```
CONTROL LOGIC TEST

 31 MS TURN-ON

    SUPPLY VOLTAGES:     -02.45
                         -16.48    62 ERRORS

00 SHUT-DOWN AT V  +01.47
51 SHUT-DOWN AT V  +00.81

ROM OUTPUT SHORTED

      04 44
RCY    0  1
TMS    1  0
 A     0  1
 P     0  1
 I     0  1
C60    1  0
C64    1  0

                            62 ERRORS
```

```
CONTROL LOGIC TEST

 31 MS TURN-ON

   SUPPLY VOLTAGES:     -02.45
                        -16.49 ** NO ERRORS **
                        -13.54 ** NO ERRORS **

00 SHUT-DOWN AT V  +01.47
51 SHUT-DOWN AT V  +00.81

                       ** NO ERRORS **
```

Fig. 3. *Typical teleprinter output for a defective assembly (left), and printout for a perfect board (right).*

scanner to each of the 29 output transistors and addressing the 64 states at each output, the 1856 bits of data are read into computer memory and compared against the correct data. If no errors are found at 10% high voltage, the sequence is repeated for 10% low voltage.

If errors are detected, the computer analyzes the error pattern to determine if an output transistor is shorted or open, if a word wire is open or exchanged, if the braid is incorrectly woven, or if various other errors exist. These diagnostics are printed as shown in the example, Fig. 3, which indicates two wires exchanged (address 04 and 44) and one output transistor shorted for a total of 62 bit errors. Also on this test the timing of the turn-on circuitry is measured and printed in milliseconds, as well as the voltage at which control is forced to the shut-down state. Typical time for the test and print-out is 50 to 60 seconds.

ROM Test by Comparison Method

The 32,768 bit main Read Only Memory is the most complex of the logic subassemblies and would require considerable input/output time to test by the method used for the braid memory. Thus a more sophisticated ROM tester interface, Fig. 4, has been built to compare the ROM assembly against a reference ROM at calculator speeds through 64 exclusive-OR gates. Each exclusive-OR output sets a corresponding buffer flip-flop, which in turn halts testing and flags the computer to input the error data. Thus the entire 32,768 bits are tested and the errors stored in milliseconds to seconds depending on the number of errors. If errors are detected at nominal operating conditions, diagnostics are performed much as with the braid memory test.

If no errors were found, the computer adjusts the sense amplifier bias in 0.25 volt increments until a specified number of failures occur or a +50% to —50% bias margin has been tested. The computer then organizes this error data in table form showing failures as a function of bias voltage. Typically 600,000 to 1,200,000 bit comparisons are performed in a margin test in about two minutes. The margin tables aid in early detection of any process change which affects operating margins, and also serve to define an optimum bias value for each ROM assembly to assure greater reliability of operation.

A variation of the ROM tester is used to test the Read Only Memory data board before the drive circuits and sense amplifier circuits are soldered to it. This tester has drive and sense circuits that make pressure contact to the ROM data board and may be used with or without the aid of the computer.

Fig. 4. *Read-Only Memory tester interface compares the ROM assembly mounted on top with a reference unit.*

The goal of minimizing total test time and increasing operating margins through use of an HP 2116B Computer test facility is fast becoming a reality. Based on data from early production runs, over 90% of fully assembled units operate perfectly at initial turn-on. By comparison, in subassembly testing about 50% of flip-flops, control logic and ROM subassemblies pass the computer test initially without the need for any further adjustment. For those problems that do arise in final test, trouble-shooting is still required, for the complexity at this level exceeds the diagnostic capabilities of the computer test facility.

Acknowledgments

I am pleased to acknowledge the contributions by Bob Watson who initially proposed use of the computer to test the 9100A Calculator, and by Jack Anderson who offered valuable suggestions for adaptation of the computer to production testing. Norm Carlson and John Scohy provided electronic tooling in construction of the test interface units; Rex James and Clair Nelson contributed to development of the read only memory tester.

Charles W. Near

Chuck Near received his B.S. degree in Electrical Engineering from Michigan State University in 1961. Upon joining Hewlett-Packard the same year, he helped design the HP Model 3440A Digital Voltmeter. In 1964 he earned his MSEE from Stanford University on the HP Honors Cooperative Program.

During development of the HP Model 9100A Calculator at HP Laboratories, Chuck had design responsibility for the Calculator Read Only Memory, and has been issued a patent on this memory device. In July of 1967 Chuck was transferred from HP Labs to the Loveland Division where he designed the computer test facility for calculator production.

Chuck is a member of Tau Beta Pi, Eta Kappa Nu and IEEE.

How the Model 9100A Was Developed

By Bernard M. Oliver
Vice President for
Research & Development

Some lab projects are endothermic: the desired reaction proceeds only with the application of considerable heat and pressure and stops the moment these are relaxed. Others are exothermic: when the proper ingredients are brought together the reaction starts automatically and it is only necessary to harness and control the power that is generated. The 9100A project was one of the most exothermic I have known.

The ingredients started coming together in the late summer of 1965 when we were shown a prototype of a calculator invented by Malcolm McMillan that had one interesting feature: it could calculate all the common transcendental functions. The machine operated in fixed point and took a few seconds to calculate a function, but it did demonstrate the feasibility of providing these functions in a small calculator, and the power of the algorithm used to compute them.

The second ingredient was Tom Osborne who came to see us carrying a little green balsa wood calculator, which he had built on his own to demonstrate the virtue of some of his design concepts. What impressed us was its millisecond speed and its ten digit floating point operation and display.

A third ingredient was the imagination of Paul Stoft and other engineers in his group. It took no genius to see the appeal of a calculator that combined the speed and dynamic range of Tom's machine with the transcendental computing ability of the other machine. But to combine them into a small machine faster than either prototype, to adapt the transcendental algorithm to floating point, to add programmability and magnetic card storage and entry of programs, to provide the flexible display with automatic roundup, and to design the whole assembly for automated production required not only imagination but engineering skill of the highest order.

Another ingredient, the read only memory, which stores all the calculating and display routines, was already under development by Arndt Bergh and Chuck Near before we began the 9100A project. By carrying printed circuit techniques beyond the existing state of the art, Chuck was able to compress the required 32,000 bits into an amazingly small space. While the 9100A uses only discrete diodes and transistors, it it fair to describe the read only memory as one large integrated circuit with extremely long life expectancy.

As soon as the development began, everyone, it seemed, had ideas for new features. Hardly a day went by without someone proposing a new keyboard with a new key arrangement or new functions. It was Dick Monnier's responsibility as section leader to steer the project through these conflicting currents of ideas. Although we went down a couple of blind alleys, for the most part the course was held true. That we arrived at such an elegant solution in a short time is a tribute to Dick's navigation.

Tom Osborne joined the development team as a consultant on the general architecture of the machine. His contributions include the basic logical design, and the details of the control logic, flip-flops, mother board gates, and the memory drive and sense circuitry. He also contributed a large measure of sound judgment that resulted in an economy of design, low power consumption, high performance, and the ability of the 9100A to interface with peripherals and systems.

Tony Lukes developed the display routines, which include the features of automatic roundup in fixed point display, choice of decimal places, suppression of insignificant zeros, and the display of program step addresses and key codes, as well as numerical data. Tony also developed the program storage, editing, and execution routines which, together with the display, make the 9100A easy to use and program.

I should at this point emphasize that the 9100A, while small, is, in a sense, much more complicated than many general purpose computers. Most general purpose computers have relatively large memories but can execute directly only certain elementary machine instructions. To compute complicated functions and indeed, in some cases, even to perform simple arithmetic, the computer must be externally programmed. By contrast the 9100A has a very sophisticated external instruction set: the entire keyboard. The 9100A is a small computer with a large amount of 'software' built in as hardware.

The task of compressing the floating point arithmetic operations and functional computations into the limited read only memory of the 9100A was accomplished by Dave Cochran. To make sure that all the calculations were accurate over the enormous range of arguments allowed by the floating point operation, to assure exact values at certain cardinal points, and above all to get so much in so little memory space was an enormous achievement.

I sometimes wonder if Dave realizes what a remarkable job he did. It took several passes. On the first pass it appeared hopeless to include all the functions. But by nesting routines and by inventing a number of space saving tricks he was able to save enough states to crowd them all in. Then various bugs were discovered and more states had to be freed to correct these. The 'battle of the states' continued for several months and the end result was one of the most efficient encoding jobs ever done.

The necessity for magnetic card program storage and entry became apparent as soon as we had an operating prototype. Don Miller, Dick Osgood and Bob Schweizer deserve praise for the speedy development of this unit, which adds so much to the convenience of the 9100A.

Clarence Studley supervised the overall mechanical design and assembly, reducing to manufacturing drawings and to the final metal the handsome cabinet styling of Roy Ozaki, Don Aupperle and others in the Industrial Design group, while Harold Rocklitz and Doug Wright handled much of the tooling.

Many other people contributed to the 9100A — too many to give proper credit to all; but I must mention the fine art work of Frank Lee on the read only memory and other printed circuit boards, Chung Tung's work on the core memory electronics and Bill Kruger's development of the short high brightness cathode ray tube. Chris Clare made many contributions to the project especially in the area of interfacing the calculator with printers and other peripherals. A special measure of recognition is due Ken Petersen whose expert technicianship and whose genius at trouble shooting saved us weeks of time and bailed us out of many tight spots. Ken also laid out the multilayer mother board with its thousand diode gates and interconnections.

The transfer of the 9100A from Hewlett-Packard Laboratories to the Loveland Division took place gradually rather than abruptly. As various portions reached the final prototype stages, responsibility for these was assumed by the Loveland group headed by Bob Watson in engineering and by Jack Anderson in production. Many visits both ways and some transfer of people to Loveland accomplished the transmission of much unwritten information. The Loveland team introduced several engineering improvements. Especially significant were the improved read only memory margins obtained by Rex James and the many contributions by Ed Olander, whose comprehensive understanding of the entire machine helped greatly. That we were able to go from an incomplete lab prototype stage in Palo Alto to a pilot run of final instruments in Loveland in only 10 months attests to the skill and dedication of the Loveland group and to the fine cooperation on both sides.

Finally, I must confess that very few projects receive as much direct attention from corporate management as this one did. Early in the spring of 1967 a skiing injury landed Bill Hewlett in the hospital. We learned about this right away when he called up to have some 9100A programming pads sent over. I found myself hypnotized by the project and unable to share my time equitably. Here was management in the unusual role of consumer, for if Bill and I did anything constructive it was mainly to assess and modify the developing product from the user's standpoint. I owe the 9100A group an apology for being constantly in their hair, and everyone else in HP Labs an apology for slighting their projects. Now that it's all over I find the 9100A as fascinating to use as it was to develop. *Caveat emptor!*

HEWLETT-PACKARD JOURNAL **SEPTEMBER 1968** *Volume 20 · Number 1*

TECHNICAL INFORMATION FROM THE LABORATORIES OF THE HEWLETT-PACKARD COMPANY PUBLISHED AT 1501 PAGE MILL ROAD, PALO ALTO, CALIFORNIA 94304
Editorial Staff: F. J. BURKHARD, R. P. DOLAN, L. D. SHERGALIS, R. H. SNYDER Art Director: R. A. ERICKSON

First in a line of solid-state display devices are these one- and three-digit numeric indicators. Compatible with integrated circuits, they need only BCD input signals and five-volt power to display any numeral from 0 to 9 in an array of bright red dots.

Solid-State Displays

By Howard C. Borden and Gerald P. Pighini

SOLID-STATE DISPLAYS ARE HERE. Developing them has taken more than six years of research and development in light-emitting materials, plus Hewlett-Packard's resources in solid-state technology, integrated-circuit design and manufacture, ceramic metallization and etching, and optoelectronic packaging. The result is the new HP Model 5082-7000 Numeric Indicator, a small, low-power, all-semiconductor module which accepts four-line binary-coded-decimal input signals and displays the corresponding digit, 0 through 9, as an array of brightly glowing red dots (Fig. 1). A similar but larger module, Model 5082-7001, displays three digits in line.

Compatibility with integrated circuits is a significant advantage of the new solid-state indicators over other types of display devices. They need only five-volt power and logic levels of 0 and 5 volts (nominal), compatible with transistor-transistor logic (TTL) and diode-transistor logic (DTL).

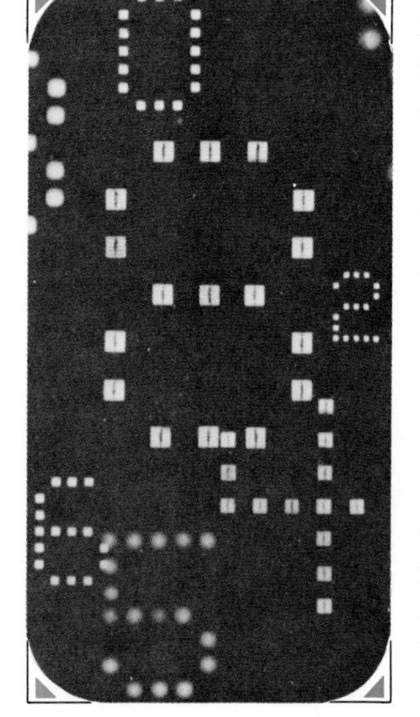

IC compatibility, in fact, was the principal motivating factor in the development of the solid-state display. However, it wasn't the only one. The list of advantages of solid-state displays is impressive. Among them are thin single-plane presentation, ruggedness, and high 'solid-state' reliability. Because they are free from fundamental degradation mechanisms, they are expected to have long life. They don't generate RFI, and they are amenable to low-cost, high-volume production using semiconductor batch fabrication techniques.

The new solid-state indicators have other advantages as well. They have a surface light distribution which gives constant brightness over wide viewing angles. They have high contrast and color purity, both of which contribute to readability. They are free from parallax because they produce all numerals in the same plane, and they respond in less than one microsecond to input-code changes; hence they are useful as readout devices when test results are being photographed with high-speed cameras. The brightness of the solid-state indicators is voltage-variable; it can be adjusted for optimum readability under widely varying ambient light conditions.

Obvious uses of the new indicators are in instrument panels, status boards, and information displays, or anywhere a need exists for a compact, IC-compatible, variable-brightness readout module that can perform both the decoding function and the readout function. The possibilities for these indicators and for future solid-state displays are intriguing; some are discussed on page 4.

Itself an instrument manufacturer, Hewlett-Packard may eventually become its own best customer for solid-state readouts, especially where a large character font is needed. Limited character font is a disadvantage of gaseous display tubes. Where greater flexibility in alpha-numeric and symbolic display is needed, as in the

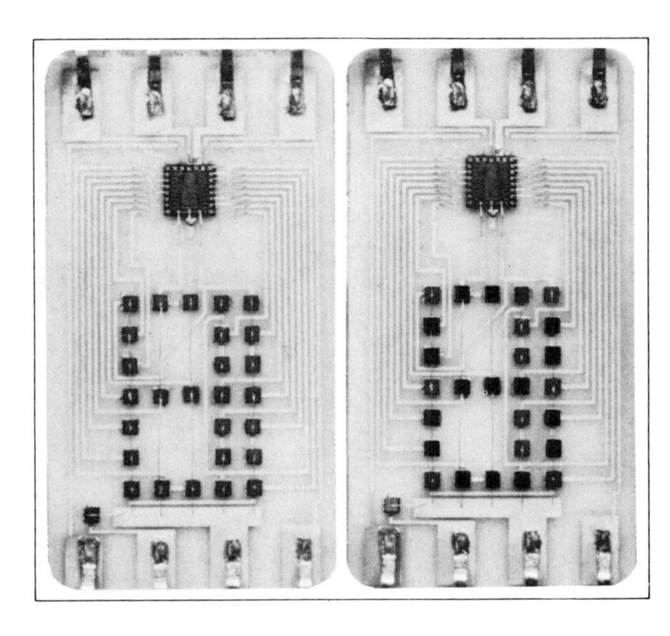

Fig. 1. *Model 5082-7000 Solid-State Numeric Indicator accepts four-line BCD and displays any numeral from 0 to 9 in an array of bright red dots. It takes five-volt power and logic levels compatible with integrated circuits. Shown here is the bare module; mounted behind red glass, only the numeral is visible.*

HP Model 9100A Calculator, HP has used cathode-ray-tube displays. CRT's are still the most economical solution to the type of display requirements found in the calculator. However, CRT's are large, need high voltages, and must have circuits to generate and recirculate the characters. Although we don't have it yet, an all-solid-state alternative to the CRT seems to be desirable. The new numeric indicators are a first step in that direction.

Optical Characteristics

A display is an interface between a machine and a man. The man is affected by the optical characteristics of the display. Character font, size, color, viewing angle, brightness, and contrast all contribute to the subjective effect of the display on the man.

Character Font. Fig. 2 shows the character font of the new numeric indicators. The numerals 0 through 9 are produced by selectively energizing a matrix of gallium arsenide phosphide [Ga(As,P)] light-emitting diodes. There are 28 diodes: 27 are arranged in a 5 x 7 rectangular array (not all of the 35 matrix locations are occupied by diodes) and the 28th is offset at the lower left to serve as a decimal point.

The characters are designed to preclude ambiguity. It is unlikely that one number will be mistaken for another, even if one or two diodes should fail to light.

Fig. 2. *The new solid-state indicators come in one-digit and three-digit modules. Here four three-digit modules display the full character font, which includes the numerals 0 through 9 and a decimal point.*

Solid-State Displays, Present and Future

The new one-digit and three-digit solid-state numeric indicators are only a beginning. Larger and smaller characters, larger character sets, more colors, and discrete light sources are under development. Different kinds of displays are in the future.

Using What's Available Now

Small size, low power, low voltage, and high brightness make the presently available ¼ inch numeric indicators suitable for a wide range of applications in commercial and military equipment. Displays with as many 250 to 275 characters per square foot can be assembled using one-digit and three-digit modules. An obvious use is for numeric readouts from instruments and electronic data processing equipment. In telemetered or computer-driven status boards, the modules' small bulk and high image definition should prove valuable. The numeric indicators have no sealed-in toxic gases and operate at low voltages, and so are particularly well suited for closed environments, or for environments where there are explosion hazards. Their red color makes them useful for displays in darkrooms or in areas where personnel must be dark-vision adapted.

Color TV Still Years Away

Although it will surely happen, perhaps another ten years will pass before solid-state display technology is sufficiently advanced to make a wall-mounted color television set practical. Red, blue, and green light-emitting diodes will be needed to produce the visible spectrum with acceptable color fidelity, and at present we lack the means for producing the blue ones. Cost is another problem. About a quarter-millior points would be needed for a standard TV presentation. At today's production costs, the chips alone would cost nearly $25,000 for a five-inch-wide screen. Power requirements are also a problem. A five-inch-wide screen, 25% saturated to a luminance of 75 footlamberts, would need about 100 watts. A square foot of diodes driven solidly to a brightness of 50 footlamberts would draw about one kilowatt, at the electroluminous efficiencies that are typical of today's materials. For all these reasons, a solid-state color TV seems to be years away.

What About the Near Future?

Somewhat nearer to reality than solid-state television displays are smaller solid-state characters and symbols which can be incorporated in probes, micrometers, and other tools, so the tools themselves will become digital measuring instruments. Small displays can also be head mounted, say in a pilot's helmet. Character densities of 50,000 to 250,000 characters per square foot are feasible, the size reduction being limited only by the human side of the man-machine interface.

Small displays are potentially useful as markers in optical instruments, too. One such application arises when recording data on film. An 18 x 32 dot pattern is sometimes used to identify frames for automatic scanning equipment. With their nanosecond turn-on and turn-off times, light-emitting diodes could easily supply the dot pattern, which could then be transmitted to the film by fiber optics.

Alphanumeric Modules Are Imminent

Almost ready for production at HP are alphanumeric indicators which use the same 5 x 7 dot matrix as the new numeric indicators, with all 35 positions in the matrix occupied by diodes. Six-bit ASCII code controls the indicators. The character font, illustrated below, includes the letters A through Z, the digits 0 through 9, and the symbols +, −,), (, ·, *, ?, =, /, and ·. Two integrated-circuit chips decode the inputs and drive the diodes. The IC's are designed so they can be modified easily and economically whenever a change of font is wanted.

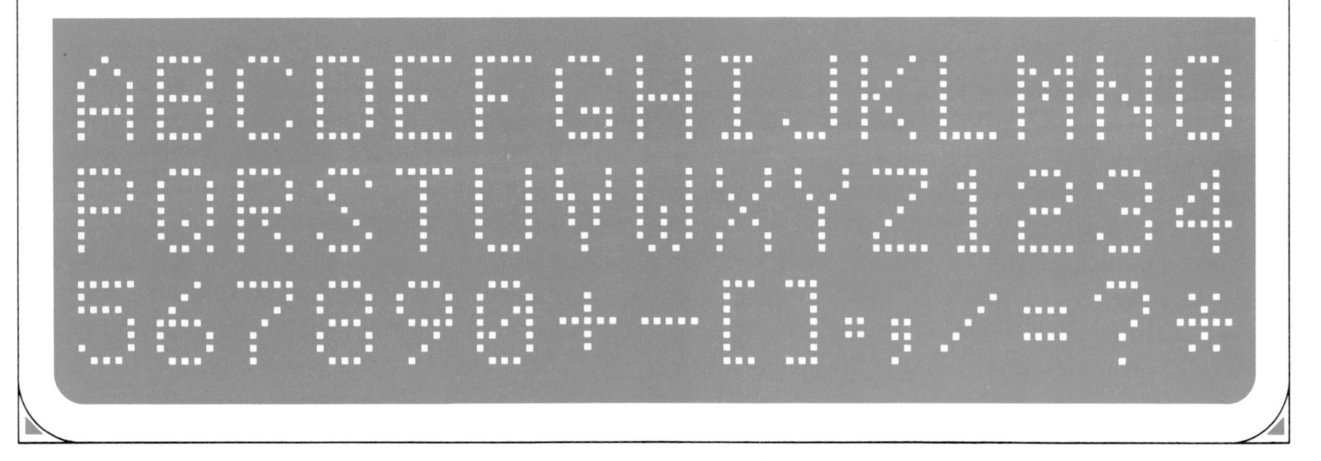

Character Size. The characters are ¼ inch high. However, because of an unexplained subjective phenomenon, the characters appear larger by at least 50%. The apparent character height is about ⅜ to ½ inch. This phenomenon isn't visible in the photographs in this article, so the reader should be aware when looking at these photographs that a 'living display' would have a substantially different subjective effect. After a study of larger and smaller sizes in various display applications, it was decided that ¼ inch characters are quite satisfactory for general instrumentation display, and this size was selected for the initial products. People with normal vision can read the characters at distances up to 8 feet.

Both the one-digit and the three-digit indicators are intended primarily for mounting in a horizontal line. The minimum center-to-center spacing between characters in separate modules is limited by the package width to 0.570 inch. Within the three-digit module the character spacing is 0.400 inch, center-to-center. The minimum vertical center-to-center spacing is slightly more than one inch. Large displays in which each character is individually replaceable can be constructed out of one-digit modules; such displays can have as many as 250 characters per square foot.

Color. The color of the light emitted by the numeric indicators is red. It has a dominant wavelength of 655 nm and is of high purity. Although alloys of gallium, arsenic, and phosphorus can be made to emit any color between infrared and green, the luminance (i.e., visual brightness) of HP's Ga(As,P) alloy is greatest, for a given input energy, when the light has a wavelength of 655 nm. Red is also well suited for use in darkrooms, ready rooms, and other dark environments, since it impairs the eye's dark-adaptation considerably less than other colors.

Brightness. Typical Ga(As,P) diodes used in the numeric indicators produce luminances of 75 footlamberts with 4.0 Vdc applied to the module. At 50 fL brightness, the characters appear quite bright under normal factory lighting. Brightness can be varied between 5 and more than 50 footlamberts by adjusting the dc voltage applied to the 'LED+' terminal of the module. As shown in Fig. 3, the luminance variation is nearly linear from three to four volts.

Reduced display brightness is desirable in dark rooms, or where power is at a premium. The power input to the light-emitting diodes varies roughly linearly with LED+ voltage, as does the power dissipated in the current-limiting circuits in the module. Power to the logic circuits in the module remains constant at about 150 mW.

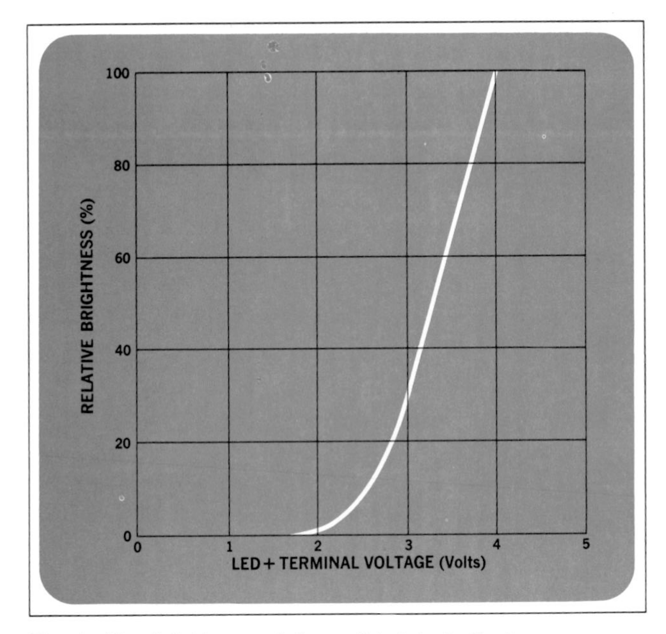

Fig. 3. *The brightness of the solid-state indicators can be adjusted by varying the voltage applied to the LED+ terminal of the module. Variable brightness and red color make the indicators useful in dark working environments.*

Contrast. Contrast is the ratio of the luminance of a lighted diode to the luminance of the surrounding area, which in this case is a white ceramic substrate with gold metallic striping. For optimum contrast, the numeric indicators should be viewed through a red 'notch' filter. (The filter isn't supplied with the indicators.) Ideally, the filter should transmit 100% of the red light from the diodes (density = 0) and 0.1% or less of the rest of the visible spectrum (density ≥ 3.0). Rohm and Haas' Plexiglas #2423 as it is presently manufactured isn't quite this good, but it is effective and inexpensive.

Viewing Angle. There are two general display situations. One is typified by bench-mounted instruments; here wide viewing angle is important, either so several instruments can be observed by one operator, or so a group of people can observe the same display. In such applications the new numeric indicators can be viewed from angles as great as 60° from the normal in a horizontal plane and as great as 70° in a vertical plane. They have a cosine, or Lambertian, surface light distribution, meaning their light is equally dispersed in all directions; this makes their brightness independent of viewing angle.

The second display situation is typified by an aircraft instrument panel, where the observer's head is in a relatively fixed position with respect to the display. In these situations it's often desirable to trade off wide-angle viewability for minimized light reflection or for the higher

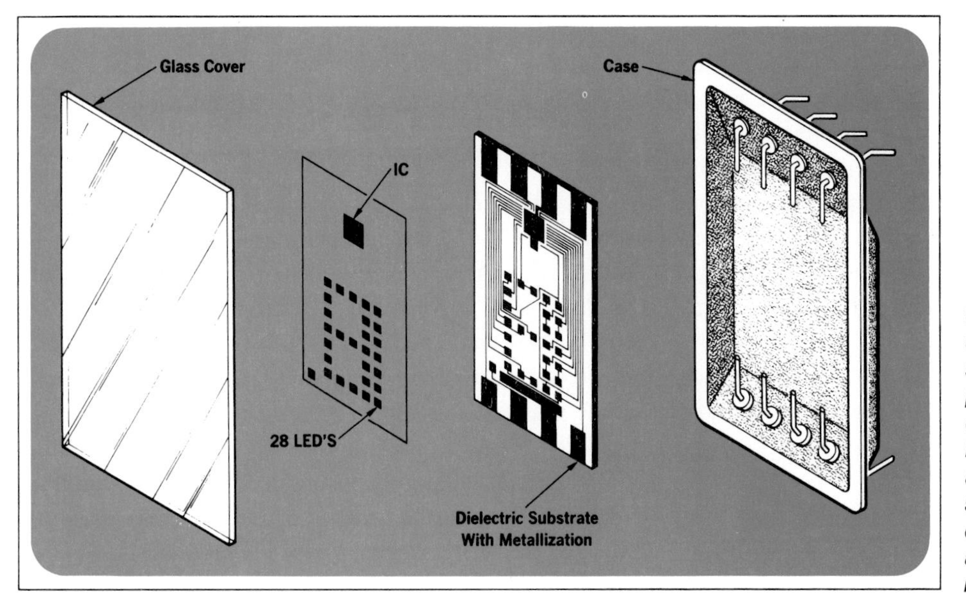

Fig. 4. *Model 5082-7000 Solid-State Numeric Indicators consist of 28 gallium arsenide phosphide red-light-emitting diodes and an IC chip containing some 400 circuit elements, all mounted on a ceramic substrate and sealed in a case designed for wide-angle viewability. On the substrate is a metal interconnection pattern.*

efficiency possible with narrow-lobe light emission. The radiation pattern of the numeric indicators can be modified by appropriate filters, lenses, and shades.

Electrical Characteristics

While the optical characteristics of the man/machine interface are important to the man, the machine is more interested in the electrical characteristics of the display.

Each digit of the solid-state indicators has eight inputs.

They are:

- one line for a 5 Vdc filtered power supply for integrated-circuit logic operation. This supply should be regulated to prevent overvoltage conditions and should be able to provide about 30 mA per digit.
- one line (LED+) for a 4.0 Vdc light-emitting-diode power supply, capable of providing up to 200 mA per digit.* If brightness variation is required, this supply should be variable from about 2.0 to 4.0 volts.
- four lines for 8-4-2-1 BCD negative logic, 3.5 V < '0' < 5.0 V, 0 V < '1' < 1.5 V. The BCD coding, Table I, conforms to ASCII coding.
- one line for decimal point control. A 10 mA, current-limited source is needed. It is turned on to illuminate the decimal point.
- one line for ground, common to all signals and power supplies.

The decoding circuitry has no memory, so the display will conform to the input code within less than a microsecond — typically less than 200 ns. The modules also have no overvoltage protection. Transients exceeding 6 V on the BCD lines or the integrated-circuit power-supply lead, or exceeding 5 V on the LED+ lead, may cause damage, so protection should be provided.

Mechanical and Thermal Characteristics

At 50 fL average diode brightness, and with the numbers 5 or 8 illuminated (17 diodes lighted), the light-emitting diodes dissipate about 250 mW. Another 250

Table I. Binary Code Truth Table

X_8	X_4	X_2	X_1	Display
0	0	0	0	0
0	0	0	1	1
0	0	1	0	2
0	0	1	1	3
0	1	0	0	4
0	1	0	1	5
0	1	1	0	6
0	1	1	1	7
1	0	0	0	8
1	0	0	1	9
1	0	1	0	Blank
1	0	1	1	Blank
1	1	0	0	Blank
1	1	0	1	Blank
1	1	1	0	Blank
1	1	1	1	Blank

Note: Negative Logic '0' = Line High: $3.5 \leq V \leq 5.0$
'1' = Line Low: $0.0 \leq V \leq 1.5$

* The LED supply may also be a full-wave-rectified unfiltered source of frequency 50 Hz or higher. Lower frequencies can be used if noticeable flicker is not objectionable.

mW is dissipated by the decoding circuitry. Hence the modules are rated at ½ watt per digit. Heat sinking should be adequate to dissipate this amount of power with a temperature rise of 10°C or less above ambient.

In mounting the display modules, care should be taken to protect their glass front windows. The indicators don't need a vacuum, so they'll continue to operate even with substantial package damage. However, the hermetic seal will be lost if the front window is broken, and this may reduce the life of the module.

The leads of the indicator modules are 0.100 inch apart, compatible with current printed-circuit-board practice.

How They're Made

Five basic elements make up the one-digit solid-state numeric indicators, as shown in the exploded diagram Fig. 4. All the parts except the case are manufactured by Hewlett-Packard. Tekform Products Company manufactures the case. The five elements are:

- A ceramic substrate. The front of the substrate is thin-film metallized and photolithographically etched to form an interconnection pattern consisting of 8 input pads and 18 output drive lines. The back of the substrate is also metallized so the thermal resistance between the substrate and the case will be low.
- Twenty-eight gallium arsenide phosphide red-light-emitting diode chips. Twenty-seven are arrayed in a 5 x 7 matrix and one is offset to serve as a decimal point. The cathodes of the diodes are bonded to the metal interconnection pattern on the substrate. The light is emitted from the anode side of the diodes. All of the anodes are wired togther and connected to the metallized substrate by ultrasonic lead bonding, using 0.001 inch aluminum wire.
- A monolithic silicon integrated-circuit chip. The chip translates standard 8-4-2-1 binary-coded-decimal input codes into 18 current-limited outputs which drive the 27 diodes in the 5 x 7 matrix. (A separate external source drives the decimal point.) The IC chip has only 18 outputs because certain combinations of the 27 diodes are always lighted together. The IC terminals are connected to the metal layer on the substrate by ultrasonic bonding, using 0.001 inch aluminum wire.
- A tin-plated Kovar® case with glass-to-metal-sealed leads. The case is designed for wide-angle readability of the characters. The ceramic substrate is soldered to the case.
- A glass window cover whose coefficient of thermal expansion is matched to that of the case. The glass is joined to the case with epoxy to form a hermetic seal.

® Westinghouse Electric Corporation

Ga(As,P) Light-Emitting Diodes

Each light-emitting diode chip is a simple mesa structure (see Fig. 5). An n-type alloy of gallium arsenide phosphide is grown epitaxially on a gallium arsenide substrate. The p region is then diffused and capped with a comb-type metal anode. The comb-type anode distributes the diode current evenly over the diode's cross-section while masking less than 25% of the light.

When forward bias is applied to the diode, the potential barrier at the junction is reduced so current can flow. Electrons are injected into the p region and holes are injected into the n region. Eventually these minority carriers recombine, and in some of the recombinations energy is given off as photons. Most of the light is generated in a space charge layer about 0.5 μm wide on the p side of the junction, where several percent of the recombinations result in near-edge photons — that is, photons whose energy is near the bandgap energy.

Since the anode surface is very close to the junction, most of the light generated internally reaches the surface. However, only a few percent of the photons escape. The loss is caused by internal surface reflection, which is a result of the difference between the refractive index of the Ga(As,P) alloy ($n \approx 3.5$) and that of its surroundings ($n \approx 1.0$).

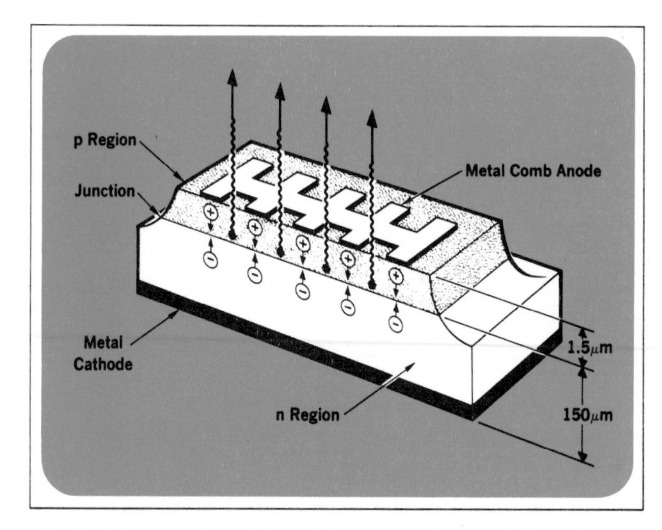

Fig. 5. *Each light-emitting diode chip is a mesa-type p-n junction diode. Light is given off in recombinations of minority carriers, predominantly on the p side of the junction. The comb anode gives even current distribution while blocking less than 25% of the light.*

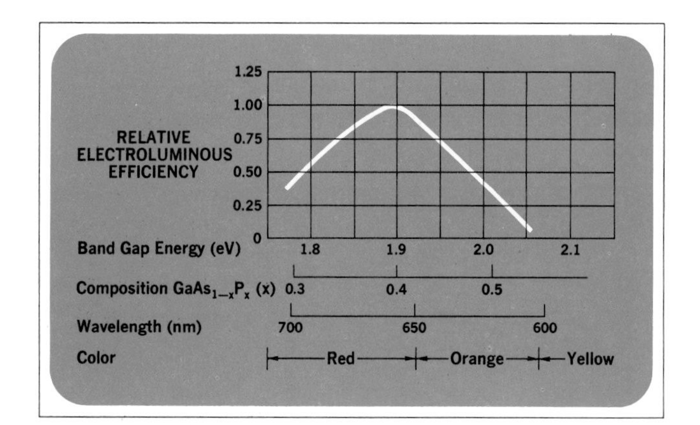

Fig. 6. *Red was chosen as the color of the light emitted by the new solid-state indicators because the electroluminous efficiency of HP's gallium arsenide phosphide alloy is highest for that color. Electroluminous efficiency is a measure of visual brightness per unit diode current. It is a function of the alloy composition, that is, the value of x in the formula* $GaAs_{1-x}P_x$.

At very low diode currents, that is, currents in the nanoampere range, only a small fraction of the total diode current contributes to light emission. As the current is increased into the milliampere range, the electroluminous efficiency also increases. Electroluminous efficiency, expressed in footlamberts per unit current density, is a measure of the perceived brightness that results from a given amount of current. Electroluminous efficiency becomes constant, and diode luminance increases nearly linearly, for diode currents of 3 mA to 100 mA. Typical diodes have junction areas of 0.002 cm^2 and electroluminous efficiencies of 15 footlamberts per ampere per square centimeter; this is equivalent to a brightness of 75 fL at 10 mA. Efficiencies as high as 100 $fL/A/cm^2$ have been observed in some diodes.

The light-emitting properties of the diodes are very stable with time. The diodes' half-life, the time required for the luminance to decrease to 50% of its original value, appears to be more than 100,000 hours. This estimate is based on presently available data, straight-line extrapolated to the 50% point.

Composition Determines Electroluminous Efficiency

Gallium arsenide phosphide has the formula $GaAs_{1-X}P_X$, where x is between 0 and 1. The value of x determines the optical bandgap energy, which in turn determines the radiation wavelength for near-edge emission (photon wavelength is inversely proportional to energy, and the energy of near-edge photons is approximately the bandgap energy). We have found that the electroluminous efficiency of the Ga(As,P) we are now producing is highest for a composition in which x is 0.4 (see Fig. 6). This corresponds to a bandgap energy of 1.9 electron-volts and a wavelength of 655 nm, and gives the diode its characteristic red color. Although the eye's response to this wavelength is only 10% of its peak response

Fig. 7. *Ga(As,P) wafers up to two inches in diameter are grown in this HP-designed vertical-flow, RF-heated, cold-wall, dual-reactor, vapor-phase epitaxial system. The light-emitting diodes are made from these wafers. All parts of the solid-state indicators except the case are made by HP.*

(which occurs at 555 nm), it isn't possible to generate a brighter light for a given current at wavelengths shorter than 655 nm. This is because the number of photons generated per unit current drops more sharply with decreasing wavelength than the eye's sensitivity increases.

The tradeoff between the eye's response and the efficiency of photon generation, which is controlled by varying the value of x in the formula $GaAs_{1-X}P_X$, was only one of many tradeoffs that had to be decided upon as the light-emitting diodes were developed. Material was the first variable. Ga(As,P) was selected because its bandgap energy is high enough to provide visible light, its doping profile can be closely controlled, and its near-edge recombination mechanism is relatively strong compared to competing energy-dissipating recombinations. Another tradeoff was how to optimize the injection of electrons into the p side of the junction. This is controlled by the dopants and the doping profiles that are used in the diodes. The six years that were required to develop the diodes were largely spent in unravelling the complex physics and determining the tradeoffs involved. [1]

Special Manufacturing Facility

It was necessary to create a source of Ga(As,P) that would be capable of maintaining sufficiently close control over composition, doping profiles, and dimensions. Therefore, a special reactor [2] was designed and built by HP (Fig. 7). It is a vertical-flow, RF-heated, cold-wall, dual-reactor, vapor-phase epitaxial system capable of growing wafers of Ga(As,P) up to two inches in diameter. The system holds the phosphorus-to-arsenic ratio constant within $\pm 1\%$ across the growing epitaxial layer. Because of this precise control, the entire wafer of Ga(As,P) can be used to make diodes, and light-emitting properties are quite uniform from diode to diode. (A typical wafer had a mean luminance of 291 fL at 10 A/cm^2 and a standard deviation of only 25 fL.)

RF induction heating was chosen for the reactor instead of resistance heating for the following reasons.

- It is possible to keep all of the glass portions of the apparatus at temperatures well below those of the reaction zone, thereby minimizing a possible source of contamination.
- The thermal mass of an induction heated system can be made small, thereby reducing the total time required for the growth process.
- Sharp temperature profiles, desirable for high deposition efficiency, are easily achieved.
- The volume of the system for a given substrate area can generally be made smaller than a comparable resistance-heated unit. This gives shorter system time constants when time-variant gas flow rates are used.

Heating is accomplished by the inductive coupling between an RF solenoid and a high-purity graphite susceptor. Since the skin depth of the RF field in graphite is a few millimeters at the operating frequency (450 kHz), the interior of the chamber is heated primarily by radiation from the outer walls of the susceptor.

GaAs substrates are placed on a horizontal pedestal within the susceptor so the growth surface is normal to the gas flow. Gas flows are measured with electronic flowmeters to ensure the required degree of control and reproducibility in the growth process. The phosphorus (in the form of PH_3) flow is automatically programmed by an electromechanical valve to achieve the desired profile of phosphorus within the epitaxial layer. Flow

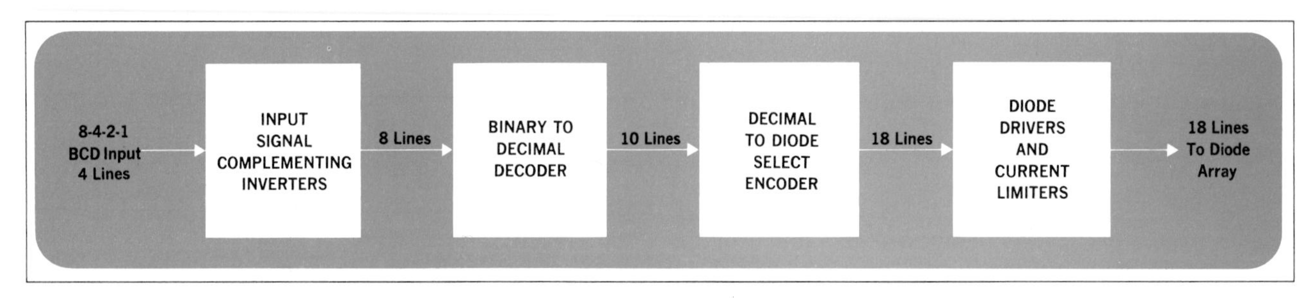

Fig. 8. *The IC chip decodes BCD input signals into ten signals representing the digits 0 through 9. These signals are then encoded into signals which drive groups of diodes. There are 18 groups, and the diodes in each group are always lighted together. A two-step decode/encode process was chosen so the input code or character font could be changed simply by changing the interconnections between the logic elements.*

Measuring Luminance

At HP, diode luminance is measured by optically imaging the diode's light-emitting surface on a fiber-optic probe. The fiber-optic probe transmits the light through a filter. From the filter the light goes to a photomultiplier tube, and the tube's output is measured by a digital voltmeter which reads directly in footlamberts. The system is calibrated using a 100 fL standard source. Matched photopic-filter/photomultiplier-tube assemblies and certified sources, traceable to N.B.S., are purchased from Gamma Scientific Corporation. The electrical response of the filter/photomultiplier assemblies to photon excitation matches the response of the human eye.

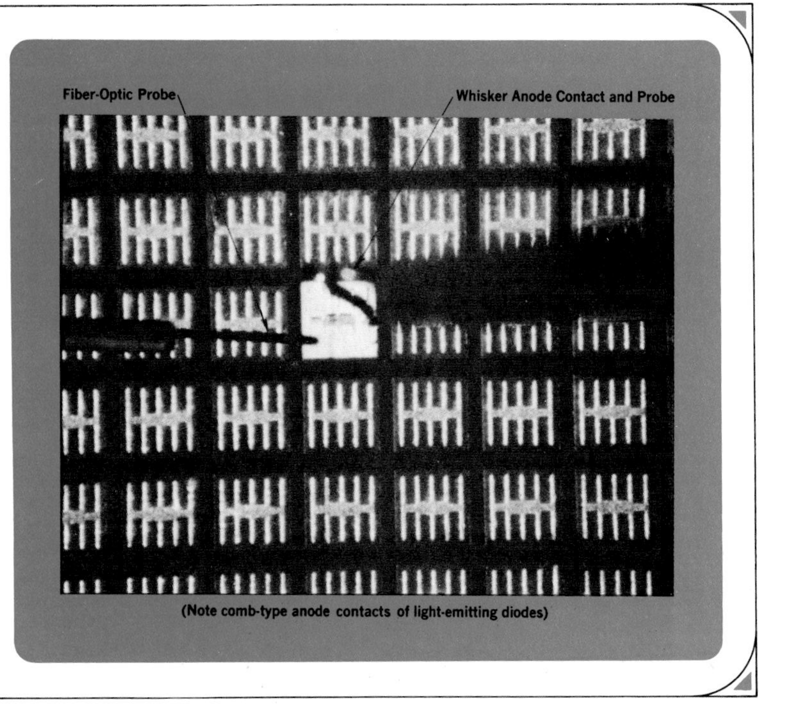

(Note comb-type anode contacts of light-emitting diodes)

rates can be changed very quickly, so a wide variety of compositional profiles can be obtained.

Substrate temperature is controlled to ±1°C or better by a closed-loop control system whose thermocouple sensing element is located within the substrate pedestal. Temperatures elsewhere in the susceptor are measured by optical pyrometry. The temperature of the arsenic (in the form of $AsCl_3$) reservoir is controlled by a thermoelectric cooling unit.

The initial capacity of the light-emitting-diode production facility is about 1.5 million diodes per year.

The integrated circuits and ceramic substrates for the solid-state numeric indicators are also manufactured by HP. An automatic machine is now being developed to sort and test the diodes and IC chips, then orient them and attach them to the substrates. Testing is done by a data acquisition and processing system controlled by an HP 2116A Computer.

IC Decode-Encode Logic

Some 400 circuit elements are contained in the IC chip used in the solid-state numeric indicator modules. Four basic functions are performed in the chip (Fig. 8). Incoming four-line BCD signals are first complemented. Then the eight signals — the four BCD input signals and their four complements — are decoded into ten mutually exclusive line signals, each of which will excite one of the ten decimal digits 0 through 9. Of the sixteen possible binary input codes, the six that don't represent a decimal digit produce blanks. Complementing the BCD inputs was done to minimize the overall complexity of the IC chip; it greatly simplifies the decoding circuitry.

The third function performed in the IC is to encode the ten mutually exclusive line signals into signals that select the proper diodes to produce each character. Diodes that are always lighted together are excited by a single output from the IC chip. The number of diode groups required to produce ten digits is eighteen, so the chip has eighteen outputs. Each of the ten mutually exclusive line signals activates a subset of the eighteen outputs. These diode drive signals then go through combination power amplifiers and current limiters, and excite the light-emitting diodes.

The two-step decode-encode organization was chosen to allow some flexibility in changing the input and output options of the chip. The chip is the electrical equivalent of a ten-position, eighteen-gang switch which has series resistors in all of its outputs and is operated by four BCD signal lines. By simple changes in one of the masks used in making the IC, either positive or negative BCD inputs can be accommodated, and any 10 of the 2^{18} possible output combinations can be selected. Therefore, changes in character font, presentation of special symbols, or changes in input code can be accomplished quite easily.

Each of the four binary input lines is connected to twenty grounded-collector pnp transistors in the decoding area of the IC chip. It is connected to the bases of ten of these directly. Between the input and the other ten transistors is an npn emitter follower, which drives an inverter; the input line is connected to the base of the emitter follower, and the inverter output is connected to the bases of the ten pnp transistors. Thus there are eighty pnp transistors in the decoding area of the IC chip. However, not all are operational. Only those needed for the desired decoding function are given emitter connections. The direct-connected pnp transistors that have emitter connections are active when the binary input signal is in the low state (0V). The emitter follower, the inverter, and the remaining pnp transistors that have emitter connections are active when the binary input signal is in the high state (5V). Emitter connections are made by etching holes in the oxide layer of the chip. When the metal interconnection layer is added, the metal comes in contact with the emitter of a transistor only where there is a hole in the oxide. Thus the chip can be made to respond to either positive logic or negative logic simply by changing the oxide cut mask used in making the IC.

The same technique allows flexibility in changing the output code of the IC chip. Each of the ten lines going from the binary-to-decimal decoder into the decimal-to-diode-select encoder is connected to eighteen npn emitter-follower OR gates. However, not all of the 180 npn transistors are connected. Oxide cuts are made for emitter connections only where connections are needed to turn on one of the eighteen diode drive lines. Any drive line is turned on when one of the ten mutually exclusive decimal lines is turned on *and* the encoding npn transistor for the drive line has an emitter connection.

Diode Drive Circuit

From the decimal-to-diode-select encoding circuitry, each of the eighteen diode drive lines goes to an npn grounded-emitter inverter which drives a pnp emitter follower. Between the pnp emitter and the output connection to the light-emitting diode is a current-limiting resistor. The resistor is connected to the n side (cathode) of the light-emitting diode, and the p side (anode) of the diode is connected to the LED+ terminal of the module. Between this terminal and the ground terminal of the module is the external light-emitting-diode power supply, its positive side connected to the LED+ terminal. When the diode is forward biased (turned on), about 1.6 volts appear across the diode, and less than one volt appears across the saturated pnp emitter follower. The remainder of the power supply voltage appears across the current-limiting resistor. As the power supply voltage is varied,

Howard C. Borden

Howard Borden is manager of solid-state displays at HP Associates. He began his 28 year career in 1939 as a mechanical engineer. Following a tour of duty with the U.S. Navy and further work as a mechanical engineer, he decided to go into business for himself. He bought an appliance store, introduced television, and saw it grow to become the largest part of his business. For the past 18 years he has been in research and development in the field of electro-optical-mechanical systems. He joined HP in 1966.

Howard studied mechanical engineering at the Massachusetts Institute of Technology and electronics at the U.S. Navy Radio Materiel School. In 1961 he received the BA degree in economics from Stanford University, where he also minored in industrial engineering. He is a member of IEEE, the American and Western Economic Associations, the Optical Society of Northern California, and the Society for Information Display.

Gerald P. Pighini

Gerry Pighini has been in the solid-state electronics field since 1955, when he received his BS degree in metallurgical engineering from Brooklyn Polytechnic Institute. He has been with HP since 1965, first as chief engineer of HP Associates and then as production manager for semiconductors before assuming his present responsibilities as manager of solid-state displays. Before coming to HP, Gerry was chief manufacturing engineer for a producer of semiconductor devices.

Gerry is co-author of a paper on growing large gallium arsenide phosphide wafers for light-emitting-diode production. He is a member of the Society for Information Display.

the diode and emitter-follower voltage drops remain nearly constant. The voltage across the resistor, and therefore the diode current, varies nearly linearly with the power supply voltage. This makes the brightness of the diode vary almost linearly with the supply voltage. All of the current-limiting resistors are closely matched (a characteristic of IC's) so the brightnesses of the individual diodes are very nearly equal.

The total time required for a change in input code to travel through the IC is about 100 ns. The response time of the light-emitting diodes is about 10 ns, so the total response time of the display is typically about 110 ns.

The IC chip is a low-resistivity p^+ substrate with a p-type epitaxial layer. The output currents flow through the substrate, thereby minimizing the current density in the aluminum interconnecting metal. This construction is a departure from the conventional IC, which is grown on an n^+ substrate.

Acknowledgments

The authors wish to acknowledge the contributions of people from several HP divisions which have made this product possible. Developmental work was done in John Atalla's solid-state laboratory, which is part of Hewlett-Packard Laboratories, the corporate research and development facility. Materials work was under Paul Greene and Robert Burmeister, Jr., device work under Robert Archer, and IC development under John Barrett. Ed Hilton's integrated-circuit department of the Frequency and Time Division is producing the IC chips, with James Grace overseeing the transition from the solid-state laboratory. The ceramic substrates are a product of George Bodway's metal-film facility in the Microwave Division. Light-emitting-diode production, component testing, and final assembly of the pieces is done by the solid-state-display group at HP Associates.

References

[1]. 'Solid state module makes for light reading,' Electronics, September 2, 1968.

[2]. R. A. Burmeister, Jr., G. P. Pighini, and P. E. Greene, 'Large Area Epitaxial Growth of $GaAs_{1-x}P_x$ for Display Applications,' to be published in Transactions of the AIME.

SPECIFICATIONS

HP Model 5082-7000
Numeric Indicator

ABSOLUTE MAXIMUM RATINGS
POWER SUPPLY: All power and logic inputs referred to common (ground).
IC LOGIC VOLTAGE: 5 Vdc
LED (LIGHT EMITTING DIODES) VOLTAGE: 4.5 Vdc
STORAGE TEMPERATURE RANGE: −65° to +95°C
OPERATING TEMPERATURE RANGE: −55° to +95°C

TYPICAL CHARACTERISTICS
LUMINANCE AT 4 V LED +: 75 fL.
POWER DISSIPATION AT 50 fL: 500 mW.
PEAK WAVELENGTH: 655 nm.
SPECIAL LINE HALFWIDTH: 30 nm.
CHARACTER RESPONSE TIME ('on' or 'off'): <1 μs
IC LOGIC CURRENT AT 5 Vdc (logic + to ground): 25 mA
LED CURRENT AT 4 Vdc (LED + to ground): 200 mA.

PIN ASSIGNMENT
1. Logic 1 — 5. Logic 4
2. Decimal Point — 6. Ground
3. LED + — 7. Logic + (5 Vdc)
4. Logic 8 — 8. Logic 2

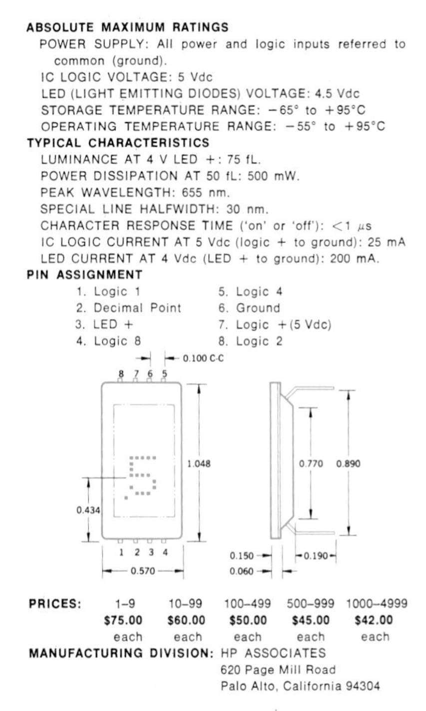

PRICES:

1–9	10–99	100–499	500–999	1000–4999
$75.00 each	**$60.00** each	**$50.00** each	**$45.00** each	**$42.00** each

MANUFACTURING DIVISION: HP ASSOCIATES
620 Page Mill Road
Palo Alto, California 94304

Reliability

The following cumulative test results have been obtained from reliability testing performed at HP Associates in accordance with the latest revision of Military Semiconductor Specification MIL-E-5400, MIL-STD-202 and MIL-STD-750. The following results were obtained with solid state displays (5082-7000) sampled from the production line.

END POINTS:
1. Generate proper character font 0–9 and decimal points.
2. No change in average unit brightness (fL) within limits of measurement accuracy (±10%).
3. Seal Hermeticity. Meets MIL-STD-883, Method 1014, Test Condition A and D (at PE = 20 P.S.I. and T_1 = 2 hours).

Test	Reference	Test Conditions	Units Tested	Failed
Humidity	MIL-STD-202C Method 106	24 hr. cycles from 25°C to 65°C @ 95% R.H., 5 cycles	38	0
Altitude (Nonoperating)	MIL-STD-202C Method 105C Condition B	30 min. @ 50,000 ft.	38	0
Temperature (Nonoperating)	MIL-STD 202C Method 107B Condition A	−65°C to +85°C 5 cycles	38	1
Shock	MIL-STD-202C Method 213 Condition C	5 drops, 6 orientations 100 g, 6 ms	37	0
Vibration Variable Frequency	MIL-E-5400 Curve IV	±10 g, 70-500-70 Hz	37	0
Thermal Shock	MIL STD-750 1056.1	0°C to 100°C 5 cycles	37	0

Test	Reference	Test Conditions	Units Tested	Total Unit Hours	Failed
High Temperature Life	MIL-STD-750 1031.1	500 hrs. storage at 85°C	10	5,000	0
Steady State Operating Life	MIL-STD-750 1026.1	1000 hrs. @ 4 V LED + cycling 0–9 at ½-s rate	10	10,000	0

A System for Automatic Network Analysis

By Douglas Kent Rytting and Steven Neil Sanders

WITH THE ADVENT OF new, highly sophisticated microwave devices and systems, there has developed a need for fast, accurate, and complete characterization of the networks that comprise them. Two popular techniques exist for characterization of microwave networks. They can be broadly classified as fixed-frequency or swept-frequency techniques. The power of the fixed-frequency technique is that the mismatch, tracking, and directivity errors of the measurement system can be minimized by 'tuning out' the residual errors, achieving high accuracy. Fixed-frequency techniques, however, are slow and somewhat tedious. Swept-frequency techniques in general offer a fast means of gathering data across broad bandwiths, and the advantage of intuitive insight into the device being tested. Normally, it is difficult to account for all errors, when using the swept-frequency technique, since they can not be completely 'tuned out' in the broadband case. The question is, can we somehow devise a microwave measurement system which will provide the advantages of the above two techniques without incurring any of the disadvantages? Before answering, let's take a broader look at the total microwave measurement field.

What should we list as desirable characteristics of a microwave measurement system?[1]

1. Complete device characterization capability
   a. linear (amplitude and phase)
   b. non-linear
   c. noise
2. Accuracy
3. Speed
4. Flexibility
5. Ease of use

A system aimed at achieving many of these characteristics is the HP Model 8542A Automatic Network Analyzer shown in Fig. 1. The effort has been to combine the advantages of the fixed-frequency and swept-frequency techniques. The Automatic Network Analyzer uses a stepped-CW sweep rather than a continuous one, so a finite number of points for measurement and error correction results. Instead of tuning-out the system errors at each frequency point, the systematic internal errors are first measured, then taken into account as the device is measured. The internal system errors are vectorially subtracted from the measurement data, correcting the measurement and leaving only the true characteristics of the device. System errors need only be characterized at the *beginning* of a set of measurements. A complete error model of the Automatic Network Analyzer can be constructed by measuring appropriate standards. The stand-

PRINTED IN U.S.A.

ards used are such devices as shorts, opens, and sliding loads, which are relatively easy to characterize and manufacture to accurate tolerances.

To recap, the system measurement procedure is as follows: 1) calibration, 2) measurement, and 3) correction of the data.

Clearly, data storage and mathematical manipulation are required. A relatively small instrumentation computer will provide this function elegantly.[2] Blending the microwave instruments, computer, and software programming will yield a powerful and flexible system which well achieves many of the desired characteristics.

1. Linear device characterization.

   What is the best way to characterize a microwave network? There are many sets of parameters available. The [z], [h], [y], [ABCD] matrices are popular low-frequency characterizations. In recent years a set has been developed that is more practical, easier to use, and easier to measure at microwave frequencies. This set is the family of s-parameters.[3] They need not be confined to microwaves, being also valid to represent networks at low frequencies.

2. Accuracy

   To insure that the error-correcting procedure is

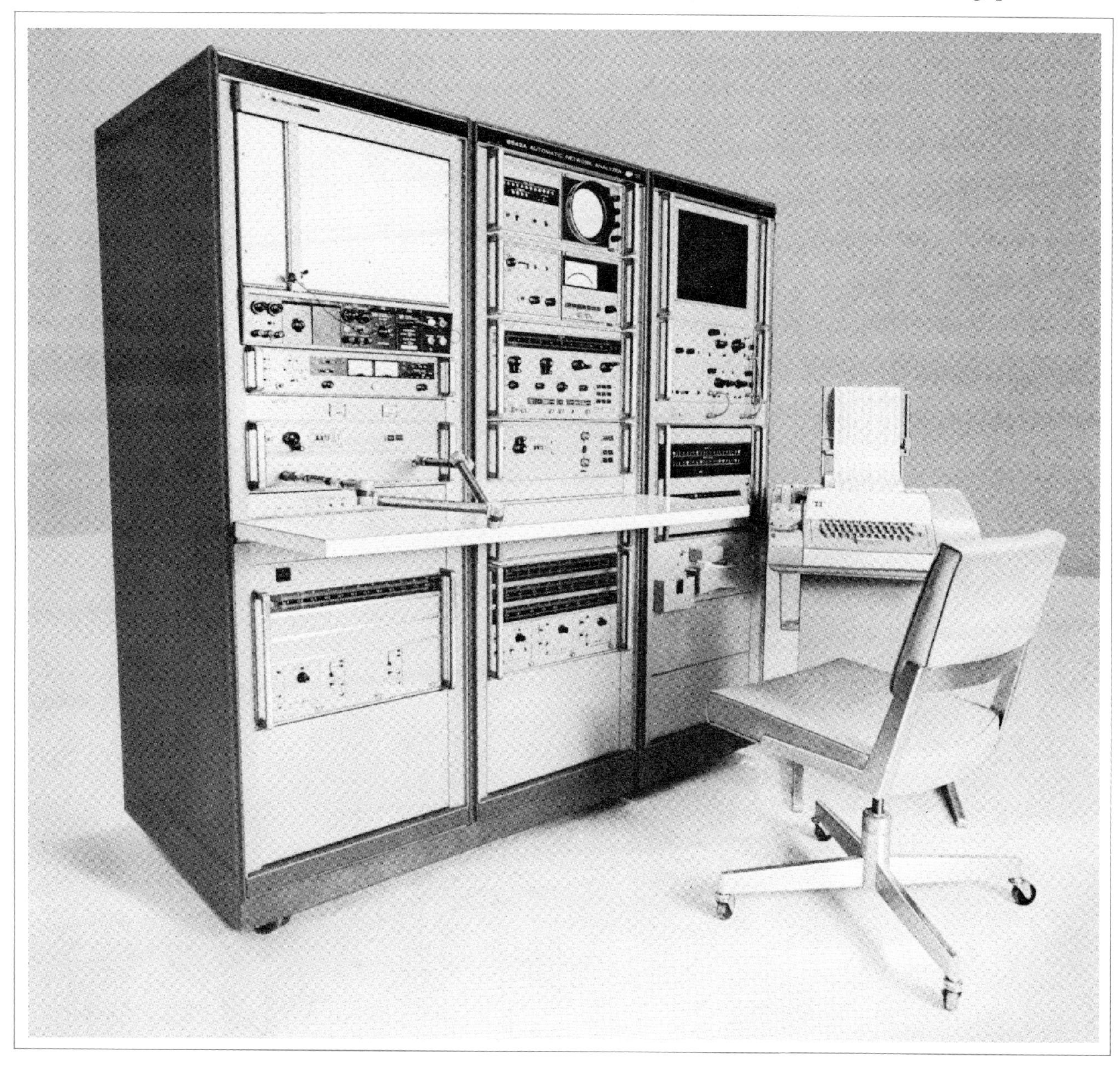

Fig. 1. *The automatic network analyzer system in this photo has an unusual number of readouts—note X-Y recorder, oscilloscope displays, and teleprinter. The number of possible configurations is almost unlimited.*

valid, accurate microwave instrumentation with good repeatability is required. Inaccuracies in the system are then minimized by the error correction technique given earlier. The residual system errors, all of second-order importance, then are only those caused by imperfect repeatability of connectors and switches, noise in the system, system drift, and errors in the standards used for calibration.

It is important to measure *both* magnitude and phase to achieve accuracy in characterizing networks at microwave frequencies. Even when the intent is only to measure magnitude, phase information is required for high accuracy. For example, when measuring the transmission of a filter, trying to correct for mismatch errors without knowing the phase of the mismatch terms can cause a large ambiguity in the amplitude measurement, no matter how accurate the amplitude detecting device.

3. Speed

   The computer can easily control all the instrument functions normally operated by the user. Then, too, the calculating power of the computer greatly shortens the time required for complete network characterization.

4. Flexibility

   S-parameters are the parameters most easily measured at microwave frequencies. However, they may not be the desired output from the system. S-parameters comprise a total characterization of the network. Therefore, the computer can transform from the s-parameter set to any other consistent parameter set one may wish. Not only can h, y, or z parameters be determined, but also group delays, VSWR, return loss, or substantially any other desired format. Transformations into other domains are also feasible, such as determining time domain response from frequency domain data. After accurately determining just one set of parameters, the system, with its software, opens an enormous range of measurement capabilities. Flexibility expands further when the Automatic Network Analyzer is augmented with computer-aided design programs. Once a device is characterized by the system, the resulting data can then be used by a computer to synthesize optimum networks for the device.[4]

   It is also important that the instrumentation and software structure of the Automatic Network Analyzer be *modular*, so the system can easily be expanded without altering its original configuration or operation. This simplifies adding computer peripherals or instrumentation options.

5. Ease of use

   Since the computer controls the instruments, makes the measurements, and manipulates the data, the user is relieved of the mundane and difficult parts of the measurement procedure. The imagination of the R&D or production engineer is not merely supplemented; it is indeed amplified by the system, and furthermore his ideas, now in software, are made usable by many people. More time becomes available and more desire is created to do the long and difficult measurements needed for imaginative design. It is important that interactive hardware and software interfaces be provided between the system and the user. This requires appropriate programming language and a good programming structure. Then, too, the user/hardware interfaces must be simple and effective.

Basic System

The HP Model 8542A Automatic Network Analyzer concept is shown in Fig. 2. The system has three main sections, source, measurement, and computer.

The signal generator provides the RF power from 0.1 to 18 GHz required to test the unknown device. The frequency is computer-controlled and can be stabilized (phase-locked) with at least 650,000 stable, repeatable frequency points across an octave band. An automatic leveling control circuit provides level power and a good source match.

RF from the signal generator is applied to the device under test via the s-parameter test set.[5] With S_1 and S_2 set as shown in the diagram, the ratio of the test to reference channels is proportional to s_{11} of the device being tested. If S_1 and S_2 are both switched, we measure s_{22} of the device. If S_1 or S_2 are switched separately, we measure s_{12} or s_{21} of the device, respectively.

The "complex ratio detector" measures the *complex* ratio, i.e. the amplitude ratio and phase difference between the reference and test channels.[6] This information is digitized and routed to the computer via the instrument interface.

The computer takes the s-parameter data and stores it as either calibration data, if measuring standards, or raw, uncorrected data if measuring a device. Output from the computer can be routed to the display, or to other computer peripherals.

There are two basic systems, phase-locked and non-phase-locked, and two possible modes of operation in each, manual or automatic. In the manual mode, the system operates as a group of standard instruments without computer control.

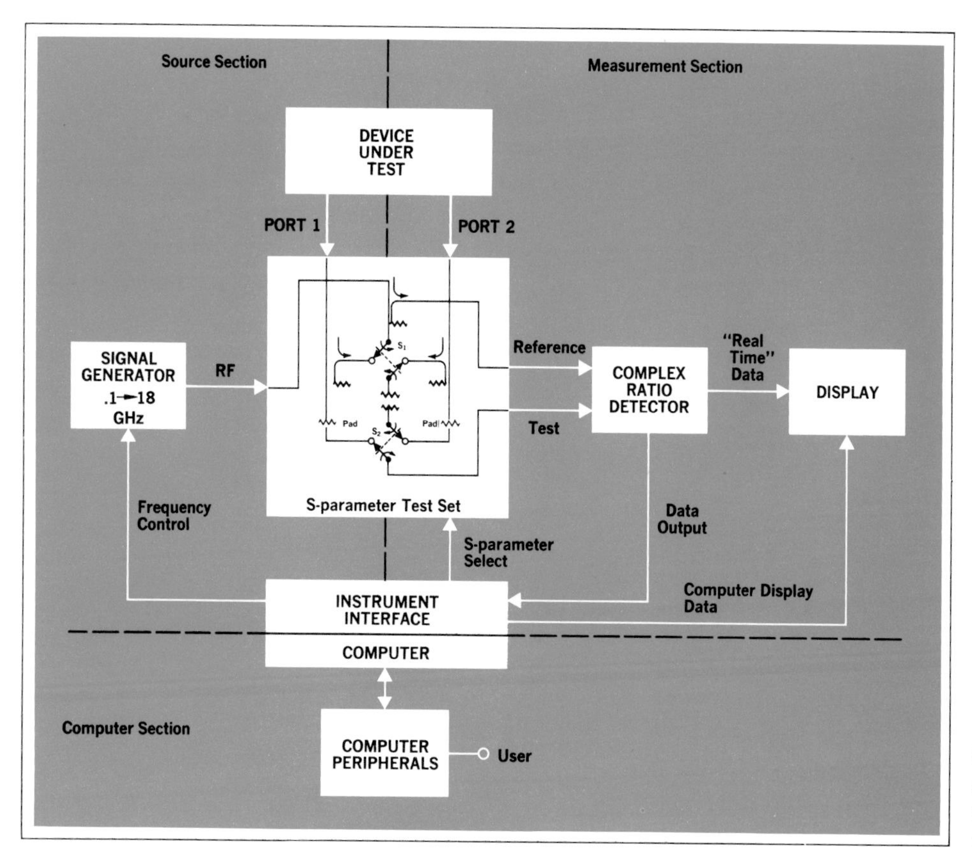

Fig. 2. *Functional diagram shows interrelations among the elements of Automatic Network Analyzer system.*

Operation

When discussing the operation of the system, we will consider only the phase-locked version operating in the automatic mode. A block diagram of the source and measurement sections of the system is shown in Fig. 3.

The total measurement sequence can be broken down into two steps. The first is to prepare the system to make a measurement. The second is to measure and digitize the high-frequency s-parameter data for storage in the computer.

Typically, this is the preparation sequence:

1. The desired s-parameter is selected in the s-parameter test set.
2. The multiplexed signal source is phase-locked to the reference oscillator via the high-frequency phase lock circuitry.
3. When the system is phase-locked, a phase-lock status indication is sent through the instrument interface to the computer. The system will not take a measurement until phase-lock has occurred.
4. The magnitude of the test signal into the detector is adjusted so as to be within an optimal 5 dB range, to enhance measurement accuracy.
5. DC offset and drift in the system are measured for later correction by the computer.

Now that the system is prepared, the high-frequency s-parameter data are converted into an equivalent digital form:

1. The s-parameter test set puts out reference and test signals proportional to the desired s-parameter.
2. The reference and test channel signals are translated to a *fixed* IF of 20.278 MHz by the harmonic frequency converter.[7] Amplitude and phase information are not altered in this down conversion.
3. The automatic gain control in the IF strip normalizes the test channel amplifier to the reference channel amplifier. The reference and test channels are then further down-converted to 278 kHz for optimum detection.
4. The synchronous detector decomposes the real and imaginary components of the test channel signal into an equivalent dc form.
5. This dc voltage is digitized by the A/D converter.

Fig. 3 shows the source section and measurement section overlapping. Actually, the same circuitry is used for both functions. The s-parameter test set and frequency converter used by the measurement section are also used by the source section for phase-locking. This eliminates using a separate phase-lock loop to establish the 20.278 MHz IF in the measurement section, and actually improves system performance.

High Frequency Phase Lock Loop

Perhaps the high frequency phase lock can be better understood with a qualitative look at its operation. The phase lock loop shown in Fig. 4 compares the phase of the two inputs into multiplier A. The two relative phase inputs are zero and θ degrees. If the signal source FM input is disconnected and re-connected, the two inputs into multiplier A, ω_{IF} and ω'_{IF} will initially be at different frequencies. After transients have disappeared $\omega'_{IF} = \omega_{IF}$, and θ is forced to a constant value, the phase offset of the loop θ_o. The locking phenomenon obeys a non-linear, second order differential equation which will not be discussed here.

age required by the signal source, to keep ω_s constrained to $|\omega_S - \omega_r| = \omega_{IF}$.

The phase lock loop described above becomes more complicated when the signal source must cover a range from 110 MHz to 18.0 GHz. The reference oscillator cannot cover a bandwidth this broad, so the signal source must phase-lock to *harmonics* of the reference oscillator. This is accomplished by replacing multiplier B with a sampler, and generating harmonics of the reference oscillator beyond 18.0 GHz as shown in Fig. 5. Case 1 transforms to locking above a harmonic by the amount f_{IF} and Case 2 transforms to locking below a harmonic by f_{IF}. In Fig. 5, the phase error θ is approximately 90° for

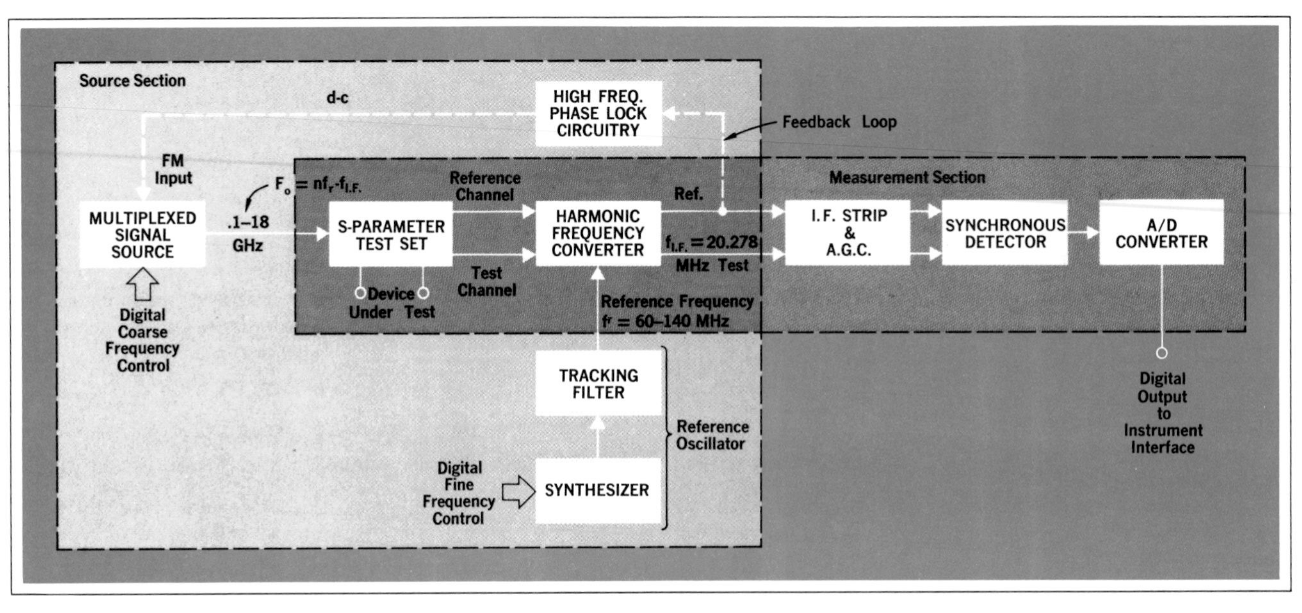

Fig. 3. *Block diagram of source and measurement sections of the Automatic Network Analyzer system.*

To get a feel for the operation of the phase lock loop in steady state, consider Case 1 of Fig. 4. Assume the steady state value of θ is $\theta_o \approx 90°$. If for some reason $\omega'_{IF} = \omega_S - \omega_r > \omega_{IF}$, θ starts to increase because $\theta = (\omega'_{IF} - \omega_{IF})\,t + \theta_o$ radians. This causes v_e to go negative, and this reduces ω_s and ω'_{IF}. A reduced ω'_{IF} causes θ to increase at a *slower* rate. This negative feedback reduces ω'_{IF} to ω_{IF}. For Case 2, notice that

$$\cos(\omega'_{IF}t + \theta) = \cos[(\omega_r - \omega_s)t - \theta],\ \omega_r - \omega_s > 0$$

which shows that phase information is reversed compared to Case 1. This phase reversal must be compensated with another phase reversal if the loop is to be stable. This is accomplished as the loop automatically shifts θ_o by 180°, which changes the sign of the slope for multiplier A's characteristics. In general v_e adjusts itself to the volt-

$f_s = nf_r + f_{IF}$ (upper sideband Case 1), or close to 270° for $f_s = nf_r - f_{IF}$ (lower sideband Case 2). n is a harmonic number of the reference oscillator. θ is delayed by 90° and applied to phase detector B. Its output will be approximately cos 180° for Case 1 and cos 0° for Case 2. This polarity change is sensed, causing S_1 to open for the upper sideband case. Thus ambiguity concerning the source output frequency is eliminated. The source now phase locks when $nf_r - f_s = f_{IF}$ as shown in Fig. 6.

The computer is given a source frequency f_s by the user. It programs the reference oscillator to

$$f_r = (f_s + f_{IF})/n\ MHz.$$

The reference oscillator then phase-locks with the accuracy and long-term stability of a crystal standard. This accurately defines the comb spectrum nf_r shown in Fig. 6. The computer also controls the coarse tuning of the signal

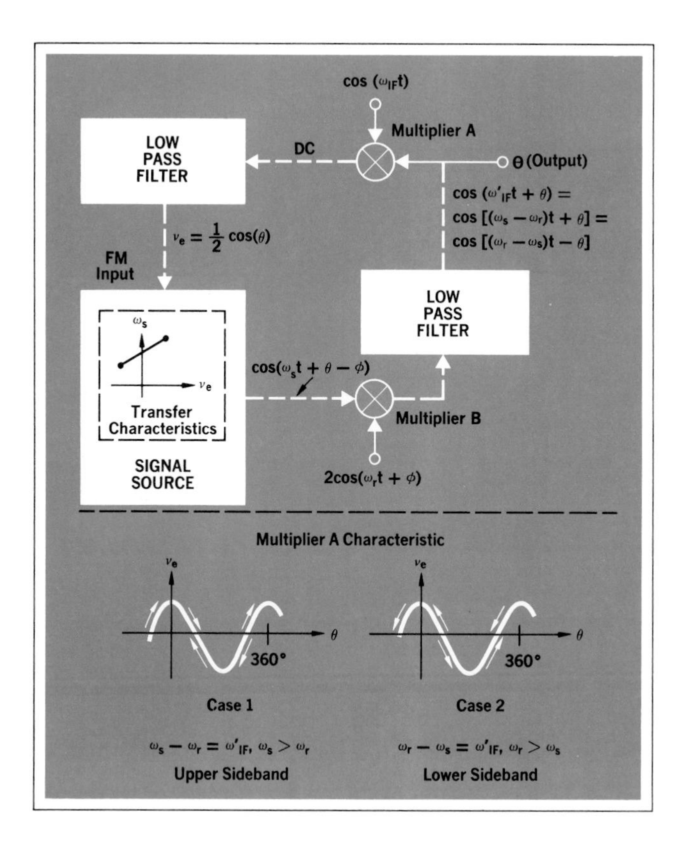

Fig. 4. *Simplified diagram of phase lock loop and multiplier characteristics, showing operational scheme.*

source frequency f_s. Because frequency errors in coarse tuning may be large, perhaps 40 MHz, a search generator is added to the phase lock loop, which increases its frequency locking range. The search generator systematically changes f_s symmetrically about the coarse tuned frequency. When f_s passes a possible lock point, phase lock occurs and the search generator is turned off. To prevent the phase lock loop from locking on the wrong harmonic (harmonic skipping), the search generater should search less than $\pm f_r/2$ about the coarse tuning frequency, and the coarse tuning must be closer than $\pm f_r/2$. These constraints are most important at the lowest reference frequency used. If harmonic skipping occurs, the source frequency f_s is offset an amount f_r, typically 120 MHz. The symmetrical clipping network and 0.2% coarse tuning accuracy eliminate the possibility of harmonic skipping.

The IF frequency, f_{IF}, could be eliminated and the signal source can still be stabilized. This would also eliminate the sideband sensing circuitry. This is not done because s-parameter information is translated to an intermediate frequency, and some means of IF stabilization must also be provided. Two phase lock loops could be used, but one will work more simply. If the reference channel IF strip is connected to the filtered output of multiplier B, the signal source and reference channel IF

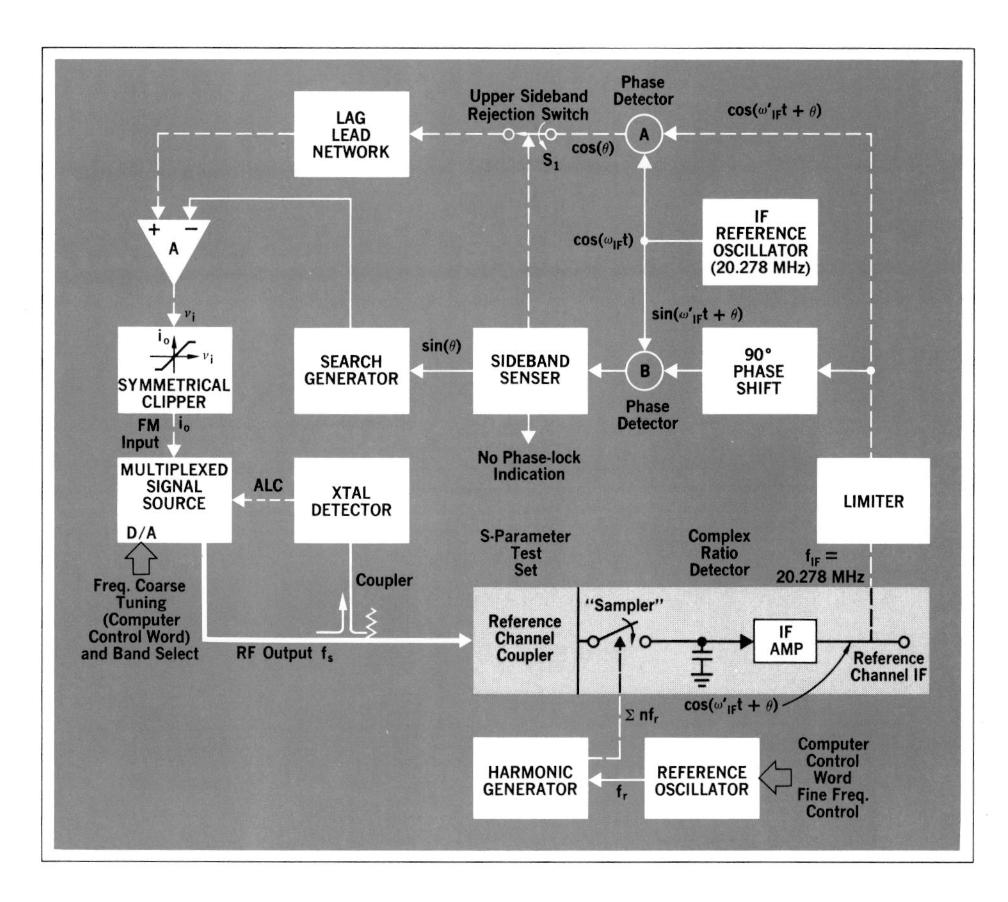

Fig. 5. *Detailed block diagram of high-frequency phase lock loop.*

About the hardware . . .

Each article in this issue of the Hewlett-Packard Journal deals with some aspect of automatic network analysis. The HP Model 8542A Automatic Network Analyzer is not a single product of fixed characteristics, but is instead a modular option system. Each system produced is tailored to its use by incorporating appropriate choices among options. There are choices of signal sources, test sets, detectors, computers, and computer peripherals.

The specific system described throughout this issue incorporates two important options for highest accuracy, a frequency-stabilized signal source, and a new precision detector. The accuracy curves given by Hand are typical of systems operating between 2.0 and 12.4 GHz with the options, but of course are not typical of others. The new precision detector will be available to update earlier HP automatic analyzers of the 8542A family.

Many Hewlett-Packard divisions contributed directly to the system's development. Some instruments could be directly placed in the system with no modifications. Others, which were not originally designed with digital interfaces, were updated. The source and measurement instruments were modified to accept the broadband phase-locking capability. The performance characteristics of other instruments were improved to meet the accuracy requirements of the system. A number of new instruments were designed especially for the Automatic Network Analyzer.

are simultaneously stabilized in frequency. The reference channel IF voltage is maintained relatively constant by the ALC feedback loop in the signal source. This provides a low-noise, high-level feedback point to connect the high-frequency phase-lock loop.

Reference Oscillator and System Noise Considerations

The reference oscillator provides the stable reference frequency (f_r) required by the high frequency phase-lock loop. The frequency range for f_r is from 60 to 140 MHz in 100-Hz steps, synthesized from a crystal standard.[8] The synthesis process is controlled by a fine-frequency control word from the computer. To achieve the high spectral purity requirement of the system, the broadband AM and PM noise out of the synthesizer is reduced by using an oscillator phase-locked to the output frequency of the synthesizer. This phase-lock loop really comprises a tracking filter. The box on page 9 explains how the phase-lock loop can perform in this way.

The tracking filter cannot respond to a phase noise input which lies outside the phase-lock loop bandwidth, as demonstrated mathematically in the box. It can also be seen that there is no steady-state frequency error caused by the tracking filter. A *bandpass* filter is thus formed, which is centered about the reference frequency (f_r), and which tracks the reference frequency over the 60 to 140 MHz range.

There are two other sources of noise which contribute to the output power spectrum of the reference oscillator besides the noise from the synthesizer. One is the noise

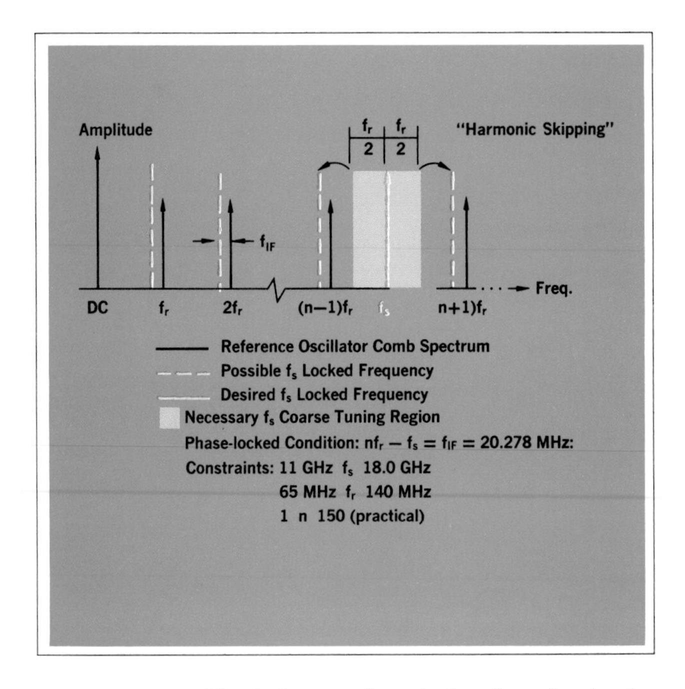

Fig. 6. *Source phase-locks when* $nf_r - f_s = f_{IF}$.

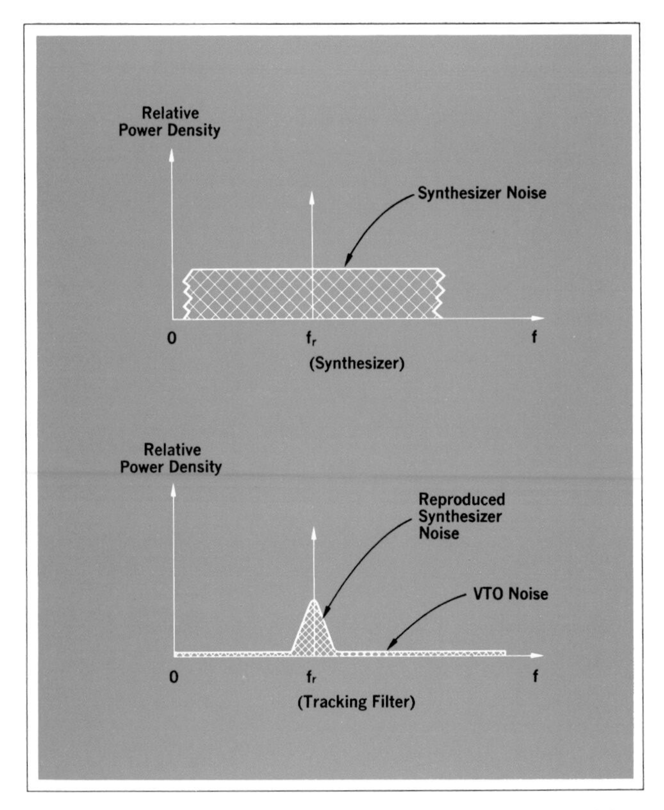

Fig. 7. *Power spectral densities.*

from the phase detector and dc amplifier in the phase-lock loop of the tracking filter. Careful low-noise design minimizes this source of noise. The other source of noise is the phase noise of the VTO. Let us see how this noise is reduced.

The loop is a *high-pass* filter to the phase noise produced by the VTO. Therefore, the VTO noise within the loop bandwidth will be reduced by the high-pass filter, but noise outside the loop bandwidth will pass. A high-quality varactor-tuned VTO was therefore designed to minimize the phase noise outside the loop bandwidth.

We thus have a typical engineering tradeoff. The noise of the synthesizer is reduced by decreasing the loop bandwidth, but decreasing the loop bandwidth increases the noise contribution of the VTO. An optimum bandwidth exists which will minimize the sum of the two noise sources. There also is a requirement for minimum lock-up time, which is a function of the loop bandwidth. In the light of these factors, it was possible to pick a best compromise for loop bandwidth. The synthesizer output spectrum and the resulting spectrum, after passing through the tracking filter, are shown in Fig. 7.

The accompanying figure will help to understand how the phase-lock loop performs as a tracking filter. A phase-lock loop acts as a filter to both amplitude and phase sidebands around a carrier. To see how this interesting function is achieved, consider the block diagram here.

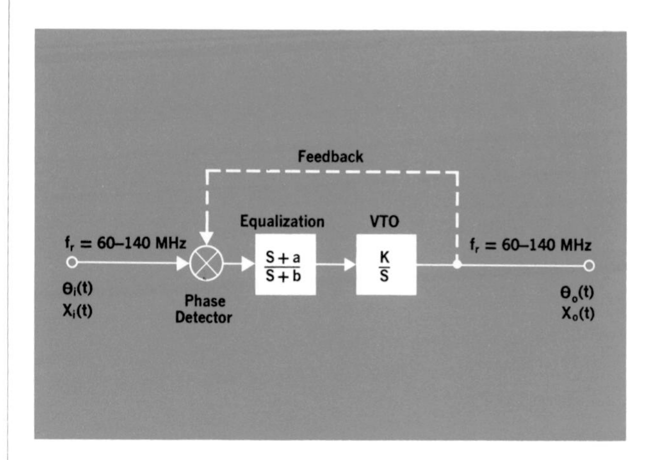

The signal input to the filter is

$$X_i(t) = A_c \cos[2\pi f_r t + \phi_i(t)],$$

where $\phi_i(t)$ is any input phase function with zero mean which has the spectrum $\Phi_i(f)$.

Let $\theta_i(t) \triangleq 2\pi f_r t + \phi_i(t)$.

The two terms on the right hand side are not correlated. The tracking filter responds only to the total argument $\theta_i(t)$, and hence to the phase modulation, but it will not respond to the amplitude modulation. There is a small amount of AM-to-PM conversion, but it may be regarded as negligible.

The transfer function for the tracking filter in the diagram is

$$\frac{\Theta_o(S)}{\Theta_i(S)} \triangleq H(S) = \frac{k(S+a)}{S^2 + (k+b)S + ka}, \quad (1)$$

which is a low-pass filter function. It follows that

$$\theta_o(t) = 2\pi f_r t + \phi_\varepsilon(t) + \phi_i(t) * h(t),$$

where

$h(t)$ = the inverse Laplace transform of [H(S)]

$\phi_\varepsilon(t)$ = transient and steady state phase error between the input and output of the filter and

$\phi_i(t) * h(t)$ = convolution of $\phi_i(t)$ with h(t).

The output signal $x_o(t)$ will be

$$x_o(t) = B_c \cos[2\pi f_r t + \phi_\varepsilon(t) + \phi_i(t) * h(t)],$$

where B_c is an arbitrary constant.

Using the Fourier transform and assuming $\phi_i(t) * h(t)$ is small, the two-sided *steady-state spectrum* is

$$X_o(f) = \frac{B_c}{2}\Big\{[\delta(f - f_r)e^{i\phi_\varepsilon} + \delta(f + f_r)e^{-i\phi_\varepsilon}] + i[\Phi_i(f - f_r)H(f - f_r) - \Phi_i(f + f_r)H(f + f_r)]\Big\}, \quad (2)$$

where

$\Phi_i(f - f_r)H(f - f_r)$ = filtered output translated to f_r,

ϕ_ε = steady state phase error, and

$\delta(f - f_r)$ = sinusoidal component at f_r.

The one-sided power spectrum of $X_o(f)$ follows from Eq. 2 and is shown below.

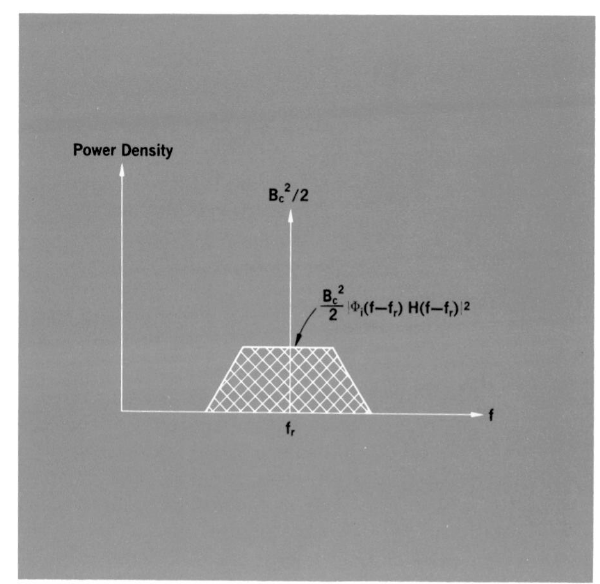

In summary, Eq. 1 gives the loop transfer function, and Eq. 2 is the resultant steady-state spectrum. This approximates the typical response seen on a spectrum analyzer.[9]

This discussion applies to the high-frequency phase-lock loop as well as to the tracking filter.

One more question remains, regarding the total phase noise measured by the computer. There are three main sources of this noise: (1) The reference oscillator noise multiplied by the harmonic number n, (2) multiplexed signal source noise, (3) noise produced by components in the high-frequency phase-lock loop, primarily the samplers. The noise from (1) and (2) is reduced by the high-pass filter characteristic of the high frequency phase-lock loop, but the equivalent noise created by the sampling process is not reduced. All three noises are then reduced by a 10-kHz low-pass filter in the detector, by the integrating A/D converter, and by computer time averaging. The resultant noise figure of the system is mainly determined by the noise figure of the samplers.

Future

Future automatic systems may become measurement terminals, much like those of the computational time-sharing services now available. In this type of system, digital data and high level commands are transferred between the measurement terminal and the central processor, linking the user with the power of a large central computer.

References

[1]. Richard A. Hackborn, 'An Automatic Network Analyzer System,' microwave journal, Vol. 11, No. 5, May 1968.

[2]. Kay B. Magleby, 'A Computer for Instrumentation Systems,' **Hewlett-Packard Journal,** March 1967.

[3]. Richard W. Anderson, 'S-Parameter Techniques for Faster, More Accurate Network Design,' **Hewlett-Packard Journal,** Feb. 1967.

[4]. Les Besser, 'Combine S-Parameters with Time Sharing and Bring Thin-film, High-frequency Design Closer to a Science than an Art,' Electronic Design, Aug. 1, 1968.

[5]. Stephen F. Adam, George R. Kirkpatrik, and Richard A. Lyon, 'Broadband Passive Components for Microwave Network Analysis,' **Hewlett-Packard Journal,** Jan. 1969.

[6]. Richard W. Anderson and Orthell T. Dennison, 'An Advanced New Network Analyzer for Sweep-Measuring Amplitude and Phase from 0.1 to 12.4 GHz,' **Hewlett-Packard Journal,** Feb. 1967.

[7]. Wayne M. Grove, 'A DC to 12.4 GHz Feedthrough Sampler for Oscilloscopes and Other RF Systems,' **Hewlett-Packard Journal,** Oct. 1966.

[8]. Alexander Tykulsky, 'Digital Frequency Synthesizer Covering 0.1 MHz to 500 MHz in 0.1 Hz Steps,' and 'Phase Noise in Frequency Synthesizers,' **Hewlett-Packard Journal,** Oct. 1967.

[9]. A. Bruce Carlson, 'Communication Systems: An Introduction to Signals and Noise in Electrical Communication,' McGraw-Hill, 1968.

Douglas Kent Rytting

Doug Rytting came to the HP Microwave Division in Palo Alto directly from Utah State in 1966. He began quickly to contribute to the Division's engineering effort, with work on broadband detection schemes, improved probes for the Vector Voltmeter, and system interfacing for early Automatic Network Analyzers. Recently he became Project Supervisor for the company's ANA developments. Doug and his wife, Sharon, have a two-year-old girl. Doug is in charge of a Cub Scout program in Cupertino, and is pursuing an MS at Stanford under the HP Honors Program. He's a member of Phi Kappa Phi and is on one of the national technical committees of the IEEE microwave group.

Steven Neil Sanders

With a BSEE from Utah State in 1967, Steve Sanders came straight to HP. He has been steadily associated with microwave automatic network analyzer systems ever since, doing new work on the high-frequency aspects of phase-locked loop circuits, among other system contributions. He is now Project Engineer for ANA systems. Steve is about to receive his Master's from Stanford under the HP Honors Program, and plans to continue toward a higher degree. He is a member of Phi Kappa Phi. Although they are new parents, Steve and his wife, Annette, are not giving up their shared outdoor activities — skiing, ice skating, and hiking.

Software for the Automatic Network Analyzer

By William A. Ray and Warren W. Williams

SUCCESS IN DESIGNING COMPUTER-OPERATED SYSTEMS depends as much on software — the set of detailed program instructions — as upon hardware. The software package is often designed and written to solve only one special problem. However, in keeping with the overall objectives of the HP automatic network analyzer systems, software objectives had to take on an unusual degree of generality:

1) Make the collection of instruments act like one very powerful instrument.
2) Give the engineer the ability to expand software flexibility in graduated steps.
3) Make it easy to add instrumentation or computing power to existing systems.

These objectives led to designing a library of building blocks or program routines, each performing some specialized function, these blocks combining together efficiently to perform complete measurements.

Standard Measurement Software

An 8542A is installed initially with a set of general purpose programs ready to make most linear network analysis tests. Table I lists such a package for a 2-port test set.

A series always begins with system calibration. The program asks the user to connect a series of known devices, such as short circuits or through connections, at the system's test ports. The program then measures these devices, storing the differences between their apparent and ideal parameters. The reflection from a short, for instance, is affected by source match and coupler tracking. After measuring the magnitude and phase of the apparent reflection, the computer can calculate these error terms. Later, measurements can be corrected automatically for these effects, so the user can concentrate on *his* device instead of on the microwave measurement problems.

Once the system has been calibrated, the other programs in Table I can be loaded to make specific tests. These all have access to the calibration data in a reserved section of the computer memory.

Fig. 1 shows the teletype print-out and plot from a typical session. The system has been calibrated and is now ready to perform as a transistor test set. It is only necessary to load the transistor test program. The system signifies it is ready by typing 'CONN DEVICE.' Once the user has inserted the transistor, he types in some logging information for his own use (underlined) and presses carriage return. The system then automatically measures, corrects and saves the s-parameter description of the transistor. This fundamental system measurement is forward and reverse reflection and transmission with 50 ohm loads and sources. The description is a complete one in the sense that the other 2-port parameters — h, y, and z — can be derived from it[1]. The listings and plot in Fig. 1 were all derived from the same measurement, requested just by selecting a task number from a table. The entire sequence, including calibration and type-out, required only a few minutes.

Thus the system acts like an instrument. The user connects his unknown device, then specifies actions through the teletype keyboard and a switch panel. He need not be concerned about the array of knobs and dials before him; the computer controls all their settings. By using a programmed test sequence, the computer system becomes easier to run than a manual set up.

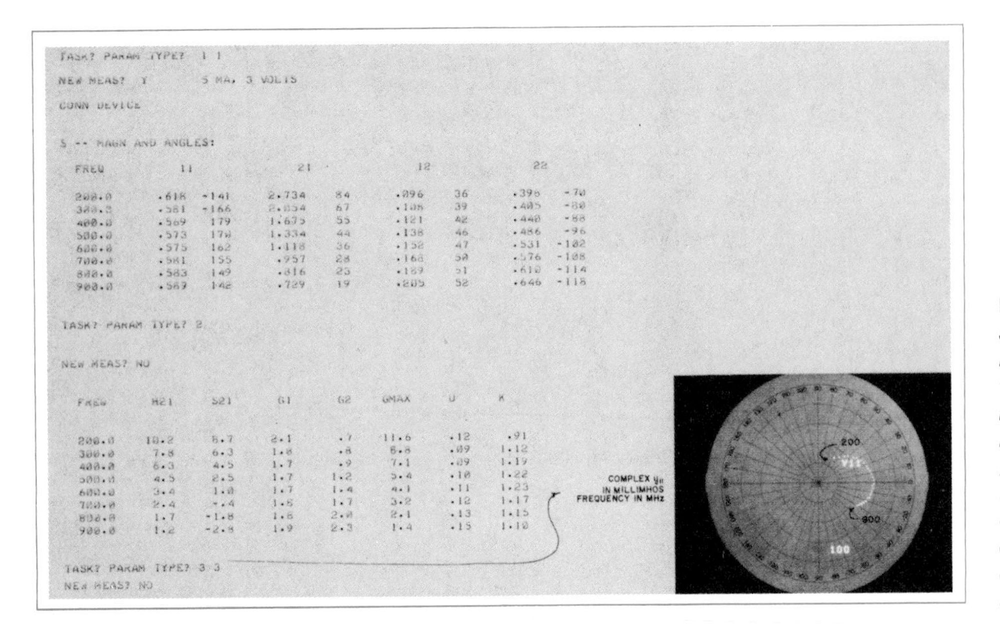
TASK? PARAM TYPE? 1 1

NEW MEAS? Y 5 MA, 3 VOLTS

CONN DEVICE

S -- MAGN AND ANGLES:

FREQ	11		21		12		22	
200.0	.618	-141	2.734	84	.096	36	.396	-70
300.0	.581	-166	2.054	67	.108	39	.405	-80
400.0	.569	179	1.675	55	.121	42	.440	-88
500.0	.573	170	1.334	44	.138	46	.486	-96
600.0	.575	162	1.118	36	.152	47	.531	-102
700.0	.581	155	.957	28	.168	50	.576	-108
800.0	.583	149	.816	23	.189	51	.610	-114
900.0	.587	142	.729	19	.205	52	.646	-118

TASK? PARAM TYPE? 2

NEW MEAS? NO

FREQ	H21	S21	G1	G2	GMAX	U	K
200.0	10.2	8.7	2.1	.7	11.6	.12	.91
300.0	7.8	6.3	1.8	.8	8.8	.09	1.12
400.0	6.3	4.5	1.7	.9	7.1	.09	1.19
500.0	4.5	2.5	1.7	1.2	5.4	.10	1.22
600.0	3.4	1.0	1.7	1.4	4.1	.11	1.23
700.0	2.4	-.4	1.5	1.7	3.2	.12	1.17
800.0	1.7	-1.8	1.6	2.0	2.1	.13	1.15
900.0	1.2	-2.8	1.9	2.3	1.4	.15	1.10

TASK? PARAM TYPE? 3 3

NEW MEAS? NO

Fig. 1. *Typical session run with standard software by the Automatic Network Analyzer. After taking a full set of s-parameter measurements the instrument has then calculated various gain parameters in dB. U is a unilateral gain factor; K is a criterion of stability. Y_{11} is one of a number of complex parameters which can be calculated and displayed at will.*

BASIC Interpreter

Eventually, the user will desire some special test not available in the standard programs supplied. Often this is simply a different way of displaying the fundamental s-parameter data, so he can take advantage of the standard 8542A calibration/corrected measurement routines. For many such purposes, he can use an interactive language, BASIC[2].

BASIC is a simplified subset of such full programming languages as ALGOL or FORTRAN. BASIC is algebraic and easy for technical people to learn, requiring only a few hours even on one's first exposure to computer programming. As each line of program is typed in, it is immediately checked for errors: misspelling, missing punctuation, etc. This is an ideal teaching method because incorrect responses are caught and corrected right away. Other types of errors which involve the logic of the program rather than its syntax only become apparent when it is run. Again, BASIC has an advantage because it is an interpreter, that is, it runs the program from BASIC statements instead of converting to machine language. So one can run the program and immediately go back to an editing mode to fix any errors.

In the 8542A, BASIC is normally used for special outputs. A standard calibration tape is used and the data transferred to a reserved area. The user then reads in BASIC plus an s-parameter measurement program supplied, finally typing in his special display coding.

Fig. 2 shows such a sequence for a reflection test, plotting return loss on the rectangular CRT. Note how the program can be run, errors fixed and then run again. Once BASIC was loaded, no additional paper tapes were required.

Statements are identified by the line numbers on the left. Lines beyond 9000 were a standardized reflection measurement routine. The only programming required was lines 100 through 240 which called the measurement routine (GOSUB 9000) and then displayed the plot (line 210).

FORTRAN

In any computer there will come a time when the memory capacity is exceeded. At this point one must trade some convenience or capability for more program space. For the automatic network analyzer, this means trading the convenient interactive programming features of BASIC for a compiler language like FORTRAN. The compiler checks for syntax errors and translates to machine-language instructions once, as a separate operation from running the program. Thus one does not have to keep editing and interpreting intelligence in the computer when making measurements; the space is available for a more elaborate measurement program. The compiled program runs faster than an interpreter program because the translation has been done previously.

Note that although one gives up editing interaction while writing the program, the resulting programs can still interact with the user. The standard measurement programs are all written in FORTRAN; they interact through switches and teletype as shown in Fig. 1.

The transition from BASIC to FORTRAN is straightforward because the two languages are similar in many ways, simply using different words to say the same thing. Fig. 3 shows a FORTRAN listing which performs the

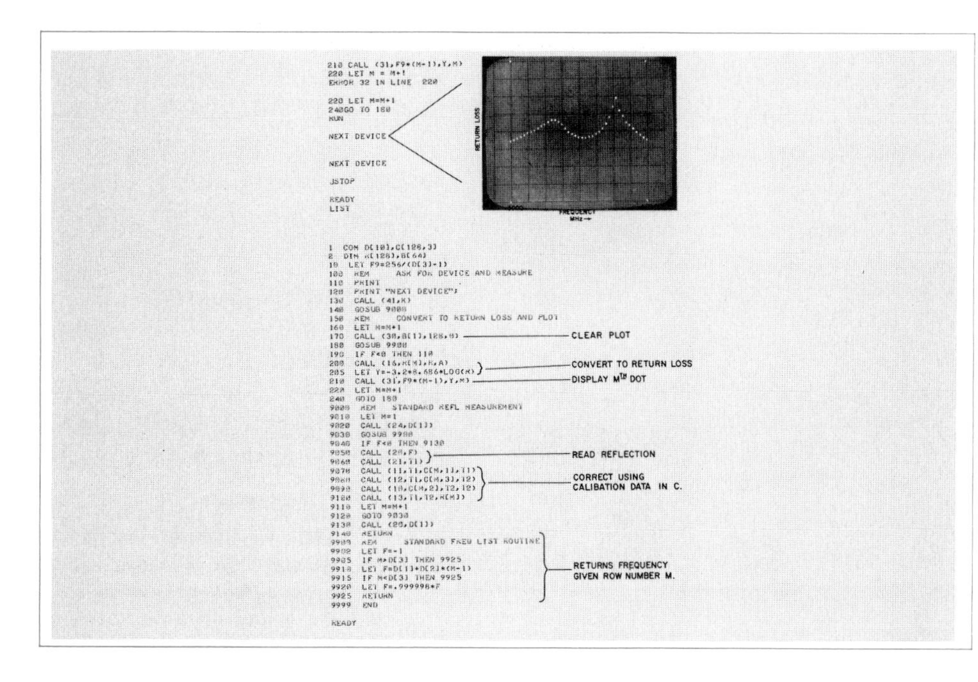

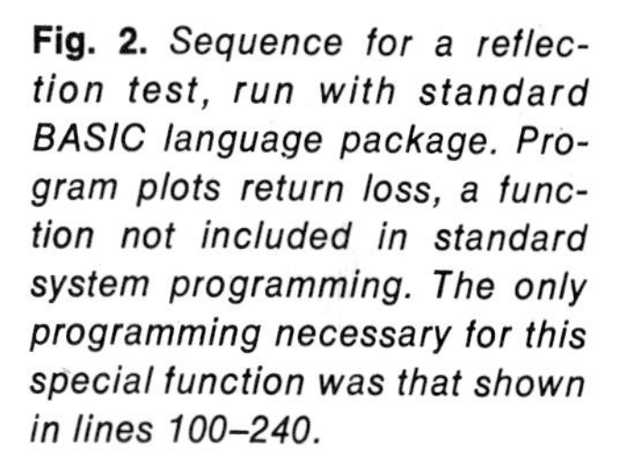

***Fig. 2.** Sequence for a reflection test, run with standard BASIC language package. Program plots return loss, a function not included in standard system programming. The only programming necessary for this special function was that shown in lines 100–240.*

same function as the BASIC program in Fig. 2. Unlike BASIC, it required a half-hour's work to convert the FORTRAN language program to a machine language program, a process which must be repeated each time a change is made in the program. Besides allowing for bigger programs, FORTRAN has two advantages over BASIC. Variable names are easier to read (CALIB instead of C for the calibration data array) and true subroutines are available (call CORR1 instead of GOSUB 9000).

The availability of true subroutines is very important because it makes modular software possible. In the BASIC program, the code between 9000 and 9140 was executed by a GOSUB from another part of the program. This is the same function as call CORR1 in the FORTRAN program. In both cases the main program can transfer to the routine from any number of places, and have it return automatically when done. The difference is that the BASIC corrected measurement routine shared the same variables as the rest of the program, while any CORR1 variables are totally isolated from any other routine. This is not very important with simple programs because one can keep all the names straight. However, programs may grow until they become too complicated to treat as one entity. Now true subroutines are needed to isolate functional modules so they can be written and verified — 'debugged' — once, then added to one's personal library. In the program shown, the user need only know how to call for the corrected measurement; he cannot affect its reliability by using one of its variable names or accidentally changing a statement.

This capability has been used in the 8542A software to build a modular hierarchical package, so constructed that an engineer need only re-write a portion to meet his special needs. Figure 4 shows the subroutine hierarchy for the example program in Figure 3. The main measurement program RPLOT has access to this package through the high level commands noted on the arrows which are the names called in RPLOT (see Fig. 3). Each of these routines calls in turn another layer of more specialized routines down to those which communicate with the instruments. Most, such as the instrument supervisor, are common to all the 8542A software.

The entire library that is needed to write the standard software is available for others writing new test programs. Of course the engineer is not restricted to writing main programs; as his needs and capabilities increase he can call, modify or replace any portion of the package. For example, he may want to change the test frequency rule from linear steps to logarithmic spacing, or perhaps to a table of critical frequencies. Since this information has been concentrated in one short routine (CALF3), he need replace only it, rather than re-writing the many routines which call it up. This is what is meant by *functional* software modularity.

Growth Capabilities

The software has been designed to grow, either by adding additional instruments (bias supplies, X-Y recorders, etc.) or adding computing power. The standard 8542A system uses a computer of 8000-word memory. This memory can be expanded with 4000-word internal core

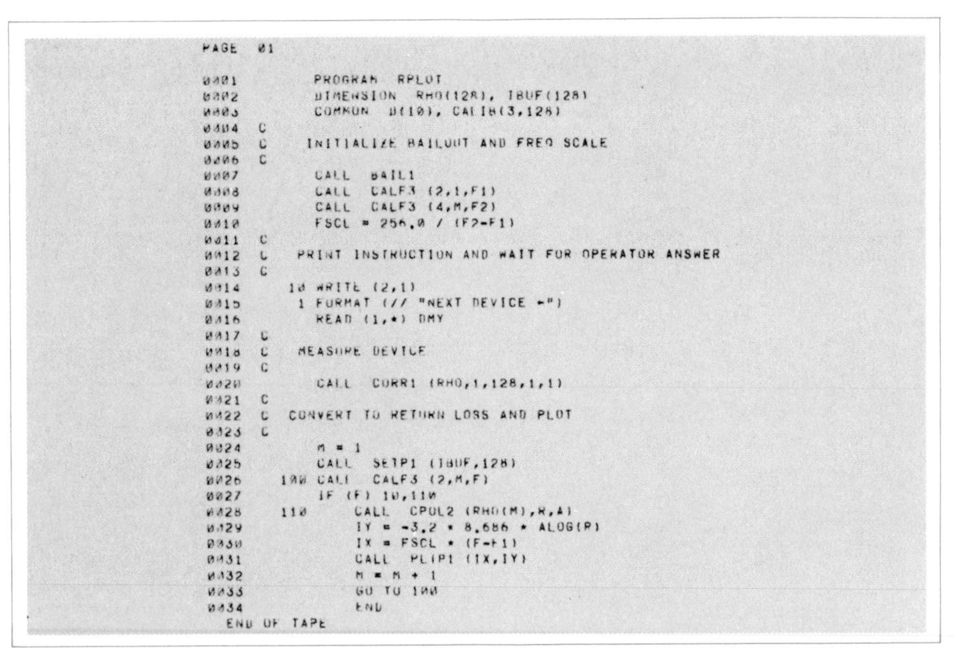

```
PAGE  01

0001          PROGRAM  RPLOT
0002          DIMENSION  RHO(128), IBUF(128)
0003          COMMON  D(10), CALIB(3,128)
0004   C
0005   C    INITIALIZE BAILOUT AND FREQ SCALE
0006   C
0007          CALL  BAIL1
0008          CALL  CALF3 (2,1,F1)
0009          CALL  CALF3 (4,M,F2)
0010          FSCL = 256.0 / (F2-F1)
0011   C
0012   C   PRINT INSTRUCTION AND WAIT FOR OPERATOR ANSWER
0013   C
0014       10 WRITE (2,1)
0015        1 FORMAT (// "NEXT DEVICE ←")
0016          READ (1,*) DMY
0017   C
0018   C   MEASURE DEVICE
0019   C
0020          CALL  CORR1 (RHO,1,128,1,1)
0021   C
0022   C  CONVERT TO RETURN LOSS AND PLOT
0023   C
0024          M = 1
0025          CALL  SETP1 (IBUF,128)
0026      100 CALL  CALF3 (2,M,F)
0027          IF (F) 10,110
0028      110       CALL  CPOL2 (RHO(M),R,A)
0029                IY = -3.2 * 8.686 * ALOG(R)
0030                IX = FSCL * (F-F1)
0031                CALL  PLTP1 (IX,IY)
0032                M = M + 1
0033                GO TO 100
0034                END
  END OF TAPE
```

Fig. 3. *FORTRAN program for same function as Fig. 2. The calls to CORR1 and CALF3 access subroutines in place of the BASIC code between lines 9000 and 9999.*

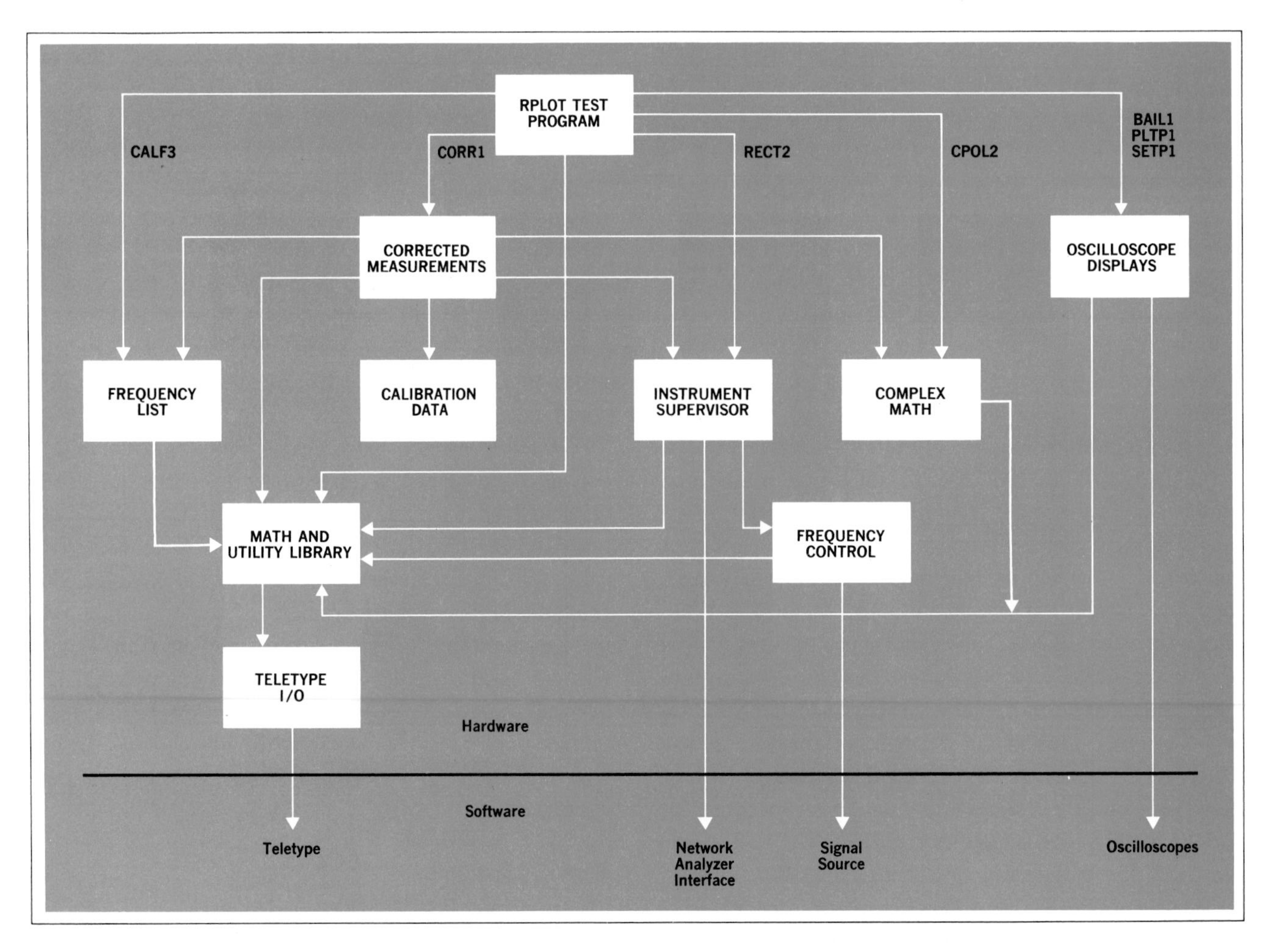

Fig. 4. *Heirarchy of routines used by the example FORTRAN program RPLOT of Fig. 3. Here the special test program used only the upper layers, through a few high-level commands. The layered structures, however, can be accessed at any level to suit the application at hand.*

memory increments or with hundred-thousand word discs, allowing much more complex programs.

In the FORTRAN discussion above, the trade-off between convenience and memory size was mentioned. With increased memory, this trade-off is altered drastically. In fact, a 16K (16,000-word) computer is roughly 5 times as powerful as an 8K version because both FORTRAN and BASIC require so much of the smaller memory for utility routines. Only 2000 to 3000 words out of 8000 are available for test programs and data, while the large machine has 10,000 words available because it does not add to the overhead. This makes it possible to combine an entire series of standard programs (such as the one in Table I) into a single program, plus doubling the data capacity. Similarly, much more sophisticated BASIC programs can be written.

A disc memory which records data magnetically on a rotating surface adds a different kind of capability. With internal core memory, the computer can access any piece of data within 2 μsec. With the disc, it may take up to 30 msec to access the first word of a data block. However, the disc economically provides much larger storage — 176,000 or more words. The 8542A network analyzer disc system saves a number of FORTRAN and BASIC programs on the disc which can be read into memory in 100 msec instead of the full minute which it takes to read a paper tape. BASIC or FORTRAN programs can call one another, so the user can add BASIC tasks to standard FORTRAN programs. A task in the standard program can call up a user's BASIC program after leaving the measured data on the disc. The BASIC program can now analyze the data and return to the standard program for the next test. The user is not even conscious of the exchange, since it happens in a fraction of a second. Now the measurements are made with the efficiency of a compiled program, but the analysis program can be written in the convenient interactive mode of BASIC.

Future

Although the 8542A software is a unique package for bringing software flexibility to microwave engineers, it does so within the framework of conventional computer languages and techniques. This experience has uncovered a number of new requirements for further instrumentation software, particularly for improvements in languages and programming systems. Implementing for these needs will soon make it even easier for the engineer to engage the computer's full capabilities.

Acknowledgments

We are grateful for the assistance of Lucienne Jackson in preparing the software, and to Jesse Pipkin and Lyle Jevons for their help in designing and refining the package.

References

[1] 'S-Parameter Techniques for Faster, More Accurate Network Design,' **Hewlett-Packard Journal,** Feb., 1967.
[2] Peterson, Gerald L., 'BASIC, The Language of Time-Sharing,' **Hewlett-Packard Journal,** Nov., 1968.

William A. Ray

After taking his Bachelor's in electrical engineering at Stanford in 1963, Bill Ray went to Stanford Research Institute for a couple of years. He did work there on antennas — measurements, analysis and theory — and that led him to acquire knowledge in depth about computer capabilities. At Hewlett-Packard, Bill first did application engineering on automatic network analyzers, then moved to the design labs to head software development for these new systems. His marriage and subsequent fatherhood caused him to give up racing sports cars and flying airplanes, but he still sails. His craft is a Javelin.

Warren W. Williams

Bill Williams is a native of Texas. His education included five years as an electronic technician in the Navy. Bill did his undergraduate work at Columbia, New York, and at San Francisco State, taking a degree in math at State in 1967. He came directly to HP from State, and has been concerned with automatic network analysis ever since. He is now the systems programmer for the design group. Married to a California girl, Bill is now a father twice over. He speaks Japanese and hopes one day his HP career will use it.

Developing Accuracy Specifications for Automatic Network Analyzer Systems

By B. P. Hand

In the usual measurement system either those few instrument specifications which directly affect accuracy are added linearly, or one predominates and is taken as the system error. In microwave measurement systems, however, the number of sources of error is so much greater that adding them linearly results in too pessimistic an error figure — one which is extremely unlikely to occur. Furthermore, automatic measurement systems remove various frequency-dependent errors which are usually the major ones. As a result, many of the more subtle errors which are usually neglected as having unimportant effect must be considered.

Establishing accuracy specifications for the various 8542A systems, then, has involved examining and including many additional sources of error and developing means to combine them in realistic fashion. A general technique has been evolved which appears to be applicable in principle to any automatic measurement system.

Reviewing The Ideal Case

Before any discussion of the technique and its results, it is in order to review the whole method of operation of the system, insofar as accuracy is concerned — the calibration process, the model set up as a result, the measurement process, and the correction of the measurement data to yield the final results. A general case will be considered, in which both reflection and transmission coefficients are measured. A representative system would involve, for example, completely characterizing a coaxial attenuator in the range from 2 to 12.4 GHz. Other cases, such as measurements below 2 GHz, or of reflection only, differ only in detail.

The test unit includes two directional couplers for sampling the incident signal (Reference) and the reflected or transmitted signal (Test). Externally, it has an 'Unknown' port, to which the device under test is connected, and a 'Return' port, to which two-port devices are also connected. The model set up in the calibration process includes the properties of this unit, plus some of the errors due to the other instrumentation.

Calibration Process

The calibration process involves making sufficient measurements with standards and conditions of known characteristics to determine all these properties. Fig. 1 is a signal flowgraph of the system model, with the various model coefficients identified.

The s-parameters of any device connected to the Unknown or between Unknown and Return ports are represented by s_{11}, s_{21}, s_{12}, s_{22} in the usual notation.

Flowgraph analysis results in the following general expressions for the ideal measured values of reflection (M_R) and transmission (M_T) coefficients:

$$M_R = e_{00} + \frac{s_{11}e_{01}(1 - s_{22}e_{22}) + s_{21}s_{12}e_{22}e_{01}}{1 - s_{11}e_{11} - s_{22}e_{22} - s_{21}s_{12}e_{11}e_{22} + s_{11}e_{11}s_{22}e_{22}}$$

$$M_T = e_{30} + \frac{s_{21}e_{32}}{1 - s_{11}e_{11} - s_{22}e_{22} - s_{21}s_{12}e_{11}e_{22} + s_{11}e_{11}s_{22}e_{22}}$$

Using D_1 for the common denominator:

$$M_R = e_{00} + \frac{s_{11}e_{01}(1 - s_{22}e_{22}) + s_{21}s_{12}e_{22}e_{01}}{D_1}$$

$$M_T = e_{30} + \frac{s_{21}e_{32}}{D_1}$$

The calibration measurements are as follows:

1. Reflection with a sliding load. The computer measures the reflection as connected, and then, three times, directs the operator to slide the load and makes another measurement. It then constructs a circle through these four values and finds the center of the circle, so that, effectively, $s_{11} = 0$, s_{21}, s_{12} and s_{22} are also zero, of course, so

$$M_1 = e_{00}$$

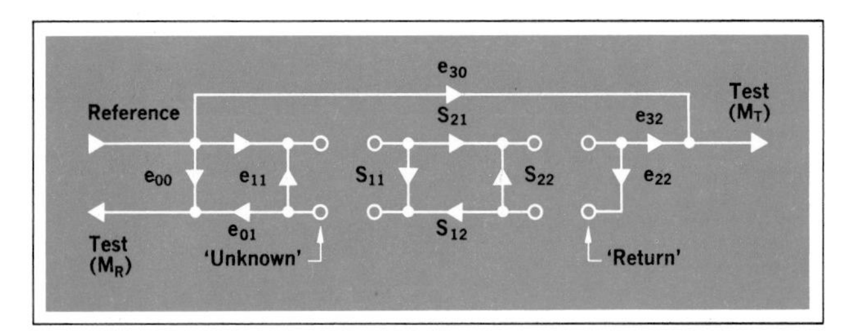

Fig. 1. *Signal flowpath of system model.*

2. Transmission without a through connection (both measurement ports terminated).

$$s_{11} = s_{21} = s_{12} = s_{22} = 0,$$

so $$M_2 = e_{30}$$

3. Reflection with a direct short.

$$s_{11} = -1;\ s_{21} = s_{12} = s_{22} = 0,$$

so $$M_3 = e_{00} - \frac{e_{01}}{1 + e_{11}}$$

4. Reflection with an 'Offset Short' (a shorted coax line one-quarter wavelength long at midband).

$s_{11} = -e^{-j2\beta l}$, where l is the length of the short. For convenience, let $-e^{-j2\beta l} = \Gamma s$. $s_{21} = s_{12} = s_{22} = 0$,

so $$M_4 = e_{00} + \frac{\Gamma s\, e_{01}}{1 - \Gamma s\, e_{11}}$$

5. Reflection with a through connection.

$$s_{11} = s_{22} = 0,\ s_{21} = s_{12} = 1,$$

so $$M_5 = e_{00} + \frac{e_{22}e_{01}}{1 - e_{11}e_{22}}$$

6. Transmission with a through connection.

$$s_{11} = s_{22} = 0,\ s_{21} = s_{12} = 1,$$

so $$M_6 = e_{30} + \frac{e_{32}}{1 - e_{11}e_{22}}$$

These six equations are solved by the computer to give the desired model parameters:

$$e_{00} = M_1$$

$$e_{11} = \frac{\Gamma s(M_1 - M_3) + (M_1 - M_4)}{\Gamma s(M_3 - M_4)}$$

$$e_{01} = \frac{(1 + \Gamma s)(M_1 - M_3)(M_1 - M_4)}{\Gamma s(M_3 - M_4)}$$

$$e_{30} = M_2$$

$$e_{22} = \frac{M_5 - M_1}{e_{01} + (M_5 - M_1)e_{11}}$$

$$e_{32} = (M_6 - M_2)(1 - e_{11}e_{22})$$

At the end of the calibration process, these six values are stored for use in correcting all subsequent measurements.

Measurement Process

On an unknown, reflection and transmission measurements are made, then either the device is reversed between ports, or internal switching reverses the direction of signal flow, depending on which test unit is used, and the two measurements are made again. The flowgraph analysis is essentially the same in either case. The measurements result in four ideal values as follows:

$$M_{R1} = e_{00} + \frac{s_{11}e_{01}(1 - s_{22}e_{22}) + s_{21}s_{12}e_{22}e_{01}}{D_1}$$

$$M_{T1} = e_{30} + \frac{s_{21}e_{32}}{D_1}$$

$$M_{T2} = e_{30} + \frac{s_{12}e_{32}}{D_2}$$

$$M_{R2} = e_{00} + \frac{s_{22}e_{01}(1 - s_{11}e_{22}) + s_{12}s_{21}e_{22}e_{01}}{D_2}$$

where D_2 is D_1 with the s-parameters interchanged:

$$D_2 = 1 - s_{22}e_{11} - s_{11}e_{22} - s_{12}s_{21}e_{11}e_{22} + s_{22}e_{11}s_{11}e_{22}$$

This corresponds to reversing the device.

Correction Process

The computer then solves the equations above for s_{11}, s_{21}, s_{12}, and s_{22}. No explicit solution is given here, since it would be very complex. The computer uses an iterative process which has been demonstrated to have negligible error.

Sources of Error

The above process takes into account only some of the possible sources of error. There are a great many others, but these may be conveniently grouped into three types:

1) Imperfections of the three calibration standards. These are carefully manufactured to very close tolerances, but the tolerances do exist and must be taken into account. Besides diameter and length variations, other factors considered are plating, skin depth, eccentricity, surface roughness, and air dielectric.

2) Noise. This, of course, enters into every measurement. Its effect is reduced by averaging multiple measurements — at least two at high signal levels and up to twelve at lower levels.

3) 'Gain Error'. This is a catch-all term to include all other sources of error, mainly in the instrumentation, which directly affect the magnitude and phase of a measured signal. Whereas noise adds, gain error multiplies. Some of the major sources are the IF attenuator, RF connectors, and switch repeatability errors.

Actual Relations

Considering now what actually happens in the whole process, it can be seen that the effect of noise and gain error must be included in every measurement, while errors in the calibration standards enter into their corresponding measurements. Referring again to the six calibration measurements, and using primes to indicate actual values:

1) Instead of $M_1 = e_{00}$, the computer gets a value

$$M_1' = (1+e_1)(e_{00} + \frac{Se_{01}}{1-Se_{11}} + N_1)$$

Here e_1 is the gain error, N_1 the noise, and S the reflection coefficient of the sliding load. S is not necessarily zero; the Z_0 of the line may not be exactly 50 ohms and the center-finding process is affected by noise.

2) No standards are involved, so

$$M_2' = (1+e_2)(e_{30}+N_2)$$

3) The errors of the direct short itself turn out to be negligible, so

$$M_3' = (1+e_3)(e_{00} - \frac{e_{01}}{1+e_{11}} + N_3)$$

4) The offset short reflection coefficient can be in error in both magnitude and phase, so

$$M_4' = (1+e_4)(e_{00} + \frac{\Gamma s' \, e_{01}}{1 - \Gamma s' e_{11}} + N_4)$$

5) and 6) No standards are involved, so

$$M_5' = (1+e_5)(e_{00} + \frac{e_{22}e_{01}}{1-e_{11}e_{22}} + N_5)$$

$$\text{and } M_6' = (1+e_6)(e_{30} + \frac{e_{32}}{1-e_{11}e_{22}} + N_6)$$

These then are the actual measured values from which the computer derives the model parameters. Re-designating the computer quantities, with ideal quantities in parentheses:

$$(e_{00}) \quad L = M_1'$$

$$(e_{11}) \quad R = \frac{\Gamma s(M_1' - M_3') + (M_1' - M_4')}{\Gamma s(M_3' - M_4')}$$

$$(e_{01}) \quad T = \frac{(1+\Gamma s)(M_1' - M_3')(M_1' - M_4')}{\Gamma s(M_3' - M_4')}$$

$$(e_{30}) \quad L1 = M_2'$$

$$(e_{22}) \quad R1 = \frac{M_5' - M_1'}{T + (M_5' - M_1')R}$$

$$(e_{32}) \quad T1 = (M_6' - M_2')(1 - RR1)$$

In the measurements on an unknown, gain error and noise enter again, so

$$M_{R1}' = (1+e_{R1})(e_{00} + \frac{s_{11}e_{01}(1-s_{22}e_{22}) + s_{21}s_{12}e_{22}e_{01}}{D_1} + N_{R1})$$

$$M_{T1}' = (1+e_{T1})(e_{30} + \frac{s_{21}e_{32}}{D_1} + N_{T1})$$

$$M_{T2}' = (1+e_{T2})(e_{30} + \frac{s_{12}e_{32}}{D_2} + N_{T2})$$

$$M_{R2}' = (1+e_{R2})(e_{00} + \frac{s_{22}e_{01}(1-s_{11}e_{22}) + s_{12}s_{21}e_{22}e_{01}}{D_2} + N_{R2})$$

The correction then is made by solving for the s-parameters, using these actual measured values and the actual stored values for the model parameters. The problem is to determine the possible error in the resulting calculated s-parameters.

Problem Solution

It can readily be seen that any attempt at a complete explicit solution is pointless since, while the maximum magnitudes of the various terms are known, their phases are, in general, quite unpredictable. Various approaches were taken involving the combining of random-phase terms in the root of the sum of the squares (RSS), or the assignment of phase in steps to different variables in turn. These were all felt very unsatisfactory, because of the interlocking and implicit relations between the terms and because there was no way of determining what the confidence level was.

Finally, it was decided to solve the problem statistically, taking advantage of a computer-driven random-number generator. This is the technique that has resulted in the current accuracy specification.

Each of the random-phase variables — the six model parameters and all the gain error and noise terms — is assigned a phase angle between 0 and 2π at random. The magnitudes of the parameters and the noise are assigned random values between an estimated minimum and the maximum allowed in production testing of the instruments, while the gain error magnitude is the RSS of all the individual contributions. The desired input magnitudes of the s-parameters of the unknown are assigned in the desired steps while their phases are also random. For convenience, s_{11} and s_{22} have the same magnitude, while s_{21} and s_{12} are equal in both phase and magnitude.

The whole process gone through by the 8542A is written into a general-purpose error-analysis program. Actual rather than ideal values are used in all calculations. This yields a set of calculated s-parameters which are compared with the input values to determine their errors. The entire calibration-measurement-correction-comparison process is carried out 100 times, new random magnitudes and angles being assigned each time. The errors in s_{11} and s_{22} and in s_{21} and s_{12} are compared with previous errors and the maximum values are stored. At the end of 100 cycles, the 9 worst values are stored. The program then discards the 8 worst and prints out the 9th worst. Since s_{11} and s_{22} have the same nominal values, in effect there are 200 calculated values of the same quantity compared. Thus the worst 4% are discarded and the printed values correspond to a 96% confidence level. The same applies to s_{21} and s_{12}.

Results

The advantages of this technique are that it gives results of predictable confidence level and that the confidence level may be set at whatever value desired. A smoother set of data can readily be obtained by taking more samples. The curves of Fig. 2 show the complete results for the 8743A in the range from 2 to 12.4 GHz. These curves were obtained with the standard software averaging of from 2 to 12 measurements, depending on signal level. To reduce the effect of noise at low level, the programs may readily be modified to average more measurements. The dotted curve in Fig. 2b was obtained by reducing the noise level by a factor of 10, corresponding to averaging 100 measurements. It shows the substantial improvement in accuracy to be expected.

The results obtained with this technique apply to all 8542A systems in general. A particular system at a given frequency may be consistently better or worse, since some errors taken as random in the general case are systematic in the individual case. However, this suggests the next logical step. Any user with the necessary equipment may determine his own system parameters and put these into the error-analysis program, substituting known magnitude and phase for random magnitude and phase wherever possible. The result after running this program will be an accuracy specification applying to that particular system. The more variables thus processed the better the resulting specification.

Acknowledgement

The constant encouragement and assistance of Richard E. Hackborn are gratefully acknowledged.

B. P. Hand

Phil Hand has been with Hewlett-Packard since 1947. His degrees include the Bachelor's in EE from Santa Clara and the Master's, also in EE, from Stanford. The list of HP products to which he has made significant contributions is long. Among other microwave products, he was associated with the 430-series of power meters, and the 370, 375, and 382 attenuators during the 1950's. In charge of the Hewlett-Packard Measurement Standards Laboratory from 1957 to 1968, he recently was named senior technical specialist to the Microwave Division. An authority and a performer on the recorder and other ancient musical instruments, Phil is also an enthusiastic wine-taster and bird-watcher.

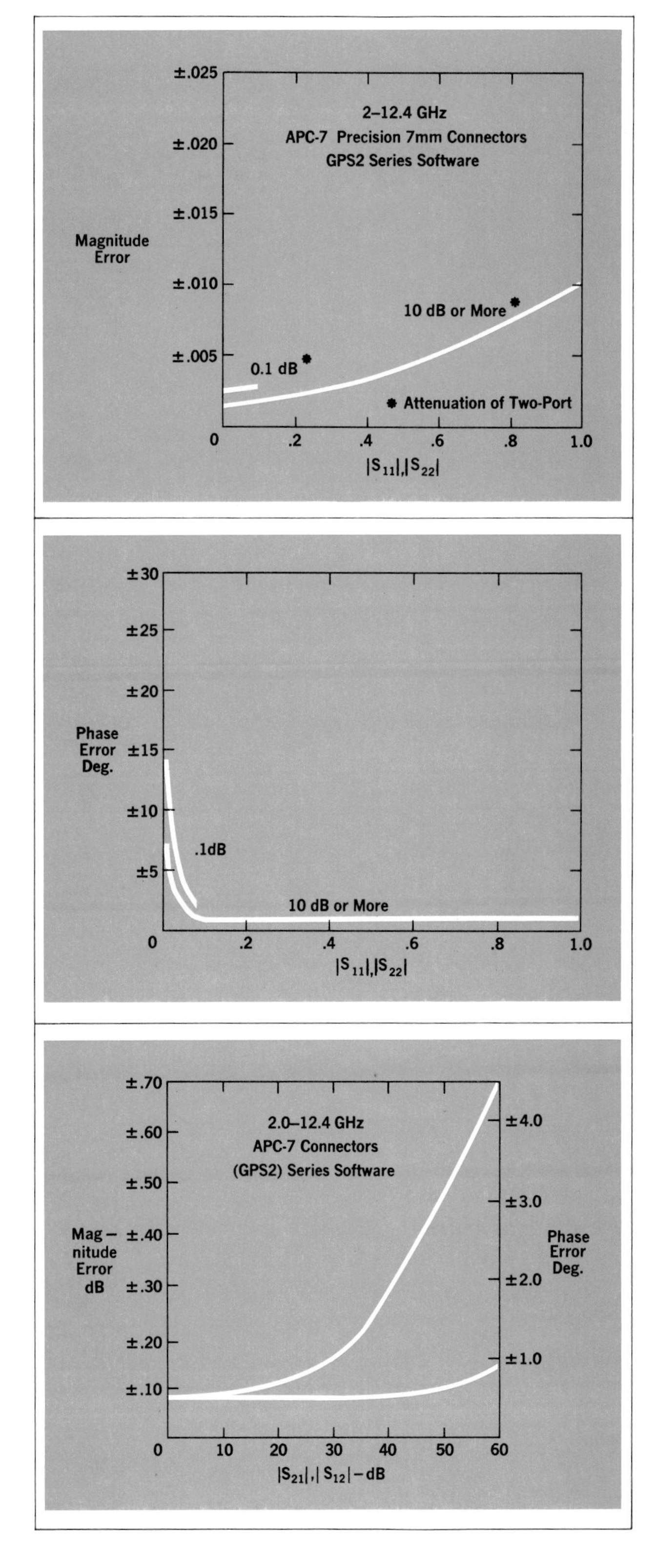

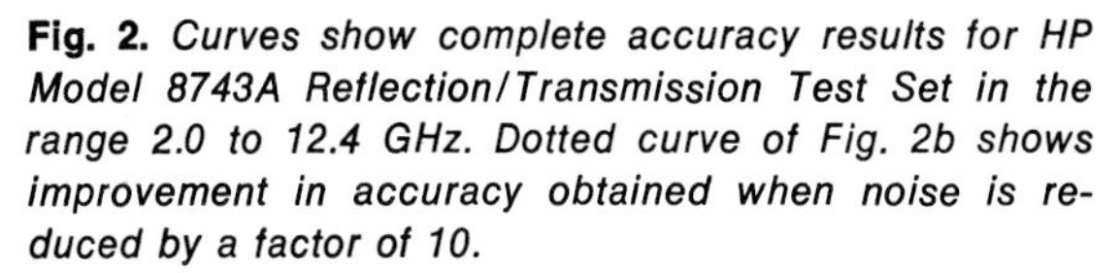

Fig. 2. *Curves show complete accuracy results for HP Model 8743A Reflection/Transmission Test Set in the range 2.0 to 12.4 GHz. Dotted curve of Fig. 2b shows improvement in accuracy obtained when noise is reduced by a factor of 10.*

Applications of the Automatic Network Analyzer

By Brian Humphries

THE POWER OF THE AUTOMATIC NETWORK ANALYZER lies in its ability to characterize RF and microwave devices completely, accurately, and rapidly, then to process and present the information in almost any way desired. The consequences of this power are perhaps best made evident with examples.

Automatic Component Pre-testing

A key ingredient in some recent advances in microwave microcircuitry has been the ability of the automatic network analyzer to fully characterize both active and passive components before they are committed to final circuits. Some new, small, mechanically tuned oscillators for X and Ku bands use negative resistance devices such as Gunn or Impatt diodes in miniature cavities. It is possible to assure they will perform as desired by analyzing both the devices and the cavities before assembly. To make best use of each diode, one wants to know the frequency range within which its impedance is negative real, and the bias current for optimum negative resistance. Measuring the diode's reflection coefficient in a 50-ohm system, as a function both of frequency and bias, gives the whole story. This fully reveals regions of potential instability, and of optimum behavior. It requires very many tests and calculations, which the automatic analyzer quickly and easily performs. The information can immediately appear on a scope face, if desired, or a teleprinter will make a tabulated record.

The analyzer also speeds the task of tuning the cavity to the desired frequency, and determining the impedance seen by the diode at resonance. The oscilloscope photograph in Fig. 1 shows how the analyzer displays the real and imaginary parts of input impedance from 12.5 to 18.0 GHz, with data displayed every 150 MHz. The instrument repeatedly sweeps the band, measuring reflection coefficient at each frequency. It continually calculates impedance and presents the information. The cavity now can be adjusted in real time, with full knowledge of the effect.

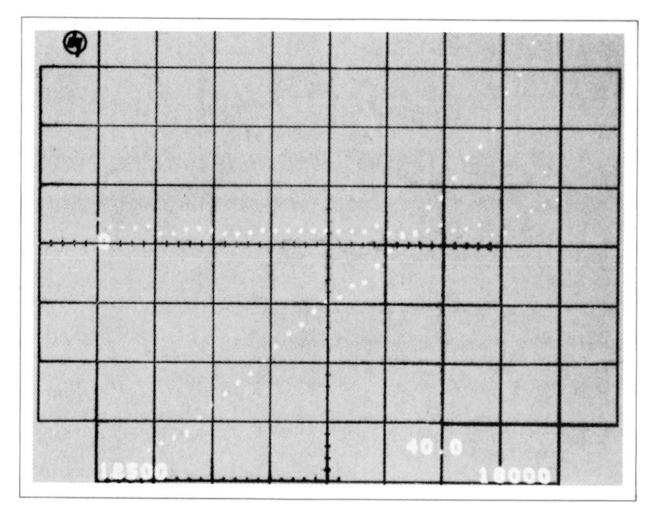

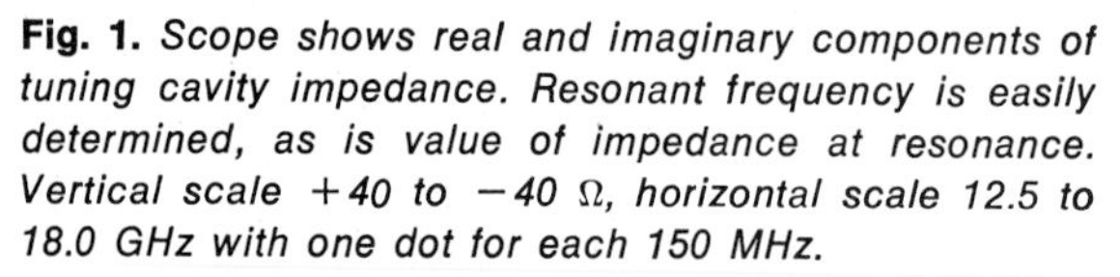

Fig. 1. *Scope shows real and imaginary components of tuning cavity impedance. Resonant frequency is easily determined, as is value of impedance at resonance. Vertical scale +40 to −40 Ω, horizontal scale 12.5 to 18.0 GHz with one dot for each 150 MHz.*

Automatic Testing for Production Control

A recent hybrid microcircuit amplifier (HP 35005A) delivers 40 dB gain, ±3 dB, across the band 0.1 to 2.0 GHz. It would have been almost impossible to put it into production without the automatic analyzer's ability to provide process control information during the actual production cycle. The amplifier consists of four sections, each of two stages. The analyzer takes a full set of scattering parameters on each unpackaged transistor. These are then compared with acceptable limits. Transistors thus pre-screened are attached to a substrate containing the passive elements of a two-stage section. The analyzer then fully characterizes this subassembly, listing its s-parameters on a teleprinter. With this information the operator accepts or rejects the device. If the section is accepted, the analyzer then generates a punched paper tape of the information. As many as 97 of these records are now analyzed together on another computer to make an optimum sort of the entire lot, so groups of four may be combined to produce optimum yield of whole amplifiers having the desired overall gain and flatness.

Automatic Finished-product Analysis

Fig. 3 shows the teleprinter output from a program that tests isolator/filters. The analyzer takes VSWR and isolation data, compares the data against preset limits and indicates an out-of-tolerance condition by printing an x in the appropriate column. The program can be instructed to select appropriately among the data, and print out only useful data for that and surrounding points. All the operator need do is answer 'Y' to the question, 'New meas?', comply with commands to connect the device, and enter the serial number of the unit to be tested.

When equipped with an optional phase-locked signal source, for highest frequency precision, the analyzer is well suited to characterize narrow-band devices. The scope photos in Fig. 4 show how minutely the analyzer can examine a sharp filter (HP Model 536A Coaxial Frequency Meter).

Frequency Domain to Time Domain Conversion

Much of the analyzer's power is in the many possible mathematical operations by the system computer. With this the analyzer often can derive data which the hardware cannot directly measure. Fig. 5 displays the output resulting from a program which measures reflection coefficient in the frequency domain, then converts the data into the time domain. The time domain information is related to distance, using known propagation velocity, and the results of an equivalent time domain impulse test are displayed on a scope. Here the measurement was of a strip-line transistor fixture connected to a 10-cm airline and terminated in a 50-ohm load. The power of the technique is such that one can realize resolution of the order of 1 cm and sensitivity of 0.001 in reflection coefficient. An added benefit of frequency domain testing is to analyze limited-band systems such as waveguide where time domain reflectometry has been impractical.

Calculating Group Delay

Group delay is given by the derivative of the transmission phase characteristic. All that is required to calculate group delay directly from measured phase information is a program which can approximate the slope of the transmission phase curve. Fig. 6 shows such a measurement

POWER AMP STANDARD #1

FREQ	S11		S12		S21		S22	
	MAG	PHASE	MAG	PHASE	MAG	PHASE	MAG	PHASE
100	.28	-45	.00	55	2.98	-125	.17	-127
416	.25	-59	.00	94	3.62	113	.17	-163
732	.29	-92	.00	173	3.87	36	.13	-160
1048	.33	-123	.00	161	3.92	-42	.16	-141
1364	.39	-148	.01	132	3.54	-113	.22	-134
1680	.42	-179	.01	101	3.17	175	.30	-151
1996	.42	145	.02	75	2.82	100	.32	161

GAINS(DB)

9.49
11.18
11.74
11.85
10.96
10.00
9.00

GMAX= 11.85 GMIN= 9.00 GVAR= 2.85

CKT #

Fig. 2. *A typical teleprinter listing gives test operator a quick look at data on two-stage section of 0.1–2.0 GHz amplifier. If gain limits are found to be within specifications, operator commands s-parameter data to be punched on tape for off-line analysis.*

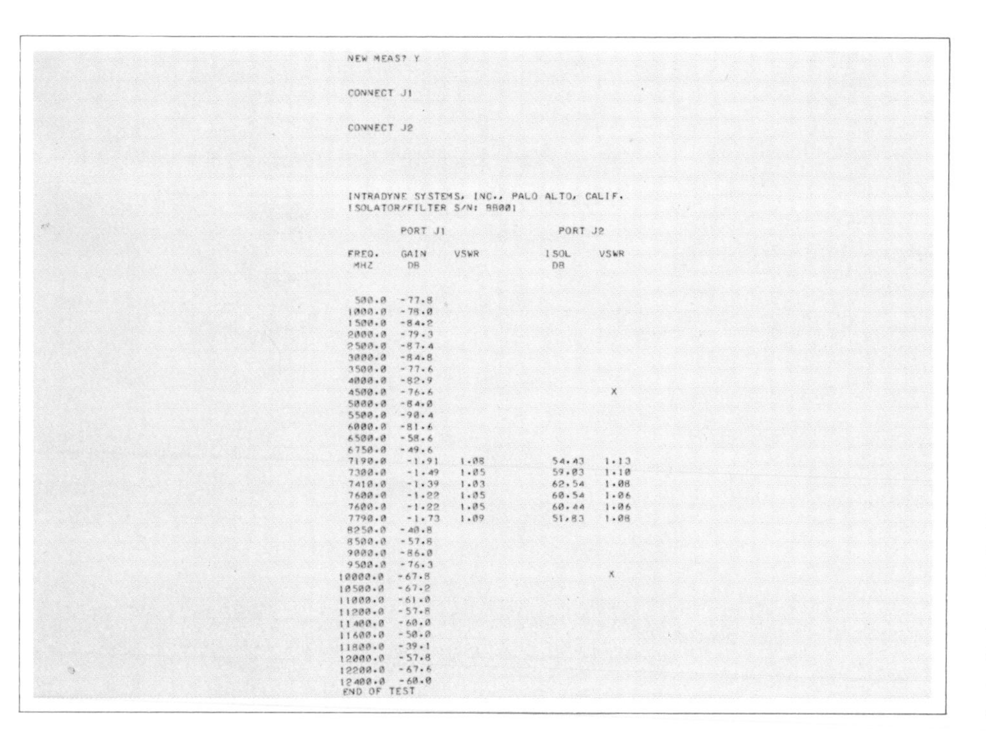

NEW MEAS? Y

CONNECT J1

CONNECT J2

INTRADYNE SYSTEMS, INC., PALO ALTO, CALIF.
ISOLATOR/FILTER S/N: 88001

	PORT J1		PORT J2	
FREQ. MHZ	GAIN DB	VSWR	ISOL DB	VSWR
500.0	-77.8			
1000.0	-78.0			
1500.0	-84.2			
2000.0	-79.3			
2500.0	-87.4			
3000.0	-84.8			
3500.0	-77.6			
4000.0	-82.9			
4500.0	-76.6			X
5000.0	-84.0			
5500.0	-90.4			
6000.0	-81.6			
6500.0	-58.6			
6750.0	-49.6			
7190.0	-1.91	1.08	54.43	1.13
7300.0	-1.49	1.05	59.03	1.10
7410.0	-1.39	1.03	62.54	1.08
7600.0	-1.22	1.05	60.54	1.06
7600.0	-1.22	1.05	60.44	1.06
7790.0	-1.73	1.09	51.83	1.08
8250.0	-40.8			
8500.0	-57.8			
9000.0	-86.0			
9500.0	-76.3			
10000.0	-67.8			X
10500.0	-67.2			
11000.0	-61.0			
11200.0	-57.8			
11400.0	-60.0			
11600.0	-50.0			
11800.0	-39.1			
12000.0	-57.8			
12200.0	-67.6			
12400.0	-60.0			

END OF TEST

Fig. 3. *Printout from a program designed for high-speed production testing of an isolator/filter combination. Note the different frequency intervals above and below the passband, and the indication of specific points of interest within the band. Software instructions selected out these significant points from the welter of other information.*

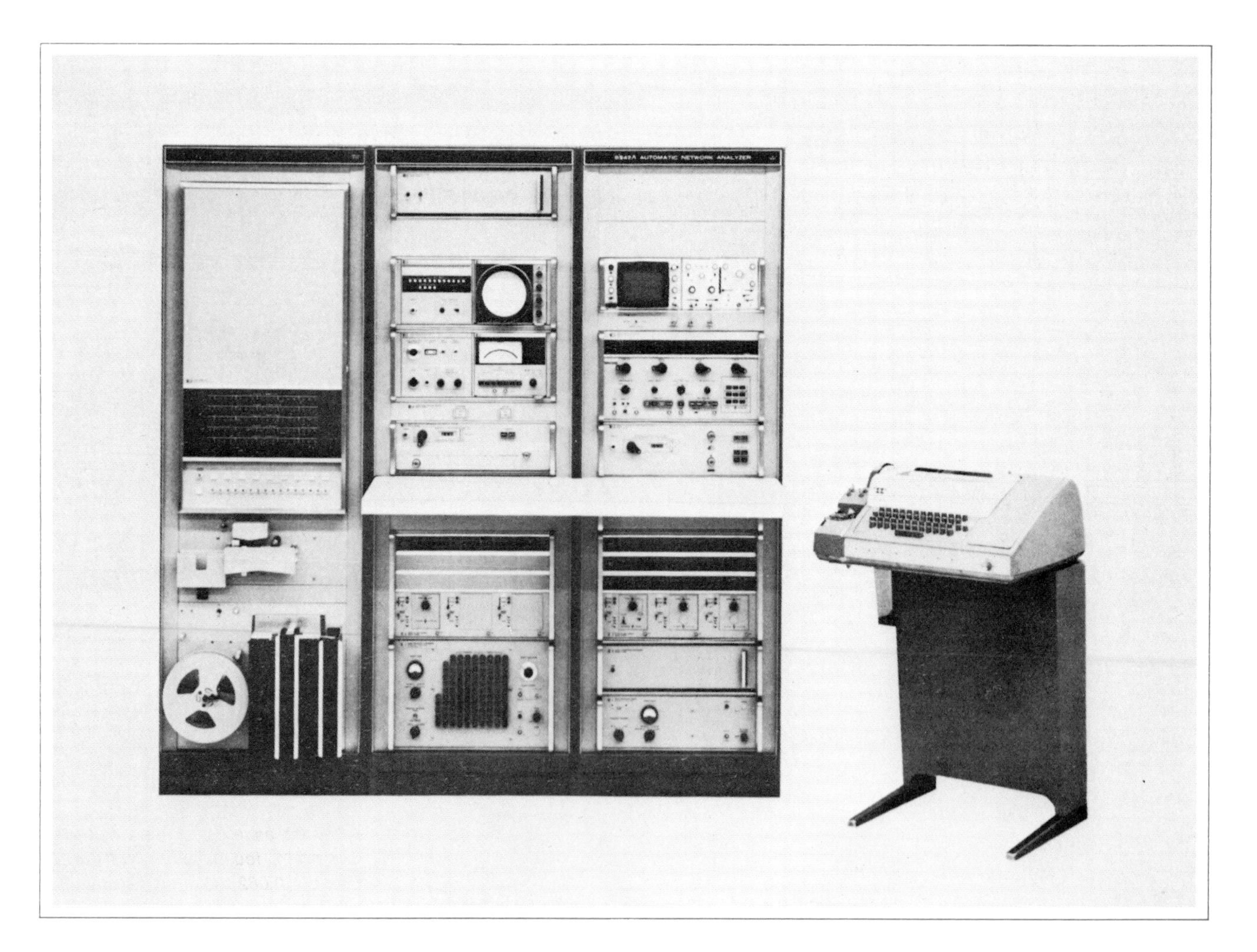

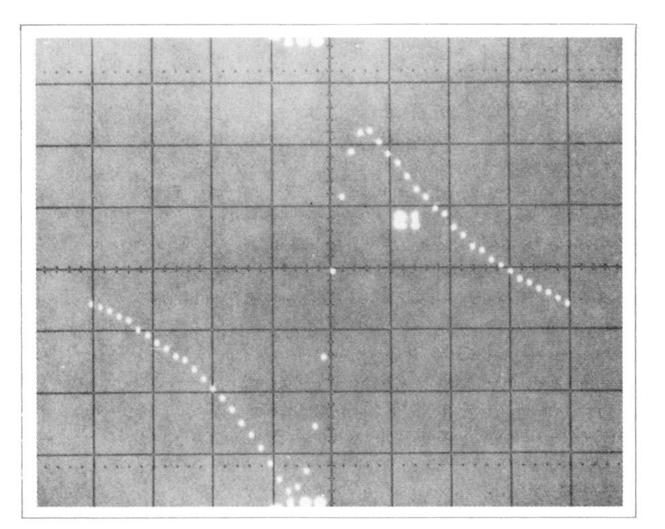

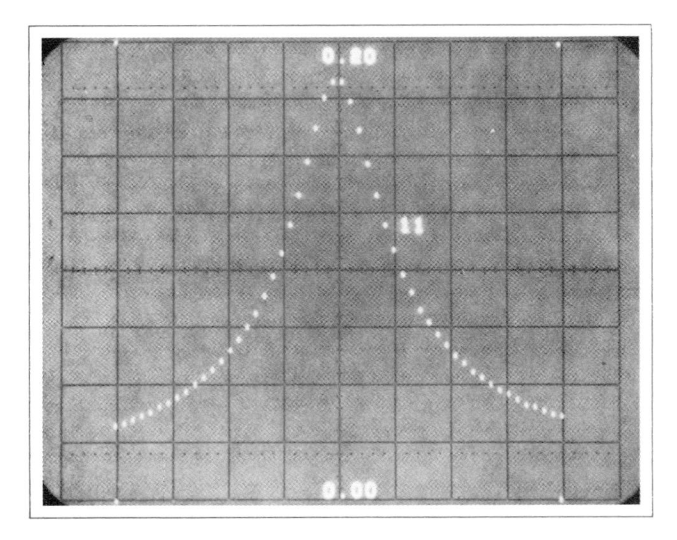

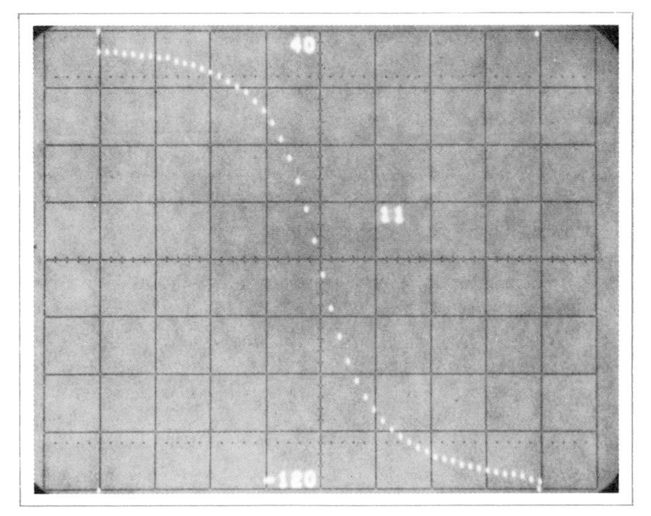

Fig. 4. *Transmission and reflection characteristics of sharp filter (HP Model 536A Frequency Meter). Curve of transmission magnitude (above, left) is centered on 3 GHz, vertical scale 0.4 dB/division. Transmission phase characteristic (above, right) is shown at 2°/div. Magnitude (lower left) and phase (lower right) of reflection coefficients are shown. Scale factor for magnitude is 0.025/div, and for phase 20°/div.*

for a 20 cm airline. Theoretical value for an ideal line is 0.66 ns. Good agreement with measured data is evident.

Conclusion

It has been possible here to indicate only broadly, by these few examples, what sort of problems the automatic network analyzer can efficiently solve. Among the many who have contributed application information to us, Mr. Ed Oxner of Intradyne Systems, Inc. is due particular thanks.

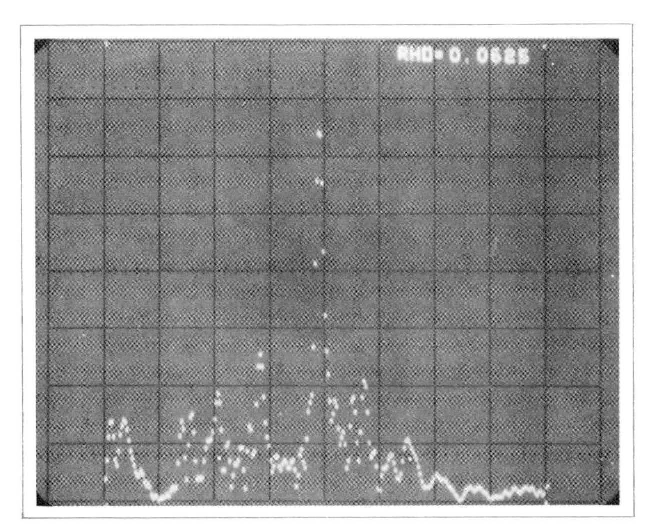

Fig. 5. *Reflection coefficient as a function of distance for a stripline transistor fixture fed through a 10 cm airline. Full scale reflection is 0.0625. Full scale distance is 60 cm.*

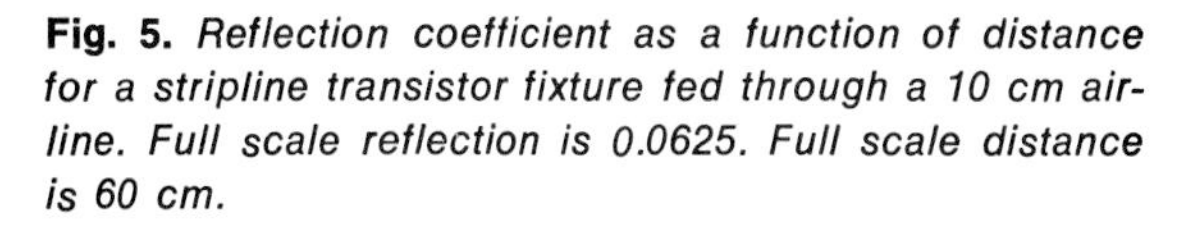

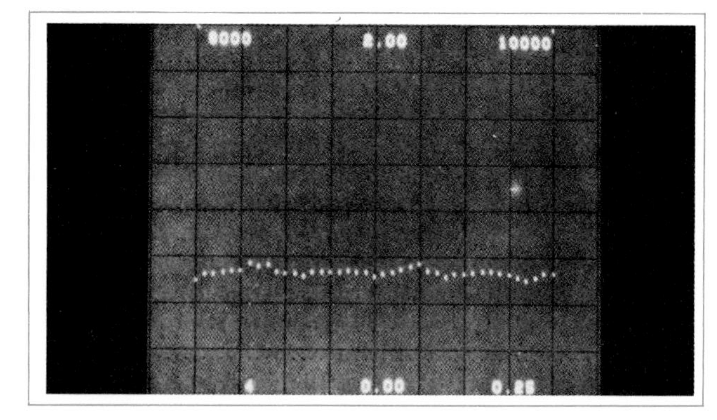

Fig. 6. *Group delay of a 20 cm airline. The measurement was made between 8000 and 10,000 MHz. The scale factor is 0.25 ns/div. The theoretical value of group delay for an ideal line is 0.66 ns.*

Brian A. Humphries

Brian Humphries is a native of England. His degree is that rarity, a BSc. Brian took it from Leeds University in 1959. After a year's graduate apprenticeship at Associated Electrical Industries (Rugby, England), he joined HP's U.K. activity, then in Bedford. Successive assignments in field engineering and sales management led him eventually to his present position as Systems Marketing Manager for the HP Microwave Division, in Palo Alto. He maintains membership in the British IEE.

SPECIFICATIONS
HP Model 8542A Automatic Network Analyzer

AVAILABLE CONFIGURATIONS AND CAPABILITIES

RF STIMULUS OPTIONS

STANDARD SIGNAL SOURCE

Plug-in oscillator covering 0.11–12.4 GHz with three rf units. Analyzer available with one, two, or three rf units to cover part or all of range. Automatic frequency setting ±0.25% + 10 MHz. Max. output power at least 0 dBm, 0.11 to 4.0 GHz, at least +10 dBm, 4 to 12.4 GHz. Broadband power leveling and automatic band selection in multi-band versions.

FREQUENCY-STABILIZED SIGNAL SOURCE

Uses modules from Standard source and incorporates Frequency Synthesizer as precision frequency reference. Resulting source has frequency accuracy ±1 part in 10^6 + 5 kHz. Frequency-stabilized source available with fourth rf unit to extend Analyzer coverage to 18.0 GHz.

DC STIMULUS OPTION

DC BIAS SUPPLY

Programmable, dual-output power supply provides ±30 V ±0.5 A output to bias transistors, diodes, solid-state amplifiers, etc.

MEASUREMENT OPTIONS

NETWORK ANALYZER

Two channels (reference and test). Makes amplitude and phase measurements from 0.11 to 12.4 GHz (18.0 GHz optional) to determine both reflection and transmission coefficients of device under test. Includes integrating analog-to-digital converter to digitize measured information for data manipulation in the Control and Digital Processor Sub-system, and CRT readout devices for both corrected and uncorrected displays.

TEST SETS

Passive instruments selectively feed proper signals to Network Analyzer to determine both reflection and transmission coefficients of two-port network under test. Contain broadband directional couplers, calibrated line stretcher, and an array of microwave switches. Test sets available: one for frequency range 0.11–2.0 GHz, one for range 2.0–18.0 GHz. Under normal system configurations, power incident on device under test can typically be set manually anywhere between −26 and −2 dBm in the range 0.11 to 2.0 GHz, and between −34 and −14 dBm in the range 2.0 to 18.0 GHz.

Transistor fixtures available to measure TO-18 (TO-72) and TO-5 (TO-12) packaged devices from 0.11 to 2.0 GHz. Accommodate standard lead configurations without need to cut leads.

Bias insertion networks, 50Ω, apply dc to test device via center conductors of input and/or output coaxial transmission lines. Two different bias tees available, one covering 0.1–3.0 GHz, another covering 1.0–12.4 GHz.

CALIBRATION EQUIPMENT

Calibration programs supplied with 8452A Automatic Network Analyzers require a set of standards for given connector type (APC-7, N, OSM, GR-900) or waveguide size. Each basic calibration kit contains standards common to entire frequency range covered by that connector or waveguide size. Additional standards required in specific frequency ranges are separately available.

CONTROL AND DIGITAL PROCESSING OPTIONS

INSTRUMENTATION COMPUTERS

Choice of Model 2114B (8,192-word, stored-program computer with seven I/O channels, expandable to 24) or Model 2116B (8,192-word stored-program computer, with 16,384-word option, sixteen I/O channels, expandable to 48. High-speed punched tape input is standard with each computer. Optional memory expansion: both magnetic tape and disc memory peripherals are available.

INPUT/OUTPUT OPTIONS

TELEPRINTERS

Model 2572A (modified ASR-33) for systems where teleprinter use is 5 hours per day or less; for heavier duty, Model 2754B (modified ASR-35) is available.

PUNCHED TAPE OUTPUT

120 character/second tape punch, recommended for systems expected to deliver considerable hard-copy output or where FORTRAN is anticipated to modify standard software. (Both teleprinters have punched tape output of more limited speed and format.)

OSCILLOSCOPE

Measurement subsystem includes CRT display of polar or rectangular plots, corrected or uncorrected, in usual size screens. Large-screen scope, readable at distances up to 10 feet, optional.

X-Y PLOTTER

Optional X-Y point plotter makes hard copy replicas, up to 11" x 17", of any oscilloscope display. Typically a 50-point display can be transferred in less than 15 seconds, including alpha/numeric labels.

HEWLETT-PACKARD JOURNAL **FEBRUARY 1970** *Volume 21 • Number 6*

TECHNICAL INFORMATION FROM THE LABORATORIES OF THE HEWLETT-PACKARD COMPANY PUBLISHED AT 1501 PAGE MILL ROAD, PALO ALTO, CALIFORNIA 94304

Digital Fourier Analysis

Some of the theoretical and practical aspects of measurements involving Fourier analysis by digital instrumentation.

By Peter R. Roth

WHEN THE CHARACTERISTICS OF A SIGNAL OR SYSTEM ARE MEASURED, the measurements most often made are the spectrum of the signal and the transfer function of the system. For example, if the transfer function of the landing gear and wing structure of an aircraft is known, and if the spectrum of the vibrations from typical runways can be determined, then the roughness of a landing can be evaluated. Or if the spectrum of the vibrations caused by typical roads can be determined, an automobile suspension system may be designed and tested to maximize ride comfort.

It is the questions of how to measure spectra and transfer functions, especially when signals more complex than simple sine waves are involved, that we will examine in this article.

The techniques to be described are based upon computation of the Fourier integral

$$S_x(f) = \int_{-\infty}^{+\infty} x(t) exp\{-i2\pi ft\} dt. \qquad (1)$$

While in principle the methods that will be examined are not new and have been partially implemented using analog instruments, their full development has waited on the availability of digital processors with sufficient speed and flexibility.*

How does computation of the Fourier integral help us make meaningful measurements? Consider the Fourier transform written in its sine-cosine form:

$$S_x(f) = \int_{-\infty}^{+\infty} x(t) \{\cos 2\pi ft - i \sin 2\pi ft\} dt. \qquad (2)$$

This equation states that the transform averages a time function input x(t) with a set of sines and cosines to determine the content of x(t) at some frequency f. Thus the transform resolves the time function into a set of components at various frequencies much as a set of analog filters would. However, it not only yields the amplitude at each frequency, but also resolves the in-phase (real, cosine) component and the quadrature (imaginary, sine) component, thereby giving magnitude and phase information which is difficult to obtain in any other way.

* The new HP Model 5450A Fourier Analyzer is one of these digital instruments. See article, page 10.

PRINTED IN U.S.A.

The Fourier transform is also valuable when it is applied to measurements on systems. The result of the operation of a linear system on any input signal in the time domain may be determined from the convolution of the system impulse response h(t) with the input signal x(t) to give the output y(t):

$$y(t) = \int_{-\infty}^{+\infty} h(\tau)\, x(t - \tau) d\tau. \qquad (3)$$

Visualizing the result of this operation is all but impossible for anything other than a simple case. But if the Fourier transform is applied to this convolution integral, a simple, easily understood relationship results. The output spectrum S_y is the product of the input spectrum S_x and the transfer function H:

$$S_y(f) = S_x(f) \cdot H(f). \qquad (4)$$

The simplest implementation of a measurement technique based on this relation is the use of a sine-wave input for x(t). Since the sine wave contains but one frequency component it provides a simple way of measuring the transfer function using voltmeters and phasemeters. However, not all systems may be measured using sine waves, either because there is no way of inserting such a signal into the system, or because the sine wave is not a realistic signal form.

A more general measurement method is to measure the input and output time series, in whatever form they may be, and to calculate H using S_x, S_y, and the Fourier transform. This method has several advantages, as I will show. But first, because the most powerful computational techniques available today are digital, it's necessary to say something about the nature of the Fourier transform when it is implemented on a digital processor.

Digital Fourier Transforms

Digital techniques make us realize very clearly that all measurements are discrete (i.e., have finite resolution) and of finite duration. All digital memories are obviously discrete and finite in size. Therefore, the equation for the Fourier transform must be changed to a finite sum for digital processing. This means, first of all, that the time function to be transformed must be sampled at discrete intervals, say Δt. It also means that only a finite number, say N, of such samples may be taken and stored. The record length T is then

$$T = N\Delta t. \qquad (5)$$

The effect of finite Δt is well known; it limits the maximum frequency that may be sampled without 'aliasing' error to

$$f_{max} = \frac{1}{2\Delta t}. \qquad (6)$$

Any components above this Nyquist frequency or its multiples are folded back onto frequencies below f_{max}. In practical measurement situations this aliasing presents little or no difficulty, since f_{max} can be chosen to include all significant components of the input signal, or a filter may be used before the sampler to eliminate any strong components above f_{max}.

The effect of finite record length T is also important. When a Fourier integral is taken over a finite record length T the result is a Fourier series, and the spectrum has discrete lines and finite resolution. A Discrete Finite Transform (DFT), which must be used whenever a Fourier transform is computed digitally, is more like a Fourier series than a transform, since it assumes that the input is periodic in the interval T and has a spectral resolution of

$$\Delta f = \frac{1}{T}. \qquad (7)$$

The DFT is written as

$$S_x(m\Delta f) = \frac{1}{N}\sum_{n=0}^{N-1} x(n\Delta t)\, exp\{-i\frac{2\pi}{N}mn\}. \qquad (8)$$

It yields in the frequency domain $\frac{N}{2}$ real (cosine) components and $\frac{N}{2}$ imaginary (sine) components from a sampled time record of N points. I will refer to this result as the *linear spectrum* to keep it sorted out from certain other spectrum forms.

While this raw form of spectrum has certain uses, it it of limited value because of its dependence on the time position of the input record. A waveform of constant shape will always have the same energy at any one frequency, but how this energy is distributed between the sine and cosine terms depends on the phase shift or time position of the waveform. Fig. 1 gives an example of this. Fig. 1a is the real part of the linear spectrum of a square pulse. It has the expected sin x/x form. However, the real part of the linear spectrum of the same pulse delayed a small amount, Fig. 1b, does not have this form. The linear phase shift given the spectrum by delaying the waveform has changed the distribution of the spectrum between its real and imaginary parts. On the other hand,

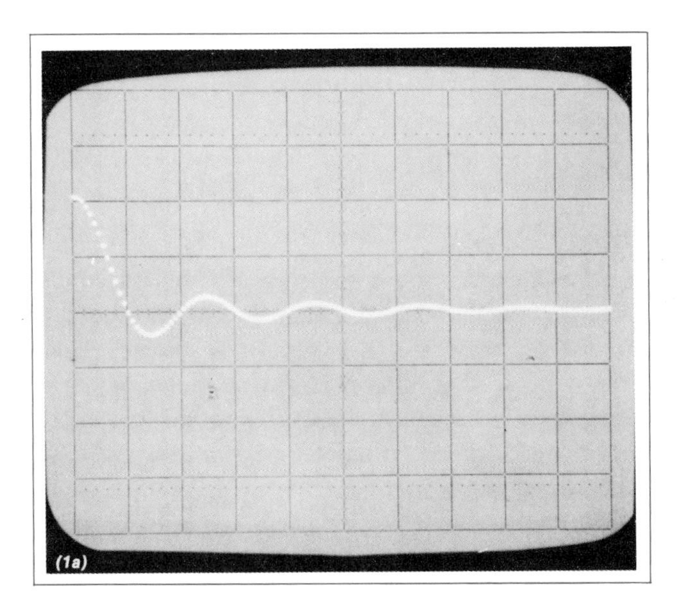

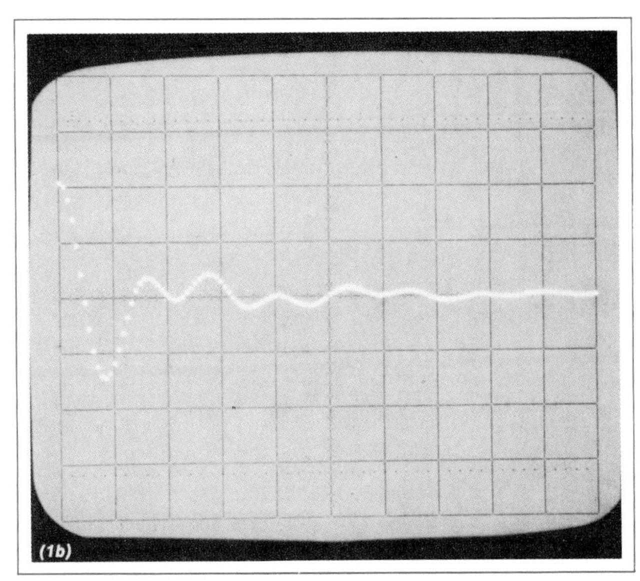

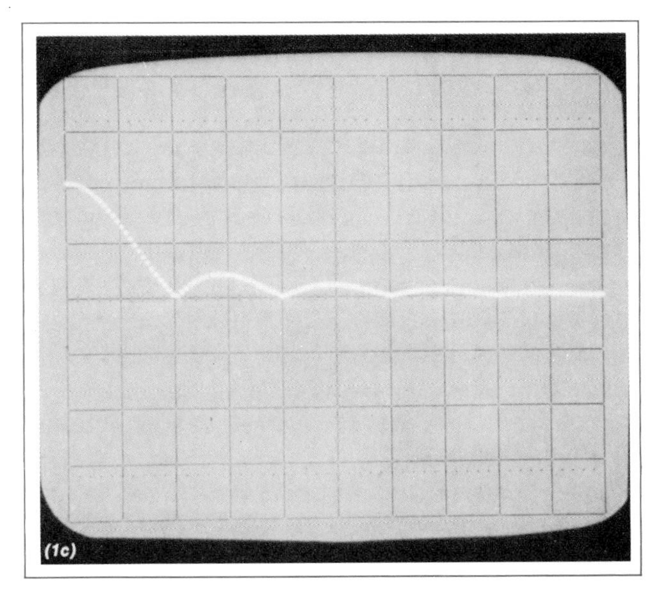

Fig. 1. *The real parts of the Fourier transforms of two rectangular pulses which are identical except for a time shift are shown in (a) and (b). (c) is the magnitude of the Fourier transforms of both pulses. Because it is independent of time position, the magnitude-and-phase form of the transform is more useful than the real-and-imaginary form. However, the square of the magnitude, or the power spectrum, is even more widely used.*

if we examine the *magnitude* of either the undelayed pulse spectrum or the delayed pulse spectrum, we see that it is constant (Fig. 1c) and that the energy in any line is the same no matter what the time position of the input waveform is.

It is clear, then, that to obtain a constant linear spectrum independent of time position it is at least necessary to convert the real and imaginary components of the spectrum into magnitude and phase. While the linear magnitude spectrum is a valid and perfectly acceptable way to achieve a useful spectrum it is cumbersome from a computational standpoint. A closely related function, the 'auto' spectrum or 'power' spectrum, gives the same basic information, is faster to compute, and can be applied to measurements which the linear magnitude spectrum cannot.

Power or Auto Spectrum

The auto spectrum, $G_{xx}(f)$, is formed by multiplying the value of the linear spectrum, $S_x(f)$, by its own complex conjugate.

$$G_{xx}(f) = S_x(f) \cdot S_x^*(f) = [A(f) + iB(f)]\,[A(f) - iB(f)] \tag{9}$$

$$G_{xx}(f) = A^2(f) + B^2(f) \tag{10}$$

Each spectral line of $G_{xx}(f)$ is proportional to the voltage squared at frequency f, or more exactly to the variance of the input waveform at frequency f. The auto spectrum is useful because it is the magnitude squared of the linear spectrum. For this reason, and because it has no imaginary part, it is independent of the time position of the input waveform. It is the square-law auto spectrum that is usually implied when the term 'spectral analysis' is used.

Analyzing Random Signals

The auto spectrum, because of its independence of time and phase, is a useful tool for analyzing signals that are deterministic, that is, for signals that do not change in spectral form from sample record to sample record, or only change in a predictable way. However, the auto spectrum is an even more useful tool for the analysis of signals that are stationary and random, that is, signals whose spectra will vary from sample record to sample record but will have a measurable mean or average value. Many processes generate signals whose spectra cannot be predicted for any single sample record, but whose spectra are stable on the average. Examples of such processes are 1/f noise in an amplifier, or the sea state noise in a sonar system. On the other hand, the process being measured may be a combination of deterministic and random spectra. For example, consider the fine-line components of the noise due to the rotating members of a turbojet hidden by the random noise of the combustion, or the tonal components of an acoustic signal hidden in the random noise of the ocean.

What is more important about a random spectrum is that for a single sample record of length T (i.e., of spectral resolution $\Delta f = 1/T$) the spectral lines are just as random as the time series that generated the spectrum no matter how long T is. In basic engineering and mathematical texts on the Fourier transform, the transition from periodic functions whose spectra are described by Fourier series to totally aperiodic functions whose spectra are described by the Fourier transform is made by making the record length T go to infinity in the limit. That this procedure does not work for the ultimate in aperiodic functions, random signals, can be intuitively demonstrated in two ways.

First consider a wave analyzer with a bandwidth Δf and a meter with very small damping. The response time of this analyzer is about $1/\Delta f$ or T seconds. If a random signal is applied to this wave analyzer an independent reading can be made about every T seconds. Now, if the bandwidth of the wave analyzer is cut to $\Delta f/2$, the response time of the filter and hence the time between independent readings becomes 2T. While the meter will move half as fast in this case, the randomness of the reading as expressed by the variance of the independent readings will be unchanged, since independent readings are twice as far apart. Thus no matter how long a record (i.e., how narrow a bandwidth) is used, no improvement in statistical certainty can be made. The only way to improve the reading is to put an integrating circuit on the meter that is much slower than the response time due to the reciprocal of the bandwidth $1/\Delta f$. Then the final reading will be the result of averaging many independent readings.

To show this effect for a DFT consider a spectrum computed from N equally spaced time samples over a sample record of length T, yielding $\frac{N}{2}$ real and $\frac{N}{2}$ imaginary frequency components. From N time points, exactly N values are obtained in the spectrum, and since no new information about the signal is added by the DFT, each spectral line will have no more statistical certainty than a sample point in the time function from which the spectrum was computed. In fact, for a spectrum of Gaussian noise of any spectrum shape, the variance of a spectral line for one sample record is equal to the expected value for the measurement. Such a measurement is so uncertain that it is no measurement at all.

However, if a number of independent samples of the spectrum are averaged, the variance of the resulting estimate of the spectrum will be reduced in a fashion analogous to integrating readings from the wave analyzer meter. Such a case is demonstrated in Fig. 2. Fig. 2a is a spectrum computed from a single sample record of a signal consisting of a sine wave plus random noise. Because the variance of one sample is equal to the expected value for each line it is impossible to tell which of the spikes is the spectrum of the sine wave and which is due to the variability of the estimate. Fig. 2b shows a spectrum computed from an average of 100 samples. Here the variability of the estimate is reduced to the point where it is perfectly clear where the single tone lies. It is also clear what the spectral shape of the Gaussian noise is. In fact, a statistical certainty for the estimate of the random spectrum is easily computed from the relationship that one standard deviation σ is

$$\sigma = \frac{1}{\sqrt{K}}, \tag{11}$$

where K is the number of sample spectra averaged.

For the case of 100 spectral averages 3σ is 1.1 dB. Thus one would expect that only one estimate in a thousand would fall farther away than 1.1 dB from the measured value in Fig. 2b. To achieve this degree of statistical stability using an analog wave analyzer with a 1 Hz bandwidth would require a 100-second integration.

Two Input Waveforms

So far we have considered measurements on one time series only. However, we often have to take measurements from two signals simultaneously so the relationship

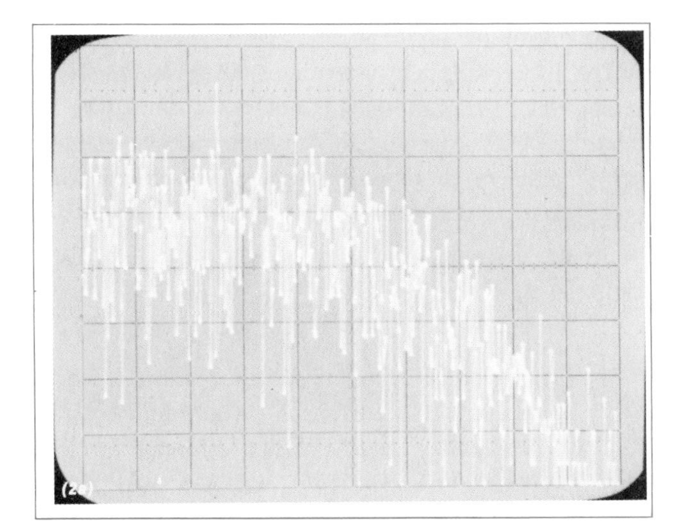

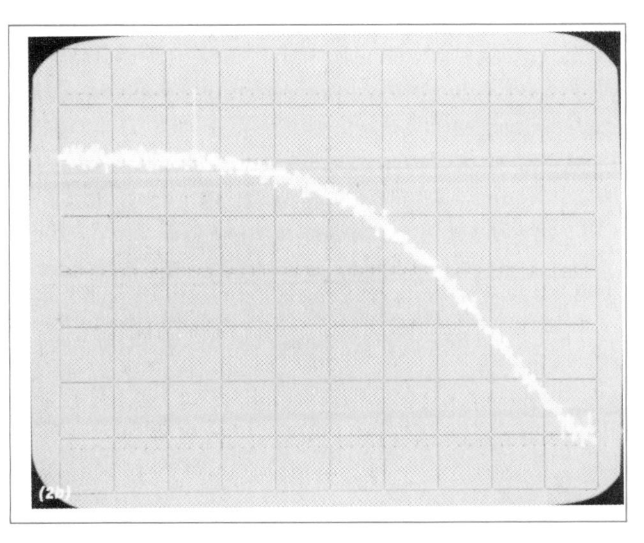

Fig. 2. *The power spectrum (a) computed from a single sample record of a random signal is as random as the signal itself. But when 100 such spectra are averaged, the result (b) shows not only the spectral shape of the random signal, but also that there was a sinusoid hidden in the signal.*

between two points in some process may be determined. For example, in the situation shown in Fig. 3 the relationship between input x(t) and output z(t) might be of interest. There are two distinct quantities that can be measured in such a situation. The first is the degree to which the output depends on the input. That is, is z(t) caused by x(t) or is z(t) due in part to some unrelated signal such as n(t)? Second, if z(t) is caused at least partly by x(t), what is the form of this relationship?

It is important to be aware that, although neither causality nor relationship can exist without the other, each contains different information about the process. It is also important to note that no single measurement of the correlation between two signals, either in the time domain

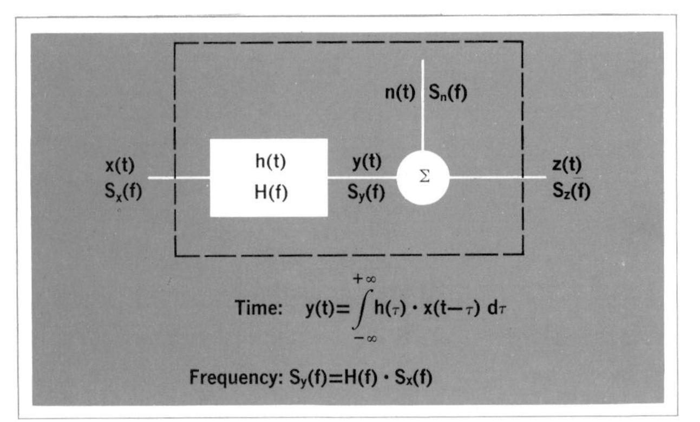

Fig. 3. *Is the output z(t) caused entirely by x(t) or is there also unrelated noise n(t)? What is the form of the relationship between x(t) and y(t)? The transfer function H(f) provides an answer to the second question. A quantity called the coherence function answers the first.*

or the frequency domain, is capable of separating these two quantities.

Cross Spectrum or Cross Power Spectrum

The cross spectrum, also known as the cross power spectrum, illustrates these points. The cross spectrum $G_{yx}(f)$ between two signals y(t) and x(t) in a process or system is formed by multiplying the linear spectrum of y(t) by the complex conjugate of the linear spectrum of x(t) measured at the same time.

$$G_{yx} = S_y S_x^* = (A_y + iB_y)(A_x - iB_x) \qquad (12)$$

$$G_{yx} = (A_y A_x + B_y B_x) + i(B_y A_x - B_x A_y) \qquad (13)$$

These relationships show that the cross spectrum is not a positive real quantity like the auto spectrum, but in general is both complex and bipolar. A physical interpretation of this function is quite straightforward. If there are components at a given frequency in both x(t) and y(t), the cross spectrum will have a magnitude equal to the product of the magnitudes of the components and a phase equal to the phase difference between the components.

While this interpretation is exactly true when x(t) and y(t) are uncontaminated by noise, an additional dimension must be added when unrelated signals are added to the process. Any single sample of the cross spectrum G_{zx} between the output z(t) of the linear system of Fig. 3 and the input x(t) will show the combined effects of x(t) and n(t) merged into z(t). However, if n(t) is unrelated to x(t) (i.e., random and uncorrelated), its contribution to the magnitude of G_{zx} will not have a constant phase from sample record to sample record as will that of x(t). If many sample records are averaged, the random phase of the contribution of n(t) will ultimately cause it to have a negligible contribution to the cross spectrum.

How many independent samples of the cross spectrum it takes to achieve a result of a given accuracy cannot be determined without some further information beyond the cross spectrum itself. Also required is information about the relative contributions of the various signals to the measurement. This makes an important point: *a simple cross spectrum measurement does not differentiate between causality and relationship.* Without more information than is contained in the simple cross spectrum it cannot be determined if a high value in a cross spectrum is due to a strong gain of the measured system at that frequency, or to a large input x(t), or to a strong contaminating signal n(t). In the time domain, the crosscorrelation function also suffers from this same inability to discriminate between causality and relationship in a measurement.

Transfer Functions

While the cross spectrum does not give a definite measurement it leads to two measurements which not only separate relationship and causality but also give quantitative results. The first of these functions measures the relationship between x(t) and z(t). It is a familiar function, the transfer function H(f) of the system (Fig. 3).

The transfer function is the ratio of the output linear spectrum for zero noise to the input linear spectrum.

$$H(f) = \frac{S_y(f)}{S_x(f)}\,. \tag{14}$$

Multiplying the numerator and denominator of this ratio by S_x^* shows that the transfer function can also be expressed as the ratio of the cross spectrum to the input auto spectrum.

$$H = \frac{S_y S_x^*}{S_x S_x^*} = \frac{G_{yx}}{G_{xx}} \tag{15}$$

There are two important points with regard to transfer functions measured in this way. The first is that this technique measures phase as well as magnitude since the cross spectrum contains phase information. Second, this measurement procedure is not limited to any particular input, such as sinusoids. In fact, the input signal may be random noise, or whatever signals are normally processed by the system being measured. For example, a telephone transmission system might be tested while in use with the normal traffic providing the test signal.

Coherence Functions

The major error in transfer function measurements develops when the output z(t) is not totally caused by the input x(t) but is contaminated by internal system noise n(t). Consider the input-output cross spectrum when there is uncorrelated noise with spectrum S_n added to the output.

$$G_{zx} = (S_y + S_n)S_x^* = G_{yx} + G_{nx} \tag{16}$$

If the noise n(t) is truly uncorrelated with x(t), and if enough averages of $\overline{G_{zx}}$ are taken, the contribution of $\overline{G_{nx}}$ to $\overline{G_{zx}}$ will approach zero, and $\overline{G_{zx}}$ will approach $\overline{G_{yx}}$. How rapidly the average of G_{zx} will approach $\overline{G_{yx}}$ depends upon how much noise there is in the output spectrum, that is, to what degree z(t) is caused by x(t).

To measure this coherence between x(t) and z(t) it is necessary to compute a new quantity, the *coherence function,* defined as[1]

$$\overline{\gamma^2} = \frac{\overline{G_{zx}}\,\overline{G_{zx}^*}}{\overline{G_{zz}}\,\overline{G_{xx}}} = \frac{|\overline{G_{zx}}|^2}{\overline{G_{zz}}\,\overline{G_{xx}}} \tag{17}$$

The horizontal bars denote ensemble averages.

After a number of records are averaged the numerator of the coherence function will reduce to $\overline{G_{yy}}\,\overline{G_{xx}}$. The denominator of the coherence function will be the auto spectrum of the normal output plus the noise, times the input auto spectrum. The output-plus-noise auto spectrum is

$$G_{zz} = (S_y + S_n)(S_y + S_n)^* = G_{yy} + G_{yn} + G_{ny} + G_{nn}. \tag{18}$$

After averaging, the cross terms in equation 18 disappear because they are uncorrelated with S_y, leaving

$$\overline{G_{zz}} = \overline{G_{yy}} + \overline{G_{nn}} \tag{19}$$

for the output auto spectrum. The coherence function then has an averaged value of

$$\overline{\gamma^2} = \frac{\overline{G_{xx}}\,\overline{G_{yy}}}{(\overline{G_{yy}} + \overline{G_{nn}})\overline{G_{xx}}} = \frac{\overline{G_{yy}}}{\overline{G_{yy}} + \overline{G_{nn}}} \tag{20}$$

Equation 20 shows that the coherence function $\overline{\gamma^2}$ has a value between 0 and 1, depending on the degree to which the output of the system is causally related to the input. This number not only defines the degree of causality, a useful quantity in itself, but it also defines the number of averages of the cross spectrum and input auto spectrum that are required to define the transfer function to a given degree of accuracy.

An Example

Fig. 4 is an example of the separation of causality and relationship in a measurement. The system under test had a second-order highly damped transfer function. The input signal was Gaussian noise band-limited to the Nyquist folding frequency (10 kHz in this case).

$\overline{\gamma^2}$ for this measurement was about 0.8 out to the point where the transmission attenuation was about 20 dB. Beyond this frequency the data had too small a value to compute $\overline{\gamma^2}$ with any accuracy and it fell off to zero. The midband value of 0.8 for $\overline{\gamma^2}$ indicates that there was uncorrelated noise added to the system at some point other than the input. This could have been due either to real noise or to nonlinearities.

The transfer function, on the other hand, is a smooth well-defined function whose 3 dB and 90° phase points are at the same frequency. This indicates a good measurement of the relationship between input and output in spite of a fairly high uncorrelated noise environment.

Fig. 5 points up even more clearly the difference between a simple cross spectrum measurement and a transfer function measurement. Here the magnitude of the cross spectrum $\overline{G_{zx}}$ and the transfer function $\overline{H}$ are pictured on the same dB scale. Twenty-five sample records were averaged to determine system response, using a white noise input. One standard deviation on the input spectrum for this measurement is 20%, and since the cross spectrum does not employ information about the input its statistical certainty is poor. However, calculating the transfer function using the input power spectrum measured simultaneously with the cross spectrum reduces the statistical variation and gives a result with a few tenths of a dB of variation rather than 3 or 4 dB. In spite of the fact that a flat noise source is used, measurement of the transfer characteristics using a cross relationship alone is both inefficient and inaccurate.

Figs. 4 and 5 also illuminate a number of advantages of calculating the transfer function from the input spectrum and the cross spectrum. Clearly a good measurement can be made in spite of system noise. Also a measurement can be made using realistic test signals such as band-limited random noise. The phase measurement is unaffected by harmonic distortion and can be accurately made over wide dynamic ranges between input and output. The measurement can be made even more rapidly when there is no contaminating noise present. Thus, digital techniques of Fourier analysis offer powerful methods for transfer-function determination that are unavailable with analog instruments.

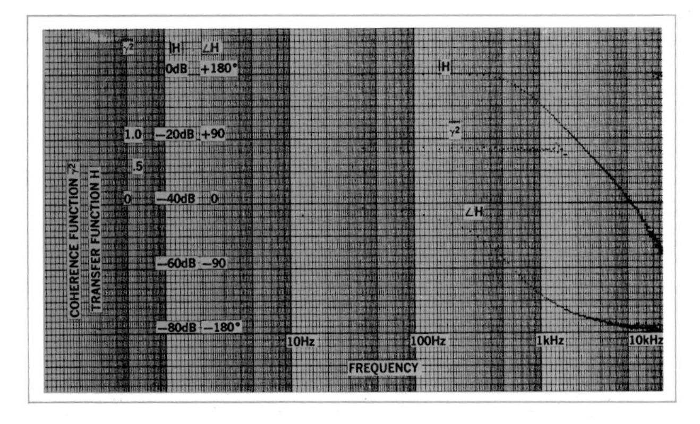

Fig. 4. *Transfer function of a second-order highly damped system measured by digital analyzer. Coherence function $\overline{\gamma^2}$ = 0.8 indicates the presence of uncorrelated noise in the system (1.0 would indicate no noise), but transfer function is smooth and well defined, indicating a good measurement in spite of the noise.*

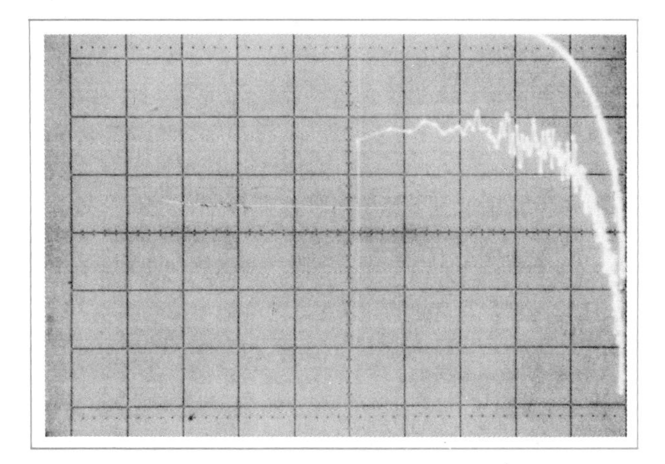

Fig. 5. *Transfer function of second-order highly damped system and cross power spectrum of input and output measured by digital analyzer. Spectra of twenty-five sample records were averaged. Not only do the magnitudes of $\overline{H}$ and $\overline{G_{zx}}$ differ, but also the statistical uncertainty in $\overline{G_{zx}}$ is much greater. This is because the computation for H takes into account the input power spectrum G_{xx}, whereas the computation for G_{zx} does not.*

Correlation Functions

So far I have described measurements that produce functions of frequency as their results. There are also functions of time which can be used in some of the same ways as spectra to clarify the nature of linear processes. These are correlation functions. The crosscorrelation function for two functions x(t) and y(t) is

$$R_{yx}(\tau) = \frac{1}{T}\int_0^T x(t)y(t-\tau)dt. \quad (21)$$

The autocorrelation function R_{xx} is the same function with $y(t) = x(t)$. Naturally, when implemented on a digital processor the integral is replaced by a sum.

The computation proceeds as follows. First the average value of the sample-by-sample product of the two functions is computed over some interval T. Then the functions are displaced relative to each other and the process is repeated for the new value of the displacement τ. This is repeated for all values of τ and the results plotted as a function of τ.

The result of all this is a function R_{yx} which peaks when the functions y and x displaced by τ match each other well. The best use of the crosscorrelation function is to determine the delay between y(t) and x(t). The autocorrelation function, on the other hand, is used to determine periodicities in a single function, since it will peak every time the displacement is equal to the period.

It is interesting to consider several alternatives to a direct calculation of correlation functions. First of all, it can be shown that the auto spectrum and the autocorrelation function are Fourier transforms of each other. The same holds true for crosscorrelation and cross spectrum.[1]

$$G_{xx} = F[R_{xx}] \text{ and } R_{xx} = F^{-1}[G_{xx}] \qquad (22)$$

$$G_{yx} = F[R_{yx}] \text{ and } R_{yx} = F^{-1}[G_{yx}] \qquad (23)$$

Thus it is possible to calculate a correlation function by transforming a waveform to find the appropriate spectrum, complex conjugate multiplying the spectrum by itself or another spectrum, and then taking the inverse transform. While this may appear to be the long way around, it actually requires fewer multiplications to find a correlation function than calculating the average displaced products directly. Certain precautions must be observed because the discrete Fourier transform always assumes the sampled function is periodic with period T. However, it is possible to calculate an *exact* correlation function of $\pm N/2$ displacements (points) if 2N time points are available.

The lower trace in Fig. 6 shows the results of a crosscorrelation and impulse-response measurement on a damped second-order system with a white random noise input.[2] The measurement is the average of 25 sample records, but it still shows considerable statistical variation. It is difficult to determine if the ripple in the waveform is due to external noise, normal statistical variation, or the characteristics of the system being measured.

The upper waveform in Fig. 6 is the inverse Fourier transform of the transfer function computed from $\overline{G_{zx}}$ and $\overline{G_{xx}}$. This result shows much less statistical variation and is a more efficient way to compute the system impulse response, although it still does not give information about the effect of uncorrelated noise. For this we still need the coherence function.

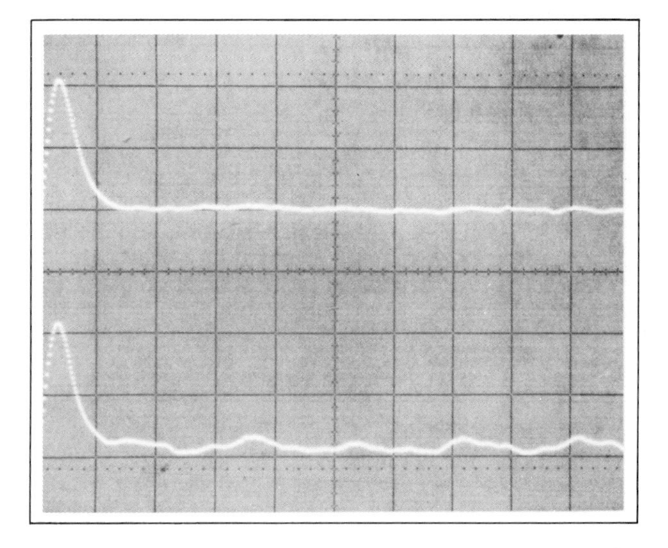

Fig. 6. *Crosscorrelation between white noise input and the output of a fourth-order linear system has the shape of the system impulse response. Lower trace is the crosscorrelation function computed directly. Upper trace was computed by inverse transforming the system transfer function, which was calculated by dividing the input-to-output cross power spectrum by the input power spectrum. Both traces are the average of 25 measurements. Upper trace is smoother as a result of taking into account the actual input power spectrum.*

References

[1]. J. S. Bendat and A. G. Piersol, 'Measurement and Analysis of Random Data,' John Wiley and Sons, 1966.

[2]. R. L. Rex and G. T. Roberts, 'Correlation, Signal Averaging, and Probability Analysis,' **Hewlett-Packard Journal,** November 1969.

Peter R. Roth

Peter Roth is project leader for the 5450A Fourier Analyzer. Before joining HP in January of 1965, Peter was an engineering officer in the U.S. Coast Guard for three years. Assigned to the Coast Guard's electronics laboratory in Alexandria, Virginia, he worked on high-frequency communications and on Loran A and Loran C equipment design. At HP, he has been responsible for development of the 5210A Frequency Meter, and has worked on the 5216A and 5221A, HP's first IC counters.

Peter received his BS and MS degrees in electrical engineering from Stanford University in 1959 and 1961. He is a member of IEEE and Tau Beta Pi.

A Calibrated Computer-Based Fourier Analyzer

This pushbutton-controlled digital measuring instrument performs complex analytical operations on input signals or time series. As a bonus, the user gets a general-purpose digital computer.

By Agoston Z. Kiss

ONE HEARTBEAT IN EVERY ¾ SECOND — 80 heartbeats per minute: these are the time-domain and frequency-domain descriptions of the same phenomenon. Neither contains more information than the other but to different people or to the same people in different circumstances one description may have more meaning or clarity than the other.

This duality between the time domain and the frequency domain is the basis of many important theorems and useful methods in signal and time-series analysis. Autocorrelation and crosscorrelation, power spectral density, cross power spectra, impulse response and transfer function, coherence, probability distribution and characteristic functions, convolution and filtering — these are examples of such methods. Since the principal theoretical bridge between the time domain and the frequency domain is the Fourier transform theorem, the methods of signal analysis that are based on the time-frequency duality are often called Fourier analysis.

Digital Fourier Analysis

Since 1965, the year of the Cooley-Tukey algorithm[1], Fourier analysis has been done more and more by digital techniques. The Cooley-Tukey algorithm, also called the fast Fourier transform, reduces the lengthy and cumbersome calculation of the Fourier coefficients by digital computer to a manageable, relatively rapid procedure. Computations that used to take hours can now be done in seconds. As a result, Fourier analysis is now becoming fashionable in many fields where it has not been used before because it took too long.

A version of the Cooley-Tukey algorithm is implemented in the new HP Model 5450A Fourier Analyzer, a calibrated, pushbutton-controlled instrument that can perform almost any Fourier-transform-based or related signal analysis (see Fig. 1). At the push of a button, the analyzer becomes a power spectrum analyzer, or a correlator, or an averager, or a digital filter, or any of a number of other instruments. *No knowledge of computer programming is required to operate it.* However, it can be converted into a general-purpose digital computer simply by moving a front-panel switch.

Model 5450A Fourier Analyzer combines a small general-purpose computer and some peripheral hardware into a flexible, user-oriented general-purpose instrument. An HP 2115A or 2116B computer with 8K memory is interfaced with a keyboard (Fig. 2), a dual-channel analog-to-digital converter (Fig. 3), a special display unit (Fig. 4), a teleprinter, and a punched-tape photoreader. An additional 8K memory can be installed to increase both the internal range and the number of peripherals. The analyzer has two basic modes of operation, i.e., as a Fourier analyzer or as a general-purpose computer. In the analyzer mode, it is either under keyboard control or under the remote control of another general-purpose computer.

The basic operations the Model 5450A can perform in the analyzer mode can be categorized as:

- data input/output
- transform related operations
- arithmetic operations
- data manipulations
- writing and editing of analysis routines.

Specific mathematical functions under keyboard control are:

- forward and inverse Fourier transform
- power spectrum
- cross power spectrum

- auto and crosscorrelation
- convolution
- histogram
- Hanning and other weighting functions
- real and complex multiplication and standard arithmetic operations
- integration and differentiation
- ensemble averaging

These can be executed separately or combined into complex routines.

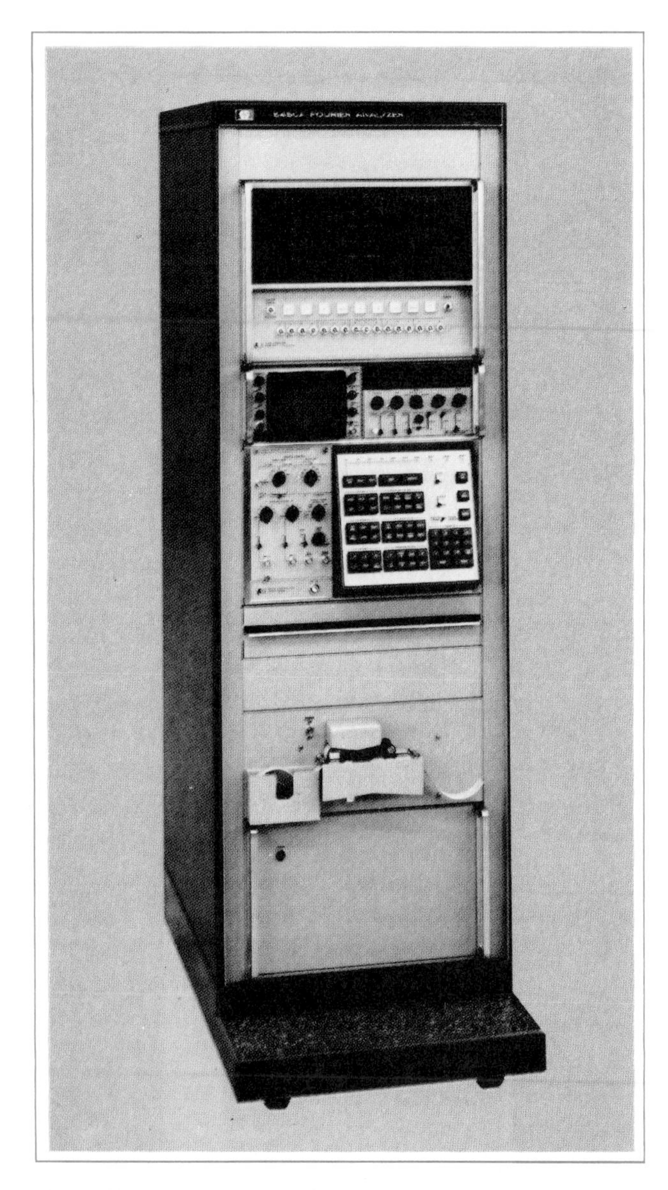

Fig. 1. *Model 5450A Fourier Analyzer is a flexible, push-button-controlled, modular, digital instrument, useful for analyzing waveforms and time series in a wide variety of systems and processes. It uses a standard HP computer for memory and computation, but requires no knowledge of computer programming. When it isn't doing Fourier analysis, the computer can be used separately.*

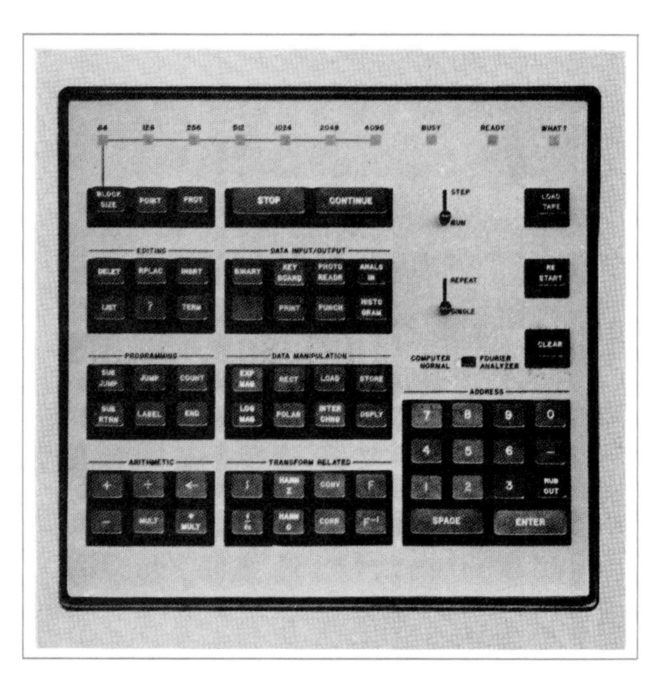

Fig. 2. *All Fourier analyzer operations are keyboard controlled. The principal operations—Fourier transforms, convolution, correlation, complex multiplication, coordinate transformations, and so on—can be called for by single keystrokes, or strung together using the programming and editing features to form routines to be run automatically later on. Typical routines can change the analyzer into a spectrum analyzer, a correlator, an averager, and many other instruments.*

Data Input/Output

There are 3K words available for data storage (8K words in the 16K version of the analyzer). This storage space can be filled up with data records; the shortest record is 64 words long and the longest is 1024 words long (4096 in the 16K version). Record lengths are pushbutton selectable in powers of two between these limits. The number of records which can be stored is the size of the data storage divided by the record length. Consequently, the 8K version can store 3 records of 1024 points each, or 6 records of 512 points each, and so on up to 48 records of 64 points each. These records are addressable as data block 0, 1, 2, . . . in every keyboard command.

Data Input

Analog data records can be read in via the analog-to-digital converter, which has two input channels with separate input attenuators. It can be switched to single-channel mode when only channel A is operational, or to dual-channel mode when channels A and B are both operational. Channels A and B are sampled simultaneously, then sequentially converted into digital values — channel

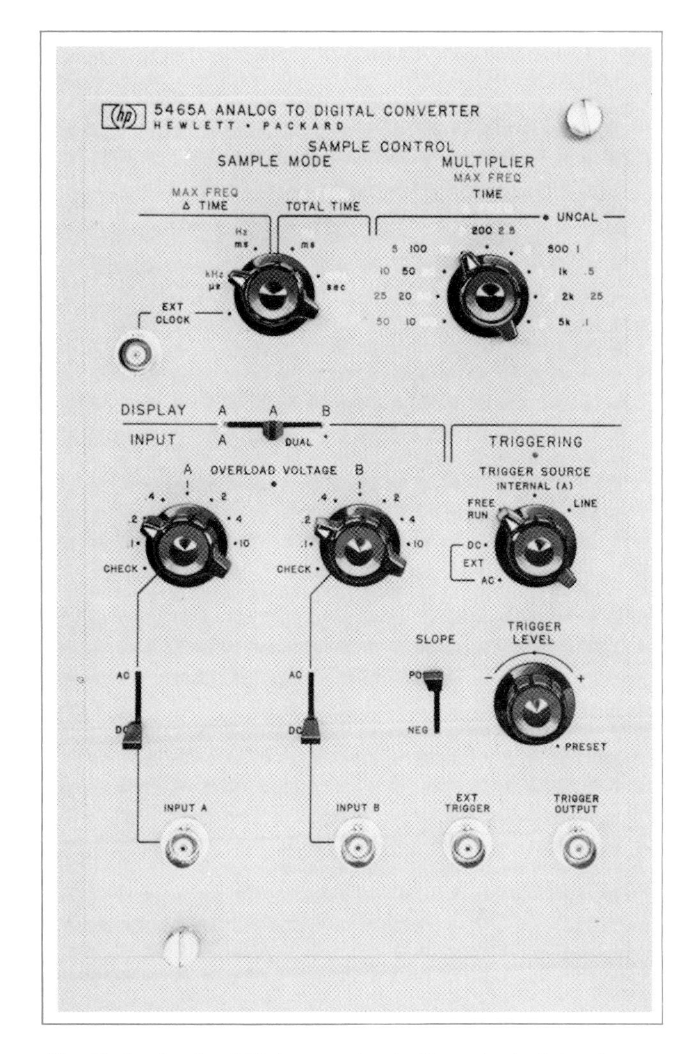

Fig. 3. *Analog-to-digital converter is the principal input device for analog signals. It can be operated as a single-channel unit or a dual-channel unit. The maximum sample rate for single-channel operation is 20μs per data point; the minimum rate is one sample in every five seconds. Data can also be read into the analyzer via peripheral devices, such as a tape reader or a teleprinter.*

A first — and stored in separate data blocks. The sample rate can be varied from 20 μs per data point (50 μs for dual-channel input) down to one sample in every five seconds.

There are some obvious but important relations between sampling time Δt, record length T, number of samples in a record (or data block size) N, frequency resolution Δf and upper frequency limit f_{max}:

$$T = N\Delta t \tag{1}$$

$$f_{max} = \frac{1}{2\Delta t} \tag{2}$$

$$\Delta f = \frac{1}{T} \tag{3}$$

Equation 1 says simply that a data record of length T seconds has been sampled N times with Δt seconds between samples.

Equation 2 is sometimes called the Shannon or Nyquist criterion of sampling, which states that to avoid loss of information, the highest frequency in a signal must be sampled at least twice per cycle.

Equation 3 really says that better frequency resolution requires a longer record. The analog equivalent of this statement is the observation that narrower-band filters take a longer time to reach steady state conditions.

A-D Converter

The analog-to-digital converter is a 10-bit ramp-type device with a 100 MHz clock. Because of its high differential linearity (3% as opposed to 25–50% for a typical successive-approximation-type A-D converter), the 60 dB dynamic range of the 10-bit converter will be appreciably improved, in some cases to as much as 90 dB, when any

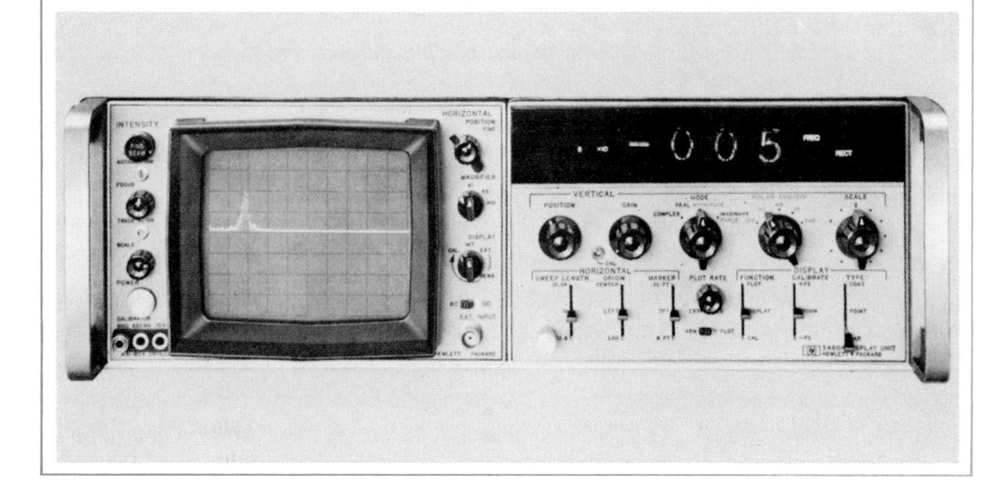

Fig. 4. *Built-in display unit is the principal analyzer output device when the recipient of the data is human. The digital display and annunciation indicate the vertical scale factor and the type of display. Data in the analyzer are always absolutely calibrated.*

averaging is done. The Fourier transform is a weighted average, of course. We have consistently observed dynamic ranges of 80 dB or more in computed transforms. How differential linearity and averaging affect dynamic range is quite a complex subject, and we hope to publish a paper on it soon.

Once the keyboard command is given for analog input, the actual record will be started by an internal or external sync signal with positive or negative slope, as selected by the user. After the last sample of the record has been stored, the analyzer calibrates the data, taking the input-attenuator setting into account, and establishes a scale factor for the record, which will follow it through all calculations. This absolutely calibrated input/output is one of the most important basic features of the analyzer.

Data can also be introduced into the analyzer through the numeric keys of the keyboard, through the teletype, through the photoreader (if they are on a punched paper tape), through the double binary I/O channels from another computer, and from digital magnetic tape (16K version only). The common feature of all the data input modes is that they can be initiated by keyboard control and that they establish calibrated data records in the analyzer.

Data Output

The most often used data output device is the display unit. Any stored data record can be displayed on the CRT by keyboard command. Also, when the analyzer is idle, it automatically reverts to a display mode, generally displaying the data record which was the subject of some I/O or analytical operation just before the idle period.

The display unit has many convenient features. It can display a time record or a frequency spectrum. When a spectrum is being displayed, its real part or imaginary part — or its amplitude or phase — can be displayed as a function of frequency, or the imaginary part can be displayed as function of the real part (Nyquist plot). Fig. 5 illustrates the possibilities. In every mode of data display, the calibration factor is also displayed as a power of 10, facilitating the readout of absolute values. Besides showing calibration, display lights also show whether the record displayed is in the time or frequency domain, whether the amplitudes are linear or logarithmic, and whether they are calculated in rectangular (real and imaginary) or polar (amplitude and phase) coordinates.

Other features of the display unit are: digital up or down scaling in ten steps, linear or logarithmic horizontal scale, markers on every 8 or 32 points, point display, continuous curve display or bars drawn from display points to the zero level horizontal axis. It also has a calibration mode, and a plotter mode in which it can drive an X-Y recorder to plot exactly what is being displayed on the CRT.

Data records can also be printed out on the teletype, punched out on paper tape either on the punch unit of the teletype or on an optional fast punch, transferred on the double binary I/O channels to another computer, plotted on a digital plotter, or stored on digital magnetic tape (the last two features on the 16K version only). Common features of all data output modes are that they can be initiated by keyboard command and that the data are always calibrated.

A final remark about the calibrated input-output feature. The analyzer, being a binary device, carries the calibration in radix two. In every output operation where the recipient is non-human (binary I/O, paper tape, digital magnetic tape), the calibration remains in radix two to retain maximum accuracy. However, in every human-related output operation (display, data printout, plotting) the calibration is changed to radix 10 for maximum user convenience.

Transform-Related Operations

The most important transform-related operations are, of course, the forward and inverse Fourier transforms. The definitions of these operations are:

$$S_x(m\Delta f) = \frac{1}{N}\sum_{n=0}^{N-1} x(n\Delta t)\, exp\left\{-i\frac{2\pi}{N}nm\right\} \tag{4}$$

and

$$x(n\Delta t) = \sum_{m=0}^{N-1} S_x(m\Delta f)\, exp\left\{i\,\frac{2\pi}{N}\,nm\right\} \tag{5}$$

where N is the number of samples (points) in the time record x(t) or frequency record $S_x(f)$.

Although the time function x(t) is always real, the spectrum, $S_x(f)$, is generally complex. In a complex spectrum, every spectral value (except dc) has to be described by two quantities, either amplitude and phase, or real (cosine or in-phase) and imaginary (sine or quadrature) components. The former is the polar-coordinate representation and the latter is the rectangular-coordinate representation. In the analyzer, all calculations are carried out in rectangular coordinates, but the results can be converted into polar coordinates by keyboard command.

Since the Fourier transform does not create new information, the Fourier spectrum of a time record with N independent data points will also contain exactly N inde-

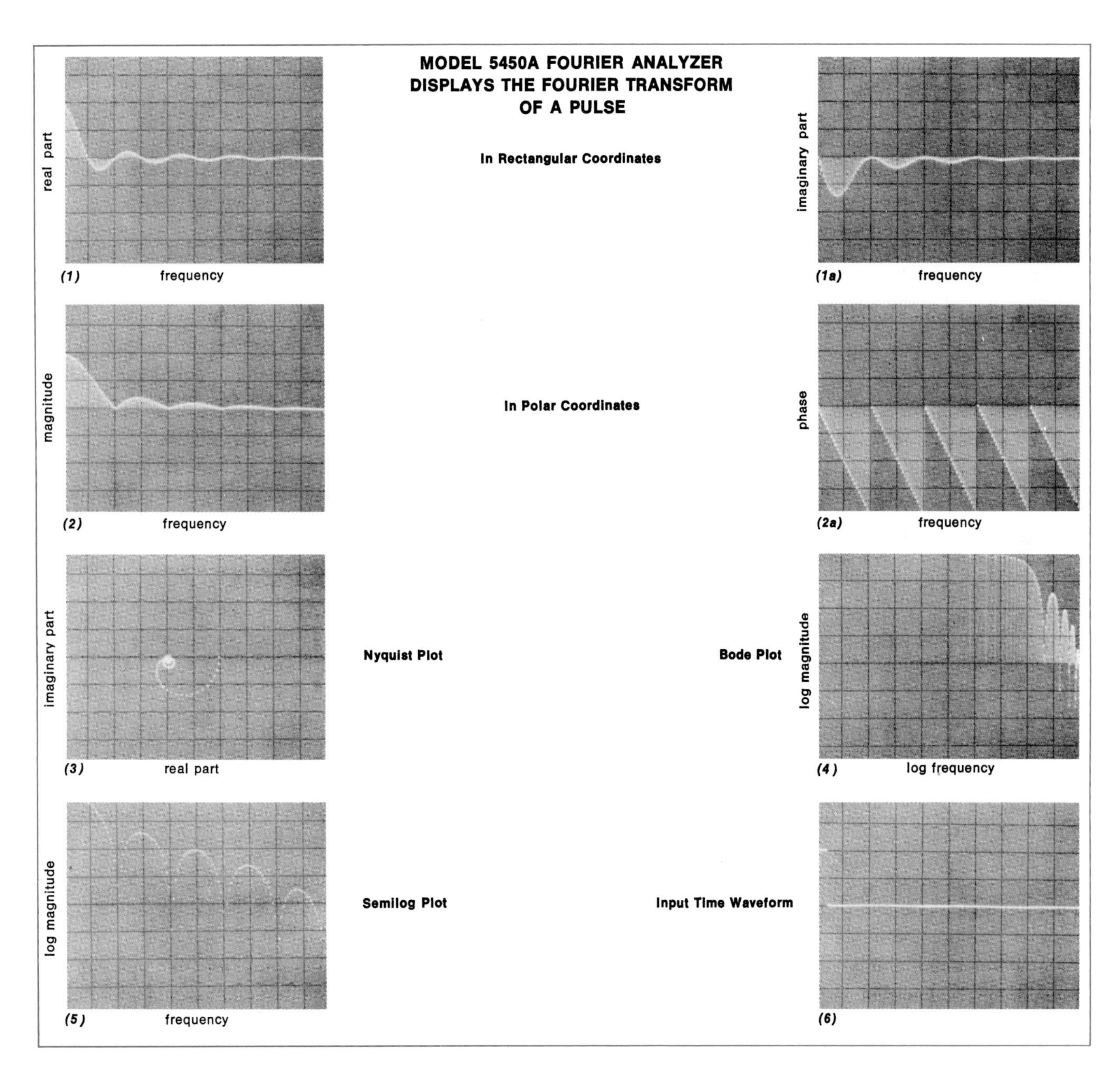

Fig. 5. *The analyzer has a display mode to suit every need, and it changes from one to another at the touch of a button or the flick of a switch. In every case the readouts on the display unit and the A-D converter indicate scale factors and type of display.*

pendent data points. But since every frequency point has to be described by two independent data values — except dc, which has no phase, and the highest frequency, which by definition has zero phase — the Fourier spectrum of a time record with N points will contain N/2 frequency-value pairs (for counting purposes dc and the highest frequency are counted as one frequency-value pair).

Equation 4 actually defines a spectrum for negative as well as positive frequencies. However, the analyzer is restricted to the analysis of physically realizable, and therefore real, time functions only. The spectra of real time functions are Hermitian (i.e., even real part and odd

A Fourier Analyzer Makes Fundamental Measurements

The measurements a Fourier analyzer makes are useful to behavioral scientists, psychophysicists, biomedical researchers, process control system designers, analytical chemists, and oceanographers, and to people working in vibration analysis, structural mechanics, acoustics, geophysics, control system design and analysis, component testing, system identification, sonar, and many other fields. The reason a Fourier analyzer is so widely useful is that, like a voltmeter, it makes fundamental measurements. For the same reason, no finite list of applications can convey a true picture of its capabilities. Here are just a few examples of applications.

Analytical chemists can use it to measure nuclear magnetic resonance (NMR) spectra, and as an averager to improve the sensitivity of their spectrum measurements.

Structural designers, e.g. of airframes, can use it to determine the transfer function and vibration modes of a structure, the spectra of vibrations induced at various points by various inputs, and the degree of coherence between vibrations at different points.

Behavioral scientists can use it to determine the transfer function of a driver, and the degree of coherence between his responses and various input stimuli.

Brain researchers can use it to measure the spectra of brain waves, and the degree of coherence between waves at different points in the brain.

Designers of process control systems and other systems—power plants, servomechanisms, etc.—can use it to determine transfer functions, impulse responses, coherence between signals, power spectra, and cross power spectra.

In application after application, the measurements are the same—transfer function, coherence function, power spectrum, cross power spectrum, and combinations of these fundamental measurements. End uses of the data differ, of course. To the designer of a structure or a control system, it's accurate information that he couldn't have obtained without the analyzer, and he uses it to optimize his design, avoid overdesign, and optimize performance adjustments. The physician analyzing an electromyogram (EMG) is looking for evidence of muscle disease. What these and other users and potential users of Fourier analyzers have in common is that they are working with time series—voltages, vibrations, sound waveforms, or perhaps just a series of data points obtained at regular intervals and punched on paper tape. On such inputs the Fourier analyzer makes measurements and computes functions that would be difficult to do by any other means. It does these things with the convenience of keyboard control, rapidly, and with great flexibility.

imaginary part), so the negative-frequency parts of the spectra of real time functions contain no additional information. Partly to increase the effective transform speed of the analyzer and partly to avoid the confusion that the mentioning of the existence of negative frequencies generally creates, we used a version of the fast Fourier algorithm that applies only to time signals that are real.

Like other versions of the fast Fourier algorithm, ours is an 'in-place' algorithm. Intermediate and final results of computations are stored in the same data block as the original data.

Correlation and Convolution

Auto and crosscorrelation are well known and widely used methods in signal analysis. They are used to improve signal-to-noise ratio, to find hidden periodicities, and so on. Convolution, on the other hand, is in most cases a mean trick nature plays on us. When we use any measuring equipment to measure an event, the result is never the phenomenon we want to observe but its convolution with the impulse response of the equipment used. Sometimes, however, even convolution can be useful. For example, smoothing a record by taking a K-point running average can be performed in the analyzer by convolving the record in question with another record containing K unit impulses.

Correlation and convolution are generally defined on the time domain, although they do depend on the frequency content of the functions in question. Since both correlation and convolution involve an enormous number of multiplications and additions — N^2 of them to be exact — to perform either of them within a reasonable time requires special hardware. But according to the convolution theorem, convolution (correlation) in one domain is multiplication (conjugate complex multiplication) in the other domain. Therefore convolution (correlation) can be reduced to two Fourier transforms and one point-by-point multiplication (conjugate complex multiplication) of the two records involved. If x(t) and y(t) are two time functions and their respective spectra are $S_x(f)$ and $S_y(f)$, the analyzer performs the following calculations: for crosscorrelation,

$$x(t) \star y(t) = F^{-1}\,[S_x(f) \cdot S_y{}^*(f)] \tag{6}$$

and for convolution,

$$x(t) * y(t) = F^{-1}\,[S_x(f) \cdot S_y(f)] \tag{7}$$

Here the superscript * stands for complex conjugate, the $*$ between two time functions for convolution, the ★ for crosscorrelation, and F^{-1} for inverse Fourier transform. These operations can be performed step by step on the analyzer, but for convenience the keyboard has a correlation key and a convolution key.

Hanning

Physically realizable devices can act only on signals which are limited in duration and in bandwidth. If infinitely long signals or signals with infinite bandwidth are passed through any physical device, they will be time and frequency-band limited by the device itself.

The simplest kind of time-limiting is the application of a square time window. If we have a function x(t) and we take a T-second long record of it, say from $t = 0$ to $t = T$, then we have really multiplied x(t) by a square pulse T seconds long with unity amplitude (see Fig. 6).

What happens to the spectrum of this function? The convolution theorem says that multiplication in one domain is convolution in the other domain. Since we multiplied x(t) by the window function $\sqcap$ (t/T), the spectrum of the time-limited function x(t) • $\sqcap$ (t/T) will be the convolution of the spectrum of the original function and the spectrum of the time window. Let us say that the spectrum of x(t) is $S_x(f)$. The spectrum of $\sqcap$ (t/T) is (Fig. 7) $\text{sinc}\pi Tf = \frac{\sin \pi Tf}{\pi Tf}$, and the spectrum of x(t) • $\sqcap$ (t/T) is $S_x(f) * \text{sinc}\pi Tf$. The maximum value of the sinc function is unity at $f = 0$, it has zero crossings at $f = 1/T$, $2/T, \ldots$, and the amplitude of the sidelobes decreases at 6 dB per octave. If $S_x(f)$ has spectral lines exactly at $f = 0, 1/T, 2/T, \ldots$, that is, if x(t) was periodic in the time window $\sqcap$ (t/T), then convolving $S_x(f)$ with $\text{sinc}\pi Tf$ will simply result in $S_x(f)$. But if x(t) was not periodic in the time window, then the spectral lines of $S_x(f)$ and the zero crossings of the window spectrum will not coincide and the convolution process will smear each spectral line of $S_x(f)$ all over the spectrum. Even if $S_x(f)$ contains one spectral line only, the result will be a series of spectral lines spaced 1/T apart and having an amplitude decay of 6 dB per octave. This phenomenon is often referred to as the leakage effect.

Leakage can be avoided only by making sure that the function x(t) is periodic in the time window. Obviously, this condition can seldom be met. Therefore, in order to reduce the effect of leakage, different window shaping ideas have been proposed. The idea of the window shaping is to make x(t) somehow 'quasi-periodic' in the time window with the least possible loss of information. Among these window-shaping methods the Hanning window has proved most popular. It is a $\frac{1}{2}\left(1 \pm \cos \frac{2\pi t}{T}\right)$ window, where both the window and its derivative approach zero at the two ends of the record. Its effect on the spectrum is that the main lobe of each line is widened by an additional 1/T, but the sidelobes decay by an additional 12 dB per octave.

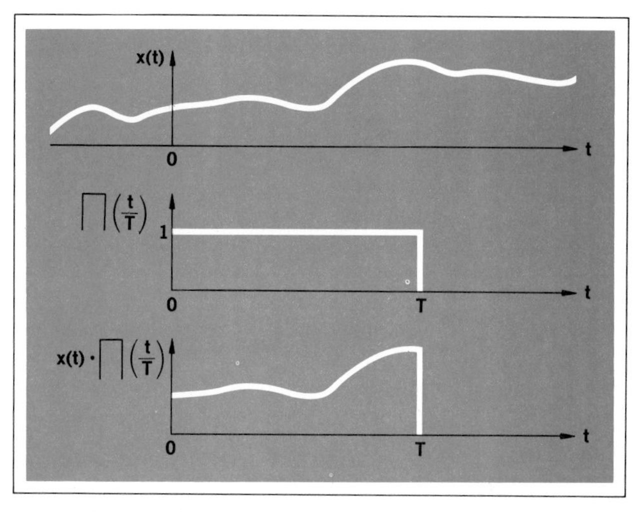

Fig. 6. *When a T-second record is taken of an analog input, the effect is to multiply the input by a square window function. If the input isn't periodic with period T, the spectral lines of the input will not be lines but will have the sin x/x shape shown in Fig. 7. To reduce this effect, Model 5450A Fourier Analyzer has built-in Hanning window-shaping functions.*

Two other window-shaping methods are the Chebyshev window and the Parzen window. The Chebyshev window achieves a faster sidelobe decay than the Hanning window but is much more cumbersome to implement. The triangular Parzen window is fairly easy to implement but not as effective as the Hanning window.

In Model 5450A two different Hanning windows can be applied by pushbutton command. The interval-centered Hanning window, HI, is used to reduce leakage as described above. The origin-centered Hanning window, HO, can be used to form a 3-point running average of records with ¼, ½, ¼ weighting.

Integration and Differentiation

There is a keyboard command to integrate any data record between any two chosen data points or to differentiate any chosen data record. The defining equations for integral and differential are:

$$\int D_k = D_{k-1} + D_k \quad, k = 0, 1, \ldots, N\text{–}1 \qquad (8)$$

$$\frac{d}{dx} D_k = D_k - D_{k-1} \quad, k = 0, 1, \ldots N\text{–}1 \qquad (9)$$

By definition, $D_{0-1} = 0$.

The integral routine is especially useful for calculating integral power spectra, cumulative probability distribu-

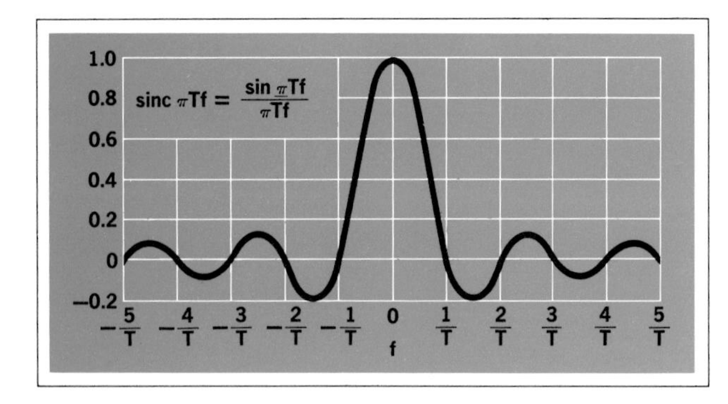

Fig. 7. *Spectral lines of a sampled function will have this sin x/x shape if the function isn't periodic in the record length T. Hanning weighting doubles the width of the main peak, but causes later peaks to fall off at 18dB per octave instead of 6dB per octave.*

tion functions or third-octave, half-octave, or full-octave filters. The differential routine can be used to calculate higher moments of probability density functions by differentiating their Fourier transforms (i.e. their characteristic functions).

Arithmetic Operations

There are keyboard commands for the addition, subtraction, multiplication, and division of data records. These operations are performed on a point by point basis. Addition is especially useful for ensemble averaging of data records either in the time domain or in the frequency domain, thereby improving the statistics of the measurement. There are separate commands for complex multiplication and conjugate complex multiplication of two selected data records. Both multiplications result in real multiplication if the records are in the time domain.

The division of a data record by another selected data record is performed as real or complex division in the time or frequency domain, respectively. In addition to these data block operations, any selected data block can be multiplied or divided by any positive or negative constant whose magnitude is less than 32767.

Data Manipulations

Since there can always be more than one data record stored in the analyzer, 'Store,' 'Load' and Interchange' keyboard commands were established to effect data transfers among them.

As I have mentioned, all transform-related and arithmetic operations are performed in rectangular coordinates. However, spectral results are often desired in polar coordinates (amplitude and phase). There are keyboard commands to change the coordinate system of any chosen data record from rectangular to polar or from polar to

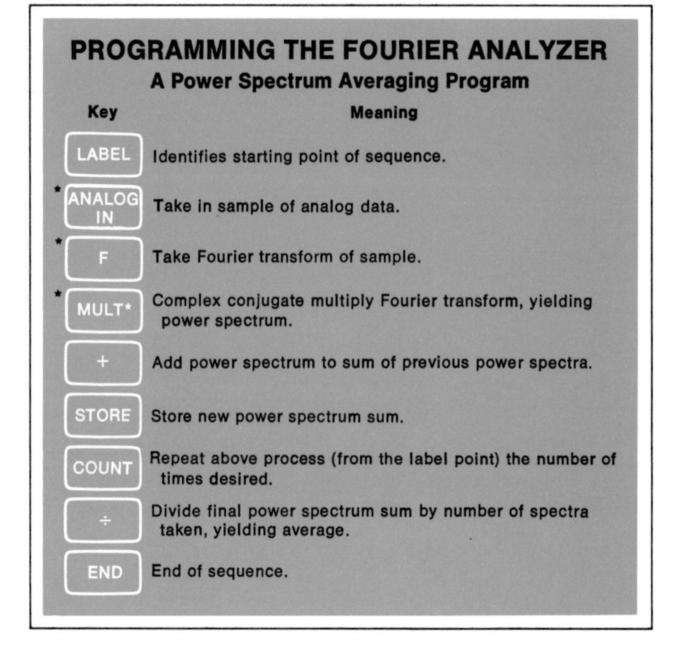

PROGRAMMING THE FOURIER ANALYZER

A Power Spectrum Averaging Program

Key	Meaning
LABEL	Identifies starting point of sequence.
* ANALOG IN	Take in sample of analog data.
* F	Take Fourier transform of sample.
* MULT*	Complex conjugate multiply Fourier transform, yielding power spectrum.
+	Add power spectrum to sum of previous power spectra.
STORE	Store new power spectrum sum.
COUNT	Repeat above process (from the label point) the number of times desired.
÷	Divide final power spectrum sum by number of spectra taken, yielding average.
END	End of sequence.

Fig. 8. *An often-used routine is the power spectrum averaging program. After the steps are entered into the analyzer's memory, the program can be listed on the teleprinter. Errors can be corrected by adding, deleting, or modifying steps. Another keystroke makes the program execute. Model 5450A will compute one 1024-point spectral estimate (the three steps marked*) in 2.4 seconds or less.*

rectangular. Further keyboard commands can change linear amplitudes to logarithmic or logarithmic amplitudes to linear. The execution of these commands ('Rectangular,' 'Polar,' 'Logarithmic Amplitude,' 'Exponential Amplitude') are based on a power-series technique in which the coefficients are calculated by Chebyshev expansion of the function desired.

Writing and Editing Routines

The power of Model 5450A Fourier Analyzer is not only in the easy access it offers to the most important basic signal analytical operations, but also, and perhaps even more so, in its capability of building automatic routines using these operations. Programming the analyzer to carry out a sequence of computations actually transforms it into a different measuring instrument — a spectrum analyzer, for example, or a signal averager, or a correlator.

Keyboard commands can be assembled into routines up to 100 steps long (200 steps in the 16K version). The routines can incorporate labels, jump instructions, subroutines, and loops, thereby providing an extremely flexible and easily learned high-level instruction set for almost any type of signal analysis. The assembled routines reside in the analyzer. They can be listed on the teletype, punched out on paper tape, re-edited using 'Delete', 'Replace', and 'Insert' edit commands, and can be run under keyboard control.

Here are some of the most often used routines.

Power Spectral Analysis. Ensemble averaging to improve the signal-to-noise ratios of power spectral estimates can be simply performed by:

1. reading in a time record
2. taking its Fourier transform
3. conjugate complex multiplying the spectrum by itself, thereby creating a power spectral estimate
4. summing the power spectral estimate into a second record
5. repeating operations 1–4 any desired number of times
6. after summing a given number of power spectral estimates, dividing the result by the number of estimates.

Fig. 8 illustrates the program, which computes

$$G_{xx}(f) = \overline{S_x(f) \cdot S_x{}^*(f)} \qquad (10)$$

Cross Power Spectra. The cross power spectrum contains the frequencies common to the individual spectra of two signals. It is the Fourier-transform of the crosscorrelation function. To create the ensemble average of cross power spectral estimates, one can follow the instructions for power spectral averaging, except in step 1 take two simultaneous records, and in step 3 conjugate complex multiply one spectrum by the other. The function computed is

$$G_{xy}(f) = \overline{S_x(f) \cdot S_y{}^*(f)} \qquad (11)$$

In Equations 10 and 11 $G_{xx}(f)$ stands for power spectrum, $G_{xy}(f)$ for cross power spectrum, $S_x(f)$ and $S_y(f)$ are the Fourier spectra of functions x(t) and y(t) respectively, the superscript * stands for complex conjugate, and the upper bar for ensemble averaging.

Digital Filtering. Let us consider a filter as a black box with one input and one output:

x(t) / $S_x(f)$	h(t) / H(f)	y(t) / $S_y(f)$

The black box can be characterized by its impulse response, h(t) or its transfer function, H(f). They are Fourier-transform pairs. The input function is x(t), and the output is y(t). $S_x(f)$ and $S_y(f)$ are their respective Fourier spectra.

The filter equation simply states that the output spectrum is the product of the input spectrum and the transfer function of the filter:

$$S_y(f) = S_x(f) \cdot H(f) \qquad (12)$$

Filtering can be easily performed in the Model 5450A by storing the filter transfer function in one of the data records and block-multiplying the spectrum of the input signal by it. Taking the inverse transform of the product results in the output function, y(t).

Inverse Filtering or Deconvolution. Equation 12 can be rewritten in the time domain using the convolution theorem (multiplication in one domain equals convolution in the other domain):

$$y(t) = x(t) * h(t) \qquad (13)$$

that is, the output of the black box is the convolution of its impulse response and the input function. Now if this black box happens to be some measuring equipment, it is x(t) that we are interested in, not y(t). The inverse operation of convolution is pretty difficult to produce, but Equation 12 can be rewritten as:

$$S_x(f) = \frac{S_y(f)}{H(f)} \qquad (14)$$

Since $S_y(f)$ and H(f) are known, the division can be performed point by point. Taking the inverse Fourier transform of the quotient results in x(t).

Transfer Function and Coherence. A method based on Equation 12 can be worked out to measure the transfer functions of unknown black boxes and to find causal relationships between inputs and outputs. This extremely important and interesting subject is discussed by Peter Roth elsewhere in this issue.

Measurement of Statistical Behavior. Random data can be characterized by their statistics: probability density functions, distribution functions, and the moments of the probability distribution. The analyzer can collect amplitude histograms or, with an optional input box, time-interval histograms. The histograms are really frequency curves; the independent variable is amplitude or (time interval) and the dependent variable is the frequency of occurrence. Histograms can be easily normalized to give probability density functions and integrated to calculate distribution functions.

The Fourier transforms of distributions are called characteristic functions; they are used mainly in theoretical work in statistics. The differentials of the characteristic functions can be used to calculate the moments and central moments of the distributions.[3]

SPECIFICATIONS
HP Model 5450A
Fourier Analyzer

ANALOG INPUT

The Analog-to-Digital Converter accepts one or two inputs. In two-channel operation both inputs are sampled simultaneously. Resolution of the ADC is 10 bits.

INPUT IMPEDANCE: 1 MΩ ±1% shunted by 45 pF max.
SENSITIVITY: 30 μV rms (sine wave).
CONVERSION GAIN (CHANNEL A):
ACCURACY (as a function of frequency):
±.2% ±1 × 10^{-4}%/Hz.
TEMPERATURE STABILITY: 0.005%/°C.
LINEARITY: Integral, ±0.05%; Differential, ±3%.
GAIN AND PHASE CHANNEL A TO B:
CONVERSION GAIN A/B: ±0.2% ±4 × 10^{-4}%/Hz.
TEMPERATURE STABILITY: 0.01%/°C.
PHASE AND DELAY A TO B: ±0.2°, ±5 μs.
SAMPLE RATE CONTROL:
MAXIMUM FREQUENCY/TIME BETWEEN SAMPLES MODE: Maximum frequency is selectable from 0.1 Hz to 25 kHz (0.1 Hz to 10 kHz in two-channel operation).
FREQUENCY RESOLUTION/TOTAL TIME MODE: Frequency resolution is selectable from 0.2 mHz to 100 Hz.

DISPLAY UNIT

Data may be displayed on the 8 x 10 cm oscilloscope or output to a plotter or remote oscilloscope in the following forms:

Y AXIS	X AXIS
Real Part Amplitude	Time
Real Part Amplitude	Frequency (Linear or Log)
Imaginary Part Amplitude	Frequency (Linear or Log)
Magnitude (Linear or Log)	Frequency (Linear or Log)
Phase	Frequency (Linear or Log)
Imaginary Part Amplitude	Real Part Amplitude

ANALOG DISPLAY ACCURACY: ±1%.
TYPES OF DISPLAY: Points, bars, or continuous (interpolation).
AMPLITUDE SCALE: Data in memory is automatically scaled to give a maximum on-screen calibrated display. The scale factor is given in volts/division, volts^2/division, or in dB offset.
LINEAR DISPLAY RANGE: ±4 divisions with scale factor ranging from 1 × 10^{-150} to 5 × 10^{+150} in steps of 1, 2, 5 and 10.
LOG DISPLAY RANGE: 4 decades with a scale factor ranging from 0 to −998 dB.
TIME AND FREQUENCY SCALE:
LINEAR SWEEP LENGTH: 10, 10.24, or 12.8 divisions.
LOG HORIZONTAL: 0.5 decade/division.
ANALOG PLOTTER OUTPUT:
AMPLITUDE: 0.5 V per oscilloscope display division.
LINEARITY: 0.1% of full scale.

BLOCK SIZES FOR TYPICAL MEASUREMENTS

The following table indicates some of the measurements made by the 5450A as well as the *maximum* block size available for these measurements.

COMPUTATIONAL SPEED

The speeds shown are based on using the 2116B Digital Processor.
FOURIER TRANSFORM:
Block Size 64: 52 ms
Block Size 1024: 1.4 s
POWER SPECTRUM ENSEMBLE AVERAGE:
Block Size 64: 110 ms/Spectral Estimate
Block Size 1024: 2.1 s/Spectral Estimate
CROSS POWER SPECTRUM ENSEMBLE AVERAGE:
Block Size 64: 210 ms/Spectral Estimate.
Block Size 1024: 4.4 s/Spectral Estimate.

DIGITAL ACCURACY AND RESOLUTION

All calculations using floating point arithmetic on a block basis. Data overflow does not occur. Amplitude resolution is 1 part in 16,000 worst case.

DATA MEMORY SIZE: 3072 words (8192 for a 16,384 word memory).
DATA BLOCK SIZE: Any power of 2 from 64 to 1024 (to 4096 with a 16,384 word memory).
DATA WORD SIZE: 16 bit real and 16 bit imaginary or 16 bit magnitude and 16 bit phase.
COMPUTATIONAL RANGE: ±150 decades.
TRANSFORM ACCURACY: 0.1% worst case error during the forward or inverse calculation.

SPECTRAL RESOLUTION

The element of spectral resolution is the frequency channel width, the maximum frequency divided by ½ the data block size.

MAXIMUM FREQUENCY: 25 kHz single channel; 10 kHz dual channel.
FREQUENCY CHANNEL WIDTH: <3.2% down to <0.2% of the maximum frequency (down to <0.05% for 16,384 word processor).
SPECTRAL RESOLUTION OF TWO EQUAL AMPLITUDE SINE WAVES: If separated by 3 frequency channel widths, there will be a null of at least 3 dB between them; if separated by 7 frequency channel widths the relative magnitudes will be correct to within 0.1%. The power spectrum for two equal amplitude sine waves separated by 5 frequency channels will have the correct relative magnitude to within 0.1%.
DYNAMIC RANGE: 4 decades over ±150 decades.
ENVIRONMENTAL CONDITIONS: 0°C to 55°C using 2116B Digital Processor (10°C to 40°C using 2115A Digital Processor).

PRICE: Systems start at approximately $50,000, depending upon choice of computer and other required options.

MANUFACTURING DIVISION: HP Santa Clara Division
5301 Stevens Creek Boulevard
Santa Clara, California 95050

MEASUREMENT	BLOCK SIZE N (Points/Ensemble)	
	8K MEMORY	16K MEMORY
Power Spectral Density — Ensemble Average	1024	4096
Voltage Spectrum — Ensemble Average	1024	4096
Cross Power Spectral Density — Ensemble Average	1024	2048
Transfer Function	512	2048
Coherence Function	512	1024
Autocorrelation of N/2 Lags	1024	4096
Crosscorrelation of N/2 Lags	1024	4096
Crosscorrelation of N/2 Lags — Ensemble Average	1024	2048
Autocorrelation of N/2 Lags — Ensemble Average	1024	2048
Power Spectral Density of One Shot Transient	1024	4096
Voltage Spectrum of One Shot Transient	1024	4096

Possibilities Unlimited

The analytical operations available in Model 5450A Fourier Analyzer can be combined in many, many ways, and only the best known were mentioned here. But the analyzer will cater to the most esoteric tastes, including, for example, *cepstrum, saphe cracking,* and *liftering.*[4]

Acknowledgments

The original idea of Model 5450A and much of the initial groundwork is due to Ron Potter. The engineering development was under the capable leadership of Peter Roth as project leader, who also developed the analog-to-digital converter. Hans-Jurg Nadig and Evan Deardorff developed the keyboard control and display units, respectively. Steve Cline wrote most of the software. Bill Katz and Dick Cavallaro gave valuable assistance in the development work and documentation. The industrial designer was Roger Lee, and the mechanical designers were Wally Mundt and Chuck Lowe. Special thanks are due to Fred London and his group for their marketing support, to Pete Schorer for the excellent manuals, and to Jim Shea and his service group—especially Al Linder—for their service support. Production is in the capable hands of Don Lien's group and Walt Noble, production engineer.

The fast-Fourier algorithm used in Model 5450A is based mainly on those of Cooley and Tukey[1] and Gentleman and Sande[5]. However, we also benefited from the ideas of I. F. Good, R. C. Simpleton, R. B. Blackman, G. C. Danielson, C. Lanczos, R. Shively, and others.

References

[1]. V. W. Cooley and J. W. Tukey, 'An algorithm for the machine calculation of complex Fourier series,' Math. Comp. Vol. 19, pp. 297-301, 1965.

[2]. R. L. Rex and G. T. Roberts, 'Correlation, Signal Averaging, and Probability Analysis,' **Hewlett-Packard Journal,** November 1969.

[3]. R. M. Bracewell, 'The Fourier Transform and Its Applications,' McGraw-Hill Book Company, 1965, Chapter 8.

[4]. B. P. Bokert, M. J. R. Healy, and J. W. Tukey, 'The Quefrency Alanysis of Time Series for Echoes: Cepstrum, Pseudo-Autocovariance, Cross-Cepstrum, and Saphe Cracking,' Proceedings of the Symposium on Time Series Analysis: Brown University, June 11–14, 1962 (M. Rosenblatt, ed.), John Wiley and Sons, Inc., 1963.

[5]. W. M. Gentleman and G. Sande, 'Fast Fourier Transform for Fun and Profit,' 1966 Fall Joint Computer Conference, AFIPS, Proceedings, Vol. 29, pp. 563-578.

Agoston Z. Kiss

Ago Kiss began his varied professional career in his native Hungary doing research in nuclear particles and electromagnetic interactions. After five years he moved to England, where he spent another five years, first as a research physicist and then as a consultant in control systems. In 1962 he came to the United States and spent three years developing bandwidth reduction and coding techniques before joining the staff of Hewlett-Packard Laboratories in 1965. At HP, Ago has worked on character recognition and gas chromatograph control, and since 1967 has been engineering group manager in charge of the development of Fourier Analyzers. Ago was responsible for defining the 5450A Fourier Analyzer and developing its algorithms.

Ago studied electrical engineering at the Polytechnical University of Budapest, and physics and math at the University of Sciences in Budapest. He has also done postgraduate work at the University of Pittsburgh. He holds several British and U.S. patents related to control circuits. For relaxation, Ago chooses the out-of-doors. Skiing and sailing are his favorite leisure-time activities.

HEWLETT-PACKARD JOURNAL *JUNE 1970 Volume 21 • Number 10*

TECHNICAL INFORMATION FROM THE LABORATORIES OF THE HEWLETT-PACKARD COMPANY PUBLISHED AT 1501 PAGE MILL ROAD, PALO ALTO, CALIFORNIA 94304

A Two-Hundred-Foot Yardstick with Graduations Every Microinch

This new and innovative laser interferometer is ready to measure distance with no warmup at all. From its specially designed two-frequency laser and heterodyning techniques it derives increased sensitivity and resistance to air turbulence. With its internal computer it can smooth jittery readings, calculate velocity, and improve resolution.

By John N. Dukes and Gary B. Gordon

A LASER BEAM makes an eminently practical working standard for measuring length. With its high degree of coherence, it is usable over distances of hundreds of feet. Its short wavelength permits resolution into the microinch range. Its wavelength can be determined to a high degree of accuracy—parts in 10^7 or better. It is locked to an atomic transition in neon and to the velocity of light, and does not require periodic recalibrations. On this yardstick, the graduations are laid down by nature herself.

The new HP Model 5525A Interferometer is a portable distance-measuring standard which takes advantage of these desirable laser attributes, and incorporates as well a number of refinements which give it significantly better performance, reliability, and ease of use than other interferometers. It measures distances from zero to more than 200 feet. It has one microinch resolution (0.4 μin or 10^{-8} m when measuring in metric units) and is accurate within five parts in 10^7. There is no need to wait an hour or more after turn-on for it to warm up; it meets its accuracy specification immediately and maintains itself in tune automatically. It has electronic averaging to make readings rock-steady even in the presence of minute vibrations, and it has unusual immunity to air turbulence, the most common cause of poor interferometer performance.

Fits In a Suitcase

The interferometer has three components (see Fig. 1): a laser head that generates a low-power laser beam, a reflector that returns the beam to the laser head, and a control box that computes and displays the readings. The entire system is so compact that it can easily be carried from place to place in a suitcase. It weighs only 44 pounds.

In practice, either the laser head or the reflector is mounted on the device whose movement is to be measured, and the other unit is mounted at a fixed point. The control box can be placed anywhere allowed by the single 15-foot cable connecting the box to the laser head.

The reflector is a glass trihedral prism, or 'cube corner,' similar to the ones recently placed on the moon. Distance is measured by electronically counting wavelengths of light. Distance *change* is measured, rather than the absolute distance between the laser head and the reflector. Thus any point may be defined as a zero reference.

Fig. 1. *Model 5525A Laser Interferometer is a rugged, easy-to-use distance-measuring system with a range of 200 feet, accuracy better than 5 parts in 10^7, and usable resolution better than one microinch. Its new two-frequency laser warms up instantly, runs cool, and doesn't mind air turbulence, the most common cause of poor interferometer performance.*

Fig. 2. *Model 5525A Interferometer has 10 microinch resolution in NORMAL and SMOOTH modes, one microinch in X 10 mode. In the SMOOTH mode, successive readings are averaged to eliminate vibration-induced jitter in the display. The system can also display the moving reflector's velocity, up to 720 in/min. Behind the door at right are thumbswitches for setting in velocity-of-light corrections.*

To convert from wavelengths of light to useful units of inches or millimeters the interferometer performs several operations. The wavelength, about 25 microinches, is first divided electronically to finer than one microinch. The result is then multiplied digitally by the wavelength of the laser light in either English or metric units, corrected for slight variations in the velocity of light due to the temperature, pressure, and humidity of air, and finally displayed to 9 digits as the distance traversed.

Two Frequencies Fight Turbulence

The interferometer derives its immunity to air turbulence from a new two-frequency system, described in detail later in this article. Turbulence has always been a serious problem for laser interferometers. The same heat waves that cause a distant image on the horizon to flutter can also affect the laser beam. The effect is equivalent to an intensity variation. Similar intensity variations are produced by absorption and dispersion from atmospheric contaminants such as smoke or oil mist, and by dirt films on the optical surfaces. A measure of an interferometer's ability to operate under these adverse conditions is the maximum loss of returned beam it can tolerate.

Where most interferometers are comfortable with 50% loss of signal, the two-frequency interferometer tolerates more than 95%. This additional margin of safety also frees the two-frequency interferometer from periodic electrical adjustments. There are no adjustments for beam intensity or triggering threshold. Furthermore, the interferometer tolerates signal variations produced when the reflector is rotated, so it can make measurements such as dynamic growth of a lathe spindle resulting from bearing self-heating.

Special Laser Designed

The heart of the interferometer is a unique single-mode helium-neon gas laser specifically designed for this application. Its output is a continuous red beam at 632.8nm (6328Å). Conventional laboratory lasers are stabilized by placing them in an oven, but for a portable distance-measuring device this presents two drawbacks. First, the time required for the oven to stabilize can approach an hour, and second, the heat of the oven can cause the object being measured to expand, thereby invalidating the results. In the new laser, the laser cavity is stabilized by using an internal zero-coefficient-of-expansion structure, combined with a servo loop for automatic tuning. The results are zero warmup time and a cool-running optical head. Also, since the tuning system doesn't have to dither the laser frequency to find line center, the display is steadier than that of other interferometers. This laser is the subject of the article beginning on page 14. Other design goals contributing to its unusual configuration were long lifetime and ruggedness.

Modes of Operation

The new interferometer has several modes of operation. In the normal mode display is nearly instantaneous, to a resolution of 10 millionths of an inch. In this region one becomes suddenly aware of how flexible even massive structures can be. Granite surface plates and large machines may be deformed by mild pressures, and vibrations produced by nearby motors produce small rapid dimensional fluctuations. When present these are seen as superimposed digits in the rightmost display tube, since the vibrations are faster than the eye can resolve.

Frequently one would like to see through such vibrations to observe other deflections, such as machine deflection under loading by the workpiece. The 'smoothing' mode of operation effectively accomplishes this. In this mode, a sequence of measured values is low-pass-filtered digitally so the display shows the average position rather than the instantaneous position. Thus the reading becomes rock-steady.

The normal and smoothed modes display distance with a resolution of 10 millionths of an inch, or about one five hundredth the diameter of a human hair. Fine as this is, there are applications in metrology, photogrammetry, and integrated-circuit mask-making for which more resolution is desirable. In the ×10 mode resolution is electronically extended by interpolating between fringes. Resultant resolution is one microinch in English units, or 10^{-8} m in metric units. 10^{-8} m (0.4 microinches) is about 25 times the atomic spacing in a crystal lattice.

An internal time reference is included for digital velocity or feed rate measurements up to 1 foot/second or 720 inches/minute. These are derived by subtracting subsequent distance measurements at precisely known intervals.

Velocity of Light Corrections

Behind a front-panel door (Fig. 2) is a vernier for making, in effect, fine and calibrated adjustments in the length of the standard. An example of its use is in calibrating a machine at an elevated temperature, say 78°, and correcting the results back to 68°. If the machine is cast iron and has a coefficient of expansion of 6.5 ppm/°F, then it will be 65 ppm too large at 78°F. Correction is effected by subtracting 65.0 from the thumbwheel switch reading.

Another use for these switches is to make slight corrections for variations in the velocity of light for very precise measurements. A simple table converts readings of barometric pressure and air temperature into the proper number to enter on the switches.

Within a few months an accessory will be available which will automatically compute the velocity-of-light correction. It will have sensors to measure air pressure, air temperature, and machine temperature. There are good arguments for both approaches; manual entry of the correction factor has the advantage of being both economical and conducive to good measurement technique, while automatic compensation saves a few operator steps but requires periodic recertification to maintain accuracy. Other accessories, which are available now, are a printer and a 90° beam-bender.

Human Engineering

Many man-hours went into the human engineering of the interferometer, and the result proves that a basic length-measuring instrument need not be complicated to be accurate. For example, the operations of starting the laser, tuning and locking it, and resetting the display are performed automatically when the power switch is turned on. Seldom-used controls such as electrical self-checks are placed behind a front panel door. Only low voltages are carried on the cable to the interferometer, so no shock hazards are present should a chip cut into the cable. For easier readability, insignificant leading zeros on the display are automatically blanked. Front-panel range changes or units changes (e.g., metric to English) do not destroy the distance information, and when the instrument is switched out of velocity or feed-rate mode, the distance reading is still valid.

The cube-corner reflector was designed for unusual versatility in mounting and ease of alignment (Fig. 3). Holes allow 90° rotation, and turning a single knurled ring clamps all axes. The base may be interchanged with standard magnetically clamped bases. The column and a rear shoulder on the housing are both ⅝ inch in diameter, for easy retention in collets or chucks. There is also a rear thread for mounting the cube-corner perpendicular to flat surfaces.

How Interferometers Work

When Apollo 11 landed man on the moon for the first time, a corner reflector was set up as part of an experiment to measure the distance to the moon very precisely. Laser pulses were bounced off the reflector, and the time of travel of the *envelope* of each pulse was a measure of the distance. A radar measures distance in the same way, but at a lower carrier frequency.

Interferometers measure distances in a different way, that is, by counting wavelengths of the *carrier* signal, rather than by measuring the travel time of the envelope of the carrier. All modern interferometers are based on techniques pioneered by A. A. Michelson in the 1890's (Fig. 4).

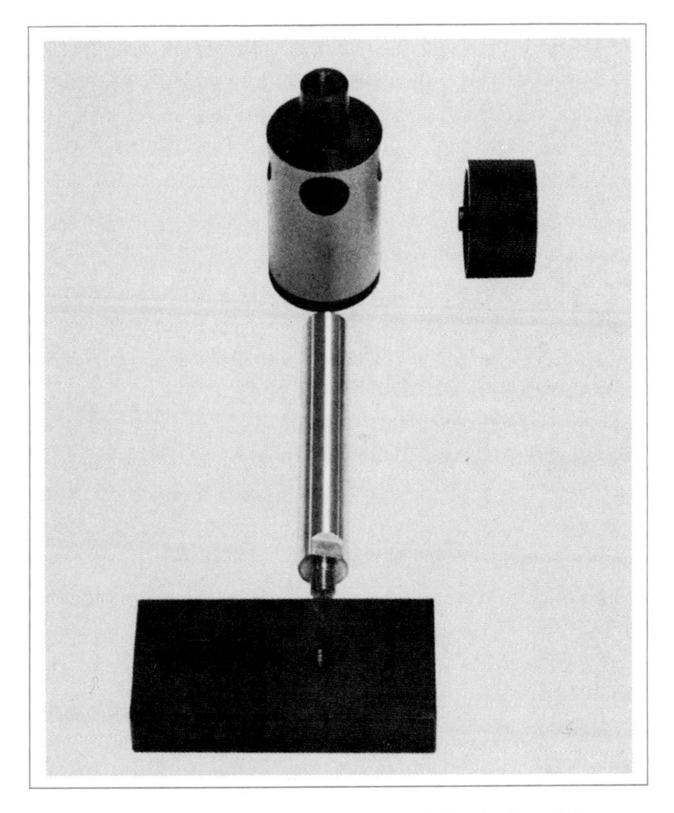

Fig. 3. *Reflector, or 'cube corner', is designed for easy mounting in a variety of ways, and for easy alignment. Turning a single knurled ring clamps all axes.*

Michelson used a half-silvered mirror to split the beam from a light source into two beams, each of which was reflected from a mirror and again recombined at the half silvered mirror. With the mirrors exactly aligned and motionless the observer sees a constant intensity of light. But if one of the mirrors is moved very slowly the observer will see the beam repeatedly increasing and then decreasing in intensity as the light from the two paths adds and cancels. Each half wavelength of mirror travel means a total optical path change of one wavelength and one complete cycle of intensity change. If the wavelength of the light is known, then the travel of the mirror can be accurately determined. It's important to note that the distance out to the moving mirror may not be known; interferometers measure the *changes of position* of the mirrors with respect to each other.

To convert Michelson's apparatus into an electronic measuring instrument basically requires only a photocell to convert beam intensity into a varying electrical signal, and an electronic counter to tally the cycles of beam intensity. To make such a device practical, however, several other improvements are necessary.

First, because mirror alignment is extremely critical, modern interferometers use cube corners instead of mirrors. Cube corners reflect light parallel to its angle of incidence regardless of how accurately they are aligned with respect to the beam. Second, modern interferometers use lasers as light sources, for two reasons: if the interferometer is to be used over any significant distance the light must be pure, i.e., single wavelength; if the interferometer is to be accurate, the wavelength must be exactly known. The laser satisfies these criteria beautifully.

A third improvement is direction-sensing electronics. A single photocell isn't sufficient to sense which way the reflector is being moved. The method used by most interferometers to sense direction is to split one of the optical beams into two portions, delay one portion in phase by 90°, and then, after recombination, detect each portion of the beam using a separate photocell. This technique gives two signals which vary sinusoidally in intensity as the reflector is moved, and they differ in phase of brightness by 90°. These two signals, after dc amplification, can be used to drive a reversible counter, and the phase separation is sufficient to inform the counter of the direction sense of the motion.

Although all commercial interferometers to date have been built this way, there is a fundamental problem with this conventional system. Fig. 5(a) illustrates the output of one of the photocells as one of the reflectors is moving. Notice that the intensity variations are centered around the triggering levels of the counter. But if the intensity of either light beam or the intensity of the source should change, the variations in intensity may not cross the triggering levels. Fig. 5(b) illustrates this condition. Thus a change in intensity can stop operation until the trigger levels are readjusted. Such a change in intensity conventionally occurs as the laser ages; it always occurs when turbulence either deflects the beam slightly or warps the wavefront, and, while trigger levels can be adjusted

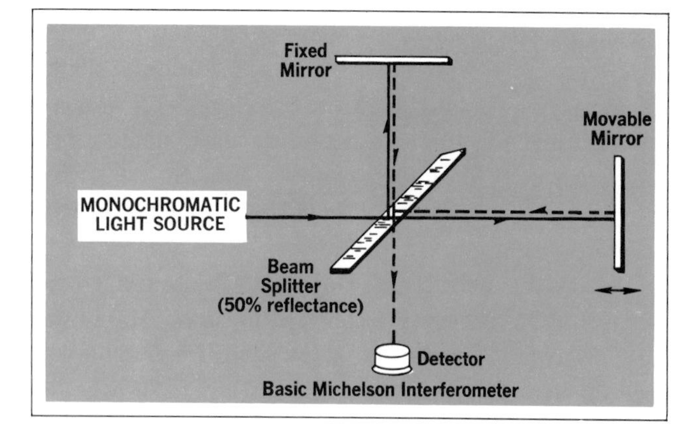

Fig. 4. *Laser interferometers are based on the Michelson interferometer of the 1890's. If one of the two mirrors moves, the observer sees the light repeatedly increasing and decreasing in intensity as light from the two paths alternately adds and cancels. Each cycle of intensity corresponds to a half-wavelength of mirror travel. If the wavelength of the light is known, intensity cycles can be counted and converted to distance traveled.*

for long-term changes, no automatic trigger-level adjustments can follow the fast changes in intensity one usually finds in a shop atmosphere where the interferometer is often used.

Two Frequencies Are Better Than One

The new HP interferometer operates on a heterodyne principle and completely avoids this problem. While conventional interferometers mix two light beams of the same frequency, the HP interferometer uses a two-frequency laser and mixes light beams of two different frequencies. Fig. 6 is a diagram of the system.

The virtue of the two-frequency system is that the distance information is carried on ac waveforms, or carriers, rather than in dc form. Unlike dc amplifiers, ac amplifiers are not sensitive to changes in the dc levels of their inputs.

The ac signals representing distance change are generated in a manner exactly analogous to the intermediate frequency carriers in the everyday FM heterodyne radio receiver. The ac signal or 'intermediate frequency' is produced by mixing two slightly different optical frequencies, near 5×10^{14} Hz, differing by only parts in 10^9. If these had to be generated by different sources, the stability requirements would be almost prohibitive. But by a fortunate circumstance, a laser can be forced to oscillate on two frequencies simultaneously, simply by applying an axial magnetic field. The two frequencies that result are very close together, but the corresponding components of the laser beam have opposite circular polarizations and can therefore be separated by polarized filtering.

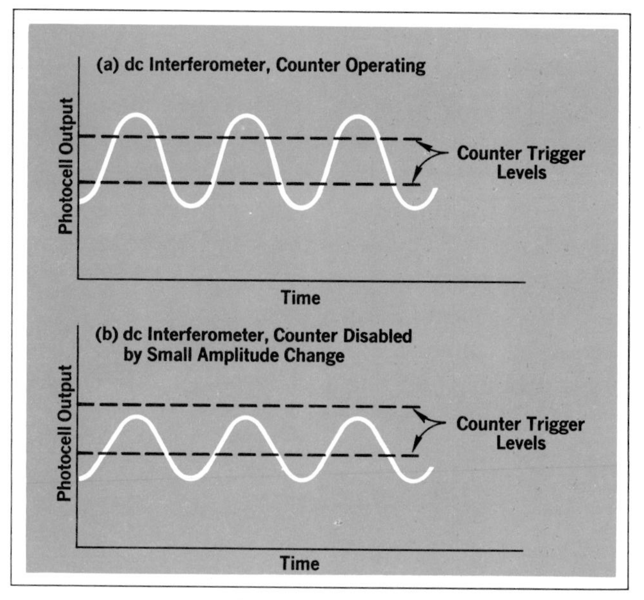

Fig. 5. *Most laser interferometers use dc systems in which intensity changes due to aging or air turbulence can interfere with triggering and cause improper counting. The new HP interferometer uses an ac system which doesn't have this problem.*

One of the two frequency components is used as the measuring beam and reflected from the cube corner. On return it is mixed with the second frequency, or 'local oscillator' in receiver language. The mixing produces the well-known fringe patterns of alternate light and dark bands caused by alternate constructive and destructive interference. The eye can't resolve these bands, however, since they flicker at a rate of several million per second. If the movable cube-corner reflector happens to be stationary, the rate will be exactly the difference between the laser's two frequencies, about 2.0 million fringes/second. Now if the reflector is moved, the returning beam's frequency will be Doppler-shifted up or down slightly, as with a passing train's whistle. A reflector velocity of one foot per second causes a Doppler shift of approximately 1 MHz. This fringe frequency change is monitored by a photodetector and converted to an electrical signal. A second photodetector monitors fringe frequency before the paths are separated, as a reference for the fringe rate corresponding to zero motion.

These two frequencies from the photodetectors are next counted in a form of reversible counter. One frequency produces up-counts, the other down. If there is no motion, the frequencies are equal, and no net count is accumulated. Motion, on the other hand, raises or lowers the Doppler frequency, producing net positive or negative cumulative counts corresponding to the distance traversed in wavelengths of light.

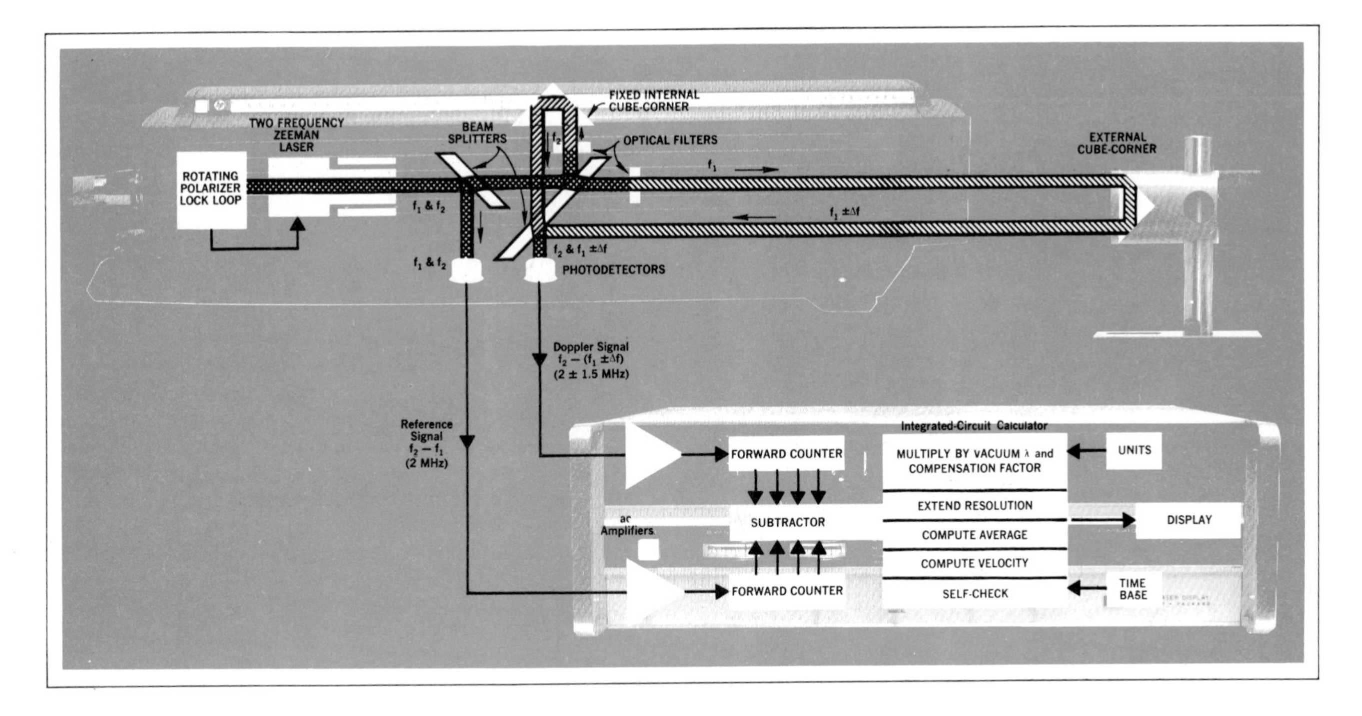

Fig. 6. *In the HP interferometer, two frequencies are generated by the laser and separated by optical filters of opposite circular polarizations. The beam of frequency f_1 is Doppler-shifted as it bounces off the moving reflector. The returned beam is optically mixed with the beam of frequency f_2 and the difference frequency is detected, amplified, counted, and converted to distance traveled.*

IC Calculator

A small integrated-circuit calculator converts these wavelengths to English or metric units of length. The low cost, small size, and high reliability of IC's make such a calculator feasible for this application. And as a bonus, having the flexible calculator present makes it economical to include other functions, such as smoothing, resolution extension, and velocity computation.

Smoothing is done by first storing the previous distance reading. Then when a new distance is computed, it is not displayed. Rather, the previous distance reading is changed slightly (0.1%) toward the new distance, and this result is displayed. Consequently the display is sluggish to rapid changes. This digital low-pass filtering is the logic designer's counterpart of the mechanical engineer's dashpots and the analog designer's RC filters. Resolution extension is achieved in much the same way, by random sampling and averaging. Velocity is computed by subtracting subsequent readings at tenth-second or thousandth-minute intervals.

Acknowledgments

Several years ago when HP engineers were first experimenting with interferometry, they found that variations in optical intensity were causing miscounts and intermittent operation. Santa Clara Division manager Al Bagley, a fast man with an analogy, suggested heterodyne operation similar to that used in frequency-measuring instruments and radio receivers. Such a system, he felt, would allow the use of ac amplification and eliminate the need for critical (and changing) dc level adjustments.

The invention moved fast from that point. Len Cutler conceived several methods for furnishing two frequencies from one single-mode laser. Don Hammond proposed optical schematics. Joe Rando contributed ideas on all fronts.

A good invention needs good treatment to be a usable instrument. This was the job of the laser measurements section of the Santa Clara Division. Ken Wayne began the project as a mechanical engineer and emerged an optical designer. The structural, thermal, and magnetic design was coordinated by André Rudé with help from product designer Roy Ingham. Later in the project they were assisted by Jobst Brandt and Jim Marrocco. Bringing five years of interferometer experience, Dick Baldwin joined mid-course as optical engineer.

The attractive industrial styling is by Roger Lee. Jon Garman made numerous digital design contributions. Analog assistance was lent by John Corcoran and Lyle Hornback. Production engineer Carl Hanson smoothed the road to replication.

John N. Dukes

Gary B. Gordon

John Dukes has degrees from Oberlin College (BA) the Universities of North Dakota (BSEE) and California (MSEE), and Stanford University (EE). As engineering section manager, he supervised development of the new laser interferometer, along the way gathering three patents pending, one on a laser locking system, another on a laser power supply, and a third on an interferometer thermal compensation technique. John previously helped design two synthesizers and led a heterodyne-converter project. In his spare time he plants trees and plays the baritone saxophone, but not, he says, simultaneously.

Gary Gordon received his degree in electrical engineering from the University of California. Before becoming project leader on the laser interferometer, he did logic design on the HP Computing Counter and conceived the Logic Probe and Logic Clip, two handy gadgets for IC logic checkout. Gary has seven patents pending, three of which cover the interferometer's error plotting, smoothing, resolution extension, and counting techniques. He received his MSEE this year from Stanford University, which means that he'll now have less studying to do and more time for sailing and designing contemporary furniture.

SPECIFICATIONS
HP Model 5525A
Laser Interferometer

ACCURACY (exclusive of velocity of light, alignment and work piece temperature): 5 parts in 10^7 ±1 count in least significant digit (±2 counts in metric units).

RESOLUTION (least count):
NORMAL MODE: 0.00001 in or 0.0002 mm.
× 10 MODE: 0.000001 in or 0.00002 mm.

OPERATING RANGE: 200 ft, 60 meters, in typical machine shop environments.

MAXIMUM MEASURING VELOCITY: 720 in/min (1 ft/s), 0.3 m/s.

VELOCITY MEASUREMENT:
RANGE:
ENGLISH: 0 to 12 in/s, 0 to 720 in/min.
METRIC: 0 to 300 mm/s, 0 to 18,300 mm/min.
ACCURACY AND RESOLUTION:
ENGLISH: ±0.0001 in/s, ±0.01 in/min.
METRIC: ±0.002 mm/s, ±0.2 mm/min.

WARMUP TIME: None

LASER TUNING: Laser tuning is automatic.

DISPLAY: 9 digits with appropriate decimal point and comma, and + or − sign.
UNITS:
NORMAL, SMOOTH and × 10 MODES: in, mm, λ/4
VELOCITY MODE: in/s, in/min, mm/s or mm/min.
Display in all modes (Normal, Smooth, × 10, and Velocity) and all units available at any time during or after a measurement without loss of any information.
Nonsignificant leading zeros are blanked for readability.

RESET: Pushbutton reset to zero.

ERROR INDICATORS: Beam interrupt, overspeed.

TEST CIRCUITS:
Front-panel pushbutton-operated test circuits verify that all computing circuits are operating properly.

ALIGNMENT TO MEASURING AXIS:
Built-in three-point kinematic suspension and precision adjustment permits accurate alignment to measuring axis. Signal strength meter on front panel makes possible easy positioning of retroreflector for maximum optical signal.

INTERCHANGEABILITY:
Any HP interferometer head will operate with any HP display unit.

VELOCITY OF LIGHT COMPENSATION:
A combined factor for barometric pressure, temperature, and humidity is derived from a supplied table. The factor, directly in parts per million with 0.1 ppm resolution is manually entered via thumbwheel switches. The range of this factor is large enough to cover any possible set of environmental conditions.

MATERIAL THERMAL EXPANSION COMPENSATION:
Thermal expansion compensation in parts per million is manually entered via thumbwheel switches.

INPUTS:
Automatic velocity of light compensation or remote manual VOL compensation.
Auxiliary: Remote front panel controls, i.e., Reset, Manual Print, Normal, Smooth, × 10, Velocity, Tuning Error, Beam Interrupt Error.

OUTPUTS:
BCD output for printer, computer, Fourier analyzer, etc. Timed contact closure for automatic NC test advance, or periodic data recording applications.

PRICE: Model 5525A (Includes Model 5500A Interferometer Head, Model 5505A Display, and Model 10550A Reflector), **$11,500.00.**

MANUFACTURING DIVISION: SANTA CLARA DIVISION
5301 Stevens Creek Boulevard
Santa Clara, California 95050

André F. Rudé

Kenneth J. Wayne

A former sports car and motorcycle racer and light airplane pilot, André Rudé settled down after his marriage and now only climbs mountains and skis. He also fixes and collects old clocks. Appropriately, when he joined HP in 1966, his first assignment was in the product design of frequency and time standards. He subsequently took on the product design responsibility for the new laser interferometer, and is now looking into new ways of using the interferometer in metrology and other fields. André is a member of ASME.

Keeping his head above water isn't one of Ken Wayne's problems; he's a scuba diver and underwater photographer with a strong interest in oceanography. Ken did the optical design and some of the mechanical design of the new laser interferometer, and has recently switched to marketing as the western region field engineer for the interferometer. Ken began his career designing jet engines, then came to HP in 1964. His BSME degree is from the University of Arizona.

A New Tool for Old Measurements – and New Ones Too

By André F. Rudé and Kenneth J. Wayne

Although Model 5525A Laser Interferometer measures distance and velocity with great range and accuracy, it is nevertheless practical and economically justifiable to use it for making gross low-accuracy measurements, since it is fast and easy to use. The diagram illustrates the spectrum of its applications.

The primary application is in **calibration of numerically controlled machine tools and coordinate measuring machines** in the machine shop and metrology lab. The laser beam can be aligned parallel to the axis of a machine tool or measuring machine in a few minutes. Because the system requires no warmup, calibration may begin immediately. Complete calibration of a three-axis machine tool with a printed record and/or a graph of errors versus command position can be accomplished in a few hours, compared with days by conventional methods. What's more, conventional methods do not automatically generate a graph. No specially trained personnel are needed, since the new interferometer is very simple to operate.

New for a laser interferometer is the ability to measure the axial growth of a rotating spindle due to frictional heat in its bearings. The cube-corner reflector is chucked and spun in the spindle and the beam is aligned parallel to the axis. This cannot be done with a conventional interferometer because, when the nonreflecting edges of the cube corner pass through the beam, the optical signal drops below the level at which the counter can safely trigger. The new interferometer, however, can tolerate a 95% loss of optical signal without error, so the small amount of light lost as the cube-corner edges go through the beam is hardly even noticed.

Since this laser interferometer also measures velocity, it is possible to **calibrate machine tool feed rate** at the same time positioning accuracy is measured.

In the metrology laboratory the new interferometer is useful for **calibration of other length standards** such as micrometer heads, tool-makers' microscopes, glass and metal scales, and even low-accuracy steel tapes over 200 feet long. With suitable fixturing it can be used for parts inspection.

It is particularly noteworthy that metrological calibrations may be performed continuously rather than in discrete steps as required with gage blocks. The HP error plotting option generates a continuous plot of leadscrew error versus position that allows the metrologist to see short-term variations in lead (leadscrew drunkenness) which otherwise might be obscured by the 'synchronous sampling' of gage blocks. Also the error incurred by transferring a measurement from a part to a stack of gage blocks can be avoided.

In keeping with the present trend toward **on-the-machine inspection,** the new interferometer is a natural for use as a length standard. The machine tool is converted to an N/C measuring machine by replacing the cutting tool with a contact probe and using the laser interferometer as the position transducer. The same tape used to machine the part is used to control the inspection process. A part may be inspected in its fixture immediately after machining.

Model 5525A Interferometer is already used as a **precision option on conventional N/C machine tools** that have their own built-in position transducers. With this arrangement it is possible to machine to normal tolerances with N/C. Then, for precision machining, the controller turns the machine over to the operator who manually positions one axis at a time using the laser interferometer. An example of this use might be precise locating of dowel holes in a large part. With the aid of beam benders, a single HP interferometer in a fixed mounting location can be used to position all axes of a machine.

From here it is only a short step to complete **closed-loop N/C or computer control** of the machine tool.

Other applications for the new interferometer are in control of step-and-repeat cameras for integrated-circuit production, in control of artwork generators, in mapmaking, and in photogrammetry. Stress-analysis and thermal-expansion-coefficient measurements of interest to the mechanical engineer and metallurgist are easily made, and physicists can make index of refraction measurements. A new and still experimental application is predicting earthquakes by accurate long-term measurements of earth movement in fault locations.

As for the future, who knows what is possible when you can measure a microinch?

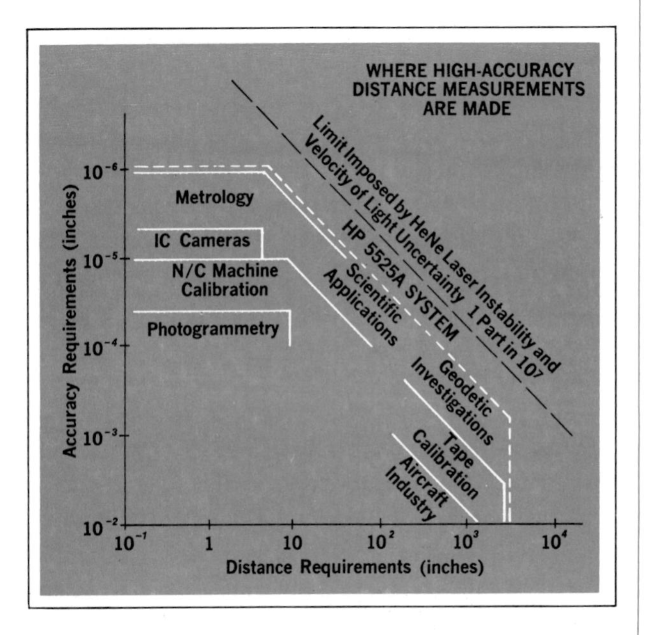

Automatic Error Plotting—a Report Card for Nonlinear Behavior

By Jonathan D. Garman

IN ALMOST EVERY CALIBRATION OF A LINEAR SYSTEM, the desired result is a plot of system error as a function of some calibration parameter. In the case of machine tool calibration, the calibration parameter is the nominal or command position of the machine tool along one of its axes, and the error is the difference between the actual position and the command position.

Although laser interferometers have greatly increased the accuracy and speed of machine-tool calibrations, methods of deriving errors and presenting them graphically have been largely unsatisfactory, ranging from tedious manual methods to complex, expensive automatic systems.

The main stumbling block in the automatic approach is that the command position of the machine tool is awkward to obtain in electrical form. In manual machines, of course, the command position is simply not available. In some automatic machines it is available, but codes are not standardized and vary from machine to machine. Even when the command position is available in electrical form, dozens of interconnections are required to make use of it.

New Error Plotting System

The HP approach doesn't solve these problems, it sidesteps them by making a few nonrestrictive assumptions about machine errors. As a result, automatic error plotting becomes a simple task, requiring only an inexpensive plug-in option for the interferometer, a single two-wire cable to the machine tool, and an X-Y recorder. The only signal required from the machine is a simple synchronization pulse to signify that the machine is in nominal position on a calibration point.

The principal assumption made in implementing the error plotting system is that the largest error to be encountered will be less than half the interval between calibration points. If the errors prove to be larger than this, the interval can be widened appropriately. The interval between calibration points can be any integral multiple of 0.010 inch and the range is 100 inches. Longer scales can be calibrated in 100-inch segments. If the interferometer is operating in the extended-resolution mode all these numbers are divided by 10. The full-scale position is selectable, and accuracy is constant (approximately 0.5% of reading) whether full scale is one inch or 100 inches.

How the system works is best explained by an example. Suppose that a positive error exists at three calibration points and the interferometer readings are 4.00012 inches, 4.10017 inches, and 4.20023 inches. The error plotting system breaks each reading into two parts and assumes the right-hand part is the error and the left-hand part is the command position. It then plots each right-hand part as a function of each left-hand part. Thus 12, 17, and 23 are assumed to be the errors in tens of microinches at the calibration points 4.00000 inches, 4.10000 inches, and 4.20000 inches, respectively. In other words, the system derives both command position and error just by inspecting the interferometer display. There is no need for complicated computations and massive interfaces.

When negative errors are encountered, the right-hand digits are complemented to obtain the correct absolute value of the error. The left-hand digits are incremented by one unit so the error will not be plotted one step too low.

Modes of Operation

The error plotting option operates in one of several modes, depending upon the type of machine being calibrated. For a full N/C machine tool, a command tape is prepared to position the machine sequentially to the desired calibration points.* The interferometer contacts are clipped in parallel with the manual-step button. The contacts step the machine through the control tape point-by-point, plotting the position and error at each calibration point. The stepping rate is front-panel selectable.

When calibrating a manually controlled machine such as a manual milling machine, the machine is placed in control of timing. A pulse must be generated each time the machine moves one calibration interval. This can be (and has been) done either by having a microswitch drop

* Some newer N/C controllers don't require the command tape.

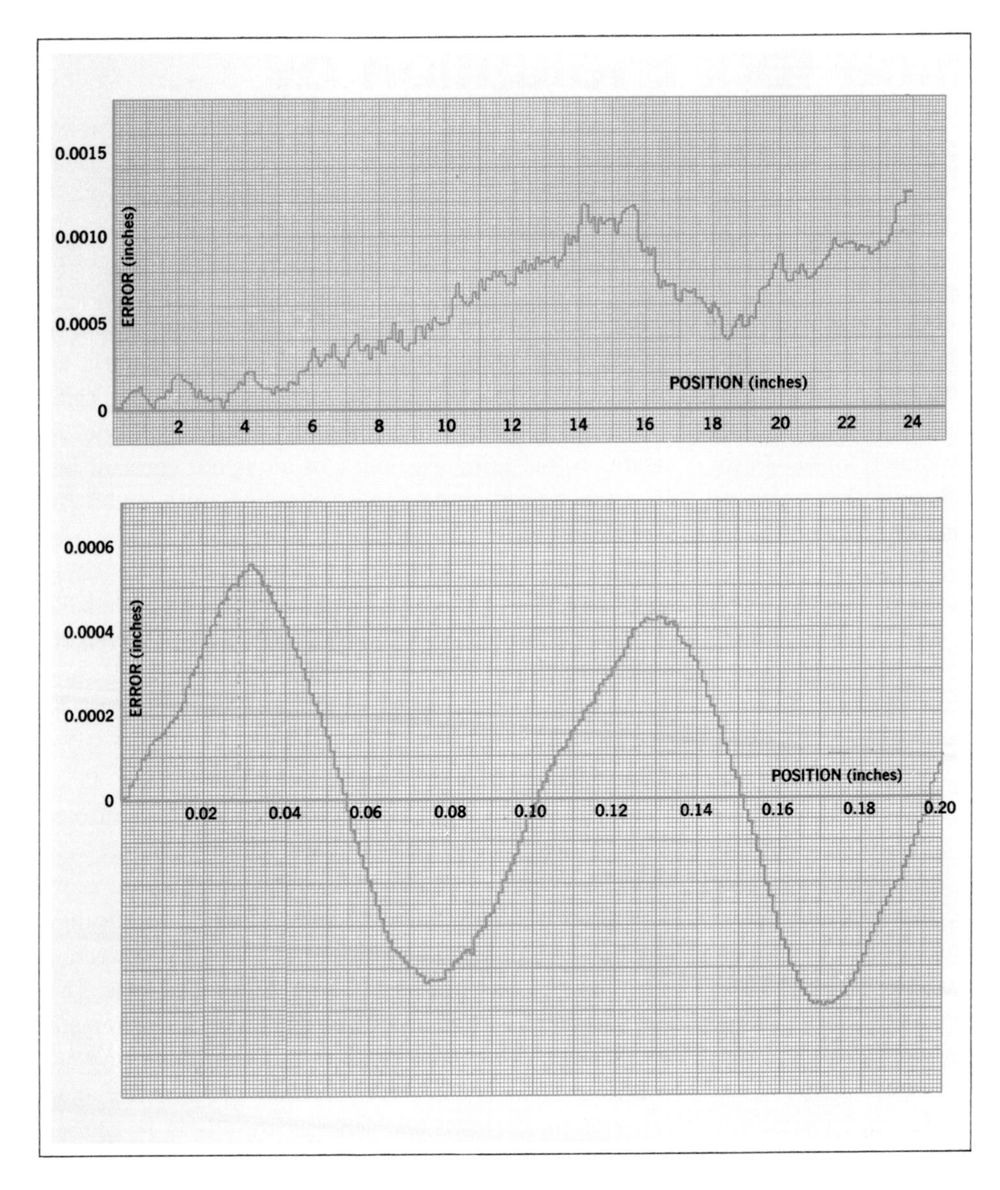

Graphic displays, like these generated by the error plotting option for Model 5525A Laser Interferometer, show machine errors in perspective. Both plots were made during a calibration of one axis of a numerically controlled milling machine that was suspected of being out of specification over large distances. The top plot shows the error at 0.1 inch intervals over the entire 24-inch axis. The upward slopes reveal that the machine's three 10-inch internal scales are all long by 50 parts per million. The rightmost scale is also 0.0005 inch out of position, as shown by the downward step, which is broadened by the effect of the machine's three-inch-wide read head. Plotting time was four minutes. The bottom plot magnifies a 0.2 inch section of the same scale, showing the error at 0.001 inch intervals. There is a cyclic error with a peak value of 0.0005 inch, indicating misadjustment of the electronic circuits which interpolate between the 0.1 inch marks of the machine's inductive scales. Similar cyclic errors are often seen on leadscrew machines and micrometers.

into teeth on a gear on the drive leadscrew, or by a lamp-photocell combination looking through a slotted disc driven by the leadscrew. The pulse commands the interferometer to make a measurement and plot the error. Measurements are taken on the fly, that is, without stopping the machine at each calibration point.

When it is impractical to generate an automatic synchronization signal, the synchronization can be done manually. This might be necessary, for example, in calibrating a precision microscope stage or a micrometer. The operator would position the instrument manually, and when ready he would press the MANUAL PLOT switch to plot the error.

Acknowledgments

The error plotting option was developed jointly with Gary Gordon, who originally proposed the technique.

Jonathan D. Garman

Jon Garman's first instrument netted him his first patent application—one on the error plotting system for the 5525A Laser Interferometer, co-authored by Gary Gordon. Jon's efforts in logic design for the interferometer also earned him the job of project leader for the forthcoming automatic velocity-of-light compensator for the 5525A. A 1967 recipient of a BSEE degree from Stanford University, Jon did graduate work at the University of Illinois before joining HP in 1968. He's an amateur trumpet player and photographer, but his first love these weekends is the old Jaguar XK140 he's restoring. That's the Jag's cylinder head in the foreground.

Machine Tool Evaluation by Laser Interferometer

By Richard R. Baldwin

INCREASING DEMAND FOR HIGH ACCURACY in mass-produced machined parts has forced the machine tool industry to rely less on the skill of the machinist and more on the accuracy of the machine tool itself. The large number of identical machining operations required for mass production has led to increasing automation of machine tools, ultimately leaving the machine tool responsible for the quality of the finished part. This, of course, has made it imperative that builders and users of machine tools continuously study and improve their tools' operating characteristics, particularly positioning accuracy.

One of the earliest problems associated with machine tool evaluation was the lack of a suitable length standard. Evaluation of positioning accuracy was commonly performed using a physical standard such as a scale or lug bar. These were available with sufficient accuracy in lengths up to about two feet, but longer standards were unwieldy and generally inaccurate. This meant that positioning accuracy of large machine tools had to be checked in short intervals by the method of 'staging' which was an extremely long and tedious process. The results so obtained were often nonrepeatable, and were not always indicative of the accuracy of the machine tool under evaluation.

The development of the laser interferometer finally provided the machine tool industry with a high accuracy length standard which could be used on machine tools of all sizes. The accuracy of the interferometer is limited by the laser wavelength, which is known to within about one part in ten million. This value compares favorably with the best physical standards available, and is certainly adequate for machine tool evaluation. In addition, the laser interferometer is extremely easy to use, allowing measurements to be made in minutes which had previously taken several hours or even days to perform.

Initial attempts at machine tool evaluation using the laser interferometer yielded results very quickly, but again the results were often nonrepeatable and not indicative of machine tool positioning accuracy.

Wavelength Variations

One reason for this involves the wavelength of the laser itself. It is the wavelength in vacuum which is known to about one part in ten million. The wavelength in air is somewhat shorter than the vacuum wavelength, since air has a refractive index slightly greater than one. In addition, the refractive index of air is not constant but is a function of air composition, temperature, and barometric pressure. To define the wavelength of the laser in air, therefore, all of these factors must be accurately determined. For this reason, all commercially available laser interferometers with fringes-to-inches conversion have provisions for determining barometric pressure, temperature, and relative humidity either automatically or via manual input.

Many users of interferometers, however, do not realize that these systems include a combination of electronic circuits and transducers which are subject to failure, and which require periodic recertification. In addition, many interferometers are built in such a way that certification of the pressure, temperature, and humidity correction factors is an extremely tedious and difficult process. This is particularly true in interferometers with automatic wavelength correction. As a result, many users do not institute an adequate recertification program for their interferometers, and this generally results in inaccurate measurements due to improper fringes-to-inches conversion.

Thermal Effects

Another and more significant source of error in interferometer machine tool evaluation is the effect of temperature on the machine tool itself. For machine tools which use a steel leadscrew to determine carriage position, this effect represents an expansion of six microinches per inch for a one-degree-Fahrenheit rise in the leadscrew temperature. If the total carriage travel were 50 inches, this effect would represent a total change in positioning accuracy of 300 microinches for each degree change in the leadscrew temperature. Further compounding the difficulty is the fact that the leadscrew operates in an extremely poor thermal environment. During operation, the leadscrew is faced on all sides with heat sources such as the driving motor, the bearings, and the drive nut.

During the first few hours of machine operation, the leadscrew temperature increases to some value well above ambient. Its final temperature, however, is not only a function of the ambient temperature but is also dependent on how the machine is operated during warmup. If the carriage is cycled on fast feed through its entire travel, for example, the leadscrew temperature will stabilize at a higher value than would result during normal operation. It is therefore important that the conditions under which a machine tool evaluation is made be well controlled and well defined. But even more important, it should be recognized that the machine tool will probably *not* show the same characteristics during actual operation. The presence of coolant, the use of different feed rates, cutting forces, and many other factors will change the positioning accuracy of a machine tool during use. A machine tool evaluation should therefore be conducted under conditions which best approximate those conditions under which the machine tool will operate during use.

An additional temperature effect occurs in the machined part itself. It cannot be assumed that the machined part will remain at ambient temperature during the machining process, and if the part is not at ambient temperature its dimensions will change due to thermal expansion or contraction when it is taken off the machine. This effect is not directly applicable to machine tool evaluation since it is due to no fault of the machine tool, but it must be understood by the machine tool user if he wishes to obtain optimum machine tool performance.

Mechanical Deflections

In addition to thermal effects, mechanical deflections of machine tool components often introduce error during a machine tool evaluation.

As an example of machine tool deflection, the author witnessed some time ago a series of interferometric machine tool tests in which the interferometer sensing head was mounted on a large tape-controlled lathe by means of a platform which replaced the tailstock. This platform provided a stable, vibration-free mounting fixture which appeared to be adequate. However, when the data obtained relative to the tailstock were compared to similar data taken relative to the spindle, certain repeatable differences became apparent. A detailed examination of the setup revealed that as the carriage approached the tailstock platform, which supported the interferometer, the platform was deflected toward the carriage by an appreciable amount. This deflection was due to bending of the entire machine bed, caused by the weight of the carriage. It would not have been valid to accept or reject the machine tool on the basis of data taken relative to the tailstock, since it is the position of the tool tip relative to the spindle which determines machining accuracy. In this case, deflections of the machine bed should have been taken into account when designing fixturing for the laser interferometer.

Despite the insidious nature of the previous example, it is one of the simpler types of machine deflection. Besides the predictable deflections caused by the weight of moving components, there exist much more complex deflections caused by thermal gradients throughout the machine tool. These effects are not constant, and are a function not only of the conditions under which the machine tool is operated, but also of the environment in which the machine tool operates. Here again, it is important that the conditions under which a machine tool is evaluated approximate those conditions under which it will be used.

In conclusion, it should be mentioned that the previously described problems are not new, nor are they unique to the laser interferometer. The same principles apply when using any measuring system. The point is that the laser interferometer does not eliminate the fundamental requirements of precision measurement. Machine tool errors are complex by nature, and can only be eliminated through intensive study. The laser interferometer represents a valuable tool for diagnosing these errors, which is the first step in improving machine tool performance.

Richard R. Baldwin

Dick Baldwin is well on the way to making a name for himself in the fields of precision optics and precision machining. During the past seven years, Dick has been involved in the design and application of optical measuring systems to meet high accuracy requirements; has been a consultant in machine tool evaluation, error mapping, and correction; has designed and fabricated physical standards for machine tool certification; and has helped develop new methods of machining with diamond tools. He is the author of several papers and patent applications related to laser interferometry and precision machining. At HP for the last year, Dick has been developing methods for evaluating high precision optical components and machine tools. Dick holds a BS degree in engineering physics from Ohio State University and is a member of the Society of Manufacturing Engineers.

An Instant-On Laser for Length Measurement

This specially developed two-frequency laser is rugged, tunes itself instantly, and runs cool.

By Glenn M. Burgwald and William P. Kruger

SOON AFTER ITS INVENTION ten years ago, some clever fellow characterized the laser as 'a solution looking for a problem.' That solution has now found problems aplenty, ranging from eye surgery through metal cutting to high-density data storage and retrieval, to name but a few. Another such problem is the precise measurement of distance, made feasible by the coherence and accurately known wavelength of laser light. When a laser is used as the light source in an interferometer, distance measurement is a simple matter of counting interference fringes, each count signifying another few millionths of an inch of distance.

For the new HP Laser Interferometer described in the article on page 2, an entirely new laser was developed. Designed specifically for interferometry, it is a single-mode helium-neon laser in which Zeeman splitting is used to divide the main spectral line into two lines separated in frequency by about 2.0 MHz. The laser is extremely rugged, can be locked on frequency without warmup, and is designed to operate reliably in industrial use for 10,000 hours or more.

Single-Mode Operation

To avoid ambiguity in translating light wavelengths into distance, the laser operates in a single transverse mode and a single longitudinal mode. It has a 'hemispheric' mirror system in which a flat and a spherical mirror face each other at opposite ends of a gas-discharge bore of predetermined length and diameter. By adjusting mirror spacing, diffraction losses are set so that only the lowest-order transverse mode, TEM_{00}, can oscillate. Fig. 1 illustrates power output versus mirror spacing for a given bore diameter, and shows typical mode bounds. The laser operates as high on the power curve as possible without danger of developing spurious transverse modes.

Single-longitudinal-mode operation is achieved by using a mirror spacing of approximately 13 cm, a value sufficiently small to keep adjacent modes in regions of low gain and thus prevent their simultaneous oscillation. Fig. 2 shows how laser gain varies with frequency for the Doppler-broadened 6328Å (632.8 nm) line of helium-neon. Mode spacing is $c/2l$, where c is the velocity of light and l is the mirror spacing. A proper choice of l is one which safely suppresses adjacent modes when some one mode is at maximum gain. But with l/R constant (R is mirror radius of curvature) smaller l means smaller output power. Optimum spacing is 13–15 cm.

Zeeman Splitting

Although single-mode operation is necessary to avoid ambiguities, a better interferometer can be built if two adjacent frequencies are available rather than a single one (see article, page 2). In the new HP laser, two frequencies are obtained by Zeeman splitting of the main spectral line. Here's what this means.

If an axial magnetic field is applied to a laser which is free from polarization anisotropy in either the mirrors or the plasma tube, the output splits into two frequencies of left and right circular polarization as shown in Fig. 3. First-order theory predicts that the frequency splitting is proportional to magnetic field strength and to the ratio of line Q to cavity Q. In the new laser, magnetic field strength is adjusted for a difference frequency of about 2.0 MHz. Line center is virtually midway between the displaced lines, so proper cavity tuning can be assured by adjusting for equal intensities of the lines.

Fig. 4 is a photograph of the interior of the laser head, showing the axial magnet.

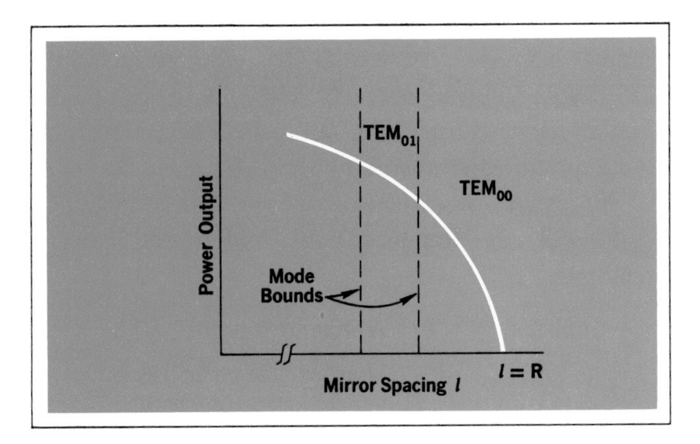

Fig. 1. *Laser power output versus mirror spacing for a given diffraction aperture (bore diameter), showing typical transverse-mode bounds. New HP laser operates in TEM_{00} mode only, but as close to TEM_{01} region as possible, so as to maximize power output.*

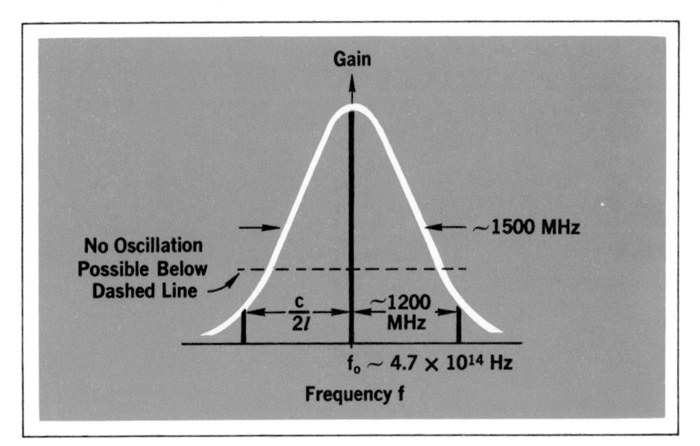

Fig. 2. This is the Doppler-broadened gain curve for the *6328Å (632.8 nm) line of helium-neon. In the new HP laser, cavity length (l) is about 13 cm, so only a single longitudinal mode can oscillate.*

Cavity Design and Tuning

The requirement of no polarization anisotropy, which must be satisfied for Zeeman splitting, precludes use of a plasma tube with Brewster-angle windows, since such windows have low reflection loss for only one kind of polarization. This in turn means that internal mirrors are obligatory. Also, it was a design objective that the laser be tuned to line center as soon as it is turned on, and this means that the plasma tube, or cavity, must include a tuning element to compensate for small changes in length.

The cavity design satisfies both these requirements. A sturdy rod, about an inch in diameter, of one of the new inorganic materials having virtually a zero coefficient of thermal expansion, is used as a combined plasma tube and mirror spacer. It has an axial hole of the proper size to control transverse-mode excitation, and the ends are precision-ground to provide proper mirror alignment.

Cavity length changes during and after warmup are small enough to be compensated by a piezoelectric wafer which forms part of an electronic servo loop. The loop monitors the intensities of the Zeeman-split lines and keeps them equal by varying the voltage on the piezoelectric element. The tuning range is adequate to compensate for all expected thermal length changes without the need for an oven.

This very stiff and stable resonator is encased in an all-glass envelope through which are brought appropriate electrical leads to generate a gas discharge and to apply voltage to the piezoelectric tuning element. Since no organic cements are used, the laser can be given high-temperature vacuum bakeout before final filling. This removes gas contamination as a factor tending to reduce the life of the laser.

Long-Life Mirrors

Because of the modest gain-per-pass in this type of laser, photons must be reflected back and forth many times through the cavity to achieve oscillation, and hence the mirrors must have extremely good reflectivity. Only tuned multilayer dielectric surfaces have the needed reflectivity at optical wavelengths. These coatings must be able to withstand evacuation, high-temperature bakeout, and exposure to all radiation from the discharge including ultraviolet. Unfortunately, the 'soft-coated' mirrors typically used externally, although excellent reflectors, deteriorate upon exposure to ultraviolet light in at most a few hundred hours.

Coating manufacturers have exerted a great deal of effort to achieve 'hard' coatings unaffected by ultraviolet light, and have found that alternate layers of silicon dioxide and titanium dioxide, for example, can provide adequate reflectivity and yet not deteriorate from ultraviolet exposure. They also withstand the thermal cycling of good vacuum practice. The new laser has mirrors of this type. Long-term life tests are under way, and some are now approaching 20,000 hours. Therefore, it seems possible to conclude that hard-coated mirrors properly made and used do not limit laser life.

Long-Life Discharge

Eliminating the most common cause of laser failure—contamination of the fill gases—does not in itself assure infinite laser life. Mechanisms exist by which the noble gases helium and neon can be lost to the discharge. For instance, helium diffuses rapidly through certain glasses. To minimize this, the new laser's envelope is made of a material which is very low in helium diffusion.

A different kind of loss involves 'capture' of noble

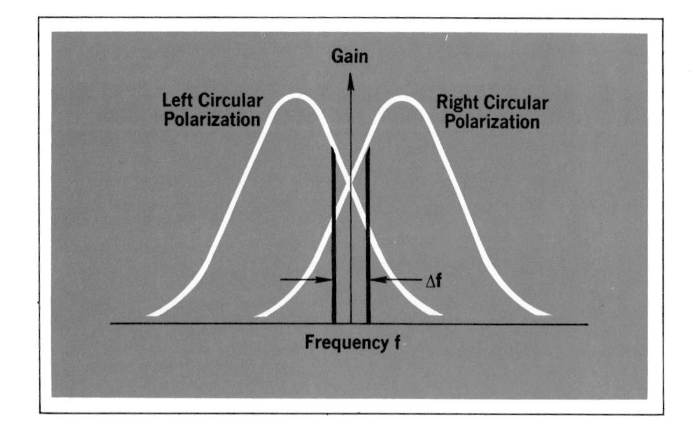

Fig. 3. *Two frequencies are derived from the single-mode laser by Zeeman splitting. An axial magnetic field causes the single line to split into two lines. The two frequencies have opposite circular polarizations.*

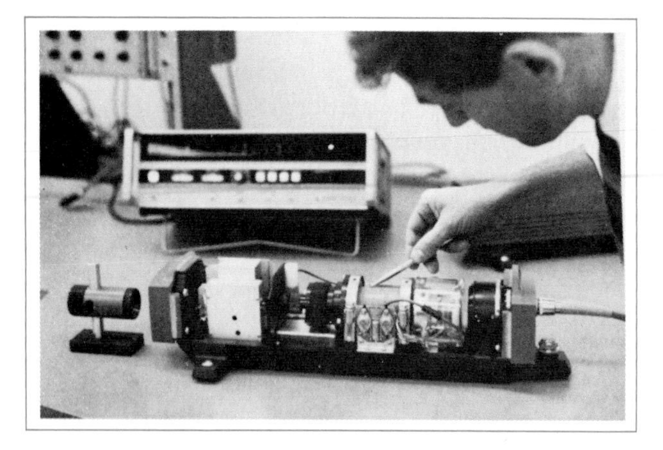

Fig. 4. *Interior of laser head, showing laser with axial magnet, telescope, and interferometer components. Pencil points to magnet.*

gases either within the metallic lattice of the cathode by ion penetration, or through a complex sputtering process which seems to plate out gas atoms when cathode material migrates to a new surface. In the new laser 'capture' factors are minimized by proper cathode design and fill pressures. The laser uses a cold cathode of rugged design whose long life has been demonstrated.

Acknowledgments

The laser described was developed in the physical electronics laboratory of Hewlett-Packard Laboratories. The two-frequency system and means for achieving suitable two-frequency operation were suggested by Len Cutler, Al Bagley, and Don Hammond. Means of drilling and grinding the zero-expansion cavities were perfected by Vas Peickii and personnel of the HP Laboratories model shop. Envelope construction fell to Bob Lorimer of the HP Laboratories glassworking shop. Innumerable ideas and suggestions were contributed by others including Hugo Fellner, Joe Rando, Wright Huntley, and Howard Greenstein.

William P. Kruger

Farmhand, newspaper editor, grocer, college professor, research physicist—Bill Kruger has been all of them. At HP since 1959, Bill was manager of cathode-ray-tube design for several years. More recently he helped invent the new two-frequency laser, and now, as a physicist on the staff of Hewlett-Packard Laboratories, he is doing research on lasers, beam-switched diodes, and spectrometers. He has five patents, mostly on gas-filled tubes. Bill holds BA and MS degrees from the University of North Dakota and has done graduate work at the University of Illinois and the University of Chicago. He is active in the Involvement Corps and other community groups, and thinks that woodworking and golf are the perfect ways to relax.

Glenn M. Burgwald

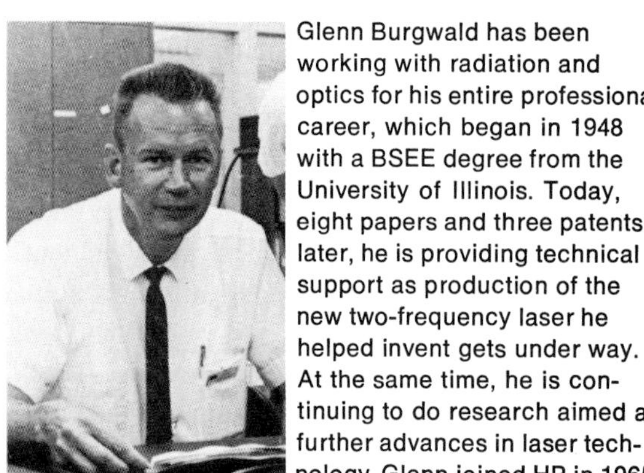

Glenn Burgwald has been working with radiation and optics for his entire professional career, which began in 1948 with a BSEE degree from the University of Illinois. Today, eight papers and three patents later, he is providing technical support as production of the new two-frequency laser he helped invent gets under way. At the same time, he is continuing to do research aimed at further advances in laser technology. Glenn joined HP in 1965, contributed to the development of the laser, then left for a while but returned in 1969 to participate in the interferometer project. Glenn's three years with the U. S. Air Force seem to have left a permanent mark—he has a passion for radio-controlled model aircraft. He keeps in shape by playing tennis.

HEWLETT-PACKARD JOURNAL **AUGUST 1970** *Volume 21 • Number 12*

TECHNICAL INFORMATION FROM THE LABORATORIES OF THE HEWLETT-PACKARD COMPANY PUBLISHED AT 1501 PAGE MILL ROAD, PALO ALTO, CALIFORNIA 94304

The 'Powerful Pocketful': an Electronic Calculator Challenges the Slide Rule

This nine-ounce, battery-powered scientific calculator, small enough to fit in a shirt pocket, has logarithmic, trigonometric, and exponential functions and computes answers to 10 significant digits.

By Thomas M. Whitney, France Rodé, and Chung C. Tung

WHEN AN ENGINEER OR SCIENTIST NEEDS A QUICK ANSWER to a problem that requires multiplication, division, or transcendental functions, he usually reaches for his ever-present slide rule. Before long, however, that faithful 'slip stick' may find itself retired. There's now an electronic pocket calculator that produces those answers more easily, more quickly, and much more accurately.

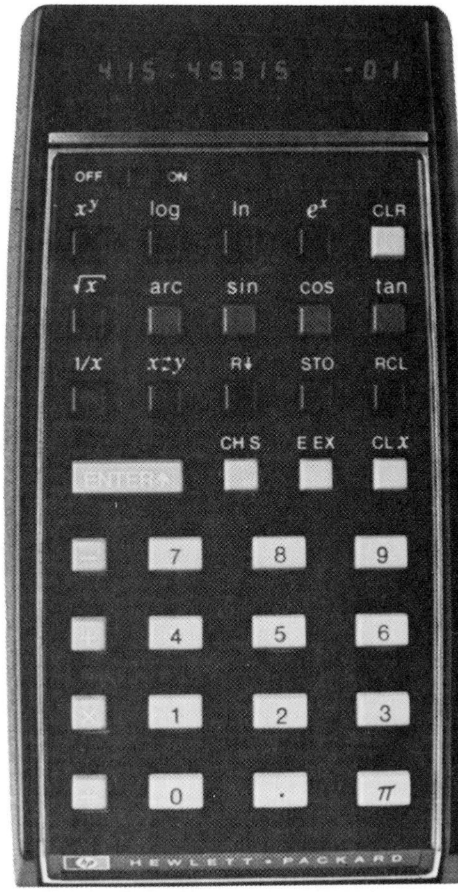

Despite its small size, the new HP-35 is a powerful scientific calculator. The initial goals set for its design were to build a shirt-pocket-sized scientific calculator with four-hour operation from rechargeable batteries at a cost any laboratory and many individuals could easily justify. The resulting nine-ounce product surprises even many who are acquainted with what today's large-scale integrated circuits can achieve.

The HP-35 has basically the same functions and accuracy of other HP calculators, and it is portable. It is a close cousin to the many four-function electronic calculators which have appeared in recent years, initially from Japan and now from many U.S. manufacturers. However, three features set the HP-35 apart from four-function calculators. First, none of the four-function calculators has transcendental functions (that is, trigonometric, logarithmic, exponential) or even square root. Second, the HP-35 has a full two-hundred-decade range, allowing numbers from 10^{-99} to $9.999999999 \times 10^{+99}$ to be represented in scientific notation. Third, the HP-35 has five registers for storing constants and results instead of just one or two, and four of these registers are arranged to form an operational stack, a feature found in some computers but rarely in calculators (see box, page 5). On page 7 are a few examples of the complex problems that can be solved with the HP-35.

Data Entry

The photograph of the calculator on this page shows how the keys are arranged. Numbers enter the display, which is also called the X register, from left to right exactly as the keys are pressed. Entry is entirely free-field, that is, digits will be displayed exactly as they are entered, including leading or trailing zeros.

The enter-exponent key, EEX, is used for entering numbers in scientific notation. For example, the number 612,000 may be entered as 6.12 EEX 5 or 612 EEX 3 or .0612 EEX 7 as well as 612000. The change-sign key, CHS, changes the sign of the mantissa or, if pressed immediately following EEX, the sign of the exponent. Mistakes during data entry can be corrected by use of the clear x key, CLx. A special key is provided for entering π.

Display

The display consists of 15 seven-segment-plus-decimal-point light-emitting-diode (LED) numerals. Answers between 10^{10} and 10^{-2} will always be displayed as floating-point numbers with the decimal point properly located and the exponent field blank. Outside this range the HP-35 displays the answer in scientific notation with the decimal point to the right of the first significant digit and the proper power of 10 showing at the far right of the display. To make the display more readable, a separate digit position is provided for the decimal point.

The display is always left justified with trailing zeros suppressed. For instance, the answer to $3 \div 4$ appears as .75 instead of 0.750000000.

Answers greater than 10^{100} (overflow) are displayed as $9.999999999 \times 10^{99}$, and answers smaller than 10^{-99} are returned as zeros. These are the HP-35's 'closest' answers. Improper operations, such as division by zero or the square root of a negative number, cause a blinking zero to appear.

Single-Operand Function Keys

The function keys operate on the number displayed in the X register, replacing it with the function of that argument. The trigonometric functions, sin, cos, tan, arc sin, arc cos, and arc tan operate in degrees only. For inverse trigonometric functions the function key is preceded by the arc key. The angle obtained will be the principal value.

The functions log x and ln x compute the logarithm of x to base 10 and base e respectively. The exponential function e^x is also provided. Other single-operand functions are the square root of x, $\sqrt{x}$, and the reciprocal of x, 1/x.

Arithmetic Keys: Two-Operand Functions

The arithmetic functions, $+$, $-$, $\times$, and $\div$, and the power function x^y, operate on the X and Y registers, with the answer appearing in X. Numbers are copied from X into Y by use of the ENTER↑ key. Thus to raise 2 to the 7th power the key sequence is 7, ENTER↑, 2, x^y and the answer 128 is displayed immediately after x^y is pressed. This is a general principle: when a key is pressed, the corresponding operation is performed immediately.

Although raising x to a power can be accomplished via the formula $x^y = e^{y \ln x}$ and in fact is done this way internally, the single-keystroke operation is much more convenient. The power, y, can be any positive or negative integer or fraction, while x must be positive.

Registers and Control Keys

Five registers are available to the user. Four of these, called X, Y, Z, and T, form an operational stack. The fifth, S, is for constant storage. (For clarity, capital letters refer to the registers and lower-case letters refer to the contents of the registers.)

Five keys are used to transfer data between registers. The ENTER↑ key pushes the stack up, that is, x is copied into Y, y into Z, z into T, and t is lost. This key is used as a separator between consecutive data entries, such as in the power example just described. Roll down, R↓, is used to view the contents of X, Y, Z, and T. Four consecutive operations of this key return the registers to the original state with no loss of data.

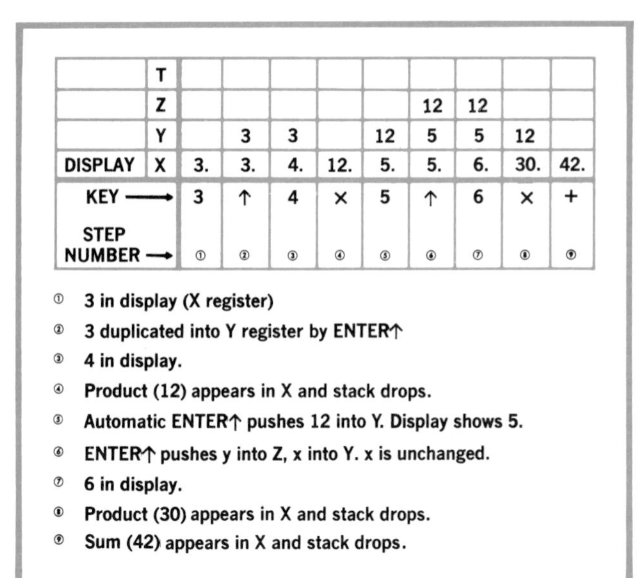

	T									
	Z						12	12		
	Y		3	3		12	5	5	12	
DISPLAY	X	3.	3.	4.	12.	5.	5.	6.	30.	42.
KEY ⟶		3	↑	4	×	5	↑	6	×	+
STEP NUMBER ⟶		①	②	③	④	⑤	⑥	⑦	⑧	⑨

① 3 in display (X register)
② 3 duplicated into Y register by ENTER↑
③ 4 in display.
④ Product (12) appears in X and stack drops.
⑤ Automatic ENTER↑ pushes 12 into Y. Display shows 5.
⑥ ENTER↑ pushes y into Z, x into Y. x is unchanged.
⑦ 6 in display.
⑧ Product (30) appears in X and stack drops.
⑨ Sum (42) appears in X and stack drops.

Fig. 1. *HP-35 Pocket Calculator has a four-register operational stack (last-in-first-out memory). Here's how the stack works to solve (3 × 4) + (5 × 6). Answers appear in display register X, in floating-point or scientific notation, to 10 significant digits.*

Fig. 2. *Instruction panel on back of calculator answers most frequently asked questions.*

The contents of X and Y can be exchanged using x⇄y; this is useful if the operands have been entered in the improper order for x^y, −, or ÷. The constant storage location, S, is accessed via the two keys store, STO, and recall, RCL.

The operational stack is automatically pushed up for data entries following any operation other than ENTER↑, STO, and CLx. This saves many uses of the ENTER↑ key. The stack automatically drops down following any two-operand function (+, −, ×, ÷, x^y). For example, to do the problem (3×4) + (5×6) the key sequence is: 3, ENTER↑, 4, ×, 5, ENTER↑, 6, ×, + (see Fig. 1). ENTER↑ is unnecessary between × and 5 because the answer to the first term, 12, is automatically transferred to Y when the 5 is entered. Also, no R↓ is necessary after the second × since the first term, 12, is automatically transferred from Z to Y after the multiplication.

The clear-all key, CLR, clears all registers including storage. Initial turn-on of the calculator has the same effect.

On the back of the calculator is an instruction panel that provides the user with quick answers to the most commonly asked 'how-to-do-it' questions (Fig. 2). The panel also shows an example of a problem solution.

System Organization

Now let's go inside and see what makes it work. The HP-35 contains five MOS/LSI (metal-oxide-semiconductor/large-scale-integration) circuits: three read-only-memories (ROMs), an arithmetic and register circuit (A&R), and a control and timing circuit (C&T). The logic design was done by HP and the circuits were developed and manufactured by two outside vendors. Three custom bipolar circuits are manufactured by HP's Santa Clara Division: a two-phase clock driver, an LED anode driver/clock generator, and an LED cathode driver. Fig. 3 is a block diagram of the calculator.

The HP-35 is assembled on two printed circuit boards (see Fig. 4). The upper board contains the display and drivers and the keyboard. The lower and smaller board has all the MOS logic, the clock driver, and the power supply.

The calculator is organized on a digit-serial, bit-serial basis. This organization minimizes the number of connections on each chip and between chips, thereby saving area and cost and improving reliability. Each word consists of 14 binary-coded-decimal digits, so each word is 56 bits long. Ten of the 14 digits are allocated to the mantissa, one to the mantissa sign, two to the exponent, and one to the exponent sign.

Three main bus lines connect the MOS circuits. One carries a word synchronization signal (SYNC) generated by a 56-state counter on the control and timing chip. On another bus, instructions (I_S) are transmitted serially from the ROMs to the control and timing chip or to the arithmetic and register

Operational Stacks and Reverse Polish Notation

In 1951, Jan Lukasiewicz' book on formal logic first demonstrated that arbitrary expressions could be specified unambiguously without parentheses by placing operators immediately before or after their operands. For example, the expression

$$(a + b) \times (c - d)$$

is specified in operator prefix notation as

$$\times + ab - cd$$

which may be read as multiply the sum a plus b by the difference c minus d. Similarly, the expression can be specified in operator postfix notation as

$$ab + cd - \times$$

with the same meaning. In honor of Łukasiewicz, prefix and postfix notation became widely known as Polish and reverse Polish, respectively.

During the following decade the merits of reverse Polish notation were studied and two simplifications in the execution of computer arithmetic were discovered. First, as reverse Polish notation is scanned from left to right, every operator that is encountered may be executed immediately. This is in contrast to notation with parentheses where the execution of operators must be delayed. In the above example, $(a + b) \times (c - d)$, the multiply must wait until $(c - d)$ is evaluated. This requires additional memory and bookkeeping. Second, if a stack (that is, a last-in-first-out memory) is used to store operands as a reverse Polish expression is evaluated, the operands that an operator requires are always at the bottom of the stack (last operands entered). For $(a + b) \times (c - d)$, the reverse Polish, $ab + cd - \times$, is evaluated as follows.

	T							
	Z					a + b		
↑	Y		a		a + b	c	a + b	
stack	X	a	b	a + b	c	d	c − d	(a + b) × (c − d)
Reverse Polish		Enter a	Enter b	+	Enter c	Enter d	−	×

These properties have made the notation a valuable tool in the computer industry. All modern computer compilers for languages such as FORTRAN and ALGOL convert statements to reverse Polish in some form before producing a program that can be executed. Some computer manufacturers have even designed their machines with special instructions to perform stack operations to facilitate execution of reverse Polish. However, the HP-35 is the first scientific calculator to fully exploit the advantages of reverse Polish and automatic stack operations to provide user convenience seldom found in calculators.

chip. The third bus signal, called word select (WS), is a gating signal generated on the C&T chip or by the ROMs; it enables the arithmetic unit for a portion of a word time, thereby allowing operations on only part of a number, such as the mantissa or the exponent.

Control and Timing Circuit

The control and timing (C&T) circuit performs the major nonarithmetic, or housekeeping functions in the calculator. These include interrogating the keyboard, keeping track of the status of the system, synchronizing the system, and modifying instruction addresses.

The keyboard is arranged as a five-column, eight-row matrix. It is scanned continuously by the C&T chip. When contact is made between a row and a column by pressing a key, a code corresponding to that row and column is transmitted over the I_A line to the read-only memory (ROM). This code is the starting address of a program in ROM to service that key. Key bounce and lockout are handled by programmed delays.

In all digital systems, status bits or flags are used to keep track of past events. In the HP-35 there are twelve status bits, all located on the C&T chip. They can be set, reset, or interrogated by microinstructions.

ROM addresses are updated on the C&T chip and sent serially to the ROMs over the I_A bus. During execution of a branch instruction, the appropriate signal—arithmetic carry or status bit—is tested to determine whether the incremented address or the branch address should be selected next.

A powerful feature of the serial organization is the ability to operate on just a single digit of a number as it flows through the arithmetic unit. On the C&T chip are a pointer register and a word-select circuit which issue a word-select signal (WS) corresponding to the time slot being operated upon. The value of the pointer register can be set, incremented, and decremented by microinstructions.

Read-Only Memory

Preprogrammed mathematical routines are stored in three ROM chips, each of which contains 256 instructions of 10 bits each. A specific select code is assigned to each ROM chip. Only one of the three ROM chips is used at any time. When a ROM selection instruction is issued, a decoder inside each ROM checks the select-code field of the instruction. In case of a match, the selected ROM turns on. Unselected ROMs turn off.

A timing circuit on each ROM is synchronized to the SYNC signal issued by the C&T chip as the calculator is turned on. This circuit then keeps the ROM chip running synchronously with the rest of the system.

A ROM address register on each ROM chip receives the address sent out by the C&T chip. The corresponding instruction is placed on the instruc-

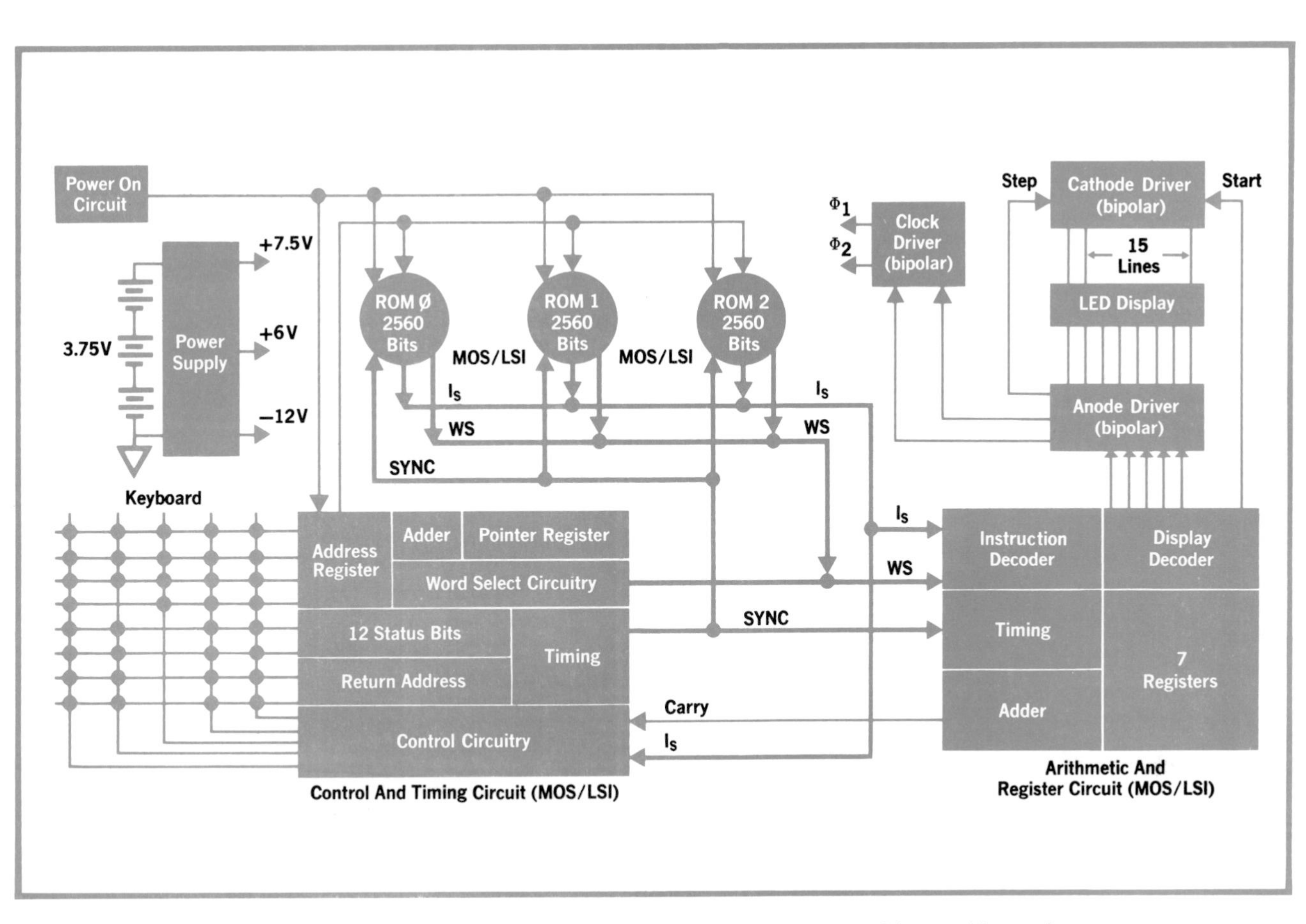

Fig. 3. *Five MOS/LSI circuits, developed by HP, are manufactured by outside vendors. Three custom HP-developed bipolar IC's are manufactured by HP. SYNC, I_s, and WS are three main bus lines connecting the MOS circuits.*

tion line, I_s, provided the ROM chip is turned on.

The ROM chip also issues word-select signals for certain classes of instructions.

Arithmetic and Register Circuit

The arithmetic and register circuit executes instructions coming in bit-serially on the I_s line. Most arithmetic instructions must be enabled by WS, the word-select signal. Data to be displayed is sent to the LED anode drivers on five lines, and one carry line transfers carry information back to the C&T chip. The BCD output is bidirectional and can carry digits into and out of the A&R chip.

The A&R circuit is divided into five areas: instruction storage and decoding circuits, a timing circuit, seven 56-bit registers, an adder-subtractor, and a display decoder.

Three of the registers are working registers. One of these and three of the remaining four registers form the four-register stack. The seventh register is an independent register for constant storage. There are numerous interconnections between registers to allow for such instructions as exchange,

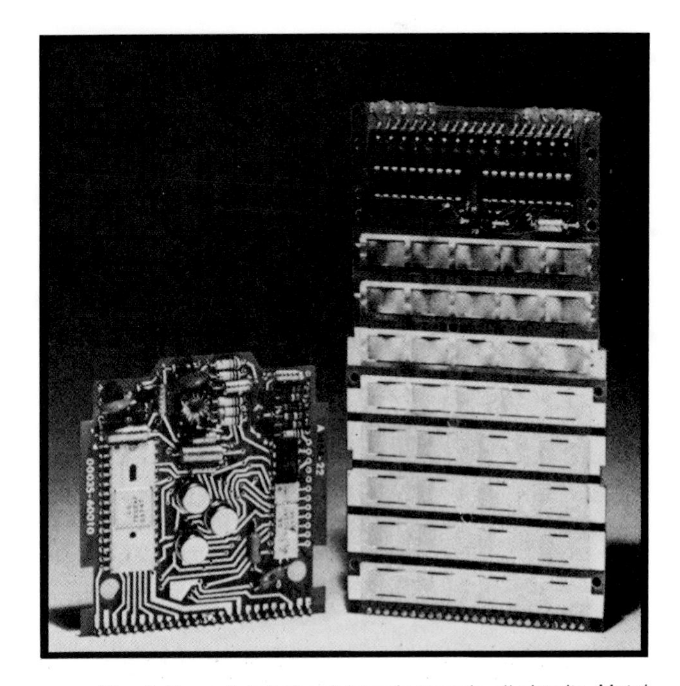

Fig. 4. *Two printed-circuit boards contain all circuits. Metal humps on the larger board are pressed down by the keys to make contact with printed traces underneath.*

How the HP-35 Compares with the Slide Rule

These comparisons were made by engineers who are not only highly proficient in slide rule calculation, but were also familiar with the operation of the HP-35. Thus, the solution times should not be taken as typical. They do, however, serve to indicate the relative time advantage of the HP-35 and to point up the still more significant advantage of its accuracy.

PROBLEM 1: COLLECTION SOLID ANGLE FROM A POINT SOURCE.

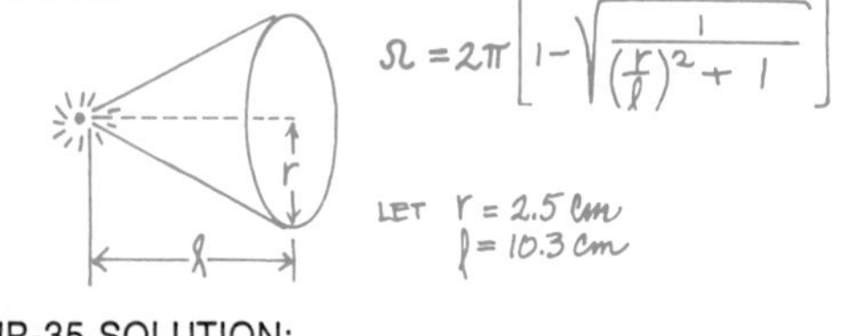

HP-35 SOLUTION:

2.5 [↑] 10.3 [÷] [↑] [×] 1 [+] [1/x] [√x] [CHS] [↑] 1 [+] 2 [×] [π] [×] → .1772825509

SLIDE RULE SOLUTION: .176

TIME ON HP-35: 20 seconds with answer to ten significant digits.

TIME ON SLIDE RULE: 3 minutes, 15 seconds with answer to three decimal places.

PROBLEM 2: GREAT CIRCLE DISTANCE BETWEEN SAN FRANCISCO AND MIAMI.

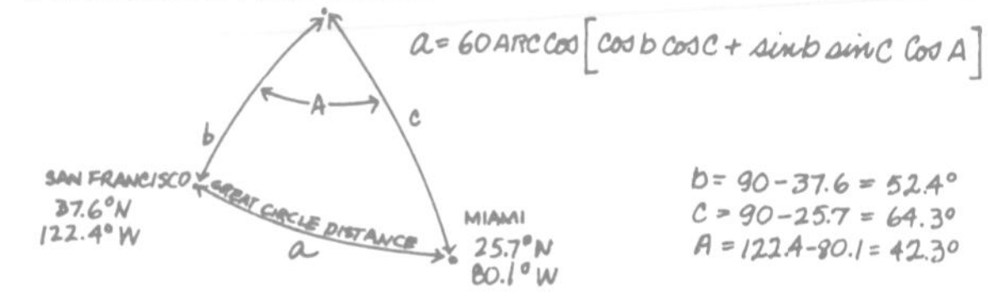

HP-35 SOLUTION:

52.4 [cos] 64.3 [cos] [×] 52.4 [sin] 64.3 [sin] [×] 42.3 [cos] [×] [+] [arc] [cos] 60 [×] → 2254.093016

SLIDE RULE SOLUTION: 2255 mm.

TIME ON HP-35: 65 seconds with answer to ten significant digits.

TIME ON SLIDE RULE: 5 minutes with answer to four significant digits.

PROBLEM 3: pH OF A BUFFER SOLUTION.

$$\alpha_H = 1 + \Sigma C_B K_A \,, \quad H^+ = \sqrt{\frac{1}{\alpha_H} \Sigma \frac{C_A}{K_A}}$$

FOR A MIXTURE OF Na_2HPO_4 @ 0.3 M/ℓ AND NaH_2PO_4 @ 8.7×10^{-3} m/ℓ

$$\alpha_H = 1 + [3\times10^{-2}](10^{7.21}) + [8.7\times10^{-3}](10^{2.16})$$

$$-H^+ = \text{LOG}\sqrt{\frac{1}{\alpha_H}\left(\frac{3\times10^{-2}}{10^{11.7}} + \frac{8.7\times10^{-3}}{10^{7.21}}\right)}$$

HP-35 SOLUTION:

7.21 [↑] 10 [x^y] .03 [×] 1 [+] 2.16 [↑] 10 [x^y] .0087 [×] [+] [STO] .03 [↑] 11.7 [↑] 10 [x^y] [÷] .0087 [↑] 7.21 [↑] 10 [x^y] [÷] [+] [RCL] [÷] [√x] [log] → −7.47877778

SLIDE RULE SOLUTION: 7.43

TIME ON HP-35: 65 seconds with answer to ten significant digits.

TIME ON SLIDE RULE: 5 minutes with answer to three significant digits.

transfer, rotate stack, and so on. An advantage of the bit-serial structure is that interconnections require only one gate per line.

Transfers into or out of the stack or the constant register are always whole-word transfers. All other arithmetic instructions are controlled by the word select signal, WS. Thus it's possible to interchange only the exponent fields of two registers, or to add any two corresponding digits of two registers.

The adder-subtractor computes the sum or difference of two decimal numbers. It has two data inputs, storage for carry or borrow, and sum and carry/borrow outputs. For the first three clock times, the addition is strictly binary. At the fourth clock time the binary sum is checked, and if the answer is more than 1001 (nine), then the sum is corrected to decimal by adding 0110 (six). The result is then entered into the last four bits of the receiving register and the carry is stored. A similar correction is done for subtraction. Carry information is always transmitted, but is recorded by the control and timing chip only at the last bit time of the word-select signal.

Simulation and Test of the MOS Circuits

In designing elaborate integrated circuits like the C&T, A&R, and ROM chips, two questions that have to be answered at the very beginning are: How is the design to be checked? and How is the final integrated circuit to be tested?

The first question has two answers. One is to build a breadboard and compare its operation with the desired operation. A second answer is a computer simulation of the circuit. When the MOS circuits (C&T, A&R, and ROMs) for the HP-35 were being designed, the computer simulation approach was chosen over a TTL or MOS breadboard. It was felt that the hardware breadboard wouldn't be an exact model of the final circuits anyway, and two or three months of development time could be saved by computer simulation because people could work in parallel rather than serially on a breadboard.

A general-purpose simulation program, HP-DABEL, had just been developed by Jim Duley of HP Laboratories. This was used to check out each gate, each circuit, each chip, and finally all the chips together. Each MOS circuit is designed as a network of gates and delay elements. For each gate output an algebraic equation was written as function of the inputs to the gate. This produced a large set of algebraic equations to be evaluated every clock time. A printout was available so the operation of any of the gates or delay outputs could be observed, as if with an oscilloscope probe. In this respect the computer simulation was much better than a hardware breadboard.

Because of the large number of equations to be evaluated each clock time the general-purpose simulation program was too slow to use for evaluating the algorithms implemented in the ROMs. For this a higher-level simulation was used, so only the input/output functions of each subsystem had to be specified. This was fast enough that all the algorithms could be checked, even the transcendental functions. If anything went wrong it was always possible to stop the program and step through it until the trouble was found. Correcting a problem was a simple matter of changing a punched card or two. These are advantages a hardware breadboard doesn't have.

The simulation approach proved very successful. It saved a lot of time not only in logic design, but also in generating the test patterns to be used for testing the final integrated circuits. After a simulation is running successfully a pattern for each input is specified such that virtually every circuit element will be exercised. By running the program and recording all the inputs and outputs a complete test pattern is generated. This is recorded on tape, ready for final test of the IC.

Fig. 5. *Light-emitting-diode cluster was specially developed for the HP-35. Magnifying lenses are built in.*

Display and Drivers

It was apparent early in the HP-35 planning that new display techniques would be required. Existing light-emitting-diode products used too much power and cost too much. HP Associates developed a magnified five-digit cluster which saves both power and cost and is packaged in a convenient 14-pin package (Fig. 5). Each digit has a spherical lens molded in the plastic over it. A slight reduction in viewing angle results, but for the handheld calculator this is not a problem.

LED's are more efficient if they are pulsed at a low duty cycle rather than driven by a dc source. In the HP-35, energy is stored in inductors and dumped into the light-emitting diodes. This drive technique allows a high degree of multiplexing; the digits are scanned one at a time, one segment at a time.

Customized bipolar anode and cathode driver circuits incorporating the required features were developed and are manufactured by HP. The anode driver generates the two-phase system clock and the segment (anode) drive signals, decodes the data from the arithmetic and register chip and inserts the decimal point, sends shift signals to the other axis of the multiplex circuitry (the cathode driver), and senses low battery voltage to turn on all decimal points as a warning that about 15 minutes of operating time remains. The cathode driver contains a 15-position shift register which is incremented for each digit position.

Keyboard

Requirements for the HP-35's keyboard were particularly difficult. The keyboard had to be reliable, inexpensive, and low-profile, and have a good 'feel'. The solution is based on the 'oilcan' or 'cricket' principle, that is, curved metal restrained at the edges can have two stable states. The larger board in Fig. 4 shows the etched metal keyboard strips running horizontally. At each key location the metal is raised to form a hump over a printed-circuit trace running underneath. Depressing a key snaps the metal down to make contact with the trace. The keys have a definite 'fall-away' or 'over-center' feel so that there is no question when electrical contact is complete.

Acknowledgments

The many people who contributed to the HP-35 did so with great energy and enthusiasm. There was a feeling throughout the project that we had a tremendous winner. Appreciation is due the particularly important contributors below:

Tom Osborne for the initial product definition and continued guidance on what a calculator was and for whom it was intended.

Paul Stoft for providing technical direction and an environment where wild ideas can flourish and for keeping our unbounded optimism in check.

Dave Cochran for initial system design, algorithm selection, and sophisticated and clever microprogramming.

Chu Yen for a super-efficient power supply and work on the recharger/ac adaptor, with the able assistance of Glenn McGhee.

Ken Peterson for the automatic logic board tester and for devising novel methods to test the elusive dynamic MOS circuitry.

Rich Marconi and Charlie Hill for the design of the display board tester and to Rich for patience through many redesigns of the PC boards.

Bill Misson and Dick Osgood for an inexpensive, reliable and producible keyboard.

Clarence Studley and Bernie Musch for creative and durable packaging and for not yelling, 'There's no more room' too often.

Jim Duley, Margaret Marsden, and John Welsch for assistance in computer programs for test patterns, simulation, and microassemblers.

Ed Liljenwall for exceptionally creative industrial design work including the improbable result that 35 easily operated keys can exist in a three-by-five-inch area.

Tom Holden and Neil Honeychurch for providing a close liaison with the manufacturing division to speed the design of the fabricated parts.

And lastly to our behind-the-scenes project leader, Bill Hewlett, who initiated the project and kept the fires burning whenever needed.

Thomas M. Whitney
Tom Whitney holds BS, MS, and PhD degrees in electrical engineering, all from Iowa State University, received in 1961, 1962, and 1964, respectively. With HP Laboratories since 1967, he has served as digital systems section leader and as section manager for the HP-35 Pocket Calculator. He's also a lecturer at Santa Clara University, currently teaching a course in microprogramming. Away from electronics, Tom spends as much time as possible outdoors, with skiing, tennis, and camping the major activities.

France Rodé
France Rodé came to HP in 1962, designed counter circuits for two years, then headed the group that developed the arithmetic unit of the 5360 Computing Counter. He left HP in 1969 to join a small new company, and in 1971 he came back to HP Laboratories. For the HP-35, he designed the arithmetic and register circuit and two of the special bipolar chips. France holds the degree Deploma Engineer from Ljubljana University in Yugoslavia. In 1962 he received the MSEE degree from Northwestern University. When he isn't designing logic circuits he likes to ski, play chess, or paint.

Chung C. Tung
Chung Tung received his BS degree in electrical engineering from National Taiwan University in 1961, and his MSEE degree from the University of California at Berkeley in 1965. Late in 1965 he joined HP Laboratories. He was involved in the design of the 9100A Calculator and was responsible for the design and development of two of the MOS/LSI circuits in the HP-35 Pocket Calculator: the control and timing chip and the read-only-memory chips. Now working for his PhD at Stanford University, Chung still manages to find time now and then to relax with swimming or table tennis.

Algorithms and Accuracy in the HP-35

A lot goes on in that little machine when it's computing a transcendental function.

By David S. Cochran

THE CHOICE OF ALGORITHMS FOR THE HP-35 received considerable thought. Power series, polynominal expansions, continued fractions, and Chebyshev polynominals were all considered for the transcendental functions. All were too slow because of the number of multiplications and divisions required. The generalized algorithm that best suited the requirements of speed and programming efficiency for the HP-35 was an iterative pseudo-division and pseudo-multiplication method first described in 1624 by Henry Briggs in 'Arithmetica Logarithmica' and later by Volder[1] and Meggitt[2]. This is the same type of algorithm that was used in previous HP calculators.

An estimate of program execution times was made, and it became apparent that, by using a bit-serial data word structure, circuit economies could be achieved without exceeding a one-second computation time for any function. Furthermore, the instruction address and instruction word could be bit-serial, too.

The complexity of the algorithms made multilevel programming a necessity. This meant the calculator had to have subroutine capability, as well as special flags to indicate the status and separations of various programs. In the HP-35, interrogation and branching on flag bits—or on arithmetic carry or borrow—are done by a separate instruction instead of having this capability contained as part of each instruction. This affords a great reduction in instruction word length with only a slight decrease in speed.

The arithmetic instruction set was designed specifically for a decimal transcendental-function calculator. The basic arithmetic operations are performed by a 10's complement adder-subtractor which has data paths to three of the registers that are used as working storage. Partial word designators (word select) are part of the instruction word to allow operating on only part of a number—for example, the mantissa or the exponent field.

Sine Algorithm

The sine routine illustrates the complexities of programming a sophisticated calculator. First, degrees are converted to radians by multiplying by $2\pi/360$. Then integer circles are removed by repeatedly subtracting 2π from the absolute value of the argument until the result is less than 2π. If the result is negative, 2π is added to make it positive. Further prescaling to the first quadrant isn't required. The resulting angle is resolved by repeatedly subtracting $\tan^{-1} 1$ and counting until overdraft, then restoring, repeatedly subtracting $\tan^{-1} 0.1$ and counting until overdraft, and so on. This is very similar to division with a changing divisor. Next the resulting pseudo-quotient is used as a multiplier. Beginning with an X vector of 1 and a Y vector of 0 a fraction of X is added to Y and a fraction of Y is subtracted from X for the number of times indicated by each multiplier digit. The fraction is a negative power of 10 corresponding to that digit position. The equations of the algorithm are:

$$\text{pseudo-division} \left\{ \theta_{n+1} = \theta_n - \tan^{-1} k \right.$$

$$\text{pseudo-multiplication} \begin{cases} X_{n+1} = X_n - Y_n k \\ Y_{n+1} = Y_n + X_n k \end{cases}$$

$$X_0 = 1,\ Y_0 = 0$$
$$k = 10^{-j} \qquad j = 0,1,2 \ldots$$

The pseudo-multiplication algorithm is similar to multiplication except that product and multiplicand are interchanged within each iteration. It is equiva-

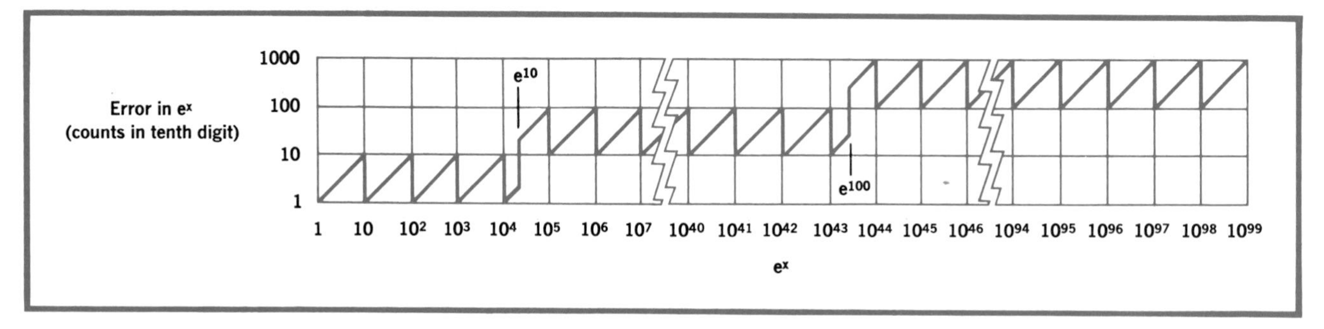

Fig. 1. *Accuracy of exponential function in HP-35 Calculator. Error bound is approximately δe^x, where δ is the error due to prescaling and the algorithm itself. δ is estimated to be equivalent to one count in the tenth significant digit of the argument x.*

lent to a rotation of axes. The resultant Y and X vectors are proportional to the sine and cosine respectively. The constant of proportionality arises because the axis rotation is by large increments and therefore produces a stretching of the unit circle. Since this constant is the same for both sine and cosine their ratio is identically equal to the tangent. The signs of each are preserved. The sine is derived from the tangent by the relationship

$$\sin\theta = \frac{\tan\theta}{(1+\tan^2\theta)^{1/2}}.$$

Accuracy and Resolution

Determination of the accuracy of the HP-35 is as complex as its algorithms. The calculator has internal roundoff in the 11th place. In add, subtract, multiply, divide, and square root calculations the accuracy is $\pm 1/2$ count in the 10th digit. In calculating the transcendental functions many of these elementary calculations are performed with the roundoff error accumulating. In the sine computation there is a divide, a multiply, and a subtract in the prescale operation, and there are two divides, a multiply, an addition, and a square root in the post-computation. Roundoff errors in these calculations must be added to the error of the basic algorithm to get the total error.

Accuracy and resolution are sometimes in conflict; for example, the subtraction of .9999999999 from 1.0 yields only one digit of significance. This becomes very important, for example, in computations of the cosines of angles very close to 90°. The cosine of 89.9° would be determined more accurately by finding the sine of 0.1°. Similarly, the sine of 10^{10} wastes all ten digits of significance in specifying the input angle, because all integer circles will be discarded.

For many functions there is no simple exact expression for the error. The exponential function is a good example. Let δ be the accumulated prescaling error and computational error in the algorithm, referred to the input argument x. Then for $\delta << 1$,

$$e^{x+\delta} - e^x = e^\delta e^x - e^x = e^x(e^\delta - 1) \approx \delta e^x.$$

Fig. 1 shows the error bound for the exponential function for various arguments, assuming that δ is equivalent to one count in the tenth significant digit of x.

References

1. Jack E. Volder, 'The CORDIC Trigonometric Computing Technique.' IRE Transactions on Electronic Computers, September 1959.
2. J. E. Meggitt, 'Pseudo Division and Pseudo Multiplication Processes,' IBM Journal, April 1962.

David S. Cochran
Dave Cochran is HP Laboratories' top algorithm designer and microprogrammer, having now performed those functions for both the 9100A and HP-35 Calculators. He was project leader for the HP-35. Since 1956 when he came to HP, Dave has helped give birth to the 204B Audio Oscillator, the 3440A Digital Voltmeter, and the 9120A Printer, in addition to his work in calculator architecture. He holds nine patents on various types of circuits and has authored several papers. IEEE member Cochran is a graduate of Stanford University with BS and MS degrees in electrical engineering, received in 1958 and 1960. His ideal vacation is skiing in the mountains of Colorado.

Packaging the Pocket Calculator

The industrial design of the HP-35 was of primary importance, often taking precedence over electrical considerations.

By Edward T. Liljenwall

THE INDUSTRIAL DESIGN OF THE HP-35 was unusual not only for Hewlett-Packard, but for the electronics industry in general. Usually, the mechanical and electrical components of a product are determined before the exterior is designed. The HP-35 took the opposite approach.

Since the calculator was to be pocket-sized, size was the overriding constraint on the design. In addition to size three other parameters were established. The calculator would have thirty to thirty-five keys, contain two or three batteries, and have a twelve-to-fifteen-digit LED display.

The industrial design began with an investigation of keyboard, packaging, and overall shape concepts. Several basic form factors were studied using sketches and simple three-dimensional models. The models were particularly valuable at this stage of development. They allowed a good evaluation of the shapes and sizes being considered.

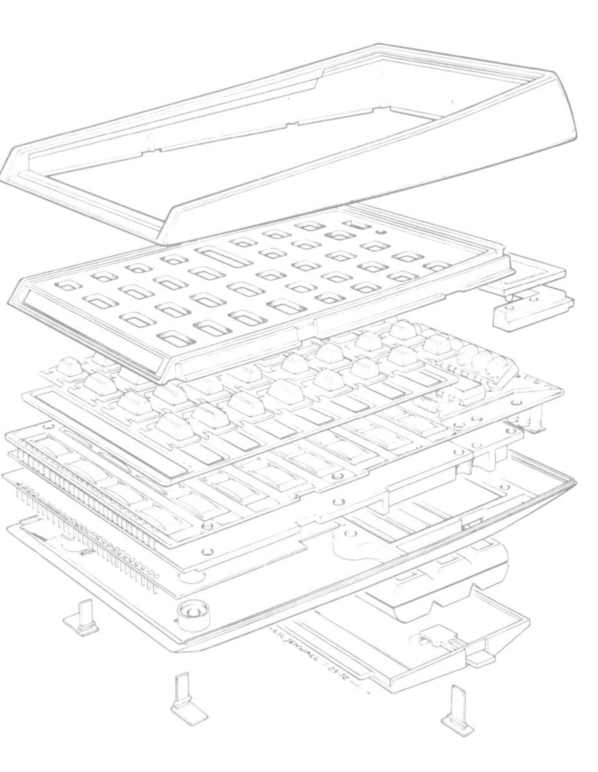

Once the preferred direction had been established a detailed model was built. The model was well accepted and approval was given to develop the concept into a reality.

Only a general idea of the electronic design was known at this point. Designing and packaging all necessary electrical and mechanical components into the tiny product became a tremendous challenge for electrical, mechanical, and industrial designers alike.

From a human-engineering standpoint, the keyboard was the most critical area of the design. The problem was to place thirty-five keys in an area approximately 2½ inches by 4½ inches and retain the ability to operate the keys without striking more than one at a time. It became apparent that the industry standard of ¾-inch center-to-center key spacing could not be maintained. A successful compromise was to use 11/16-inch center-to-center spacing for the numeric keys, and ½-inch spacing for all others. This was made possible by reducing the size of each key, thereby increasing the space between the keys.

The keys are divided into groups according to functions. The groups are separated by size, value contrast, color, and placement of nomenclature. The numeric keys, which are most frequently used, are larger and have the strongest value contrast. They have their nomenclature directly on the keys. The next group of keys according to frequency of use are identified by their blue color. The ENTER↑ key and arithmetic keys are separated within this group by placement of nomenclature on the keys. The less frequently used keys have the least value contrast, and the nomenclature is placed on the panel just above the key.

The keys have an over-center or snap feel when they are pressed. This comes from special spring contact developed by HP. The electrical and mechanical parts of the keyboard are less than 1/8 inch high.

The spring contacts are mounted on a printed-circuit board along with several other components including the fifteen-digit LED display. The display is tilted toward the operator for optimum viewing.

A second printed-circuit board carries the majority of the electronic components. The two boards are attached by a series of pin connectors.

The external package was developed from a human-engineering approach, with aesthetic appeal of major importance. The sculptural wedge shape permits the calculator to be comfortably held in the palm of one hand. It also allows the product to slide easily into a pocket. The keyboard and display slope upwards for a better viewing angle in desk-top use. The sculptural sides visually break up the total mass of the package. The top half of the case is highlighted while the bottom half is in shadow. This gives the product the appearance that it is thinner than it actually is. The product appears to be floating when viewed from a normal operating position in desk-top use. The use of textures that complement each other contribute significantly to the overall finesse and appearance. The texture on the case provides a non-slip surface, important when the calculator is being hand-held.

Viewed from the bottom, the calculator retains a clean appearance. There are no exposed screws or other fastening devices. This is aesthetically important to a product that is hand-held and viewed from all sides.

The polyvinyl-chloride feet prevent the calculator from sliding during desk-top use. The rear feet also serve as the battery-door latches, aiding in the overall cleanness of the product.

The HP-35 couldn't have been developed without an outstanding working relationship between laboratory, industrial design, manufacturing, and tooling. Everyone involved in the project shared a common desire to retain the original size and shape, and many innovative engineering concepts resulted. Many of the problems encountered during development could have been easily solved by using more conventional methods, but the result would have been a larger package. Now that the calculator is a reality, everyone feels that the extra efforts required were worthwhile.

SPECIFICATIONS
HP-35
Pocket Calculator

FUNCTIONS:

ARITHMETIC: Add, Subtract, Multiply, Divide and Square Root.

TRIGONOMETRIC: Sin x, Cos x, Tan x, Arc Sin x, Arc Cos x, Arc Tan x.

LOGARITHMIC: $\log_{10}$ x, $\log_e$ x, and e^x.

OTHER FUNCTIONS: x^y, 1/x, π and data storage and positioning keys.

SPEED OF OPERATION (typical):

Add, Subtract	60 milliseconds
Multiply, Divide	100 milliseconds
Square Root	110 milliseconds
Logarithmic & Exponential	200 milliseconds
x^y	400 milliseconds
Trigonometric	500 milliseconds

POWER:

ac: 115 or 230 V ±10%, 50 to 60 Hz, 5 watts.

Battery: 500 mW derived from Nickel-Cadmium rechargeable Battery Pack.

WEIGHT:

Calculator: 9 oz., Recharger: 5 oz.

Shipping weight: approx. 2 lbs.

TEMPERATURE RANGE:

OPERATING: 0°C to 40°C (32°F to 104°F).

STORAGE: −40°C to +55°C (−40°F to 131°F).

PRICE IN USA: $395, including battery pack, 115/230 V adapter/recharger, carrying case, travel case, name tags, operating manual.

Note: At press time, orders for the HP-35 have exceeded expectations to such an extent that a waiting list has been established. Deliveries should improve in the next few weeks.

MANUFACTURING DIVISION: ADVANCED PRODUCTS DEPARTMENT
10900 Wolfe Road
Cupertino, California 95014

Edward T. Liljenwall
Ed Liljenwall, industrial designer of the HP-35 Pocket Calculator, is a 1967 graduate of the Art Center College of Design in Los Angeles. He holds a BS degree in industrial design. Ed joined HP's corporate industrial design group in 1960 after two and one-half years as an automotive designer. As a result of his work on the HP-35, he's filed for several design patents. Away from HP, Ed likes to ski or water-ski, and is currently putting another hobby, woodworking, to good use in redecorating his home.

A Practical Interface System for Electronic Instruments

Connecting instruments into a digitally-controlled system now becomes a matter of plugging in cables. This article describes the interface system that makes this possible.

By Gerald E. Nelson and David W. Ricci

AS MODERN-DAY TECHNOLOGY gives rise to increasingly complex electronic devices, engineers are finding that to test these devices economically, they must resort to automatic systems. And, in automating their test procedures, engineers become more and more involved in instrument interconnection.

To overcome many of the problems experienced in interconnecting instruments and digital devices, a new interface system has been defined. This system gives new ease and flexibility in system interconnections. Interconnecting instruments for use on the lab bench, as well as in large systems, now becomes practical from the economic point of view.

The interface system evolved as part of the trend towards designing digital processing circuits into low-cost instruments. Now the functions formerly performed by the I/O circuit cards used in computers, calculators, and couplers are performed by the instruments themselves. Passive cabling ties the system components all in parallel into a common communications structure in much the same way that a computer or coupler backplane bus ties I/O cards together. The cabling is the only external part of the system. For reasons to be described later, the system is informally referred to as an ASCII-compatible interface bus system.

Interconnecting a variety of instruments, recorders, display devices, and calculators or computers with this new system is simply a matter of plugging in cables. Cable connectors can be stacked in parallel (Fig. 1) or the cables can be daisy-chained in any way that best fits the physical arrangement. System "overhead" is drastically reduced. Changing the system configuration is reduced to adding or replacing devices and reconnecting the cables; no modification of the existing instrument hardware is required.

A significant feature of the system is that there is complete flexibility of information flow. With all devices connected in parallel, any device can talk directly to any other (Fig. 2). A computer is not a necessity for managing the flow of information. One instrument can control another, or an

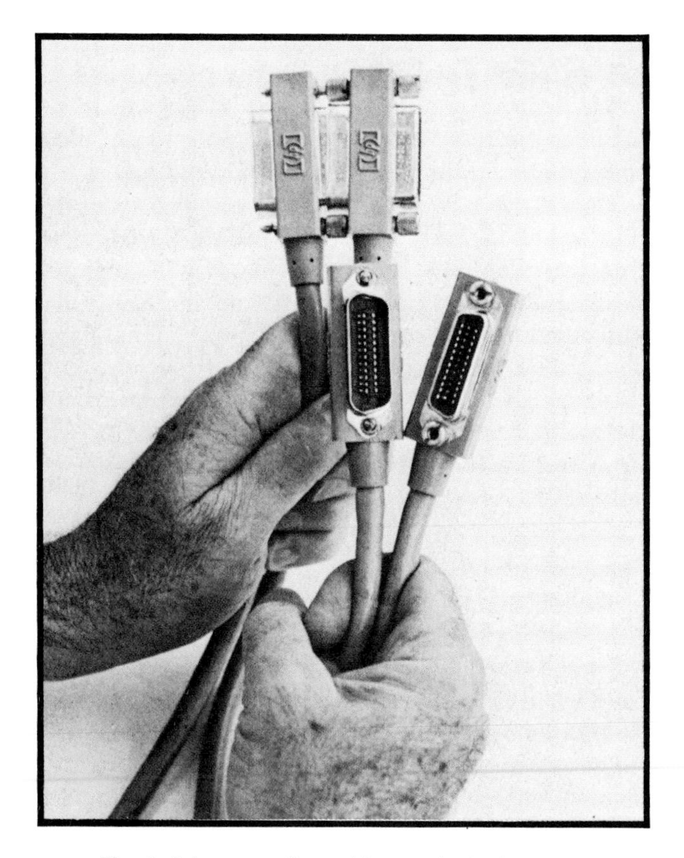

Fig. 1. *Interconnecting cables used with interface system have dual connectors. These can be stacked to accommodate variety of physical layouts by allowing more than one cable to be attached to any device.*

inexpensive card reader can control a group of instruments, making it economically feasible to automate instruments for bench or production use where cost has been a deterrent. Incorporating a calculator or computer into the system for more complex operations is also much simpler than ever before.

How Many Lines?

The interface system uses fifteen lines in the bus. Only one line might have been used with all data and control information in serial form, but this would have been far too complex to implement. On the other hand, enough lines to transmit everything in parallel, besides being incompatible with character-serial devices like teleprinters, would have been too cumbersome. For example, the HP Model 3330B Automatic Synthesizer would require about 100 lines for full parallel control.

The new interface system uses the byte-serial approach in which each multiple-bit byte of information is transmitted in parallel, and bytes are transmitted serially. The number of lines is not large enough to make the cabling cumbersome while the circuit requirements for byte-serial data transfer are modest.

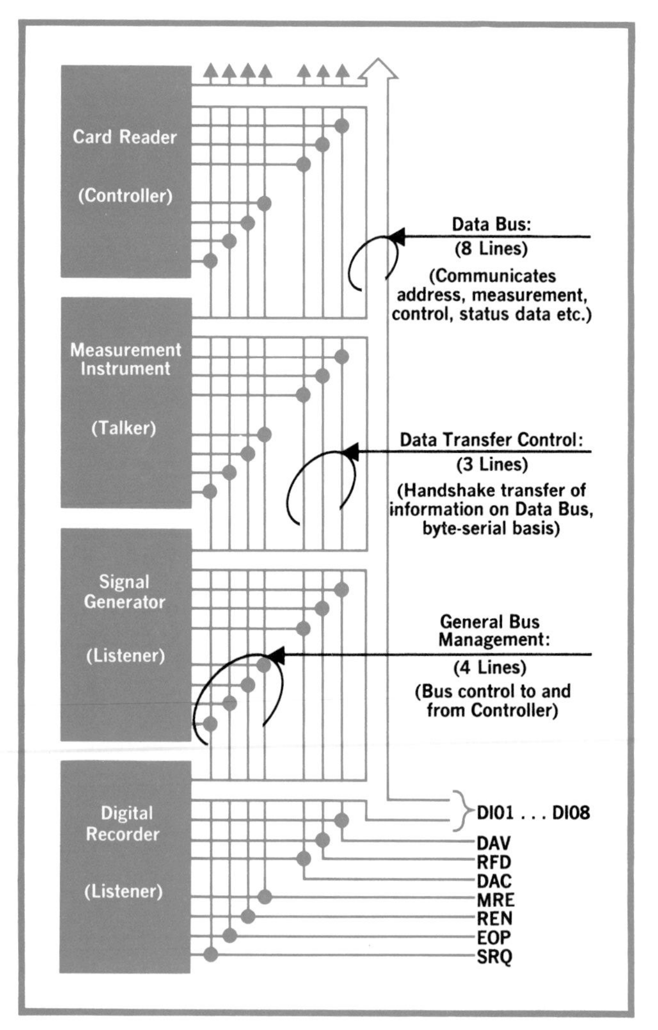

Fig. 2. *Typical interface bus system communications network. Information flow is bi-directional. Because of parallel connection, any device is potentially able to communicate directly with any other.*

The byte-serial approach has other advantages too. For example, with only 15 lines dc isolation of the lines to avoid ground loops is not an expensive undertaking. Furthermore, translation hardware and software costs are minimized since many devices, particularly those involved with the human interface, process data on a character-serial basis.

Line Assignments

The lines are summarized in Table I. Eight of the 15 lines are reserved for data input/output (DIO1 through DIO8), allowing data to be transferred in 8-bit bytes. Eight lines accommodate the 8 data bits of the standard teletype 11-unit code and they accommodate the widely-used 7-bit ASCII code, leaving one bit that can be used as a parity bit if

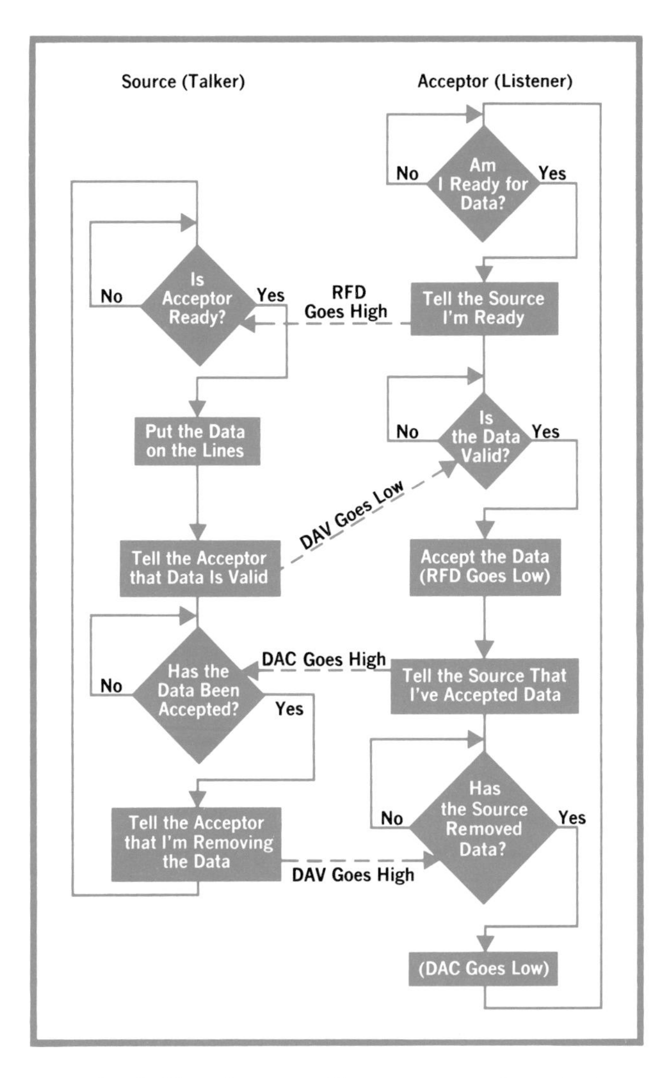

Fig. 3. *Flow diagram outlines sequence of events during transfer of data byte. More than one listener at a time can accept data because of logical-AND connection of RFD and DAC lines (see Fig. 4).*

it should be necessary to transmit data through a noisy channel. Also, most computers use word lengths that are multiples of 8 and there are software advantages in having bus bytes and computer words related in an integral ratio.

The data lines are also used for establishing the role of each device in the activity that is to follow. Each device can play one of three active roles: (1) talker, (2) listener, and (3) controller. A digital voltmeter is a talker, for instance, when it transmits data to a recorder but it becomes a listener when a card reader, acting as a controller, instructs it to change ranges. Unless a device is assigned a role, it remains inactive.

Some means of "ringing up" a device therefore is needed to establish its role. Presumably, one could set aside some of the data line code words as role addresses but with 8 bits there are only 256 different words. Because data transmitted by some devices might include the same words set aside as address words, it is desirable to not place restrictions on the codes carried by the bus.

Hence the new interface system has a control line called Multiple Response Enable (MRE). This line is driven only by the device that is currently designated as the Controller. When the Controller pulls the MRE line low, all other devices must listen to the data lines and only the Controller may then transmit codes. When MRE goes high again, only those devices that were addressed while MRE was low can use the data lines. All others remain inactive.

Responding to the Call

The MRE line, then, is an address mode/data mode selector for the data lines. When MRE is low, the codes on the data lines have the same meanings for all devices connected to the bus. When it is high, there are no restrictions on the codes used.

The codes used when MRE is low, which include certain commands as well as addresses, fall into four classes:

Code form	Meaning
X 0 0 A_5 A_4 A_3 A_2 A_1	Universal commands
X 0 1 A_5 A_4 A_3 A_2 A_1 (except X01 11111)	Listen addresses
X 1 0 A_5 A_4 A_3 A_2 A_1 (except X10 11111)	Talk addresses
X 0 1 1 1 1 1 1 and	"Unlisten" command
X 1 0 1 1 1 1 1	"Untalk" command

Note that the 6th and 7th digits indicate the meaning to be assigned to the information in the first 5 digits. The 8th digits (X) is not used when MRE is low so devices that use the ASCII 7-bit code are able to control the bus.

To allow more than one device at a time to function as a listener, any device becomes a listener when its listen address is placed on the bus (while MRE is low) and it remains a listener until the "unlisten" command is transmitted. Talkers, on the other hand, stop functioning as talkers whenever another talk address is put on the data lines (with MRE low). This prevents more than one device from talking at a time. (Talking is also terminated by the "untalk" command.)

Data Transfer

Information—whether addresses, commands, measurement results, or other data—is transferred on the data lines under control of a technique

Table I. Interface line assignments.

Line	Name	Abbreviation	Assertive State
1	Data Input/Output 1	DIO1	Low
↓	↓	↓	↓
8	Data Input/Output 8	DIO8	Low
9	Multiple Response Enable	MRE	Low
10	Data Valid	DAV	Low
11	Ready for Data	RFD	High
12	Data Accepted	DAC	High
13	End Output	EOP	Low
14	Service Request	SRQ	Low
15	Remote Enable	REN	Low

known as the three-wire handshake.

This technique gives a flexibility to the interface system that significantly simplifies the interconnection of a variety of instruments and digital devices. This flexibility results from three important characteristics. First of all, data transfer is asynchronous, that is to say there are no inherent timing restrictions. Data can be transferred at any rate suitable to the devices operating on the bus (up to a practical limit of 1 megabyte per second imposed by circuit response, cable delays, etc.).

Second, the handshake enables the bus to interconnect devices of different input/output speeds. Data transfer automatically adjusts to the slowest active device without interfering with the operation of faster devices.

Third, more than one device can accept data at the same time. Thus several devices can listen simultaneously. This capability enables all devices to accept addresses and commands when MRE is low.

The handshake sequence is diagrammed in Fig. 3. A listener indicates that it is ready to accept data by letting the Ready for Data (RFD) line go high. Listeners are connected to the RFD line in a logical-AND configuration, however, so that the line does not go high until all active listeners are ready for data (Fig. 4).

RFD going high signals the talker that all listeners are ready for data. The talker indicates that it has placed a data byte on the data lines by setting the Data Valid (DAV) line low, but it cannot do so unless RFD is high. When DAV does go low, listeners are enabled to accept the information on the data lines.

When a listener has accepted the data (stored it in a register), it lets the Data Accepted (DAC) line go high. Here again, all active listeners are connected in a logical-AND configuration so DAC does not go high until all listeners have accepted the information. DAC remains high until the talker sets DAV high again. The sequence may then repeat

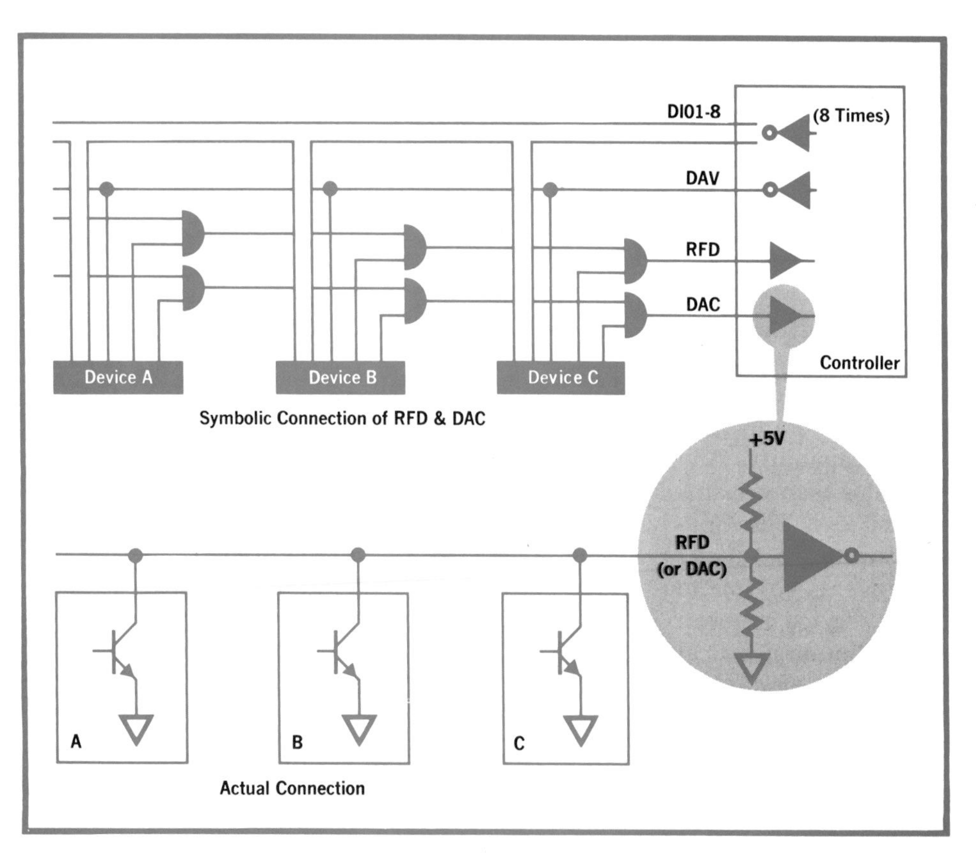

Fig. 4. *Logical-AND connection of RFD and DAC lines prevents these lines from going high until all active devices have set their outputs high (because of AND-gate configuration, assertive state for these two lines only is high).*

after DAC goes low.

No step in this sequence can be initiated until the previous step is completed. Thus, information can proceed as fast as devices can respond, but no faster than the slowest device that is presently addressed as active.

Assuming Command

The interface allows a system to have more than one controller but only one controller at a time can manage the system. For example, two instruments may work together to perform certain measurements but they might be connected to a calculator to give more sophisticated results. The calculator could instruct one of the instruments, addressed as a listener, to execute a certain subroutine. The calculator would then pass control to that instrument by using an appropriate command word. When the subroutine was completed, the controlling instrument would then pass control back to the calculator with another command word. This arrangement reduces the number of programming steps needed for the calculator.

One of the controllers must be able to gain absolute control of the bus without being addressed itself, a capability that is needed to allow an operator to initiate action in the system. This controller, designated as System Controller at the time of system configuration, has control of another line called End Output (EOP), and it is the only device in the system to have control of this line. When the System Controller pulls EOP low, all bus activity stops immediately and all devices unaddress themselves. Besides being useful for system start-up, EOP can be used if the operator wishes to abort an automatic sequence.

Asking for Attention

Suppose a listener or a non-participant wants to become a talker or otherwise gain access to the bus —a voltmeter, for example, that wants to indicate an overrange, alarm, or error condition. A status line, Service Request (SRQ), gives this capability. Any device can indicate to the Controller that it wants attention by pulling the SRQ line low. The Controller then responds at its discretion, most likely permitting the current cycle of events to run to completion before acting on the service request. Devices are connected to the SRQ line in a logical OR configuration so that any number of devices can pull down on the line at the same time.

When the Controller responds to SRQ, it enters a service-request identification cycle. It begins by setting MRE low and then transmitting an identify-service-request command. All devices accept this command and when MRE goes high again (following the handshake cycle), those that have service requests pull down on one of the data lines, each device being assigned a particular line during system configuration. At the same time, the devices let the RFD line go high. When RFD goes high, the Controller reads the data lines to determine which devices asked for service and it takes whatever action it has been programmed to take.

Remote/local

The bus has one more control line: Remote Enable (REN). When this line goes low, control of each instrument's functions is transferred from the front-panel controls to the interface bus. This line can be pulled low only by the System Controller and gives the operator the ability to switch all instruments to remote control from a central location. Setting REN high restores local control.

The System Controller can also switch instruments to bus control by transmitting a switch-to-remote code, one of the universal commands. With this arrangement, a front-panel pushbutton, if the instrument is so equipped, can restore local control to that instrument, allowing the operator to take over control at any time. The Controller can return all devices to local control with a switch-to-local code.

Preferred Codes and Formats

Although any code of 8 bits or less can be used on the data lines when MRE is high, it is highly desirable that devices which input or output decimal numbers do so in a manner compatible with calculators or computers. The use of a common code also minimizes the need for code translation.

An example of the problems that arise with nonstandard codes or formats concerns one of the early instruments designed for use with the new interface system. To simplify internal logic, this instrument used the ASCII colon (:) for a radix point instead of the usual period (.). Then it was found that a calculator would have to output a string such as <12:62> by breaking the number into an integer part and two separate digits, increasing the programming time. The instrument has now been modified to accept a period as the radix point.

A code set easily generated and understood by standard software in computers and calculators is the "printing ASCII," so-called because the 64 characters used, out of the possible 128 words available with 7 bits, are those used by teletypes. Hewlett-Packard instruments designed for use with the new interface system use this code, hence the reference to the bus as ASCII-compatible.

Hardware

Driver and receiver circuits connected to the bus operate at typical TTL voltage levels (high state greater than 2.4 V and low state less than 0.4 V). Driver circuits are typically open-collector NPN transistors capable of sinking more than 48 mA. Tri-state drivers may be used on the data input/output lines where high data rates are desired. Receiver circuits are standard TTL gates.

A 24-pin connector accommodates all lines, using single conductors for data input/output lines, twisted pairs for control lines, and a single shield surrounding all lines.

Interface System Summary

Number of Devices:
15 per system maximum

Signal Lines:
15 assigned (8 data, 7 control)

Data Rate:
1 Megabyte per second maximum

Data Transfer:
Byte serial, bidirectional using interlocked handshake technique

Transmission Path:
50 feet total accumulated cable length maximum

Address Codes:
31 Listen addresses
31 Talk addresses

Control:
May be delegated, never assumed; maximum of 1 talker and 14 listeners.

Hewlett-Packard instruments that can be used with the new interface system include (although these instruments are available now, in some cases interfaces will not be available until dates shown):

Signal Sources:	3320A/B Frequency Synthesizers 3330A/B Automatic Synthesizers 8660A Synthesized Signal Generators (Spring '73)
Measuring Instruments:	3490A Digital Voltmeter (Winter '72) 3570A Network Analyzer 5340A Automatic Counter
Controllers:	3260A Marked Card Reader 9820A Calculator (with 11144A Interface Kit) 2100A Computer (with 59310A Interface Kit—Summer '73)

Gerald E. Nelson
Jerry Nelson, fresh out of the University of Denver with a BSEE, joined the HP Loveland Division in 1964. Initially he worked on the 675A Sweeping Signal Generator. He then became project leader for the 3330A/B Automatic Synthesizer for a time before moving on as group leader for the 3570A Network Analyzer and related systems. These activities involved him in the definition of the new interface system.

Along the way, Jerry earned his MSEE degree at Colorado State University in the HP Honors Co-op Program. During leisure hours, he likes to sail but he also cares for a 2½-acre cherry orchard. What to do with all those cherries? He makes wine.

David W. Ricci
Dave Ricci has been involved in digitally interfacing instruments almost from the moment he joined HP's Santa Clara Division in early 1965. Once that work on his first project, the 5431A Display Plug-in for HP's Multichannel Analyzer, was completed Dave started on digital peripheral interfaces for the Multichannel Analyzer. Of late he has been coordinating interface efforts for all of HP's Counters.

A third-generation Californian, Dave obtained his BSEE at California State Polytechnical College and, following a stint with another company working in communications, earned an MSEE degree at the University of California at Berkeley. In his spare time, Dave and his wife like to ski and sail. They also like to backpack, taking along their dog, a friendly Samoyed, who carries her own pack.

A Common Digital Interface for Programmable Instruments: The Evolution of a System

HP's corporate interface engineer describes the trends, philosophy, and ancestors that have helped define the new HP instrument interface system.

By Donald C. Loughry

INSTRUMENTATION SYSTEMS CAN'T WORK WELL UNLESS their separate elements—instruments, computers, peripherals—can communicate effectively with one another. There must be a way to tell an instrument (perhaps a digital voltmeter) what to do (go to the 10 Vdc range) and when to do it. There must also be a way for the instrument to tell what it has accomplished (DVM reading is +9.765 Vdc). That these mainly digital messages be communicated clearly and conveniently serves the common interests of design engineers, instrument manufacturers, and system users alike.

Recognizing this, representatives of various Hewlett-Packard divisions began about eight years ago to discuss what techniques might be appropriate for a digital interface system that would be applicable to most HP programmable instruments. One such general-purpose system has now been developed. It is described in detail in the article beginning on page 2. In this article I will summarize what has happened in the last few years to influence the evolution of this system.

Where We've Been

The earliest programmable instruments were programmed by contact closures. Frequently this was accomplished on a line-per-function basis where each program input line was dedicated to a specific task (i.e., one program line for each range or function). Program input lines typically did not have storage. Inputs were usually generated by means of simple, economical, mechanical switches, or later, solid-state switches. In either case, the logic convention or assertion state was "ground true."

Unlike program inputs, measurements results were usually output in coded form (e.g., two's complement binary or binary-coded decimal). The mathematical perspective of engineers, along with ease of interpretation, usually led to the "positive true" assertion state for measurement data. Thus, input and output logic conventions were frequently dissimilar. This presents no particular problem when separate input and output lines are used, but the interface circuitry becomes costly when there is a direct connection between inputs and outputs and each uses a different logic convention.

Separate unidirectional input and output lines for each instrument, the absence of program storage capability, inconsistent codes and formats, unique control techniques for each instrument, and liberal use of uncoded line-per-function program control signals are a few of the parameters that characterize the instrument interface environment from which we are now emerging.

Tracing the Evolution

Looking back at specific hardware, one can see the changes in interface techniques that were implemented as problems were recognized and experience was gained.

The history of today's digital interface methods at Hewlett-Packard begins with the "Adam and Eve" of instrumentation systems, a frequency counter output coupled to a digital recorder input. Digital information was presented in parallel form, that is, all data bits for a complete frequency measurement were presented simultaneously. Information flow was unidirectional: the first frequency counters did not have programmable inputs, and it wasn't necessary to output data from the digital recorder. The marriage of these two products back in the mid-1950's proved highly successful. The event sparked much activity, and digitally coupled instrumentation systems were born at HP.

Within five years digital interfaces were commonplace. Digital voltmeters, scanners, power supplies, remote displays, and other products were added to the growing family. By this time some products, digital voltmeters in particular, required both digital inputs and outputs. The voltmeter not only generated measurement data but was programmable as well. This was another significant event in the interface world.

Along with this new capability came the need to program many products within one system. The solution was to create a special purpose controller, backed up by a paper tape or punched card memory. The interface network formed a radial or star structure as shown in Fig. 1. Most of these systems were custom-designed mixtures of standard instruments and special products. While the number of system elements expanded rapidly, each digital interface remained unique with little similarity to others.

The next interface milestone saw the introduction of small, stand-alone, relatively low-priced, standard instrument systems with program control capability internal to one of the products. These systems forced the start of more commonality among the interface characteristics. Programmable inputs needed to have common signal levels and logic conventions. Attention was focused on common code structure and data format.

Fig. 2 illustrates the typical standard data acquisition system in vogue at this time. Information flow was still essentially unidirectional and each signal line was dedicated to carry only one type of information.

To correct some of the format and code translation problems, additional coupler units were developed to interact with various recording media. A different coupler was required for each output recording medium (e.g., magnetic tape, punched cards, punched paper tape, hard copy, displays, etc.).

A breakthrough occurred in the late 1960's as the marriage of computers and instrumentation systems became a reality. Overall system capability soared, data rates climbed significantly, and message traffic volume increased dramatically over the system interfaces.

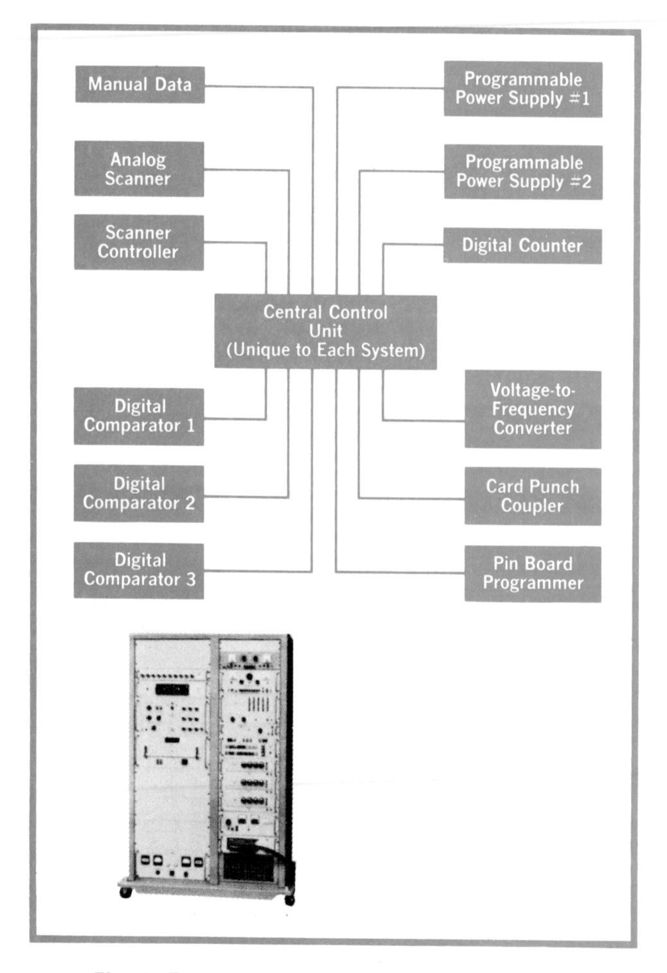

Fig. 1. *Early instrumentation systems had a special control unit and a "star" interface pattern.*

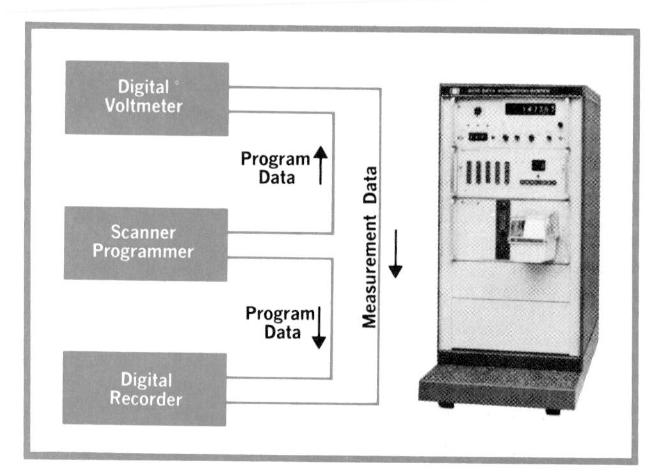

Fig. 2. *Later, small stand-alone systems had the control function within one of the instruments.*

In these systems, control was vested in one location, the central processor mainframe. The interface hardware, particularly the input/output circuits, approached uniformity as a result of the economic demands of the CPU design and the wide availability of solid-state microcircuits. The demands of system software led to a reduction in the variety of instrumentation program codes and data formats.

Bus techniques were frequently used to interconnect internal sections of the CPU mainframe, and

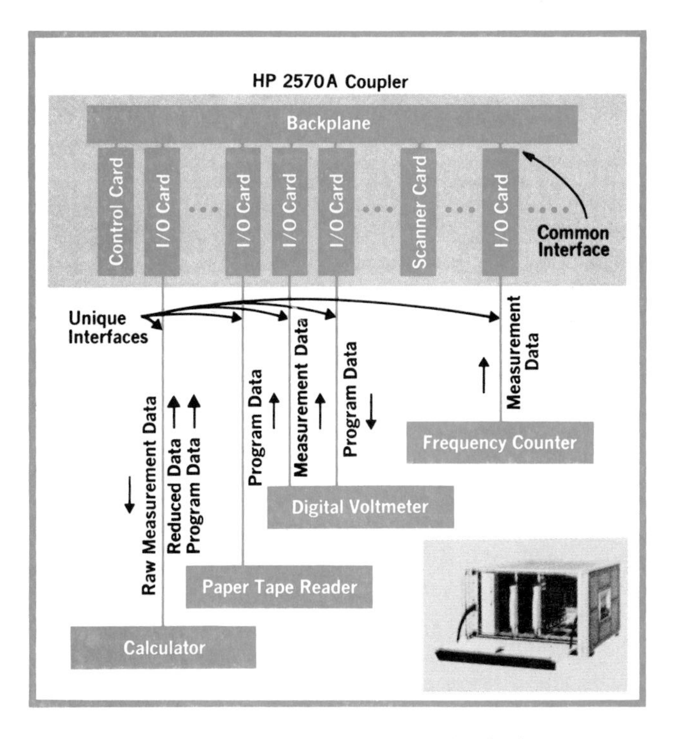

Fig. 3. *A recent instrumentation coupler is the forerunner of the new general-purpose interface system now being introduced into HP instruments. Individual instruments communicate in character-serial format via a bidirectional backplane bus. Control can be delegated to different devices at different times.*

some of these same capabilities were extended to the instrument interface in the form of word serial (16-bit parallel) data transfers and more bidirectional interfaces. The overall interface network, however, remained a star pattern with each instrument connected to the mainframe processor by its own cable and plug-in I/O card.

The plug-in I/O card and backplane structure used in processor mainframes gave rise to a new coupler philosophy that came into existence around 1970. Fig. 3 shows the coupler as the data interchange hub for a data acquisition system. The backplane of the coupler is a bus structure for a closely interrelated collection of I/O cards, one to each instrument or device.

This system concept represented the beginning of an instrument bus structure. Looking at the interface backplane, data rates were high, information flow was bidirectional, control could be delegated to different elements of the system, codes and formats were standardized. At the backplane side of the I/O cards the interface circuits were identical. However, the individual instruments still retained their unique interface characteristics.

In many respects this collected bus system was the immediate predecessor to what has been described in the preceding article: a distributed bus interface system where similar functions and capability exist without being concentrated on one backplane.

Where We're Headed

Turning from the past to the present and future, we can identify several trends. For one, serial interface techniques are in common use throughout the data processing field in devices such as teleprinters, line printers, and magnetic tape units. These products, as well as calculators and computers, are now frequently used in instrumentation systems. Programmable instruments are therefore required to interface with many devices which use some form of byte-, character-, or word-serial interface. Even simple program input devices like paper tape and mark-sense card readers are character-serial by nature.

Remote data collection stations, interaction with time-shared systems, and the frequent use of common-carrier communication facilities are also on the increase. All of these trends seem to point to a serial interface.

Another trend is toward intelligent instruments, for which a few bytes of input data define the limits for an entire series of events. No longer is it necessary to program each discrete action.

These trends help create an environment conducive to a common interface system. There are other trends that will influence the character of that system.

One of these is towards decreasing instrument response times as instrument capability responds to the need for greater data rates in automatic measurement systems. Greater instrument data rates tend to increase the amount of message traffic that must be carried by the interface.

While the trend toward smart instruments is decreasing the need for programming data, the number of programmable parameters is increasing. It is not unusual to have all front-panel controls available for digital remote programming. This means that any interface system must provide for an increasing repertoire of messages.

Now Is the Time

With these trends both demanding and helping to create more effective interproduct communications, is there reason to believe that a common interface system is feasible at this time? The answer is yes, there are many reasons.

One key fact is the availability, capability, and variety of inexpensive digital integrated circuits.

Interface circuit costs no longer inhibit the addition of storage (20¢ per bit), conversion between serial and parallel data (10¢ per bit), interface control logic, and the driver/receiver circuits necessary in any interface system.

The advent of intelligent instruments has decreased interface message requirements, thereby increasing the feasibility of a serial interface system. General acceptance of a common message convention (polarity or logic assignment of binary data), now a virtual reality within HP, has appreciably aided interface feasibility.

Not so obvious is the experience gained over the last few years. The last five years in particular have taught us many hard lessons and made us better able to develop an effective interface system.

Perhaps the most significant factor that makes an effective interface system feasible today is acceptance of the fact that no single interface system either can or needs to solve all problems! Instrumentation systems tend to be in the middle of the spectrum in terms of such interface parameters as data rates and path widths, data transmission lengths, number of interconnected devices, and message traffic volume. Defining an interface system for this middle range now appears possible. Interfaces for special environments such as extremely long transmission paths and electrically noisy environments are indeed needed, but less frequently. Overall costs are reduced and system flexibility is increased if individual instruments are not required to meet all of these special needs.

Interface System Objectives

As a first step toward developing an effective interface system it is helpful to identify the most important objectives and design criteria. Of many objectives, four stand out. The interface system must be:

1) capable of interconnecting small, low-cost instrument systems by means of simple passive cable assemblies without restricting individual instrument performance and cost.
2) capable of interconnecting a wide range of products (measurement, stimulus, display, processor, storage) needed to solve real problems.
3) capable of operating where control and management of the message flow over the interface is not limited to one device but can be delegated in an orderly manner among several.
4) capable of interfacing easily with other more specialized interface environments.

Additional criteria to consider would include at least these: operation under asynchronous conditions; communication with two or more devices simultaneously; limitation of the total number of signal lines to no more than can be accommodated in one computer word; compatibility with the most widely used codes for information interchange; ability to alter codes and data rates to achieve system flexibility; ability to alter the communication network to optimize system performance.

The byte-serial, bus-oriented interface system described in the article on page 2 is one that meets these criteria. This interface system is now being used in many new HP products.

Acknowledgments

The present interface system developments are the result of corporate-wide vision, commitment, diligence, and enthusiasm. Though the list of engineers is long, Gerald E. Nelson and David W. Ricci made major contributions. The patience, understanding, and constructive commentary from customers has had its positive influence as well. The efforts of all are hereby acknowledged.

Donald C. Loughry

Don Loughry started out at HP in 1956 as a production test department manager. In 1958 he moved to the development laboratory to work on custom and standard systems, and in 1963 became engineering manager of the Dymec Division. He's held his present position of corporate interface engineer since 1968. In that capacity, he's responsible for companywide interface guidelines and services. He's a member of IEEE and is active in various groups working on interface standards, including IEC, ISO, and ANSI.

Don is a 1952 B.S.E.E. graduate of Union College. When he's not busy interfacing, his interests vary widely, they include gardening, photography, classical music, skiing, camping, and star gazing (he grinds his own telescope lenses). He's also active in his church and in interfaith groups.

An Economical Full-Scale Multipurpose Computer System

This is the first 16-bit computer system to have a hardware stack architecture and virtual memory. It handles time-sharing, batch processing, and real-time operations in several languages concurrently.

by Bert E. Forbes and Michael D. Green

THE HP 3000 COMPUTER SYSTEM is Hewlett-Packard's first full-scale multipurpose computer system. Its primary objective is to provide, at low cost, a general-purpose* computer system capable of concurrent batch processing, on-line terminal processing, and real-time processing, all with the same software. Many users can access the system simultaneously using any of several programming languages and applications library programs.

The HP 3000 (Fig. 1) is an integrated software-hardware design. It was developed by engineers *and* programmers to provide a small computer capable of multiprogramming. Unlike many computers of the past, it was not built by the engineers and turned over to the programmers to see what they could do with it.

Helping define the objectives for the HP 3000 was HP's long experience with both customer and internal use of 2100-Series Computers and 2000-Series Time-Shared Systems. These computers and systems have been widely used in educational, instrumentation, industrial, and commercial applications. These are also expected to be the primary applications areas for the HP 3000 (see page 7).

A comprehensive set of software and the hardware to support it has been developed for the HP 3000. Software includes the Multiprogramming Executive operating system, several programming language translators, and an applications library.

Architectural Features

The scope of the software for the HP 3000 requires certain capabilities in the computer on which the software is to run. Among these are efficient program segmentation, relocation, reentrancy, code sharing, recursion, user protection, code compression, efficient execution, and dynamic storage allocation. All are provided in the HP 3000 design.

Efficient program segmentation makes it possible to run programs which are much larger than the available memory without incurring a large overhead. Much of the power and flexibility of the HP 3000 comes from the *virtual memory* that results

*"General-purpose" means that a user is not restricted to a single application, but can readily write programs to fit his own application, whatever it might be.

Fig. 1. *HP 3000 Computer System has multilingual and multiprogramming capabilities usually found on much larger systems.*

from the segmentation capabilities of the system.

Swapping of programs and data is made easier by an automatic **relocation** technique that is part of the addressing structure of this multiprogramming computer. The operating system doesn't have to take time to adjust all addresses in a program, nor does it have to put something in the same physical location every time.

Reentrancy is a property of HP 3000 code. It makes it possible for a given sequence of instructions to be used by several processes without having to be concerned about the code being changed or temporary variables being destroyed by the other processes.

Automatic relocation and reentrancy make **code sharing** possible. It would be extremely wasteful of main memory to keep multiple copies of programs in memory. In the HP 3000, one copy of a program can be shared by many processes.

Another consequence of reentrancy is **recursion**, or the ability to have a routine call itself. The hardware stack architecture of the HP 3000 plays an important role in recursive calls.

One of the key items in designing a multiprogramming operating system is that of **user isolation and system protection.** If the operating system and the users are not completely protected from the intentional or unintentional destructive actions of another user, the system will crash so often as to be unusable. HP 3000 protection covers programs, data, and files that exist in the system.

In a small-word-size machine, the amount of addressable memory is limited. To take full advantage of it, the HP 3000 has **dynamic storage allocation.** All temporary and local variables are assigned physical memory only when needed at procedure or block entry and are deallocated upon exit.

The HP 3000's unified real-time, terminal-oriented, and batch environment is accomplished without the use of fixed or variable memory partitions. Instead, priorities are used to control system resources. Partitioning, it was felt, places arbitrary restrictions on memory, the most valuable resource in a multiprogramming system.

Multiprogramming Executive Operating System

The HP 3000 has a single operating system called the Multiprogramming Executive (MPE). MPE is a general-purpose system that can handle three modes of operation concurrently. In time-sharing, one or more users can interactively communicate with the system via computer terminals. In batch processing, users can submit entire jobs to be performed by the system with no interaction between the system and the user. In real-time operations, tasks are dependent upon the occurrence of external events and must be performed within critical time periods.

MPE also provides many services to users, such

as input/output handling, file management, memory management, and system resource allocation and scheduling.

There are many advantages to an operating system of this sort. For example, subsystems (compilers, applications programs, etc.) need not be customized for different operating systems or configurations. In fact, the same subsystem can be used concurrently by an interactive user and by a batch user. Another advantage is that software can be generated much more efficiently because the operating system already performs many of the more difficult tasks. Also, with a single operating system, it is easier to attain consistency throughout the various software subsystems, thereby simplifying the user/system interface.

Although there is only a single operating system on the HP 3000, it can be adapted to operate in a number of different hardware configurations, each tailored to the needs of its users (Fig. 2). Thus, one installation may run only small batch processing jobs using a card reader and a line printer, while another may add a number of time-sharing terminals. The same software is used by all these installations.

Programming Languages

Several different programming languages have been developed for the HP 3000. Most important of these is SPL, the Systems Programming Language. This is a higher-level programming language designed specifically for systems programming. Almost all of the HP 3000 software has been developed using SPL, the few exceptions being some of the applications programs.

The reasons for using a higher-level language rather than an assembly language for systems programming are much the same as those for using a higher-level language for applications programming. It's possible to write and debug programs more quickly, to modify them more easily, and to make them more reliable and easier to read and understand. Furthermore, programs often perform better because more time can be spent on general methods than on coding details. Improving programs by rewriting substantial sections of code is not distasteful, as it often is when programs are written in assembly language. In general, in a given period of time much more software can be developed by using SPL than by using a lower-level assembly language.

Since SPL is designed for HP 3000 systems programming, it was necessary to give SPL programmers easy access to all the features of the central processor. For this reason, much of the syntax is based on these features, and the machine code generated by the compiler is related in an obvious way to the higher-level statements in the language.

Other programming languages which have been developed for the HP 3000 are FORTRAN, COBOL, and BASIC. SPL, BASIC, and FORTRAN are all recursive, that is, programs, procedures, and subroutines can call themselves. HP 3000 software also includes scientific and statistical applications program libraries, and text editing and formatting facilities.

Program Environment

Traditionally, 16-bit computers have been von Neumann-like machines with little or no distinction between program code and data. In a multiprogramming environment there is much to be gained from separating the two. In the HP 3000, a typical user's environment consists of one or more program code segments and a data segment (Fig. 3). All code is nonmodifiable while active in the system. Overlay techniques can therefore be used (that is, new code can be written over old code) without having to write the old code back out on the swapping disc, since an exact copy already exists there. The data area consists of global data (data common to several procedures) and a push-down stack that is handled automatically by the hardware.

Code Segmentation

In the HP 3000, code is grouped into logical entities called segments, each consisting of one or more procedures. Each segment may be up to 16K words long. Programs are normally broken into segments by the user, although he may choose not to do this and his program will run as a single segment unless it is too large, in which case an error message will be generated.

There is a master directory, the Code Segment Table (CST), that contains one entry for each segment that is currently active on the system. The CST is maintained by the operating system and is used by the central processor for procedure entry and exit. The table doesn't occupy a fixed position in memory but its address is always stored in absolute location 0.

Each two-word CST entry contains the beginning address and the length of the code segment. There are also four bits that are used by the central processor. One of these, the reference bit, is used to implement a software least-recently-used overlay algorithm. Another, the trace bit, causes a procedure call to the trace routine if set. The mode bit specifies whether the segment will be run in privileged or user mode. The absent-from-main-memory

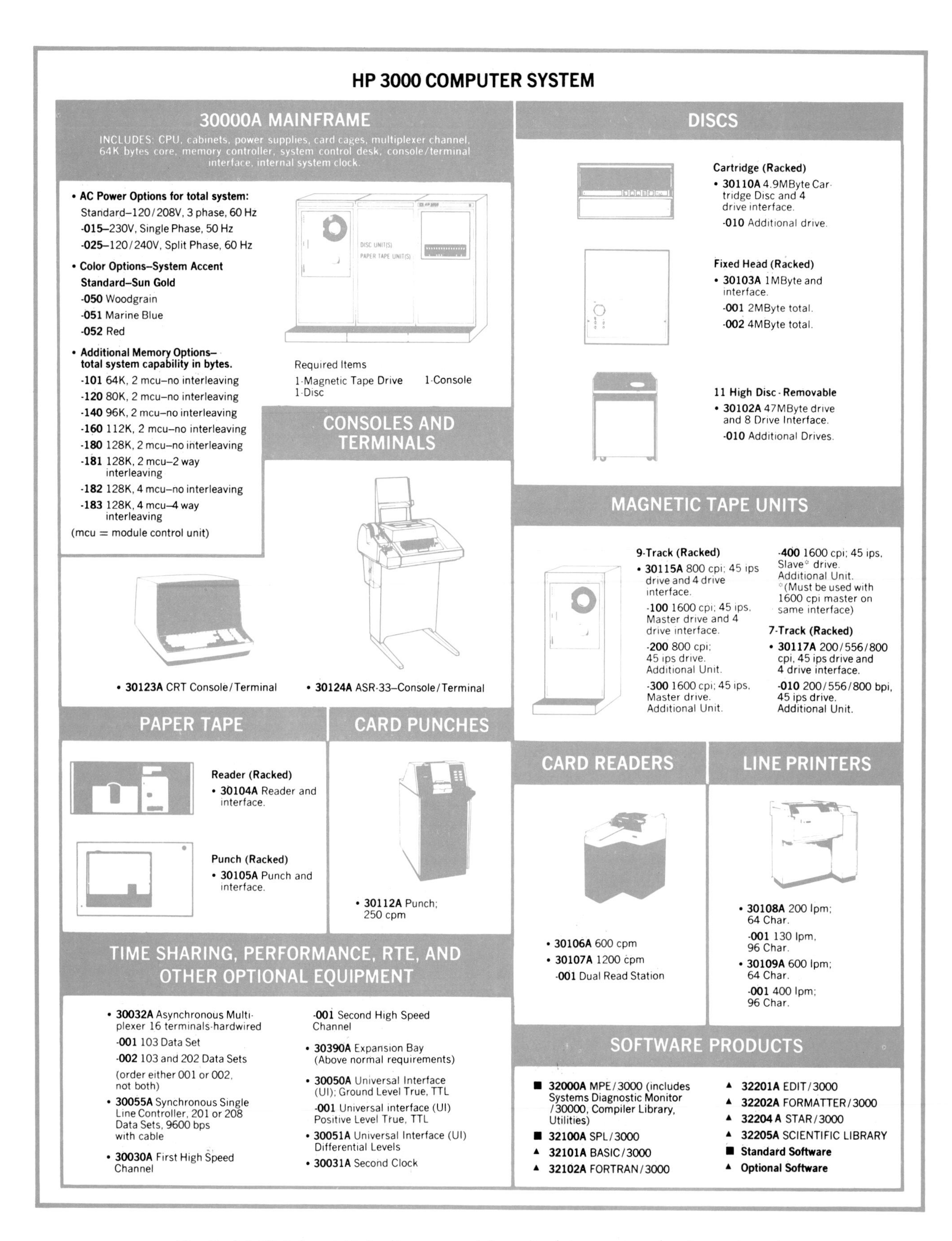

Fig. 2. *HP 3000 Computer Systems are modular and can be configured to fit a variety of applications. The same software is used by all configurations.*

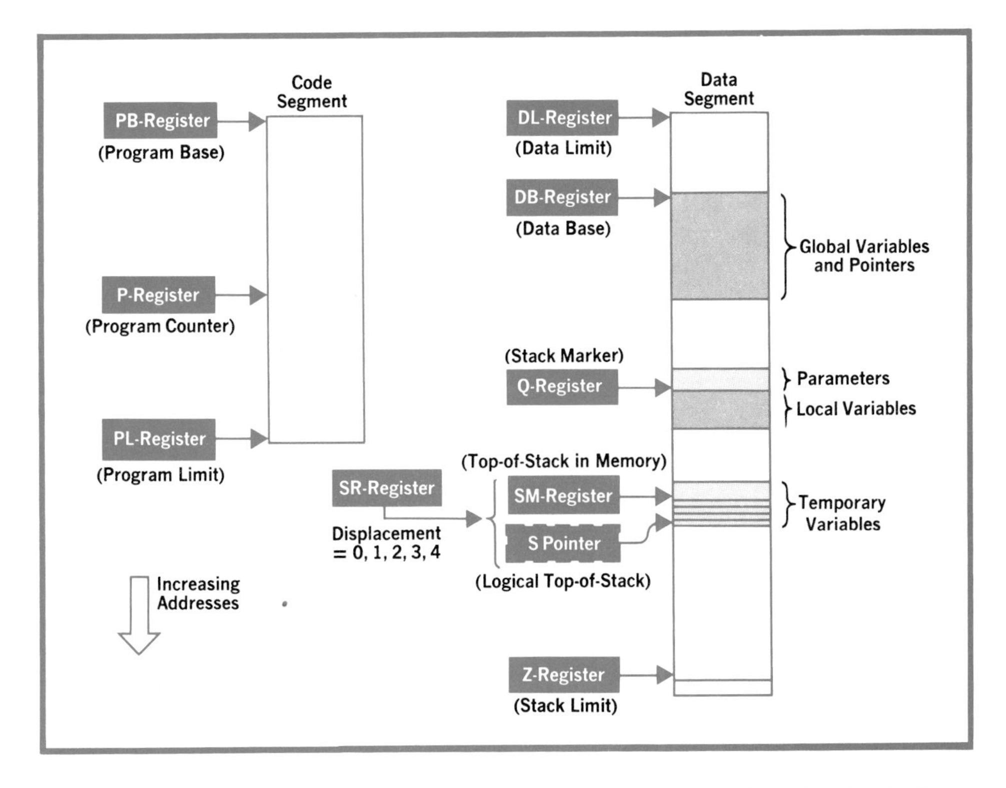

Fig. 3. *Code and data are kept separate in the 3000 Computer System. Code is never modified and can be shared by several users. Code segmentation gives the system virtual memory capability. Data segments are organized as pushdown stacks.*

bit causes a procedure call to the make-present routine if set, and it implies that the second word of the CST descriptor is a disc address. The maximum number of entries in the CST is 255.

Every procedure call must go through the Code Segment Table and must check the absent-from-main-memory bit. This is part of the virtual memory implementation of the HP 3000. If the procedure called is in a program segment that isn't in main memory, the required segment is automatically brought in. When a segment is given control of the central processor, the program base (PB) and program limit (PL) registers are set from the CST entry of that segment.

While code segmentation is normally specified by the user, data segmentation is handled by the MPE/3000 operating system. A normal user has only one data segment, which is limited to 32K words. Additional data segments may be requested.

Relocatable Code

Relocation is the normal mode of operation in the HP 3000 because of its relative addressing capability. All addressing is relative to hardware registers.

Fig. 4 shows the memory reference instruction format. The address mode bits have been Huffman-coded to give the maximum displacement range on the most frequently used codes.

In the code segment, normal addressing is relative to the program counter register (P). Indirect addressing is similar except that the content of the indirect cell is assumed to be relative to its own location.

In the data segment, the addressing modes are designed to match the types of data encountered in a procedure-oriented language. Fig. 3 shows the organization and common use of the data area. Global variables and pointers are stored relative to the data base (DB) register. The DB+ mode has a direct range of up to 255 words without indexing or a 64K word range with indexing.

Stack Operation

The stack concept,* which on the HP 3000 is fully used for the first time in a 16-bit machine, allows dynamic storage allocation on a procedure level. The stack is the area of a user's data segment between the DB register and the stack pointer (S).

Local stack storage in a procedure is allocated only upon entry and is automatically freed upon exit. This allows reuse of that area of memory by other parts of the program. The stack also provides automatic temporary storage of intermediate results until they are needed later in a computation. This is transparent to the programmer, and the compiler doesn't have to be concerned with saving and restoring registers.

Parameters that are passed to procedures are

*A stack is a linear collection of data elements which is normally accessed from one end on a last-in-first-out basis. An everyday example is a stack of plates in a plate warmer in a cafeteria.

A Computer for All Reasons

Education and Instrumentation are traditional fields for HP, and the HP 3000 Computer System significantly enhances the company's capabilities in these areas. The new system is also well suited to advanced industrial and commercial applications.

Education

HP computers entered the education field in 1968. The HP 2000A Time-Shared BASIC System, along with its successors, provided cost-effective computer aided instruction (CAI), problem-solving, and computer science education. A math drill and practice program, an instructional dialog facility, and an instructional management facility are available to the teacher for use on the HP 2000. These programs are written in HP BASIC and are therefore upward compatible with the HP 3000. In addition to these programs, there are other CAI packages available for use on both HP systems.

In addition to the time-shared CAI use of the HP 3000, the Multiprogramming Executive operating system allows simultaneous batch mode computation. This permits a school to use the computer for administrative tasks concurrently with CAI, giving a more cost-effective solution to the needs of school systems. Many secondary schools will also be able to teach programming and other computer science concepts using the multiprogramming capability of the HP 3000.

Junior colleges and small four-year colleges, to keep costs down, often find it necessary to have only one computer for all their activities. The HP 3000, with its simultaneous multilingual time-sharing and batch operating modes, has the ability to tackle diverse computing needs. In addition to these two modes, real-time experiments may also be handled by the operating system, and this makes the system useful to university scientific departments.

Because the HP 3000 was designed around the latest concepts in computer science, it has many features in hardware and software that university professors have been teaching in recent years. The 3000 should provide a computer science department with a machine that can be used not only as case study of advanced architecture, but also as a vehicle for operating system study. The modular structure of the software allows students to rewrite small portions of the system as projects and then try them in the operating system. It would be too large a task to write a whole system in a semester, but a small self-contained module is the right size for a term project.

Instrumentation

While the education field is mainly interested in the time-sharing and batch modes of operation under the Multiprogramming Executive (MPE), the instrumentation field makes extensive use of the compatible real-time capability. Previous systems generally had one or the other, but not all three in a unified environment.

MPE provides the ability to collect data and control processes in real time while allowing the data so generated to be accessed through the common file system by terminal-oriented and batch mode programs. This multi-mode capability is a natural extension and combination of the real-time executive and time-sharing systems that use the HP 2100 family of computers.

Industrial/Commercial

The HP 3000 will find many industrial and commercial applications. One reason is that it is designed to support hierarchical computing systems. The data-base handling capabilities of MPE, along with a powerful and wide-bandwidth I/O structure, make the 3000 a good middleman computer. It will have extensive data communication facilities for connection to a large general-purpose computer and will be able to control several minicomputers on the other end of the hierarchy.

There will be many instances of this computer-to-computer connection in the future. Standard software protocols and hardware interfaces are being developed for the HP 3000 to support these systems. The HP 2100 family provides compatible minicomputer facilities in systems where the 3000 is the host computer. Intercomputer links may be by direct connection or by modems over common carrier facilities.

Commercially oriented languages and data base management systems currently in development will give HP the ability to develop and support commercial applications such as on-line inventory management, order entry and production control. The hierarchical computing capability combined with this business data-processing software will make the HP 3000 more and more useful in industry and commerce, particularly if the strong trend toward distributed processing continues as expected.

---

pushed onto the stack before the procedure call. When the procedure call occurs the status of the presently executing code segment is stored on the stack and the Q register is set to point at the top of the stack (S). Parameters are then accessed by Q− addressing, while the local variables used by the procedure are accessed by Q+ addressing, as shown in Fig. 3.

Upon exiting from a procedure the operating system retrieves the status of the previously executing code segment from the stack and returns control to the instruction following the procedure call.

Addressing in the negative direction with respect to the stack pointer (S) register is useful for accessing temporary results left on the stack during processing. The area between the data limit (DL) register and DB may be addressed only indirectly and is used for such purposes as storing symbol tables and the like.

Reentrant Code

The separation of code and data, the use of a pushdown stack with Q+ and Q− addressing modes and the nonmodification of code make reentrant code the natural way to write HP 3000 programs. Reentrant code, in conjunction with the

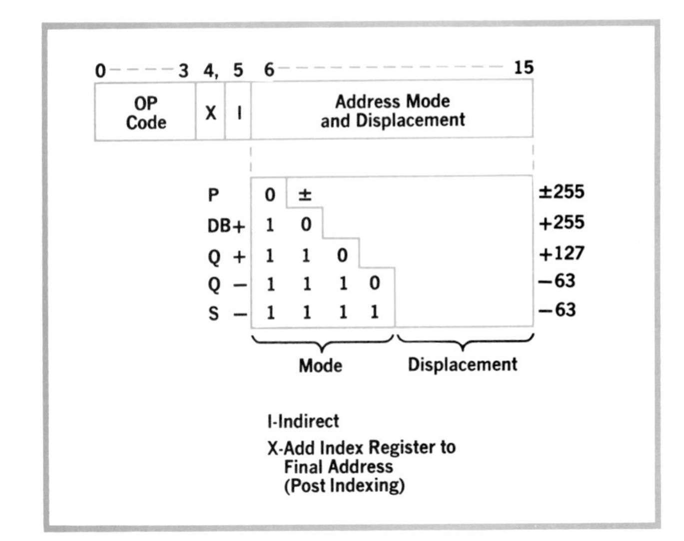

Fig. 4. *Memory reference instruction format. All addressing is relative to hardware registers, making code and data easily relocatable.*

use of the Code Segment Table as the master directory of all active segments, allows code segments to be shared between users. Control is transferred through the CST to the proper segment number of the shared code as determined by the loader when the segment was made active. Thus only one copy of a compiler or a library or the operating system intrinsics need be available, saving valuable space in main memory.

Protection Features

User isolation and protection takes several forms on the HP 3000. Programs may execute in one of two modes: privileged or user. In privileged mode no bounds checking is done except for stack overflow ($S > Z$), and all instructions are available for use. All system interrupts including external (I/O) interrupts are handled on a separate interrupt control stack so the user running when the interrupt occurs is fully protected. In user mode, access is limited to within the user's own code and data areas.

In addition to the hardware memory protection, files are protected by the MPE/3000 file management system. Access to files may be controlled at several levels which range from unrestricted access by anyone to controlled access available only to the creator of the file.

Modular Hardware Organization

HP 3000 hardware is organized on a modular basis. A major feature is the central data bus, which can service up to seven independent and asynchronous modules. These can be central processors, memory modules, and/or various types of input/output channels including a high-speed selector channel capable of transferring data at a rate of 2.8 megabytes per second.

Modules attached to the bus are technology-independent. Thus the memories may be magnetic core, semiconductor, or anything else. Up to four memory modules can be attached to the bus, and these can be interleaved (two-way or four-way).

Bert E. Forbes

Bert Forbes has been designing computers for HP since 1967. He was project manager for the HP 3000 CPU and has several patents pending as a result of that project. He's a member of ACM and the author of articles on computer architecture and integrated circuits for minicomputers. Now at HP's Geneva, Switzerland, data center, he's supporting the European introduction of the HP 3000. Bert received his B. S. degree in electrical engineering from Massachusetts Institute of Technology in 1966 and his M. S. E. E. degree from Stanford in 1967. He's also done work towards the Ph. D. degree. He's an amateur photographer and a connoisseur of fine wines, and is active in church youth and social-action groups.

Michael D. Green

Mike Green came to HP in 1966. He's been project manager for ALGOL/2116, 2000A Time-Shared BASIC, and BASIC/3000, and he's currently project manager for MPE/3000. Mike graduated from Columbia University in 1964 with a B.S. degree in mathematics, then got his M.S. degree in computer science at Stanford University in 1966. He's a member of ACM. For relaxing away from the world of computers, he favors bicycle touring and chess.

Central Bus Links Modular HP 3000 Hardware

Sharing the bus can be one or more CPU's, I/O processors, memory modules, high-speed I/O channels, and special devices. The microprogrammed CPU's have a procedure-oriented stack architecture.

by Jamshid Basiji and Arndt B. Bergh

ON THE HARDWARE LEVEL, the HP 3000 Computer System consists of independently functioning modules communicating over a high-speed multiplexed central data bus (Fig. 1). The modules may include one or more central processing units (CPUs) and input/output (I/O) processors, one to four memory modules, one or more selector channels for high-speed input/output, and one or more special-purpose modules. Hardware modularity makes the system flexible and expandable, and leaves the door open for future performance improvements through new technologies such as faster memories.

The memory now available is a magnetic core memory that has a cycle time of 960 nanoseconds. Optional is an interleaved addressing capability that places sequential addresses in different memory modules. Memory modules can operate concurrently. With interleaving, the system can support a 5.7 megahertz byte data rate.

The 3000 CPU is a microprogram-controlled processor. It has a stack architecture and special hardware to make procedure execution very efficient. Instructions are implemented in microprogrammed read-only memories, making possible a powerful instruction set with some instructions resembling those of higher-level languages.

The data for each user is organized as a data stack. In general, a stack is a storage area in core memory where the last item stored in is always the first item taken out. The stack structure provides an efficient mechanism for parameter passing, dynamic allocation of temporary storage, efficient evaluation of arithmetic expressions, and recursive subroutine or procedure calls. In addition, it enables rapid context switching—21 microseconds to establish a new environment when an interrupt occurs. In the HP 3000, all features of the stack (including checking for overflow and underflow) are implemented in hardware.

Bus Operation

The central data bus is a high-speed synchronous bus that can service up to seven modules. The transfer cycle time of the central data bus is equivalent to the cycle time of the system master clock, 175 nanoseconds. During each transfer cycle sixteen bits of data plus parity and eight bits of source-destination addresses and operation code are transmitted from the source module to the destination module.

Control of the bus is distributed among the modules; there is no central control. The bus control and interface logic for a given module is in the module control unit (MCU) for that module.

Bus cycles are granted to a transmitting module when two conditions are met. First, the transmitting module must request a bus cycle from its MCU and the destination module must be willing to accept the message in the next cycle. The willingness of a module to accept a message is indicated by the logical state of its "Ready" line. There are seven "Ready" lines in the central data bus, one for each module.

The second condition that must be met before a module is granted a bus cycle is that there must not be any higher priority module seeking to obtain the next bus cycle. Module priority is a function of data transfer urgency. Memory modules have the highest priority, and the high-speed selector channel has a higher priority than the CPU or input/output processor (IOP). A module, when ready to transmit a message, blocks lower priority modules by lowering its "Enable" line. There are seven ded-

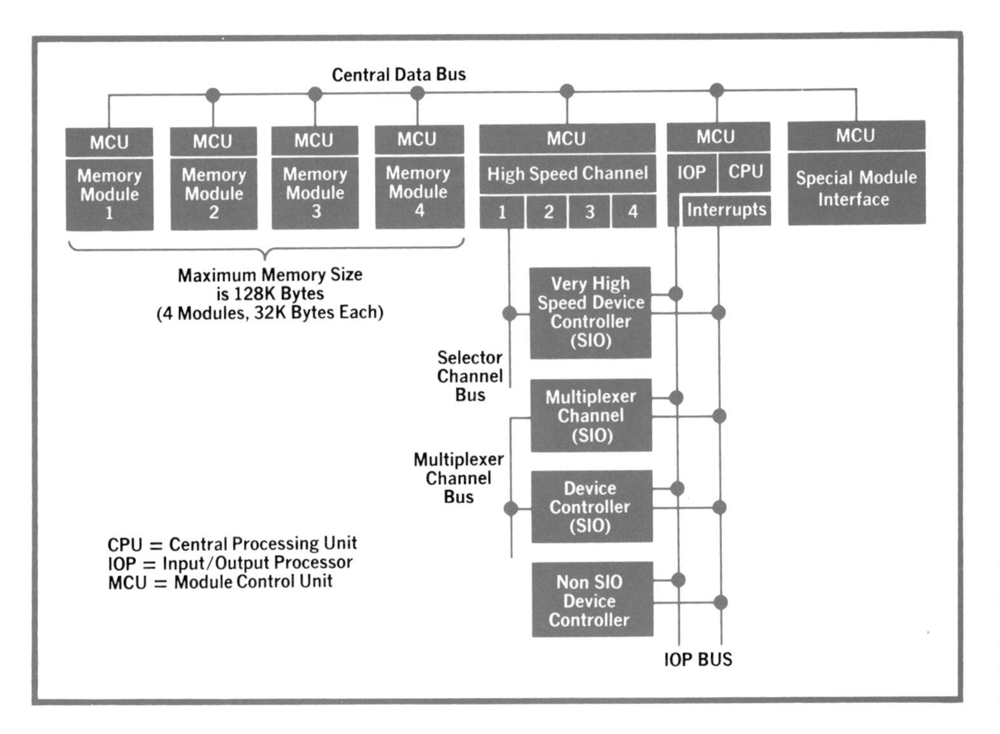

Fig. 1. *Central data bus of HP 3000 serves up to seven independent modules. 16 bits of data plus parity and eight bits of address and operation code are transferred in 175 nanoseconds.*

icated "Enable" lines, one for each module, in the central data bus. Each MCU checks the status of all higher priority modules prior to granting the next bus cycle to its host module.

With this bus-cycle allocation scheme, the "handshaking" mode of operation is not necessary, so data transfer speed is improved.

The central processing unit and the input/output processor share a module control unit. Thus the CPU and IOP share a single port on the central data bus. The IOP has a higher priority for bus access than the CPU, although both have independent access to the bus. The IOP provides the I/O devices with a direct path to memory through a buffered connection between the central data bus and the I/O bus.

The Central Processor

Because it provides a great deal of instruction power very economically, the microprogrammed read-only memory (ROM) method of logic control was chosen for the HP 3000. The central processing unit, Fig. 2, has a general-purpose microprocessor structure with some special features to aid the stack architecture. The 170 individual instructions are implemented by sequences of microinstructions stored in the control ROM.

In the CPU are approximately 30 registers. Those of most interest to the user are the four top-of-stack data registers (A, B, C, D), three code-segment registers (PB, P, PL), a status register, an I/O mask register, an index register (X), and six stack pointer registers (DL, DB, Q, SM, SR, Z). The DB register is the base of the stack, and the S register, defined as SM + SR, is the top of the stack. The area between Q and S is for local variables of the current procedure or routine. The top-of-stack registers are logical extensions of the stack area in core and their use greatly improves instruction execution time. The SR register tells how many of these registers are filled.

To improve the efficiency of handling data in the CPU, a two-stage "pipelined" data path structure is used. In the first stage, data is selected from the source registers and fed onto the two data buses (R and S) and into the bus storage registers shown in Fig. 2. These storage registers are the pipeline holding registers and serve as the data source for the second stage. In the second stage this data is processed through the arithmetic logic unit and a shift network, and the result is optionally tested and stored in selected destination registers. New data is entered into the stream on each clock pulse to keep the pipeline full and maximize throughput. The 175 ns clock time achieved with this structure is much lower than would have been possible if the whole source-to-destination processing were done in one clock period.

Communication paths from the CPU to outside modules include a path to memory through the MCU and central data bus, a path to device controllers through the I/O processor and I/O bus, and a path to the control panel through a special panel interface.

CPU Operation

The CPU performs tasks by sequentially enabling the appropriate logic to pass data through the processing structure and to perform other non-data-path functions. For each sequential step a 32-bit ROM word, divided into seven coded control fields, enables the required functions. Each 32-bit ROM word constitutes a microinstruction. As shown in Fig. 3, the seven fields in each microinstruction are the R and S bus source register fields, the operation or function field, the shift field, the register store field, the test field, and a special field for executing non-data-related tasks.

Because each control field can, in general, select only one meaningful field option at a time, it was possible to encode them with little loss of capability. For a slight reduction in speed, field encoding, or "vertical microprogramming," offers considerable ROM cost savings over the one-bit-per-option method.

Branching capability is provided by redefining the R bus, shift, and special fields to be interpreted as a branch address when a Jump or Jump Subroutine instruction occurs in the function field. Constants also are generated by redefining fields when a function field designator occurs.

Programs and Microprograms

As the CPU executes a user program, it sequentially fetches software instructions from main memory. From the binary pattern of each instruction, a combination ROM lookup table and decoding logic generates a ROM address and stores it in a presettable indexing ROM address register. This register is used first to access and then to step through the sequence of microinstructions, or microprogram, that causes the software instruction to be executed. There is a microprogram in ROM for each of the 170 machine instructions.

The CPU executes a software program in the normal sequence of phases, that is, instruction fetch, data fetch, and execute. In the HP 3000 these phases are more accurately described as instruction pre-fetch, optional data address computation or hard-

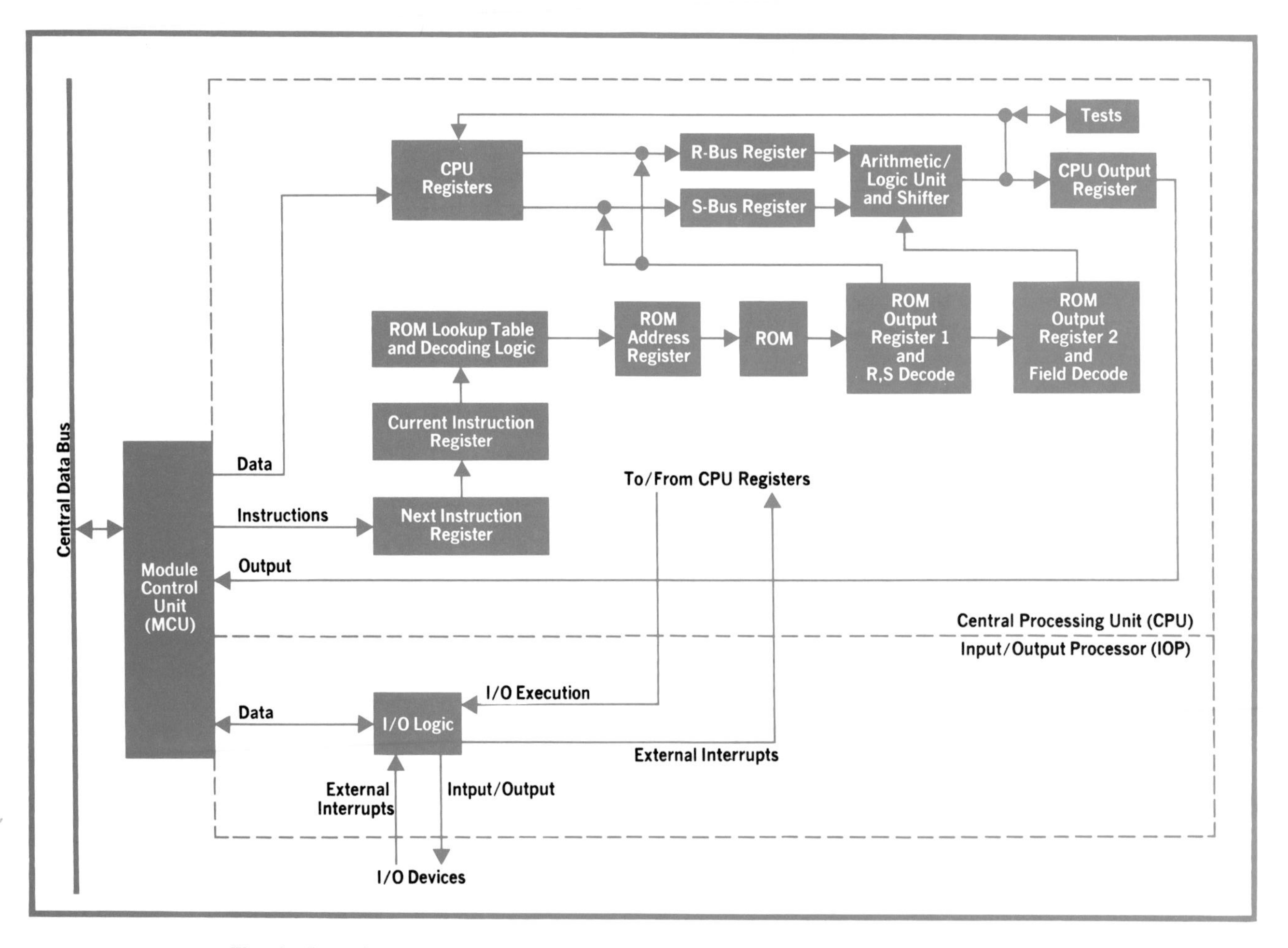

Fig. 2. *Central processor has a general-purpose microprocessor structure with special features to aid stack operation.*

ware stack register preadjust, and instruction execution. Instruction prefetch is an automatic hardware activity that gets the next instruction during the execution of the present instruction, thus avoiding the normal instruction fetch time. For memory reference instructions, hardware has been provided to compute the absolute memory address, that is, to add the displacement and index to the appropriate base register. A general bounds-testing routine in ROM then checks the computed address for validity before the individual instruction microprogram is used. Instructions that use only top-of-stack data normally (90% of the time) don't require a data fetch, but if necessary, these instructions are first routed through a microprogram that fills the appropriate number of hardware stack registers from the equivalent logical locations in core.

Interrupts

As the execution of each instruction is completed a microprogram control signal is issued that starts the execution of the next instruction unless an interrupt is requesting service. If an interrupt has occurred, a force to an interrupt microprogram takes place. This causes the status of the present user program to be stored on the stack. Then if the interrupt is not directly user related, the microprogram transfers the status to a system interrupt stack and calls the first instruction of the software program serving that interrupt. After the interrupt has been serviced control is returned to the MPE operating system.

TOS Hardware

To achieve faster execution of instructions that reference the top elements of the stack, special hardware has been provided. Up to four of the top elements of the stack can be kept in four top-of-stack hardware registers, and manipulation of these registers by the microcode has been made as easy as possible. The TOS hardware includes the four registers and renaming logic that allows each of the four registers to assume any of the four positions relative to the top of the stack. Thus, the stack can be logically shifted up or down by simply renaming the registers, without moving the contents of one register to another. The number of stack elements that currently reside in the TOS hardware registers is kept in the TOS register pointer, SR.

Memory

Memory modules on the HP 3000 are designed to be self-contained asynchronous units of up to 64K bytes each. The maximum memory limit is 128K bytes in up to four modules. The modules interface with the system through an MCU port on the central data bus. Only data transmissions to the system have to be synchronized with the system clock; all other memory timing and control is contained within each module. Since no fixed response time is required, faster memories can be interfaced as they become available.

Memory commands include read, write, and a special multiprocessor semaphore function: read and write all 1's within one memory cycle.

The present memory is a 960-nanosecond three-wire 3D magnetic core memory using the same core stack and phased X-Y drive current arrangement as is used in the HP 2100A Computer.[1] A basic module consists of one timing, control, and MCU interface card, one X-Y switch and inhibit-current load card and one to four 8K word stack cards. Because the sense amplifiers, X-Y drivers and inhibit drivers all are on the stack card, memory expansion only requires the addition of one stack card for each additional 16K bytes.

Input/Output Processor

The functions of the I/O processor have been distributed between a kernel processor attached to the CPU and one or more multiplexer channels on the I/O bus. The kernel processor controls the I/O bus, which is the data path from external devices to memory and the communication path between external devices and the CPU. The multiplexer channel does the bookkeeping for block transfers of data to and from memory for up to 16 devices. When needed, additional multiplexer channels may be added to the system.

Input/output operations in the HP 3000 are divided into three categories: direct I/O, programmed I/O and interrupt processing. Programmed I/O operations have priority on the I/O bus over other types.

Direct I/O operations take place as a result of the execution of an I/O instruction by the CPU. These operations either exchange a word of information between the top-of-stack register (TOS) in the CPU and the I/O device controller, or cause a control function to take place in the I/O system. During the execution of I/O instructions the CPU microprocessor performs the basic control functions such as assembling the I/O command, checking the status of the I/O device controller, and exchanging a word of information between the TOS register and the I/O device via the I/O bus.

Programmed I/O operations are aimed at transferring blocks of data between I/O devices and the memory. This type of operation begins for an I/O device when the CPU issues an SIO instruction for that device. The device controller in conjunction with the multiplexer channel then executes the I/O control program for that device without further CPU intervention. This allows the CPU and I/O processing to carry on in parallel.

The interrupt structure is a multilevel priority network that allows the processing of CPU programs or lower-level interrupts to be preempted by higher-level interrupts. This assures a prompt response to critical external processes. A "polling" scheme is used in the priority network. Up to 253 devices are allowed on the interrupt poll line, and the interrupt priority of a device is determined by its logical proximity to the CPU on the interrupt poll line. A 16-bit mask register is provided for the purpose of masking off groups of interrupts. Any number of devices can be assigned to any particular mask group.

I/O bus transfer cycles are granted to multiplexer channels based on their priorities. A polling scheme similar to the interrupt polling is used to resolve priority among the multiplexer channels. However, the data poll line is separate from the interrupt poll line, so the data priority of a channel can be different from its interrupt priority.

Selector Channel

High-speed devices may communicate directly to the central data bus through a selector channel. Unlike the multiplexer channel, the selector channel is designed to service one device at a time for the duration of the execution of the I/O control program for that device. This eliminates the time-slice multiplexing overhead, thereby allowing the SEL channel to achieve higher data transfer rates than are possible with the MUX channel. The selector channel is a part of the SEL module, which is an independent system module that contains up to four selector channels and has an independent port to the central data bus (see Fig. 1). This port enables the selector channels to fetch and execute their own I/O command words and transfer data between the memory and the I/O devices independently of the I/O processor. Each selector channel has its own SEL bus and can interface up to eight devices through this bus.

Special Devices

Ports on the central data bus are not device-dependent. Therefore, they can be used for special custom devices should the system application warrant their use. An example of such a device might be a communications processor.

Acknowledgments

By Richard E. Toepfer
Engineering Section Manager,
Multiprogramming Computer Systems

The design of a system like the HP 3000 requires the contributions of a large group of people. The following list represents the members of the Data Systems Development Laboratory who were principally concerned with the design and realization of the hardware and its associated diagnostic software. **Mainframe Electronics Design:** Harlan Andrews, Jim Basiji, Arne Bergh, Bill Berte, Wally Chan, Ken Check, John Dieckman, Mauro DiFranceso, Bert Forbes, Gordon Goodrich, Barney Greene, John Grimaldi, Jim Hamilton, Marty Kashef, Jim Katzman, Walt Lehnert, Frank McAninch, Joe Olkowski, Mike Raynham, Gene Stinson, Tak Watanabe, Steve Wierenga, Dennis Wong. **Mainframe Mechanical Design:** George Canfield, Bob Dell, Joe Dixon, Bill Gibson, Gary Lepianka, Larry Peterson, Bob Pierce, Don Reeves, Fred Reid. **Mass Storage Subsystems:** Naresh Aggarwal, Ole Eskedal, Karl Helness, Ed Holland, Jake Jacobs, Earl Kieser, Harry Klein, Stan Mintz, Malcolm Neill, Cliff Wacken. **I/O Subsystems:** Mitch Bain, Oty Blazek, Vince Emma, Ron Kolb, Tom Kornei, Rick Lyman, Al Marston, Joe Mixsell, Bill Murrin, Jack Noonan, Jim Obriant, Ken Pocek, Willard Reed, Willis Shanks, Elio Toschi, Lloyd Summers. **Diagnostic Software:** Bob Bellizzi, Gary Curtis, Hank Davenport, Dan Gibbons, Pete Graziano, Tony Hunt, Walt Wolff, Tom Ellestad.

Particular credit must be given to the following individuals and groups whose special talents greatly contributed to the success of our development effort. Coordinators—Karl Balog and Ollie Saunders. Printed circuit layout—Bob Jones and staff. Industrial Design—Gerry Priestly and staff. Material and Reliability Engineering—Bernie Levine and staff. Publications—Joe Kintz and staff. System Management—Dave Crockett and staff.

Reference

1. Hewlett-Packard Journal, October 1971.

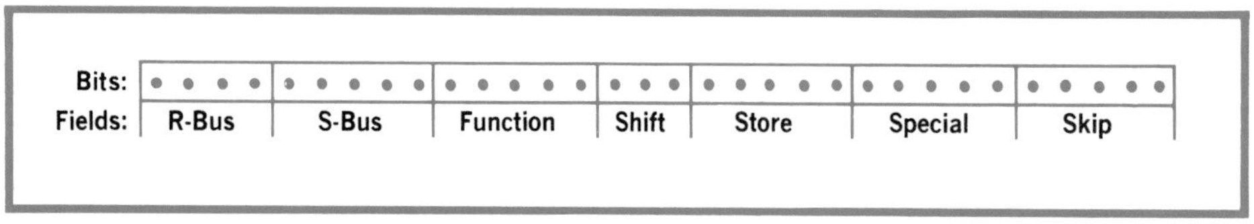

Bits:	• • • •	• • • • •	• • • • • •	• • •	• • • • •	• • • • • •	• • • • •
Fields:	R-Bus	S-Bus	Function	Shift	Store	Special	Skip

Fig. 3. *170 HP 3000 instructions are implemented by sequences of microinstructions stored in read-only memories. Each 32-bit microinstruction has seven coded fields.*

SPECIFICATIONS
HP 3000 Computer System

DESCRIPTION
Multiprogrammed general-purpose computer system implemented with complementary hardware and software providing for concurrent real-time, batch, and time sharing processing.

CENTRAL PROCESSOR
ARCHITECTURE
- Hardware-implemented stack
- Separation of code and data
- Nonmodifiable, reentrant code
- Variable-length code segmentation
- Virtual memory
- Dynamic relocatability of programs

IMPLEMENTATION
- Microprogrammed CPU
- 175 nanosecond microinstruction time
- Memory protect, parity checking, power-fail/auto restart
- Protection between users
- Central data bus
- Concurrent I/O and CPU operations

INSTRUCTIONS
- 170 instructions
- 16 bits per word
- 16-bit and 32-bit integer; 32-bit floating point hardware arithmetic
- Triple-word shifts to aid 48-bit floating point software

MEMORY
- Technology independent, speed independent
- Up to four modules
- Interleaving
- Addressable to 65K words (131,072 bytes)
- 17 bits includes parity bit

I/O AND PERIPHERALS
GENERAL
- Privileged control of I/O
- Concurrent I/O operations
- Three ways to implement I/O
- Direct memory access by all channels
- Device-independent I/O program execution
- Up to 253 devices

I/O SYSTEM
- Multiplexer channel
- Selector channel
- Direct I/O

INTERRUPT SYSTEM
- Up to 253 external interrupts
- Independent masking and priority structures
- Microprogrammed environment switching
- Common stack for interrupt processing
- 17 internal interrupts plus 7 traps

PERIPHERALS
- Mass Storage:
  - Fixed Head Disc: 1, 2 or 4 megabyte, 496 kHz byte transfer rate
  - Removable Media: 47 megabytes, 320 kHz byte transfer rate; 4.9 megabytes, 245 kHz byte transfer rate
- Magnetic Tape: 7 Track—45 ips, 200, 556 or 800 bpi; 9 Track—45 ips, 800 or 1600 bpi
- Card Readers: 600 or 1200 cards per minute
- Card Punches: 35 or 250 cards per minute
- Line Printer: 200 or 600 lines per minute, 64 or 96 character sets; 132 columns
- Paper Tape Reader: 500 cps
- Paper Tape Punch: 75 cps
- Consoles: CRT or ASR-33

SOFTWARE
MULTIPROGRAMMING EXECUTIVE
- Batch processing
- On-line terminal processing
- Real time processing
  - Production control
  - Automatic testing
  - Process control
  - Information retrieval
  - Data acquisition

SYSTEMS PROGRAMMING LANGUAGE (SPL)
- High-level syntactic structure
- Ability to address hardware registers explicitly
- Bit manipulation
- Branches based explicitly on hardware status
- Use of all hardware data types and operators

PROGRAMMING LANGUAGES
- FORTRAN: extended version of ANSI Standard FORTRAN (x3.9-1966)
- BASIC: most powerful currently available
- COBOL: highest level of Federal Government Standard
- Compiler Library: provides common compiler functions

DATA MANAGEMENT
- IMAGE: comprehensive information management system
- QUERY: interactive language interface to data base via IMAGE

SUPPORT SOFTWARE
- EDIT: editing of source or text files
- SORT: sort and/or merge of multiple files
- TRACE: debugging tool for FORTRAN and Systems Programming Languarge

SCIENTIFIC SOFTWARE
- Scientific Library
- STAR—Statistical Analysis Routines

DIAGNOSIS OF HARDWARE
- System Diagnostic Monitor—runs on-line diagnostics
- Stand-alone diagnostics
- Microdiagnostics

PRICE IN USA: $125,000 for small system to over $500,000 for large system.

MANUFACTURING DIVISION: HP Data Systems
11000 Wolfe Road
Cupertino, California 95014

Jamshid Basiji
Since coming to HP in 1969, Jim Basiji has worked on high-speed I/O processing techniques, developed the architecture of the I/O system and central data bus for the HP 3000, and helped design the I/O processor and central processor modules for the HP 3000. A graduate of the University of California at Berkeley, Jim received his B.S.E.E. degree in 1965 and his M.S.E.E. degree in 1966. Before joining HP, he worked on computer development and advanced computing techniques for IBM. His idea of a fascinating way to spend his free moments is with a good book.

Arndt B. Bergh
With HP since 1956. Arne Bergh has had a variety of research and development responsibilities as a member of HP Laboratories and various operating Divisions. His projects have included instruments, magnetic devices, memories, and computers, the latest being the hardware design of the HP 3000. He holds four patents and has others pending on the HP 3000 and on a 1024-bit bipolar ROM. Arne received the A.B. degree in chemistry from St. Olaf College in 1947 and the M.S. degree in physics from the University of Minnesota in 1950. He's a member of ACM and IEEE. He has a private pilot's license, but his real passion is flying over the water, racing his Daysailer sailboat in local, regional and national competition.

Software for a Multilingual Computer

SPL is a high-level language that produces code that's as efficient as other systems' assembly-language code. Other 3000 languages are FORTRAN, BASIC and COBOL.

by William E. Foster

PROGRAMMING LANGUAGES NOW AVAILABLE FOR THE HP 3000 USER are FORTRAN, BASIC, and SPL (Systems Programming Language). COBOL will be available in summer 1973. The system will support all these languages simultaneously.

Systems Programming Language

SPL is an ALGOL-like language. Its objective is to provide systems programming capability from a high-level language rather than the traditional assembly language. The benefits are faster coding and easier debugging. Virtually all the HP 3000 software is written in SPL.

It's imperative, of course, that a systems programming language produce efficient object code, and this was another major objective of SPL. Code optimization has been achieved through the logic of the compiler and through close correlation between the SPL syntax and the 3000 instruction set.

A significant aspect of SPL is that it may be used as either a machine independent or a machine dependent programming language. At the machine independent level, the syntax of SPL closely resembles that of ALGOL. It isn't necessary for the programmer to understand the architecture of the 3000 to program at this level.

The machine dependent programmer is one who has some knowledge of the 3000 architecture (instruction set, stack, status register, etc.); the greater his knowledge, the more he is able to make use of the machine dependent features of SPL. The effect of using these features of SPL is improved object code.

Fig. 1 illustrates the two levels of SPL applied to the same programming problem. Fig. 2 is an example of a more typical SPL program.

FORTRAN/3000

FORTRAN is one of the most widely used and oldest programming languages. Initial specifications for the language date back to 1954. FORTRAN/3000 is an ANSI-standard compiler with extensions that enhance the capability of the language and use the features of the HP 3000. Among these features are CHARACTER variables, which were added to the language to provide the capability of string manipulation. Additionally, a great deal of power is provided in the area of input/output operations.

- Free-field I/O. Variables may be input and output in a free-field manner, without the specification of a FORMAT statement.
- Output expressions. Expressions may be included in the output list (Fig. 3). For example,

  WRITE (3,10) I*S,A+B

  is a legal FORTRAN/3000 statement.
- Logical unit table. A global table is created by the compiler and built by the loader that is used to associate FORTRAN logical unit (storage device) numbers with internal file numbers. The FORTRAN programmer has the capability, with the use of library routines, to tailor this table to his own needs. For instance, he may explicitly open a file through a call to the file system intrinsic FOPEN, then set the returned file number to correspond to a particular FORTRAN unit number (say unit #7). Subsequent READ or WRITE statements using unit #7 would, in fact, be referencing this file.
- FORMAT specification. Two important specifications have been added to the FORMAT statement: the T-specification, which positions the format scanner to specific locations in the record, and the S-specification, which outputs character data with a field width that corresponds to the

Machine independent method

The conventional approach, used in most programming languages, would be to use a temporary variable in making the exchange:

SPL statement	Purpose	Generated Code
TEMP: = A;	Store the value of A in TEMP.	LOAD A STOR TEMP
A: = B;	Store the value of B in A.	LOAD B STOR A
B: = TEMP;	Store the original value of A in B.	LOAD TEMP STOR B

Machine dependent method

A more efficient approach would be to use the special SPL symbol TOS. When used in place of an identifier, this symbol denotes the current top of stack.

SPL statement	Purpose	Generated Code
TOS: = A;	Push the value of A onto the stack.	LOAD A
A: = B;	Store the value of B in A.	LOAD B STOR A
B: = TOS;	Store the current value that is on the top of the stack into B, then pop the stack.	STOR B

Fig. 1. *An SPL program to swap the values of two integer variables, A and B, illustrating the machine dependent and machine independent levels of the language.*

length of the associated list element (Fig. 3).

- Direct-access I/O. Disc files may be referenced as direct access devices. For example, the statement

  READ (3@RECNUM) A,B

  reads from logical unit #3 the record specified by RECNUM, and transmits the data to the list elements A and B.

Other extensions of standard FORTRAN are mixed-mode arithmetic, free format program entry for more convenient usage from terminals, removal of restrictions on indexing and DO-loops, and an interactive debugging facility.

Machine dependent characteristics of FORTRAN/3000 are that programs are recursive and reentrant. In the HP 3000, code and data are stored separately, and code is never altered. This means that programs can be shared by several jobs. If one job is using a program and is interrupted by another job that uses the same program, the first job can later reenter the program and continue from the point of interruption. Thus programs are reentrant. Recursive means that programs can call themselves. Another machine dependent feature is that storage for local variables is allocated on the stack dynamically when functions or subroutines are entered, and deallocated upon exit.

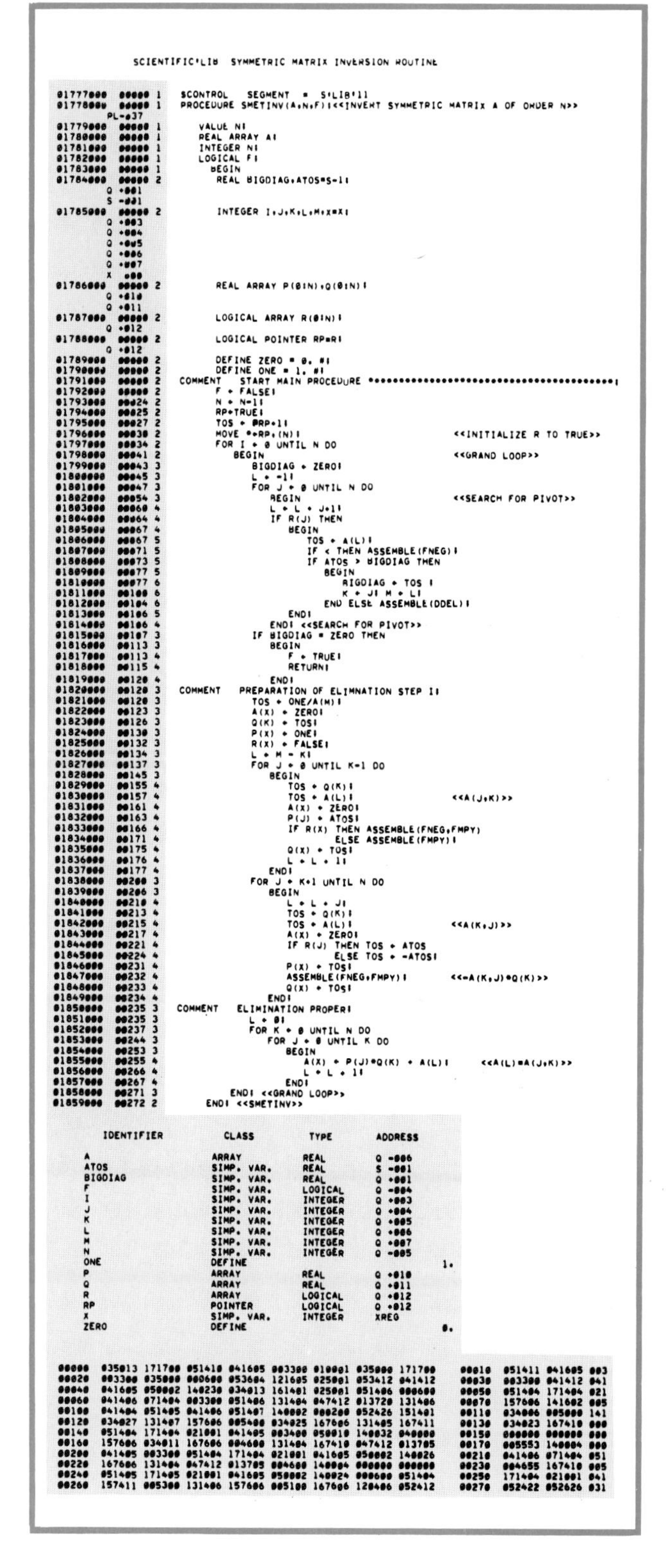

```
                SCIENTIFIC'LIB  SYMMETRIC MATRIX INVERSION ROUTINE

01777000  00000 1   $CONTROL   SEGMENT  =  S'LIB'11
01778000  00000 1   PROCEDURE SMETINV(A,N,F);<<INVERT SYMMETRIC MATRIX A OF ORDER N>>
          PL-037
01779000  00000 1      VALUE N;
01780000  00000 1      REAL ARRAY A;
01781000  00000 1      INTEGER N;
01782000  00000 1      LOGICAL F;
01783000  00000 1        BEGIN
01784000  00000 2          REAL BIGDIAG,ATOS=S-1;
          Q +001
          S -001
01785000  00000 2          INTEGER I,J,K,L,M,X=X;
          Q +003
          Q +004
          Q +005
          Q +006
          Q +007
          X  00
01786000  00000 2          REAL ARRAY P(0:N),Q(0:N);
          Q +010
          Q +011
01787000  00000 2          LOGICAL ARRAY R(0:N);
          Q +012
01788000  00000 2          LOGICAL POINTER RP=R;
          Q +012
01789000  00000 2          DEFINE ZERO = 0. #;
01790000  00000 2          DEFINE ONE = 1. #;
01791000  00000 2   COMMENT   START MAIN PROCEDURE ****************************************;
01792000  00000 2          F ← FALSE;
01793000  00024 2          N ← N-1;
01794000  00025 2          RP←TRUE;
01795000  00027 2          TOS ← @RP+1;
01796000  00030 2          MOVE *←RP,(N);                          <<INITIALIZE R TO TRUE>>
01797000  00034 2          FOR I ← 0 UNTIL N DO
01798000  00041 2             BEGIN                                <<GRAND LOOP>>
01799000  00043 3                BIGDIAG ← ZERO;
01800000  00045 3                L ← -1;
01801000  00047 3                FOR J ← 0 UNTIL N DO
01802000  00054 3                   BEGIN                          <<SEARCH FOR PIVOT>>
01803000  00060 4                   L ← L + J+1;
01804000  00064 4                   IF R(J) THEN
01805000  00067 4                      BEGIN
01806000  00067 5                         TOS ← A(L);
01807000  00071 5                         IF < THEN ASSEMBLE(FNEG);
01808000  00073 5                         IF ATOS > BIGDIAG THEN
01809000  00077 5                            BEGIN
01810000  00077 6                               BIGDIAG ← TOS ;
01811000  00100 6                               K ← J; M ← L;
01812000  00104 6                            END ELSE ASSEMBLE(DDEL);
01813000  00106 5                      END;
01814000  00106 4                   END; <<SEARCH FOR PIVOT>>
01815000  00107 3                IF BIGDIAG = ZERO THEN
01816000  00113 3                   BEGIN
01817000  00113 4                      F ← TRUE;
01818000  00115 4                      RETURN;
01819000  00120 4                   END;
01820000  00120 3   COMMENT   PREPARATION OF ELIMNATION STEP I;
01821000  00120 3                TOS ← ONE/A(M);
01822000  00123 3                A(X) ← ZERO;
01823000  00126 3                Q(K) ← TOS;
01824000  00130 3                P(X) ← ONE;
01825000  00132 3                R(X) ← FALSE;
01826000  00134 3                L ← M - K;
01827000  00137 3                FOR J ← 0 UNTIL K-1 DO
01828000  00145 3                   BEGIN
01829000  00155 4                      TOS ← Q(K);
01830000  00157 4                      TOS ← A(L);                 <<A(J,K)>>
01831000  00161 4                      A(X) ← ZERO;
01832000  00163 4                      P(J) ← ATOS;
01833000  00166 4                      IF R(X) THEN ASSEMBLE(FNEG,FMPY)
01834000  00171 4                              ELSE ASSEMBLE(FMPY);
01835000  00175 4                      Q(X) ← TOS;
01836000  00176 4                      L ← L + 1;
01837000  00177 4                   END;
01838000  00200 3                FOR J ← K+1 UNTIL N DO
01839000  00206 3                   BEGIN
01840000  00210 4                      L ← L + J;
01841000  00213 4                      TOS ← Q(K);
01842000  00215 4                      TOS ← A(L);                 <<A(K,J)>>
01843000  00217 4                      A(X) ← ZERO;
01844000  00221 4                      IF R(J) THEN TOS ← ATOS
01845000  00224 4                              ELSE TOS ← -ATOS;
01846000  00231 4                      P(X) ← TOS;
01847000  00232 4                      ASSEMBLE(FNEG,FMPY);        <<-A(K,J)*Q(K)>>
01848000  00233 4                      Q(X) ← TOS;
01849000  00234 4                   END;
01850000  00235 3   COMMENT   ELIMINATION PROPER;
01851000  00235 3                L ← 0;
01852000  00237 3                FOR K ← 0 UNTIL N DO
01853000  00244 3                   FOR J ← 0 UNTIL K DO
01854000  00253 3                      BEGIN
01855000  00255 4                         A(X) ← P(J)*Q(K) + A(L);     <<A(L)=A(J,K)>>
01856000  00266 4                         L ← L + 1;
01857000  00267 4                      END;
01858000  00271 3             END; <<GRAND LOOP>>
01859000  00272 2          END; <<SMETINV>>

       IDENTIFIER          CLASS          TYPE        ADDRESS

     A                     ARRAY          REAL        Q -006
     ATOS                  SIMP. VAR.     REAL        S -001
     BIGDIAG               SIMP. VAR.     REAL        Q +001
     F                     SIMP. VAR.     LOGICAL     Q -004
     I                     SIMP. VAR.     INTEGER     Q +003
     J                     SIMP. VAR.     INTEGER     Q +004
     K                     SIMP. VAR.     INTEGER     Q +005
     L                     SIMP. VAR.     INTEGER     Q +006
     M                     SIMP. VAR.     INTEGER     Q +007
     N                     SIMP. VAR.     INTEGER     Q -005
     ONE                   DEFINE                                 1.
     P                     ARRAY          REAL        Q +010
     Q                     ARRAY          REAL        Q +011
     R                     ARRAY          LOGICAL     Q +012
     RP                    POINTER        LOGICAL     Q +012
     X                     SIMP. VAR.     INTEGER     XREG
     ZERO                  DEFINE                                 0.

00000  035013 171700 051410 041605 003300 010001 035000 171700    00010  051411 041605 003
00020  003300 035000 000600 053604 121605 025001 053412 041412    00030  003300 041412 041
00040  041605 050002 140230 034013 161401 025001 051406 000600    00050  051404 171404 021
00060  041406 071404 003300 051406 131404 047412 013720 131406    00070  157606 141602 005
00100  041404 051405 041406 051407 140002 000200 052426 151401    00110  034006 005000 141
00120  034027 131407 157606 005400 034025 167606 131405 167411    00130  034023 167410 000
00140  051404 171404 021001 041405 003400 050010 140032 040000    00150  000000 000000 000
00160  157606 034011 167606 004600 131404 167410 047412 013705    00170  005553 140004 000
00200  041405 003300 051404 171404 021001 041605 050002 140026    00210  041406 071404 051
00220  167606 131404 047412 013705 004600 140004 000000 000000    00230  004655 167410 005
00240  051405 171405 021001 041605 050002 140024 000600 051404    00250  171404 021001 041
00260  157411 005300 131406 157606 005100 167606 120406 052412    00270  052422 052626 031
```

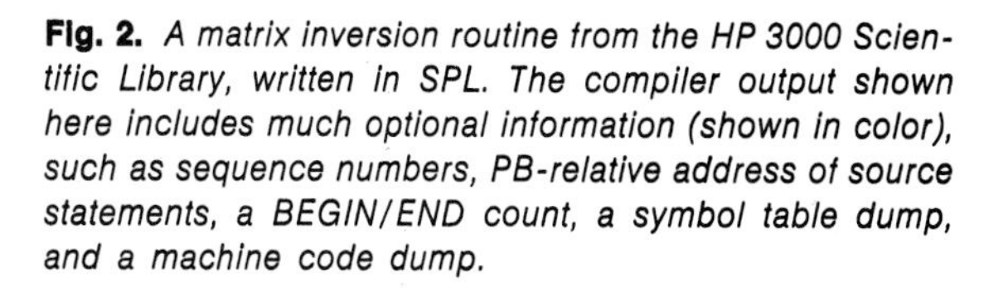

Fig. 2. *A matrix inversion routine from the HP 3000 Scientific Library, written in SPL. The compiler output shown here includes much optional information (shown in color), such as sequence numbers, PB-relative address of source statements, a BEGIN/END count, a symbol table dump, and a machine code dump.*

BASIC/3000

The HP 3000 BASIC subsystem runs as an interpreter rather than a compiler, which means that programs are not translated into machine code that is directly executable, but into an intermediate language that is executed by control routines.

The primary reasons for having an interpreter instead of a compiler are faster development and greater debugging facilities. The interactive debugging mode in BASIC provides the following capabilities:

- Tracing of the path of execution through a program and changes in the values of variables
- Interactively displaying the dynamic nesting structure of a program, that is, the order in which programs and functions are called
- Displaying and modifying the values of variables
- Altering the execution sequence of a program.

One aspect of an interpreter is that programs are really data to the interpreter. Therefore, BASIC programs do not execute as code segments and so are not sharable. For this reason, HP is currently developing a BASIC compiler that accepts the internal file generated by the interpreter and generates executable code. In this way, BASIC programs will not only run as sharable code segments, but will also execute faster.

The BASIC/3000 language is a superset of HP 2000 BASIC, incorporating many extensions:

2000

26 numeric arrays
26 string variables
one data type (32-bit real)

3000

286 numeric arrays
286 strings or string arrays
four data types (16-bit integer, 32-bit real, 48-bit real, 64-bit complex)

Other extensions include compound statements (Fig. 4), mixed-mode arithmetic, multiple-line functions, string-valued functions, access to all MPE files and peripheral devices, capability of calling SPL procedures, many additional predefined string and numeric functions, string arrays, program overlays, picture I/O formatting, statement execution frequency reporting, dynamic array redimensioning, handling of non-BASIC files, and additional file commands.

```
      WRITE(6,10) "PRESSURE", P
      WRITE(6,10) "TEMPERATURE", 2*T
   10 FORMAT (" THE VALUE FOR ", S, "IS", F7.3)

Result: (assume P=1.0339 and T=55.87)

      THE VALUE FOR PRESSURE IS 1.034
      THE VALUE FOR TEMPERATURE IS 111.740
```

Fig. 3. *FORTRAN/3000 program illustrating the use of an expression in an output list, and the "S" specification in the FORMAT statement.*

```
                10 IF A>B THEN 60
             ┌─►20 ELSE DO
             │  30 IF B <=C THEN B=C+1
DO-          │┌►40 IF C # D THEN DO
DOEND        ││ 50 C = C + FNK(D,D*A,C)
Pairs        ││ 60 D = Z + A
             │└►70 DOEND
             │  80 ELSE 110
             └─►90 DOEND
```

Fig. 4. *An example of a BASIC/3000 compound statement.*

SPL, BASIC, and FORTRAN are all recursive, that is, programs, procedures, and subroutines can call themselves. Fig. 5 illustrates this property.

COBOL/3000

COBOL (COmmon Business Oriented Language) is the result of an effort to establish a standard programming language for business processing. The original specifications were drawn up in 1959 by CODASYL (the COnference on DAta SYstems Languages). COBOL/3000 conforms to the highest level of Federal Government Standard COBOL and has the added capability of interprogram communications.

COBOL is a structured language that consists of Indentification, Environment, Data, and Procedure divisions. A feature of COBOL that makes it attractive in commercial applications is that it provides fixed-point arithmetic up to 18 digits; this eliminates the problem of round-off error which exists in "floating-point" formats.

Switching Languages Made Easy

HP 3000 languages share many common attributes that aid the user in switching from one language to another. Among the areas of compatibility are:

- Program-to-program communication. SPL, FORTRAN, and COBOL programs can all call programs written in either SPL, FORTRAN, or COBOL. BASIC programs can call SPL, FORTRAN, or COBOL programs as well as other BASIC programs. Files written in one language are accessible by other languages.
- Compiler construction. The command languages for all of the compilers are consistent. For ex-

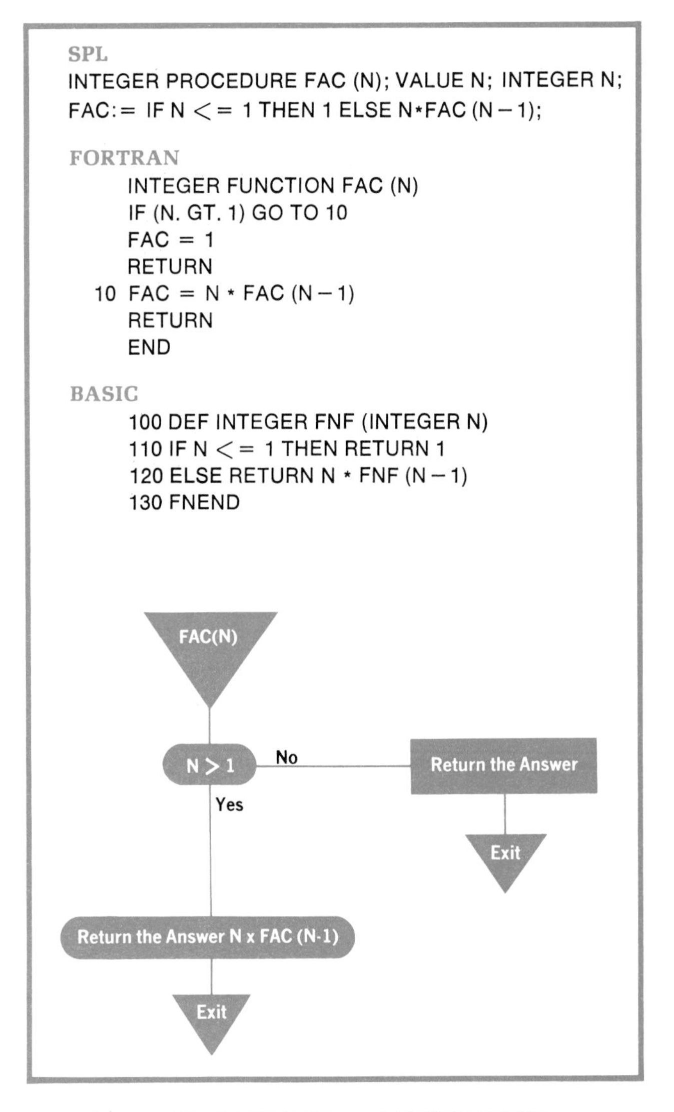

Fig. 5. *SPL, BASIC/3000, and FORTRAN/3000 programs to calculate integer factorials. All three languages have recursive capabilities.*

ample, the commands that tell the compiler to merge a source file with an update file are identical for each compiler. Also, the language translators share the same system library routines. These library routines are used both during compilation and as run-time routines to implement the language features. For example, the program that converts a character string into an internal binary number is used both by SPL at compile time and by the FORTRAN formatter at execution time. This modularity not only simplifies the task of making changes to common programs, but also reduces the development cost by eliminating duplication of effort. The steps in compiling and executing programs are as follows (Fig. 6):

1) The source program (main program plus subroutines) is compiled into relocatable modules that are stored in the user's subprogram file (USL). If the programmer decides to change any part of his program, he can recompile any subroutine, or the main program, into the USL file and the old copy of that subroutine will be deactivated. (It will still exist in the file, and could later be reactivated.) The relocatable modules can be added, deleted, activated, or deactivated from the USL. Also, these modules can be copied from one USL to another.
2) Next, the USL file is prepared into a Program File. Preparation consists of segmenting the code and defining the initial stack size.
3) Now, the Program File can be allocated/executed. The segments are allocated into virtual memory, external references are satisfied from the libraries, and the program is scheduled for execution according to its priority.

General-Purpose Applications Software

Several general-purpose software packages are now available for the HP 3000. There is a scientific library, an interactive statistical package, a text editor, and a text formatter. Other packages will be available in the future.

Scientific Library. The scientific library consists of a collection of SPL procedures that reside in the system library. The initial capabilities include: error function/complimentary error function, gamma and $\log_e$ gamma functions, exponential, sine-cosine, Fresnel integrals, elliptic integrals and elliptic functions, Bessel functions, and statistical procedures including elementary statistics (kurtosis, means, etc.), one-way frequency distribution, correlation, and multiple linear regression. This library

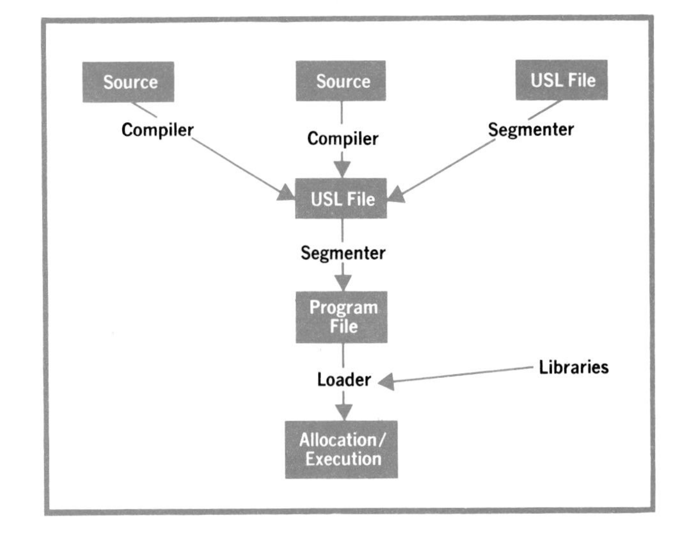

Fig. 6. *HP 3000 compilation/execution process.*

will be kept open for future enhancement.

Interactive Statistical Analysis Package (STAR). This subsystem provides the user with the capability of performing various kinds of statistical analysis in an interactive (question-answer) mode. This package may also be used in a batch mode. All of the statistical capabilities that exist in the scientific library are available to the STAR user, along with the following additions: data file manipulation (creation, editing, etc.), scatter diagrams, histograms, and variable transformation.

The output from STAR may be to the user's terminal, or to a line printer. All results are displayed in an easily readable, tabular form. The data may be input directly from the terminal, or from the batch input device, or from a file created by a FORTRAN, SPL, or BASIC program.

In keeping with the modular structure of the HP 3000 system, STAR makes use of the scientific and compiler libraries in performing its functions. As new capabilities are added to the scientific library, these capabilities will be easily extendable to STAR merely by adding the necessary input/output routines and calling on the scientific library to perform the calculations.

Text Editor. EDIT/3000 is a general-purpose utility that provides the user with the capability of easily creating and manipulating files of upper and lower case ASCII characters. Lines and characters can be inserted, deleted, replaced, searched for, and so on. The files to be edited can be FORTRAN, SPL, BASIC, or COBOL source files, or textual material such as reports.

One feature of this program not usually found in text editors is its ability to selectively modify text depending on conditions found within the text itself. When this is done, the "edit language" has an ALGOL-like structure with the metacommands WHILE, NOT, and OR acting upon statements that can be compound statements (groups of statements enclosed by a BEGIN-END pair). These commands and statements can be nested indefinitely. Interactive users can write an edit program to send messages to the terminal and place input from the user in appropriate places within the text file. Together, these features make the editor a powerful tool for many applications other than simple program editing.

Text Formatter. This program lists ASCII files under the control of format records imbedded in the text file. FORMAT/3000 may also be used with the text editor. The formatter provides the capability of preparing simple documents to be listed on line printers or other ASCII devices.

Acknowledgments

The following people were directly involved in the implementation of the languages and general-purpose products:

SPL: Doug Jeung, Gerry Bausek, Tom Blease.
FORTRAN: Jerry Smith, Terry Hamm, Jim Hewlett, John Couch
BASIC: Mike Green, Terry Opdendyk, John Shipman
COBOL: Steve Ng, Waldy Haccou, John Welsch, John Yu, Paul Rosenfeld, Gerry Bausek
STAR, Scientific Libraries: Paul Rosenfeld, Dave Johnson
Editor/Formatter: Fred Athearn.

Credit is also due the many people in software QA and publications who have done such a great job.

William E. Foster

As section manager for systems software, Bill Foster is responsible for programming languages and operating systems for 2100, 2000, and 3000 Computer Systems. Bill received his B.A. degree in mathematics from California State University at San Jose in 1966, then spent the next three years developing satellite orbit prediction, tracking, and reentry software. In 1969 he got his M.S. degree in applied mathematics from the University of Santa Clara and joined HP as a software project manager. He became a section manager in 1971 and assumed his present job in 1972. A member of ACM and the American Management Association, Bill is now a candidate for the M.B.A. degree at Santa Clara. He enjoys golf, tennis, bicycling, basketball, hydroplaning (he built his own boat), and exploring the Bay Area by motorcycle, and he's now taking flying lessons.

Single Operating System Serves All HP 3000 Users

The Multiprogramming Executive operating system takes care of command interpretation, file management, memory management, scheduling and dispatching, and input/output management for time-sharing, batch, and real-time users.

by Thomas A. Blease and Alan Hewer

MULTIPROGRAMMING EXECUTIVE (MPE/3000) is a general-purpose disc-based operating system that supervises the operation of the HP 3000 Computer System and its variety of users.

MPE/3000 allows users to access the system concurrently in three distinct but compatible modes: batch processing, time sharing, and real-time processing. MPE is designed to take maximum advantage of system resources, to make the system easy to use, and to relieve the user of the need for detailed knowledge of the internal hardware or direct interaction with it. Each user's environment is protected; program protection is provided by hardware and data protection by any of several software facilities depending on the degree of security desired.

MPE/3000 has a modular organization that makes it more convenient to check out and maintain, and provides a flexible base on which additional capabilities may later be developed. Users interact with the 3000 System through the command interpreter, one of the functional units of MPE. Programming access to the hardware is provided by system routines called MPE intrinsics. Uniform access to disc files and input/output devices is provided by the MPE file system. MPE also has memory management, an input/output system, and scheduling for dynamic allocation of resources.

Process Structure

Underlying the modularity of MPE/3000 and its ability to support three kinds of users concurrently is its process structure. Except for a few specialized system controls such as the dispatcher and interrupt structure, all operating-system and user functions are performed as a series of processes.

A process is the basic entity that can be executed by the central processor. While a *program* identifies a static sequence of instructions and data, a *process* denotes the dynamically changing sequence of states of an executing program. Under MPE/3000, a process consists of:

- A unique process control block which describes and controls the process,
- A private (stack) data segment, accessible only by the process, for data operation and storage, and
- An instruction in a code segment which may be private to the process or may be shared with other processes.

Processes are organized hierarchically in a tree structure as shown in Fig. 1. Each process has only one immediate ancestor, but may have several immediate descendants. Control and information flows are restricted to proceed only along branches of this logical tree structure. The primary interactions which are provided for are creation, deletion, control, and intercommunication.

The root process is the *progenitor*. All immediate descendants of the progenitor are *system processes*. They include:

- I/O system controller processes, which queue, initiate, and complete all input/output requests for all devices configured under the operating system.
- The make-a-process-present (MAPP) process, which schedules the allocation of memory resources to data segments belonging to active processes.
- The device recognition (DREC) process, which performs the administrative tasks of allocating input/output devices and also verifying and initiating new users under the operating system.
- The user controller (UCOP) process, which is

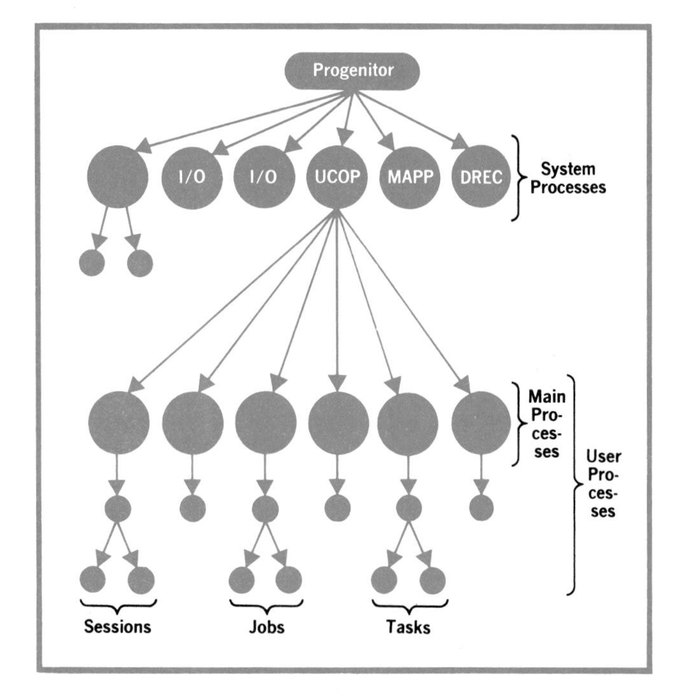

Fig. 1 *Multiprogramming Executive (MPE) operating system for HP 3000 has a process structure. All functions are performed as a series of processes.*

defined as the ancestor of all user processes currently running under MPE/3000. The primary responsibility of UCOP is to create, supervise, and delete user process tree structures.

Of these system processes, the most important is UCOP, the root process of the user structure. An immediate descendant, created by UCOP, is called a *main process*, and the code executing under it is normally the command interpreter. The process tree structure originating at a main process defines a *job* (job/session/task). A basic feature of a job is its complete independence from all other jobs currently existing.

Apart from the progenitor and several specific system processes which together constitute the operating system and which must exist, the process tree structure is completely dynamic, expanding and contracting as operating system and user requirements change.

Memory Management

The primary function of MPE/3000 memory management is the allocation of main memory to meet the demands of users. "Main memory" is core memory as opposed to disc memory. The memory management module is also responsible for code segment table entries, data segment table entries, and overlay disc storage for data segments.

Main Memory Organization

Main memory is organized into three contiguous areas (Fig. 2). The first area contains system tables, interrupt procedures, and MPE intrinsics which must be core resident, that is, always present and accessible in main memory.

The second area is of variable length and is used to satisfy requests from users for core resident storage. This area is dynamically expanded and contracted and can be of zero length.

The remaining main memory is referred to as *linked* memory. Linked memory is composed of free (not currently being used) and assigned (allocated for a code or data segment) areas of varying sizes. Areas not currently in use are linked together and form the free space list. Similarly, the assigned areas are linked together and form the assigned space list. Each area contains an information header defining its size. If the area is assigned, the header also contains information about disposition (I/O pending, etc.), segment type (code or data with in-

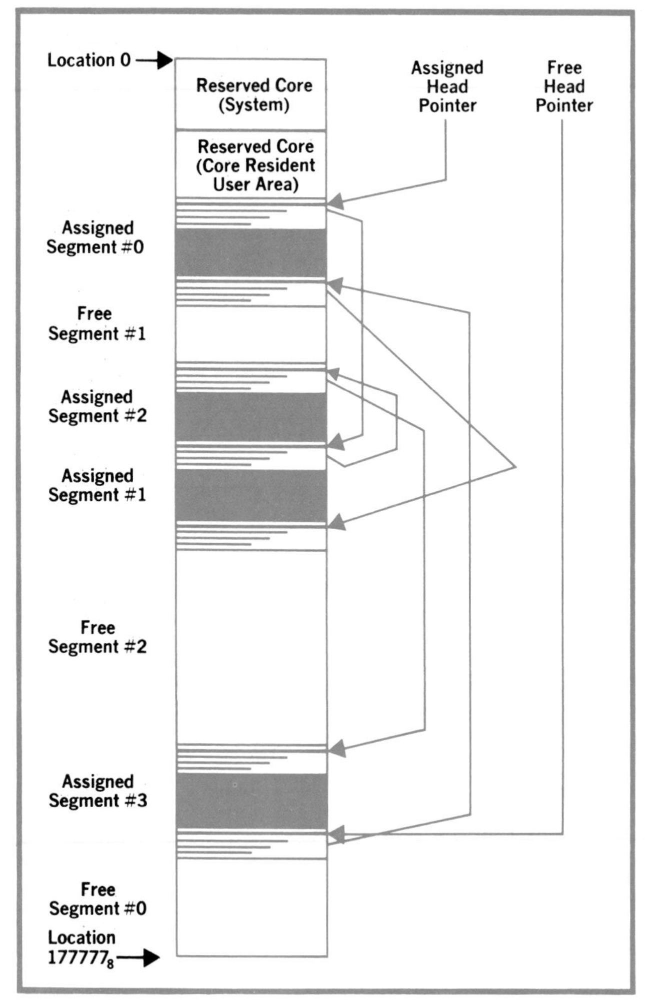

Fig. 2 *Main memory is organized into reserved and linked memory. Linked memory consists of free and assigned areas.*

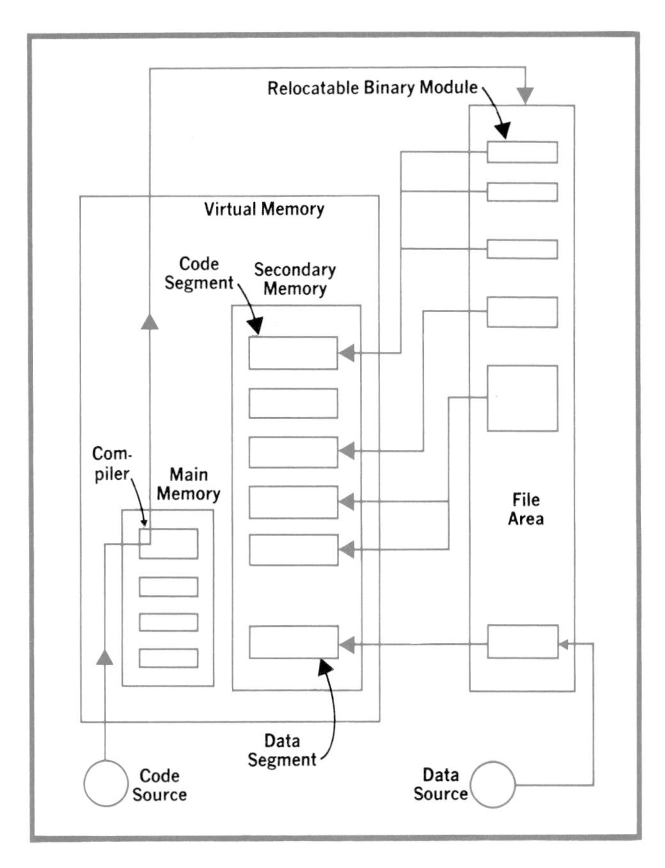

Fig. 3 *Program segmentation gives the HP 3000 virtual memory. MPE automatically brings into main memory only those code segments that are currently needed. Thus a user's program may be much larger than main memory.*

dex into code segment table or data segment table), disc address, priority, and frequency of access. This additional information is used in the selection of assigned areas to overlay when a request cannot be satisfied from the free area list.

Virtual Memory

Virtual memory consists of main memory plus an area of mass storage called secondary memory, or the swapping area (Fig. 3). The swapping area is on disc or drum memory, although not necessarily on a single device; it may include areas of several devices. In the swapping area is a collection of pieces of code or data defined as segments. As a program executes, segments are swapped in and out of main memory by the operating system. Whether a segment is in main memory or absent, it is nevertheless part of virtual memory. Thus from the point of view of a user, he is working with a memory that appears to be many times larger than the actual physical size of main memory. His own program may exceed the 65K-word maximum main memory capacity and still allow space for many other users on the same machine.

As shown in Fig. 3, code is entered into the computer in some source language, is translated to binary form by a compiler, and is stored in the file area. Each compiled program or subprogram exists in the file area as a relocatable binary module.

When the user is ready to execute his program, the appropriate command is given and the operating system loads the binary modules of his program into the swapping area of virtual memory. Simultaneously with this transfer, the binary modules are formed into segments as specified by the user. In some cases no actual change takes place; for example, a small program may consist of just one segment and the loader will probably not move it from a file disc onto the system disc unless the user wants this done.

Data segments are allocated dynamically when a program is loaded, and are always on the system disc.

Scheduling/Dispatching

To accommodate the different modes of operation which may coexist under MPE/3000, the scheduling system is based upon a priority structure. All processes are logically organized into a linear master scheduling queue in order of their priority.

The dispatcher is responsible for allocating the central processor to the active processes in the scheduling queue. A process is considered active if it requires access only to the central processor. Otherwise, it is considered inactive, awaiting some other resource.

The basic organization of the scheduling queue is shown in Fig. 4. System processes are scheduled directly onto the master queue. Subqueues are used to schedule processes belonging to users. Note that since processes are scheduled independently, not all processes in a job are necessarily entered in the same subqueue.

There are five standard subqueues. Three are linear in structure. In a linear (sub)queue, the highest priority active process is given access to the central processor by the dispatcher, and it maintains this access until it becomes inactive or until it is preempted when a higher priority process becomes active. The three linear subqueues are for core-resident processes, real-time processes, and low-priority (idle) processes.

The other two subqueues are circular subqueues. These are for time-share processes and batch processes. In a circular subqueue, all processes are considered to be of equal priority and each active process accesses the central processor for a certain time interval. At the end of this time interval, the process releases the CPU and the next active process in the subqueue is dispatched. This continues in a

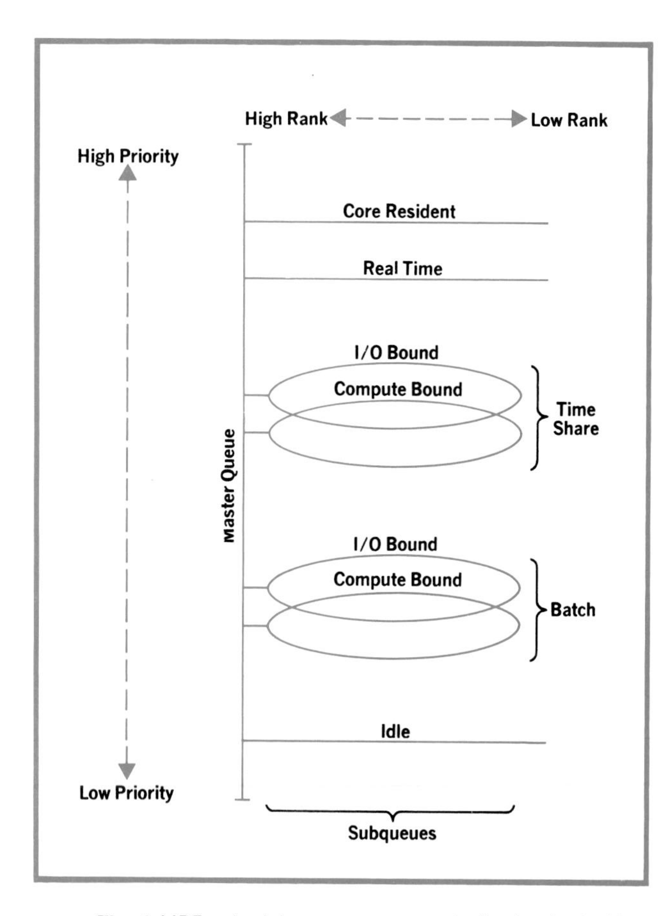

Fig. 4 *MPE schedules processes on the basis of priorities. Processes are organized into a linear master queue and five subqueues.*

round-robin manner.

Each of the two circular subqueues is composed of two subqueues—a higher priority subqueue containing I/O-bound processes and a lower priority subqueue containing compute-bound processes. The dynamic rescheduling of processes between the dual subqueues is performed by MPE/3000. In the case of highly interactive time-share processes, this arrangement provides quicker response at the terminal.

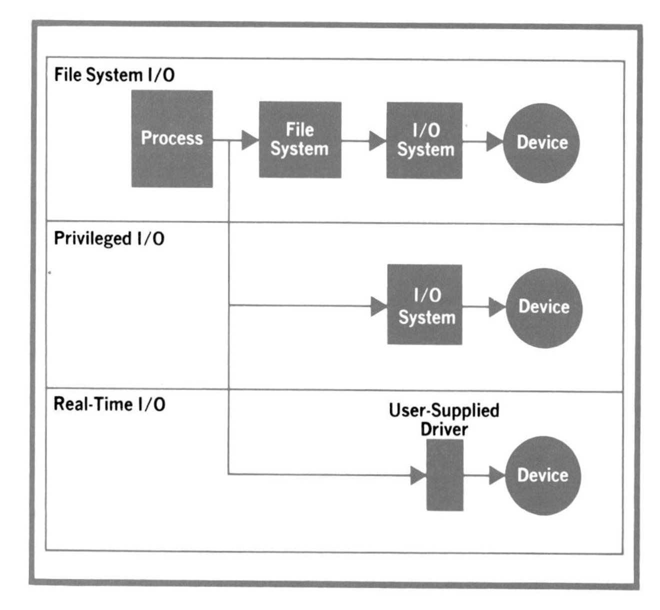

Fig. 5 *Basic HP 3000 input/output access methods.*

I/O System

The purpose of the MPE/3000 I/O system is to perform input/output operations for the file system. The user doesn't interact directly with the I/O system, but indirectly via the file system. However, privileged users may access the I/O system directly, and users with real-time capability may bypass both the file system and the I/O system for direct access to specific devices. Fig. 5 shows the basic I/O access methods.

In a typical I/O operation the sequence of operations is as follows. An executing user process generates a file request to the file system. The file system calls the attach-I/O intrinsic. Attach-I/O allocates an I/O queue entry and links it into the queue for the device specified. When all earlier requests for the device have been completed and the I/O monitor process has the highest priority among all other processes, the I/O monitor process begins execution of this request. There is one I/O monitor process for each device controller.

The I/O monitor process first assures that the data buffer is frozen in memory. The initiator section and the I/O program issue an SIO instruction to the device controller and return control to the I/O monitor process. Data is then transferred between the I/O device and the data buffer.

When the I/O monitor process is again dispatched, it recognizes that an interrupt has occurred and calls the completion section of the device driver. The completion section checks for successful completion and returns the results of the I/O operation to the file system via the I/O control block. The user's process is activated upon I/O completion.

When the user process is again dispatched, return is made to the point following the file request.

Acknowledgments

The following people were directly involved in the design and implementation of MPE/3000: Harlan Andrews, Larry Birenbaum, Terry Branthwaite, Jean-Michel Gabet, Jack MacDonald, Bob Miyakusu, Chris Larson, Tom Ellestead, Paul Rosenfeld, Steve Brown, and Myron Zeissler.

Thomas A. Blease

Tom Blease's career in software design and implementation got its start in 1960 when he received his B.A. in mathematics from the University of California at Berkeley. In the ensuing years he held positions in that field with several organizations in Florida and California. At HP since 1969, he participated in the design and implementation of SPL and MPE for the HP 3000. He's a member of ACM and he enjoys a good hike on his days off.

Alan Hewer

Alan Hewer received his B.A. and M.A. degrees in mathematics from Christ's College, Cambridge University, England in 1960 and 1963, respectively. Between 1960 and 1970 he worked on software design and implementation with various companies in England and the United States. When he joined HP in 1970, he was first involved in the hardware design of the HP 3000. Later in the project he took on his recent responsibilities in the design and implementation of MPE/3000.

Hewlett-Packard Company, 1501 Page Mill Road, Palo Alto, California 94304

HEWLETT-PACKARD JOURNAL

JANUARY 1973 Volume 24 • Number 5

Technical Information from the Laboratories of Hewlett-Packard Company

Hewlett-Packard S.A., CH-1217 Meyrin 2
Geneva, Switzerland
Yokagawa-Hewlett-Packard Ltd., Shibuya-Ku
Tokyo 151 Japan

The Logic Analyzer: A New Instrument for Observing Logic Signals

Designed specifically to solve digital design and trouble-shooting problems, this new instrument provides a digital display with storage, positive and negative digital delay, combinatorial triggering, and digital sequence comparison.

by Robin Adler, Mark Baker, and Howard D. Marshall

DIGITAL CIRCUITS confront designers and troubleshooters with measurement problems that differ in many ways from those in analog circuits. As a result, instruments developed for analog measurements are often less than optimum for the new digital measurements, so there is a need for new kinds of instruments designed specifically to solve digital problems.

Model 5000A Logic Analyzer, Fig. 1, is just that. An entirely digital instrument for displaying logic signals, it has a digital display, digital functions, and digital controls, and it operates in a manner that is intuitive to digitally oriented users.

A good example of the kind of problem that is easily solved by the Logic Analyzer is observation of long and infrequent logic sequences in calculators and other ROM-controlled systems. With conventional instruments, seeing these sequences is quite difficult and identifying individual bits is virtually impossible. With the Logic Analyzer and its digital storage, once-per-keystroke calculator sequences are easily captured. The Analyzer's digital delay makes it possible to observe any section of a thousand-bit sequence with no uncertainty as to which bits are displayed and no need to count clock pulses. These abilities also make it easy to see and analyze the long, non-repetitive signal sequences that occur in serial data transmission, such as between remote terminals and a computer.

Another problem arises in disc drives and other motor-driven computer peripherals. Continuous variations in drive-motor speed may cause so much jitter in the data waveforms that they are impossible to interpret when observed by conventional means. The Logic Analyzer's digital delay removes the jitter, so the display is stable and easily interpreted.

Other Logic Analyzer features are useful in a variety of digital applications. In its SPIKE mode, the Analyzer captures and stores the short, randomly occurring noise pulses that often cripple entire systems while escaping detection by conventional means. Negative delay helps the user analyze causes of errors in single-shot data sequences by displaying data that occurred prior to a trigger point. And a trigger point need not be defined simply as an edge occurring at a single node; the Analyzer can be set to trigger on coincidences of logic HIGH's or LOW's at two or three nodes.

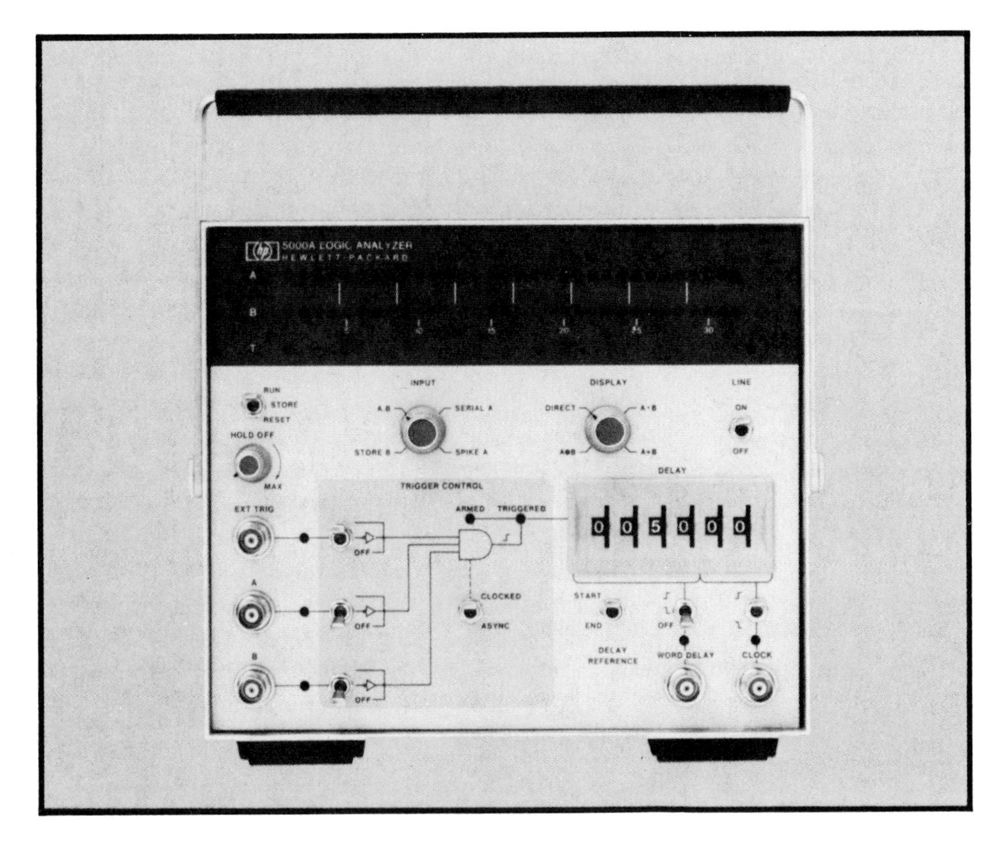

Fig. 1. *Model 5000A Logic Analyzer's two rows of 32 light-emitting diodes display the sequences of logic states at the A and B inputs. Positive and negative delay, combinatorial triggering, storage, spike detection, and several display modes help solve a multitude of digital measurement problems in design and troubleshooting.*

Clocked Light-Emitting-Diode Display

The Logic Analyzer displays 32 bits of digital data on each of its two rows of red light-emitting diodes (LED's). Each LED row displays data from one of the Analyzer's two data inputs, Channels A and B. The LED's turn on to indicate logic HIGH's and turn off for logic LOW's (Fig. 2).

The horizontal parameter of the display is time, proceeding from left to right. Time is quantized into discrete intervals according to the clock signal of the circuit or system under test. Thus the LED's represent the logic states of the data inputs during each of 32 successive clock cycles. The system clock signal is applied to the Analyzer's CLOCK input and serves to synchronize the Analyzer with the system under test.

How does this display compare to the more familiar oscilloscope display of logic waveforms? They are very similar, the Analyzer deleting some oscilloscope information and emphasizing other parameters more pertinent to digital needs. Oscilloscopes display voltage versus time; the Analyzer display quantizes each of these parameters, voltage into logic state and time into clock cycles.

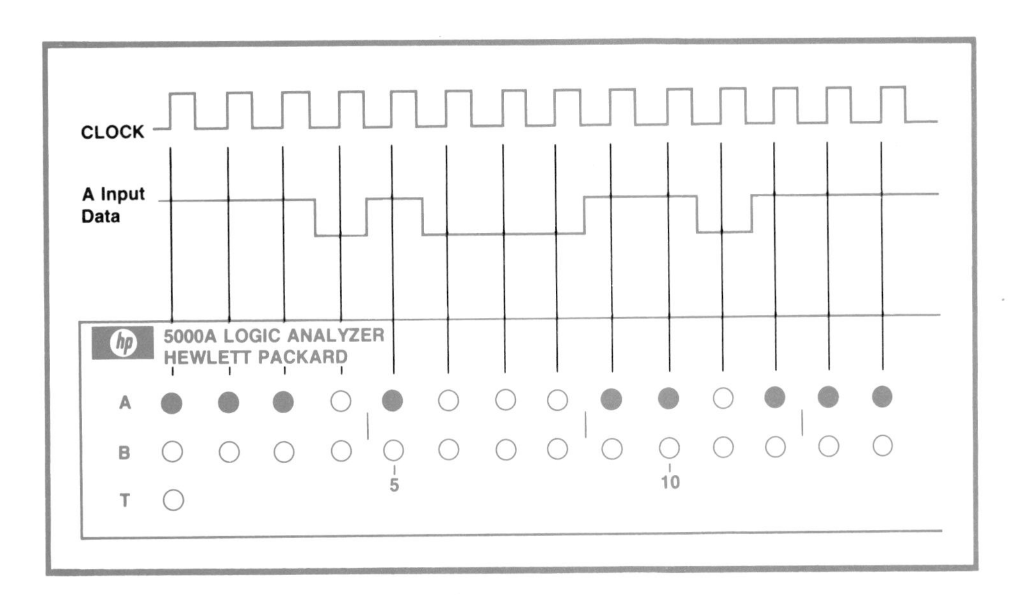

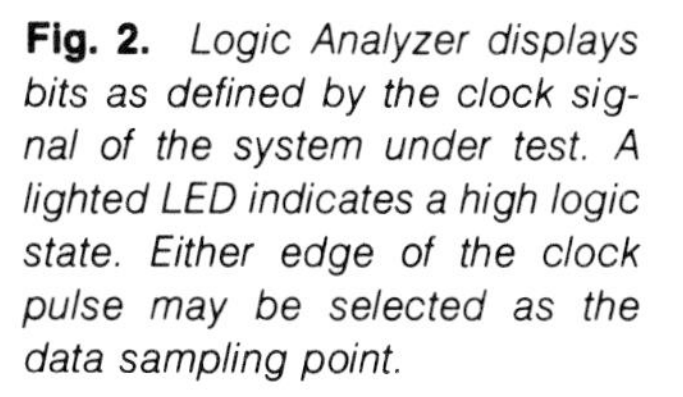

Fig. 2. *Logic Analyzer displays bits as defined by the clock signal of the system under test. A lighted LED indicates a high logic state. Either edge of the clock pulse may be selected as the data sampling point.*

Omitted from the Analyzer's display is analog voltage and timing information. These of course, are the signal parameters over which the digital designer has least control. Once he has selected a particular IC family, he cannot, for example, adjust high-state output levels or propagation delays beyond the limits specified on data sheets.

What the user does control is what the Analyzer is designed to display, that is, the procession of logic states as a function of clock cycle, or in other words, the function of the circuit. On and off LED's correspond naturally and intuitively to HIGH and LOW logic states. Using the test system's clock for the horizontal quantization means the data is displayed in terms of bits—just the information the user needs to compare circuit response to device truth tables or system timing diagrams.

Oscilloscopes, on the other hand, force the user to make the quantization into bits. To do this, he must display the clock along with the data and visually divide the continuous data stream into the bits of interest. At the same time the user must mentally impress the threshold voltage of the logic family on the waveforms to make decisions about HIGH's and LOW's.

This is not to say that either instrument precludes use of the other. Both are essential, although for different things. The oscilloscope is most important early in the design cycle, when questions about ringing, race conditions, fan-outs, and power distribution are important. These are questions that need to be answered before the system begins to run. Later in the design cycle, when the sequences of logic states are most important, and in production test and field service, the Logic Analyzer is likely to be more useful. There are also many measurements (often involving the Analyzer's digital delay) that can be made only by using both instruments together.

Digital Combinatorial Triggering

Triggering begins the Analyzer's data input and display processes. The trigger point is the reference point to which all following events are related.

The Analyzer's controls provide for triggering on either single events or multiple events occurring simultaneously. This capability is necessary in digital applications because digital systems commonly function in response to parallel data patterns occurring simultaneously on several circuit nodes. A device, for example, might be designed to accept data only when two control lines are HIGH concurrently. Therefore, synchronizing a measurement with the start of a process often requires the ability to recognize multiple simultaneous events. This the Analyzer can do. It triggers on an AND combination of up to three signals or their complements.

A major Logic Analyzer design consideration was that all controls be simple to use and intuitively understood by digital users. The trigger section is a good example. All trigger switches are grouped in a shaded area on the front panel (Fig. 3). The algorithm describing trigger requirements is drawn on the front panel as a simple schematic diagram using common digital symbols.

Triggering requires a high-going edge at the output of the front-panel AND gate. Trigger data can originate at three inputs, A, B, and EXT TRIG, reaching the AND gate via the slope-control toggle switches in each of the three paths. When any of these three-position switches is set to the upper position, data is transmitted directly from the associated data input to the AND gate. With the switch in the middle position opposite the inverter, data is inverted before application to the gate. In the lower or OFF position, the input is removed from the trigger process.

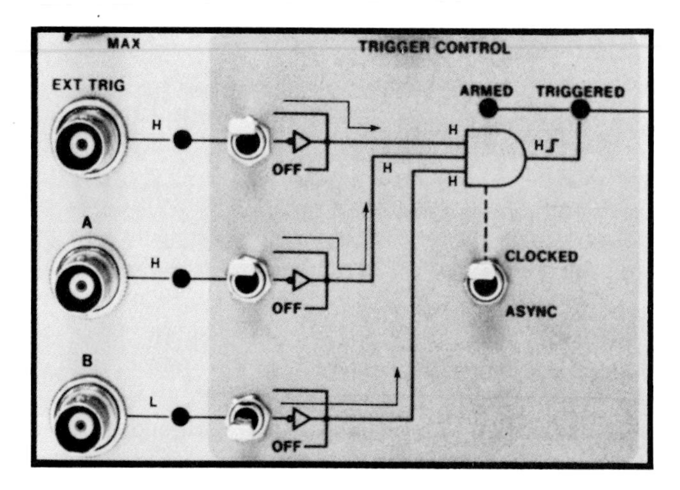

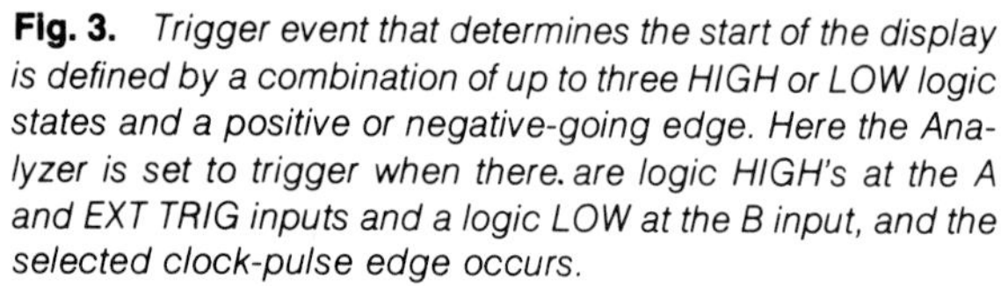
Fig. 3. *Trigger event that determines the start of the display is defined by a combination of up to three HIGH or LOW logic states and a positive or negative-going edge. Here the Analyzer is set to trigger when there are logic HIGH's at the A and EXT TRIG inputs and a logic LOW at the B input, and the selected clock-pulse edge occurs.*

With this arrangement the user can select any three-bit combination of HIGH's and LOW's as his trigger word. With the slope control switch up, the input must be HIGH for triggering; with it opposite the inverter the input must be LOW.

In ASYNC mode, when the AND gate pulses HIGH, the TRIGGERED light flashes immediately to indicate triggering. In CLOCKED mode, the AND output must remain HIGH until sampled by the clock.

Displayed data is always retained until new information is written over it after a subsequent trig-

Logic Analyzer Applications in Digital System Waveform Measurements

Wherever oscilloscopes are used to display digital information, the Logic Analyzer will be a valuable addition to the user's test equipment. Specific areas include computers, magnetic tape and disc drives, card readers, teleprinters, paper tape readers, point-of-sale readers and terminals, calculators, digital data transmission, ROM-controlled instruments, and others.

Disc Drives

Magnetic disc drives are fast-access, mass storage devices that record information on rotating magnetic platters or discs. Data is stored and read serially. The address of each data record on the disc is contained in a preamble that precedes the record and is part of the serial data stream. When the contents of a given record are desired, its address is entered and the data stream is monitored for that address. When the address is detected, a comparison circuit generates a pulse and data transmission begins. If this pulse is used to trigger an oscilloscope or Logic Analyzer, subsequent data is viewable. However, the preamble data has already passed by; seeing it requires a negative delay. The 5000A's END mode provides negative delay and makes it possible to see the address and other information stored in the preamble.

Looking farther into the data record following the trigger creates other difficulties. The timing of the waveforms depends on the rotational speed of the disc, which continually varies. Hence the waveforms jitter with respect to the trigger. Jitter is cumulative and becomes quite large several thousand bits into the record. Stable presentation of jittering data requires digital delay. A particular data bit always occurs a fixed number of clock cycles into the record, even though the time delay between the trigger and the bit of interest varies widely. Dialing this delay into the Logic Analyzer's thumbwheel register will produce a stable display of any desired bit.

The Logic Analyzer's single-shot storage simplifies finding and analyzing intermittent faults, which are common in disc drives and other computer systems. When a system fails once per hour or even less often, the 5000A will, whenever the error occurs, capture a sweep of data and hold it indefinitely. With negative delay, data that preceded the error can be displayed so the cause of the error can be analyzed.

Microprogrammed Devices

Microprogrammed, read-only-memory-controlled devices commonly generate long non-repetitive data sequences that are difficult to analyze. Calculators are a good example of such devices. A program to calculate the square root of a number might contain over ten thousand bits and repeat only as often as the square-root key is pressed.

In troubleshooting a ROM-controlled instrument, it may be possible to single-step through long sequences. However, this is often inconvenient, and the problem may go away at slow speeds. The Logic Analyzer allows the system to run at its normal rate while the test is being made. It can display not only data sequences but also spikes or transients on critical control lines. In cases where ROMs control arithmetic operations involving several shift registers, the Analyzer can be used to display in detail the complex operations that take place in these shift registers. This analysis capability goes far beyond merely observing the final result of a complex operation.

The Logic Analyzer's digital display, digital delay, and single-shot storage also simplify calculator measurements. The digital clocked display presents data and program instructions in the form of bits, just as they appear on timing diagrams or ROM truth tables. Storage captures the transient waveforms. If RZ (return-to-zero) encoding is used, SPIKE A mode will produce the desired display.

Because the serial instructions or data are usually grouped into words, digital delay by both bits and words is useful. With once-per-word pulses applied to the Analyzer's WORD DELAY input, any word can be displayed without counting pulses to find it. Bit delay within the selected word is helpful when word length exceeds the display capacity.

Digital Data Transmission

In digital data transmission, characters, words, and blocks of data take the form of fast single-shot bursts of ones and zeros. This is true whether the data is being transmitted between a terminal and a computer, between two data banks via modems and telephone lines, or between a frequency counter and a calculator in a small automatic test system. The Logic Analyzer easily displays these bursts. The Analyzer display is locked to the data by the clock and the Analyzer triggers only when data is received. Successive characters can be displayed one after another and stored as long as desired.

To access a specific part of a data block, such as a preamble, a postamble, a control bit, or a parity bit, the Analyzer's positive and negative digital delay can be used. If a trigger can only be found at the end of the preamble, engaging negative delay by selecting DELAY REFERENCE "END" will display the data that leads up to the trigger, namely the preamble. With exactly the same trigger point, the end of the data block can be observed by dialing in the correct amount of positive delay and retransmitting the data.

Analyzer and Oscilloscope

The synergism that exists between the Logic Analyzer and the oscilloscope is most evident when either instrument is used as a trigger source for the other. When the Analyzer's countdown of bit and word delays is complete, a pulse that can be used to trigger an oscilloscope is generated at the rear panel TRIG OUT connector. The 5000A's combinatorial triggering and digital delay combine with the oscilloscope's voltage and timing information to display analog parameters of hard-to-locate data bits. Stable oscilloscope displays of disc data or calculator program words are possible with this technique.

With a delayed-sweep oscilloscope, the reverse arrangement is often fruitful. Triggering the Analyzer from the delayed gate output of the oscilloscope begins the 5000A data input at the start of the intensified region on the oscilloscope CRT. The oscilloscope delay vernier can be used to scan rapidly through the waveform searching for suspect data. The Analyzer simultaneously displays the precise bit structure of the scanned data and indicates any spikes that are present. Searching for a particular control character in a serial transmission to or from a data terminal is a good example of the type of problem that can be solved using this technique.

(Continued on page 6)

(Continued from page 5)

Everyday Uses

While the examples given so far have concentrated on dramatic applications of the Logic Analyzer in difficult system measurements, the Analyzer will probably be used most often as a general-purpose bench instrument in common digital troubleshooting situations. For example:

- With the CLOCK input connected to the clock line of a flip-flop and the channel A input connected to the Q output of the flip-flop, the flip-flop can be checked for proper divide-by-two operation, indicated by a display of alternately off and on LED'S.
- An AND gate has pulses at its two inputs but the output remains LOW. Combinatorial triggering can determine if the inputs are ever HIGH simultaneously.
- With digital delay determining the number of clock pulses between overflows, proper division ratio in a programmable frequency divider is easily verified.
- The output pulse of a monostable multivibrator (one-shot) may be viewed using asynchronous triggering and SPIKE mode, no matter what the pulse relationship of the clock to the one-shot output.
- Data bits dropped in a recirculating shift register can be detected by comparing the circulating data with a stored reference.

ger. If only a single trigger occurs, the data is held indefinitely. The result is automatic, single-shot storage. There is no need to realize beforehand that data is non-recurring and engage special storage controls.

Even if a data sequence is repetitive, the Analyzer can be commanded to store it by placing a front-panel switch in the STORE position. With this same switch in the RUN position the display will change each time a trigger occurs. An adjustable HOLDOFF control prevents triggering for up to five seconds in the RUN mode so new data can be displayed at a convenient rate. At the end of the holdoff period an ARMED light on the front panel turns on to indicate that the Analyzer is ready to trigger.

Digital Delay

Normally, once the Analyzer has triggered, the next 32 bits of data at the A and B inputs are displayed. Other sections of the data sequence can be displayed using the positive and negative delay capabilities of the Analyzer. Positive delay, expressed as a number of clock pulses (or a number of word pulses and a number of clock pulses), is dialed into front-panel thumbwheels, and delay equal to the indicated number of pulses is inserted between triggering and the beginning of the display (Fig. 4). Thus the 32-bit "display window" is movable up to 999,999 bits (or 9999 words and 99 bits) downstream from the fixed trigger point, and there is no doubt as to which bits are being displayed.

Why *digital* delay? In synchronous systems, the clock signal is the prime mover; all signal sequencing is in response to it. Systems are designed so timing is in terms of numbers of clock pulses rather than their duration. An example is a system that might be set to generate an output pulse 1024 cycles after a start command. That this interval might happen to be 0.509 ms long is almost unimportant; the 1024 clock cycles are the significant point.

When the clock frequency is continually changing, as in a disc drive, this is even more significant. Here the time delay between the start of data transmission and a particular data bit is continually varying. The number of clock periods in the interval, however, is fixed, so digital delay will locate the desired data bit without jitter.

Digital delay is also useful when data streams are extremely long. Calculator sequences are an example. When trying to find a particular bit in a several-thousand-bit sequence, digital delay to the desired bit is the only way to get there and be certain of it.

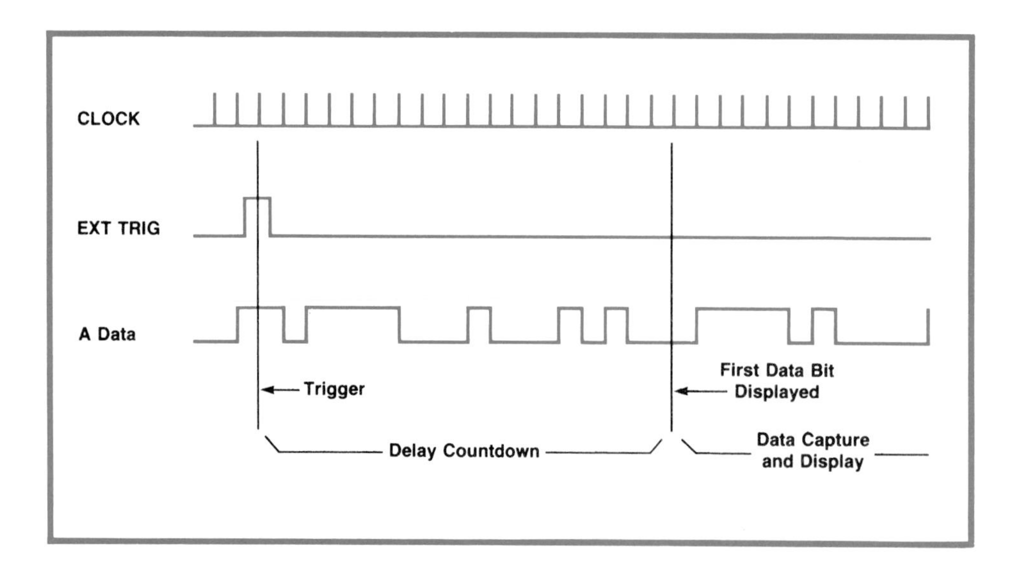

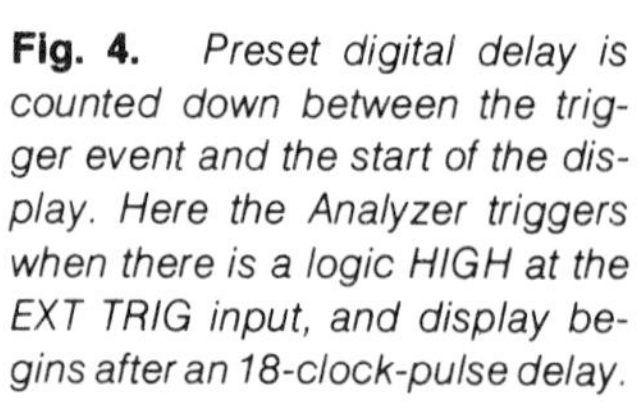
Fig. 4. *Preset digital delay is counted down between the trigger event and the start of the display. Here the Analyzer triggers when there is a logic HIGH at the EXT TRIG input, and display begins after an 18-clock-pulse delay.*

Negative Delay

Some oscilloscopes use an analog delay line to preserve information close to the trigger long enough for the CRT to display it. The 5000A extends this concept using digital storage to permit display of events up to 32 clock cycles before the trigger.

Negative delay is engaged by setting the DELAY REFERENCE switch to the END position. The trigger then occurs at the end of the display registers and the data displayed is the data that occurred on the 32 clock cycles prior to the trigger.

Uses of negative delay are numerous. For example, when analyzing serial trigger circuits in disc drives or data terminals, the serial word leading up to and causing the trigger can be displayed. Or, triggering from an error condition, the pre-error data may be examined to locate the cause of the error.

Input and Display Modes

In its most common mode of operation the Logic Analyzer is a two-channel device with a row of LED's for each channel. The various INPUT and DISPLAY modes use the two inputs and displays in other ways to perform additional functions. SERIAL A mode, for example, cascades the B display register onto the end of the A display. The result is a single 64-bit display loaded from the A input, and a single-channel 64-bit negative delay capability, as well. STORE B provides selective use of the Analyzer's storage. B-channel data is stored while the A display reloads with each trigger.

SPIKE A mode captures short, intra-clock-period pulses caused by noise or other problems (Fig. 5). These would ordinarily be missed by the Analyzer's normal once-per-clock-cycle sampling of the data inputs. High-going pulses occurring at the A input are displayed in the A register and low-going pulses in the B register. One reason SPIKE mode is necessary is that without it RZ (return-to-zero) data would either be undisplayable or undistinguishable from NRZ (non-return-to-zero) data.

The settings of the DISPLAY mode switch select various Boolean combinations of A and B input data for display. AND (A • B), OR (A + B), and EXCLUSIVE-OR (A ⊕ B) combinations are possible. Results appear in the A display; the B display is blanked.

A ⊕ B mode in conjunction with STORE B is especially useful in production test applications. EXCLUSIVE-OR is a comparison function; the result A ⊕ B is HIGH only when A and B are different. A ⊕ B mode allows rapid digital comparison of two supposedly identical data streams with any differences appearing as lighted LED's. Using

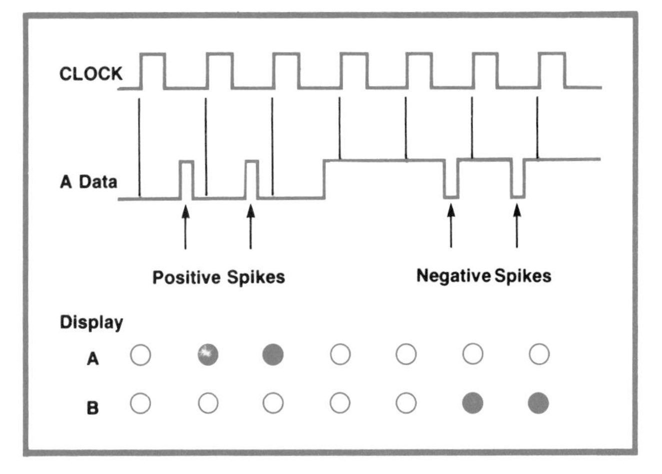

Fig. 5. *Spikes occurring between clock pulses may be desired signals or noise. In SPIKE A mode, the Analyzer displays spikes and shows whether they are positive-going or negative-going.*

STORE B mode, the B data may be loaded once and retained indefinitely, a useful capability when data from a reference device is being compared with data from one or more test devices (Fig. 6).

Getting Into the Analyzer

To minimize loading of circuits under test regardless of their IC family, each of the Logic Analyzer's five input channels (A, B, EXT TRIG, WORD DELAY, and CLOCK) has high input impedance: 1 MΩ in parallel with 25 pF. This also permits use of

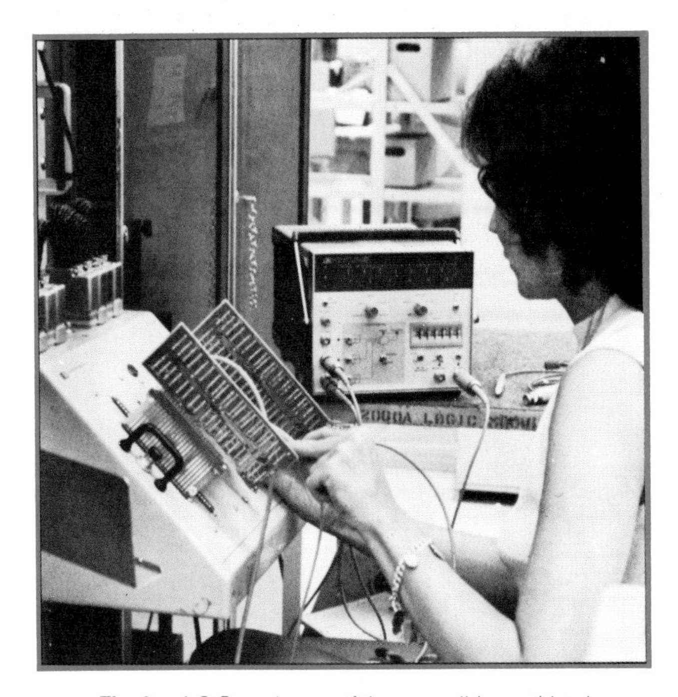

Fig. 6. *A ⊕ B mode, one of three possible combined-waveform display modes, is useful for comparing test and reference circuit boards in production testing. Only differences are displayed.*

the Analyzer with standard Hewlett-Packard high-impedance probes and their accessories to further minimize circuit loading and to maximize probing flexibility.

The input channels are essentially identical. Each acts as a threshold detector to compare a high-impedance, low-level input to a dc threshold level and produce a logic signal indicating whether the input is HIGH or LOW. The inputs share a common threshold voltage, which is set by rear-panel controls. Two selectable threshold ranges are provided, each with its own voltage control. The high range covers ±1.4 volts and the lower one ±0.14 volts. The actual threshold voltage can be monitored at the rear panel for precise threshold adjustment. With 10:1 divider probes, effective thresholds are 10 times the values indicated on the rear panel. Thus ±14 volts of threshold variation is possible, sufficient to cover requirements of all types of digital integrated circuits.

Although the Analyzer is specified for clock repetition rates up to 10 MHz, the actual equivalent bandwidth of the signal inputs is greater than 50 MHz. This relatively high bandwidth is necessary to minimize skew between input channels and to provide high sensitivity to narrow input pulses.

The basic circuit of each input amplifier is illustrated in Fig. 7. Q1A and Q1B, a matched FET pair, are used in a totem-pole voltage-follower configuration to form a unity-gain amplifier. CR1 and CR2 clip excessive input signals to provide protection against overloads of ±200 V dc or ±400 V transient. U1 is a high-speed, high-gain analog comparator IC with complementary TTL outputs.

As an added convenience to the user, each input channel drives an LED annunciator that indicates the logic state of that input. Each annunciator functions as a logic probe, turning on to indicate logic HIGH's and off for logic LOW's, and flashing at a rate of a few Hz for pulse-train inputs. The annunciator circuit (Fig. 8) includes a pulse stretcher so the user can easily see the presence of single-shot or low-repetition-rate pulses, often very difficult to detect. The threshold voltage can be quickly set by connecting one of the inputs to a voltage equal to the desired threshold and adjusting the threshold control until the annunciator barely turns on or flickers.

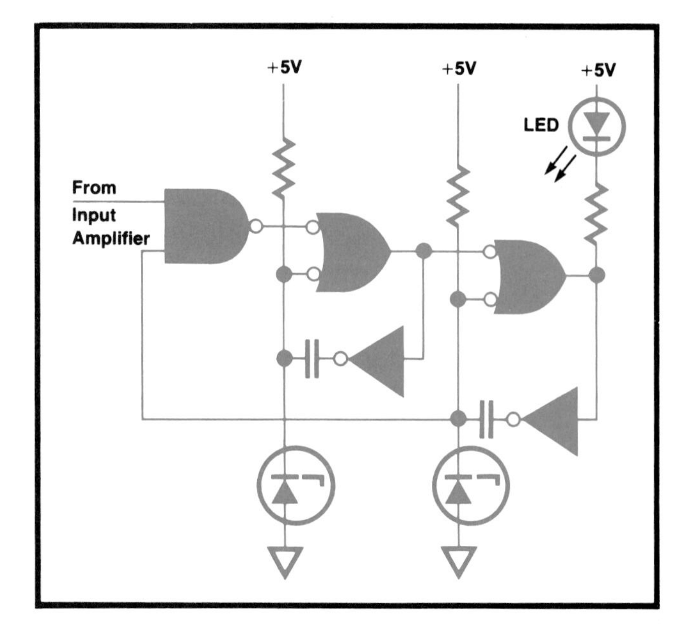

Fig. 8. *At each input is an LED annunciator that acts like a logic probe. LED turns on when the input is HIGH and off when it is LOW. Pulses are stretched to give a visible flash.*

Data Capture

The data on the A, B, and EXT TRIG channels goes from the comparator in the input amplifiers to the input flip-flops, where it is sampled and

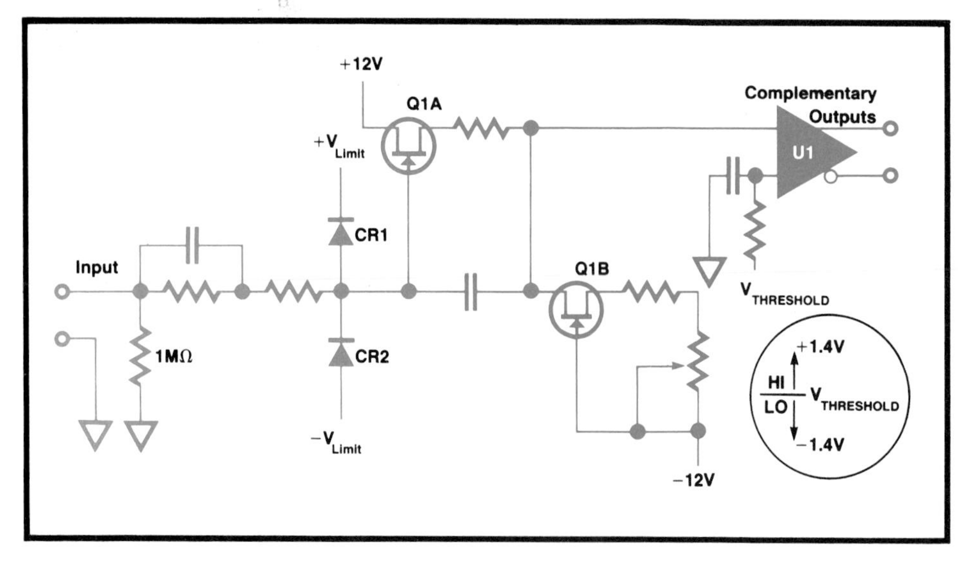

Fig. 7. *High-impedance input amplifiers minimize circuit loading. Variable threshold assures compatibility with all IC logic families.*

quantized into bits. Data at the inputs to these flip-flops is transferred to their outputs at the clock pulse edge selected by the slope control on the front panel. This takes place whenever there is a clock signal present. However, the display is not loaded until the Analyzer is triggered and all digital delay is counted to zero. When this occurs, data is transferred to the input registers a short time (50 ns) after it is loaded into the sampler flip-flops. (See block diagram, Fig. 9.)

For easier use and interpretation of the Analyzer and its display, the first bit in each display row represents the data at the A or B input when the trigger event occurs. It is also necessary that these first and subsequent bits be displayed immediately, as soon as they are sampled, without waiting (e.g., for the next clock pulse). To accomplish this the contents of the input samplers are continuously monitored for the trigger event. When it occurs, a delayed version of the input clock is gated to the input registers to load the contents of the samplers for display. On each succeeding clock pulse data is loaded into the samplers and 50 ns later is transferred to the input registers by the delayed clock. Loading continues in this way until the two 32-bit channels are filled.

Meanwhile, data has also been flowing from the input registers to the display registers. Details of this process are described in the section headed "Display Considerations."

Once loaded, data is frozen for a minimum of 50 ms before a new trigger and input sequence can begin. This guarantees that even an intermittent bit is displayed long enough to be recognized.

Triggering

One of the most difficult problems to overcome in the display of any waveform is the definition of a trigger or sync point to which the data can be referenced. It is necessary that the trigger, however defined, be a unique event, one that occurs only once in the cycle of the machine whose waveforms are being analyzed. When signals are repetitive, this guarantees that triggering always occurs at the same point in the sequence and, most importantly, that the display will be stable.

Examination of trigger requirements in synchronous systems reveals several possibilities. First, the trigger may be well defined as an event on a single line—for example, start-program pulses, reset pulses, or gating signals. In this case, the familiar single-channel, edge-sensitive triggering is sufficient and the positive or negative edge of the pulse provides a unique data reference. It is crucial that an edge rather than a level serve as the trigger since the pulse may extend over many clock periods. The edges are unique; the level is not.

If the trigger cannot be derived from a single line, several signals can be ANDed together to provide the unique reference. This is controlled using the Analyzer's three trigger control switches.

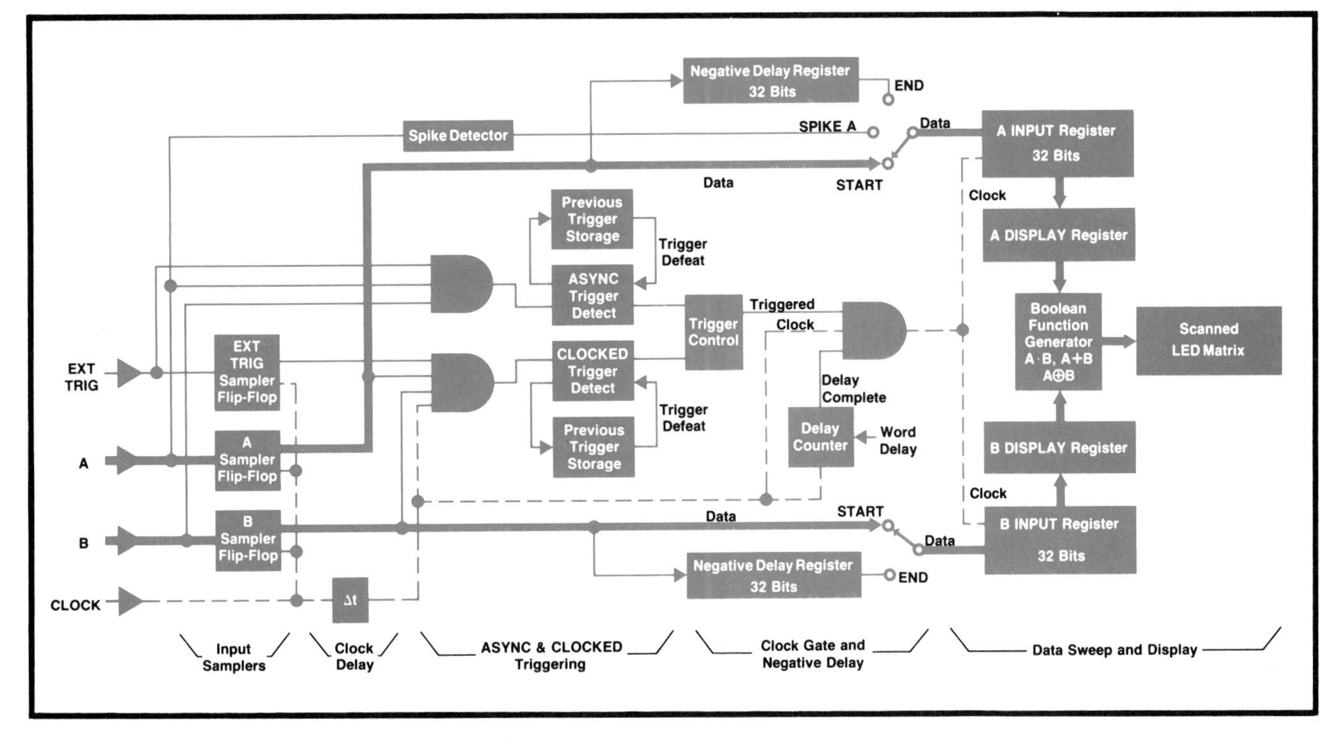

Fig. 9. *Logic Analyzer block diagram.*

Edge sensitivity of the Analyzer's trigger is achieved by an algorithm performed in the blocks labeled "Previous Trigger Storage" in Fig. 9. In words, the algorithm is simply that for a given event to trigger the Analyzer, the previous clock cycle must not have contained a trigger event. Thus the Analyzer does not trigger, for example, if its inputs are such that the front panel AND gate output is continuously HIGH.

In many cases it is desirable to ignore spikes and transients when triggering. This can be done by selecting the CLOCKED mode of operation, in which the inputs are monitored for the trigger condition only at the selected clock edge (Fig. 10). The converse is also possible, in case the desired trigger is a spike. When this happens—that is, trigger information exists only between sampling points—ASYNC mode should be selected. The spike will be captured and stored until the clock occurs to enter it as a trigger condition. Both of these situations are common in digital testing.

Digital Delay Countdown

Once the Analyzer has been triggered, it counts down any digital delay that has been entered in the thumbwheel switch register. The contents of the thumbwheel switches are loaded into six down counters, which are then decremented.

The chain of six down counters may be decremented solely by clock pulses or by a combination of clock pulses and WORD DELAY pulses. When the word delay function is engaged, the left four decades of the thumbwheel switch register are counted down by pulses of lower repetition rate than the clock. These might correspond, for example, to pulses that occur once per serial word in a calculator program. The four digits of word delay in combination with the remaining two digits of clock delay provide coarse and fine digital delays between the trigger and the start of the display.

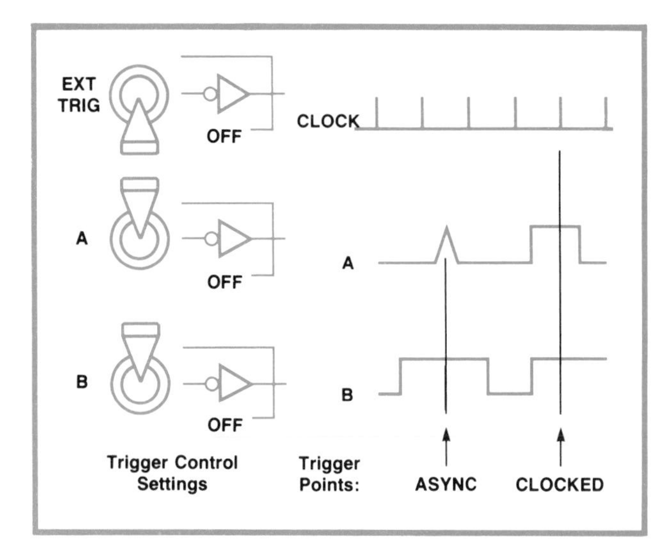

Fig. 10. *In CLOCKED mode, the Logic Analyzer monitors its inputs for the trigger condition only at the selected clock edge. In ASYNC mode, triggering occurs as soon as the trigger condition appears at the inputs. Triggering is always edge-sensitive.*

The Analyzer's negative delay is implemented by running all incoming data through 32-bit digital delay lines (shift registers) prior to loading it into the input registers. When a trigger occurs, therefore, the data that is displayed actually occurred 32 clock periods earlier and has spent the interim traveling through the shift register. The result is a display whose trigger event appears displayed at the right side, or end, of the display. Hence the START and END positions of the DELAY REFERENCE switch.

In END mode the thumbwheels still cause delay in the positive sense. The display ends the indicated number of pulses after the trigger. By selecting the END mode and less than 32 bits of positive delay, pre-trigger and post-trigger events can be seen in the same display.

Display Considerations

In addition to the input registers, which provide the sixty-four bits of information storage, there is also a second set of shift registers that provide an interface to the LED display. The function of these display registers is threefold.

First, the display registers are necessary to position the displayed data bits in the proper place. It was desired that data should always be displayed with the oldest data on the left and the newest data on the right, just as on an oscilloscope display.

The input shift register places each new bit in the rightmost position and pushes older data to the left. Thus if the contents of the input register were displayed directly, data would shift in from the right. To prevent this, as each new bit is entered, the contents of the input register are loaded in parallel into the display register. The display register then quickly shifts so the oldest bit always appears at the left side of the display. This occurs on each clock cycle (Fig. 11). With 32-bit displays, the number of bits that data must be shifted on each clock cycle is thirty two minus the number of bits entered since the start of display.

This data positioning circuitry is required only for low-frequency data, because the movement of data apparent at slow rates cannot be seen above about 1 kHz. Therefore the data positioning circuit is disabled above 1 kHz and data is transferred to the display registers only after all 32 bits of the

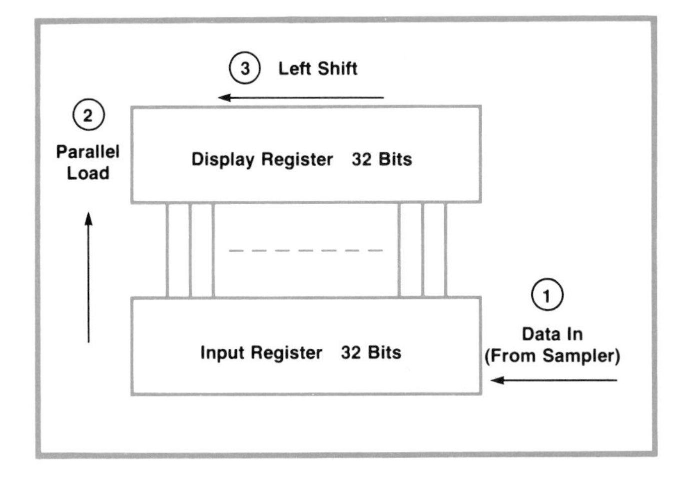

Fig. 11. *Logic Analyzer displays data with oldest information at the left. At clock rates below 1 kHz the three steps shown occur on each clock cycle. The result is that each bit is displayed, properly positioned, as soon as it is entered. Data does not appear to shift in and is viewable while the display is being loaded.*

input register have been reloaded. The advantage is that no high-speed circuitry is needed in the display register.

The second purpose of the display register is to perform the multiplexing function necessary for transferring the data to the scanned display. After the display register is loaded and the data is positioned in the correct orientation, each 32-bit shift register is broken into four eight-bit closed loops. The data in each loop is circulated and one bit of the eight is monitored for the multiplexed data signal. The eight lines corresponding to the eight closed loops of circulating data in the two channels supply the drive (through buffers) to the anodes of the LED's in the 8 × 8 scanned matrix. To properly decode the multiplexed data, the cathodes of the LED's are scanned in synchronism with the shifting of data in the eight-bit loops.

The third function of the display registers is as a source of data to the circuits that perform the Boolean operations (A + B, A • B, A ⊕ B) on the A and B inputs. These Boolean operations take place as the data is scanned into the LED matrix and therefore do not affect the data that is stored in the input registers. This means that any Boolean function can be performed without disturbing the stored data. Thus if it is desired to see this data in its original form it is readily available.

Spike Detection

Selection of the SPIKE A mode makes possible the detection of asynchronous events occurring between clock pulses. A spike is detected whenever more than one logic transition occurs within a single clock period. If the positive-going transition occurs first, the spike is defined as positive and is displayed on the A channel; conversely, negative spikes are displayed on the B channel.

The spike detector consists of two flip-flops (Fig. 12). One is sensitive to the positive-going edge of the data input and the other is sensitive to the negative-going edge. If during any given clock period both flip-flops are set, a spike has occurred. The detected spike is entered into the input register by the next clock pulse.

A positive spike is differentiated from a negative one by a detector and a latch which determine which of the two flip-flops was set first. At the

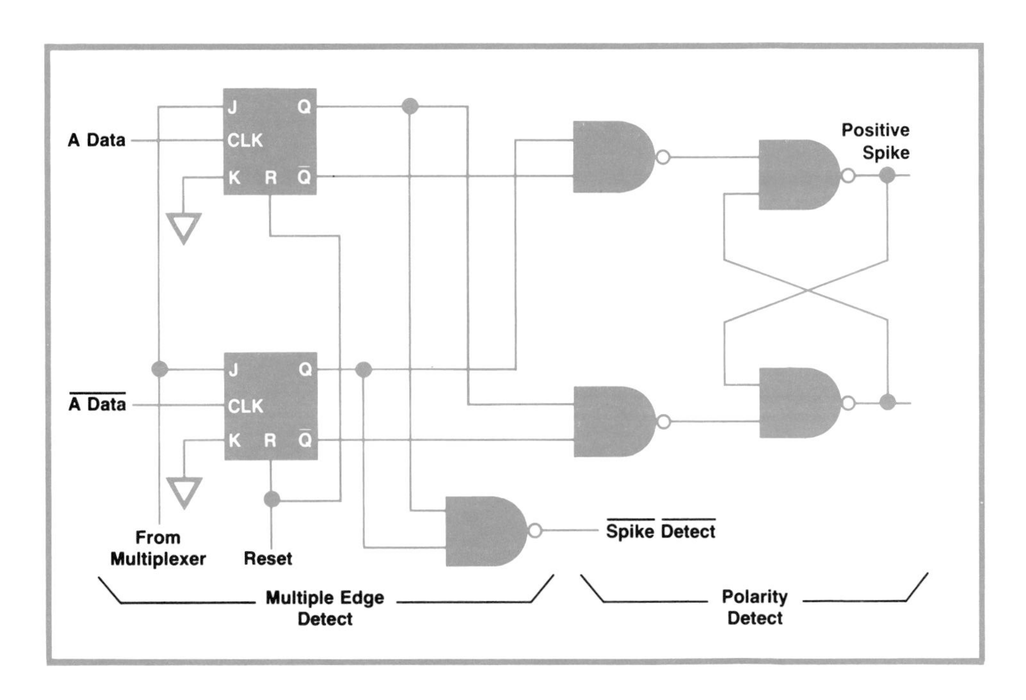

Fig. 12. *Spike detection circuitry detects multiple edges within a clock cycle.*

The IC Troubleshooters

Logic testing needs similar to those for which the 5000A Logic Analyzer was developed had earlier prompted development of a group of handheld instruments that were forerunners of the 5000A. Beginning in 1968 with the now ubiquitous Logic Probe, this family of instruments has grown to include a TTL/DTL Logic Probe, a Logic Clip, a Logic Pulser, a Logic Comparator, and Logic Probes for testing ECL and HTL/MOS circuits.

Logic Probe

The Logic Probe was the first test instrument designed and optimized strictly for digital applications. It has a digital readout, a lamp near the probe tip, that displays logic levels and pulses occurring on the circuit node being probed. The lamp glows brightly to indicate logic HIGH's, goes off for logic LOW's and glows at half brilliance to indicate open circuits or voltages between the preset logic thresholds. Continuous pulse trains cause blinking of the lamp at a 10 Hz rate and single pulses are stretched to provide a visible flash of the lamp: on for HIGH-going pulses and off for LOW-going ones.

Logic Probes are the quickest, surest way of detecting the presence or absence of single or infrequent pulses. No adjustments are needed, and signals can be rapidly traced through circuits by monitoring only two points: the schematic and the probe tip. There is no chance the probe will slip off the intended node while the user turns his head to read a remote display.

The Probe's value derives from the greater speed with which design and troubleshooting faults that result in bad nodes can be found. In these cases the Probe by itself will rapidly isolate the failure. When more detailed analysis is required, such as that of a Logic Analyzer or oscilloscope, the Probe is a useful adjunct. When the user is unable to identify the specific cause of a fault, the Probe can usually localize the search to a small group of suspect IC's and thus more quickly focus the power of the analyzer or oscilloscope on the problem.

The 5 volt probe, Model 10525T for TTL and DTL integrated circuits, has now been joined by two other models. Model 10525E has ECL logic thresholds and a −5.2 volt power input voltage for compatibility with all types of emitter-coupled logic. It's also the fastest of the three probes: pulse detection is guaranteed down to 5 ns. Model 10525H is designed for high voltage logic families such as HTL and HiNIL. A built-in power supply voltage regulator accepts input voltages anywhere between +12 and +25 volts. Logic thresholds are preset to +2.5 volts and +9.5 volts. The "H" probe is also useful with many types of MOS, discrete component, and relay logic systems.

Logic Clip

The second handheld logic tester, introduced in 1970, is Model 10528A Logic Clip. IC's are multi-pin devices, and it's often of interest to see information at several pins at the same time. The Logic Clip was developed to display the logic states of all 14 or 16 pins of a DIP IC simultaneously. The Clip attaches directly to TTL/DTL IC's, automatically seeks the V_{CC} and ground pins, and connects its power and ground buses to the proper pins. The Clip's 16 light-emitting diodes then display the logic states of all pins of the IC, one LED corresponding to each pin. The LED's light to indicate logic HIGH's and remain off for logic LOW's.

The Clip is handy when analyzing sequential circuits such as those containing IC counters or shift regisers. These typically have four or more outputs. With a slow stimulus (about 1 Hz) the progression of logic states can be followed on the Clip. Other applications include monitoring the output states of a ROM or displaying static input-output relationships in combinatorial IC's such as inverters or NAND gates.

Logic Pulser

The Logic Pulser, Model 10526T, was the fourth of the IC Troubleshooters to be developed. The Probe and Clip are response monitors; they depend on the circuit under test to supply stimulus to IC's while they display the responses. The Logic Pulser contributes in-circuit stimulus, thereby making possible the same kind of stimulus-response testing that has long been invaluable in analog troubleshooting.

The problem of stimulating digital IC's in a circuit is more difficult than it might at first appear. Digital outputs have very low output impedances (less than 5 ohms). They are designed this way to provide wide immunity to spurious noise. Connected in circuits, each relatively high-impedance IC input is driven by a low-impedance output that clamps it either HIGH or LOW at all times. Stimulus is possible by overriding the driving IC output with large amounts of current, but then destruction of the driving stage is a real possibility.

The Logic Pulser solves this problem by generating a very narrow pulse, briefly overriding the driving output. The Pulser will source or sink up to 0.65 amperes each time its pulse button is pressed. Narrow pulse width of 0.3 microseconds, coupled with the manual activation, make the duty cycle very small, so there is negligible added power dissipation in the driving IC and no danger of damage. Automatic selection of the polarity of the output pulse geatly simplifies operation; HIGH nodes are pulsed LOW, and LOW nodes HIGH with no adjustments required.

The Pulser provides stimulus at rates appropriate for moni-

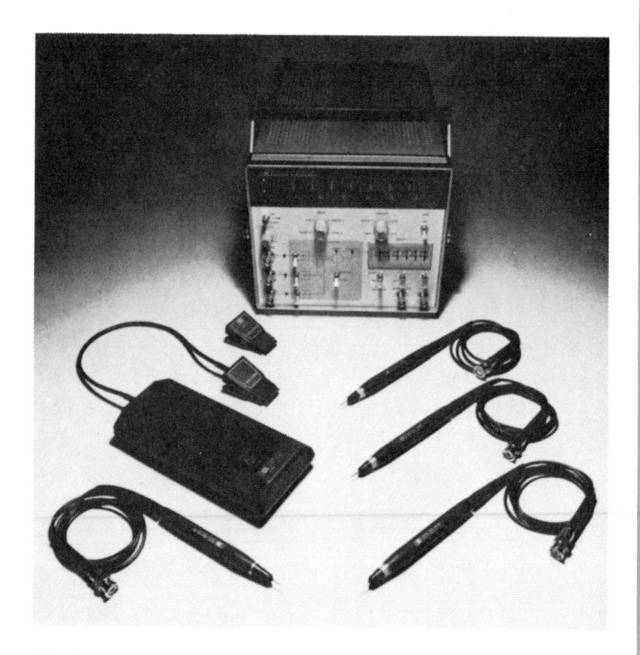

Clockwise from lower left: 10526T Logic Pulser, 10529A Logic Comparator, 10528A Logic Clip, 5000A Logic Analyzer, 10525T Logic Probe, 10525E Logic Probe, 10525H Logic Probe.

toring responses with the Logic Clip. For example, the proper progression of logic states in an IC counter may be verified using Pulser and Clip as a stimulus/response team. The Logic Probe, with its pulse stretcher, is a good companion for the Pulser when testing combinatorial logic such as gates and inverters. The Logic Pulser's 0.3 μs pulse, injected at a gate input, should appear at the gate output and be displayed by the Probe. If it doesn't, the gate is defective.

The Logic Probe, Pulser, and Clip are available in a single package as the 5015T Logic Troubleshooting Kit.

Logic Comparator

The Pulser, Clip and Probe leave most of the task of analyzing results to the user, who must interpret the circuit responses that the instruments display for him. Through his knowledge of circuit operation he decides if his new design is functioning the way he intended or if a particular IC in the instrument he is troubleshooting has failed.

Model 10529A Logic Comparator, an in-circuit IC tester, goes further, analyzing the detected signals to display logical faults rather than HIGH's and LOW's. The comparator functionally tests TTL and DTL IC's in their normal circuit environment without removing them from their printed circuit boards. Failures of the test IC are displayed on the Comparator's 16 LED's, each of which corresponds to a pin of the test device.

The 10529A compares the operation of the test IC to a reference IC of the same type that is inserted in the Comparator. Power and input signals are borrowed from the test IC. Outputs of the two IC's are compared and whenever a logic difference exists the LED corresponding to the differing pin is lighted. Brief or intermittent errors are stretched to provide a visible flash of the LED.

Blank reference boards are supplied with the Comparator, ready for the user to load with the IC's he wants to test. After inserting the IC into the board the user bends power and ground pins to contact the buses that supply the Comparator's power, and solders the IC into place. The user then breaks a trace on the reference board to identify each output pin; this permits the Comparator to differentiate between inputs and outputs. The reference board is then ready for use. An accessory kit of reference boards pre-programmed with 20 commonly used TTL IC's is available.

In a troubleshooting situation, a suspect IC is selected for testing and the corresponding reference board is placed in the Comparator's drawer. The Comparator is then clipped onto the selected test IC and the display is checked for lighted LED's.

The Logic Probe, Pulser, and Clip also complement the Logic Comparator in troubleshooting applications. Many variations are possible. For example, the Probe can be used to isolate the failure to a specific board or group of IC's. Then the Comparator can be brought in to test the smaller number of possibilities. Once a bad node has been located by the Comparator, the Probe and Pulser can analyze the cause. Simultaneously probing and pulsing the suspect node will identify a short to ground or to the power supply: the Pulser can't pulse its own supply buses, so if no signal is registered on the Probe a solder short is likely.

The Pulser is handy when the Comparator is used to test sequential logic. If the reference IC turns on to a different state than the test IC, an error may be indicated. Pulsing the reset input of the test IC synchronizes test and reference IC's to the same state and the test becomes valid. This external synchronization is necessary whenever it is not performed automatically by the circuit or by a manual reset control.

Model 5011T Logic Troubleshooting Kit combines all four TTL and DTL fault-finders: Probe, Clip, Pulser, and Comparator.

References

1. R. Adler and J. Hofland, "Logic Pulser and Probe: A New Digital Troubleshooting Team," Hewlett-Packard Journal, September 1972.
2. M. Baker and J. Pipkin, "Clip and Read Comparator Finds IC Failures," Hewlett-Packard Journal, January 1972.
3. G.B. Gordon, "IC Logic Checkout Simplified," Hewlett-Packard Journal, June 1969.

end of each clock period, the contents of the spike circuit are cleared. To assure that there are no dead times for spike detection, two identical spike detection circuits have been included. Each is active on alternate clock cycles. Thus even when the spike detector is being cleared, a new spike can be detected.

Special Applications

Additional capabilities built into the Logic Analyzer greatly extend its flexibility in certain applications.

First, external access is provided to the data stored in the A and B registers. This data is valuable when using the Analyzer as a serial-to-parallel converter or whenever computer analysis of displayed data is desired. Computer-aided fault isolation in production test is one example.

The top edges of the two register boards that store displayed information are designed to mate with flat-cable connectors. Holes and trace patterns already exist on the boards for IC's to interface the Analyzer's TTL levels to those of other logic families. Layout is for the 7404 pin configuration, and any similar devices (7405, 7407, etc.) may be used to tailor the outputs to specific requirements. Once interface devices are selected and soldered in place, data is available at the connectors at the top of the card. A rear-panel signal, TRIG OUT, functions as the strobing command, signaling with a logic LOW when data is valid.

Second, if it is desired to use the 5000A as a digital trigger and delay generator for an oscilloscope, the 50 ms data hold time (during which the display is frozen) may be reduced to 3μs. To do this one simply moves the programming plug on the Control A board to the TEST position. The Analyzer is then retriggerable 3μs after a display sweep is completed. The 50 ms delay is necessary for proper operation of the Analyzer's display section, so all displayed information must be ignored with the plug in the TEST position. Data at the top of the register boards, however, is still valid.

Finally, if special triggering capability beyond that provided by the Logic Analyzer is required, inclusion of user-designed add-ons has been facilitated. All necessary data and trigger control signals in addition to +5V power are available at the top of the Trigger Board, which also mates with a card-edge connector. Serial triggering, the ability to trigger when a particular serial bit pattern occurs, could be added in this manner, for example.

Mechanical Design

The 5000A Logic Analyzer is the first instrument to be packaged in the new HP cabinet system. This system features die-cast front and rear frames and removable aluminum side rails that connect the front and rear frames. With top, bottom, and both side covers removed, access to all sides of the instrument is possible (Fig. 13). The covers slide into slots on the frame; each is quickly removed by loosening only a single screw.

The internal design of the Logic Analyzer emphasizes ease of assembly from both manufacturing and service viewpoints. Its chassis is a single printed-circuit mother board containing a series of card-edge connectors. All electronic components are mounted on printed-circuit modules that plug into the mother board, which contains all inter-assembly connections. The printed-circuit modules are easily extendible or removable for trouble-shooting purposes, and hand wiring is kept to a minimum. Even the front and rear-panel controls are printed-circuit mounted; the only hand wiring in the instrument is the IEC-required power and transformer wiring.

The Logic Analyzer cabinet is a standard half-rack module, 15 inches deep. Its handle, 15 pound weight, and form factor make it light and convenient to carry (Fig. 14).

Acknowledgments

The authors wish to acknowledge the following people whose help was greatly appreciated throughout the design and production introduction phases: Jim Marrocco for mechanical design, Jack Nilsson for his inputs on serviceability, Chuck Taubman and Gary Gordon for their guidance and management, Roy Criswell for a smooth, on-time production introduction, and Jesse Pipkin for his constant efforts to configure an instrument in concert with market needs.

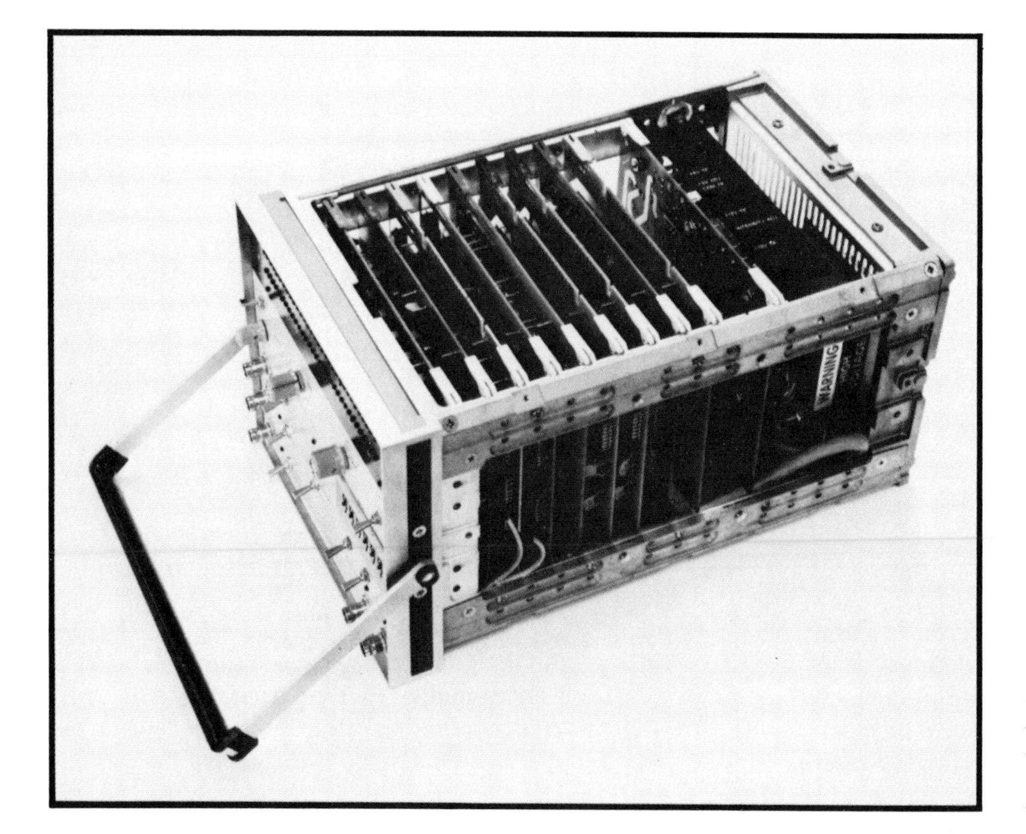

Fig. 13. *Logic Analyzer is mechanically designed for ease of assembly and repair. Top, bottom, and sides are easily removed to gain access to all sides of the instrument.*

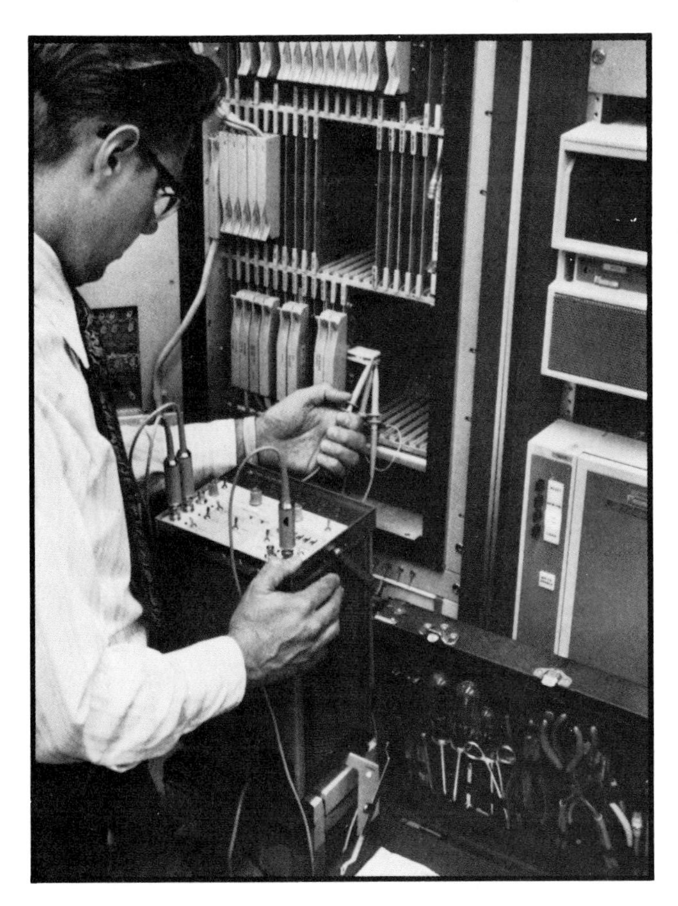

Fig. 14. *Light weight, handle, and form factor make the Logic Analyzer suitable for field service applications.*

Mark Baker

Logic test product marketing engineer Mark Baker assumed his present position after serving as engineering project leader for the 5000A Logic Analyzer. A native of Oklahoma, Mark received his BS degree in electrical engineering from Oklahoma State University in 1969 and his MSEE degree from Stanford University in 1970. His first product for HP was the 10528A Logic Clip, which he designed during the summer of 1969. He's also contributed to the design of the 5525A Laser Interferometer and the 10529A Logic Comparator. For recreation, Mark likes water skiing, tennis, and tinkering with cars. He and his wife live in Cupertino, California.

Howard D. Marshall

Howard Marshall received his BS degree in engineering from California Institute of Technology in 1970, and came to HP the same year. After a brief exploration into high-speed test systems, he helped invent the 10526T Logic Pulser and made major contributions to the design of the 5000A Logic Analyzer. He's now a project leader in the logic test laboratory. A native of Nampa, Idaho who now lives in Woodside, California, Howard enjoys backpacking, bicycling, and making his own wine.

Robin Adler

Robin Adler joined HP in 1970 after receiving his BS degree in electrical engineering from California Institute of technology. After two years of digital design he became project leader for the 10526T Logic Pulser. When that project was completed in 1972 he moved into logic test marketing and now has marketing responsibility for the 5000A Logic Analyzer. Robin spends a good deal of his spare time restoring old cars, mostly foreign ones. Weekends and vacations he's likely to spend in the mountains, camping or skiing, depending on the season.

SPECIFICATIONS

HP Model 5000A Logic Analyzer

Inputs

NUMBER: 5 (Channel A, Channel B, External Trigger, Word Delay, Clock).

INPUT IMPEDANCE: 1 MΩ shunted by approximately 25 pF.

INPUT THRESHOLD VOLTAGE: Continuously variable over ±1.4 V (±14 V with 10:1 divider probe).

MAXIMUM INPUT VOLTAGE: ±200 V continuous, ±400 V transient.

ANNUNCIATORS: One per input to display logic state; pulse stretching for single pulses.

DATA AND TRIGGER INPUTS (Channel A, B, External Trigger)
- MINIMUM SETUP TIME: 15 ns (8 ns typical).
- MINIMUM HOLD TIME: 0 ns.

CLOCK INPUT
- MAXIMUM PULSE REPETITION RATE: 10 MHz.
- MINIMUM PULSE WIDTH: 15 ns (10 ns typical with 100 mV overdrive).
- HYSTERESIS: Approximately 10 mV.

WORD DELAY INPUT
- MAXIMUM PULSE REPETITION RATE: ½ of Clock input repetition rate (input is synchronized to Clock).
- MINIMUM PULSE WIDTH: 15 ns.
- WORD DELAY SLOPE CONTROL SWITCH: Permits selection of high or low-going edges of word pulse or disabling of function.

Input Modes

A, B: Two-channel operation.

SERIAL A: A and B display registers cascaded into a single 64-bit display loaded from Channel A input.

SPIKE A: Detects multiple transitions at A input during a clock period. Positive spikes are displayed in A display register, negative spikes in B register.
- MINIMUM SPIKE WIDTH: 15 ns (10 ns typical with 100 mV overdrive).

RUN-STORE-RESET MODES
- RUN: Normal operation, A and B registers loaded each time Analyzer triggers.
- STORE: Analyzer triggers once (or finishes current sweep) and retains data, rearming inhibited.
- RESET: Clears A and B display registers (except in Store B mode) and rearms Analyzer when released. Also aborts sweep or delay countdown if in progress.

STORE B MODE: Data in Channel B display register is retained (sweep is completed if in progress). Channel A loads each time Analyzer triggers. Reset control does not affect data stored in B register.

Trigger Controls

FUNCTION: Initiate display or delay countdown (also see Digital Delay section).

ARMING: Analyzer must be armed in order to trigger. Arm LED lights to indicate arming.
- MINIMUM SWEEP REARMING TIME: Approximately 50 ms after last clock pulse of sweep.
- HOLDOFF CONTROL: Increases rearming time up to approximately 5 seconds.

TRIGGER CONDITIONS: Triggering may be from either Data input, External Trigger input, or logical AND combinations of two or three inputs. Slope selection is provided by 3-position slope control switch for each input.

TRIGGER SLOPE CONTROL SWITCHES
- SWITCH UP: Input must be HIGH for triggering.
- SWITCH IN CENTER POSITION: Input must be LOW for triggering.
- SWITCH DOWN: Input does not affect triggering.

TRIGGERING MODES
- CLOCKED MODE: Analyzer triggers on first clock pulse after all input conditions defined by slope control switches are met. Trigger condition must remain until clock pulse occurs.
- ASYNCHRONOUS MODE: Analyzer triggers when trigger conditions are met. Conditions need not remain until clock pulse occurs.
  - MINIMUM PULSE WIDTH: 40 ns.
  - MINIMUM SETUP TIME: 60 ns.

TRIGGER LED: Indicates Analyzer is triggered. Remains on until completion of sweep.

Digital Delay

POST-TRIGGER DELAY RANGE: Display begins 0 to 999,999 clock periods after trigger event.

PRETRIGGER (NEGATIVE) DELAY RANGE: Display begins 0 to 32 clock periods (64 in Serial A mode) before trigger event.

WORD DELAY: When enabled, permits 2 levels of digital delay.
- DELAY RANGE: 0 to 9,999 pulses at Word Delay input plus 0 to 99 pulses at Clock input.

Display

TYPE: Red light-emitting diodes.

LENGTH: 32 LED's per register.

LOGIC CONVENTION
- ON LED: Logic HIGH (input more positive than threshold voltage).
- OFF LED: Logic LOW (input less positive than threshold voltage).

DISPLAY MODES
- DIRECT: Data at A and B inputs displayed by A and B registers.
- A • B: Logical AND of A and B inputs displayed in A register, B register blanked.
- A + B: Logical OR of A and B inputs displayed in A register, B register blanked.
- A ⊕ B: Logical EXCLUSIVE-OR of A and B inputs displayed in A register, B register blanked.

DISPLAY RELATIONSHIP TO CLOCK: Display advances horizontally one LED per clock pulse.

Rear Panel

THRESHOLD ADJUSTMENT
- THRESHOLD RANGE SWITCH: Selects low range (approximately −0.14 to +0.14 V) or high range (approximately −1.4 to +1.4 V) for input amplifier threshold voltage.
- VERNIER CONTROLS: Permit continuous adjustment over either range.
- THRESHOLD OUTPUT: Enables measurement of threshold voltage, $Z_0 \simeq 5.6$ kΩ.

TRIGGER OUT: Goes HIGH when delay countdown is completed; remains HIGH until display sweep has completed.

5 V OUT: Supplies 5 V ±5% at 200 mA.

CHECK-OPERATE-COMPENSATE
- CHK: Provides alternating HIGH/LOW display when Check Out output is connected to Channel A or B input. No external clock required.
- OPER: Normal Analyzer operation.
- COMP: Used for adjusting compensation of divider probes.

CHECK OUT: Provides signal at ½ clock input repetition rate (except in Check mode); TTL compatible.

PRICE IN U.S.A.: $1900.00

MANUFACTURING DIVISION: SANTA CLARA DIVISION
5301 Stevens Creek Boulevard
Santa Clara, California 95050 U.S.A.

The "Personal Computer": A Fully Programmable Pocket Calculator

This 11-ounce battery-powered marvel has the computing power of an advanced scientific pocket calculator and is programmable as well, so it can adapt to any number of specialized uses.

by Chung C. Tung

AN ENGINEER OR SCIENTIST in need of an on-the-spot answer in the laboratory, a pilot making an in-flight course correction, a surveyor running a traverse in the field, a businessman estimating returns-on-investment during a conference, a physician evaluating patient data—it isn't difficult to think of everyday examples of people whose professions require certain types of calculations over and over again. Were one available, a computer or programmable calculator would obviously be of great assistance to such people. Unfortunately, you can't carry one of those around in your pocket, can you?

You can't, unless it's battery-powered, weighs only 312 grams, and measures only 8.1 × 14.7 × 3.4 cm, like the HP-65 Pocket Calculator. This new computing device is a combination of electronic calculator and general-purpose small computer. It offers the convenience and easy operation of a calculator, but its programmability makes it versatile enough to fit the needs of a wide variety of disciplines, including science, engineering, finance, statistics, mathematics, navigation, medicine, surveying, and many others.

Although the HP-65 is designed to operate in a logical, easy-to-learn way, it is capable of sophisticated computations. It has 51 built-in mathematical functions and data-manipulation operations, a four-register operational stack, nine addressable data registers, and five user-definable keys. It can run programs that have as many as 100 steps. There are two program flags and four comparison tests for program branching.

Perhaps most significant is that within its small package the HP-65 contains a tiny magnetic card reader and recorder. Users can store their programs on magnetic cards for later use, or they can take advantage of hundreds of preprogrammed cards containing programs commonly used in various disciplines.

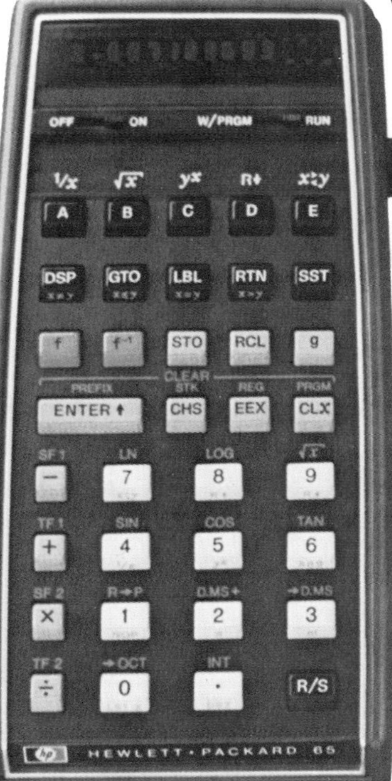

Fig. 1. *The first fully programmable pocket calculator has 51 preprogrammed operations, a 100-step program memory, five user-definable keys, and a tiny magnetic card reader-recorder. After a program card is read it can be placed in the window above the keyboard to show the new functions of the user-definable keys.*

Keyboard Design

The HP-65 inherits its size and shape from its ancestor, the HP-35. However, its keyboard layout is quite different. Thirty-five keys control more than eighty operations (see Fig. 1).

In the interests of logical operation and simplicity, many different techniques were used in designing the keyboard layout. Keys of the same nature are grouped into clusters. Some nomenclature has been placed on the lower side of the keys to reduce busyness. Nomenclature for multiple-keystroke operations is color-coded to make the keystroke sequences associative. All functions are classified as immediate (+,−,×,÷), direct, inverse, or miscellaneous, and are grouped and color-coded accordingly. Key sizes, colors, value contrast, and nomenclature have all been chosen to guide the user.

Arithmetic operations (+,−,×,÷) and data entry keys are in the same locations as on the

HP-35. Data entry keys include the digits 0 through 9, the decimal point, the ENTER↑ key used for entering values into the four-register operational stack, and the EEX (enter exponent) and CLx (clear display) keys.

Keys in the top row, labeled A, B, C, D, and E, stand for user-definable functions or subroutines. In the second row, the GTO (go to), LBL (label), RTN (return), and SST (single step) keys are used for programming. The DSP (display) key is used for formatting the display. Numbers can be displayed either in fixed-point notation or in scientific notation with a selected number of digits after the decimal point.

Keys in the third row, labeled f, f^{-1}, STO, RCL, and g, are all prefix keys. They have to be followed by one or two more keystrokes to complete a command.

At the lower right corner of the HP-65 keyboard is the R/S (run/stop) key. Pressing it begins program execution. Execution of a running program halts if an R/S step is encountered in the program.

User-Definable Keys

When power is first applied to the HP-65 the top-row keys (A,B,C,D,E) call for the functions 1/x, $\sqrt{x}$, y^x, x⇆y, and R↓ (roll down the operational stack). However, the functions of these keys can be changed, either by keyboard programming or by inserting a previously recorded magnetic card into the HP-65's reader slot. The new functions of the keys can be written on the card and the card inserted into the window above the top row of keys to show the new functions.

With the five definable keys, the user can readily change the HP-65 into a specialized calculator tailored to his needs. For example, Standard Pac program 11A, a compound interest program, changes these keys to n (number of periods), i (interest rate), PV (present value), FV (future value), and COMPUTE, similar to keys on the HP-80 financial calculator (see Fig. 2). Standard Pac 06A, a π-network impedance-matching program, converts the HP-65 into an electrical engineer's calculator. Given input and output resistance, frequency, and Q, the program computes the values of the two shunt capacitors and the

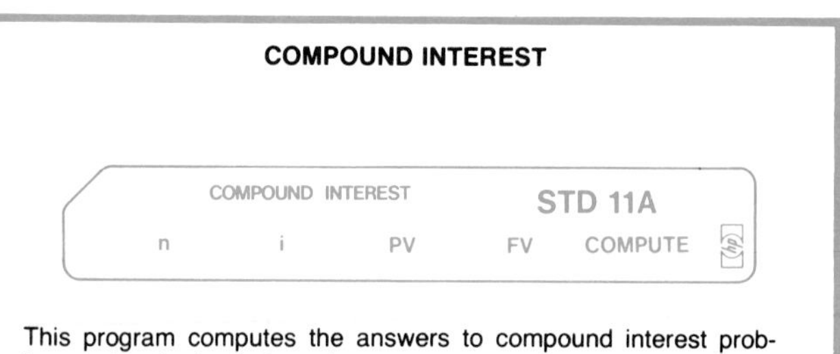

This program computes the answers to compound interest problems using the basic formula:

$$FV = PV\,(1 + i/100)^n$$

where:
n = Number of time (compounding) periods.
i = Interest rate per time period (in percent).
PV = Present value (value at the beginning of the first time period).
FV = Future value (value at the end of n time periods).

Any three of the variables (n, i, PV, FV) can be inputs. The program computes and stores the fourth variable. Input variables can be entered in any order and need not all be reentered to change one variable.

HP-65 User Instructions

STEP	INSTRUCTIONS	INPUT DATA/UNITS	KEYS		OUTPUT DATA/UNITS
1	Enter program				
2	Initialize		RTN	R/S	
3	Input three of the following				
	n·	n	A		n
	and/or i	i(%)	B		i(%)
	and/or PV	PV	C		PV
	and/or FV	FV	D		FV
4	Compute the remaining variable				
	n		E	A	n
	or i		E	B	i(%)
	or PV		E	C	PV
	or FV		E	D	FV
5	To modify problem go to 3 and				
	change appropriate input(s).				

Fig. 2. *Users may write their own programs and store them on blank magnetic cards, or may take advantage of many preprogrammed cards. This compound interest program converts the calculator into a financial calculator.*

PI NETWORK IMPEDANCE MATCHING

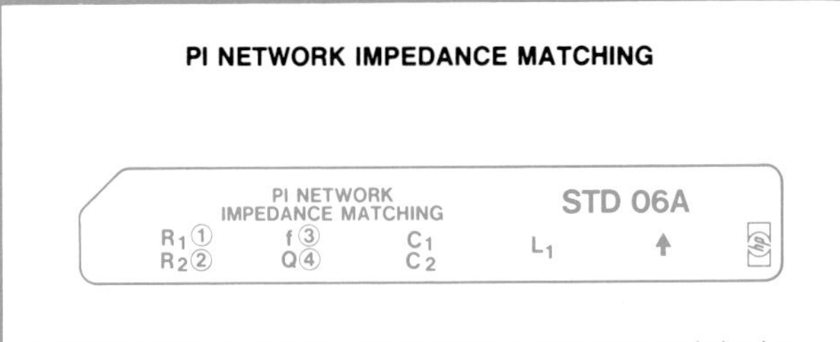

A lossless network is often used to match between two resistive impedances, R_1 and R_2, as shown.

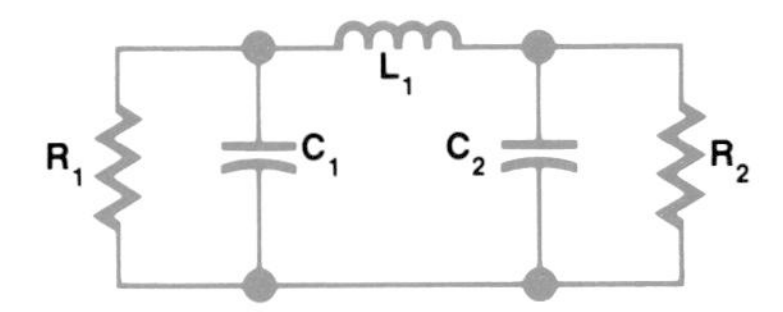

Given the values of R_1 and R_2, the frequency f, and the desired circuit Q (center frequency/desired half-power bandwidth), the values of C_1, C_2, and L_1 can be found.

Notes:

1. R_1 must always be greater than R_2 and

$$Q > \sqrt{R_1/R_2 - 1}\,.$$

2. Circled numbers on the magnetic card designate the register in which a variable is stored.

HP-65 User Instructions

STEP	INSTRUCTIONS	INPUT DATA/UNITS	KEYS		OUTPUT DATA/UNITS
1	Enter program				
2	Initialize		RTN	R/S	0.0000
3	Input R_1	$R_1(\Omega)$	E	A	$R_1(\Omega)$
	and R_2	$R_2(\Omega)$	A		$R_2(\Omega)$
	and f	f(Hz)	E	B	f(Hz)
	and Q	Q	B		Q
4	Compute C_1		E	C	C_1(F)
5	Compute C_2		C		C_2(F)
6	Compute L_1		D		L_1(H)
7	Recall inputs (optional)				
	R_1		RCL	1	$R_1(\Omega)$
	and/or R_2		RCL	2	$R_2(\Omega)$
	and/or f		RCL	3	f(Hz)
	and/or Q		RCL	4	Q
8	For new case change appropriate				
	input in step 3.				

Fig. 3. *This program converts the HP-65 into an electrical engineer's calculator.*

series inductor (see Fig. 3).

For more information on HP-65 keys and programming, see the article beginning on page 8.

Magnetic-Card Reader/Writer

The HP-65's built-in magnetic card reader/recorder uses mylar-based ferrite-oxide-coated cards 0.95 cm wide and 7.1 cm long. Each card can store 100 program steps or 600 bits of information. A two-track self-clocking recording scheme is used to maximize the system's tolerance to head skew and motor speed variations (see box below).

When the right-hand switch just below the display is in the RUN position, insertion of a card into the reader slot in the right side of the calculator triggers the motor and read circuits. All 100 program steps on the card are read into the calculator's memory.

When the same switch is in the W/PRGM position, insertion of a card triggers the motor and writing circuits, and the contents of the calculator's memory are written on the card. A card that has the file-protect tab (upper left corner) clipped off will not trigger the writing circuits; the program on the card is thereby protected against accidental erasure.

More information on the design of the tiny card reader is contained in the article beginning on page 15.

System Organization

Like its ancestor, the HP-35, the HP-65 contains an arithmetic and register circuit, a control and timing circuit, and three read-only-memory (ROM) circuits. In the HP-65, each ROM is actually a quad ROM containing the equivalent of four single ROMs. Like its more recent predecessor, the HP-45, the HP-65 also

A Self-Clocking Two-Track Recording Technique

When the HP-65 records a program on one of its small magnetic cards, it places two side-by-side tracks of varying magnetic flux on the card. One track represents the logical 1's in the binary data stream coming from the program memory, and the other track represents the logical 0's. The 1 track contains a flux reversal for each 1 in the data sequence and no flux change for each 0. The 0 track contains a flux reversal for each 0 and no flux change for each 1. The diagram shows an example.

The technique is self-clocking because there is a flux transition for each bit. Thus no separate clock track is needed.

The scheme also maximizes the system's tolerance to head skew and motor speed variations. The data can be recovered correctly even if a transition in one track almost overlaps a transition in the other. By contrast, other two-track schemes usually have one clock track and one data track, and may misread data when there is only minor misalignment of clock and data transitions.

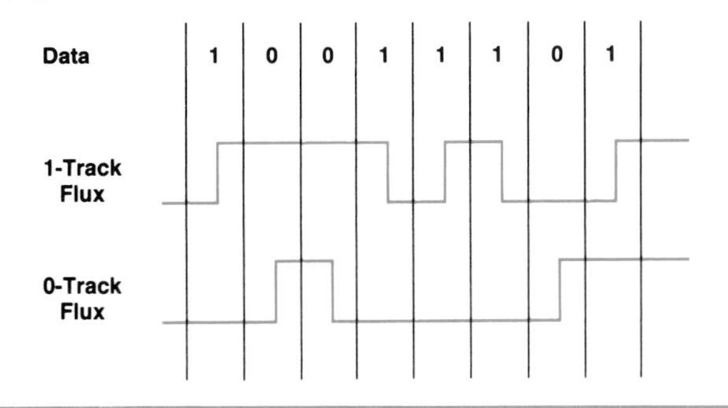

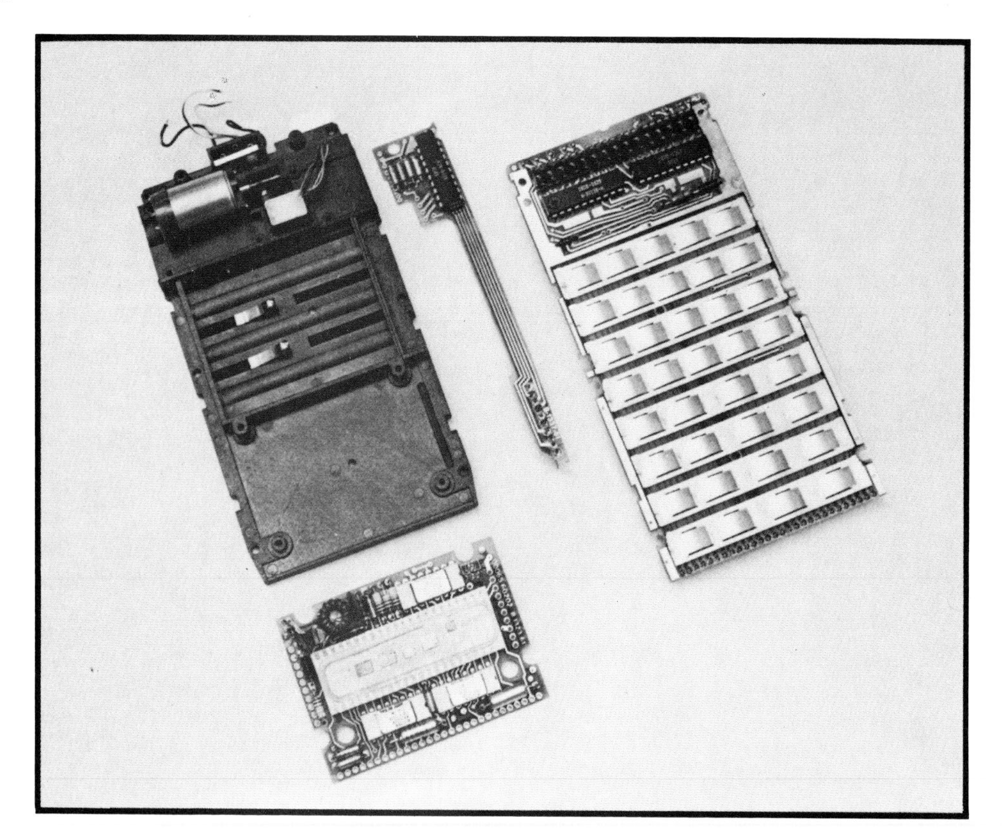

Fig. 4. *Inside the HP-65.*

has a data-register circuit. New in the HP-65 are a program storage circuit, a card-reader control circuit, and a card read/write circuit. Fig. 4 is a photo of the components and Fig. 5 is a block diagram.

The *data register circuit* is made up of ten addressable data registers of 56 bits each, enough to store 14 digits each. Data register 0 is used to implement a LSTx (last x) function and always contains the previously displayed number. Registers 1 through 9 are addressable from the keyboard. Addresses and data are transferred between the data register circuit and the arithmetic and register circuit by way of a bidirectional BCD bus.

The *program storage circuit* stores the user's program. Each program step is stored as a six-bit word. There are 100 words of storage.

The program memory is designed to act like a carousel. The memory is implemented in a dynamic shift register. Program words, a pointer, and a beginning-of-memory marker circulate continuously in this register, a complete circulation taking approximately 3½ milliseconds. The pointer always points to the next program word and can move freely within the memory. Thus program steps can be inserted into the program or deleted from it at any point, without re-keying the other steps. Addressing is symbolic rather than absolute, and label-searching hardware is built into the program storage circuit.

When the calculator is in use, every keystroke places a corresponding keycode on the I_A (instruction address) bus. This code is stored in the buffer of the program storage circuit. If the calculator is in W/PRGM mode, the code is then inserted into the program memory. If the calculator is in RUN mode, the keystroke is executed.

When a stored program is executed, the pointer points to the next executable memory word and the buffer contains a copy of that word. The buffer contents are decoded and placed on the I_S bus as a microinstruction that causes the calculator to enter a subroutine in the ROM to service the key that was pressed.

Card Reader Circuits

The *card-reader control circuit* and the program storage circuit together form a complete memory circulation path. When the HP-65 is used as an ordinary calculator the card-reader control circuit merely short-circuits the data-in and data-out lines of the program storage circuit.

When a card is being read, the card-reader control circuit places the outputs of the head sense amplifier onto the data-in line of the program storage circuit. When a card is being written the card-reader control circuit takes the memory words in sequence and transforms them into recording signals for the head write amplifiers.

When two unsynchronized sequential memories like the program memory and the magnetic card are working together, timing and control are critical.

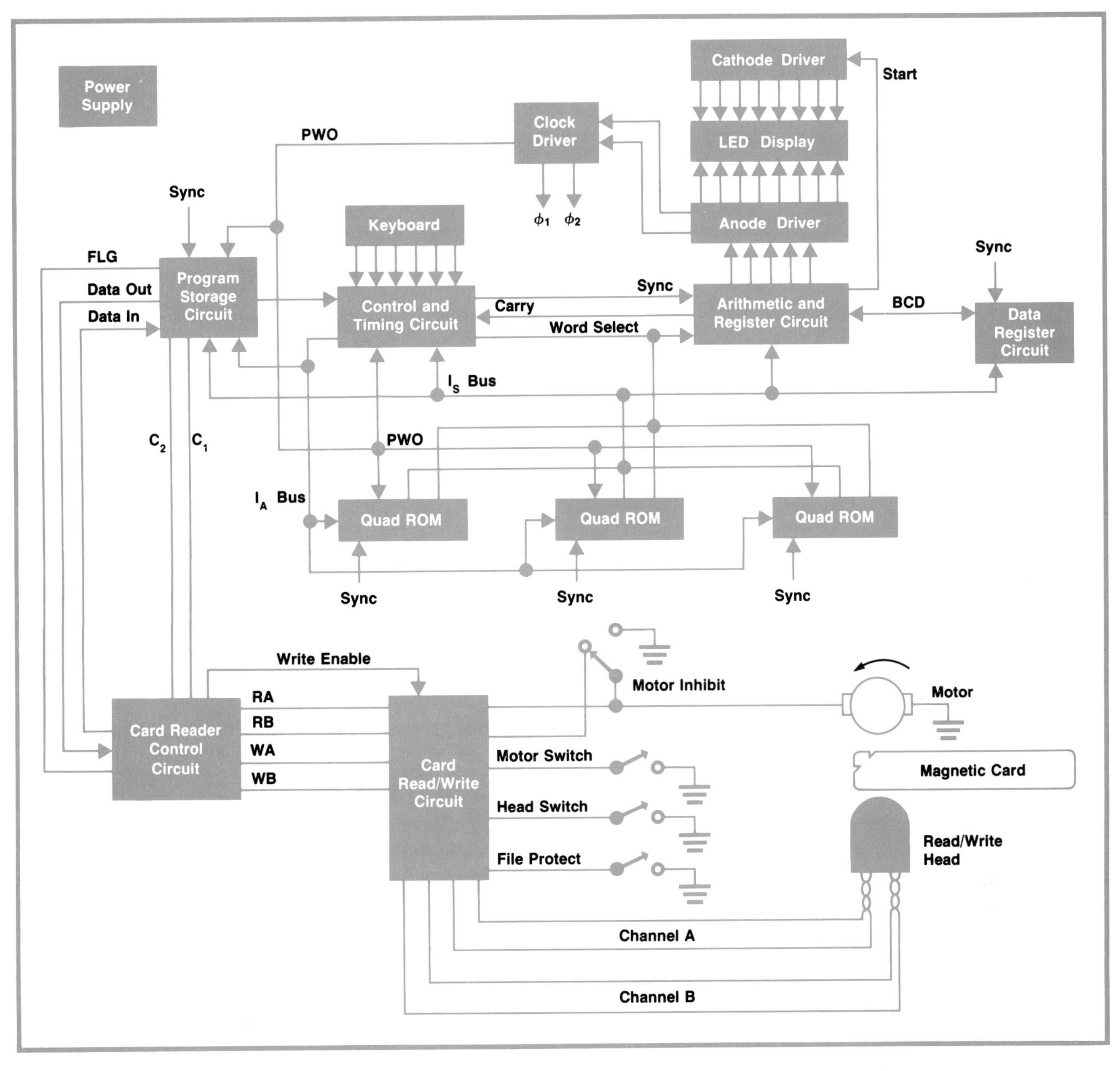

Fig. 5. *Architecture of the HP-65 is stack and bus oriented, like earlier HP pocket calculators. New circuits are the program storage circuit, the card-reader control circuit, and the card read/write circuit.*

Therefore, during reading or writing the card-reader control circuit assumes control of the program memory and controls the movement of the pointer. Thus only limited handshaking is needed between the card-reader control circuit and the program storage circuit.

The *magnetic-card read/write circuit* is a bipolar LSI (large-scale-integrated) circuit. It serves as the interface between the magnetic read/write head and the card-reader control circuit. There are two identical channels on the chip, each consisting of a read amplifier, a threshold detector, a latch, and a write amplifier.

The read amplifier takes a one-millivolt signal from the head, amplifies it approximately 200 times and feeds it into the threshold detector. The detector removes noise and triggers the latch to provide a logic signal to the card reader control circuit. The write amplifier takes logic information from the card-reader control circuit and transforms it into a recording current that flows through the inductive head and generates the flux that magnetizes the card.

The read/write circuit also provides a dc motor control circuit that regulates the widely varying battery voltage so the card-reader motor runs at constant speed.

Acknowledgments

The HP-65 is the creation of a large number of people who contributed their efforts with great enthusiasm and engineering professionalism. The support of the IC departments of the Loveland and Santa Clara Divisions, and of the Loveland hybrid circuit department, was indispensable. Special acknowledgments are due Bill Hewlett for his encouragement; Tom Whitney and Paul Stoft for their direction and help in administration and coordination between a number of departments; Tom Osborne for valuable discussions and suggestions; Allen Inhelder, Darrel Lauer, and Ed Liljenwall for industrial design—their consistent human engineering consideration is particularly valuable to a simple, elegant keyboard design; Dave Cochran for card read/write circuit design; Rich Whicker for the design of the card-reader control circuit; Bob Schweizer for valuable suggestions on the design of the read/write principle and circuits and for help on system breadboarding and simulator design; Clarence Studley and Bernie Musch for packaging and mechanical part design; Bob Taggart for his creative design of the card reader mechanism, with the able assistance of Dick Barth; John Bailey for help on the motor and head switches; Bill Boller for his close liaison with Manufacturing Division; Homer Russell and his team for applications software support. Special thanks are also due the tool designers of Manufacturing Division, whose creative, outstanding tools made the state-of-the-art plastic parts for the HP-65 possible.

Chung C. Tung

Chung Tung received his BSEE degree from National Taiwan University in 1961 and his MSEE degree from the University of California at Berkeley in 1964. As a member of HP Laboratories from 1965 to 1972, he was involved in the design of the 9100 Calculator and the HP-35 and HP-45 Pocket Calculators. He subsequently joined the Advanced Products Division as manager of the HP-65 project and is now engineering section manager there. Chung has done work towards his PhD degree at Stanford University, although he isn't presently enrolled there. He is married, has a son and a daughter, and lives in Santa Clara, California. Music, reading, and table tennis are his principal modes of relaxation.

SPECIFICATIONS

HP-65 Programmable Pocket Calculator

PRE-PROGRAMMED FUNCTIONS:

TRIGONOMETRIC: sin x • arc sin x • cos x • arc cos x • tan x • arc tan x

LOGARITHMIC: log x • lnx • e^x • 10^x

OTHER: y^x • $\sqrt{x}$ • 1/x • π • x^2 • n! • conversion between decimal angle, degrees/minutes/seconds, radians or grads • rectangular/polar coordinate conversion • decimal/octal conversion • degrees(hours)/minutes/seconds arithmetic • integer/fraction truncation

OTHER FUNCTIONS:

REGISTER ARITHMETIC: Addition, subtraction, multiplication or division in serial, mixed serial, chain or mixed chain calculations

FEATURES:

DISPLAY: Up to 10 significant digits plus 2-digit exponent and appropriate signs

DYNAMIC RANGE: 10^{-99} to 10^{99}

Primary functions activated by single keystroke; alternate functions use prefix keys

Five user definable keys

Four-register operational stack

Program memory for storage of up to 100 steps

Single step running and/or inspection of a program

Insert/delete editing features

Nine addressable memory registers

"Last X" register for error correction and number reuse

Two flags for skip or no-skip programming or branching to another part of program

$x \neq y$, $x \leq y$, $x = y$, $x > y$ relational tests

Magnetic card reader/writer

Built-in counter

Automatic decimal point positioning

Selective round-off; range: 0-9 decimal places

Two display modes: fixed point and scientific

Indicators for improper operations and low battery condition

Operates on rechargeable batteries or ac

Light-emitting diode (LED) display

APPLICATION PROGRAM PACS:

Standard Pac (17 miscellaneous programs, 2 diagnostics, 1 head cleaning card, 20 blank cards)

Math Pac I (40 mathematical programs)

Math Pac II (33 mathematical programs)

Stat Pac I (37 statistical programs)

Survey Pac I (34 surveying programs)

Medical Pac I (33 medical programs)

EE Pac I (35 electrical engineering programs)

POWER:

AC: 115 or 230 V, ±10%, 50 to 60 Hz, 5 watts.

BATTERY: 500 mW derived from nickel-cadmium rechargeable battery pack.

WEIGHT: 11 ounces (312 g) with battery pack. Recharge: 5 ounces (155 g).
SHIPPING WEIGHT: approx. 3 lbs (1.4 kg).

DIMENSIONS:

LENGTH: 5.8 in (14.7 cm).

WIDTH: 3.2 in (8.1 cm).

HEIGHT: 0.7 to 1.4 in (1.8 to 3.4 cm).

TEMPERATURE RANGES:

OPERATING: 32°F to 104°F (0°C to 40°C).

CARD READER: 50°F to 104°F (10°C to 40°C).

PRICE IN U.S.A.: $795. Includes rechargeable battery pack, 115/230V ac adapter/recharger, soft carrying case, safety travel case, owners handbook and quick reference guide, program forms, standard pac of prerecorded cards, HP-65 Newsletter, and 1-year subscription to Catalog of Contributed Programs.

MANUFACTURING DIVISION: ADVANCED PRODUCTS DIVISION
19310 Pruneridge Avenue
Cupertino, California 95014 U.S.A.

Programming the Personal Computer

Wherein are revealed the functions of the keys, how problems are solved, and a bit of what goes on inside.

by R. Kent Stockwell

THE HP-65 CALCULATOR uses the same reverse Polish keyboard language, the same four-register operational stack, and the same architecture as its predecessors, the HP-35,[1] the HP-45, and the HP-80.[2] It also has two important features that are new to hand-held calculators. One is its greatly expanded function set, and the other is programmability, complete with conditional and unconditional branching, user-definable functions, and magnetic-card program storage.

Function Set

Thirteen HP-65 keys are for data entry. These are the digits 0 to 9, the decimal point, CHS (change sign), and EEX (enter exponent). Numbers may be entered with or without a power-of-ten exponent.

Keyed-in digits set the value of the X register, which is also the display, in the four-register operational stack.* The CLx (clear x) key allows corrections. Any other key except SST and R/S terminates entry of a number.

The four arithmetic functions (+, −, ×, ÷) operate on x and y, the contents of the X and Y registers. Operands are loaded into the stack with the ENTER↑ key; they may then be operated upon by the function keys. Operations execute immediately and results appear in X.

Thirty-three other functions derive from using three prefix keys (f, f^{-1}, g) to condition eleven suffix keys (digits 0-9 and decimal point). The two gold-colored prefix keys, labeled f and f^{-1}, access the functions printed in gold above the suffix keys and the inverses or complements of these functions. The blue prefix key, g, accesses the functions printed in blue on the angled lower side of the suffix keys. (The no-prefix meanings of the suffix keys appear on their top faces.) All of these functions execute immediately, operating on x, or x and y, or the entire operational stack. Thus, for example, the key sequence f 4 obtains sin x in the display, f^{-1} 4 obtains $\sin^{-1}$ x, and g 4 obtains 1/x.

*Capital letters are names of registers and lower-case letters are register contents.

Computations requiring more data storage than is provided by the operational stack may use any of nine data storage registers. For example, pressing STO 4 stores x into register four, leaving x unchanged. Pressing RCL 4 recalls r_4 to X, leaving r_4 unchanged. Arithmetic accumulation to any storage register is accomplished by inserting the desired operation key between STO and the digit key that addresses the register. Thus the key sequence STO <arithmetic operator><digit n> gives r_n<arithmetic operator>x in register R_n and leaves x in the display.

The user can change the display format as required by the particular problem. The key sequence DSP <digit n> rounds the display value to n digits after the decimal point in scientific notation,* while DSP . <digit n> results in an absolute display rounded to n digits following the decimal point. For example, 12.366 gives 1.24 01 in DSP 2 mode and 12.37 in DSP . 2 mode. Display rounding does not affect internal values.

All functions involving angles, that is, sin, cos, tan, R→P (rectangular to polar conversion), → D.MS (conversion to degrees, minutes, seconds), and the inverses of these functions accept arguments or produce results in degrees, radians, or grads, set by the key sequence g ENTER↑ or g CHS or g EEX, respectively. These settings remain in effect until changed.

On the theory that users should be able to correct key-sequence errors with minimal effort, any prefix key overrides any previous prefix key, and the sequence f ENTER↑ clears any prefix keys. Thus, for example, the key sequence STO + f g g 4 gives 1/x,

*One digit to the left of the decimal point with power-of-ten exponent, e.g., 2.54 × 10^{12}.

while g f ENTER↑ 4 gives the value 4 in the display.

By now it must be clear how key conditioning with color-coded keys and legends has been used to provide access to many functions with a limited number of keys on a small keyboard. Although another level of conditioning would further expand the function set (e.g., f g 4 or f^{-1} g 4 or g f 4 could possess functional meanings), this would greatly increase keyboard complexity, keyboard busyness, and internal control programming. For these reasons, most of the key conditioning remains at the one-prefix level.

HP-65 functions are listed on page 14. Fig. 1 shows an example of a problem solution.

Programming

All operations described so far apply when the switch in the upper right-hand corner of the HP-65 keyboard is in the RUN position. When this switch is in the W/PRGM position, the keystrokes are stored in the 100-step program memory instead of being executed. Twenty-five frequently used two-keystroke sequences merge into a single memory step; thus the program memory may actually contain more than 100 keystrokes.

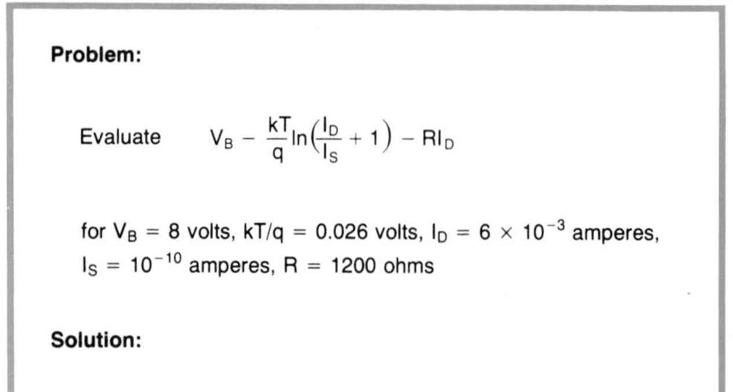

Problem:

Evaluate $V_B - \frac{kT}{q}\ln\left(\frac{I_D}{I_S} + 1\right) - RI_D$

for V_B = 8 volts, kT/q = 0.026 volts, $I_D = 6 \times 10^{-3}$ amperes, $I_S = 10^{-10}$ amperes, R = 1200 ohms

Solution:

	Stack Registers			
Keystrokes	Display X	Y	Z	T
8	8.			
ENTER↑	8.00×10^{0}	8		
.026	.026	8		
ENTER↑	2.60×10^{-2}	.026	8	
.006	.006	.026	8	
ENTER↑	6.00×10^{-3}	.006	.026	8
EEX 10 CHS	10^{-10}	.006	.026	8
÷	6.00×10^{7}	.026	8	8
1	1	6×10^{7}	.026	8
+	6.00×10^{7}	.026	8	8
f ln	1.79×10^{1}	.026	8	8
×	4.66×10^{-1}	8	8	8
−	7.53×10^{-1}	8	8	8
1200	1200	7.53×10^{-1}	8	8
ENTER↑	1.20×10^{3}	1200	7.53×10^{-1}	8
.006	.006	1200	7.53×10^{-1}	8
×	7.20×10^{0}	7.53×10^{-1}	8	8
−	3.34×10^{-1}	8	8	8

Calculator in DSP 2 Mode

Fig. 1. *An example of HP-65 use as a scientific calculator.*

STO 1	RCL 1	g R↓
STO 2	RCL 2	g R↑
STO 3	RCL 3	g x⇌y
STO 4	RCL 4	g LSTx
STO 5	RCL 5	g NOP
STO 6	RCL 6	g x≠y
STO 7	RCL 7	g x≤y
STO 8	RCL 8	g x=y
		g x>y

Fig. 2. *User programs may have as many as 100 steps. These twenty-five keystroke sequences merge into a single step. Thus programs may contain more than 100 keystrokes.*

The memory itself contains no absolute addresses. Instead, it is a circulating shift register organized into six-bit words. One word is a marker that denotes the boundary between the beginning and the end of the memory. Another word is a pointer which denotes the last step executed in run mode, and the last step filled in program mode. As a program runs, this pointer is moved down through memory. Branching is accomplished by moving the pointer to the location of the destination label. User-defined function calls are implemented by leaving the main pointer at the call and activating a second pointer at the function location (see Fig. 3). When the return to the calling location occurs, the second pointer is deactivated and the first pointer reactivated. Neither the marker nor the pointers subtract from the 100 user steps.

Programs may contain three types of tests to allow conditional execution of all operations. These are x-y comparisons (x≠y, x≤y, x=y, x>y), four flag tests

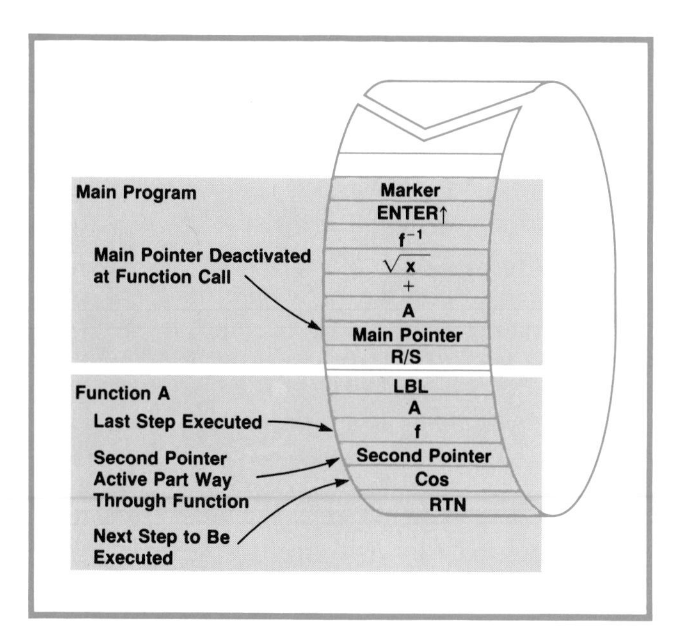

Fig. 3. *The program memory circulates continuously, its beginning and end denoted by a marker. The main pointer moves as programs are entered or executed. A second pointer is activated when a user-defined function is called.*

Problem:

Find the diode current I_D in the circuit shown. Also find its sensitivity with respect to V_B and R, i.e., $\partial I_D/\partial V_B$ and $\partial I_D/\partial R$.

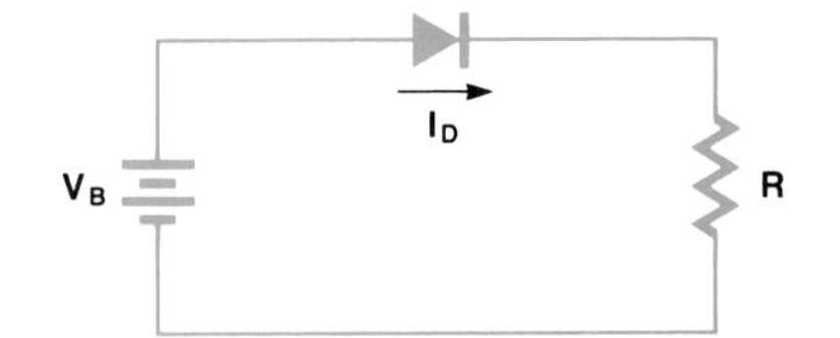

Equations:

$$V_B = \frac{kT}{q}\ln\left(\frac{I_D}{I_S}+1\right) + RI_D$$

$$\frac{\partial I_D}{\partial V_B} = \left[\frac{kT}{q}\left(\frac{1}{I_D + I_S}\right) + R\right]^{-1}$$

$$\frac{\partial I_D}{\partial R} = -I_D\left[\frac{kT}{q}\left(\frac{1}{I_D + I_S}\right) + R\right]^{-1}$$

I_S = diode saturation current in amperes
R = resistor value in ohms
V_B = battery voltage in volts
kT/q = thermal voltage in volts

Algorithm:

For Newton-Raphson iteration,

$$I_D(n+1) = I_D(n) - \frac{f[I_D(n)]}{f'[I_D(n)]}$$

where $I_D(n)$ = nth guess
$f[I_D(n)]$ = function evaluated for nth guess
$f'[I_D(n)]$ = first derivative of function, evaluated for nth guess
$I_D(n+1)$ = (n + 1)st guess

Let $$f(I_D) = V_B - \frac{kT}{q}\ln\left(\frac{I_D}{I_S}+1\right) - RI_D$$

Then $$f'(I_D) = -\left[\frac{kT}{q}\left(\frac{1}{I_D + I_S}\right) + R\right]$$

Specify convergence criterion: if $|I_D(n+1) - I_D(n)| < C$ the algorithm halts.

Program halts after ten iterations. The user may then start ten more iterations.

Example:

$I_S = 10^{-10}$ A
R = 1.2 kΩ
V_B = 8 V
kT/q = 0.026 V
C = 10^{-9} A

Load card and follow user instructions.

Results:

I_D = 6.278 A
$\partial I_D/\partial V_B$ = 0.8305 mA/V
$\partial I_D/\partial R$ = −5.213 μA/Ω

Time required to compute I_D (step 3): 11 seconds.

Flow Chart for Iteration:

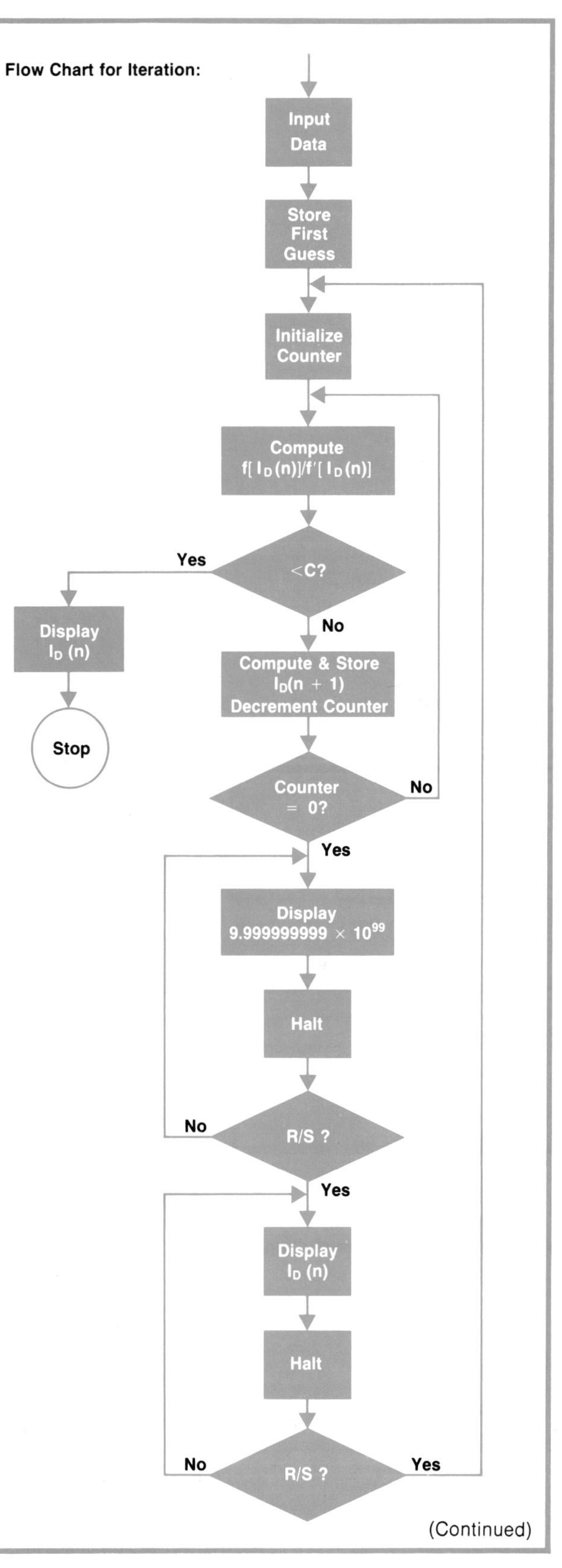

(Continued)

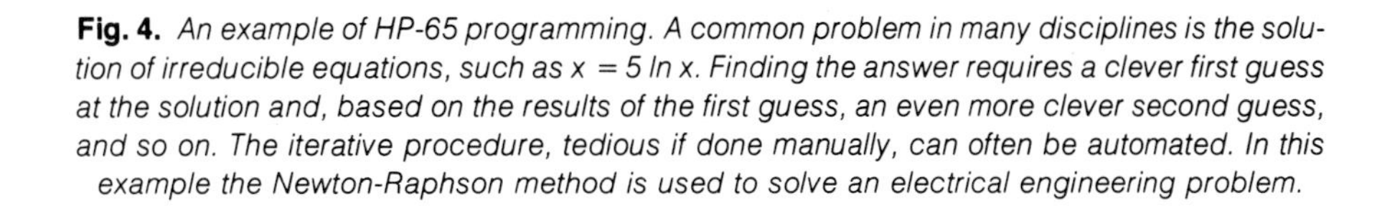

Fig. 4. *An example of HP-65 programming. A common problem in many disciplines is the solution of irreducible equations, such as x = 5 ln x. Finding the answer requires a clever first guess at the solution and, based on the results of the first guess, an even more clever second guess, and so on. The iterative procedure, tedious if done manually, can often be automated. In this example the Newton-Raphson method is used to solve an electrical engineering problem.*

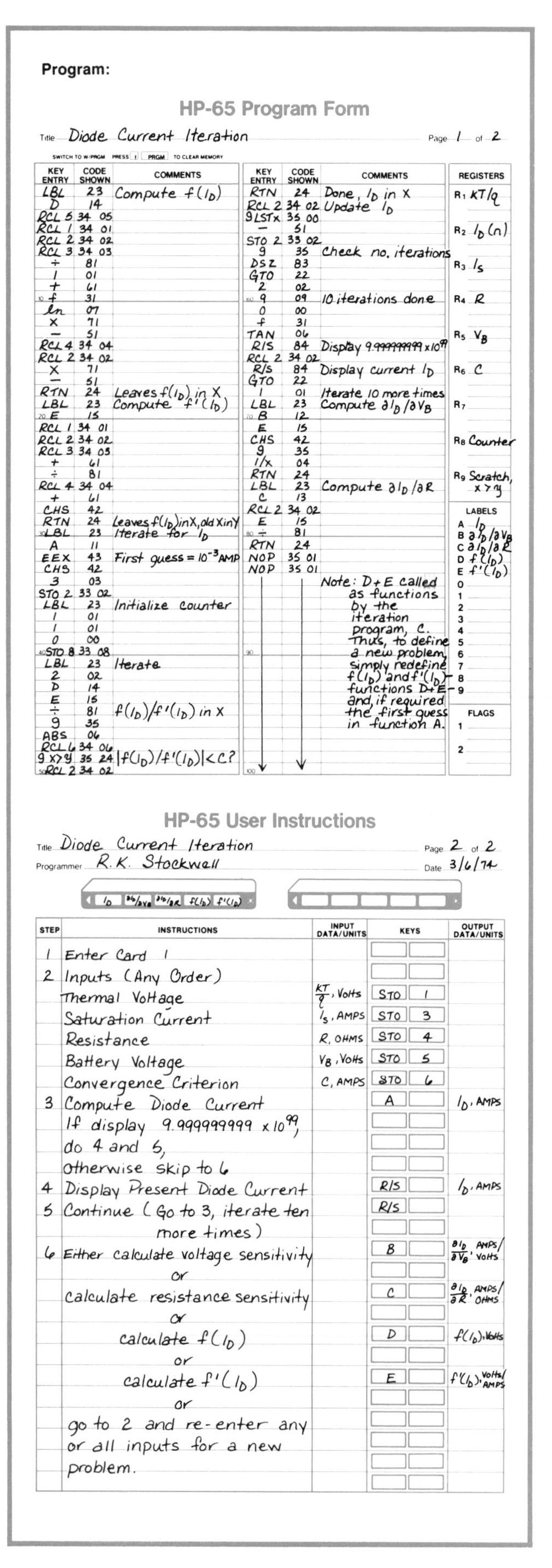

Program:

HP-65 Program Form

Title: Diode Current Iteration — Page 1 of 2

SWITCH TO W/PRGM PRESS f PRGM TO CLEAR MEMORY

KEY ENTRY	CODE SHOWN	COMMENTS	KEY ENTRY	CODE SHOWN	COMMENTS
LBL	23	Compute $f(I_D)$	RTN	24	Done, I_D in X
D	14		RCL 2	34 02	Update I_D
RCL 5	34 05		g LSTx	35 00	
RCL 1	34 01		−	51	
RCL 2	34 02		STO 2	33 02	
RCL 3	34 03		g	35	Check no. iterations
÷	81		DSZ	83	
1	01		GTO	22	
+	61		2	02	
f	31		9	09	10 iterations done
ln	07		0	00	
×	71		f	31	
−	51		TAN	06	
RCL 4	34 04		R/S	84	Display $9.999999999 \times 10^{99}$
RCL 2	34 02		RCL 2	34 02	
×	71		R/S	84	Display current I_D
−	51		GTO	22	
RTN	24	Leaves $f(I_D)$ in X	1	01	Iterate 10 more times
LBL	23	Compute $f'(I_D)$	LBL	23	Compute $\partial I_D/\partial V_B$
E	15		B	12	
RCL 1	34 01		E	15	
RCL 2	34 02		CHS	42	
RCL 3	34 03		g	35	
+	61		1/x	04	
÷	81		RTN	24	
RCL 4	34 04		LBL	23	Compute $\partial I_D/\partial R$
+	61		C	13	
CHS	42		RCL 2	34 02	
RTN	24	Leaves $f(I_D)$ in X, old X in Y	E	15	
LBL	23	Iterate for I_D	÷	81	
A	11		RTN	24	
EEX	43	First guess = 10^{-3} AMP	NOP	35 01	
CHS	42		NOP	35 01	
3	03		↓	↓	Note: D+E called as functions by the iteration program, C. Thus, to define a new problem, simply redefine $f(I_D)$ and $f'(I_D)$ — functions D+E — and, if required the first guess in function A.
STO 2	33 02				
LBL	23	Initialize counter			
1	01				
1	01				
0	00				
STO 8	33 08				
LBL	23	Iterate			
2	02				
D	14				
E	15				
÷	81	$f(I_D)/f'(I_D)$ in X			
g	35				
ABS	06				
RCL 6	34 06				
g x>y	35 24	$\lvert f(I_D)/f'(I_D)\rvert < C$?			
RCL 2	34 02				

REGISTERS

Register	Contents
R_1	KT/q
R_2	$I_D(n)$
R_3	I_s
R_4	R
R_5	V_B
R_6	C
R_7	
R_8	Counter
R_9	Scratch, x>y

LABELS

Label	Function
A	I_D
B	$\partial I_D/\partial V_B$
C	$\partial I_D/\partial R$
D	$f(I_D)$
E	$f'(I_D)$

FLAGS: 1, 2

HP-65 User Instructions

Title: Diode Current Iteration — Page 2 of 2

Programmer: R.K. Stockwell — Date: 3/6/74

Card: I_D | $\partial I_D/\partial V_B$ | $\partial I_D/\partial R$ | $f(I_D)$ | $f'(I_D)$

STEP	INSTRUCTIONS	INPUT DATA/UNITS	KEYS	OUTPUT DATA/UNITS
1	Enter Card 1			
2	Inputs (Any Order)			
	Thermal Voltage	KT/q, Volts	STO 1	
	Saturation Current	I_s, AMPS	STO 3	
	Resistance	R, OHMS	STO 4	
	Battery Voltage	V_B, Volts	STO 5	
	Convergence Criterion	C, AMPS	STO 6	
3	Compute Diode Current		A	I_D, AMPS
	If display $9.999999999 \times 10^{99}$, do 4 and 5, otherwise skip to 6			
4	Display Present Diode Current		R/S	I_D, AMPS
5	Continue (Go to 3, iterate ten more times)		R/S	
6	Either calculate voltage sensitivity		B	$\partial I_D/\partial V_B$, AMPS/Volts
	or			
	calculate resistance sensitivity		C	$\partial I_D/\partial R$, AMPS/OHMS
	or			
	calculate $f(I_D)$		D	$f(I_D)$, Volts
	or			
	calculate $f'(I_D)$		E	$f'(I_D)$, Volts/AMPS
	or			
	go to 2 and re-enter any or all inputs for a new problem.			

(there are two flags, each of which may be set or cleared and then tested for set or clear), and decrement and skip if zero (DSZ). Except for DSZ, each test, if false, causes program control to skip the next two memory steps; otherwise, execution continues normally. The DSZ operation decrements data-storage register R_8 by one, using integer arithmetic, and if the result is zero, program control skips the succeeding two steps.

Literal labels with the GO TO function implement branching. Thus LBL<n> is the destination for GTO <n>, where n is a digit or a key A-E in the top row.

The HP-65 user may store two types of programs in the program memory. First, he may precede a section of memory containing various functions with LBL <m>, where m is A, B, C, D, or E, and terminate the section with RTN (return). Thereafter, pressing key A, B, C, D, or E in the RUN mode causes that memory section to execute immediately. Any or all of keys A to E may be defined but the sum of memory steps for all functions cannot exceed 100. These user-defined functions behave exactly like the preprogrammed functions described earlier, yet the user may create the functions to fit his special needs.

The user's second option is to precede a block of code with a label definition and terminate it with the R/S (Run/Stop) key. In RUN mode this key stops an executing program; if no program is running, pressing the R/S key starts execution. Pressing GTO<label name>R/S then starts program execution, and the program halts at the R/S in memory. If the program starts at the beginning of memory no label is needed; in RUN mode control can be transferred to the beginning of memory by pressing RTN. Programs defined in this way may call any of the functions A through E; the desired key is simply entered into the program definition.

The SST (single step) and DEL (delete) functions implement debugging and editing. In W/PRGM mode, each depression of SST advances the memory pointer one step and displays each memory step as a two-digit key code. These codes represent digit keys by their values and all other keys by a row-column index of the key position referenced to the upper left-hand corner of the keyboard. For example, the decimal-point key is in the eighth row, third column, so its code is 83. In the RUN mode, each depression of SST advances the memory pointer one step and executes the adjacent memory step.

The key sequence g CLx in W/PRGM mode deletes the displayed memory step and moves up the next step to fill the gap. Any keys entered in W/PRGM mode are automatically inserted following the displayed memory step. Thus the replacement operation consists of a delete operation followed by the desired key.

The sequence f CLx clears the entire memory.

Programs can be stored on magnetic cards for later use. Cards can be recorded and rerecorded as many times as desired. To protect a recorded program on a card, further recording can be prevented by clipping the notched tab on the upper left corner of the card. Users may write on the card and place it in a slot above the keys **A** through **E**, thereby labeling any specially defined keys.

Fig. 4 shows an example of HP-65 programming.

Firmware

To direct the various computational and control functions of the HP-65, 3072 words of read-only memory (ROM) are used. Each ROM word contains ten bits and constitutes a calculator microinstruction. Microinstructions grouped together in blocks perform the various external functional tasks of the calculator. A task may require one block of words or several blocks woven together. For example, the **CLx** function requires only a few words, while the sin function uses the tan function, which uses the add function, and so on.

Although production of efficient microcode is an iterative process, the first step is the choice or design

HP-65 Normalization Routine

User input form is stored in two processor registers, A and B.

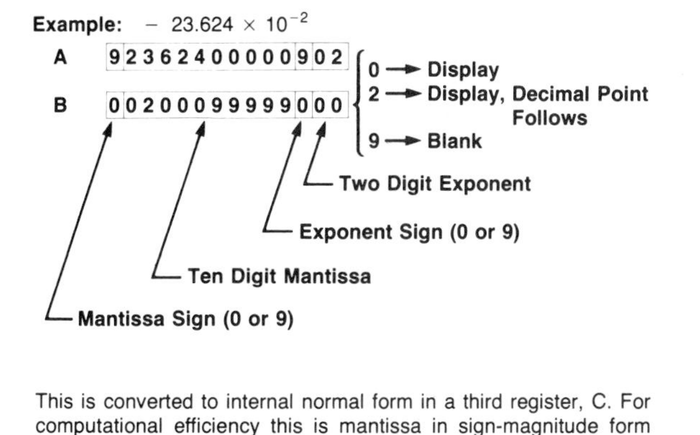

This is converted to internal normal form in a third register, C. For computational efficiency this is mantissa in sign-magnitude form with one digit to the left of an implied decimal point and exponent in ten's complement form (see ref. 4). In the mantissa sign position, 9 represents minus, 0 represents plus. Thus, -23.624×10^{-2} would store internally as

9 2 3 6 2 4 0 0 0 0 0 9 9 9

Program Listing:

Step Number	ROM Address	ROM Code	ROM Subroutine Addresses	Labels	Program Statements
51	L01063:	.....111..		FIX5 :	P - 1 -> P
52	L01064:	.1111.1.1.			C + 1 -> C[X]
53	L01065:	..111.11.1	-> L1073		JSB FIX7
		⋮			⋮
56	L01070:	..11.11.1.		FIX3 :	0 -> C[XS]
57	L01071:	11.1..11..		FIX4 :	13 -> P
58	L01072:	.1.11.1.1.			C - 1 -> C[X]
59	L01073:	........1.		FIX7 :	IF B[P] = 0
60	L01074:	..11..1111	-> L1063		THEN GO TO FIX5
61	L01075:	11....11..			12 -> P
62	L01076:	1..11...1.		FIX6 :	IF A[P] >= 1
63	L01077:	.11111..11	-> L1174		THEN GO TO FIX2
64	L01100:	.1.....11.			SHIFT LEFT A[M]
65	L01101:	.1.11.1.1.			C - 1 -> C[X]
66	L01102:	..11111..1	-> L1076		JSB FIX6
		⋮			⋮
102	L01146:	.111111.1.		FIX1 :	C + 1 -> C[XS]
103	L01147:	..111...11	-> L1070		IF NO CARRY GO TO FIX3
104	L01150:	..1.1.1.1.			0 - C -> C[X]
105	L01151:	..111..1.1	-> L1071		JSB FIX4
		⋮			⋮
119	L01167:	111.1.111.		FIX0 :	A EXCHANGE C[W]
120	L01170:	.11...111.			C -> A[W]
121	L01171:	1..11..11.			IF A[M] >= 1
122	L01172:	.11..11.11	-> L1146		THEN GO TO FIX1
123	L01173:	1.111..11.			0 -> A[M]
124	L01174:	111.1..11.		FIX2 :	A EXCHANGE C[M]

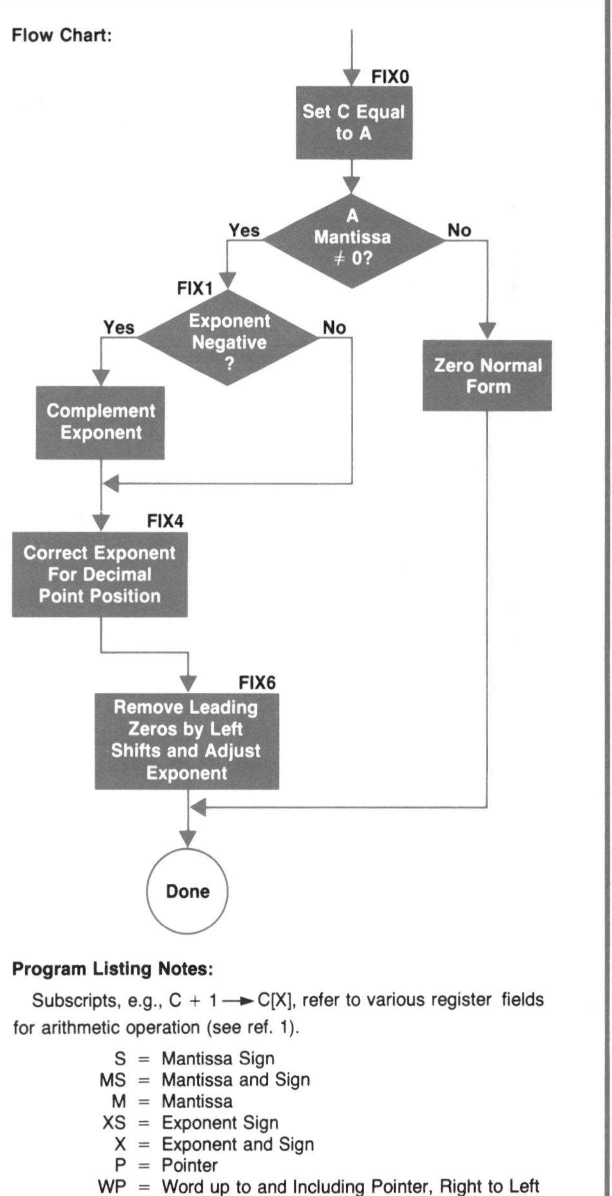

Fig. 5. *An example of the HP-65's internal microprogramming. Even such a seemingly trivial operation as digit entry requires careful design so it seems trivial to the user. Values must be displayed as keyed in, yet be normalized to a standard internal form. This is the normalization routine and the flow chart and ROM listing for it.*

of an algorithm. This may involve such constraints as accuracy, execution speed, microinstructions required, or even available design time. Next, a functional flow chart is drawn to outline the sequence of various operations and any conditional operations. This flow chart is then expanded to sufficient detail that it can be translated to microinstructions and implemented on a calculator simulator. More often than not there are implementation errors to correct; sometimes the entire algorithm is faulty, requiring a new design. When the design is complete, integrated-circuit read-only memories are produced.

Where possible, the HP-65 uses the proven algorithm implementations from the HP-35 and HP-45 (trigonometric, logarithmic, and exponential routines). This saved development time and reduced implementation error probabilities.

Many HP-65 algorithms would provide interesting descriptions here, but one that demonstrates appreciable complexity is the digit-entry routine. Designing this seemingly trivial function so as to seem trivial to the user required considerable patience and careful thought. Usually, any entry will produce an undesirable result unless the designer specifically accounts for it. Values must be displayed as keyed in, yet they must be normalized to some internal form. The table below lists some of the design constraints on this algorithm.

USER ACTION	DESIRED RESULT
More than ten mantissa digits	Ignore all digits after tenth
First key of new entry	Overwrite existing x if key follows ENTER↑ or CLx; otherwise do automatic ENTER↑
Extra digits after EEX	Shift exponent left; new digit becomes least significant digit of exponent.
Multiple decimal point	Ignore all decimal points after first
Decimal point after EEX	Ignore
Leading zeros keyed in	Accept and display leading zeros, zero normal form.
EEX first key of new entry	Enter one in mantissa; following digits enter exponent.
Decimal point first key of new entry	Display only decimal point; zero normal form.
Digits after decimal point	Continue appending digits; no effect on internal exponent
Digits before decimal point	Continue appending digits, increment internal exponent.
CHS following EEX	Complement exponent sign
Multiple CHS	Complement mantissa sign, or exponent sign if EEX has been pressed.

Such an algorithm was explained in a previous issue.[3] Fig. 5 shows the flow chart and ROM listing for the normalizing routine.

Acknowledgments

Many people of course, contributed ideas to this effort. Particular acknowledgment is due the following: Paul Stoft and Tom Whitney for bringing together the necessary technical resources and people; Dave Cochran, for the trigonometric and exponential routines used in the HP-35, and for help in understanding the HP-35 architecture; Francé Rodé for further explanations of the HP-35 architecture; Peter Dickinson for suggestions and criticisms concerning algorithm implementations, particularly the extension of the HP-35 algorithms; Tom Osborne for helpful advice and suggestions regarding the function set and the external behavior of the HP-65; Homer Russell and Wing Chan for helpful suggestions and criticisms for the function set, and for

patiently keeping up with numerous daily changes; Steve Walther for providing the microinstruction language compiler; Darrel Lauer and Al Inhelder for crystallizing the keyboard layout from a myriad of suggestions; Ed Heinsen and Lynn Tillman for extending the simulation software to accommodate the increased complexity of the HP-65.

References

1. T.M. Whitney, F. Rodé, and C.C. Tung, "The 'Powerful Pocketful': an Electronic Calculator Challenges the Slide Rule", Hewlett-Packard Journal, June 1972.
2. W.L. Crowley and F. Rodé, "A Pocket-Sized Answer Machine for Business and Finance", Hewlett-Packard Journal, May 1973.
3. D.S. Cochran, "Internal Programming of the 9100A Calculator", Hewlett-Packard Journal, September 1968.
4. M.M. Mano, "Computer Logic Design", Prentice-Hall, 1972, chapter 1.

R. Kent Stockwell

Kent Stockwell joined HP four years ago. As a member of HP Laboratories for most of that period, he's done program development, modeling, and numerical analysis for computer-aided circuit design and, more recently, the firmware development for the HP-65. Kent studied electrical engineering at Massachusetts Institute of Technology, graduating in 1970 with SB and SM degrees. A native of Kalamazoo, Michigan, he now lives in Palo Alto, California, where he's currently remodeling his house and putting his woodworking skills to good use. He also plays trombone and baritone horn, and enjoys backpacking in the mountains of California and Colorado.

APPENDIX

HP-65 Programmable Pocket Calculator Functions and Operations

Arithmetic
- add
- subtract
- multiply
- divide

Logarithmic
- natural logarithm (base *e*)
- natural antilogarithm (base *e*)
- common logarithm (base 10)
- common antilogarithm (base 10)

Trigonometric
- set operating mode (degrees, radians, or grads)
- sine
- arc sine
- cosine
- arc cosine
- tangent
- arc tangent
- add or subtract degrees/minutes/seconds
- convert angle from degrees, radians, or grads to degrees/minutes/seconds and vice versa
- convert polar coordinates to rectangular coordinates and vice versa

Exponential
- square
- square root
- raising a number to a power (y^x)
- reciprocal (can be used with y^x function to extract nth roots)

Other Preprogrammed Functions and Operations
- extract integer or decimal portion of a number
- factorial
- recall value of π to 10 significant digits
- convert decimal-base integers to or from octal-base integers
- "roll down" or "roll up" numbers in operational stack
- clear display
- clear operational stack
- clear all nine addressable memory registers
- recall last input argument from separate "last-x" storage register
- store or recall numbers from any of the nine addressable memory registers
- register arithmetic
- display formatting

Program Structure and Edit Functions
- clear program memory
- user-definable keys (A-E)
- label
- go-to
- return
- run/stop
- no-operation
- set flag 1
- test flag 1
- set flag 2
- test flag 2
- $x = y$
- $x \neq y$
- $x \leq y$
- $x > y$
- decrement and skip on zero
- delete program step
- single-step

Designing a Tiny Magnetic Card Reader

Here's how it was designed and how it works.

by Robert B. Taggart

ONE THING WE HAD WAS an abundance of ideas for tiny card readers. The HP-65 card reader project began in the electronic research laboratory of HP Laboratories before the introduction of the HP-35. The basic design goal for the card reader was to propel a magnetic card the size of a piece of chewing gum at a constant speed of 3¾ to 6½ cm/s. Of course, it also had to fit inside the HP-35 package.

Many schemes were tried, including music box mechanisms, hand feed, gravity feed, and dashpot systems. Motor-driven schemes didn't generate much enthusiasm at first. But one day while digging through a file on motors that was about to be discarded, I found a brochure describing a Swiss-made motor less than 1½ cm in diameter. Within a day we obtained two samples. They had ample torque and their unique construction provided very low brush noise. The motors had a tested lifetime of over one thousand hours.

These tiny motors turn at speeds in excess of 10,000 r/min. The problem of reducing this high speed down to 6 cm/s was solved by using a worm gear, which provides a large speed reduction in one stage. Other schemes were tried but the worm and wheel combination proved best.

Gripping of the card under all conditions was another problem that had to be solved. The card is very small and would be handled extensively. Even grease could not be allowed to stand between a card and good gripping. Plastics and rubbers of various kinds and textures were tried. We tried to put tire-like treads into the rubber and even put ridges in the card for better grip. But finally we found a polyurethane rubber with the right texture that grips even when the card is coated with oil.

The major problems we faced were caused by trying to squeeze so much information onto such a short card. If the card could have been longer the design would have been much easier. However, having a short card offers the user the convenience of labeling the top row of keys with the program card. This restricted the length of the card to less than the width of the calculator. All kinds of things become critical in trying to read and write on short magnetic cards in such a small machine at 300 or more bits per inch. Azimuth alignment of the magnetic head at 400 bits per inch must be accurate within ±¼ degree. The best way to align heads with this precision is under a microscope. But can you align a head to this kind of accuracy in a molded plastic part? Based on the HP Manufacturing Division's confidence in the plastic we decided to try. We succeeded, thereby achieving a significant cost saving over using a metal frame. Using such a short card required bunching the magnetic head very close to the drive roller and gear train. This unusual geometry combined with the tight tolerances of ±0.001 inch on some dimensions made the reader frame a complex challenge.

The short card and higher bit density created numerous problems related to keeping the vibration of the drive train to a minimum. Of all the problems we faced this proved to be the most difficult. Finding the right process for making the worm gear and refining that process to a high degree made it possible. A method was developed to couple the worm gear to the motor and all these ideas combined to provide the necessary precision and smoothness. The bit-to-bit speed variation was held to less than 10%.

Another challenging problem involved inventing the set of switches that turn on the motor when the card is inserted, then turn on the magnetic head when the card is over the head, and finally provide file protection when the card corner is cut off. Three switches are provided. Some of these serve double duty in that they help wrap the card over the magnetic head for better head contact. The geometry of the switches is unusual and many new ideas were required to fit three switches, a gear train, a drive roller, a backup roller, and magnetic head pole tips in a volume of $1\frac{1}{4} \times 1\frac{1}{4} \times 2\frac{1}{2}$ cm. This very tight spacing was made necessary by the short card length and high bit density.

The short card and the fact that the recording rate is controlled by the clock in the calculator requires that the speed of the card be kept nearly constant under conditions of varying temperature and humidity. For instance, with a slow clock and a fast card reader it is conceivable that the card might go through the machine before all the program could be recorded on it. Conversely, if the clock were fast and the card slow the bit density could exceed the maximum permitted by the head alignment.

To solve this problem the voltage across the motor is regulated and each machine's card speed is set within ±2% by an external trimming resistor on the bipolar circuit that sets the voltage across the motor. We had decided at the outset to use as large a motor as would possibly fit to maximize the amount of available torque. The greater the available stall torque the less sensitive the card reader will be to changes in load. This eliminates the need for feedback speed control.

Speed control turned out to be a very nasty problem particularly at low temperature. It was discovered that at freezing temperatures the polyurethane rubber becomes hard as a rock. This increases the current drain on the motor enormously and reduces the card speed to almost zero. Numerous other types of rubber were tried, many of which remained softer at low temperature, but none of which gripped as well as one polyurethane composition. Eventually by increasing the thickness of the polyurethane we reduced the magnitude of the problem. It was at this point that we appreciated the unique gripping properties of this type of polyurethane.

How the Card Reader Works

Fig. 1 is a diagram of the card reader. As the card is inserted into the right side of the machine, it is forced against one edge of the card slot by one of two tiny leaf springs. This helps align the card with the magnetic head. Pushing the card farther into the machine causes it to activate the motor start switch when the card approaches the rubber drive roller. This turns the motor on.

Each of the three switches in the card reader is activated in the same way. The card displaces a nylon ball resting on the bottom of the card slot. Movement of the ball forces a tiny finger of copper to move upward. The end point of this copper switch finger makes contact with a contact pad on the underside of the keyboard printed-circuit board. The contact point of the switch moves a distance several times the thickness of the card to provide a reliable contact. Each switch is adjusted to the proper contact position during assembly.

When the bipolar motor circuit is turned on a precise voltage is fed to the motor terminals to establish the motor/card speed to within a few percent. The motor turns at a speed near 10,000 revolutions per minute. The motor is directly coupled by a tiny polyurethane sleeve to a miniature worm gear. The end of the worm gear rests against a thrust ball bearing and drives a helical gear. The helical gear is pressed onto the hub of the polyurethane rubber drive roller, which grips the card.

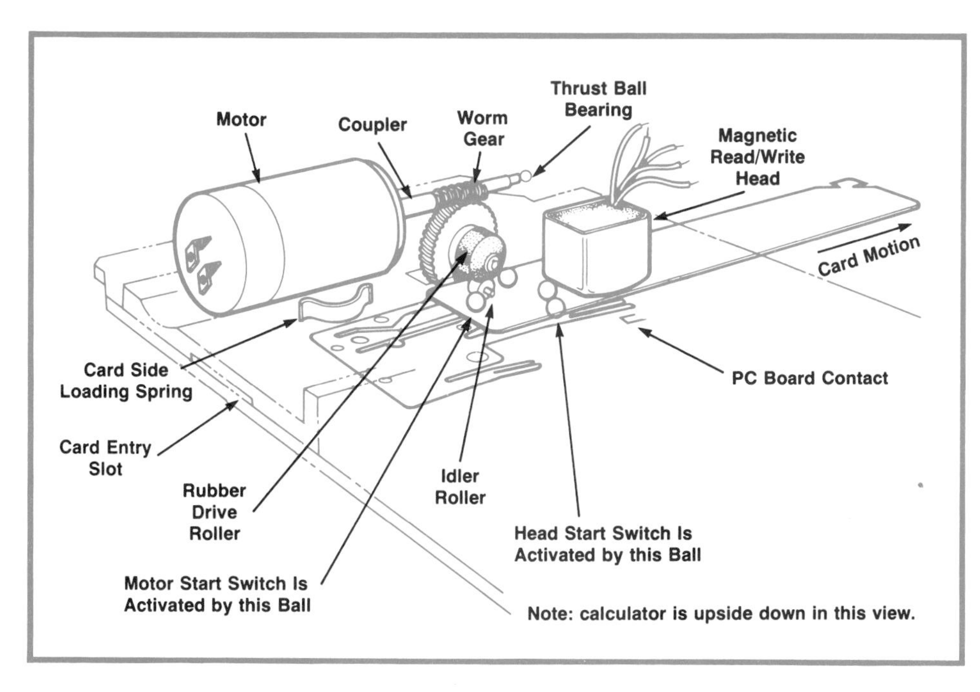

Fig. 1. *The HP-65 card reader.*

Once the motor is turned on by the motor start switch the user must push the card another small fraction of a centimeter so the rubber roller will grip the card. The card is then pinched and driven between the rubber roller and a fixed tiny idler roller made of nylon. The card proceeds through the machine toward the magnetic read/write head. Little bumps in the plastic support plate combine with the switch balls to wrap the card over the gap of the head.

As the card passes the magnetic gap of the head the leading edge of the card activates a second switch. This switch starts the write or read circuitry, depending on the position of the W/PRGM-RUN switch. Activation of this second switch lets the circuits know that the card is over the head and is assumed to be moving at the proper speed.

At nearly the same time that the second switch is activated, a third switch may or may not be activated depending on whether the corner of the card is cut off. This is the file protect scheme, which prevents the user from writing over a previous program. When the second switch is activated the third switch is interrogated. If the third switch is not activated (the corner being cut off) the machine will not write over that card when the calculator is in the W/PRGM mode. The data is written/read by a two-track recording scheme which is described elsewhere. As the card proceeds out the left side of the calculator it is held against the side of the card guide by a second side-loading leaf spring. When the trailing edge of the card passes the motor start switch the motor and read/write circuits are shut off. The card may then be removed from the machine.

Fig. 2 shows the card-reader parts disassembled.

Acknowledgments

The author would like to thank the following people for their contributions during the design stage: Dick Barth, Bob Schweizer, Bernie Musch, Clarence Studley, John Bailey, Craig Sanford, Bill Boller and the HP Manufacturing Division, Fred Rios and his model shop, the Advanced Products Division model shop, and the APD production tooling groups.

Robert B. Taggart

Bob Taggart is an engineering group leader at HP's Advanced Products Division. He holds a 1967 BS degree in mechanical engineering and a 1968 MS in biomedical engineering, both from Northwestern University, and the degree of Engineer from Stanford University (1970). In 1970 he joined HP Laboratories and for three years was involved in biomedical and pollution studies before moving to APD to develop the HP-65 card reader. Author of three papers on microwave antenna design and five patent applications (two granted), Bob is originally from Pompton Lakes, New Jersey. He skis, plays tennis, and enjoys white-water river running. He and his wife welcomed their first child, a boy, on March 31.

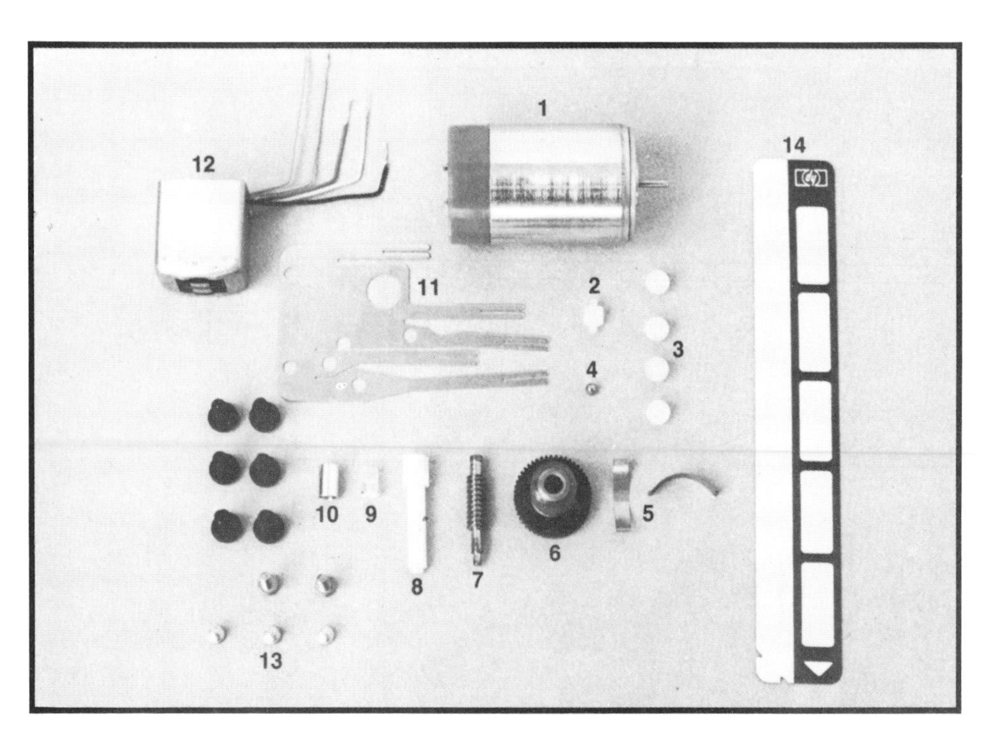

Fig. 2. *HP-65 card reader parts: 1. Motor 2. Idler roller 3. Switch activation balls 4. Thrust ball bearing 5. Card side loading springs 6. Rubber drive roller and helical gear 7. Worm gear 8. Drive pin 9. Coupler 10. Coupler sleeve 11. Switch contacts 12. Read/write head 13. Self-tapping switch adjustment screws 14. Magnetic card, 7.1 cm by 1 cm.*

Testing the HP-65 Logic Board

The board and its automatic test system are designed for rapid production testing and troubleshooting.

by Kenneth W. Peterson

DESIGNING AN INSTRUMENT for minimum cost means not only that parts and assembly costs are minimized. The time required for testing, troubleshooting, and repair is also critical and must be held to an absolute minimum. Yet too many products are designed without giving adequate consideration to the time that can be spent analyzing and locating any faults that may show up in production testing.

For a high-volume, complex product like the HP-65, computerized testing is the only feasible method. To aid the computer in diagnosing failures and isolating them to the responsible components, it's essential that access be provided to all of the product's pertinent nodes. The HP-65 logic board was designed with this in mind.

The layout of the HP-65 logic board is shown in Fig. 1. Input and output lines are brought out to two edges of the board. Additional test points are routed to a third edge, giving access to a total of 45 test points.

Test System

A block diagram of the computerized tester designed for the HP-65 logic board is shown in Fig. 2. The tester functionally compares the operation of a known-good unit with that of a test unit. All outputs from the test logic board are interfaced to voltage-comparison circuits to check for proper logic-level thresholds.

As the tester exercises the reference and test units with identical inputs (either on the key lines or on the card reader lines), the units' outputs are captured in two pattern storage registers on each clock time.

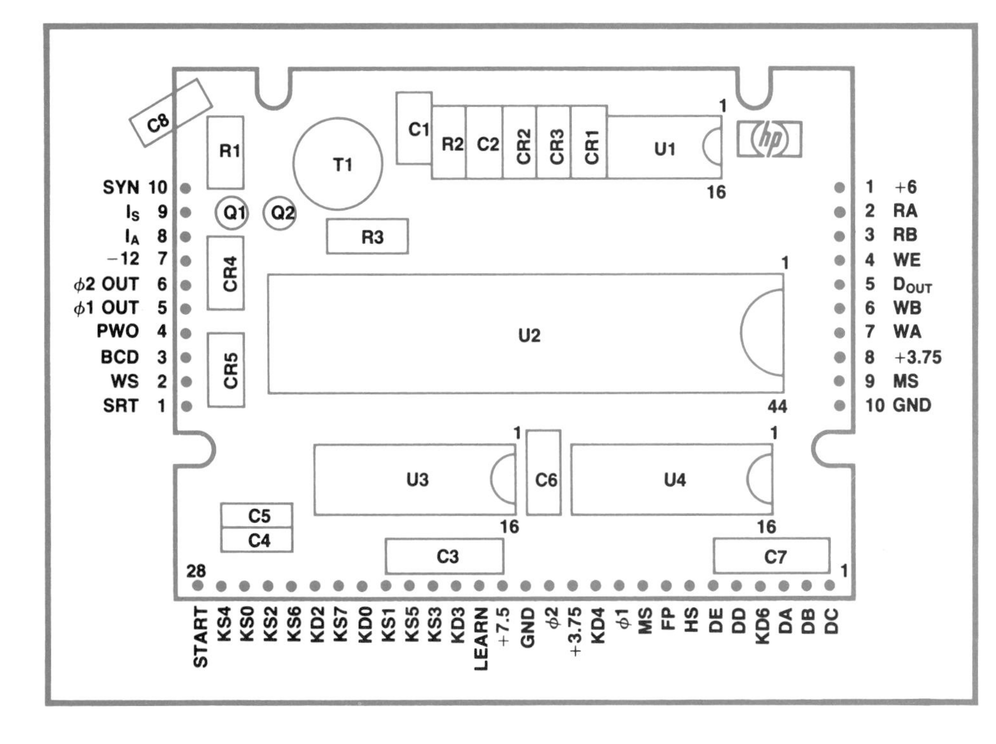

Fig. 1. *The HP-65 logic board is designed for automatic testing. Access to 45 test points is provided.*

The patterns are continually compared. If they ever fail to compare, their error condition is frozen in the pattern storage registers and a flag is sent to the computer. The computer can then request all pertinent information from the tester, which keeps a record of events leading up to a failure. For example, microinstructions are saved and decoded to show which chip was talking on the calculator buses when the failure occurred. Other information retained by the tester is the present ROM address, the previous ROM address, the identity of the active ROM (one of twelve), the bit count at the time of failure, and the number of word times since the start of the test sequence.

All this information and the contents of the two pattern storage registers are then sent to the computer. The computer sorts through the data with the aid of a diagnostic table and prints out the nature of the error and the component most likely to have caused it.

Tester Architecture

The HP-65 logic board tester has six main parts.

The *controller* is a 32-state ROM machine with 16 qualifiers and two-way branching on test conditions. There are eighteen instruction lines. Each state issues instructions that control the condition of other hardware in the tester. The controller also contains the master clock, a crystal-controlled oscillator set to the specified limit of the MOS circuits, 200 kHz.

The *interface circuits* consist of a voltage comparator and buffer amplifier between each calculator's MOS circuits and the input to the pattern storage register, which is a TTL circuit. The voltage comparator checks the upper and lower limits of the logic levels.

The *pattern storage and comparator circuit* consists of two 16-bit registers of D-type flip-flops with a 16-bit parallel comparison circuit between them to check for parity. If the patterns do not compare after each clock period (5μs) an error flip-flop is set. The clock is then inhibited to each of the 16-bit pattern storage registers and the error condition is saved.

Input/output information is received by the tester and transmitted to the computer over 16 parallel lines. Inputs are stored in a 16-bit buffer register which receives commands and data from the computer. Three of the input lines from the computer are assigned as output select control lines to specify which data to send back to the computer. The other 13 input lines are used to control the logic board: six lines specify the key code (function), four lines simulate switches to the card reader, one line varies the power-supply limits, and two lines are used for status information.

A 16-bit, eight-channel multiplexer controlled by the three output-select lines is used to return data to the computer. A full handshake between the tester and the computer is done; this assures proper data-transmission timing.

Inputs to the logic board are either through the key lines or the card reader control lines. The key lines are made through an 8 × 5 matrix. A six-bit code

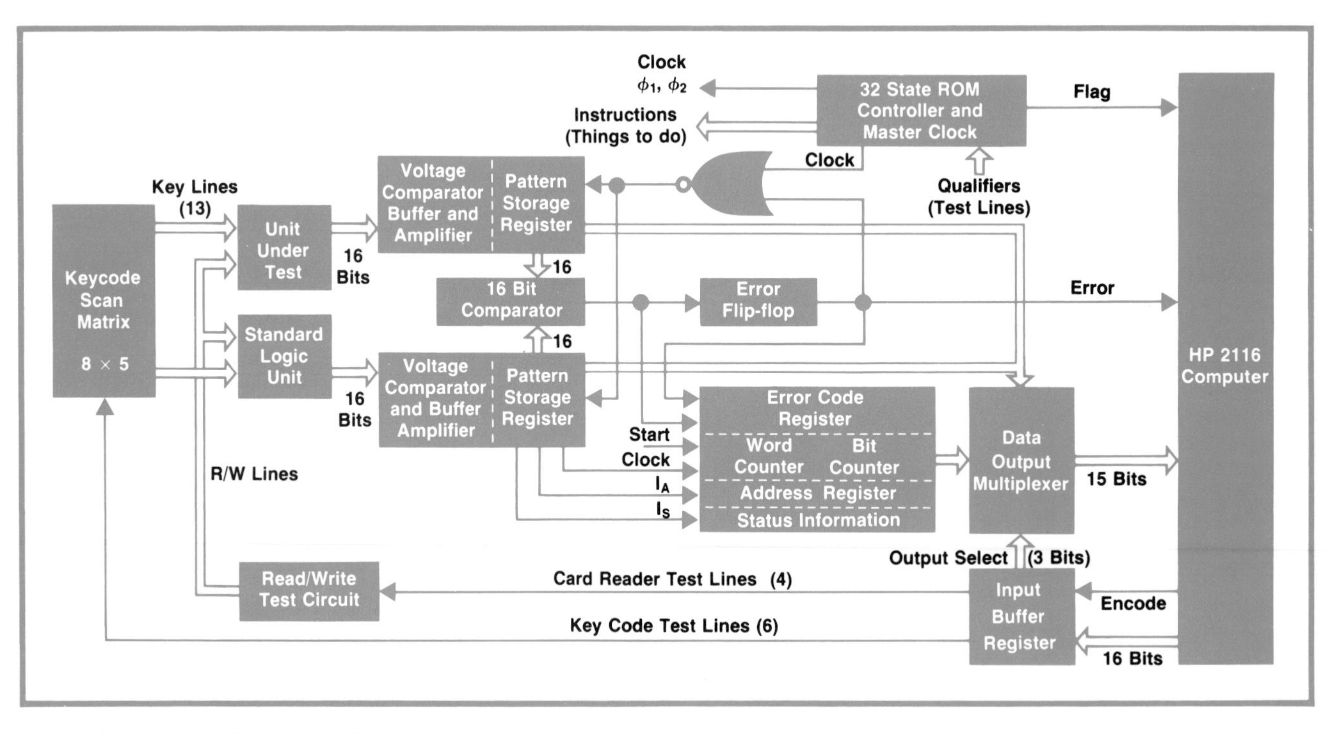

Fig. 2. *Logic board tester compares a reference board with a test board. Failures are analyzed by the computer, which prints out diagnostic messages.*

from the computer specifies the key code. There are four lines to the card reader: learn mode switch, motor switch, head switch, and file protect switch.

The *read/write test circuit* is a 15-state controller with instruction decoding for determining when to shift the read/write data pattern in and out. To simulate the magnetic card, reference write patterns are stored in two 600-bit shift registers. A third 600-bit shift register is used as a counter to keep track of how many shifts have been done.

Card Reader Testing

To make the card reader circuit read or write it is necessary to close the motor switch and the head switch. Then the condition of the learn mode switch line specifies whether the unit will read or write. The file protect line specifies the condition of the write enable line.

The test sequence consists of writing a test pattern from each of the two logic boards and comparing the pattern of the reference board to that of the board under test. The reference patterns are stored in two 600-bit shift registers. The next step is to read the same pattern back into the calculator and then write it back out again. The second write sequence is used to verify that the read circuit is functioning properly. The write enable line is tested by making the file protect signal a logical zero on the first write sequence and a logical one during the second write sequence.

Overall Test Sequence

The steps in the test sequence are as follows:

- Power off, push start button.
- Turn power on, set time delay (this allows the power supply to rise to the proper level).
- Test power supply and clock for proper high and low limits. Also test if power-on pulse was given properly. If not go to error routine.
- If everything is correct give a pseudo power-on (PWO) signal. This sets the starting address to zero.
- Test if both systems are in synchronism and if not, slow the clock to the unit under test until both systems are in synchronism.
- Release power-on pulse and start comparison test. When display has turned on and no error has occurred, flag computer for first key code sequence.
- After receiving first key code, enable key code matrix and continue to enable until display turns off.
- Wait until calculator display has turned back on again.
- Test if calculator is ready for its next test sequence, which can be through the key lines again or through the card reader lines.
- The test sequences are continued until the computer gives an end code. Then the "good" light turns on.
- If an error occurs the tester generates an error code corresponding to that fault. The computer is then flagged with the error line high.
- The computer then sends back a series of output select codes to specify which information is to be sent back to the computer. After the computer has received all information from the tester, an end code is sent to the tester. The tester turns on a "bad" light.
- The computer then prints out diagnostic messages.

Test Time

The time required for testing a logic board is determined by the number of test sequences and the length of time of each function. Typical test time is about one minute for the complete test sequence, that is, for a good board.

Acknowledgments

I would like to thank Larry Gravelle and Harry Griffin of Advanced Products Division's electrical tooling department for all the support they gave, Marlin Schell for writing the software package, and Rich Whicker for some suggestions and ideas on testing the card reader chip.

Special thanks to Ron Bernard, Sue Gross and Carl Forsyth for their help putting the prototype together.

Ken Peterson

Ken Peterson, a member of HP Laboratories since 1965, has worked on the 2116 Computer, the 9100 Calculator, and all of HP's pocket calculators. Ken was a radar technician in the U.S. Air Force and later studied electronics for two years at a community college. His home is in Fremont, California, just south of his native Oakland, and he frequently attends Oakland Raiders' professional football games, especially enjoying it when the Raiders beat their rivals, the Kansas City Chiefs. He's a softball player and a skier, too. He and his wife Carmen have five daughters ranging in age from four months to fourteen years.

High-Power Solid-State 5.9-12.4-GHz Sweepers

Two new RF plug-ins for the 8620C Sweep Oscillator produce more than 50 mW of output power, thanks to a new gallium arsenide field-effect transistor.

by Louis J. Kuhlman, Jr.

IN MICROWAVE MEASUREMENTS, high-power sweep oscillators are often needed to increase dynamic range, improve accuracy, saturate amplifiers, or overcome system losses.

- Measurement dynamic range is limited by either detector saturation or sweeper output power. In measurements involving high insertion loss, dynamic range can benefit directly from increased output power.
- Measurement accuracy can be improved by using attenuators to reduce mismatch errors and resistive power separation devices to reduce frequency response errors. No loss in dynamic range results if the sweeper can compensate for the resistive loss with additional power.
- When measuring the distortion and power handling capability of an amplifier, sufficient power is required to compress gain to the desired level. Also, mixer conversion loss measurements are a function of local oscillator drive, so extra sweeper output power allows a wider range of devices to be measured.

Fig. 1. *Compact Model 8620C Sweep Oscillator produces over 50 mW of output power with either the 86242C RF Plug-In for the 5.9-to-9.0-GHz range or the 86250C RF Plug-In for the 5.9-to-12.4-GHz range.*

High-power swept measurements have normally been made using bulky and expensive backward-wave-oscillator (BWO) sweepers or traveling-wave-tube (TWT) amplifiers. The BWO or TWT tubes in these units require periodic replacement. Also, the user has to buy another system in addition to his modern solid-state sweeper.

Two new high-power plug-ins for the 8620C sweep oscillator solve this problem by using a recently designed 100-mW GaAs FET amplifier covering the entire 5.9-12.4-GHz range. The 86242C RF Plug-In has a frequency range of 5.9-9.0 GHz and the 86250C has a frequency range of 8.0-12.4 GHz. The compact 8620C/86242C and 8620C/86250C (Fig. 1) are the highest-power microwave solid-state sweepers currently available.

Designed to meet the requirements of all major sweeper applications, the new sweepers provide more than 17 dBm of power, leveled within ±0.5 dB

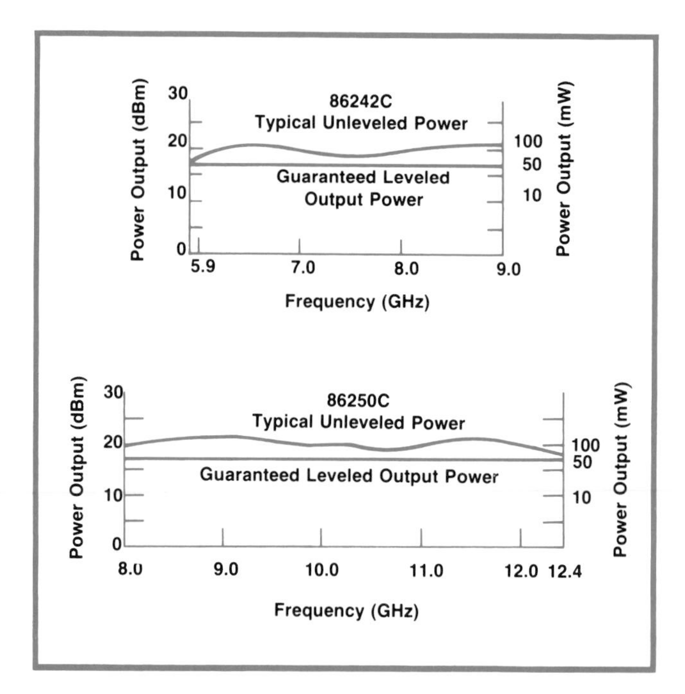

Fig. 2. *Typical output power of 86242C and 86250C RF Plug-Ins. Internal leveling option levels power within ±0.5 dB.*

with the internal leveling option (see Fig. 2). When used with the HP 8755 Frequency Response Test Set, no external modulator is required; therefore, full output power is available for measurements.

Plug-In Design

As Fig. 3 shows, the basic elements of the new plug-ins are the YIG-tuned oscillator (YTO) and the GaAs FET amplifier/modulator. The YTO is tuned by a current-induced magnetic field. The YTO driving circuit has been described in detail elsewhere.[1]

Power leveling in the plug-ins is achieved by sensing the output power with a broadband directional coupler and detector, comparing this signal to a level control voltage, and applying the error signal to a PIN-diode modulator. This modulator is built into the GaAs FET amplifier.

As Fig. 4 shows RF power leveling is accomplished by controlling the operation point of the PIN modulator as a function of V_{sense} and V_{set}. V_{sense} is a voltage derived by detecting and amplifying a small portion (−20 dB) of the output signal. V_{set} is a reference voltage; its magnitude is determined by the front-panel power level control.

In the loop, V_{sense} and V_{set} are summed, and their sum is compared to zero volts at the preamplifier. Any deviation from zero is amplified and used to shift the modulator operating point, changing V_{sense} so that its sum with V_{set} is returned to zero.

In externally leveled applications the ALC loop gain must be adjusted to compensate for variations in RF coupling ratio and detector speed of response.

A 5.9-12.4-GHz GaAs FET Power Amplifier

The GaAs FET power amplifier in the 86242C and 86250C RF Plug-Ins provides 10-dB gain at 100-mW output power and 0-60 dB of attenuation with an input PIN modulator. A single amplifier design covers the full 5.9-12.4 GHz range, using five 1-μm-gate GaAs FETs produced by HP. Up to now, most GaAs FET design efforts have concentrated on either low-noise broadband performance[1] or narrow-band high-power (>1 watt) output.[2] This amplifier appears to be the first application of GaAs FETs in broadband medium-power amplification. The high carrier mobility and power efficiency of the GaAs FET make reliable solid-state replacements for backward-wave-oscillator (BWO) sweepers possible.

Typical input-to-output isolation of a GaAs FET at 10 GHz is 25 dB or greater. This property allows very successful application of unilateral design approaches;[3] that is, input and output matching structures are designed with knowledge that they do not interact significantly.

A simplified input and output equivalent circuit for the GaAs FET is shown in Fig. 1. Matching circuits transform the input and output impedances to obtain broadband matching for gain and power. The input and first interstage matching circuit are designed for maximum small-signal gain. The second and third stages, which employ two GaAs FETs in parallel, are designed for maximum large-signal power gain.

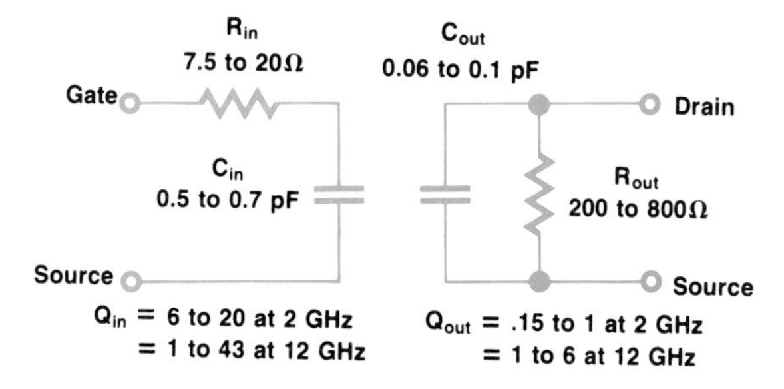

Fig. 1. *GaAs FET input and output simplified equivalent circuits.*

Impedances presented to the drain for maximum large-signal gain were measured with external tuners and large-signal RF conditions. Results of these large signal measurements and other design considerations have been described elsewhere.[4]

Assembly

Assembly procedures have been established to protect the GaAs FET from mechanical and electrical damage during handling. The GaAs chips are first pulse-soldered to a gold-plated copper pedestal for good heat sinking, input-to-output isolation, and ease of handling during later assembly steps. Sapphire matching circuits are soldered onto a single carrier, leaving space for the FET/pedestal assemblies. Epoxy attachment of the FET/pedestal assembly to the microcircuit makes it easier to replace faulty active devices. Temporary grounding bonds are attached to the FET contacts to eliminate the possibility of electrical damage by static discharge and transients during subsequent processes. An assembled 100-mW 5.9-12.4-GHz GaAs FET amplifier is shown in Fig. 2.

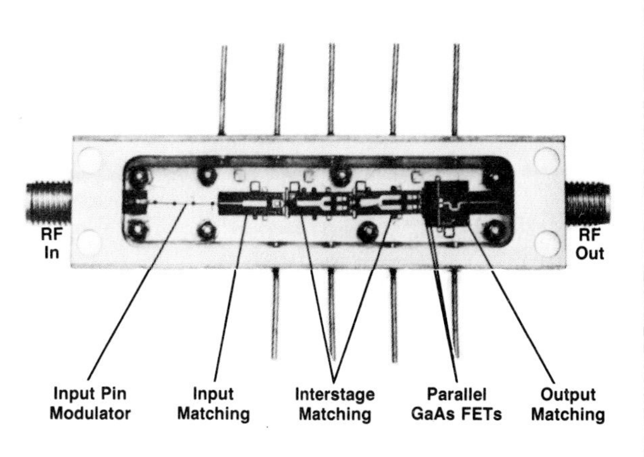

Fig. 2. *100-mW 5.9-to-12.4-GHz GaAs FET amplifier.*

References

1. C.A. Liechti and R.L. Tillman, "Design and Performance of Microwave Amplifiers with GaAs Schottky-Gate Field Effect Transistors," IEEE Transactions on Microwave Theory and Techniques, Vol. MTT-22, No. 5, pp. 510-517, May 1974.
2. R. Camisu, I. Drukier, H. Huang and S.Y. Narayan, "High-Efficiency GaAs MESFET Amplifiers," Electronics Letters, Vol. II, No. 21, pp. 508-509, October 16, 1975.
3. "S-Parameter Design," Hewlett-Packard Application Note 154.
4. D. Hornbuckle and L. Kuhlman, "Broadband Medium-Power Amplification in the 2-12.4 GHz Range with GaAs MESFET's," IEEE Transactions on Microwave Theory and Techniques, Vol. MTT-24, No. 6, June 1976.

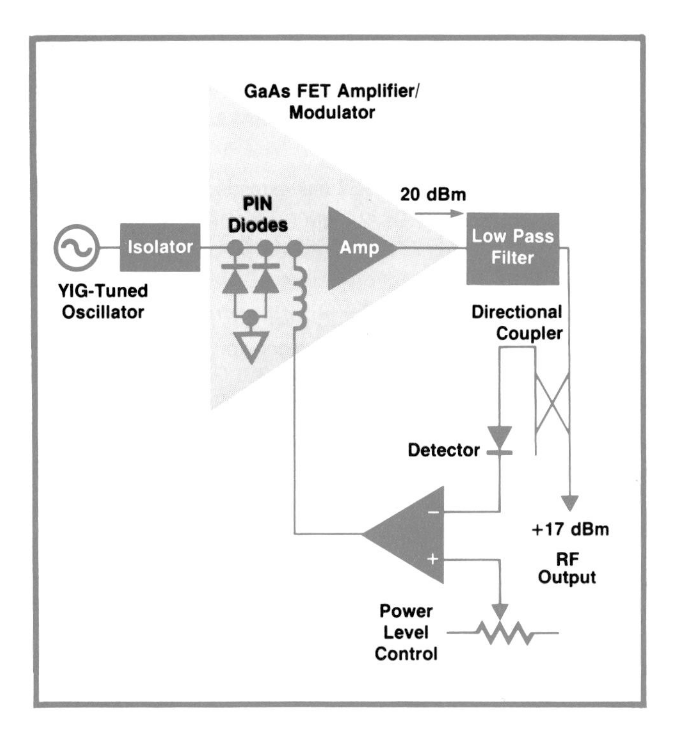

Fig. 3. *Major elements of the new RF plug-ins are the YIG-tuned oscillator and the GaAs-FET amplifier modulator.*

With many commercial sweepers the power level changes when the ALC gain is adjusted, making gain and power adjustment an iterative procedure. This problem has been solved in the 86242C and 86250C by summing V_{sense} and V_{set} before the gain control. Both voltages are attenuated equally, so they maintain a constant ratio and produce a constant output power.

Direct 8755 Modulation

A primary design goal was to allow direct modulation of the instrument when used with the 8755 Frequency Response Test Set. This eliminates the need for the external modulator normally used in this application, and as a consequence, makes the full +17 dBm available to the device under test.

Satisfying this requirement meant that the RF power would be modulated by the 27.8-kHz square wave modulation signal produced by the 8755A. The RF power would have to have a minimum on/off ratio of 40 dB and maintain 40/60 symmetry over the 20-dB range of the power level control.

The design solution makes the loop become a sample-and-hold circuit during the time the RF power is blanked. When the RF power is enabled, the output of the ALC main-amplifier/sample-and-hold will be the same as before the RF power was blanked. The PIN modulator current is the same, the RF power is the same, and the sum of V_{set} and V_{sense} is zero. Thus the loop can stabilize without having to slew fast enough to respond directly to the 27.8-kHz modulation.

FM Driver

The FM driver block diagram is shown in Fig. 5. The FM input signal is split into high and low-frequency paths. The low-frequency signals are sent to the YTO driver to allow frequency deviations of up to ±150 MHz.

The YTO FM coil is driven by a differential cascode amplifier. This achieves ±5-MHz frequency deviation at a 5-MHz modulation rate, as well as a ±1.5-dB small-signal frequency response from dc to 10 MHz.

Acknowledgments

The authors wish to thank the other contributors to this team effort. Our thanks go to Doug Fullmer for

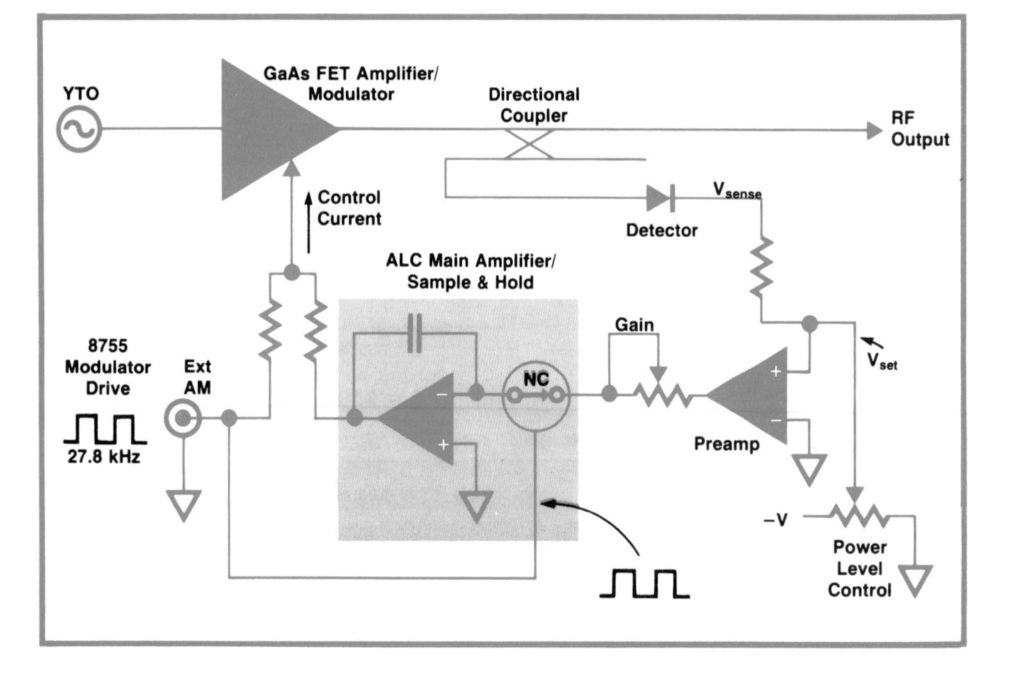

Fig. 4. *Automatic level control diagram for the 86242C and 86250C RF Plug-Ins. Design allows direct modulation by the 8755 Frequency Response Test Set, eliminating external modulators and making full output power available.*

The GaAs FET in Microwave Instrumentation

by Patrick H. Wang

Ever since the GaAs Schottky-barrier-gate FET (GaAs MESFET) was proposed by C.A. Mead[1] in 1966, it has been generally considered the device of the future for microwave applications. Initial development of the GaAs FET was slow, but significant breakthroughs in fabrication technology and material handling in 1970 and 1971 allowed realization of the device's promise. One-micrometre-gate-length MESFETs with an f_{max} of 50 GHz and useful gain up to 18 GHz became available from various research laboratories.

Today we are seeing what might be called a "GaAs FET revolution" in the microwave semiconductor industry. GaAs FETs with a 3.5-dB noise figure at 10 GHz for low-noise amplifier applications and power GaAs FETs with one-watt output power beyond 8 GHz are readily available commercially. Besides serving as microwave amplifiers, GaAs FETs are beginning to be used in many novel applications such as microwave oscillators, mixers, modulators, and digital switching.

As a three-terminal microwave device, the GaAs MESFET has two fundamental advantages over silicon bipolar transistors. First, the GaAs FET is a unipolar device; its operation depends upon majority carriers only. Electron mobility in GaAs is six times higher than in silicon and peak drift velocity is two times higher. This gives the GaAs FET a larger transconductance and a shorter electron transit time, and hence a high gain at higher frequencies. Second, GaAs FETs are fabricated on a semi-insulating substrate; this greatly reduces parasitic capacitances, again improving high-frequency performance. The silicon bipolar transistor is still very popular in applications up to 8 GHz, but the GaAs FET has extended the microwave solid-state frequency range to beyond 18 GHz, while providing an alternative to the bipolar transistor from 2 to 8 GHz.

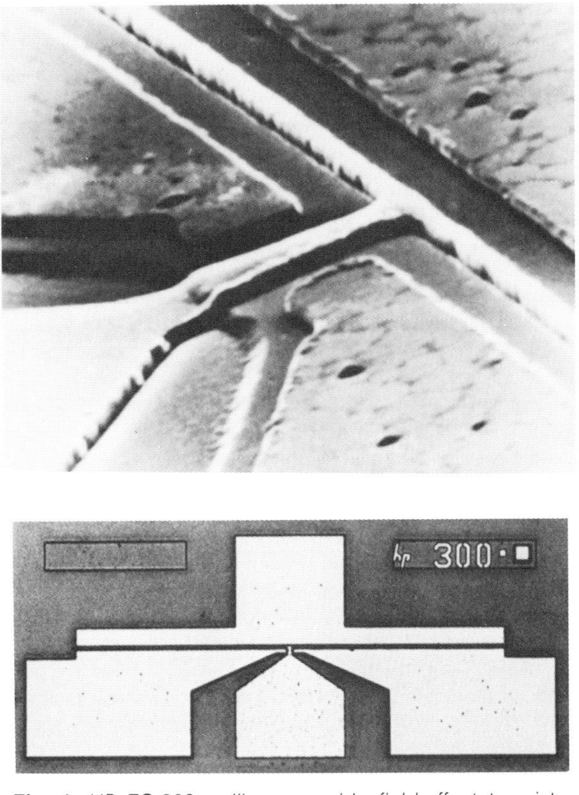

Fig. 1. *HP TC-300 gallium arsenide field-effect transistor chip. Scanning electron microscope photo shows 1-μm-wide gate stripe.*

Device Design

The HP 86242C and HP 86250C sweepers mark the first use of the GaAs FET in microwave instrumentation. Each of the new plug-ins uses five Hewlett-Packard TC-300 GaAs FETs.

Fig. 1 is a photograph of the HP TC-300 GaAs FET used in the 5.9-12.4-GHz amplifier. The gate is a 1×500-μm aluminum strip. The gate pad is located on the semi-insulating substrate with negligible capacitance. The layout is optimized for common-source operation, with both the drain and source pads maximized to allow convenient use of mesh and ribbon bonding.

Fig. 2 shows the vertical structure of the FET. A thin (0.3-μm), highly doped (10^{17} cm^{-3}) n-type epitaxial layer was grown, using liquid-phase epitaxial techniques, on a semi-insulating substrate. The drain and source ohmic contacts were formed by allowing Ge-Au into the heavily doped GaAs layer. The 1-μm Schottky-barrier gate was formed in deposited aluminum of 0.7-μm thickness. A dielectric passivation layer was applied over the active area for protection.

Fig. 3 is an equivalent circuit of the HP TC-300 FET. At 10 GHz, the TC-300 has typical maximum available gain of 10 dB and output power of 17 dBm at the 1-dB gain-compression point.

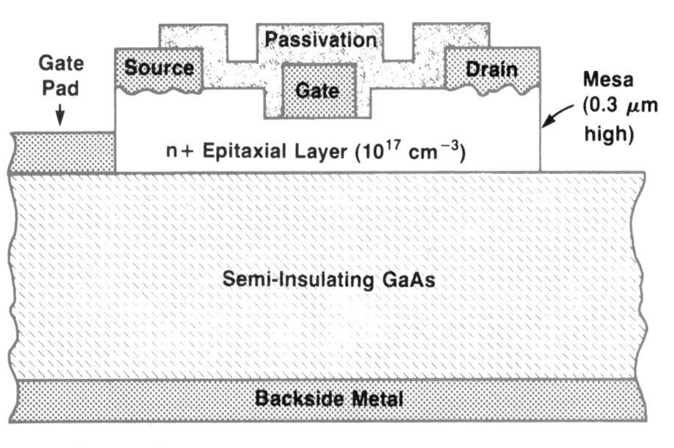

Fig. 2. *Cross section of FET geometry.*

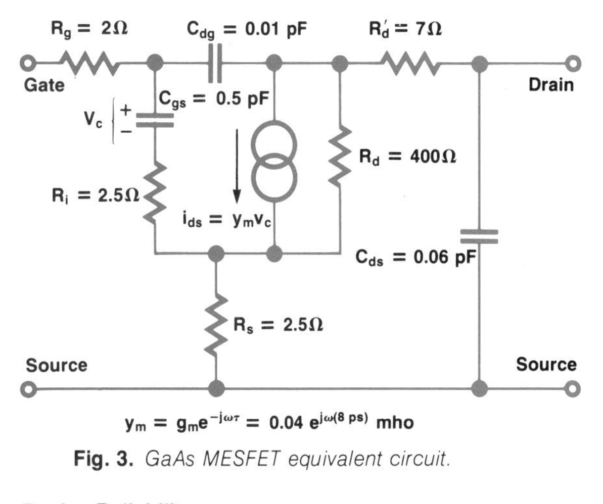

Fig. 3. *GaAs MESFET equivalent circuit.*

Device Reliability

The III-V compound materials are still relatively new in the semiconductor industry. Therefore, many people are suspicious of GaAs FET reliability. From our extensive reliability test results and from reports from other leaders in this field[2,3], the device itself is very reliable and an MTBF (mean time between failures) of 10^6 to 10^8 hours is expected. Even though the gate is made of aluminum with fine geometry (1 μm × 0.7 μm), metal migration should not be a problem because the gate is not supposed to carry any appreciable current. Even under heavy RF driving conditions, the gate current can be reduced by a proper negative bias on the gate. However, because the GaAs FET has a very high impedance at low frequencies, it is more vulnerable to static charge or transient spike damage. Extreme care must be exercised during device handling and testing, and spike protection circuits are recommended.

Acknowledgments

GaAs FET development has taken a long time and much team effort. Many people contributed to the project during the many different phases of the program. Their combined efforts made the end product possible. Special acknowledgment should go to C. Liechti and E. Gowen of HP Laboratories for their initial device and process development; to C. Li for his continuous effort in LPE work; to M. Marcelja, R. Tillman and C.C. Chang for their final state process development work; and to A. Chu, P. Froess, P. Chen and D. Lynch for their contributions in device characterization, testing, and reliability evaluations.

References

1. C.A. Mead, "Schottky-Barrier Gate Field Effect Transistor," Proceedings of the IEEE, Vol. 54, p. 307, 1966.
2. D.A. Abbott and J.A. Turner, "Some Aspects of GaAs MESFET Reliability," IEEE Transactions on Microwave Theory and Techniques, Vol. 24, p. 317, 1976.
3. T. Irie et al., "Reliability Study of GaAs MESFETs," IEEE Transactions on Microwave Theory and Techniques, Vol. 24, p. 321, 1976.

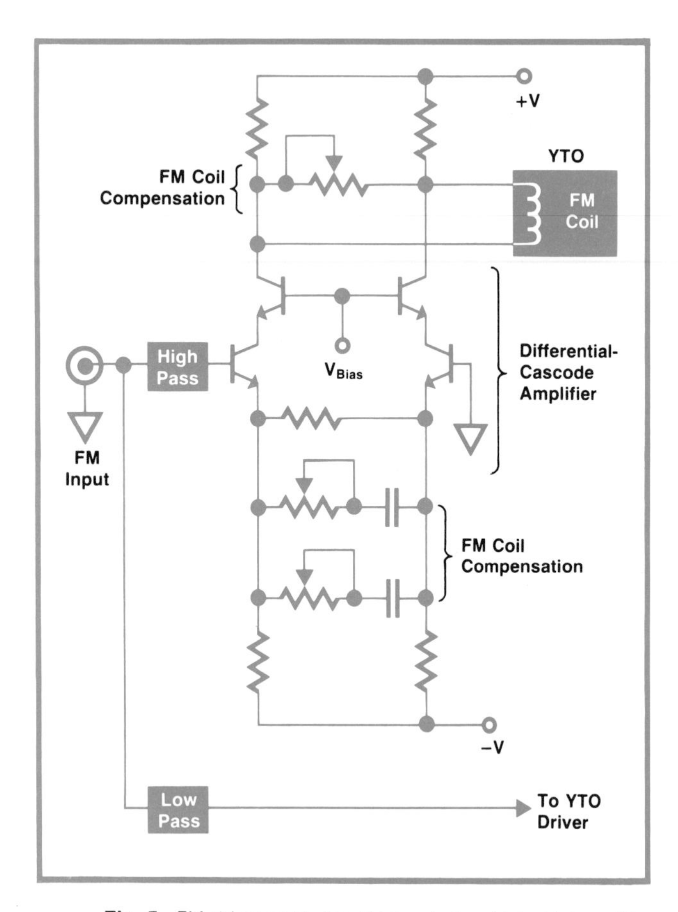

Fig. 5. *FM driver splits the FM input signal into high and low frequencies. Low frequencies go to the YTO driver, high frequencies to the YTO FM coil.*

his ALC circuit design work and coordination of the 86242C/250C production introduction. The original idea of a differential cascode FM driver was provided by Steve Sparks. Roger Stancliff provided useful advice on ALC circuitry. Mechanical design by Bill Misson proved cost-effective and easy to assemble. Dave Eng provided the industrial design. Microcircuit package design and fabricated part processes were provided by Rich Swain and Al Willits. GaAs FET microcircuit assembly process development was provided by Floyd Bishop, Tom Jerse, and Pete Planting. Production engineering support came from Ken Wong at the instrument level and Val Peterson on the microcircuits. The operation and service manual was authored by Steve Williams. Tom Babb and Jim Arnold developed the instrument field support plan. We are also grateful for the leadership of Jack Dupre during this development project.

Reference

1. P.R. Hernday and C.J. Enlow, "A High-Performance 2-to-18-GHz Sweeper," Hewlett-Packard Journal, March 1975.

New Low-Noise GaAs FET Chip has 25-Milliwatt Output at 10 GHz

Although the GaAs FET transistor used in the 86242C and 86250C RF Plug-Ins is not available commercially, a similar device is available from HP. The first of a new family of gallium arsenide Schottky-barrier field-effect transistors from Hewlett-Packard, the HFET-1000 chip has a 14.5-dBm linear power output at 10 GHz. Noise figure at 10 GHz is 3.6 dB typical with 6.9 dB associated gain; maximum available gain at that frequency is 11.0 dB.

Designed for general use in telecommunications, radar, and other low-noise amplifiers in the 2-to-12-GHz range, this new GaAs FET has a 1-by-500-micrometre gate. The chip is rugged both mechanically and electrically for easy wire bonding and die attaching. A scratch and dust resistant layer covers the active device area.

The HFET-1000 chip shows very good consistency from one production run to the next. The devices undergo 100% visual inspection and dc testing.

SPECIFICATIONS

HP Model 86242C or 86250C RF Plug-Ins in 8620C Sweep Oscillator

Frequency

	86242C	86250C
FREQUENCY RANGE:	5.9 to 9.0 GHz	8.0 to 12.4 GHz
FREQUENCY ACCURACY: (25°C)		
CW MODE:	±35 MHz	±40 MHz
ALL SWEEP MODES (Sweep Time >0.1 sec):	±40 MHz	±50 MHz
FREQUENCY STABILITY:		
TEMPERATURE:	±750 kHz/°C	±1.2 MHz/°C
10% LINE VOLTAGE CHANGE:	±40 kHz	±40 kHz
10 dB POWER LEVEL CHANGE:	±1.5 MHz	±1.5 MHz
RESIDUAL FM IN 10 kHz BANDWIDTH (CW MODE)	<15 kHz peak	<15 kHz peak

Power Output

(Applies to both models)

MAXIMUM LEVELED POWER: (25°C)	>17.0 dBm
INTERNALLY LEVELED (Opt. 001):	>17.0 dBm
POWER VARIATION:	
UNLEVELED:	<±3.0 dB
INTERNALLY LEVELED (OPT. 001):	<±0.5 dB
EXTERNALLY LEVELED (excluding Coupler and Detecor variation):	<±0.1 dB
SPURIOUS SIGNALS (below fundamental at 17 dBm):	
HARMONICS:	>30 dB
NONHARMONICS:	>60 dB
RESIDUAL AM: (AM noise in 100 kHz bandwidth below maximum power output):	>50 dB
SOURCE VSWR: 50Ω nominal impedance	
LEVELED (internal leveling only):	>1.6
UNLEVELED TYPICALLY:	<2.5

Modulation

(Applies to both models)

EXTERNAL FM:

MAXIMUM DEVIATION FOR MODULATION FREQUENCIES: dc - 100 Hz, ±150 MHz; dc - 1 MHz, ±7 MHz; dc - 10 MHz, ±1.5 MHz

FM FREQUENCY RESPONSE: ±1.5 dB (dc - 10 MHz)

SENSITIVITY: (nominal) FM mode −20 MHz/V; Phase-lock mode −6 MHz/V

INTERNAL AM:

1 kHz Squarewave, RF Blanking, and Marker ON/OFF Ratio: >40 dB

EXTERNAL AM: Specific requirements* guaranteeing HP 8755 operation with ±6V, 27.8 kHz square wave mod drive connected to external AM input. *(Symmetry: 40/60; ON/OFF Ratio: >40 dB)

PRICES IN U.S.A.:

86242C RF PLUG-IN: $3850.

86250C RF PLUG-IN: $3850.

Opt. 001 (Internal leveling): $450.

Opt. 004 (Rear panel RF output): $80.

MANUFACTURING DIVISION: SANTA ROSA DIVISION
1400 Fountain Grove Parkway
Santa Rosa, California 95404 U.S.A.

SPECIFICATIONS

HP Model HFET-1000 Microwave GaAs FET Chip

DESCRIPTION: The HFET-1000 is a Gallium Arsenide Schottky Barrier Field Effect Transistor Chip designed for low noise figure, high gain and substantial power at 10 GHz. The chip is provided with a dielectric scratch protection layer over the active area. The gate width is 500 micrometers resulting in a typical linear output power greater than 25 mW.

ELECTRICAL SPECIFICATIONS AT T_A = 25°C

Symbol	Parameters and Test Conditions	Min.	Typ.	Max.
I_{DSS}	Saturated Drain Current (mA) V_{DS} = 4.0 V, V_{GS} = 0 V	40		120
V_{GSP}	Pinch Off Voltage (V) V_{DS} = 4.0 V, I_{DS} = 100 μA	−1.5		−5.0
g_m	Transconductance (mmho) V_{DS} = 4.0 V, ΔV_{GS} = 0 V to −0.5 V	30	45	
$G_{a(max)}$	Maximum Available Gain (dB) V_{DS} = 4.0 V, V_{GS} = 0 V			
	f = 8 GHz		13.0	
	10 GHz		11.0	
	12 GHz		9.5	
F_{min}	Noise Figure (dB) V_{DS} = 3.5 V I_{DS} = 15% I_{DSS} (Typ. 12 mA)			
	f = 8 GHz		2.9	
	10 GHz		3.6	
	12 GHz		4.1	
G_a	Associated Gain At N.F. Bias (dB)			
	f = 8 GHz		8.9	
	10 GHz		6.9	
	12 GHz		4.3	
P_{1dB}	Power at 1 dB Compression (dBm) V_{DS} = 4.0 V, I_{DS} = 50% I_{DSS} f = 10 GHz		14.5	

MAXIMUM RATINGS AT T_A = 25°C

Symbol	Parameter	Limits
V_{DS}	Drain to Source Voltage −5V≤V_{GS}≤0 V	5V
V_{GS}*	Gate to Source Voltage 5.0 V≥V_{DS}≥0.0 V	−5V
T_{CH}**	Max. Channel Temperature	125°C
	Storage Temperature	−65°C to +125°C

*Max. Forward Gate Current should not exceed 1 mA.

**Θ_{CB}-Thermal resistance, channel to back of chip = 100°C/W.

PRICE IN U.S.A.: 1-9, $135 ea. 10-24, $105 ea.

MANUFACTURING DIVISION: MICROWAVE SEMICONDUCTOR DIVISION
3172 Porter Drive
Palo Alto, California 94304 U.S.A.

Patrick H. Wang

Pat Wang is section manager for GaAs device technology in HP's Microwave Technology Center. With HP since 1966, he's been concerned with the development of microwave integrated circuits, microwave IC amplifiers, and microwave transistors. An IEEE member, he's authored six papers on microwave transistors for IEEE publication and conferences. Pat received his BA degree in literature from National Taiwan University in 1960 and his MSEE degree from Stanford University in 1966. He's married, has three daughters, lives in Santa Rosa, California, and is interested in photography.

Louis J. Kuhlman, Jr.

Jack Kuhlman received his BSEE and MSEE degrees from the University of Washington in 1969 and 1971. With HP since 1971, he's done production and development engineering on YIG-tuned oscillators, served as project leader for the 86242C and 86250C RF Plug-Ins, and designed the GaAs FET amplifier used in the plug-ins. Jack was born in Snohomah, Washington, and now lives in Santa Rosa, California. He's married and has three children. He enjoys carpentry and woodworking—he and his wife have designed and built two houses—and outdoor activities like skiing, camping, hunting, and gardening.

Hewlett-Packard Company, 1501 Page Mill Road, Palo Alto, California 94304

HEWLETT-PACKARD JOURNAL

NOVEMBER 1976 Volume 28 • Number 3

Technical Information from the Laboratories of Hewlett-Packard Company

Hewlett-Packard Central Mailing Department
Van Heuven Goedhartlaan 121
Amstelveen-1134 The Netherlands
Yokogawa-Hewlett-Packard Ltd., Shibuya-Ku
Tokyo 151 Japan

Signature Analysis: A New Digital Field Service Method

In a digital instrument designed for troubleshooting by signature analysis, this method can find the components responsible for well over 99% of all failures, even intermittent ones, without removing circuit boards from the instrument.

by Robert A. Frohwerk

WITH THE ADVENT OF MICROPROCESSORS and highly complex LSI (large-scale integrated) circuits, the engineer troubleshooting digital systems finds himself dealing more with long digital data patterns than with waveforms. As packaging density increases and the use of more LSI circuits leaves fewer test points available, the data streams at the available test points can become very complex. The problem is how to apply some suitable stimulus to the circuit and analyze the resulting data patterns to locate the faulty component so that it can be replaced and the circuit board returned to service.

The search for an optimal troubleshooting algorithm to find failing components on digital circuit boards has taken many directions, but all of the approaches tried have had at least one shortcoming. Some simply do not test a realistic set of input conditions, while others perform well at detecting logical errors and stuck nodes but fail to detect timing-related problems. Test systems capable of detecting one-half to two-thirds of all possible errors occurring in a circuit have been considered quite good. These systems tend to be large, for factory-based use only, and computer-driven, requiring program support and software packets and hardware interfaces for each type of board to be tested. Field troubleshooting, beyond the logic-probe capability to detect stuck nodes, has been virtually neglected in favor of board exchange programs.

The problem seems to be that test systems have too often been an afterthought. The instrument designer leaves the test procedure to a production test engineer, who seeks a general-purpose solution because he lacks the time to handle each case individually.

Obviously it would be better if the instrument designer provided for field troubleshooting in his original design. Who knows a circuit better than its original designer? Who has the greatest insight as to how to test it? And what better time to modify a circuit to accommodate easy testing than before the circuit is in production?

New Tools Needed

But here another problem arises: what do we offer the circuit designer for tools? A truly portable test instrument, since field troubleshooting is our goal, would be a passive device that merely looked at a circuit and told us why it was failing. The tool would provide no stimulus, require little software support, and have accuracy at least as great as that of computer-driven factory-based test systems.

If a tester provides no stimulus, then the circuit under test must be self-stimulating. Whereas this seemed either impossible or at best very expensive in the past, a self-stimulating circuit is not out of the question now. More and more designs are microprocessor-oriented or ROM-driven, so self-stimulus, in the form of read-only memory, is readily available and relatively inexpensive.

By forcing a limitation on software, we have eliminated the capability to stop on the first failure and must use a burst-mode test. Another restriction we will impose is that the device under test must be synchronous, in the sense that at the time the selected clock signal occurs the data is valid; not an unfair condition by any means, and it will be justified in the article beginning on page 15.

There are only a few known methods for compressing the data for a multiple-bit burst into a form that can be handled easily by a portable tester without an undue amount of software. One method used in large systems is transition counting. Another method, a much more efficient data compression technique borrowed from the telecommunications field, is the cyclic redundancy check (CRC) code, a sort of checksum, produced by a pseudorandom binary sequence (PRBS) generator.

A troubleshooting method and a portable instrument based on this concept turns out to be the answer we are seeking. We call the method signature analysis and the instrument the 5004A Signature Analyzer. The instrument is described in the article on page 9. Here we will present the theory of the method and show that it works, and works very well.

Pseudorandom Binary Sequences

A pseudorandom binary sequence is, as implied, a pattern of binary ones and zeros that appears to be random. However, after some sequence length the pattern repeats. The random-like selection of bits provides nearly ideal statistical characteristics, yet the sequences are usable because of their predictability. A PRBS based upon an n-bit generator may have any length up to 2^n-1 bits before repeating. A generator that repeats after exactly 2^n-1 bits is termed maximal length. Such a generator will produce all possible n-bit sequences, excluding a string of n zeros. As an example, let us take the sequence: 000111101011001. This is a fifteen-bit pattern produced by a four-bit maximal-length generator ($15=2^4-1$). If we were to wrap this sequence around on itself, we would notice that all possible non-zero four-bit patterns occur once and only once, and then the sequence repeats.

To construct a PRBS generator we look to the realm of linear sequential circuits, which is where the simplest generators reside mathematically. Here there exist only two types of operating elements. The first is a modulo-2 adder, also known as an exclusive-OR gate. The other element is a simple D-type flip-flop, which being a memory element behaves merely as a time delay of one clock period. By connecting flip-flops in series we construct a shift-register as in Fig. 1, and by taking the outputs of various flip-flops, exclusive-ORing them, and feeding the result back to the register input, we make it a feedback shift register that will produce a pseudorandom sequence. With properly chosen feedback taps, the sequence will be maximal length. The fifteen-bit sequence above was produced by the generator in Fig. 1, with the flip-flops initially in the 0001 state since the all-zero state is disallowed. The table in Fig. 1 shows the sequence in detail. The list contains each of the sixteen ways of arranging four bits, except four zeros.

If we take the same feedback shift register and provide it with an external input, as in Fig. 2, we can overlay data onto the pseudorandom sequence. The overlaid data disturbs the internal sequence of the generator. If we begin with an initial state of all zeros and supply a data impulse of 1000..., the result is the same sequence as in Fig. 1 delayed by one clock period.

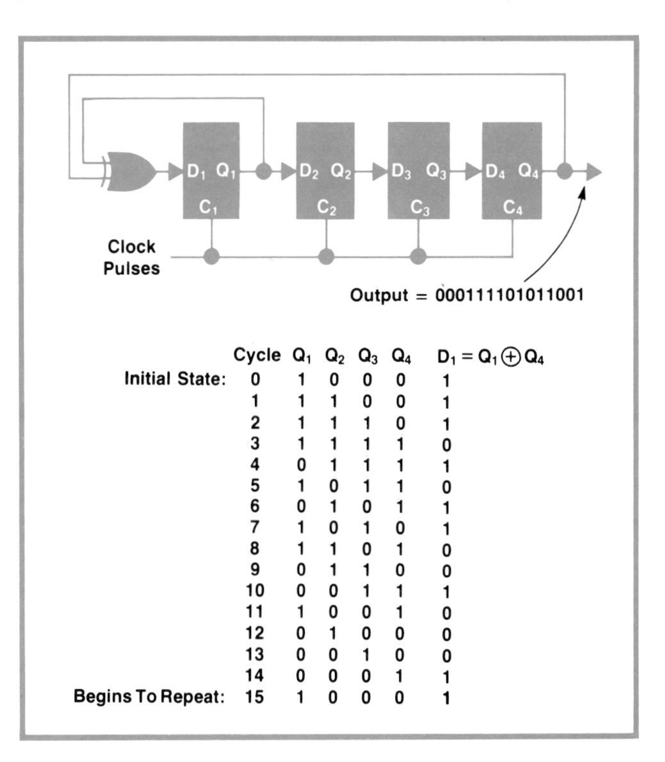

	Cycle	Q_1	Q_2	Q_3	Q_4	$D_1 = Q_1 \oplus Q_4$
Initial State:	0	1	0	0	0	1
	1	1	1	0	0	1
	2	1	1	1	0	1
	3	1	1	1	1	0
	4	0	1	1	1	1
	5	1	0	1	1	0
	6	0	1	0	1	1
	7	1	0	1	0	1
	8	1	1	0	1	0
	9	0	1	1	0	0
	10	0	0	1	1	1
	11	1	0	0	1	0
	12	0	1	0	0	0
	13	0	0	1	0	0
	14	0	0	0	1	1
Begins To Repeat:	15	1	0	0	0	1

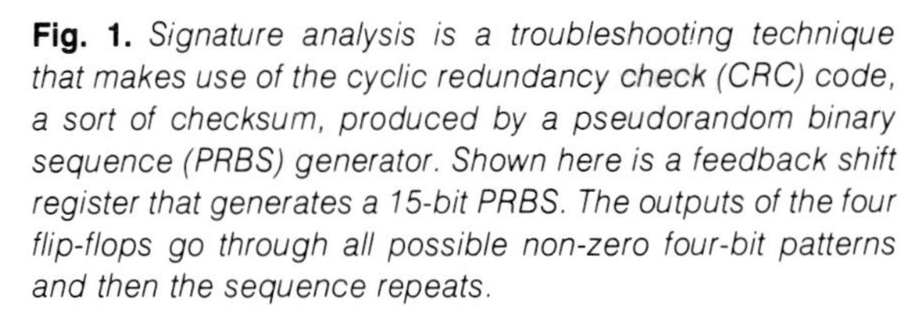

Fig. 1. *Signature analysis is a troubleshooting technique that makes use of the cyclic redundancy check (CRC) code, a sort of checksum, produced by a pseudorandom binary sequence (PRBS) generator. Shown here is a feedback shift register that generates a 15-bit PRBS. The outputs of the four flip-flops go through all possible non-zero four-bit patterns and then the sequence repeats.*

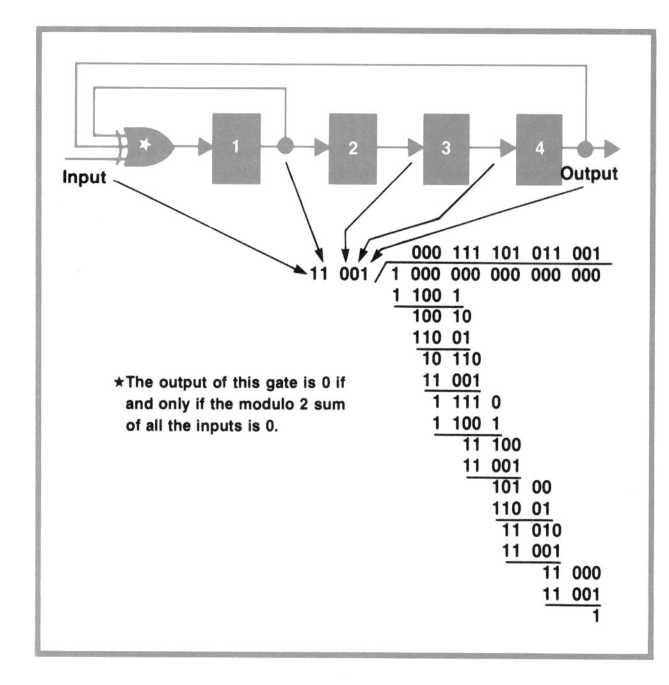

Fig. 2. *When the feedback shift register of Fig. 1 is provided with an external input, data can be overlaid on the PRBS generated by the circuit. Feeding data into a PRBS generator is the same as dividing the data by the characteristic polynomial of the generator.*

Shift Register Mathematics

A shift register may be described using a transform operator, D, defined such that $X(t) = DX(t-1)$. Multiplying by D is equivalent to delaying data by one unit of time. (Recall that we are concerned only about synchronous logic circuits.) In Fig. 2 the data entering the register is the sum of samples taken after one clock period and four clock periods along with the input data itself. Thus, the feedback equation may be written as $D^4X(t) + DX(t) + X(t)$ or simply X^4+X+1.

It happens that feeding a data stream into a PRBS generator is equivalent to dividing the data stream by the characteristic polynomial of the generator. For the particular implementation of the feedback shift register considered here the characteristic polynomial is X^4+X^3+1, which is the reverse of the feedback equation. Fig. 2 shows the register along with longhand division of the impulse data stream (100...). Keep in mind that in modulo-2 arithmetic, addition and subtraction are the same and there is no carry. It can be seen that the quotient is identical to the pattern in Fig. 1 and repeats after fifteen bits (the "1" in the remainder starts the sequence again).

Because the shift register with exclusive-OR feedback is a linear sequential circuit it gives the same weight to each input bit. A nonlinear circuit, on the other hand, would contain such combinatorial devices as AND gates, which are not modulo-2 operators and which would cancel some inputs based upon prior bits. In other words a linear polynomial is one for which $P(X+Y) = P(X) + P(Y)$. Take the example of Fig. 3, where the three different bit streams X, Y, and X+Y are fed to the same PRBS generator. Notice that the output sequences follow the above relationship, that is, $Q(X+Y) = Q(X) + Q(Y)$. Also, notice that Y is a single impulse bit delayed in time with respect to the other sequences and the only difference between X and X+Y is that single bit. Yet, Q(X+Y) looks nothing like Q(X). Indeed, if we stop after entering only twenty bits of the sequences and compare the remainders, or the residues in the shift register, they would be: $R(X+Y) = 0100$, $R(X) = 0111$.

Error Detection by PRBS Generator

Looking at this example in another manner, we can think of X as a valid input data stream and X + Y as an erroneous input with Y being the error sequence. We will prove later that any single-bit error, regardless of when it occurs, will always be detected by stopping the register at any time and comparing the remainder bits (four in this case) with what they should be. This error detection capability is independent of the length of the input sequence. In the example of Fig. 3, R(X+Y) differs from the correct R(X), and the effect of the error remains even though the error has disappeared many clock periods ago.

Let us stop for a moment to recall our original goal. We are searching for a simple data compression algorithm that would be efficient enough to be usable in a field service instrument tester. As such it was to require only minimal hardware and software support.

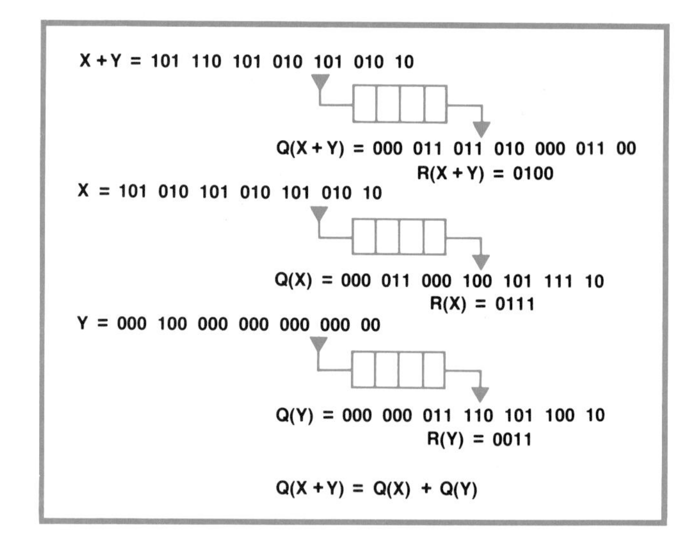

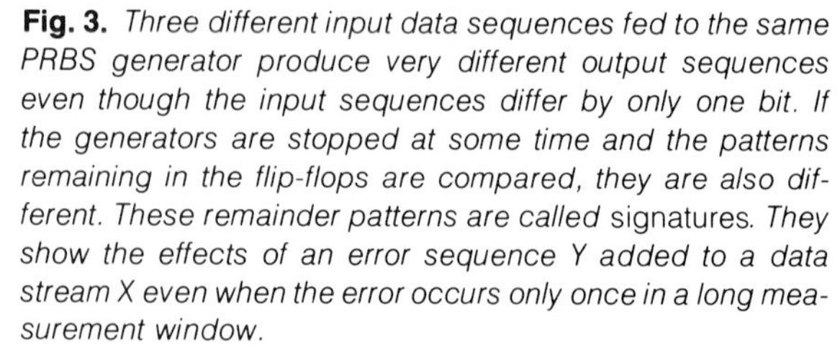
Fig. 3. *Three different input data sequences fed to the same PRBS generator produce very different output sequences even though the input sequences differ by only one bit. If the generators are stopped at some time and the patterns remaining in the flip-flops are compared, they are also different. These remainder patterns are called* signatures. *They show the effects of an error sequence Y added to a data stream X even when the error occurs only once in a long measurement window.*

We have now found such an algorithm. If the circuit designer arranges his synchronous circuit so as to provide clock and gate signals that produce a repeatable cycle for testing, then the feedback shift register is the passive device that we need to accumulate the data from a node in the instrument under test. By tracing through an instrument known to be good, the designer merely annotates his schematic, labeling each test point with the contents of the shift register at the end of the measurement cycle, and uses this information later to analyze a failing circuit. Because this PRBS residue depends on every bit that has entered the generator, it is an identifying characteristic of the data stream. We have chosen to call it a *signature*. The process of annotating schematics with good signatures as an aid in troubleshooting circuits that produce bad signatures has been termed *signature analysis*.

Errors Detected by Signature Analysis

We have claimed that any single-bit error will always be detected by a PRBS generator. But how about multiple errors? Also, our goal was to maintain error detection capability at least as good as existing methods. Earlier mention was made of transition counting, which appears to be the only other method that could easily be made portable. To show how signature analyis stands up against transition counting requires a mathematical discussion of the error detection capabilities of these methods. Take first the PRBS.

Assume X is a data stream of m bits, P is an n-bit PRBS generator, P^{-1} its inverse ($P^{-1}P = 1$), Q is a quotient and R the remainder.

$$P(X) = Q(X)\cdot 2^n + R(X). \qquad (1)$$

Take another m-bit sequence Y that is not the same as X and must therefore differ by another m-bit error sequence E such that

$$Y = X + E.$$

Now,

$$P(Y) = Q(Y)\cdot 2^n + R(Y)$$

so,

$$P(X+E) = Q(X+E)\cdot 2^n + R(X+E).$$

But all operators here are linear, so

$$P(X) + P(E) = Q(X)\cdot 2^n + Q(E)\cdot 2^n + R(X) + R(E).$$

Subtracting (or adding, modulo 2) with equation 1 above,

$$P(E) = Q(E)\cdot 2^n + R(E). \qquad (2)$$

However, if Y is to contain undetectable errors,

$$R(Y) = R(X).$$

It follows that

$$R(Y) = R(X+E) = R(X) + R(E) = R(X),$$

$$R(E) = 0.$$

Substituting into equation 2,

$$P(E) = Q(E)\cdot 2^n,$$

and all undetectable errors are found by

$$E = P^{-1}Q(E)\cdot 2^n. \qquad (3)$$

For a single-bit error

$$E = D^a(1)$$

where D is the delay operator, a is the period of the delay, and "1" is the impulse sequence 1000... Substituting into (3),

$$D^a(1) = P^{-1}Q(D^a(1))\cdot 2^n.$$

D commutes with other linear operators, so

$$D^a(1) = D^aP^{-1}Q(1)\cdot 2^n$$

$$1 = P^{-1}Q(1)\cdot 2^n$$

$$P(1) = Q(1)\cdot 2^n.$$

But by the original assumptions,

$$P(1) = Q(1)\cdot 2^n + R(1)$$

and by addition

$$R(1) = 0.$$

However, it has been shown by example that $R(1) \neq 0$. Therefore, $E \neq D^a(1)$ and the set of undetectable errors E does not include single-bit errors; in other words, a single-bit error is always detectable. (An intuitive argument might conclude that a single-bit error would always be detected because there would never be another error bit to cancel the feedback.)

To examine all undetectable errors as defined by equation 3, it helps to consider a diagrammatical representation, Fig. 4, of:

$$E = P^{-1}Q(E)\cdot 2^n.$$

Since X, Y, and E are all m-bit sequences, it follows that $Q\cdot 2^n$ must be an m-bit sequence containing n final zeros. Q therefore contains (m−n) bits. Hence, there are 2^{m-n} sequences that map into the same residue as the correct sequence, and there are $2^{m-n}-1$ error sequences that are undetectable because they leave the same residue as the correct sequence. 2^m sequences can be generated using m bits and only one of these is correct, so the probability of failing to detect an error by a PRBS is

$$\text{Prob (PRBS, fail)} = \frac{\text{Undetectable Errors}}{\text{Total Errors}} = \frac{2^{m-n}-1}{2^m-1}.$$

For long sequences, large m,

$$\text{Prob (PRBS, fail)} \approx 1/2^n.$$

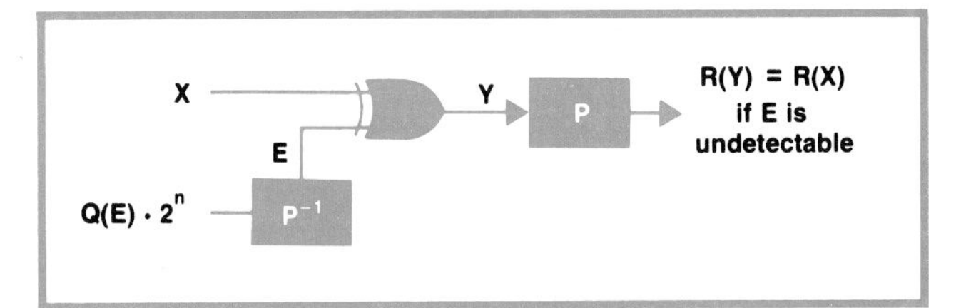

Fig. 4. *A diagrammatical representation of errors undetectable by signature analysis. For long data sequences the probability of not detecting an error approaches $1/2^n$, where n is the number of flip-flops in the feedback shift register.*

In summary, a feedback shift register of length n will detect all errors in data streams of n or fewer bits, because the entire sequence will remain in the register, R(X) = P(X). For data streams of greater than n bits in length, the probability of detecting an error using a PRBS is very near certainty even for generators of modest length. The errors not detected are predictable and can be generated by taking all m-bit sequences with n trailing zeros and acting upon such sequences by the inverse of the n-bit PRBS generator polynomial P, that is

$$E = P^{-1}(Q \cdot 2^n).$$

Furthermore, such error detection methods will always detect a single-bit error regardless of the length of the data stream. It can also be proved that the only undetectable error sequence containing two errors such that the second cancels the effect of the first is produced by separating the two errors by exactly 2^n-1 zeros.[1] The one sequence of length n+1 that contains undetectable errors begins with an error and then contains other errors that cancel each time the original error is fed back.

Errors Detected by Transition Counting

It appears that signature analysis using a PRBS generator is a difficult act to follow, but let us give transition counting a chance. A transition counter assumes an initial state of zero and increments at each clock time for which the present data bit differs from the previous bit. With a transition counter the probability of an undetected error, given that there is some error, is:

$$\text{Prob (Trnsn, Fail)} = N_u/N_t,$$

where N_u = number of undetected errors and N_t = total number of errors. But

$$N_u = \sum_{r=0}^{m} p_{ur}$$

where p_{ur} = Prob (undetected errors given r transitions). However,

$$p_{ur} = N_{ur} \cdot p_r$$

where N_{ur} = number of undetected errors given r transitions, and p_r = Prob (counting r transitions). Reducing further,

$$N_{ur} = N_r - N_c,$$
$$p_r = N_r/N_s,$$

where N_c = number of ways of counting correctly (=1), N_s = total number of m-bit sequences, and N_r = number of ways of counting r transitions:

$$N_r = \binom{m}{r} = \frac{m!}{r!(m-r)!}$$

The binomial coefficient $\binom{m}{r}$ expresses the number of ways of selecting from m things r at a time. Looking back to the original denominator,

$$N_t = N_s - N_c.$$

Putting all of this together,

$$\text{Prob (Trnsn, Fail)} = \frac{\sum_{r=0}^{m} (N_r - N_c)(N_r/N_s)}{N_s - N_c}.$$

Or,

$$\text{Prob (Trnsn, Fail)} = \frac{\sum_{r=0}^{m} \left[\binom{m}{r} - 1\right] \binom{m}{r}/2^m}{2^m - 1}$$
$$\approx 1/\sqrt{m\pi}.$$

This is the probability of a transition counter's failing to detect an error in an m-bit sequence.

A similar argument finds the probability of the specific case where a single-bit error is not detected by a transition counter. There are 2^m sequences of m bits and any one of the m bits can be altered to produce a single-bit error, so that there are $m \cdot 2^m$ possible single-bit errors. To determine how many undetected single-bit errors exist, we must look at how to generate them.

Upon considering the various ways of generating single-bit errors that are undetectable, a few observations become obvious. We can never alter the final bit of a sequence, because that would change the transition count by plus or minus one, which would be detected. The only time we can alter a bit without getting caught is when a transition is adjacent to a double bit; that is, flipping the center bit in the patterns 001, 011, 100, or 110 will not affect the transition count. In other words, the transition count for ...0X1... and ...1X0... does not depend on the value of X.

Since our transition counter assumes an initial 0 state, the first bit of the sequence, regardless of its state, can be flipped without affecting the transition count, provided that the second bit is a one. In this case only the second of m bits is predetermined, i.e., $b_2 = 1$, and there are 2^{m-1} ways of completing the sequence. Any bit other than the first or last, that is, the $m-2$ bits from b_2 through b_{m-1}, can be altered without affecting the transition count if the bit in question is flanked by a zero on one side and a one on the other. For a given bit b_i we have free choice of $m-1$ bits, since as soon as we select b_{i-1} then b_{i+1} is forced to the opposite state. There are $(m-2) \cdot 2^{m-1}$ of these midstream errors. Adding the 2^{m-1} sequences where b_1 can be changed we have a total of $(m-1) \cdot 2^{m-1}$ sequences containing single-bit errors that cannot be detected by a transition counter. But earlier we showed that the total number of single-bit errors was $m \cdot 2^m$, hence the probability of failing to detect a single-bit error is

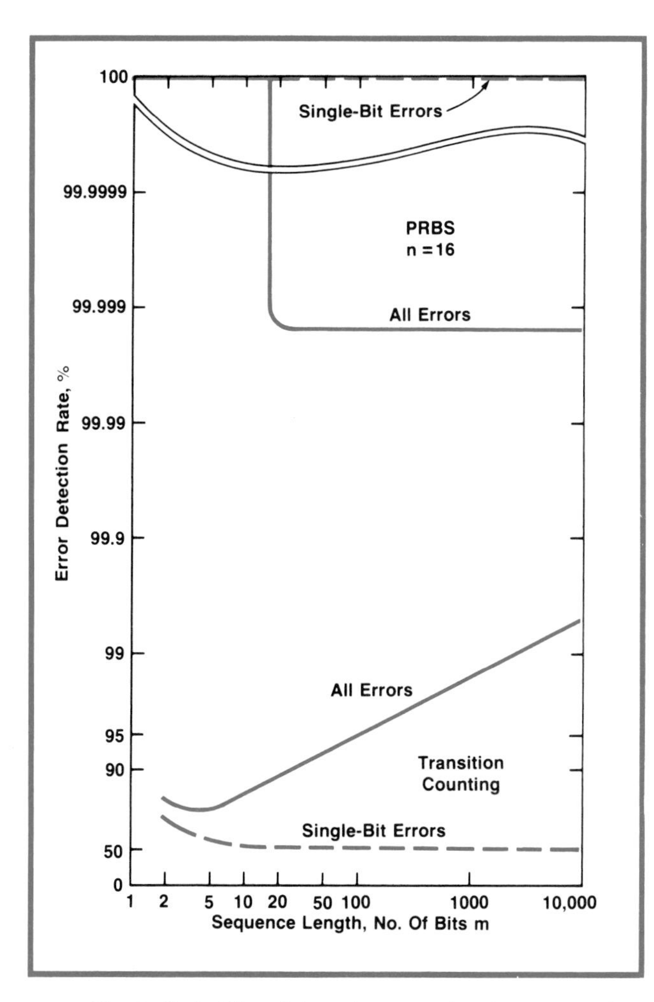

Fig. 5. *Probability of detecting errors for signature analysis and transition counting as a function of the length of the data sequence. n=16 for the PRBS generator.*

$$\text{Prob (Trnsn, Fail, single-bit)} = \frac{(m-1)\cdot 2^{m-1}}{m\cdot 2^{m}} = \frac{m-1}{2m} \approx 1/2.$$

It may be noted that the failure rate is actually somewhat higher, because a counter of limited length will overflow for long sequences, leaving some ambiguity. It can be shown that because of this overflow an n-bit transition counter will never detect more than $1/2^n$ of all errors.

Signature Analysis versus Transition Counting

We can now plot the probabilities of detecting any error using a transition counter versus a PRBS generator (see Fig. 5). It is interesting to note that the transition count method looks worst on single-bit errors, exactly where the feedback shift register never fails. Overall the transition counter looks pretty good, detecting at least half of all errors, but even a one-bit shift register could do that. The four-bit PRBS generator used in earlier examples will always detect better than $(100-100/2^4)=93\%$ of all errors. It seems conclusive that the PRBS method puts on a good performance, and if we want it to do better we merely add one more bit to the register to halve the rate of misses.

How Close Do We Want to Get?

We set out to find a means of instrument testing at least as good as present computer-based methods. These existing systems generally perform as well as the engineer who adapts them to the circuit under test. The task of adapting a circuit to be tested by signature analysis is very much the same as adapting to any other tester—engineering errors are assumed constant. If the PRBS technique is used for backtracing to find faulty components in field service, then the largest remaining block of human error is the ability of the service person to recognize a faulty signature.

It seems that a four-character signature is easily recognized, while the incidence of correct pattern recognition falls off with the addition of a fifth character. We tried this on a statistically small sample of people and found it to be so. Electronically, four hexadecimal characters is sixteen bits. A few bits more or less is not likely to complicate a shift register, but it would have an adverse effect on the user. Sixteen bits gives a detector failure rate of less than sixteen parts per million (one in 65,535), adequate for most purposes, so we settled on a four-character signature.

Since the signature offers no diagnostic information

Last In →A	B	C	D← First In	Display
0	0	0	0	0
1	0	0	0	1
0	1	0	0	2
1	1	0	0	3
0	0	1	0	4
1	0	1	0	5
0	1	1	0	6
1	1	1	0	7
0	0	0	1	8
1	0	0	1	9
0	1	0	1	A
1	1	0	1	C
0	0	1	1	F
1	0	1	1	H
0	1	1	1	P
1	1	1	1	U

Fig. 6. *In the HP 5004A Signature Analyzer, n=16 and the remainder, or signature, is displayed as four non-standard hexadecimal characters. Each character represents the outputs of a group of four flip-flops as shown here.*

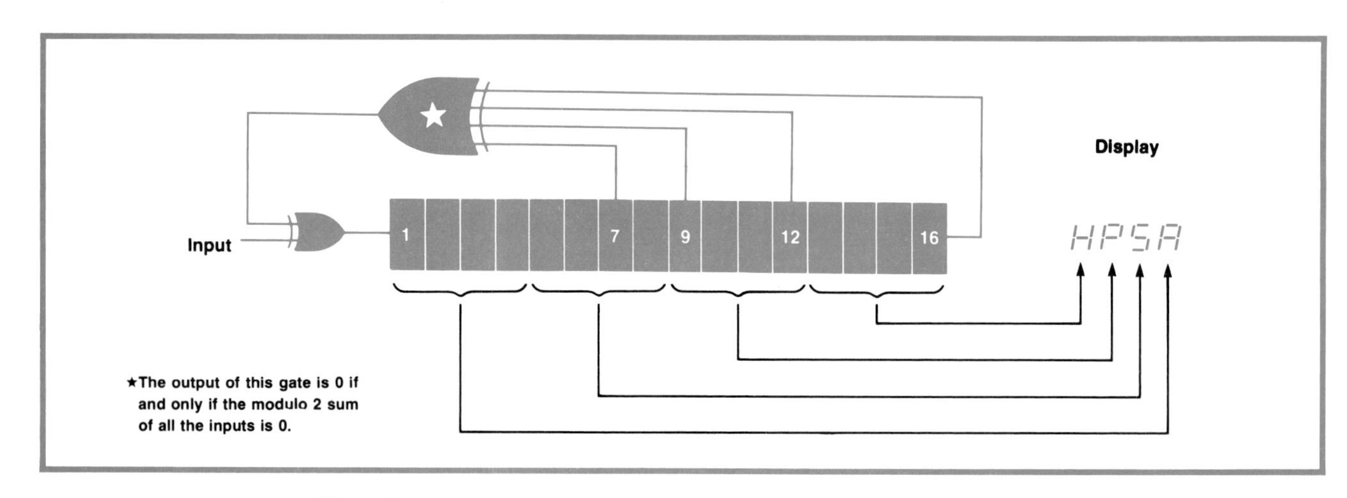

Fig. 7. *The 16-flip-flop PRBS generator used in the 5004A Signature Analyzer.*

whatsoever, but is purely go/no-go, the character set was not restricted, except to be readable. Numbers are quite readable but there are not enough of them. Another consideration was that for an inexpensive tool, seven-segment displays are desirable. The chosen character set (Fig. 6) is easily reproduced by a seven-segment display and the alpha characters are easily distinguishable even when read upside down. A further psychological advantage of this non-standard ("funny hex") character set is that it does not tempt the user to try to translate back to the binary residue in search of diagnostic information.

Register Polynomial

We have decided on a four-character display for a sixteen-bit register, but it remains to select the feedback taps to guarantee a maximal length sequence. It happens that this can be done in any of 2048 ways.[2] The computer industry uses two:

$$\text{CRC-16} = X^{16}+X^{15}+X^2+1,$$

and

$$\text{SDLC(or CCITT-16)} = X^{16}+X^{12}+X^5+1.$$

But each of these is reducible:

$$\text{CRC} = (X+1)\,(X^{15}+X+1),$$

and

$$\text{SDLC} = (X+1)\,(X^{15}+X^{14}+X^{13}+X^{12}+X^4+X^3+X^2+X+1).$$

The $X+1$ factor was included in both to act as a parity check; it means that all undetectable error sequences will have even parity. This is apparent by looking at the original polynomials and noting that they each have an even number of feedback taps, so an even number of error bits is required to cancel an error. For our purposes this clustering of undetectable errors seems undesirable. We would like a polynomial that scatters the missed errors as much as possible. For this reason we would also like to avoid selecting feedback taps that are evenly spaced or four or eight bits apart because the types of instruments, microprocessor-controlled, that we will most frequently be testing tend to repeat patterns at four and eight-bit intervals. The chosen feedback equation is:

$$X^{16}+X^{12}+X^9+X^7+1,$$

which corresponds to the characteristic polynomial

$$P(X) = X^{16}+X^9+X^7+X^4+1.$$

This is an irreducible maximal length generator with taps spaced unevenly (see Fig. 7). Our relatively limited experience with this PRBS generator has shown no problems with regard to the selection of feedback taps. The test of time will tell; even the CRC-16 generator seems to have fallen out of favor with respect to that of SDLC after having served the large-computer industry for well over a decade.

References

1. W.W. Peterson, and E.J. Weldon, Jr., "Error-Correcting Codes," The MIT Press, Cambridge, Massachusetts, 1972.
2. S.W. Golomb, "Shift-Register Sequences," Holden-Day, Inc., San Francisco, 1967.

Robert A. Frohwerk

Bob Frohwerk did the theoretical work on signature analysis. He received his BS degree in engineering from California Institute of Technology in 1970, and joined HP in 1973 with experience as a test engineer for a semiconductor manufacturer and as a test supervisor and quality assurance engineer/manager for an audio tape recorder firm. He's a member of the Audio Engineering Society. Bob was born in Portland, Oregon. Having recently bought a home in Los Altos, California, he and his wife are busy setting up a workshop for Bob's woodworking and metal sculpture, putting in a large organic garden, and planning furniture projects and a solar heating system. Bob also goes in for nature photography, sound systems, and meteorology (he built his wife's weather station).

Easy-to-Use Signature Analyzer Accurately Troubleshoots Complex Logic Circuits

It's a new tool for field troubleshooting of logic circuits to the component level.

by Anthony Y. Chan

THE NEW HEWLETT-PACKARD Model 5004A Signature Analyzer (Fig. 1) was designed to meet the need for field troubleshooting of digital circuits to the component level. The basic design goal was to implement the signature analysis technique described in the preceding article in a compact, portable instrument with inputs compatible with the commonly used logic families (TTL and 5V CMOS).

The 5004A is a service tool. It receives signals from the circuit under test, compresses them, and displays the result in the form of digital signatures associated with data nodes in the circuit under test. The signature analyzer does not generate any operational signal for circuit stimulus, depending instead on the circuit being tested to have built-in stimulus capability. The analyzer is capable of detecting intermittent faults. Its built-in self-test function increases user confidence and its diagnostic routine allows quick, easy troubleshooting with another 5004A if the instrument fails.

The signature analyzer's data probe is also a logic probe similar to the HP 545A Logic Probe.[1] The lamp at the probe tip turns bright for a logic 1, turns off for a logic 0, and goes dim when the input is open-circuited or at a bad level (third state).

What's Inside

Fig. 2 is a block diagram of the signature analyzer. During normal operation, the level detectors receive trains of start, stop, and clock control signals from the circuit under test and transmit them to the edge select switch. The edge select switch allows the user to

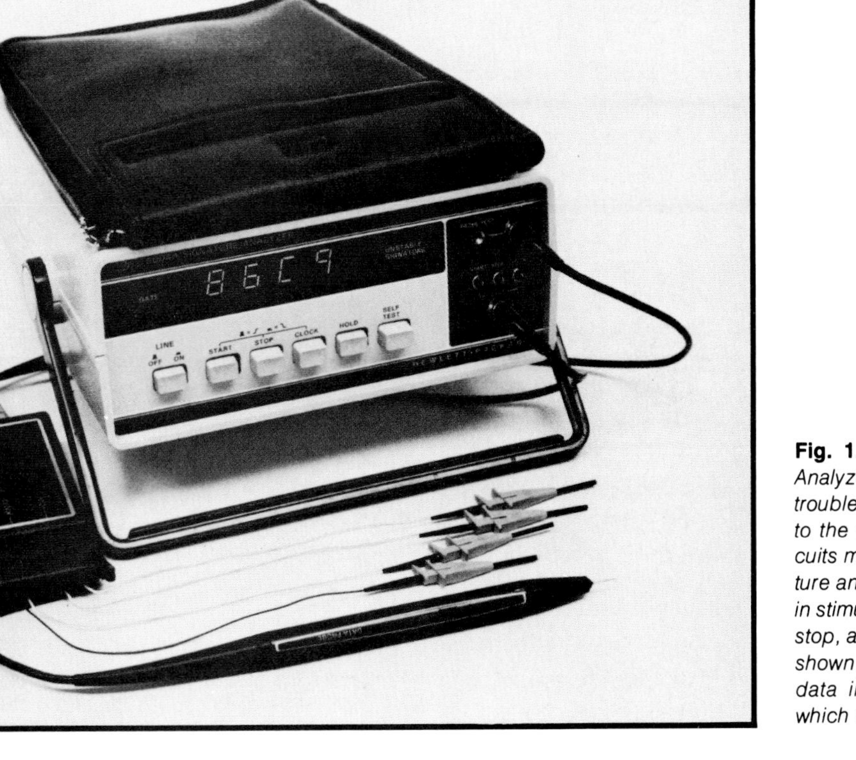

Fig. 1. *Model 5004A Signature Analyzer is a new tool for field-troubleshooting of digital circuits to the component level. (The circuits must be designed for signature analysis and must have built-in stimulus.) The 5004A gets start, stop, and clock inputs via its pod, shown here in the foreground, and data inputs via its data probe, which doubles as a logic probe.*

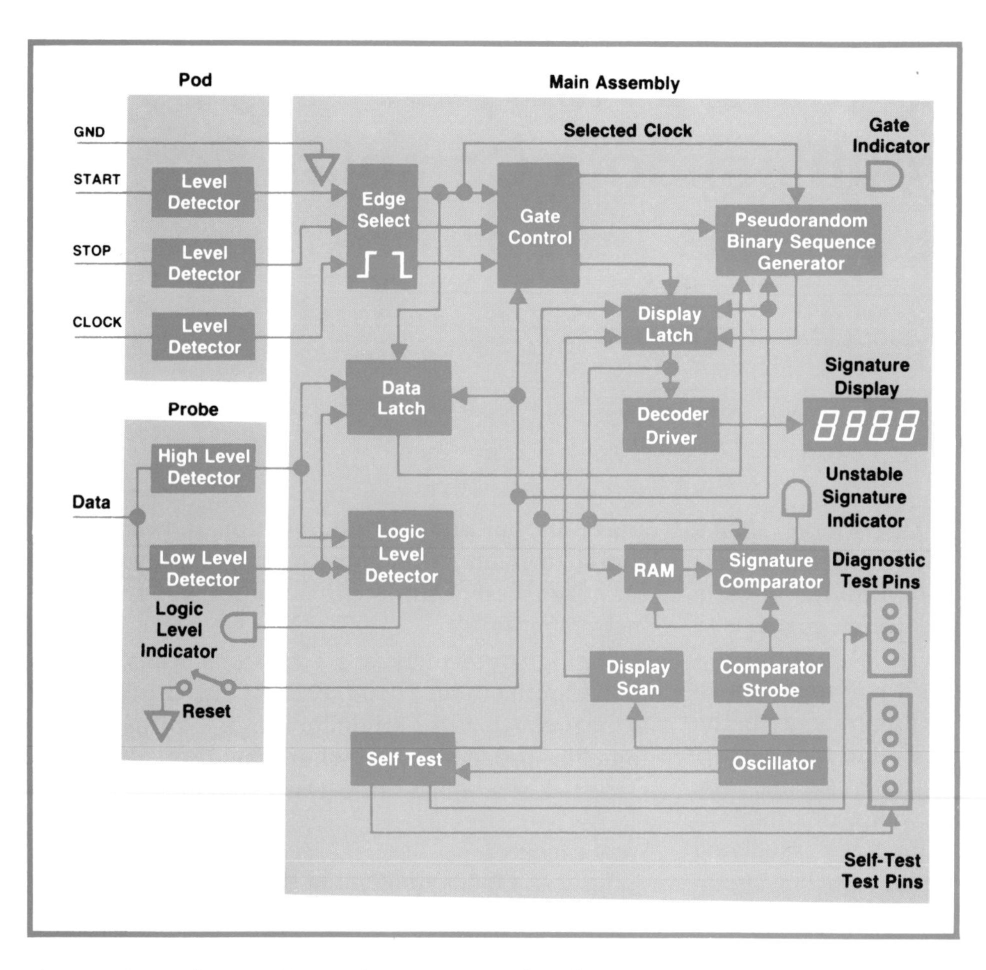

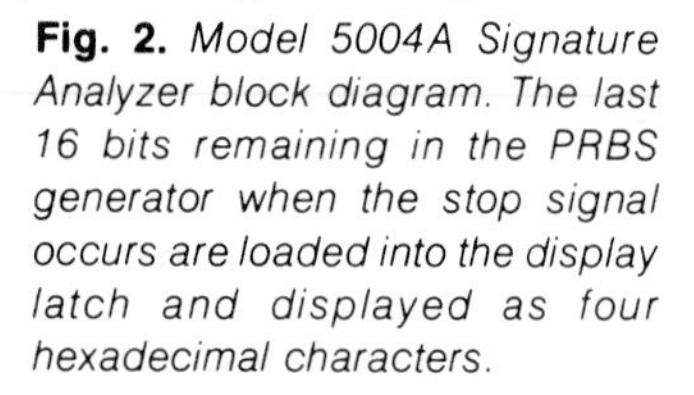

Fig. 2. *Model 5004A Signature Analyzer block diagram. The last 16 bits remaining in the PRBS generator when the stop signal occurs are loaded into the display latch and displayed as four hexadecimal characters.*

choose the polarity of signal transitions that the instrument will respond to. The gate control receives the selected control signals from the edge select switch and generates a gated measurement window (gate on) for the pseudorandom generator; it also turns the gate light on. The measurement window is the period between valid start and stop signals, and its length is measured in clock cycles (see Fig. 3). The minimum possible window length is one clock cycle.

The data probe translates voltages measured at circuit nodes into three logic states (logic 1, logic 0, and bad state) and transfers them to the data latch. The latch further translates the data, from three logic levels to two, at selected clock edges.* At each clock time, the data latch will pass a 1 or 0 level but will remain latched to the previous state if the input is in the bad state. The data latch may be the end of the road for some data because the pseudorandom generator accepts data only during the measurement window (gate on). Once data enters the pseudorandom generator, it is shifted in synchronism with the clock until the end of the measurement window. The last 16 bits remaining in the generator at gate-off time are loaded into the display latch and then output in the form of four non-standard hexadecimal characters—the signature. The display latch keeps the signature on until the end of the next measurement window, when the display is updated with new information. The signature will be stable as long as the measurement window and the data received within the window are repeatable.

*The probe recognizes three logic states instead of only two because of its logic-probe function.

Importance of Setup and Hold Times

Frequency response is one of the most important parameters in a test instrument. In the case of the

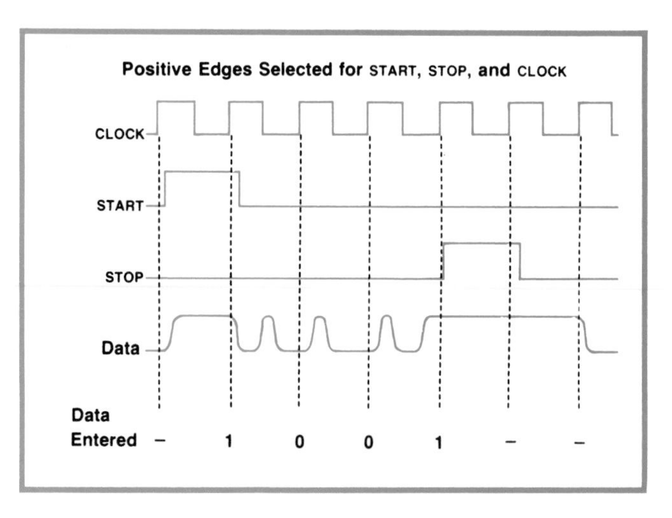

Fig. 3. *The measurement window is an integral number of clock cycles. One cycle is the minimum length.*

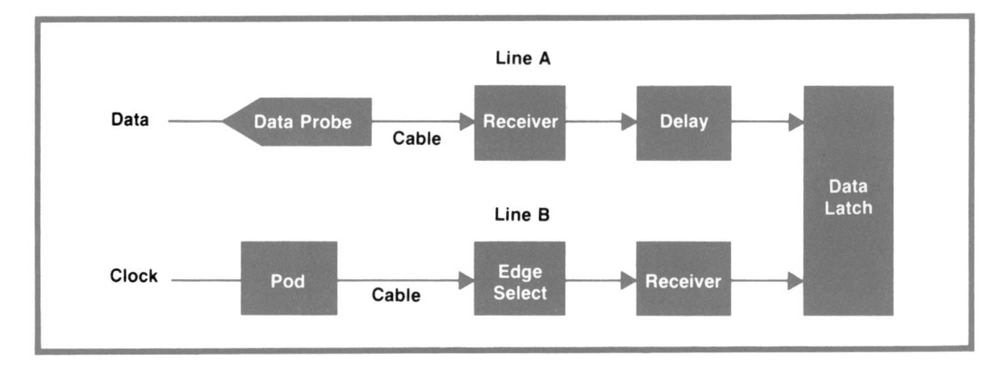

Fig. 4. *Delays through probe and pod channels are matched so that hold time (the time that input data must remain stable after a clock edge) is non-positive. Setup time (the time that input data must be stable before a clock edge) is typically 7 ns.*

signature analyzer, two other factors, data setup time and hold time, are very important as well. Data setup time is the interval for which data must be stable before the selected clock edge occurs. Hold time is the interval for which data must remain stable after the selected clock edge. Assuming a signature analyzer requiring 30 ns setup time and 10 ns hold time is used to test a circuit, then the logic of the circuit under test must be stable for at least 30 ns before the active clock edge and the logic must remain stable for 10 ns after the clock edge; otherwise, ambiguous readings may result. The setup and hold times limit the speed of the analyzer.

It is not easy for a high-speed circuit to guarantee that its logic will remain stable for some period of time after every active clock edge. The 5004A design goal was to be able to operate with reasonably short data setup time and non-positive hold time to minimize ambiguities.

Data and clock signals are received and transmitted to the data latch through the data probe, receiver, edge select switch, and cables (Fig. 4). There is one time delay for the data signal going through the data probe, cable, and receivers (line A), and another time delay for the clock pulses going through the wire and edge select switch into the data latch (line B). Every component, and therefore the time delays, may differ from unit to unit because of manufacturing tolerances. To guarantee a non-positive hold time, eliminate race conditions, and be reproducible in a production environment, the minimum delay of line A must be equal to or longer than the maximum delay of line B ($t_{Amin} \geq t_{Bmax}$). Also desired is a minimum setup time $T_s = t_{Amax} - t_{Bmin}$.

One way to achieve short setup time is to have identical circuitry in the data and clock channels, so propagation delays cancel each other. Circuitry in the data probe is very similar to that in the pod. The receivers in both channels are identical and share the same IC chip. The edge select switch in the clock line has very little delay. The symmetry results in a good match between the two signal lines, but to insure that $t_{Amin} \geq t_{Bmax}$, there is a delay circuit in line A, and the cable length of line B is shorter. Thus it is possible to guarantee hold time less than 0 ns and setup time less than 15 ns (7 ns typical).

Input Impedance

Since the 5004A is a test instrument, it is important that its inputs do not load or condition the circuit under test. It is generally true that high input impedance reduces loading. But, how high can the input impedance be before other effects cause problems?

Let's study a few cases of high-impedance input in a synchronous device (Fig. 5). Fig. 5a shows the result of the input data's changing from a logic 1 to a third state at clock 1. The input voltage is pulled toward ground by the pull-down resistance. The difference between the solid line and the dotted line is the difference of node capacitance (C) and pull-down resistance (R) tolerances. Depending on the clock rate and the RC difference, the result at clock 2 can be either a logic 0 or a third state. The same thing in reverse happens in case B. In case C, the input data is changing to an intermediate level (~1.4V). When the data changes from a logic 1 to the third state, the input is pulled towards 1.4V. The result at clock 2 is in the third state no matter what the RC time constant is.

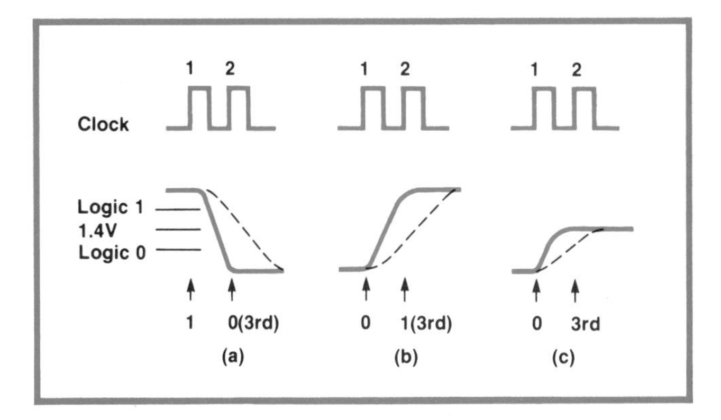

Fig. 5. *5004A input impedance is returned to 1.4V to eliminate ambiguities caused by input RC tolerances. Here are three possible results of the input data's changing from a 0 or 1 at clock 1 to a third state at clock 2. The solid and dotted lines are for different values of input RC. In (a), the input at clock 2 can be seen as a logic zero or a third state, depending on the value of RC. (b) is the reverse of (a). In (c), with the input returned to 1.4V, the result is always a third state.*

There are two other advantages to returning the input impedance to 1.4V. First, there is less voltage swing, and almost equal swings for both logic 1 and logic 0 states. Second, the logic probe open-input requirement is met.

Very high input impedance may cause problems even for a non-clocked device. It introduces threshold errors because of the offset bias current of the input amplifier, and the leakage current of the three-state bus might change the measured voltage. After study and calculation, we chose 50 kΩ to 1.4V as the input impedance and return voltage for the 5004A inputs. 50 kΩ is large enough not to load TTL and most 5V CMOS logic families, and small enough not to cause excessive offset voltage with typical leakage currents on a three-state bus.

Construction

The 5004A Signature Analyzer is constructed in a lightweight, rugged case. A hand-held data probe and a small rectangular pod are connected to the instrument by cables (Fig. 1). Inside the main case are the edge select circuit, gate control, data latch, pseudorandom generator, display latch, signature displays, signature comparator, self-test stimulus generator, and power supply shown in Fig. 2. All the electronics and mechanical components are mounted on a single printed circuit board assembly sandwiched inside two shells held together with four screws. On the front panel are four large seven-segment displays. A light to the left of the display shows gate (measurement window) activity while one on the right indicates unstable signature. Six pushbutton switches control power on/off, start, stop, and clock edge polarities, a hold mode for single cycle events or freezing the signature, and a self-test mode. Start, stop, clock, and data test sockets on the right-hand side of the front panel are for self-test and diagnostic setup. A soft pouch mounted on top of the instrument stores the data probe, pod, and necessary accessories when not in use.

Data Probe

The active data probe is a hand-held probe. Its main function is to accept tip logic information with minimum tip capacitance. The input signal is connected to two comparators through voltage dividers and an RC network (see Fig. 6). The voltage divider R1-R2 is terminated at 1.4V, which guarantees an open input at a bad level and eliminates the potential ambiguity, discussed earlier, resulting from RC tolerances.

Input overload protection is provided by on-chip clamp diodes and the external network R1 and CR1. C1 provides a bypass for fast transitions. R3, R5, and R6 set up voltage references for comparators A and B. Two comparators are needed to measure the three logic states—1, 0, and bad level. The high-speed differential-in/differential-out comparators translate the input voltage into digital signals and transmit them to the main instrument through twisted pairs. A single-contact pushbutton switch on the data probe resets the pseudorandom generator, state control, and displays.

Pod

The thin, rectangular pod houses three identical

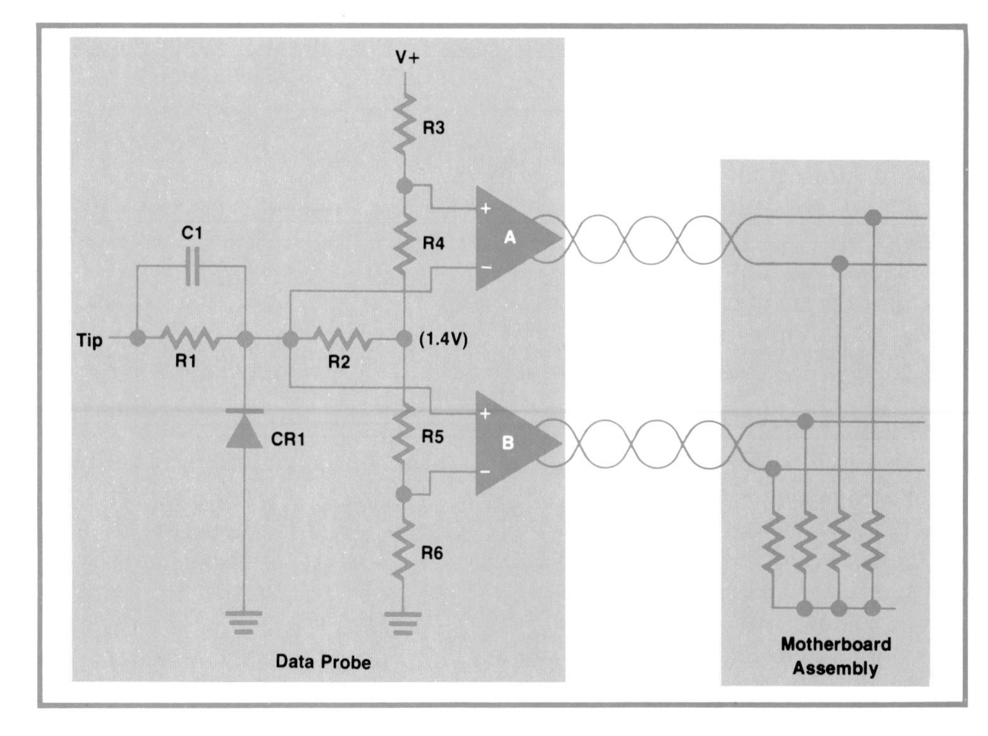

Fig. 6. *Simplified 5004A data probe schematic.*

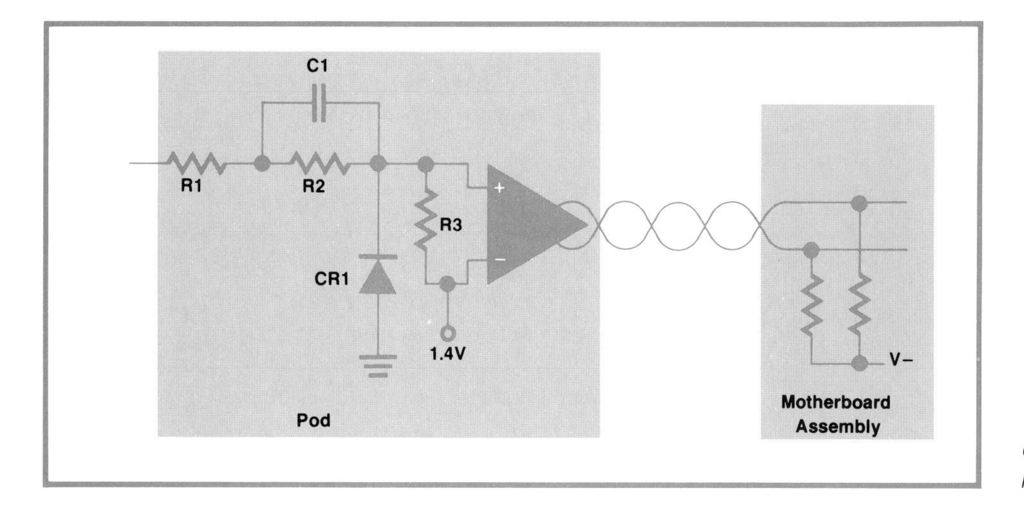

Fig. 7. *Simplified schematic of one channel of the 5004A input pod.*

channels for start, stop, and clock control inputs. The input wires can be directly plugged into any 0.03-inch round socket or connected to a "grabber" that can hook onto almost any test pin and is particularly good for IC pins. Each of the control channels is very similar to the circuitry in the data probe to match propagation delays. In fact, delay match is the major function of the pod.

In each channel of the pod is a comparator of the same type as those in the data probe. One of the comparator inputs is connected to voltage divider R1-R2-R3 while the other input is connected to 1.4V along with R3; this increases input hysteresis and sets the input threshold (Fig. 7). The impedance of R1-R2-R3 is the same as the impedance of the data probe (50 kΩ); termination at the same voltage level further improves matching.

Protection against input overload is provided by internal clamp diodes in the comparator IC and by external network R1, R2, and CR1. R1 damps the ringing generated by the inductance of the input wire. C1 provides a bypass for fast transitions.

Unstable Signature

An intermittent fault is one of the biggest problems in electronic repair. The fault comes and goes, and in most cases does not stay long enough for positive detection. Signature analysis can detect such faults if they occur within a measurement window. However, the operator may not receive the message if the measurement cycle time is too short.

The random-access memory (RAM) in the main assembly of the 5004A continuously writes and reads the display information from the display latch at the display scan rate. During each scan cycle, the signature comparator compares the signature stored in the RAM with the one in the display latch, and turns on the unstable signature light on the front panel when any difference exists. This light is stretched for 100 ms to allow recognition. The comparison is done on a sampled basis and not each time a new signature is developed, so the unstable signature detector works most of the time, but not 100%. Errors occurring in a very short measurement cycle may not always be detected by the relatively slow-scanning comparator.

Hold Mode

The hold mode works closely with the stop signal. If the hold switch on the front panel is pushed in, the hold mode will be entered at the end of the measurement window, freezing the signature display and preventing the gate control from starting a new cycle. Hold mode is particularly useful for testing single-shot events like the start-up sequence of a system.

Self Test

It is important for a user to know that a test instrument is in good working condition before it is used to test anything. The 5004A has a built-in self-test function that gives a quick, accurate check of the instrument. Pressing the self-test switch on the front panel energizes the self-test ROM, which interrupts the display update and generates a special programmed stimulus of start, stop, clock, and data signals to the test sockets on the front panel. With the start, stop, and clock control inputs connected to the corresponding test sockets and the data probe to the data test socket on the front panel, and with positive edges selected for the start, stop, and clock inputs, the signature analyzer performs the self test. When a good working 5004A is tested, its gate light flashes, the unstable signature light blinks, the logic light at the data probe tip flashes, and the signature displays 3951, 2P61, 8888 and then repeats. Pushing the hold switch in turns the gate light off and the signature displays 8888, 3951, or 2P61. The self-test routine tests the entire instrument except the clock edge select circuit and the ground wire at the pod input.

Fig. 8. *The 5004A Signature Analyzer is designed for troubleshooting with another 5004A should the self test reveal a fault. Sliding the* NORM/SERVICE *switch to the* SERVICE *position opens the three feedback loops in the instrument.*

Diagnostic Routine

When an unexpected result during self test indicates a fault, troubleshooting and repair are required. The 5004A was designed with signature analysis in mind. It can be tested with another 5004A. The instrument's top cover can be easily removed by removing four hold-down screws on the bottom and loosening two heat sink mounting screws on the back. All the components in the instrument's main case are then exposed for testing. The failing instrument is placed in self-test mode and the start, stop, clock, and ground inputs of a known good 5004A are connected to the test sockets located on the left side of the printed circuit board in the main case. Probing the circuit nodes with the data probe, reading the signatures on the analyzing 5004A, and comparing them with those printed on the schematic is an easy and almost error-free way of determining the quality of a circuit node. Once a faulty node is found, the source of the problem can be easily located with standard backtracing techniques.

When the fault is in a feedback loop, any single fault will cause all the nodes within the loop to appear bad. To pinpoint the fault, the loop must be opened. There are three feedback loops in the signature analyzer, and a slide switch (NORM/SERVICE) on the left of the main printed circuit board is provided for opening them (Fig. 8). Sliding the switch to the SERVICE position opens all three loops.

The diagnostic routine works on the entire instrument except the power supply, the ECL circuits in the data probe and pod, and their interface circuits.

SPECIFICATIONS
HP Model 5004A Signature Analyzer

DISPLAY:
SIGNATURE: Four-digit hexadecimal.
Characters 0,1,2,3,4,5,6,7,8,9,A,C,F,H,P,U.
GATE UNSTABLE INDICATORS: Panel lights. Stretching: 100 ms.
PROBE-TIP INDICATOR: Light indicates high, low, bad-level, and pulsing states. Minimum pulse width: 10 ns. Stretching: 50 ms.

PROBABILITY OF CLASSIFYING CORRECT DATA STREAM AS CORRECT: 100%.

PROBABILITY OF CLASSIFYING FAULTY DATA STREAM AS FAULTY: 99.998%.

MINIMUM GATE LENGTH: 1 clock cycle.

MINIMUM TIMING BETWEEN GATES (from last STOP to next START): 1 clock cycle.

DATA PROBE:
INPUT IMPEDANCE:
50 kΩ to 1.4V, nominal. Shunted by 7 pF, nominal.
THRESHOLD:
Logic one: 2.0V +.1, −.3
Logic zero: 0.8V +.3, −.2
SETUP TIME: 15 ns, with 0.1V over-drive. (Data to be valid at least 15 ns before selected clock edge.)
HOLD TIME: 0 ns (Data to be held until occurrence of selected clock edge.)

GATING INPUT LINES:
START, STOP, CLOCK INPUTS:
Input Impedance: 50 kΩ to 1.4V, nominal. Shunted by 7 pF, nominal.
Threshold: 1.4V ±.6 (.2V hysteresis, typical).

START, STOP INPUTS:
SETUP TIME: 25 ns. (START, STOP to be valid at least 25 ns before selected clock edge.)
HOLD TIME: 0 ns. (START, STOP to be held until occurrence of selected clock edge.

CLOCK INPUT:
MAXIMUM CLOCK FREQUENCY: 10 MHz.
MINIMUM CLOCK TIME IN HIGH OR LOW STATE: 50 ns.

OVERLOAD PROTECTION: All inputs ±150V continuous, ±250V intermittent, 250Vac for 1 min.

OPERATING ENVIRONMENT: Temperature: 0-55°C; Humidity: 95% RH at 40°C; Altitude: 4,600 m.

POWER REQUIREMENTS: 100/ 120 Vac, +5%, −10%, 48-440 Hz.
220/ 240 Vac, +5% −10%, 48-66 Hz.
15 VA Max.

WEIGHT: Net: 2.5 kg, 5.5 lbs. Shipping: 7.7 kg, 17 lbs.

OVERALL DIMENSIONS: 90 mm high × 215 mm wide × 300 mm deep (3½ in × 5½ in × 12 in). Dimensions exclude tilt bale, probes, and pouch.

PRICE IN U.S.A.: $990.

MANUFACTURING DIVISION: SANTA CLARA DIVISION
5301 Stevens Creek Boulevard
Santa Clara, California 95050 U.S.A.

Anthony Y. Chan

Tony Chan received his BSEE degree from the University of California at Berkeley in 1969 and his MSEE degree from California State University at San Jose in 1974. With HP since 1973, he developed the IC chip for the 546A Logic Pulser and designed the 5004A Signature Analyzer. Before joining HP, he designed linear and digital ICs for four years. A native of Hong Kong, Tony now lives in Sunnyvale, California. He's married and has two children. Having just finished remodeling his home, he's now taking on landscaping and furniture-building projects. He likes working with his hands, especially on wood and automobiles, and enjoys an occasional game of tennis.

Signature Analysis—Concepts, Examples, and Guidelines

Guidelines for the designer are developed based on experience in attempting to retrofit existing products for signature analysis and the successful application of signature analysis in a new voltmeter.

by Hans J. Nadig

THE POWER OF SIGNATURE ANALYSIS as a field troubleshooting technique is amply demonstrated by the analysis presented in the article on page 2 of this issue. The technique can even pinpoint the 20% or so of failures that are "soft" and therefore difficult to find, taking 70-80% of troubleshooting and repair time. Soft failures include those that occur only at certain temperatures or vibration levels. They may be related to noise performance or marginal design, such as race conditions that occur only when the power supply voltage is low but still within specifications. Or they may occur only when the user gives the machine a certain sequence of commands.

Signature analysis is applicable to complex instruments using microprocessors and high-speed algorithmic state machines. Yet it is simple enough so that the user of a product may be able to apply it nearly as well as more highly trained field service personnel.

Having recognized the power of signature analysis, we first attempted to apply it to existing products, including computers, CRT terminals, and the digital portions of microwave test equipment. We soon recognized that either the circuits had to be altered or the signature analysis approach would be no better than earlier methods. After some experience we were able to define rules for making a product compatible with signature analysis. These rules, summarized on page 18, are guidelines for the designer. Following them helps assure that a product will be simple and inexpensive to troubleshoot by the signature analysis method.

How We Got Started

A good way to demonstrate the advantages of signature analysis and the requirements for applying it successfully is to describe what happened when we first tried it a few years ago.

With a prototype signature analyzer we set out to apply the technique to various Hewlett-Packard instruments. We first attacked a CRT terminal with microprocessor control and ROM and RAM storage, including some dynamic memories.

The built-in self-test mode of the terminal displayed a certain test pattern on the CRT and flashed the cursor at a 2½-Hz rate. Taking advantage of this self test as a stimulus, and using the most significant address bit to start and stop the measurement, we soon recognized that these signals did not provide a stable measurement window. Some portions of the terminal operated on an interrupt basis, so the number of clock periods varied within the START-STOP window. Needless to say, the data stream changed, too. Next we concentrated our efforts on one section, the memory. To test it, we wanted to force the microprocessor into a mode in which all the memory locations were addressed, but to do this, we were forced to cut the data bus. Fortunately, we could separate the microprocessor from the data bus by using an extender board and cutting the lines there. Grounding a few lines and pulling some other lines high caused the microprocessor to repeatedly execute one instruction that automatically incremented the address each cycle, effectively stepping up through the whole address field. Fig 1 shows how this can be done for an 8080 microprocessor.

At this point we realized that the microprocessor, the clock, and the power supply were the heart of the product. We decided to call this the "kernel" (Fig. 2). By verifying the proper operation of these parts first we could then expand and test additional portions of the circuitry. With the free-running microprocessor exercising the control and address lines, we were able to test the address bus, the ROMs, and the data bus. As a START and STOP signal the most significant bit of the address bus was used, allowing us to check all the ROM locations. Since a number of RAMs were also affected, we applied a grounded jumper wire to force the enable line to the RAMs low; this was necessary to get stable signatures, since the RAMs did not contain a defined data pattern, and if addressed, randomly altered the information on the data bus.

Here, then, were our first lessons: *feedback loops cause problems unless opened; circuits not related to*

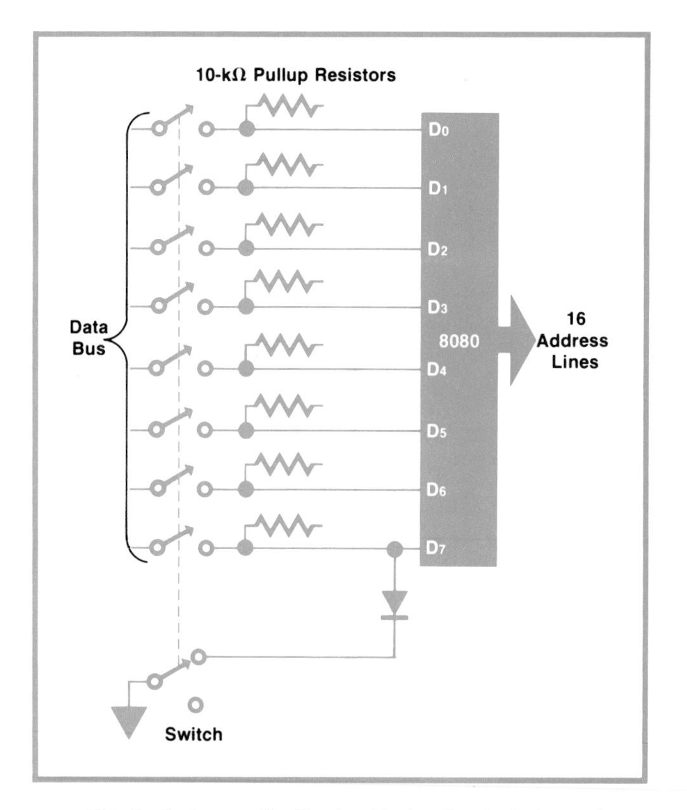

Fig. 1. *To be an effective troubleshooting technique, signature analysis must be designed into a product. For example, for a test of the address lines of a microprocessor, there should be a switch that opens the data bus and forces the microprocessor to free run. The address lines can then be checked and can also be used as control inputs to the signature analyzer.*

the test must be disabled.

Synchronous Operation Necessary

It would be ideal if one setup would allow troubleshooting most parts of an instrument. The synchronization signal with the highest rate would be connected to the clock input of the signature analyzer.

Our display terminal uses a number of different frequencies, from 21 MHz down to 1¼ Hz. A ripple counter divides the frequencies down. Trying to characterize the divider chain showed us unstable signatures for every node after the first stage. The reason was that the circuits operated asynchronously with as much as 500 ns skew from the first to the last stage. Lowering the clock frequency to about 2 MHz by removing the crystal from the oscillator, we were able to define stable signatures for the counter chain. However, one measurement lasted 10 seconds, and to verify whether it was stable or not we had to have at least two complete measurements. An alternative to reducing the clock frequency was a new test setup for the slower parts of the divider. However, it is always wise to minimize the number of necessary setups.

So we learned that *synchronous operation is essential for high-speed testing.* Fortunately, this is easy to accomplish in most microprocessor designs, even those with the newer types of microprocessors that use asynchronous handshake lines to gate information in and out. Although it might seem at first that signature analysis is not applicable in such a case, the problem of asynchronous operation can be eliminated if the handshake lines are used to clock the data into the analyzer. Also, when this is done third-state conditions are no longer a problem because at the time of the "data valid" signal the data is either high or low. Thus a seemingly asynchronous system can behave as a synchronous system as seen by the troubleshooting tool, the signature analyzer.

Need for Designed-In Capability

An interesting possibility is that of measuring all the possible fault conditions at a central node by inducing faults into a good circuit and recording the corresponding signatures at the central node in a signature fault table. Testing the central node then tells the whole story of whether the instrument is in working condition or not. If it fails, the fault table indicates where the error is and sometimes even which part has to be replaced.

In the case of the terminal, an ideal central node seemed to be the video signal. Every data and control line is ultimately concentrated in one node containing all the information to scan a dot across the screen. Using the terminal's self-test feature as a stimulus, we chose the new-frame trigger signal as our START and STOP inputs. But for some reason we could not get stable signatures, which meant that the data stream between the two gate signals was not the same for each frame.

The culprits were two signals that occurred at a much slower rate than the 60 Hz for the frames. The blinking signals for the 2½-Hz cursor and for the 1¼-Hz enhancement were changing the characteristic data stream for the frames. Not until we disabled the signal generators for the blinking did we get stable signatures.

If parts of the circuit are being disabled the comprehensiveness of the test is reduced. In this case the designer could have provided the necessary setup to do a complete test. But after the design is frozen without signature analysis in mind it is hard to apply it successfully. If the window for signature analysis is selected so that the slowest blinker is the trigger for START and STOP, it is possible to create a stable signature or, in other words, a repeatable data stream.

The characterization of the RAM required special attention. A defined pattern had to be loaded into the memories before useful signatures were obtained at the outputs. By using several jumpers to enable the write cycle and the ROM outputs, then switching into

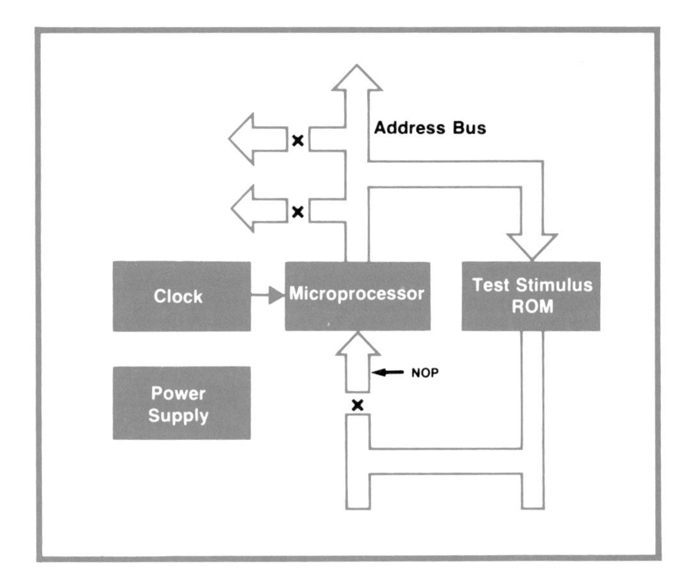

Fig. 2. *Signature analysis test procedures should verify the operation of key portions—the "kernel"—of a product first, then use the kernel to test other circuits. A typical kernel might consist of microprocessor, clock, power supply, and one or more read-only memories.*

a read mode for the RAM and disabling the ROM, we effectively loaded the content of the ROMs into the RAMs. After that, valid readings were obtainable and it was possible to trace down a bad RAM component.

Having tested and characterized about 30% of the digital boards, we next concentrated our efforts on the large display memory. Testing this dynamic memory was not easy, because it went through an asynchronous refresh cycle every 2 ms. Even adding more jumper wires, we had to admit finally that without cutting leads or altering the circuit we would not get any satisfactory results.

Looking at the CRT terminal with the oscillator crystal removed, with a cut-up extender board and jumper wires clipped into the circuit here and there, we learned the most important lesson: *signature analysis capability has to be designed into the circuit.*

After that a number of additional products were tested and the message remained the same: retrofitting is not an effective approach. On the other hand, it became clear that the additional effort to make the circuit signature-analysis-compatible is indeed very small if done at an early stage of the product development. Early, in this case, means the breadboard stage.

Thoughts on Implementation

The versatility of the signature analysis concept is impressive. As long as the data is valid at the selected clock edge, and the stimulus is repeatable, many parameters can be selected. The window length, or the number of bits in the data stream, can be of any value (100,000 bits is not unusual). Any suitable signal can be selected as the clock input, enabling the designer to make seemingly asynchronous circuits look as if they were synchronous. A major advantage is that everything happens at normal speed.

The implementation of signature analysis into a product is similar to designing a microprocessor into a product. In the latter case, the designer has to learn the instructions, and has to understand the advantages and the limitations of the microprocessor. Because of the learning curve, the first application will most likely take more time than later designs. Also, there is no cookbook approach to a microprocessor design because there are no two situations alike. The designer makes decisions based on the evaluation of power consumption, cost, size, reliability, and so on.

The same is true for the implementation of signature analysis. The design engineer must understand the function of each component and create a test stimulus that tests each function totally. Simply exercising a node may not be enough. Even a component as simple as an AND gate may have stable and correct input and output signatures and still be bad, as shown in Fig. 3. Similar cautions apply to any test method, of course. The designer must be careful to *test completely the function of the smallest replaceable part*.

Serviceability is an additional algorithm. If the service algorithm is taken into consideration at an early stage of the development, the application will be easier, and the additional cost for hardware, test program memory space, and development time will be offset by shorter test times in production. Also, the warranty service and repair costs will be much lower. Later in an actual example, we will see how even the

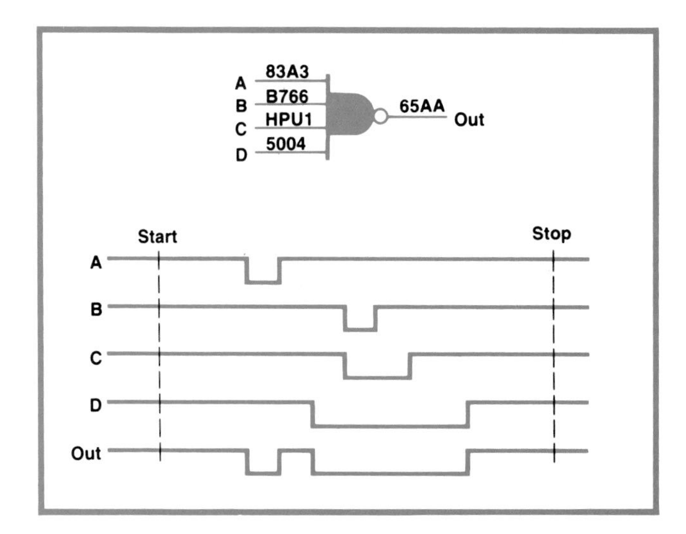

Fig. 3. *The test stimulus should test each component thoroughly. Otherwise a circuit, such as this AND gate, can have stable and correct signatures at each input and output node and still be bad. For example, input B or C might have an open bonding wire inside the IC, but in this case the error is masked by input D. Careful test stimulus design avoids this.*

Designer Guidelines for Applying Signature Analysis to Microprocessor-Based Products

General Guidelines

- Make a full commitment to use signature analysis at the definition stage of the product.
- Evaluate the trade-offs, such as increased factory costs versus lower test time in production and lower warranty and repair cost in the field. Other factors that might influence the decision are warranty cost goal, profit, cost of field repair, acceptable downtime, the cost of alternative service procedures like board exchange programs, the topography of the service organization, and the extra cost for customs if the parts have to be shipped back and forth across country borders.
- Familiarize yourself with the signature analysis service philosophy and allow some extra time for the design.
- Start to prove the basic working of signature analysis at the breadboard stage, before laying out printed circuit boards.
- Team up with the service engineer who will write the manual for the product. Do it at an early stage, before the first prototype is finished.
- If you hope for some benefit for production testing, get the production engineer involved during your definition of the test stimulus and the method of connecting the signature analyzer.
- As a design engineer, be aware that the volume of the necessary documentation can be minimized by selecting the appropriate partitioning of the tested sections in the product.

Technical Rules

- The stimulus for troubleshooting comes from within the product. The self-test stimulus can frequently be used.
- Provide if possible a free-running repeatable stimulus for continuous cycling.
- Tested nodes are to be in a valid and repeatable state at the time of the selected clock edge for triggering.
- Provide easy access to the START, STOP, and CLOCK test points.
- Feedback loops must be capable of being opened. Only an open-ended test allows backtracing.
- The test program or stimulus should exercise within the START-STOP window all the functions that are used in the instrument, although it is not necessary to perform a meaningful operation.
- Provide a controlled test stimulus to interrupt lines, open connectors, and signals that are normally asynchronous.

Additional Guidelines

- Verify the heart or kernel of the instrument first. The kernel may consist of the power supply, the clock generator, and the microprocessor. Then, use this central part to create the stimulus for the peripheral circuits.
- ROMs may be used to write the stimulus program.
- Divide the circuit into well defined portions. Several test setups may be necessary.
- Avoid the use of circuits with non-repeatable delays (e.g. one-shot multivibrators) within the test loop.
- Avoid, if possible, the third-state condition of a three-state node during the measurement cycle.

factory cost can be lowered in spite of needing some extra components, because the whole circuit could be placed on a single large board, while for the traditional board swap service approach, the circuit would have been divided into a number of easy-to-replace subassemblies, which would have required more connectors and hardware to hold the boards in place.

Signature Analysis and the Service Engineer

How would a service engineer use signature analysis if a product failed? The assumption is made that the signature analysis method is designed into the product. Instructions on the schematic or in the service manual show how to switch the product into the diagnostic mode and how to connect the signature analyzer to the device under test. Each node on the schematic is marked with a signature (Fig. 4). With the aid of the schematic the service engineer first reads the output signatures of the device under test. If they are bad, he traces back to a point in the circuit where a good signature appears at the input side of a component and a bad one at the output side. This is called backtracing.

Some understanding of the components in a digital schematic is essential. The direction of the data flow is important but no special knowledge about the actual function of the assembly is required. So, one advantage of the signature analysis service method is that less training is needed to learn to do fault tracing. We can even go a step further and develop a troubleshooting tree without the use of a schematic. A picture of the physical board with signatures at the pins of IC's or components may be used instead (Fig. 5). This way the technician is not required to know whether the circuit he is testing contains a complex storage device or simply a gate. One suggestion is to print the signatures onto the printed circuit board itself, with arrows indicating the signal flow. Another is to print a test template that is attached to the component side of the circuit board when service is required (see Fig. 6). Holes in the appropriate locations, signatures, and other instructions printed on the template guide the service person to the faulty node.

The 3455A Voltmeter—an Example

The first HP instrument using signature analysis is the 3455A Digital Voltmeter[1] (Fig. 7). The digital portion of this instrument is quite extensive. It is microprocessor controlled and contains an elaborate self-test program stored in ROM. If the self test fails, a jumper inside the enclosure is removed, breaking feedback loops and enabling the self-test program, which is then used to troubleshoot the instrument.

Signature analysis influenced other factors that make this voltmeter easier to troubleshoot down to the component level. The entire digital portion is on a

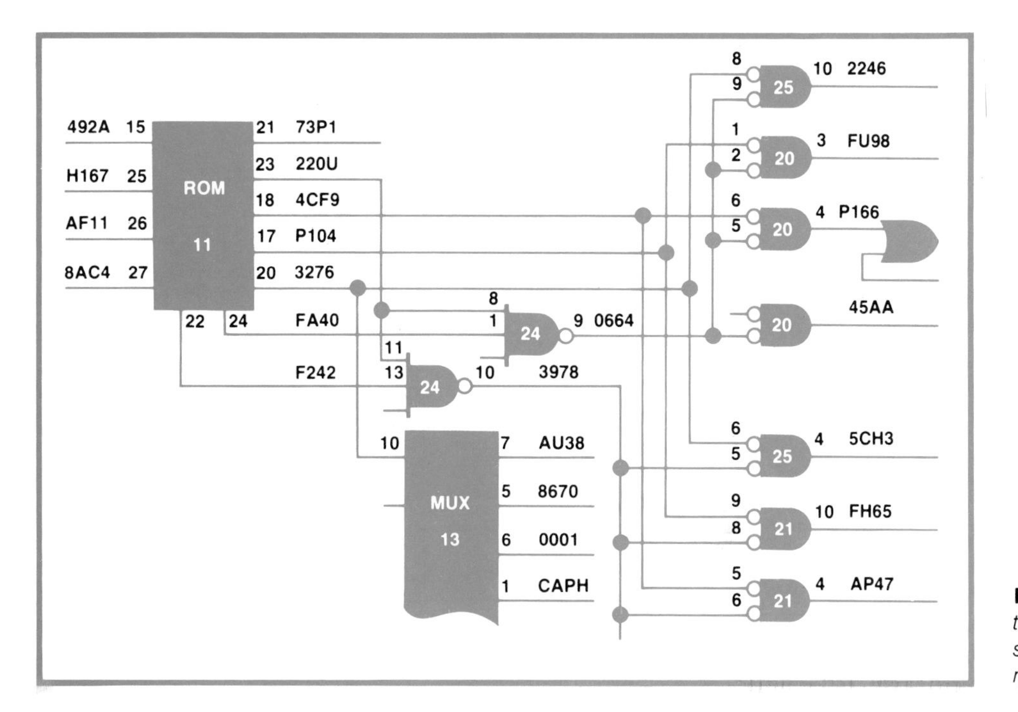

Fig. 4. *An example of an annotated schematic, showing correct signatures at various circuit nodes.*

single board. The elimination of connectors and a multitude of smaller subassemblies not only reduced production cost, but also made all the parts easily accessible for testing without special extender boards.

Some extra design time, a few more ROM locations, and the extra jumper wire were the price paid for serviceability. A cost evaluation verified that the production cost was lowered. The extra design time amounted to approximately 1% of the overall development time.

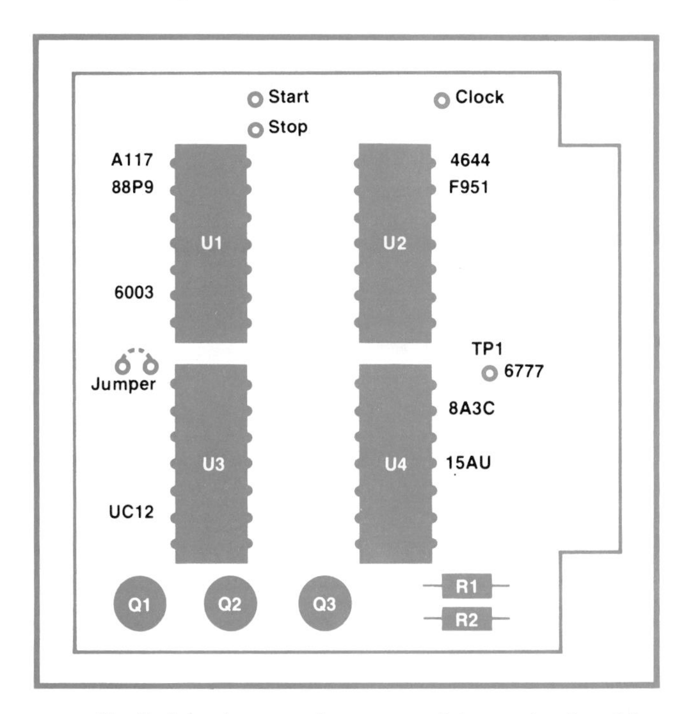

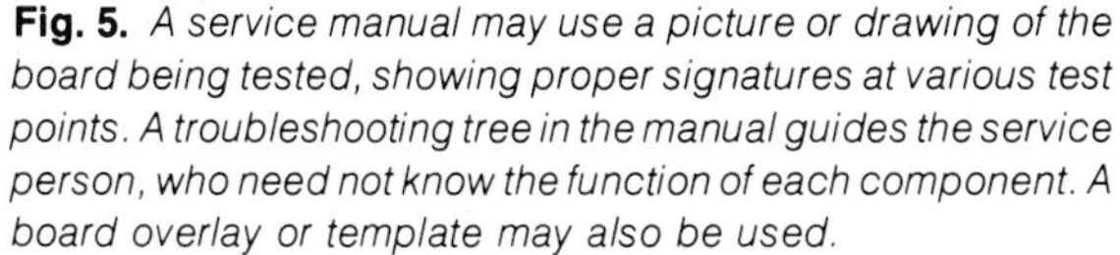
Fig. 5. *A service manual may use a picture or drawing of the board being tested, showing proper signatures at various test points. A troubleshooting tree in the manual guides the service person, who need not know the function of each component. A board overlay or template may also be used.*

Besides the design engineer, the service engineer who wrote the service manual made an important contribution to the successful application of signature analysis. He learned the internal algorithms of the product almost as well as the designer. Because there was no precedent to fall back on, he used a number of innovative ideas, which have been well accepted by the field engineers.

The service manual guides the service person to the fault within a very short time. The manual contains a troubleshooting tree that, combined with annotated schematics and graphs of board layouts, leads directly to the bad node. In some cases the manual gives instructions as to which IC to replace. In other cases the use of a logic probe, which may be the 5004A Signature Analyzer's data probe, may be required. A current sensor helps to find short circuits between traces or to ground and is particularly helpful if a long bus line should fail. A portion of the 3455A Voltmeter troubleshooting tree is shown in Fig. 8.

The first test checks the kernel, which consists of the microprocessor, the clock circuit, the power supply, and a number of external gates. After proper functioning of the kernel is verified, the test setup is changed (one control input of the signature analyzer is moved to another pin) and the remaining portions of the circuit are tested.

A special portion of the ROM control loads and reads the RAMs. Some asynchronous portions require a third test setup. Again, the connection of the START wire is simply moved to the next pin designated for

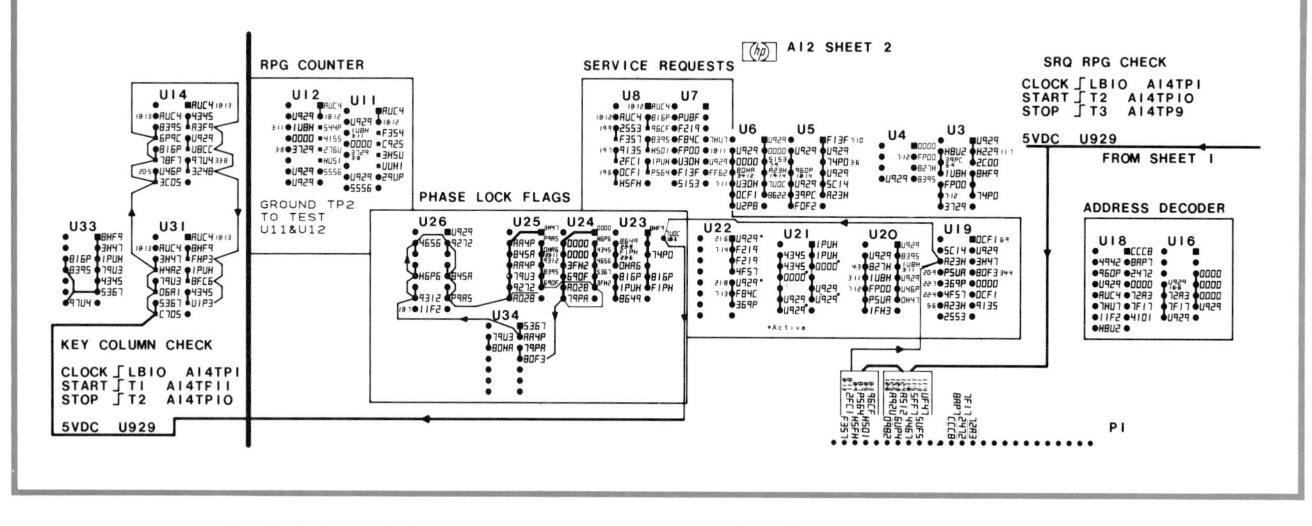

Fig. 6. *A template for signature analysis troubleshooting. The template is attached to the component side of the circuit board. Holes allow probe access to the test points. If the test point is not a source, the origin of the signal (IC and pin number) is listed next to the correct signature.*

this purpose and troubleshooting can continue.

It is obvious that proper documentation is essential. The 3455A manual shows, for each test setup, a picture of the board. Only the signatures related to that particular test are given. This helps to direct the effort towards the important areas on the board. Interrupt signals are simulated by the ROM program so they occur repeatedly at the same spot within a window and stable signatures result.

When all the signatures seem to be bad, the question arises whether the test setup itself is correct. The 5004A Signature Analyzer's self-test feature can be used to check it for proper operation. Each test setup can then be tested by touching V_{cc} with the 5004A probe. If this characteristic signature is correct, it means that the START and STOP channels are triggered at the correct moments and that the number of clock pulses within the measurement window is correct. It also tells the user that the switches on the signature analyzer are set correctly, and that all the jumpers, switches, and control buttons in the voltmeter are set to the right position. Thus the confidence level is very high at the beginning of a test routine.

A conclusion drawn from this application is as follows: success is assured if the service engineer works closely with the design engineer. This also saves time at the end of the development phase because the service engineer is fully aware of the new product's internal operation. It also forces the designer to think about serviceability.

The fact that signature analysis is built into the 3455A Voltmeter not only made serviceability but also final testing on the production line much easier. The signature analyzer is now a standard piece of equipment on the production line.

Fig. 7. *Model 3455A System Voltmeter is the first HP instrument designed for troubleshooting with the 5004A Signature Analyzer.*

Acknowledgments

The 5004A Signature Analyzer is a result of the effort of many engineers in HP's Colorado Springs and Santa Clara Divisions. Loveland Instrument Division and Santa Rosa Division added a great deal to definition of the needs and conditions of the new

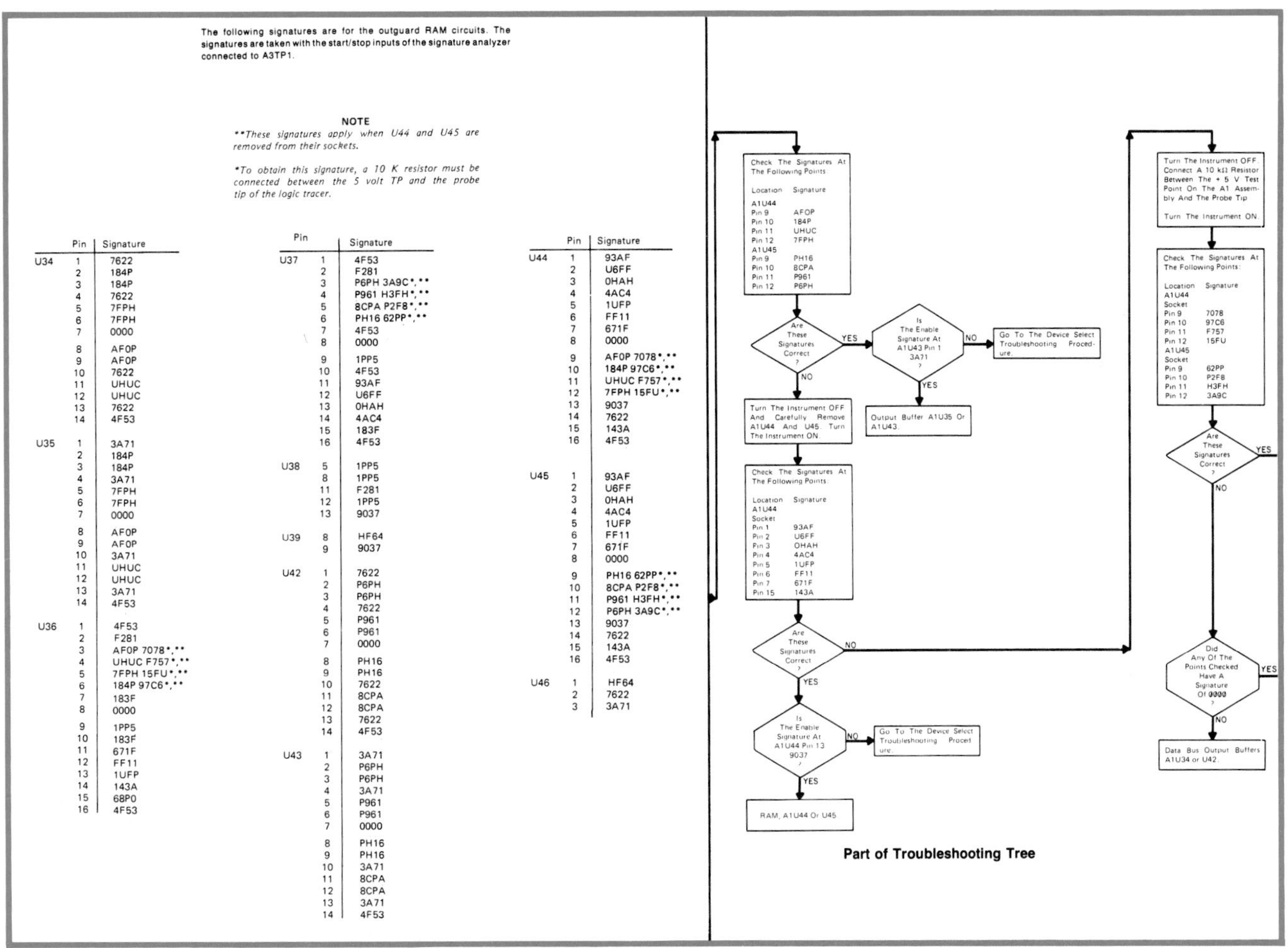

The following signatures are for the outguard RAM circuits. The signatures are taken with the start/stop inputs of the signature analyzer connected to A3TP1.

NOTE

**These signatures apply when U44 and U45 are removed from their sockets.*

**To obtain this signature, a 10 K resistor must be connected between the 5 volt TP and the probe tip of the logic tracer.*

	Pin	Signature
U34	1	7622
	2	184P
	3	184P
	4	7622
	5	7FPH
	6	7FPH
	7	0000
	8	AF0P
	9	AF0P
	10	7622
	11	UHUC
	12	UHUC
	13	7622
	14	4F53
U35	1	3A71
	2	184P
	3	184P
	4	3A71
	5	7FPH
	6	7FPH
	7	0000
	8	AF0P
	9	AF0P
	10	3A71
	11	UHUC
	12	UHUC
	13	3A71
	14	4F53
U36	1	4F53
	2	F281
	3	AF0P 7078*,**
	4	UHUC F757*,**
	5	7FPH 15FU*,**
	6	184P 97C6*,**
	7	183F
	8	0000
	9	1PP5
	10	183F
	11	671F
	12	FF11
	13	1UFP
	14	143A
	15	68P0
	16	4F53

	Pin	Signature
U37	1	4F53
	2	F281
	3	P6PH 3A9C*,**
	4	P961 H3FH*,**
	5	8CPA P2F8*,**
	6	PH16 62PP*,**
	7	4F53
	8	0000
	9	1PP5
	10	4F53
	11	93AF
	12	U6FF
	13	0HAH
	14	4AC4
	15	183F
	16	4F53
U38	5	1PP5
	8	1PP5
	11	F281
	12	1PP5
	13	9037
U39	8	HF64
	9	9037
U42	1	7622
	2	P6PH
	3	P6PH
	4	7622
	5	P961
	6	P961
	7	0000
	8	PH16
	9	PH16
	10	7622
	11	8CPA
	12	8CPA
	13	7622
	14	4F53
U43	1	3A71
	2	P6PH
	3	P6PH
	4	3A71
	5	P961
	6	P961
	7	0000
	8	PH16
	9	PH16
	10	3A71
	11	8CPA
	12	8CPA
	13	3A71
	14	4F53

	Pin	Signature
U44	1	93AF
	2	U6FF
	3	0HAH
	4	4AC4
	5	1UFP
	6	FF11
	7	671F
	8	0000
	9	AF0P 7078*,**
	10	184P 97C6*,**
	11	UHUC F757*,**
	12	7FPH 15FU*,**
	13	9037
	14	7622
	15	143A
	16	4F53
U45	1	93AF
	2	U6FF
	3	0HAH
	4	4AC4
	5	1UFP
	6	FF11
	7	671F
	8	0000
	9	PH16 62PP*,**
	10	8CPA P2F8*,**
	11	P961 H3FH*,**
	12	P6PH 3A9C*,**
	13	9037
	14	7622
	15	143A
	16	4F53
U46	1	HF64
	2	7622
	3	3A71

Fig. 8. *An example of a troubleshooting chart from the 3455A Voltmeter service manual. The chart tells which part to replace under certain conditions.*

microprocessor-based instruments relating to troubleshooting and service.

In addition to the authors of the articles in this issue, key contributors include David Kook, who managed to pack the 5004A into an existing plastic case, and Kuni Masuda, who designed the front panel to give it a well-balanced appearance. Gary Gitzen did the breadboarding and designed the unstable signature feature.

It is also a pleasure to acknowledge the many valuable inputs from Dan Kolody and George Haag in Colorado Springs, Kamran Firooz and David Palermo in Loveland, Jan Hofland who is now with the Data Systems Division, Ed White in our own marketing department, and Dick Harris who brought the instrument into production. Gary Gordon as section manager was instrumental in getting the algorithm implemented in a service tool.

Reference

1. A. Gookin, "A Fast-Reading, High-Resolution Voltmeter that Calibrates Itself Automatically," Hewlett-Packard Journal, February 1977.

Hans J. Nadig

Hans Nadig is signature analysis project manager at HP's Santa Clara Division. Holder of an MS degree in electrical engineering from the Federal Institute of Technology in Zürich, he's been with HP since 1967, contributing to the design of instruments for Fourier analysis and serving as a project manager in the IC laboratory. Born in Berne, Switzerland, Hans served two years in the Swiss Army. He's a rock climber, a qualified instructor and guide in high-altitude mountaineering, and a campaigner for environmental protection, especially through power conservation, in local government circles. He also participates in HP's career counseling program for high-school students, and enjoys photography and gardening. Hans is married, has two daughters, and lives in Saratoga, California.

A Fully Integrated, Microprocessor-Controlled Total Station

Here's a new instrument that measures angles and distances, combines these readings, and yields true three-dimensional position information.

by Alfred F. Gort

THE MEASUREMENT OF DISTANCE by electronic techniques has become well established in recent years. This technology has been combined with existing theodolites—optomechanical angle-measuring instruments—to form three-dimensional measurement systems. Fully integrated total stations that measure both distance and angle electronically have been developed by Hewlett-Packard[1] and several other companies.

The performance of the optomechanical theodolite with an optical micrometer has been difficult to match with electronic encoder systems. Until recently the encoder systems have been considerably larger than optomechanical angle measurement systems and have not reached second-order accuracy in most cases. Because of the increasing demand for speed and accuracy, more interest now exists in combining the distance and angle measuring functions into a single instrument with arc-second accuracy for angles and accuracy to several millimetres for distance.

Hewlett-Packard's answer to this need is Model 3820A Electronic Total Station, Fig. 1. To make it a reality, several new subsystems had to be developed:

- An optical system that functions as the transmitting and receiving optics for the distance meter and as the sighting telescope for the theodolite.
- An electronic angle-measuring system comparable in size and accuracy to a second-order theodolite.
- A two-axis gravity-sensing device to provide vertical index and horizontal angle correction.
- A miniaturized distance-meter module with greatly reduced power consumption and a five-kilometre range.
- An electronic system and microprocessor to control the instrument, perform necessary computations, and output data to a peripheral for recording or processing.
- A structural frame and bearing system with the stability required for a second-order theodolite.

Optical System

The layout of the optical system for the 3820A is shown in Fig. 2. A *catadioptric* telescope with a 66-mm clear aperture is used. The design provides sufficient area for the distance meter's transmitting and receiving beams in a short telescope length. Since the majority of the telescope's power is in the reflector, the system has excellent color correction and exhibits no *secondary spectrum*. The *Mangin mirror* and corrector lens form an objective that is well corrected for *spherical aberrations* and *coma* over a 1.5° field. The 30× telescope uses a *Pechan prism* to erect the image. A symmetrical eyepiece gives a sharp field at full angle and a 12-mm eye relief. The *reticle* is illuminated for night work.

The optics system acts as an eight-power *Galilean telescope* for the distance meter. The distance meter incorporates a double heterojunction GaAs lasing diode, a chopper system, and a reference path. A beam splitter is used to reflect the infrared light into the distance-meter module while transmitting the visible spectrum to the eyepiece.

Electronic Measurement and Control System

Four transducers feed measurement data to the central microprocessor. The transducers are the distance-measuring module, the horizontal-angle encoder, the vertical-angle encoder and the tilt meter. The microprocessor controls the transducers via the I/O module, which has an eight-bit control bus and an eight-bit data bus (see Fig. 3).

The two angle encoders are optically and electronically identical and each one consists of three analog interpolation circuits plus an eight-bit digital sensor. The analog signals from the angle encoders, tilt meter, and distance meter use a common phase detector. Angle, tilt, and distance interpolation are accomplished by phase measurement. The desired transducer is selected for input to the phase detector by a control gate from the I/O module.

The microprocessor is a 56-bit serial processor with a ten-bit instruction word. A masked ROM contains 4096 of these ten-bit instructions. Ten 56-bit words can be stored in the data storage chip (RAM). This RAM stores the last measurement of each function in a dedicated location.

The instrument has two identical keyboard and display units (Fig. 4), one on the front and the other on the back of the telescope mount. This was done for user convenience when taking direct and reversed telescope readings *(plunging)* for high-accuracy measurements.

Measurement data can be transmitted via a special interface, the digital-output module, to a peripheral. This interface includes two-way handshake signals and transmits via five sliprings to the fixed base. The 38001A HP-IB* Distance Meter Interface converts these signals to an HP-IB-compatible format to facilitate interfacing to data processing systems.[2]

Angle Measurement System

Angles are electronically read from a glass disc with a metal-film pattern deposited on it. Since the *zenith* angle

Note: All terms in italics in this article are defined in the glossary on page 11.

*Compatible with IEEE Standard 488-1978.

must be an absolute value referenced to gravity, an absolute reading system was chosen instead of an incremental one. The identical system is used for both the horizontal and vertical angles.

The encoder disc used in the 3820A is shown in Fig. 5. The optical sensing system for reading this disc is illustrated in Fig. 6. The angular position on the disc is found by combining three separate measurements:

1. The instrument first measures an eight-bit *Gray-code* pattern that determines position to one part in 256. This is similar to reading the degree mark on a theodolite circle.
2. Next, the instrument interpolates a sinusoidal-track pattern of 128 cycles by dividing each cycle into 1000 parts. Thus, the circle is divided into 128,000 increments, each corresponding to an angular variation of approximately ten arc seconds.
3. Finally, the instrument interpolates the position on a track comprised of 4096 radial slits. Again, the period for each slit is subdivided into 1000 parts. This divides the circle into 4,096,000 parts, resulting in a true one-cc-grad (0.32 arc-second) resolution.

The 4096-radial-slit pattern is read at diametrically opposed points on the circle to eliminate eccentricity errors. The reading of the sinusoidal track interpolator is also corrected for the eccentricity sensed by the 4096-slit track. The microprocessor combines the readings of the three sensor systems to produce an absolute reading.

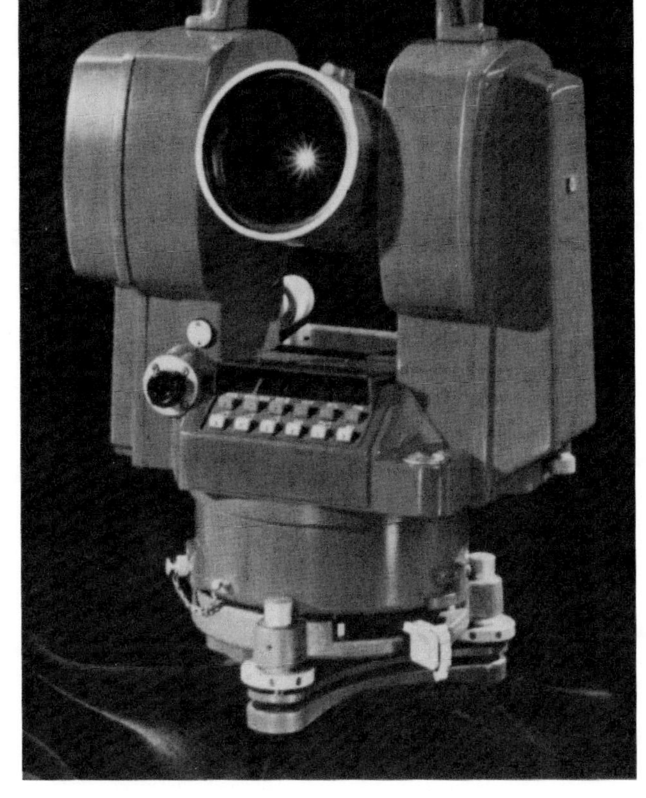

Fig. 1. *Model 3820A Electronic Total Station. This instrument combines electronic distance and angle measurement capabilities into a compact package. The values obtained with the 3820A are accurate to a few millimetres for distances up to five kilometres and to a few arc seconds for horizontal and vertical angles.*

To illustrate the principle of interpolation, consider the sinusoidal track. Fig. 7a shows the track, which varies sinusoidally in width. The wavelength of the pattern is 1080 μm and the maximum amplitude is 600 μm. Four photodiodes are placed at 90° intervals with respect to the sinusoidal period. Each diode senses the collimated illumination through the pattern. The photocurrent generated depends on the illuminated area of the diode, which in turn is dependent on the diode's position with respect to the pattern.

The relationship between photocurrent and position is derived as follows (see Fig. 7b):

$$I_1=I_0+I\sin(\phi) \quad (1)$$
$$I_2=I_0+I\sin(\phi+90)=I_0+I\cos(\phi) \quad (2)$$
$$I_3=I_0+I\sin(\phi+180)=I_0-I\sin(\phi) \quad (3)$$
$$I_4=I_0+I\sin(\phi+270)=I_0-I\cos(\phi) \quad (4)$$

The illumination amplitude I is modulated as a function of time: $I(t)=I\sin(\omega t)$. Substituting into (1) through (4) and then subtracting (3) from (1) and subtracting (4) from (2) yields:

$$I_1-I_3=2I\sin(\phi)\sin(\omega t) \quad (5)$$
$$I_2-I_4=2I\cos(\phi)\sin(\omega t) \quad (6)$$

These two signals are processed in such a manner that (I_1-I_3) is phase-shifted 90° with respect to (I_2-I_4). The resulting signals are then summed in an operational amplifier (see block diagram, Fig. 3).

$$2I\sin(\phi)\sin(\omega t)+90^\circ\text{ shift}=2I\sin(\phi)\cos(\omega t)$$
$$2I\cos(\phi)\sin(\omega t)+0^\circ\text{ shift}=2I\cos(\phi)\sin(\omega t)$$
$$2I\sin(\phi)\cos(\omega t)+2I\cos(\phi)\sin(\omega t)=2I\sin(\omega t+\phi) \quad (7)$$

The signal (7) is compared in a phase detector with the modulating signal $\sin(\omega t)$ to determine ϕ. The phase meter has a resolution of 0.36 degrees, resulting in an interpolation of one part in 1000 for the period.

The above derivation assumes zero sensor width. However, it can be shown that the equations hold for a finite-width sensor. This is because the convolution integral of a sine function with a rectangle function is still a sine or cosine function although the amplitude is changed.

A typical error graph for the encoder system is shown in Fig. 8. Fourier analysis of the graph shows an interpolation error of 3.1 cc grad, a once-around error of 2.6 cc grad, and a twice-around error (graduation error) of 4.2 cc grad. The once-around error can be eliminated by plunging and the twice-around error may be averaged out by incrementing the circle. Overall, the total standard deviation of the error is below one arc second, making the 3820A an excellent tool for second-order measurement work.

Gravity Sensing System

Traditionally, a theodolite has been equipped with a twenty-arc-second level vial parallel to the *trunnion* axis for precise leveling and a vertical compensator to correct zenith angles for residual tilt. Precise leveling is necessary to maintain horizontal-angle accuracy for steep vertical angles. In the 3820A, the vertical compensator and trunnion-axis-plate level are combined in a two-axis tilt sensor. This device eliminates the need for precise leveling.

The system is basically a two-axis electronic *autocol-*

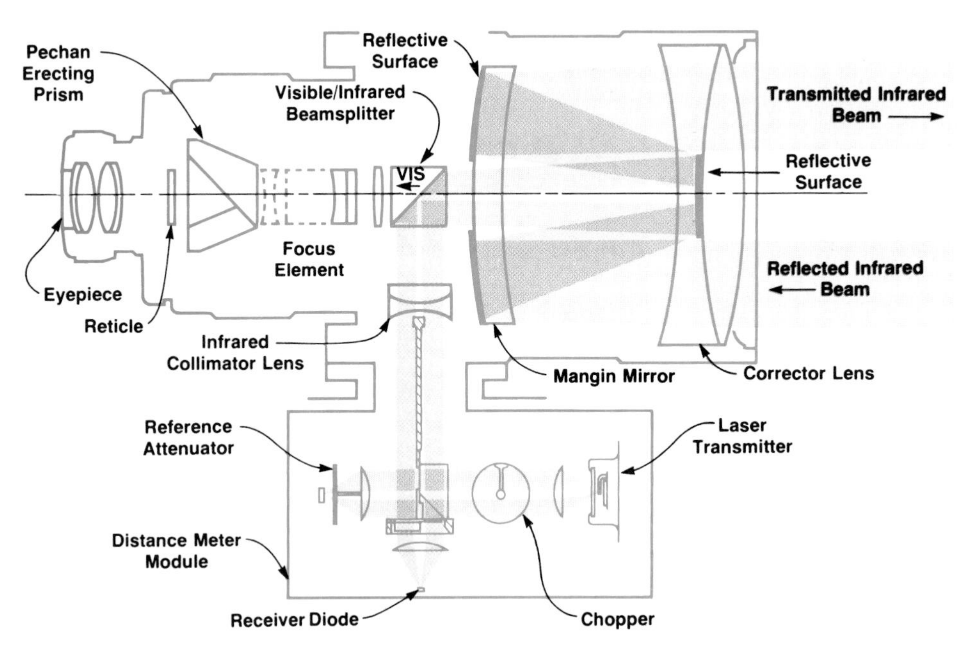

Fig. 2. *The 3820A optical system uses a compact telescope design for target sighting. An infrared laser distance measurement system shares the use of the primary optical elements.*

limator. Fig. 9 shows the optical layout of the tilt sensor. A mercury pool damped with silicone oil is used to establish the vertical reference. The surface of the mercury pool serves as a reflector that is always perpendicular to the direction of gravity. The enclosure consists of an anti-reflection-coated optical flat joined to a metal cup. This arrangement accommodates thermal expansion. The three-element illuminator lens provides highly uniform illumination of the transparent sinusoidal slits (Fig. 10). A negative lens, a positive lens, and the mercury reflector form the imaging system. The effective focal length of the system is 163 mm.

To determine the level within one cc grad, the pattern position is read to 0.7 μm accuracy. The interpolation technique is the same as used for angle measurement. Two combinations of a transparent sinusoidal slit and its four photodiode sensors are arranged orthogonally for the two axes. To prevent erroneous readings when the interpolators exceed their range, a limit sensor is incorporated. When in range, the limit sensor remains illuminated. If the range is exceeded, the limit sensor is no longer illuminated, and a display indicator flashes to inform the operator. Cross-axis

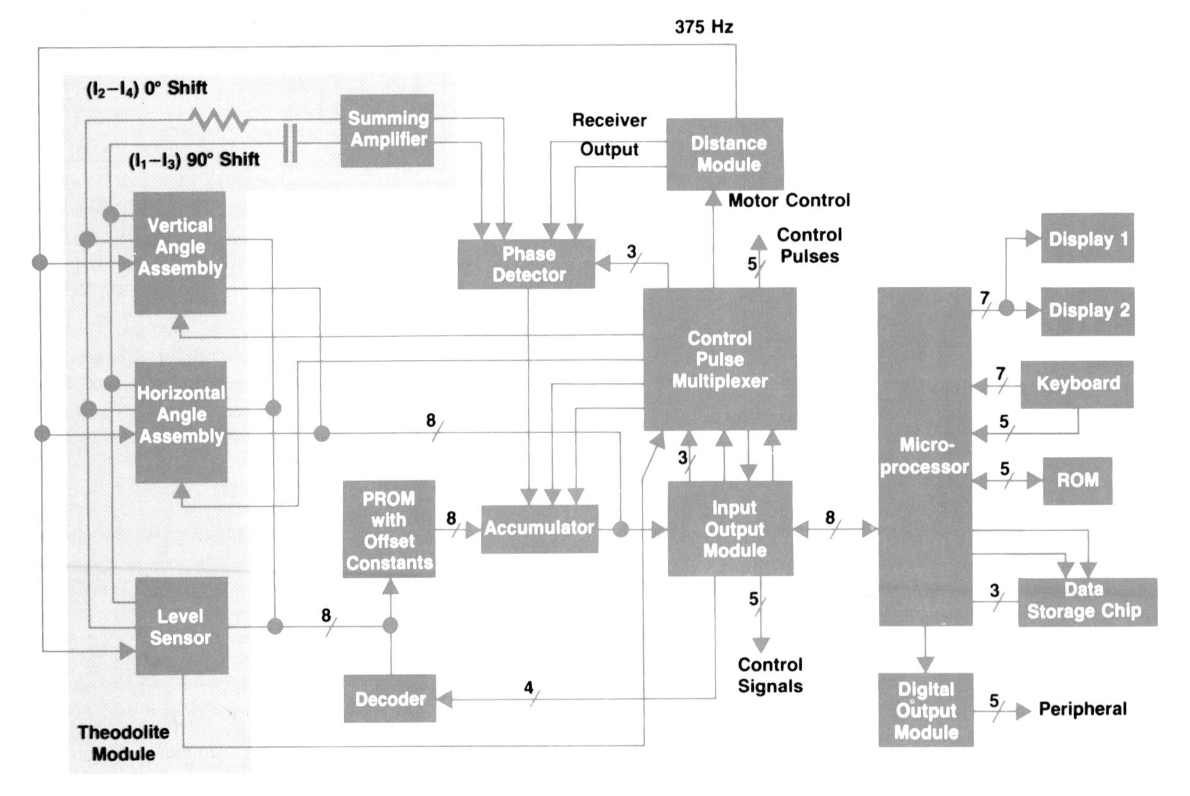

Fig. 3. *Electronic block diagram for 3820A Electronic Total Station.*

Fig. 4. *Close-up view of 3820A keyboard and display. There are two of these units, one on each side of the instrument, for user convenience.*

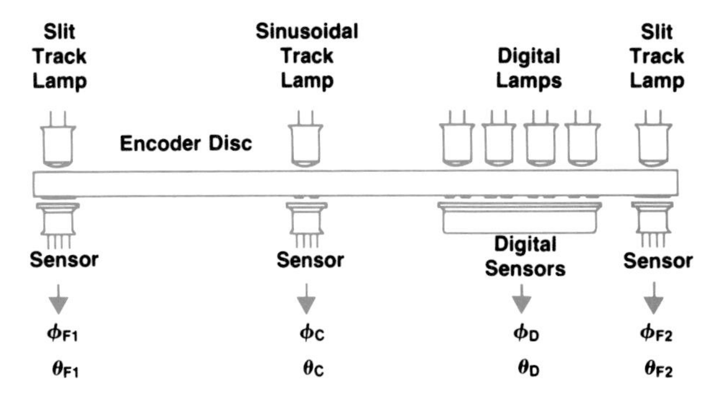

Fig. 6. *Arrangement of the optical sensors for reading the three concentric patterns on each angle encoder disc (ϕ=vertical, θ=horizontal disc components). Eccentricity errors are eliminated by reading the outermost pattern at two points located across the circle from each other (F1, F2 values).*

movement combined with the one-millimetre height of the photodiodes limits the range to ±370 cc grad (±120 arc seconds). A typical error graph for the tilt sensor is shown in Fig. 11.

Distance Meter System

The main parts of the distance meter are shown in block diagram form in Fig. 12. Although similar in principle to recent HP instruments,[3] the hardware differs greatly on the following points:

- The optical system and sighting telescope are combined.
- Physical size is greatly reduced.
- Distance meter power consumption is reduced to 1.5 W.
- A heterojunction continuous-wave GaAs lasing diode is used.

As in previous models, the distance meter contains an automatic sampling system with an internal path length and an automatic balance system to match the energy through the reference path to the energy received from the target. This matching method enables the instrument to handle a wide range of return-signal strengths—a requirement for longer-range instruments.

The infrared energy is modulated so that the phase difference between the internal reference beam and the portion of the transmitted beam that is reflected back from the target

Fig. 5. *The angle encoder disc has three concentric metal film patterns. The combination of the readings from each of these patterns allows for the determination of angular position with one-cc-grad (0.32-arc-second) resolution.*

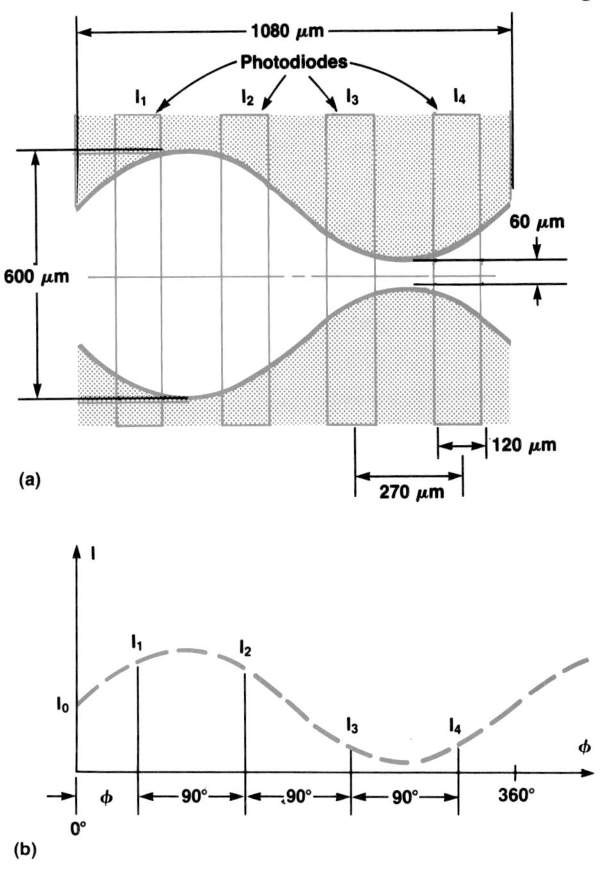

Fig. 7 *(a) Section of the sinusoidal track pattern on the angle encoder disc. The amount of illumination passing through the pattern to each of the regularly spaced photodiodes generates different photocurrents (b) that enable the determination of the pattern position relative to the diode array.*

ccgrad arcsec
+10 +3.24
Error 0 0
−10 −3.24
0 50 100 150 200 250 300 360
Angle (degrees)

Total Rms Error: 2.7 ccgrad (0.9 arcsec)

Fig. 8. *Typical interpolation error versus angular position of encoder disc.*

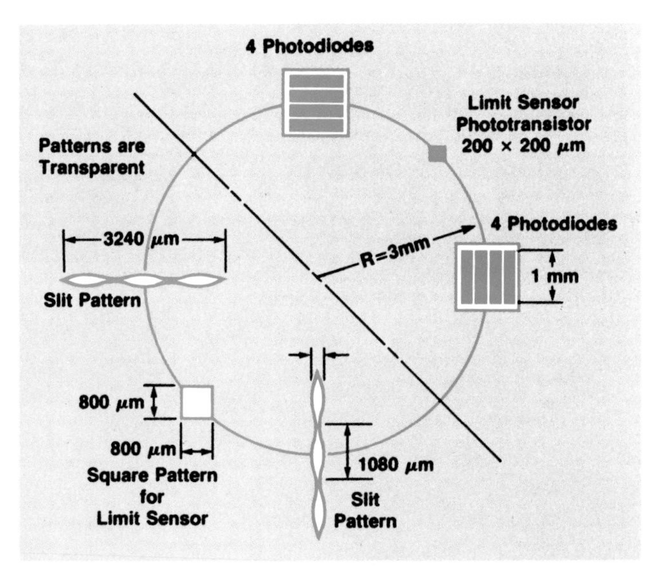

Fig. 10. *Arrangement of slit patterns and optical sensors for tilt meter. The tilt position measurement technique is the same as that used for angle measurement.*

can be measured to determine the distance to the target. Three modulation frequencies are used—15 MHz, 375 kHz and 3.75 kHz. A 360° phase shift for these frequencies corresponds to distance variations of 10, 400, and 40,000 metres, respectively. The readings for each frequency are then combined to give the absolute distance.

These modulation frequencies and the instrument readout give unambiguous distance displays for exceptionally long distances. Distances over five kilometres may be measured under ideal atmospheric conditions.

The receiver detects the returning infrared radiation with a photo-avalanche diode. The diode also functions as a mixer and provides a gain of 75. The transmitting diode is a GaAs lasing diode developed by Hewlett-Packard for the purpose of distance measurement. Fig. 13 shows the laser

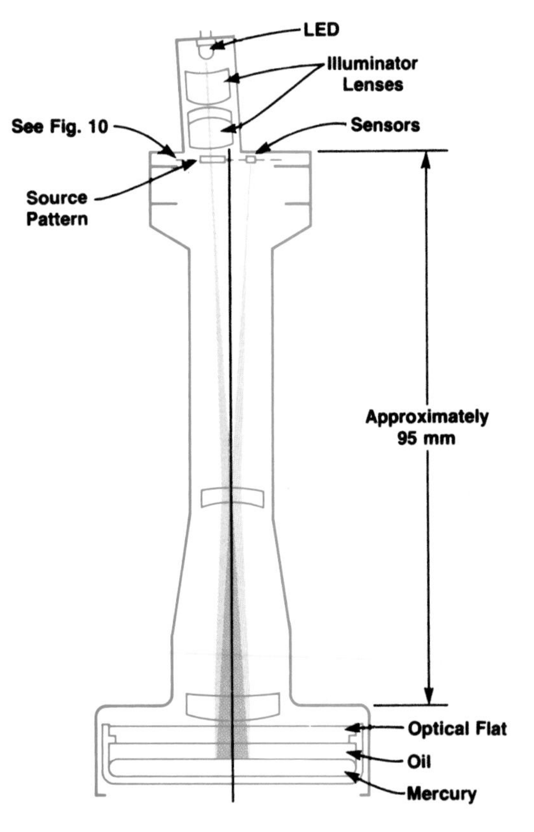

Fig. 9. *Optical system for 3820A tilt sensor. The mercury pool at the bottom establishes a reflective plane that is perpendicular to the force of gravity.*

modulation and control system. An optical feedback loop stabilizes the laser operating point over a wide temperature range. The laser and its control loop are housed inside a hermetically sealed metal package with an optically flat window. Because of the high sensitivity of the receiver and the high radiance of the laser beam, the instrument range is five kilometres with a six-prism retroreflector assembly. To prevent input circuit overload, it is necessary to use an attenuator on the telescope objective for distances less than 250 metres. Alternatively, less efficient reflectors may be used. Since the long-range accuracy depends largely on the accuracy of the modulation frequency, a low-temperature-coefficient crystal is used. The crystal stability specification is ±4 ppm from −10°C to +40°C.

The group refractive index has been derived for standard air (15°C and 760 mmHg) and a laser radiation wavelength of 835 nm using two different references (see page 9):

$n_g - 1 = 279.34 \times 10^{-6}$ where n_g is the group index (8)

$c = 299{,}792.5$ km/s (9)

resulting in a modulation frequency with zero-ppm correction of 14.985439 MHz for standard air. In the 3820A, the frequency is set at +110 ppm, or 14.987087 MHz.

Microprocessor Functions

The 3820A uses a microprocessor derived from HP

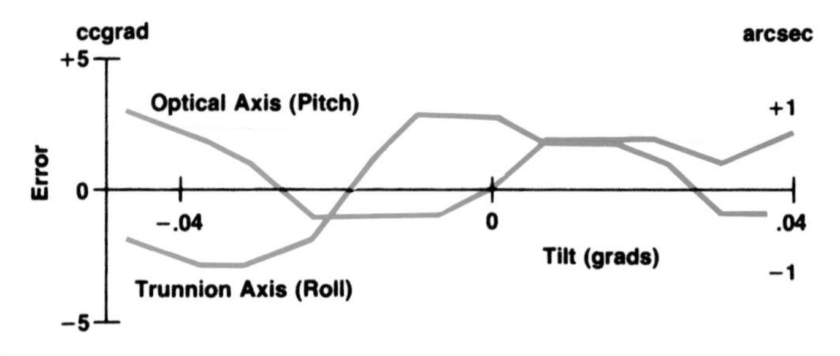

Fig. 11. *Typical axis tilt interpolation errors versus tilt angle for tilt sensor.*

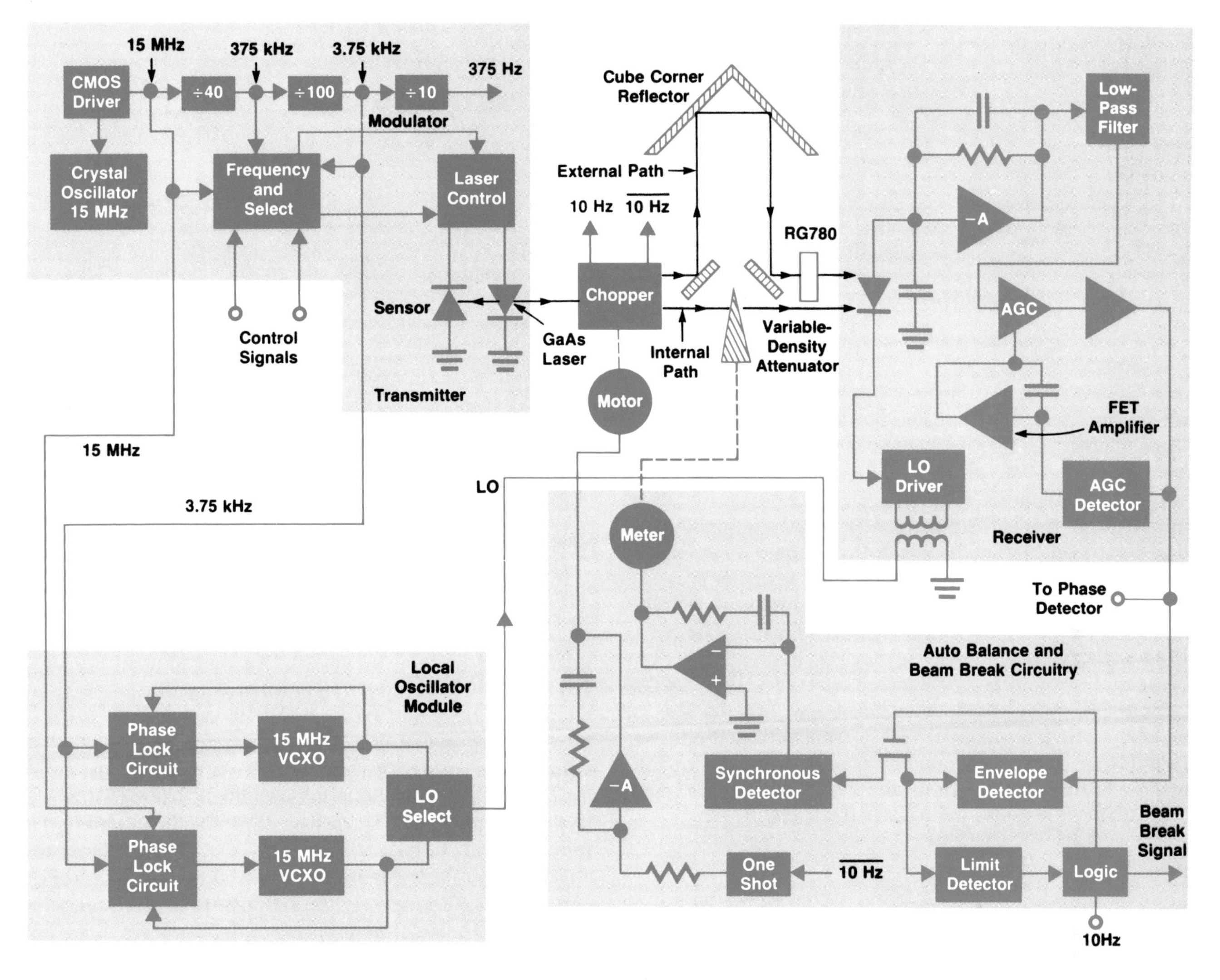

Fig. 12. *Distance meter block diagram. This assembly can measure distances up to five kilometres with ±(5 mm + 5 mm/km) accuracy over a temperature range of −10° to 40°C.*

pocket-calculator technology to perform various functions within the instrument. Among these functions are the following:

- Control various measurement sequences and the display.
- Process the intermediate results from the angle encoders, tilt meter, and distance meter.
- Perform numerous calculations and corrections such as compensating angles for instrument tilt and computing projected distances.
- Provide an internal self-test sequence, which checks for the presence of many internal signals in the angle encoders, tilt meter, and distance meter.

The routines needed for control and computations are stored in a 4K-byte ROM which is roughly divided into four parts of 1K bytes each, corresponding to:

- Distance measuring routines
- Angle measuring routines
- Self-test and service tests
- Keyboard, display and control routines.

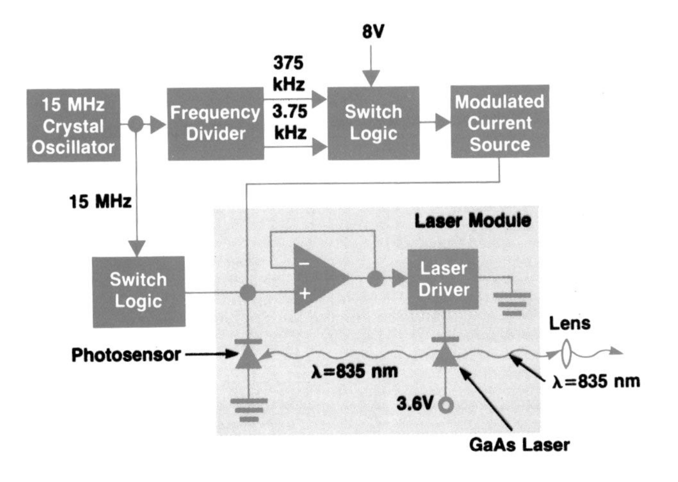

Fig. 13. *Laser control and modulation system. The system uses a custom GaAs lasing diode developed by Hewlett-Packard for this purpose.*

Multifunction Sequence

An example of the control function of the microprocessor

(continued on page 10)

Distance Correction for Variations in Air Temperature and Pressure

In an electronic distance meter the distance is derived from the amount of time required for the transmitted infrared beam to make the round trip to the retroreflector and back. Therefore, it is necessary to accurately know the velocity of the infrared radiation in air. In general, an unmodulated plane wave will have a velocity

$$v = c/n \qquad (1)$$

where c is the velocity of light in a vacuum and n is the index of refraction for the medium. The refractive index for air depends on the radiation wavelength as well as the density of the air. The wavelength dependence can be approximated by[1]

$$(n-1) \times 10^8 = 27259.9 + \frac{153.58}{\lambda^2} + \frac{1.318}{\lambda^4} \qquad (2)$$

where λ is the radiation wavelength in micrometres. This expression is for standard air, which is dry air at 15°C and 760 mmHg pressure.

For an amplitude modulated wave the amplitude maxima propagate with a group velocity u. The group velocity is related to the phase velocity v by[2]

$$u = v - \lambda \frac{dv}{d\lambda} \qquad (3)$$

Now, define the refractive index for the group velocity, n_g, as c/u. For air it can be assumed that

$$\frac{\lambda\, dv}{v\, d\lambda} << 1$$

Thus, the following approximation can be made.

$$\frac{1}{u} = \left[v\left(1 - \frac{\lambda dv}{v d\lambda}\right)\right]^{-1} \approx \frac{1}{v}\left[1 + \frac{\lambda dv}{v d\lambda}\right] \qquad (4)$$

From (1)

$$dv = -\frac{c}{n^2} dn \qquad (5)$$

Then, using (1), (4) and (5) we can derive the following result

$$n_g = \frac{c}{u} = \frac{c}{v}\left[1 - \frac{\lambda c}{vn^2}\frac{dn}{d\lambda}\right] = n - \lambda \frac{dn}{d\lambda} \qquad (6)$$

Combining (2) and (6) yields

$$(n_g - 1) \times 10^8 = 27259.9 + \frac{460.74}{\lambda^2} + \frac{6.59}{\lambda^4} \qquad (7)$$

For the laser wavelength used in the 3820A, 0.835 μm, this becomes

$$(n_g - 1) \times 10^8 = 27934.28 \qquad (8)$$

Therefore,

$$n_g = 1.0002793428$$

Since c = 299792.5 kilometres/second, the group velocity is then c/n_g = 299708.8 kilometres/second. The maximum modulation wavelength, λ_m, desired for the instrument is 20 metres so that the modulation frequency, f_m, will exhibit 360° of phase shift for every 10 metres of distance between the instrument and the retroreflector. Therefore, the highest modulation frequency is

$$f_m = \frac{u}{\lambda_m} = 14985439 \text{ Hz}$$

The influence of water vapor in the atmosphere on group velocity has been neglected since this effect is believed to be less than one ppm for near-infrared radiation.

Temperature and Pressure Correction

It is assumed that the refractive index for air varies from the ideal value of 1.0 for vacuum in proportion to the density of the air, ρ, such that

$$(n_g - 1)\Big|_{\rho_2} = \frac{\rho_2}{\rho_1} \times (n_g - 1)\Big|_{\rho_1} \qquad (9)$$

for small variations in density. Since it is also assumed that air behaves according to the ideal gas law, pV = RT, and knowing that the density is inversely proportional to the volume, V, of the air, then the refractive index for any temperature T_2 and pressure p_2 can be found by

$$(n_g - 1)\Big|_{p_2, T_2} = \frac{V_1}{V_2} \times (n_g - 1)\Big|_{p_1, T_1} \qquad (10)$$

given the value at a temperature T_1 and pressure p_1. Solving for V_1 and V_2 using the ideal gas law and using (8) for the refractive index in standard air (p_1 = 760 mmHg, T_1 = 273.2 + 15°K), expression (10) becomes

$$(n_g - 1)\Big|_{p,T} = 279.34 \times 10^{-6} \times \left[\left(\frac{p}{760}\right) \times \left(\frac{273.2+15}{237.2+T}\right)\right]$$

$$= \frac{105.92p}{(273.2+T)} \times 10^{-6}$$

where p is the pressure in mmHg and T is the temperature in °C. From this, it can be derived that the distance correction is

$$\text{Correction in ppm} = 279.34 - \frac{105.92p}{273.2+T} \qquad (11)$$

This correction can be entered into the 3820A via a front-panel control. The microprocessor then applies this correction to all subsequent distance measurements.

References

1. Froome and Essen, *The Velocity of Light and Radio Waves*, Academic Press, 1969, pp 24-28.
2. Born and Wolf, *Principles of Optics*, 2nd Ed., MacMillan, 1964.

is the routine for the **MULTI** key. This key activates a measurement sequence that combines angle and distance readings for a three-dimensional measurement result. The sequence is:

- measure and output the horizontal angle
- measure and output the zenith angle
- measure and output the slope distance.

To accomplish this complex control function the transducers have to be designed for control by a processor and also have to contain the hardware necessary for communication with the processor. Examples of these requirements are the eight control lines for the circle interpolators and the accumulator that functions as the common analog-to-digital converter for all systems. The control and reception of data from the transducers is done via the I/O module. This unit is the communication link between the processor and the measurement systems.

With the multifunction sequence it is also possible to make repeated measurements at a rate of one full sequence every 2.7 seconds. With the 3820A, the 38001A HP-IB interface and a data processor (e.g., 9825A) one can assemble a system that can track the position of slow moving targets, assuming that one can keep the telescope aimed at the target manually.

Angle Correction

The 3820A Electronic Total Station is the only instrument presently available that compensates both horizontal and vertical angles for instrument tilt. It is only necessary to level the instrument within 120 arc seconds to maintain full angular accuracy. However, leveling to one arc minute is recommended to avoid positioning errors using the optical *plummet*. This is easily and quickly accomplished using the circular bubble level on the *alidade* of the instrument. The corrections for the remaining tilt of the vertical axis with respect to gravity are:

$$\text{Zenith angle correction} = \Delta\phi = \beta + \tfrac{1}{2}\delta^2 \cot(\phi) \quad (10)$$

$$\text{Horizontal angle correction} = \Delta\theta = \delta\cot(\phi) - \tfrac{1}{2}\beta\delta\left[1 + 2\cot^2(\phi)\right] \quad (11)$$

in which: β = tilt along the telescope axis
δ = tilt along the trunnion axis
ϕ = zenith angle

Analysis of (10) shows that the second order term for $\Delta\phi$ can be neglected. For example: if $\beta = \delta = 120$ arc seconds (the limit of the tilt meter), the second order term for $\Delta\phi$ is only three arc seconds at a zenith angle of 1°. At a zenith angle of 45°, the term is less than 0.1 arc second.

Both terms of equation (11) may be significant. In the 3820A, only the first-order correction is applied when the line of sight is within 45° of horizontal. Beyond this range, both terms are used. To illustrate the importance of the horizontal-angle correction, consider that at a zenith angle of 75°, every four arc seconds of tilt introduces one arc second of error. By automatically making this correction, the 3820A eliminates the need for precise leveling.

Distance Consistency Check

As in previous HP distance meters, the 3820A incorporates a consistency check on the distance data. While distance data is being accumulated, the processor keeps a running total of the mean and variance of the readings. If the

Development of the 3820A

The development of an instrument of this complexity can only be accomplished when many favorable factors coincide. Existing technology in light-emitting devices, photodiodes, photolithography and electronic microcircuits at the start of the development project indicated that it was possible to design a compact fully-integrated instrument measuring both angles, tilt and distance.

Initally a small group was formed to investigate methods for angle measurement and design the basic optical system of the telescope and distance meter. This early group included Charles Moore who contributed much to both the optical and electronic designs, Walt Auyer who designed the mechanical elements of the angle transducers, Billy Miracle, our mechanical design leader, who designed the temperature-compensated objective cells for the telescope, Jim Epstein who designed most of the digital control system, including the software design for the microprocessor control, and Ron Kerschner who designed the mechanical systems for the distance meter and level sensor.

This team later became the nucleus of a much larger team and was aided by many specialists. Tom Christen and Hal Chase took over the design of the thin-film microcircuits needed to construct the angle transducers, tilt meter and distance meter. The industrial design of the instrument was being firmed up during this time by Arnold Joslin who made significant contributions to keeping the instrument compact and portable.

When the group was increased to full strength, Dave Daniels-Lee and Sanford Baran took on the majority of the analog circuit design while Craig Cooley joined the team to design the main structural frame and side covers of the instrument. Towards the end of the design phase Craig Cooley also designed the leveling base. With Billy Miracle taking on a larger part of the project management in the mechanical area, Dave Sims joined the group to design the mechanical parts for the main telescope. Dave also designed the carrying case for the 3820A. This case provides a high degree of protection in a compact size.

Besides the central design team, many support groups helped to realize this design. The tooling effort, involving several people, was coordinated and guided by Wilbur Saul. Administrative help was coordinated by Rod Lampe and Vicki Worden who controlled parts supply, material lists and specification drawings. Market research was performed by the marketing group in the Civil Engineering Division. Fritz Sieker and Tony Robinson contributed greatly in the areas of keyboard definition and software routines to be used for data reduction and correction. Corporate engineering and HP Labs assisted in the angle-encoder investigation and also developed the solid-state GaAs heterostructure laser that gave the distance meter its 5-km range. Special recognition is deserved by the group in HP Labs who improved the laser characteristics to make it suitable for the demanding application of electronic distance measurement. The encouragement received from Barney Oliver and Bill Hewlett helped the design teams overcome difficult hurdles.

During the production effort, a large group of new people was involved on the project. Mike Bullock served during the transition period as project manager until the production staff under the leadership of Mike Armstrong and Jim White took charge of production.

Finally special mention should go to Bill McCullough, Division Manager, and Bill Smith, Lab Manager, who believed in the project's ultimate promise and supported the effort with their guidance during the development period.

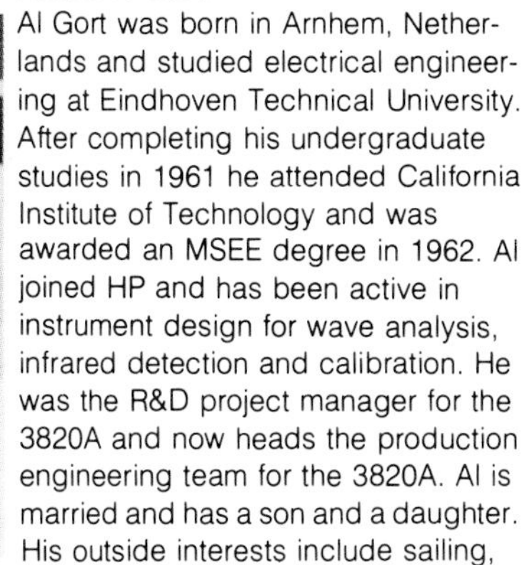

Alfred F. Gort

Al Gort was born in Arnhem, Netherlands and studied electrical engineering at Eindhoven Technical University. After completing his undergraduate studies in 1961 he attended California Institute of Technology and was awarded an MSEE degree in 1962. Al joined HP and has been active in instrument design for wave analysis, infrared detection and calibration. He was the R&D project manager for the 3820A and now heads the production engineering team for the 3820A. Al is married and has a son and a daughter. His outside interests include sailing, hiking, backpacking, cross-country skiing and growing cherries.

Glossary

ACHROMATIC. Transmits light without separating it into its constituent colors.
ALIDADE. A sighting device used for the measurement of angles. Also the mechanical structure of a THEODOLITE. See page 12.
ASPHERIC. A mirror or lens surface that varies slightly from a true spherical surface. This is done to reduce lens aberrations.
AUTOCOLLIMATION. A process of aligning a telescope's line of sight perpendicular to a mirror's surface. The telescope is used to project an image of a pattern toward the mirror. By superimposing the image reflected back by the mirror onto the original pattern in the telescope, the mirror and telescope are properly aligned.
BOUWERS. An optical design, named after the originator, in which reflections take place at two silvered concentric spherical surfaces. Since both surfaces have a common focal point, SPHERICAL ABERRATIONS are greatly reduced.
CASSEGRAIN. A reflecting telescope in which the long optical path is folded by reflecting the incoming light from a paraboloidal primary mirror onto a small hyperboloidal secondary mirror that in turn reflects the light back through a hole in the center of the primary mirror to an eyepiece.
CATADIOPTRIC. Optical processes using both reflection and refraction of light.
CHROMATIC ABERRATION. An optical lens defect that causes light color separation because the optical material focuses different light colors at different points. A lens without this defect is said to be ACHROMATIC.
COMA. A symptom of the presence of optical errors, so that a point object has an asymmetrical image (looks like an egg-shaped spot).
DIOPTER. A measure of lens power equal to the reciprocal of the lens focal length in metres.
GALILEAN TELESCOPE. A telescope using optical refraction. Its primary lens is convex and converges the incoming light. The eyepiece is a concave lens that diverges the beam from the primary lens and presents an erect image.
GRAY CODE. A modified binary code. Sequential numbers are represented by binary expressions in which only one bit changes at a time; thus errors are easily detected.
MANGIN MIRROR. A mirror in which the shallower surface of a negative MENISCUS LENS is silvered to act as a spherical mirror. The light traveling through the other surface and the glass to the mirror is then corrected by the glass for the SPHERICAL ABERRATION of the mirror.
MENISCUS LENS. A thin lens with one convex and one concave surface. The surface with the greatest radius of curvature is the convex surface for a positive lens and the concave one for a negative lens.
PARAXIAL. Lying near the axis.
PECHAN PRISM. A prism using two glass elements that shortens an optical path by reflecting a light beam internally five times. The exiting image is erect. Also known as a Schmidt prism.
PLUMMET. A device for centering a THEODOLITE over a specific location. In the 3820A this is done optically by looking through a sight in the base and centering the internal crosshairs on the location required.
PLUNGING. A technique for canceling some of the mechanical errors in a THEODOLITE. Plunging involves measuring the angles to a target twice. The target is sighted and the angles are measured. Then the ALIDADE is rotated 180°, the telescope is flipped over and the angles are measured again. The sign of the error changes between the two readings, but not the magnitude. By averaging the two readings, the error is eliminated.
RAMSDEN EYEPIECE. An eyepiece assembly using two plano-convex lenses of identical power and focal length. They are mounted with their planar surfaces facing out at each end and are separated by a distance equal to their common focal length.
RETICLE. A pattern of intersecting lines, wires, filaments, or the like placed in the focus of the objective element of an optical system. This pattern is used for sighting and alignment of the system.
RETROREFLECTOR. A device using prisms or an arrangement of mirrors to reflect light radiation back in a path parallel to the incident path.
SECONDARY SPECTRUM. The remaining CHROMATIC ABERRATION for an ACHROMATIC lens. The corrective techniques used for the lens are not equally effective for the entire color spectrum so that some regions will exhibit some color errors.
SPHERICAL ABERRATION. The optical error introduced by the fact that incident rays at different distances from the optical axis are focused at different points along the axis by reflection from spherical mirror surfaces or refraction by spherical lenses.
THEODOLITE. An optical instrument for measuring vertical and horizontal angles from a specific location to a distant target.
TRUNNION. An axle or pivot mounted on bearings for tilting or rotating the object it supports. See Fig. 1 on page 12.
ZENITH. A point directly overhead. Zenith angles are angles measured from this point.

variance is within an internal limit, the mean is displayed as the result. If the variance exceeds this limit, the mean is displayed and flashed to indicate a marginal result. Finally, if no reading can be made, a flashing zero is displayed. A more complete description of the basic process is given in the paper by White.[4] While the process in the 3820A differs in some details, it is essentially the same.

References

1. M.L. Bullock and R.E. Warren, "Electronic Total Station Speeds Survey Operations," Hewlett-Packard Journal, April 1976.
2. D.E. Smith, "A Versatile Computer Interface for Electronic Distance Meters," Hewlett-Packard Journal, June 1980, p. 5.
3. D.E. Smith, "Electronic Distance Measurement for Industrial and Scientific Applications," Hewlett-Packard Journal, June 1980.
4. J.W. White, "The Changing Scene in Electronic Distance Meters," paper presented at the ACSM Annual Meeting, St. Louis, Missouri, March 1974.

SPECIFICATIONS
HP Model 3820A Electronic Total Station

DISTANCE
RANGE (under good conditions—those found during the day when minimal heat shimmer is evident):
1 km (3300 ft) to a single-prism assembly
3 km (9800 ft) to one triple-prism assembly
5 km (16,400 ft) to two triple-prism assemblies.
ACCURACY (rms slope distance):
± (5 mm + 5 mm/km) for −10° to 40°C.
± (10 mm + 10 mm/km) for −20° to −10°C and 40° to 55°C.
UNIT OF DISPLAY (switch selectable):
0.001 m or 0.001 ft in accurate mode
0.01 m or 0.01 ft in track mode
DISPLAY RATE (track mode):
1.5 s/reading-slope distance, minimum
2.5 s/reading-projected distance, minimum
LIGHT SOURCE:
Type: Solid-state GaAs lasing diode (non-visible)
Wavelength: 835 nm, nominal value
Power Output: Complies with DHEW radiation performance standards, 21 CFR, subchapter J.
Beam Divergence: 370 cc grad (2 arc-minutes) full angle, nominal value.

ANGLE
RESOLUTION:
Degree Mode: 1 arc-second
Grads Mode: 1 cc grad
ACCURACY (rms of direction with telescope in direct and reversed position for −20° to 55°C):
HORIZONTAL: ±6 cc grad (±2 arc-seconds)
VERTICAL: ±12 cc grad (±4 arc-seconds)
UNIT OF DISPLAY: (Switch selectable).
Degrees, minutes, seconds to 1 arc-second.
Grads to 1 cc grad.
DISPLAY RATE:
2 s/reading with automatic level compensation
0.5 s/reading without automatic level compensation
AUTOMATIC LEVEL COMPENSATION:
Type: Dual-axis liquid-surface reflection
Range: ±340 cc grad (±110 arc-seconds), approximately

DIGITAL OUTPUT
TYPE: One-way-handshake bit-serial-data transfer from 3820A to peripheral.
DATA WORD: 14-digit (56-bit) BCD word.

TELESCOPE
MAGNIFICATION: 30×.
IMAGE: Erect.
OBJECTIVE APERTURE: 66 mm.
FIELD OF VIEW: 1.67 grad (1.5°).
FOCUS RANGE: 5 m (16 ft) to ∞.
Illuminated cross hairs.
Two sighting collimators for rapid target acquisition.

POWER SUPPLY (Internal, rechargeable battery):
TYPE: 3.6 Vdc nickel-cadmium (HP 11421A).
OPERATING TIME: 3 hours typical, ~400 measurements.
CHARGING TIME: 16 hours (full charge).

OPERATING CONTROLS
Switch panel for controlling operating modes of instrument.
Two function-switch-panels for convenient selection of measurement in either direct or reversed position.
Concentric lock and tangent screws.
Two-speed tangent screws.
Two-speed circle indexing screws.

MECHANICAL INTERFACING
INTERFACE: HP 11426A Leveling Base.
ROUGH LEVELING: Circular bubble on alidade with a sensitivity of 2.5 arc-minutes/mm.
CENTERING: Optical plummet in alidade with magnification of 5× and focus range of 0.6 m (2 ft) to ∞.

DIMENSION AND WEIGHT
INSTRUMENT: 162×239×298 mm (6.4 × 9.4 × 11.7 in).
WEIGHT: 9.9 kg (21.9 lb) (including battery).

PRICE IN U.S.A.: $34,000.00

MANUFACTURING DIVISION: CIVIL ENGINEERING DIVISION
815 Fourteenth Street, S.W.
Loveland, Colorado 80537 U.S.A.

Mechanical Design Constraints for a Total Station

by Ronald K. Kerschner

THE ACCURACY of a distance and angle measurement system is greatly dependent on its mechanical design. The geometrical constraints imposed on the 3820A Total Station required careful analysis and design for its fabrication.

The principal axes of the 3820A are shown in Fig. 1. The trunnion axis establishes an axis of rotation for the telescope and should be perpendicular to both the optical and vertical axes. The vertical axis should define an axis of rotation for the instrument that does not change as the alidade is rotated on its base. In addition, the vertical axis must pass through the intersection of the optical and trunnion axes for accurate angular measurement to targets at short distances.

Vertical Axis

The horizontal angle error introduced when the vertical axis does not intersect the optical axis can be expressed as

Vertical-axis centering error =

$$\tan^{-1}\left[\frac{\text{Vertical-axis offset}}{\left(\begin{matrix}\text{apparent target}\\ \text{distance}\end{matrix}\right)\times\sin\left(\begin{matrix}\text{zenith}\\ \text{angle}\end{matrix}\right)}\right]$$

This equation shows that when the target distance increases to large values, the error decreases to zero.

The vertical-axis centering is controlled to within 0.25 mm (0.01 in) by the mechanical tolerances and by adjustment of the optical axis. Axis wobble is a measure of the imperfections in the vertical axis. It is measured by taking readings for both horizontal-level-sensor axes versus the horizontal angle for the alidade. The level-sensor readings are converted to polar coordinates. The horizontal angle is subtracted from the level-sensor polar angle. Given a perfect vertical-axis bearing, the polar level vector will rotate in the same direction and at the same rate as the alidade horizontal angle. Variances in direction and rate are due to

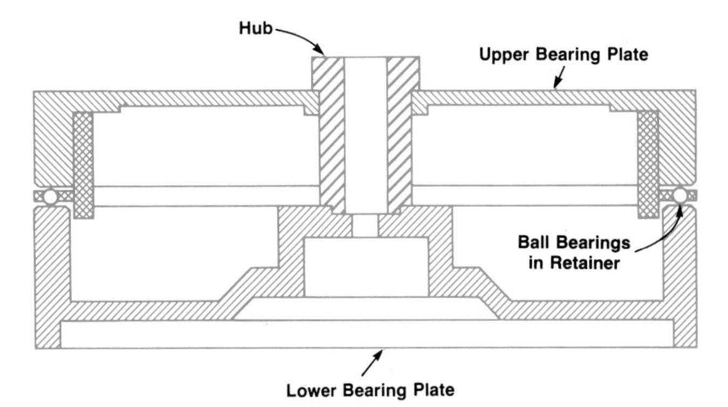

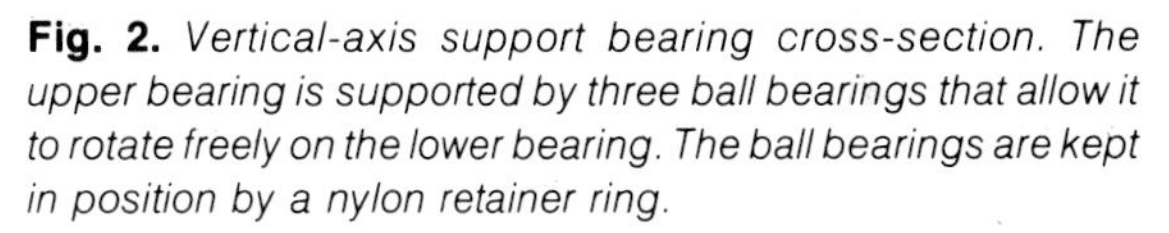
Fig. 2. *Vertical-axis support bearing cross-section. The upper bearing is supported by three ball bearings that allow it to rotate freely on the lower bearing. The ball bearings are kept in position by a nylon retainer ring.*

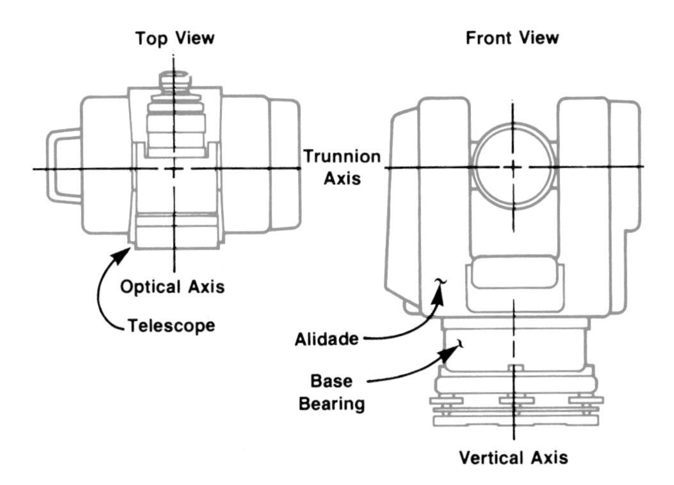

Fig. 1. *Principal axes of the 3820A. The instrument rotates horizontally about the vertical axis, the telescope rotates vertically about the trunnion axis, and the optical axis is the line of sight for the telescope.*

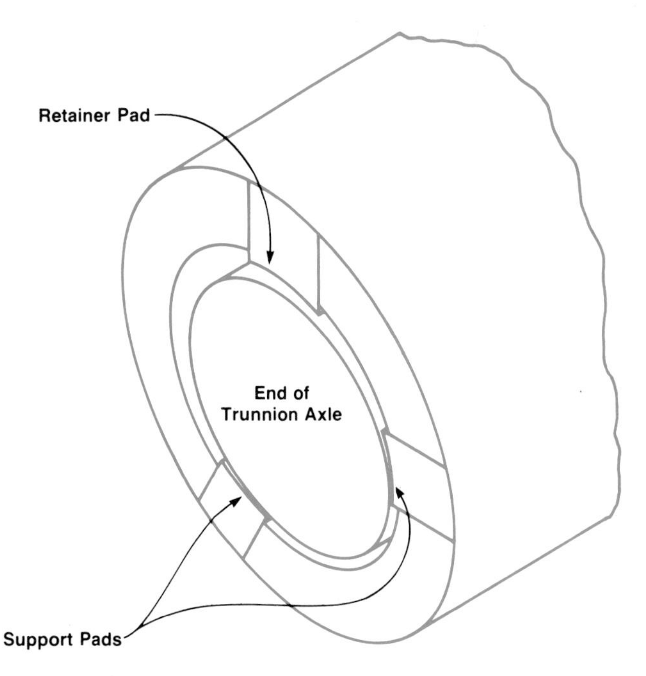

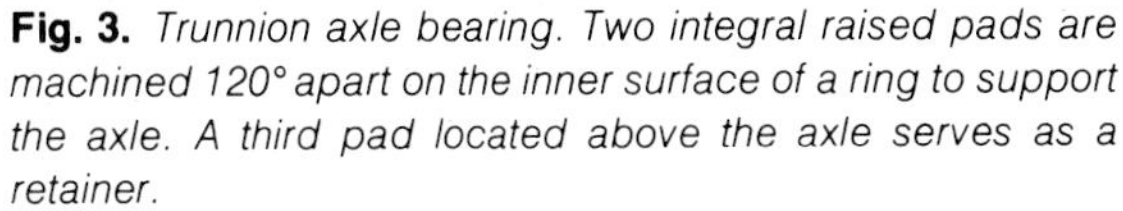
Fig. 3. *Trunnion axle bearing. Two integral raised pads are machined 120° apart on the inner surface of a ring to support the axle. A third pad located above the axle serves as a retainer.*

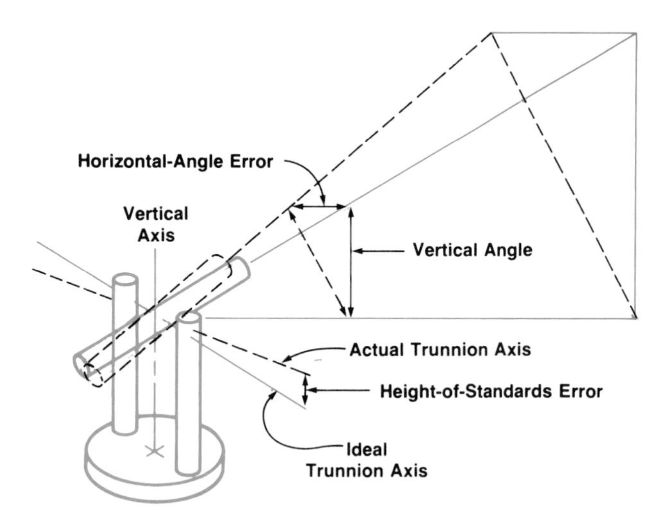

Fig. 4. *Exaggerated illustration of height-of-standards error. This error contributes to errors in horizontal angles when measuring targets at various vertical angles.*

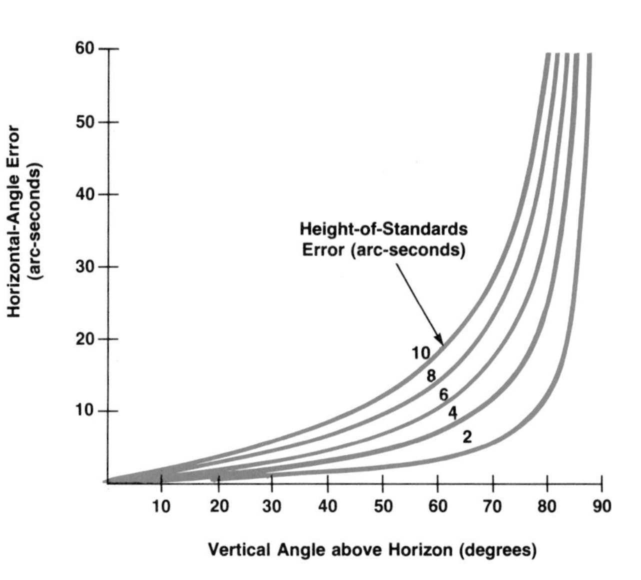

Fig. 5. *Family of curves showing horizontal-angle errors versus vertical-angle differences for varying degrees of height-of-standards error.*

imperfections in the vertical-axis bearing.

Another way to explain vertical axis wobble is to imagine a fixed screen being placed above the instrument perpendicular to the vertical axis and in a plane parallel to the earth's surface. The axis wobble is a plot of the movement of the intersection of the vertical axis with this screen as the alidade is rotated on its base. Vertical-axis wobble complicates the determination of the horizontal level-sensor indexes. Thus, the wobble is tightly controlled to minimize the problems encountered in determining the indexes.

The vertical-axis bearing system is shown in Fig. 2. The clearance between the hub and the top bearing plate controls the centering uncertainty. The lower bearing is lapped flat to within 0.25 μm (10 μin). The balls are made with a diameter tolerance of 0.5 μm (20 μin). The upper bearing is machined to have three equidistant high points around its perimeter. The surface peak-to-valley variation for this bearing is 2.5 μm (100 μin). Thus, the lower bearing establishes a plane, the balls serve as a rolling element to reduce friction, and the top bearing provides three-point kinematic contact. The bearing plates and the hub are made from hardened steel to minimize wear.

Trunnion Axis

Since the trunnion axis is always close to being perpendicular to gravity, a simple V-block bearing can be used. This type of bearing provides line contact on the axle. V-blocks are used extensively in mechanical measurement for establishing a reference on cylindrical parts. Because the instrument will be shipped and handled in positions other than those perpendicular to gravity, the bearing requires a third retaining point. The actual bearing used is shown in Fig. 3. The two lower pads provide an approximation to a V-block. The upper pad is there to retain the axle in abnormal positions and is not in contact with the axle during normal use.

Height-of-standards error results when the trunnion axis is not perpendicular to the vertical axis (Fig. 4). If a height-of-standards error exists, horizontal-angle errors are introduced when two targets are at different vertical angles. A plot of the horizontal-angle error between two points, one on the horizon and the other at varying angles above the horizon, versus their difference in vertical angle for various values of height-of-standards error is given in Fig. 5.[1] The height-of-standards error is controlled to less than five arc seconds in the 3820A.

An error in horizontal angles can also be generated by collimation error, which results if the optical axis is not perpendicular to the trunnion axis. This error is corrected to less than five arc seconds by adjusting the telescope optics and fine-tuning electronically at the factory by use of programmable read-only memories (PROMs).

Like the vertical axis, the trunnion axis also suffers from axis wobble. Wobble of the trunnion axis is caused by the profiles of the left and right ends of the axle not having the same geometry, or by scope imbalance. This wobble will create collimation and height-of-standards errors.

All of the instrument geometry errors can be cancelled by plunging, except for wobble of the trunnion axis (for a description of the plunging technique, see the glossary on page 11). For this reason wobble of the trunnion axis is tightly controlled to less than 1.5 arc seconds. However, the

Ronald K. Kerschner

Ron Kerschner received a BSME degree from Colorado State University in 1972. He joined HP in the same year and has worked on mechanical design for the level sensor and distance meter module used in the 3820A Total Station. Then he was involved with the production of the 3820A and is currently working on new products. Ron was born in Sterling, Colorado. He and his wife and two children now live in Loveland, Colorado. When Ron isn't landscaping their new home or busy with his studies for an MSEE at Colorado State, he enjoys photography, running, bicycling, hiking and cross-country skiing.

other errors are also controlled so that the 3820A can be used for single-shot readings when only moderate angle accuracy is needed.

Acknowledgments

In view of the mechanical tolerances required by the 3820A, a great deal of credit must be given to the Civil Engineering Division precision fabrication and parts inspection groups. Craig Cooley designed the alidade and bottom bearing base. Walt Auyer designed the vertical and trunnion axis systems.

Reference

1. M.A.R. Cooper, Modern Theodolites and Levels, Crosby Lockwood and Sons, London, 1971. (page 62).

A Compact Optical System for Portable Distance and Angle Measurements

by Charles E. Moore and David J. Sims

THE COMBINED DISTANCE METER and telescope optical system of the 3820A Total Station provides two substantial performance advantages. It is smaller than the alternative of using two or more separate optical systems, which is an advantage for a portable field instrument. Also, it uses a single sighting, directly at the center of the cube-corner target, for both distance and angle measurements throughout the entire range of the instrument. A combined optical system such as this provides a difficult task for the lens designer. The designer must reduce the optical aberrations to very low order to achieve good aiming for angular measurements, provide large enough optics to give good distance-meter range, and still design the shortest possible telescope.

The conventional refractive triple used in most *theodolites* proves to be unsatisfactory in two ways. First, if the telescope is balanced about the *trunnion axle*, the beam is quite large at the beam splitter. This requires a large, expensive beam splitter and an undesirably large axle. Second, the *spherical aberrations* and *secondary spectrum* of the telescope are too large for a good theodolite. The second problem can be dealt with by stopping down the telescope with an aperture located behind the beam splitter where it will not affect the distance-meter optics. This reduces the spherical aberrations to below the Rayleigh limit of $\frac{1}{4}\lambda$ optical-wavefront distortion and limits the secondary spectrum to an acceptable level while retaining the same clear aperture as most one-arc-second theodolites. The problem of the large beam splitter remains.

The solution to this problem uses a *catadioptric Cassegrain* structure as shown in Fig. 1. The folded optics combine with the telephoto effect provided by the negative-power secondary mirror to permit use of a small beam splitter. Spherical aberration can be reduced because the curved mirror provides most of the magnification for the system and has an inherently small spherical aberration due to the large effective index-of-refraction difference at the mirror. All *chromatic aberration*, including secondary spectrum, can be eliminated by having no net magnification in the refractive surfaces.

The catadioptric telescope used is not without problems. First, a catadioptric telescope is difficult to focus onto

NOTE: All words in italics in this article are defined in the glossary on page 11.

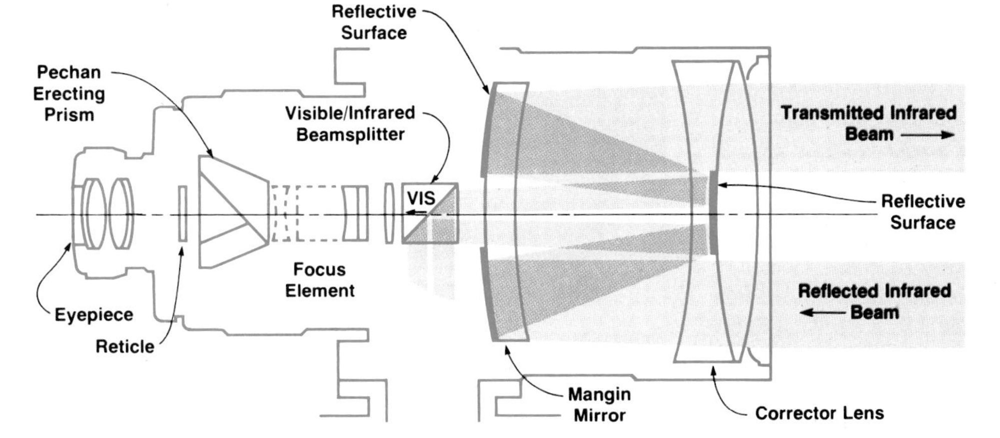

Fig. 1. *Optical system design for 3820A. Visual sighting and projection and reception of the infrared distance measurement beam are shared by the optical elements on the right. The beam splitter deflects the infrared portion down to the distance meter module while allowing the visible portion to continue to the eyepiece and focusing elements on the left.*

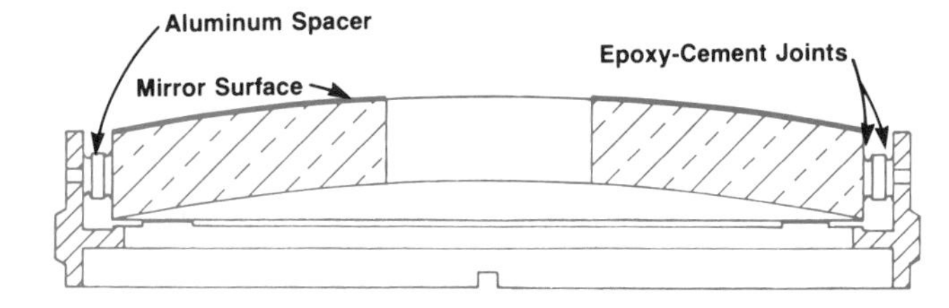

Fig. 2. *The mirror rests on three pads providing stress-free kinematic support. It is secured by cementing at six points around the edge. The difference between the thermal expansion of the cell and the mirror is compensated for by aluminum spacers in the cement joints, and by selection of the cement bond thickness.*

nearby targets. The telephoto effect that was so useful in permitting the use of a small beam splitter also produces a long effective focal length. The travel distance needed in a focusing system depends on the square of the effective focal length. This is reduced by adding a positive-power element behind the beamsplitter to reduce the effective focal length.

Second, this catadioptric Cassegrain structure is difficult to design. Many designs, such as the *achromatic* objective, the classical Cassegrain with two *aspheric* reflective surfaces, and the *Bouwers*, have been analyzed and simplified design procedures have been worked out for them. No previous work has been done for the principles of the 3820A telescope and there are few readily apparent simplifications. In the objective of the telescope there are ten curved surfaces that can be varied to control ten aberrations or *paraxial* characteristics—spherical aberration, *coma* and chromatic aberration at far and near focus, overall length, effective focal length, distance to nearest focus, and size of the secondary reflector. Each curvature affects all, or almost all, of the parameters one wishes to control. Fortunately, modern computer-aided lens design techniques make such complex designs much more manageable.

Telescope Assembly

A mechanical problem is that mirrors are very sensitive to mounting. They must be firmly mounted because any small rotation of a mirror causes twice as much change in angle for a reflected ray. To achieve two-arc-second pointing accuracy with the 3820A the main mirror should be stable to within one arc second, corresponding to a 0.00025-mm variation across the mirror. At the same time the mount should not exert any appreciable force on the mirror, since any distortion in the mirror will result in aberrations in the image. As little as 0.00006-mm distortion of the mirror can cause noticeable loss in resolution.

To avoid these problems the mirror is placed into the cell so that the front surface rests gently on three small pads. A cross-section of this cell is shown in Fig. 2. Since three points locate a sphere, this provides an unambiguous location for the front surface, while minimizing stress to the mirror. This principle of three-point support mounting is used many times in the design of the 3820A.

The mirror has aluminum spacers cemented to its edge. After the mirror has been centered in the cell by three removable leaf springs, it is permanently attached to its cell by cementing the aluminum spacers to the cell. The individual thicknesses of the aluminum spacers and of the epoxy-cement layers were chosen so that the combined thermal expansion of the glass mirror plus the cement-aluminum-cement stack matches that of the stainless-steel cell. This provides a solid mount for the mirror that does not distort the mirror over the temperature range the 3820A experiences in service. The cement attaching the pads to the mirror has a closely controlled compliance, which allows for the difference in expansion of glass and aluminum, but does not allow the mirror to move appreciably.

The cell is screwed into the telescope housing and rests on three pads. The threads fit loosely to allow the cell to rest securely on the pads without having to bend. If the cell were bent to fit both the pads and tight threads, it would distort the firmly attached mirror. The threads are cemented to help secure the cell.

The front element is mounted in a similar manner except that before the element is cemented to its cell, the centering is adjusted in an optical test fixture to correct aberrations that might be introduced by slight miscentering or tilting of other elements. Also, the front cell is shimmed to give the correct fixed focus for the distance meter. These adjustments allow the parts of the telescope to be built to achievable tolerances.

Roelof's Prism Adaptor

The 11429A Roelof's Prism Adaptor (Fig. 3) is a mechanical mount that lets the user view the sun for use in determining the azimuth of a line in surveying work. The Roelof's prism divides the sun into a four-quadrant pattern that is easily centered on the *reticle* crosshair pattern.

Acknowledgments

All of the optical parts for the 3820A, except for a few very specialized parts, are made in Civil Engineering Division's own optical fabrication shop. Billy Miracle and Ron Kerschner participated in the design of the mechanical housing and lens mounts for the 3820A telescope. Alfred Gort, the project leader, provided advice and encourage-

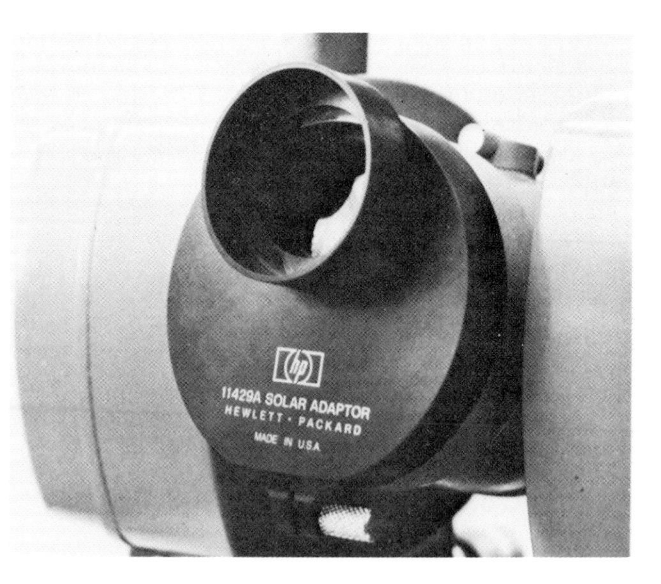

Fig. 3. *Roelof's Solar Prism Adapter. This accessory for the 3820A allows measurement of the angle of the sun above the horizon.*

David J. Sims

Dave Sims is a native of Salt Lake City, Utah and attended Brigham Young University where he was awarded BSME and MSME degrees in 1970. Dave then joined HP and has worked on mechanical designs for the 3805A and 3820A. Currently he is the mechanical design leader for the 3820A optical accessories. Dave is a co-inventor for a patent on the temperature-compensated lens mount described in this article. He speaks Norwegian and has served eight years with the Utah National Guard as a linguist interrogator. Dave met his wife in Norway during a mission for his church there and they now live with their seven children in Loveland, Colorado. Dave is involved with church leadership and enjoys boating and camping.

Charles E. Moore

Charles Moore received a BSEE degree from the University of California at Berkeley in 1966 and an MS degree in optics from the University of Rochester in 1978. With HP since 1966, he has worked on wave analyzer and voltmeter projects in addition to his work on the receiver and optics for the 3800A Distance Meter and the 3820A Total Station. Charles is a co-inventor for three patents, including one for the 3820A telescope optics. He is a member of the Optical Society of America. Charles was born in Santa Fe, New Mexico and served in the U.S. Army from 1960 to 1963. He is married, has five children, and lives in Loveland, Colorado. His interests include bicycling, running, hiking, cross-country skiing, reading, and chess. Charles has programmed an HP computer to pair opponents in the chess tournaments he enjoys directing.

ment for the optical designs. Norm Rhoads of the optics shop made contributions to improving the manufacturability of the optical designs.

Elton Bingham contributed to the mechanical design of the optical accessories after joining the program in June of 1979. Gary Gates provided manufacturing inputs to the program. Gib Webber's model shop created the early mechanical prototypes in the accessory development, and Bennett Stewart's optics shop provided prototype lenses.

An Approach to Large-Scale Non-Contact Coordinate Measurements

by Douglas R. Johnson

A COMMON PROBLEM in the manufacture of large products is the quality control of critical dimensions. The problem is easily solved on a small item by means of coordinate measuring machines. These machines use a delicate manipulator arm to gently contact the point to be measured. The x, y, and z coordinates are determined by a series of vernier scales and sensitive pressure transducers. Newer machines are computer controlled and motor driven. Resolution approaches 0.5 μm (20 μin) on the best of these machines.

When the item to be measured becomes larger, the cost of a coordinate measurement machine increases dramatically. Alternative methods become attractive as soon as any dimension of the item to be measured exceeds one metre. However, these alternate solutions often have serious drawbacks such as excessive pressure during contact, delay in obtaining results, or highly technical operator requirements. Properly configured, a 3820A system can be an ideal solution for large-scale coordinate determination.

The coordinate determination works on the principle of triangulation—an old, yet effective, solution. The digital theodolite portion of the 3820A Total Station becomes the workhorse of the system. Its high angular accuracy and resolution insure reliable results accurate to better than ten ppm without need for mechanical contact. The data output capabilities of the 3820A provide an effective way of transferring measured angular data to a small computer for real-time analysis and comparison. At least two total stations are required per installation. Improved performance may be realized by adding additional instruments.

Principle of Operation

The 3820A Coordinate Determination System (Fig. 1) works on the principle of triangulation. Two digital theodolites mounted at known points are used to measure angles. They both observe the same set of unknown points and perform accurate angle measurements. The 3820A's level compensator insures that the horizontal plane is indeed horizontal. Because the digital theodolite can measure both horizontal and vertical angles, a three-dimensional solu-

Fig. 1. *The 3820A Coordinate Determination System allows accurate dimensional and positional information to be readily obtained for large objects.*

tion is possible. Four angles are recorded to each unknown point—a vertical (zenith) angle referenced to gravity and an included horizontal angle from each 3820A. These four angles may be readily combined to yield the three coordinates X, Y, and Z of the unknown point.

To further understand the triangulation concept, consider the example shown in Fig. 2. In triangle ABP, the length (r) of one side (AB) and two angles (θ_1 and θ_2) are known. If A and B are in the same horizontal plane ($\phi_t = 90°$), then the law of sines and some elementary trigonometry yields:

$$X_p = \frac{r}{2}\left(1 + \frac{\sin(\theta_2-\theta_1)}{\sin(\theta_2+\theta_1)}\right)$$

$$Y_p = r\left(\frac{\sin(\theta_2)\sin(\theta_1)}{\sin(\theta_2+\theta_1)}\right) \quad (1)$$

The 3820A horizontal-angle measurement capability permits determination of θ_1 by subtracting the angle reading along line AB from line AP. θ_2 is similarly determined. To obtain the elevation or Z coordinate, vertical (zenith) angles are used. If ϕ_1 is the zenith angle (an angle of 0° is straight up) from the 3820A at A to P, the unknown point, ϕ_2 is the angle from the 3820A at B to P, and ϕ_t is the zenith angle between the 3820As (if they are at uneven elevations); then the following relationships may be derived.

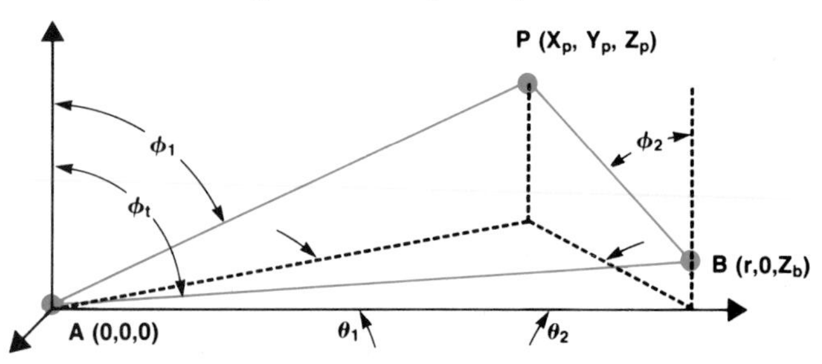

Fig. 2. *By placing a 3820A at position A, another at position B and then measuring the distance r between A and B and the angles θ_1, θ_2, ϕ_1, ϕ_2 and ϕ_t, the unknown position of point P can be determined using simple trigonometry.*

Aircraft Inspection

In the assembly of large aircraft, commercial or military, passenger or cargo, dimensional control plays a key role. Various subsections such as tail assemblies, wing assemblies, engines and their mounts are precisely constructed on rigid manufacturing fixtures. These fixtures are routinely inspected. In addition, once the aircraft is completely assembled, another dimensional inspection must be performed. This insures that critical dimensions are within tolerances, subassemblies have been properly mated, and abnormal stress is not present.

Current inspection methods involve many man-hours of plumbing, taping, and various geometric calculations. The need to perform many individual steps to obtain one measurement complicates the procedure. To obtain the coordinates of a control rivet on a wing fuel-cover latch, a measurement crew must typically:

- Tape the horizontal distance from wing rivet to wing trailing edge using a hand level, invar tape, and two plumb bobs.
- Establish the elevation difference between the rivet and trailing edge using transit and stadia.
- Plumb the trailing-edge point to the ground.
- Tape the trailing-edge height above the ground.
- Tape the distance from the ground point to the coordinate origin.
- Determine the angle from the control axis at the origin to the trailing-edge ground point and from the control axis at the origin to the wing rivet.
- Geometrically solve for the x and y coordinates of the wing rivet from the taped distance and measured angles.
- Mathematically obtain the elevation from the series of individual elevation measurements.

The 3820A Coordinate Determination System will solve the same dimensional measurement problem in less than one-tenth the man-hours while significantly improving coordinate accuracy as a side benefit. A final benefit is the reduction of errors through simple operation, automatic data transfer, and computer control.

Interfacing the 3820A via the HP-IB

by Gerald F. Wasinger

For the 3820A distance meter to talk to a controller via the HP-IB* the data has to go through some conversions. We must convert from the binary-coded-decimal (BCD) bit-serial data of the distance meter to the ASCII byte-serial data of the HP-IB. This process is accomplished through the 38001A Distance Meter Interface.**

As seen in Fig. 1, the digital output of the 3820A consists of fourteen digits of data—nine digits of measurement data and five mode annunciators. Since each digit is represented by four bits the complete data string contains 56 bits. This data is positive-true and is valid on the falling edge of the clock. The frequency of the clock is 180 kHz so that a complete data transfer from the 3820A to the 38001A occurs in 311.1 μsec.

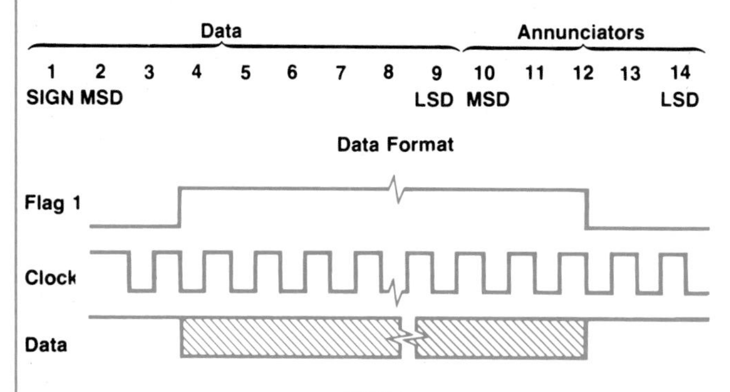

Fig. 1. *Data format and timing for the output of the 38001A. The annunciators indicate to the controller what type of data is being sent.*

Referring to Fig. 2, it is seen that the serial data is loaded into a shift register with four parallel outputs. After every fourth bit a negative pulse is generated that on the falling edge stores the four bits of the shift register into the data RAM and then, on the rising edge, increments the RAM's address counter. This process is repeated until all fourteen digits have been stored in the RAM.

*Hewlett-Packard's implementation of IEEE Standard 488-1978.
**See page 19 of June 1980 issue for a list of the 38001A specifications.

After the 14th digit, Flag 1 goes low telling the 3820A that the 38001A is not currently prepared to accept data. Also, the quad multiplexer selects the sequence for the RAM outputs. Finally, a service request is sent out on the HP-IB to tell the controller that the 38001A has data.

When the controller reads the data from the interface the output may not be in the same sequence as the one generated by the distance meter. The output sequence is programmable via the sequence RAM and gives the user the option of obtaining some or all of the data, in any order.

When the BCD data flows into the four least-significant bits of the bus transceivers, the number three is placed in the four most-significant bits. The result is the ASCII representation of the data in a form suitable for the HP-IB.

After all of the digits have been placed on the bus, a carriage return and line feed are generated to terminate the data string. Also, Flag 1 is set high to allow another data transfer from the 3820A and the multiplexer selects the now cleared 4-bit counter. Thus, the cycle is complete.

A simple program for the 9845B Computer/Controller to read 3820A multifunction data from the 38001A is listed below.

```
 10  CLEAR 717
 20  OUTPUT 717 USING "#,K";"MLKJIHGFENCO"
 30  TRIGGER 717
 40  ENTER 717; Datum, Annun
 50  IF Annun<>3 THEN GOTO 80
 60  Dist=Datum/1000
 70  GOTO 120
 80  IF Annun<>6 THEN GOTO 110
 90  Dir=INT (Datum/10)/1E4
100  GOTO 120
110  Zenang=INT (Datum/10)/1E4
120  DISP "Direction:"; Dir; "Distance:"; Dist; "Zenith Angle:"; Zenang
130  GOTO 40
140  END
```

The first three lines set the output format of the 38001A (HP-IB address 717). The rest of the program simply reads the data from the 38001A, determines whether the information is distance, direction, or

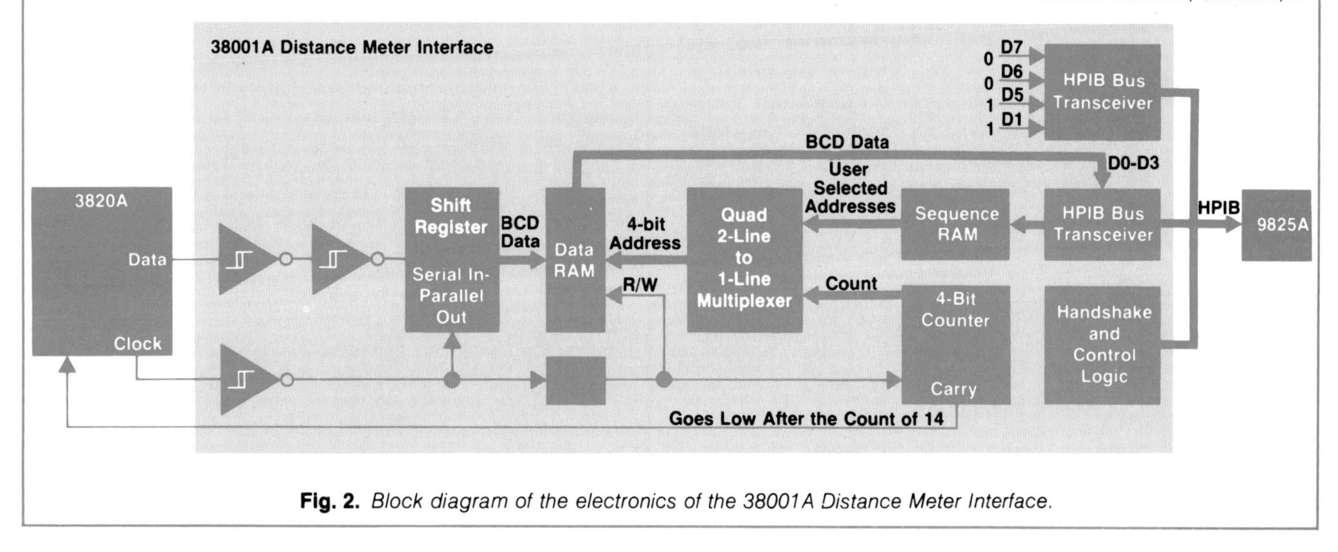

Fig. 2. *Block diagram of the electronics of the 38001A Distance Meter Interface.*

zenith angle, and then displays the data on the CRT of the 9845B. The angles are assumed to be in units of grads.

The above process is not peculiar to the 3820A. The 38001A also translates data from the 3808A, 3810B, and 3850A distance meters to a form suitable for use on the HP-IB.

Gerald F. Wasinger

Jerry Wasinger received the BSEE degree from the University of Oklahoma at Norman in late 1977. He joined HP in early 1978 and has worked on the 3808A DM and the 38001A. Jerry is a native of Oklahoma City, Oklahoma. He and his wife live in Loveland, Colorado and are expecting their first child this fall. Outside of work and his studies for the MSEE degree at Colorado State University, Jerry enjoys bicycling, hiking, electronics, photography, and computer art—he was an art major before taking up engineering as a career.

$$X_p = \tfrac{1}{2} r \sin(\phi_t)\left(1 + \frac{\sin(\theta_2 - \theta_1)}{\sin(\theta_2 + \theta_1)}\right)$$

$$Y_p = r \sin(\phi_t)\left(\frac{\sin(\theta_2)\sin(\theta_1)}{\sin(\theta_2 + \theta_1)}\right) \qquad (2)$$

$$Z_p = r \sin(\phi_t)\left(\frac{\sin(\theta_2)\cot(\phi_1)}{\sin(\theta_2 + \theta_1)}\right)$$

Since r, θ_1, θ_2, ϕ_1, ϕ_2, and ϕ_t may all be measured without contact in a matter of seconds, X_p, Y_p, and Z_p may be determined almost instantaneously.

The preceding mathematics calculates a single solution for X_p, Y_p, Z_p. In actual practice, four angles—θ_1, θ_2, ϕ_1, and ϕ_2—are measured and used to calculate the three unknown points—X_p, Y_p, and Z_p—through a least-squares reduction. This reduction finds the best fit for the four knowns into the three unknowns. The least-square residual provides a convenient check on coordinate determination accuracy.

A 9845T Computer/Controller is recommended as the computer for the coordinate determination system. The 9845T was chosen on the basis of its large memory, ease of programming, and graphics capability. The computer communicates with the 3820A through the 38001A HP-IB Interface.

If it is unsatisfactory to have the 3820A at point A as the system origin or the horizontal projection of AB as the X axis, the 9845T may three-dimensionally rotate and translate the calculated coordinates. Thus complete coordinate system flexibility can be maintained. Even the baseline distance r need not be measured. If two control points exist the 9845T may inverse triangulate to define both the length and direction of r.

The 9845T also performs the functions of data transfer, angle averaging (in the event that additional sightings are made), coordinate calculation, graphics display of measured points, and operator prompting.

Error Analysis

In most dimensional measurement systems, there are three error-related items of interest—resolution, repeatability, and accuracy. For the 3820A-based coordinate-determination package, each quantity is first a function of geometry. The magnitude of this geometric influence varies with triangle strength (relative errors are greater for combinations of large and small values of θ_1 and θ_2).

System resolution referenced to the baseline distance is at least one part in 650,000 for most measurement applications. The figure of one part in 650,000 may be expressed as a dimension in the form of 1.5 μm per metre (18 μin per foot) of baseline. This figure assumes that the 3820A is outputting in the grads mode. If measurements are output in the degrees mode, resolution suffers by a factor of over three to become one part in 200,000.

Coordinate-determination repeatability is primarily a function of 3820A angle accuracy. Extracting partial derivatives of the least-square equations (equivalent to those presented in equation set (2)) yields error terms for X_p, Y_p, and Z_p as a function of each of the measured angles. Since most of the errors in measured angles are random variables, they may be squared, summed, and rooted to

Antenna Assembly

In the parabolic antenna industry, new techniques are constantly being sought to improve manufacturing ease and versatility. One problem is to design and certify lightweight antennas suitable for space applications. A technique recently developed uses thin aluminized-mylar films that are creatively shaped into a paraboloid. Because of the thin, delicate nature of the mylar film, dimensional measurements cannot be made by normal contact measurement procedures.

The 3820A Coordinate Determination System becomes a useful alternative measuring tool. Because the system is portable, it is brought to the antenna site instead of the opposite. The X, Y, and Z coordinates are measured to the surface without contact. The resulting coordinates are compared to an ideal parabolic surface for antenna profile accuracy determination. In this application, the scale factor of the baseline is not critical since the whole antenna may be scaled larger or smaller in size without affecting the parabolic calculations.

A Deflection Measurement Example and Error Considerations

A short example may serve to clarify the usefulness of the 3820A Coordinate Determination System. Suppose we are to measure the bulges of the center of a metal tank before and after being filled with liquid (see Fig. 1). We select two instrument setups approximately 10 metres apart located in the same horizontal plane ($\phi_t = 90°$).

Measurements are made to P yielding:

$$\begin{aligned} \theta_1 &= 45° \ 19' \ 12'' \\ \theta_2 &= 46° \ 24' \ 18'' \\ r &= 9.929 \text{ m} \\ \phi_1 &= 91° \ 18' \ 01'' \\ \phi_2 &= 91° \ 19' \ 28'' \end{aligned} \tag{1}$$

The 9845T calculates coordinates for P in a matter of milliseconds:

$$(5.05855, 5.11537, -0.16330) \text{ metres} \tag{2}$$

If r is known to an accuracy of ± 0.005 m, θ_1 and θ_2 are known within 0° 00′ 02″ and ϕ_1 and ϕ_2 are known within 0° 00′ 04″ (3820A rms error specification), the absolute uncertainties of the coordinates for P due to the uncertainties in r, θ_1, θ_2, ϕ_1 and ϕ_2 are:

$$(\pm 0.00255, \pm 0.00258, \pm 0.00010) \text{ metres} \tag{3}$$

This is not an exceptionally precise determination of coordinates. However, since we are trying to determine the bulge in the tank, we are more concerned with the relationship of P-after to P-before.

After the tank is filled, the following measurements are made to P (P-after).

$$\begin{aligned} \theta_1 &= 45° \ 17' \ 14'' \\ \theta_2 &= 46° \ 22' \ 19'' \\ r &= 9.929 \text{ m} \\ \phi_1 &= 91° \ 18' \ 03'' \\ \phi_2 &= 91° \ 19' \ 30'' \end{aligned} \tag{4}$$

The new set of coordinates for P=P-after is

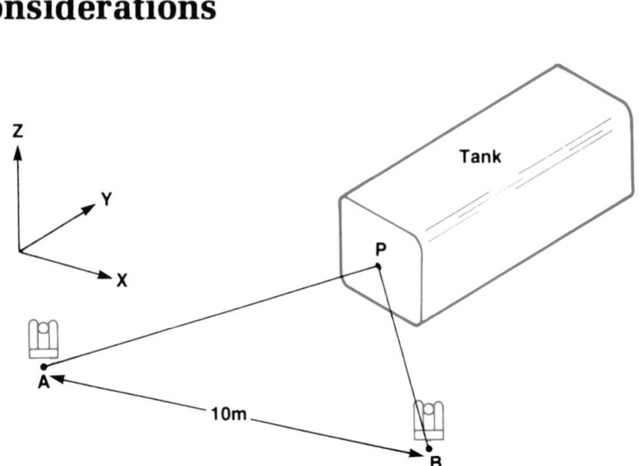

Fig. 1. *A small change in large object dimensions such as the bulge in a tank wall before and after being filled can be easily measured with the 3820A Coordinate Determination System.*

$$(5.05852, 5.10949, -0.16327) \text{ metres} \tag{5}$$

These coordinates also have absolute uncertainties of (3) above.

Since the measure of r is unchanged we may subtract the one set of coordinates from the other and get a much improved result. The difference is P-after minus P-before equal to

$$(-0.00003, 0.00588, 0.00003) \text{ metres}$$

with relative uncertainties of (using the rms error equations with r = 9.929 m)

$$(\pm 0.00006, \pm 0.00006, \pm 0.00010) \text{ metres}$$

Thus we can state that the bulge occurred in the Y direction with a magnitude of 5.88 ± 0.06 mm (0.231±0.002 in).

yield expected errors in coordinates. The magnitude of the partial derivatives varies with triangle strength as previously discussed.

The 3820A angle measurement system has a two arc-second horizontal rms error and a four arc-second vertical rms error when two sightings (one direct, one reverse) of an unknown point are taken. These angular errors are composed of sighting, 3820A circle, and 3820A geometric errors. Many of these errors are discussed elsewhere in this issue. However, when these angular error values are substituted into the partial-derivative equations for coordinate determination, extremely good results are obtained. For the triangle where $\theta_1 \approx \theta_2 \approx 45°$ and $\phi_1 \approx \phi_2 \approx \phi_t \approx 90°$ the rms error values for coordinate determination are:

E_x = 1:160,000 or 6.2 μm per metre (75 μin per foot) of r
E_y = 1:160,000 or 6.2 μm per metre (75 μin per foot) of r
E_z = 1:100,000 or 10 μm per metre (120 μin per foot) of r.

Once again, these values change as θ_1, θ_2, ϕ_1, ϕ_2 and ϕ_t vary.

Acknowledgments

Ken Frankel of Boeing Company and Bill Haight and Bob Hocken of National Bureau of Standards assisted in the preliminary understanding of the triangulation methodology, and Elton Bingham of HP verified the concepts for software development.

Douglas R. Johnson

Doug Johnson joined HP in 1978 after several years experience in semiconductor marketing and manufacturing. He is currently program manager for EDM instruments. Doug received the BSEE degree from Rensselaer Polytechnic institute in 1971 and the MBA from Arizona State University in 1975. A native of Rahway, New Jersey, he now lives in Fort Collins, Colorado. Doug enjoys skiing, sailing, hiking, jeeping, and backpacking. He is a member of the ski patrol and supports Junior Achievement. Recently, Doug and other HPites built a duplex at a ski resort using the 3820A to survey the land and stake out the construction.

Development of a High-Performance, Low-Mass, Low-Inertia Plotting Technology

A new vector plotter technology makes possible small, inexpensive graphics products that provide high-quality plots quickly.

by Wayne D. Baron, Lawrence LaBarre, Charles E. Tyler, and Robert G. Younge

THIS AND ACCOMPANYING ARTICLES in this and next month's issues will trace the evolution of a new plotting technology from muse to market. This article describes the authors' early thinking and experimentation in Hewlett-Packard's corporate research laboratories. We will also describe our interaction with colleagues in other parts of the company, and how this focused our work, helping us produce the tools they needed to carry out their product development programs.

Our interest was captured by the success of the pocket calculator. Large scale integration (LSI) of electronic circuits had taken the calculator from the desktop to the pocket. Could a similar reduction in size and cost be achieved for a calculator peripheral? A calculator is an electronic device, a peripheral is an electromechanical device. Could digital control techniques be used to put mechanical cost and complexity into LSI as well? These ideas could be debated endlessly. We decided to test them by casting a tangible objective: a vector plotter for the (other) pocket.

Initial Concept

How do you plot over the surface of a piece of notebook paper with a device that will fit into your shirt pocket? Our solution (Fig. 1) had the plotter grasp the paper along an edge and move it up and down. A steel tape, similar to a common measuring tape, moved a pen back and forth across the paper, so that (almost) any point on the paper could be marked.

Fig. 1. *This initial mechanism moved the paper up and down by its edge and moved the pen left and right by extending and retracting the steel tape on which it was mounted. A piezoelectric element was used to raise and lower the pen.*

Two rubber rollers pressed the margins of the paper against two drive rollers coated with grit. A guide bar kept the paper edge straight and a minute amount of reversible tilt in the rubber rollers insured that the paper edge remained pressed against the guide bar. The reversible tilt is obtained by containing the outside ends of the roller shafts in holes that are slightly elongated in the two directions of paper travel. During the first pass of the paper through the rollers, the grit rollers (pinions) impressed a minute pattern (rack) on the back of the paper along the margin. Subsequent passes then followed this same pattern, so that paper registration to high precision was realized. The steel tape was driven with high precision by convolute pin gears such as those used in motion-picture film drives.

At this point, our prototype moved the paper in and out and the tape back and forth with hand cranks fashioned from bent wire. How could we replace these with motor drives having adequate speed and control but little more cost? This implied a very simple, all digital control system, without any position encoders, tachometers, or accelerometers. One solution is the step motor, but its power supply and drive electronics are expensive. A dc motor can have a simple drive system, but an encoder to measure shaft angle is necessary for its control. This led us to explore whether a digital shaft encoder could be made for less cost than a step motor drive. Early experiments were encouraging, except for the resolution required to reduce quantization error and thereby eliminate grinding noises at low speeds. Attempts to purchase a satisfactory encoder for less than $150 (US) failed, and this was already much more than we were willing to spend. Fortunately, HP's Optoelectronics Division became interested in this situation, and they were able to build encoders with satisfactory specifications and much lower cost by using their light sensing and emitting, integrated circuit, and precision optics expertise. These devices are described in the article on page 10.

If position encoders are located out on the pen as in earlier slidewire plotters, all system compliance enters as a resonance in the servo system. Putting an encoder directly on the drive motor shaft eliminated this resonance and any deadband caused by the drive linkage. This simplified modeling and understanding. We proposed to keep the outboard mechanisms light, tight and simple so that any compliance there would not degrade graphic quality. Moving only the separate masses of a sheet of paper and a pen made this a practical approach. This is equivalent to making the motor rotor the dominant inertia, which in turn allows identical servo controllers for both axes.

We needed a pen lift at the end of the tape, but weight and battery power argued against a solenoid. We reduced the energy requirement manyfold by rotating the pen about its center of mass, and using a piezoelectric bimorph to drive it.

Our first graphic record (Fig. 2) improved nicely with effort and the mechanism did indeed fit inside an attractive, calculator-sized enclosure (Fig. 3).

First Product Application

As the initial design approached completion, we had the good fortune to be contacted by HP's Andover Division. They wanted to replace galvanometers, which consist of precision parts and skilled labor (both already expensive and increasing rapidly) with some form of inexpensive graphics. Their overall objective was an annotating all-digital electrocardiograph with superior graphic fidelity and quality. We quickly agreed that the initial design was not satisfactory in a hospital environment, and Steve Koerper at Andover made certain we understood how severe that environment could be: if a mechanism could be knocked off its cart repeatedly, stepped on, have serum poured into its works, and still work after months of such abuse, it would be a serious candidate for their attention. Very high pen and paper accelerations would be helpful to reduce the time necessary to produce a record so the cardiologist would know quickly if a faulty electrode connection existed. Again, high graphic fidelity is essential to proper medical interpretation.

To retain the advantages of the initial low-mass, low-inertia design in this new application, we still wanted to move only the paper and the pen. We decided to close the pen drive tape on itself to form a belt and to pass the paper through the belt, again with rollers. This allowed placing the mechanism entirely within a compact, protective structure. The pen could be attached to a precision guide. Grit wheel and rubber roller pairs pinched the paper on opposite edges, causing the resultant force to pass through the center of mass of the paper. This prevented paper buckling by high accelerations. Each track of impressions in the back of the paper was now made by only one grit wheel, enhancing the rack and pinion effect to prevent slippage under the anticipated higher acceleration.

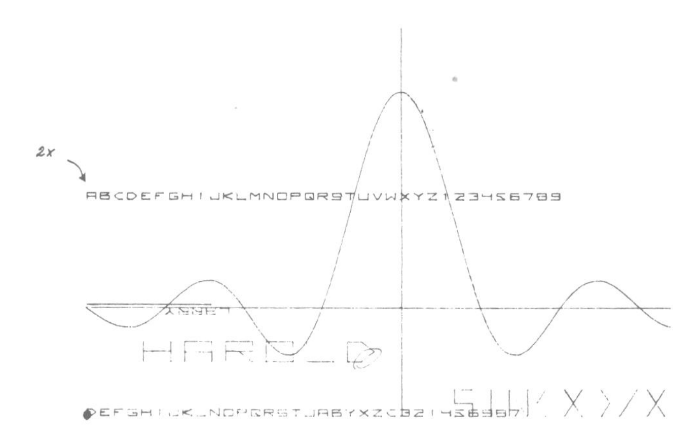

Fig. 2. *First graphic record produced by the mechanism shown in Fig. 1.*

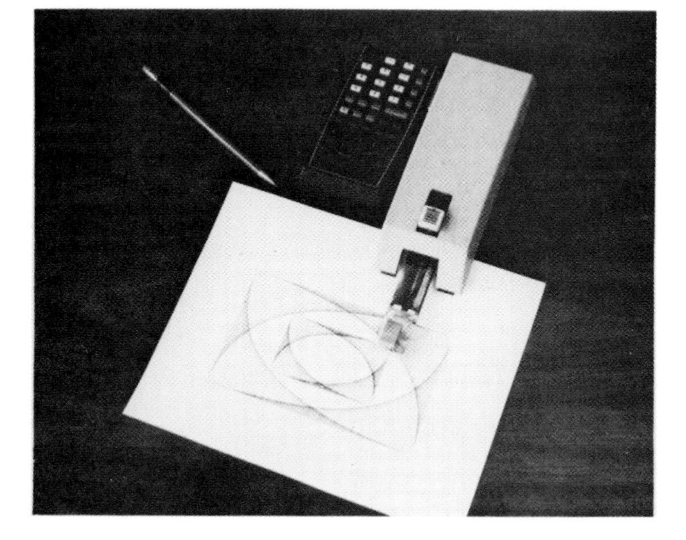

Fig. 3. *Additional effort and refinement of the initial design led to an attractive package capable of producing a quality plot as shown above.*

The control electronics were receiving a similar measure of attention in preparing for higher speed, acceleration, and quality. Some higher-order feedback would be necessary, but adding transducers would compromise our spartan digital approach that promised such attractive low cost. We decided to convert the shaft-encoder period into its inverse and use it as a velocity feedback signal. This required no additional transducers but still provided the feedback we needed for higher performance.

Mechanics and electronics were designed, built, and hooked together. We turned on the switch only to throw the pen out of its holder and three metres off the bench! After

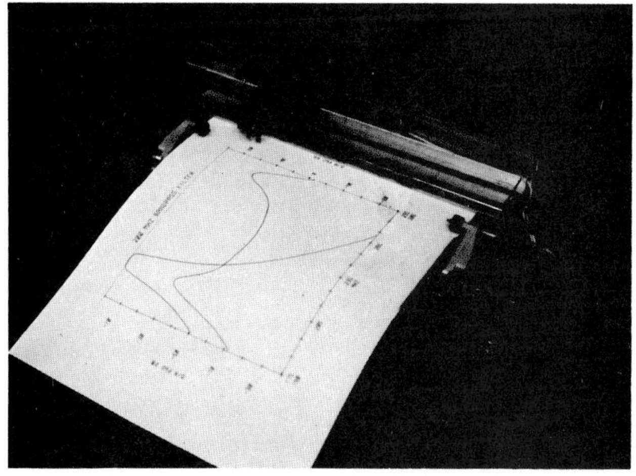

Fig. 4. *Basic mechanism using grit-wheel drive on opposite edges of the paper and a pen driven along a precision guide by a belt loop. Refinements of this basic design are used in the 4700A PageWriter Cardiograph and the 7580A Drafting Plotter.*

Digital Control Simplifies X-Y Plotter Electronics

The dominant theme of the new low-mass, low-inertia X-Y plotter technology is simplicity. This concept is not only prevalent in the mechanism, but is the basis of the electronic architecture (Fig. 1). The job of the electronics is to accept a series of vector and pen lift commands from an external source such as a calculator, computer, or another microprocessor, and then to convert these commands into the appropriate movements of the pen and paper.

The first stage of this process takes place in a custom Z80-based microcomputer. The data received with the commands is converted from user coordinates to the coordinates of the mechanism. From this data the microprocessor determines a velocity profile for the drive motor of each axis that will do the task in minimal time while keeping the line being drawn straight. The output of the microprocessor is a pulse train to each of the servo controllers. Each pulse causes the motor to move a prescribed amount. By commanding the servo with the longest move to perform at its maximum speed and acceleration and synchronizing the movement of the other axis at a slower velocity to it, a straight line is achieved (Fig. 2). This is accomplished by controlling the longer axis with a table-driven velocity profiling scheme and then using the algorithm outlined in Fig. 3 for synchronization. Making the algorithms interrupt-driven assures that the processor does not limit the ultimate speed of the plotter.

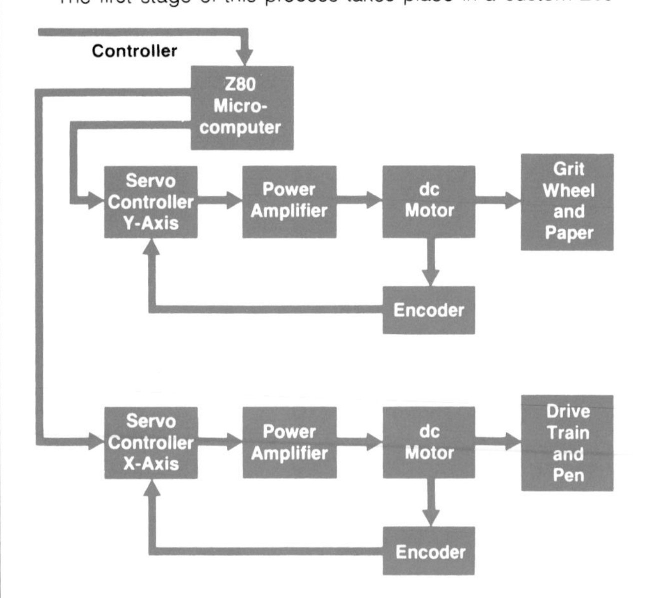

Fig. 1. *Block diagram of low-mass, low-inertia plotter electronics. Note that each axis is controlled by a separate, but identical servo control loop. Coordination between the two axes is handled by the microcomputer upon instructions from a controller.*

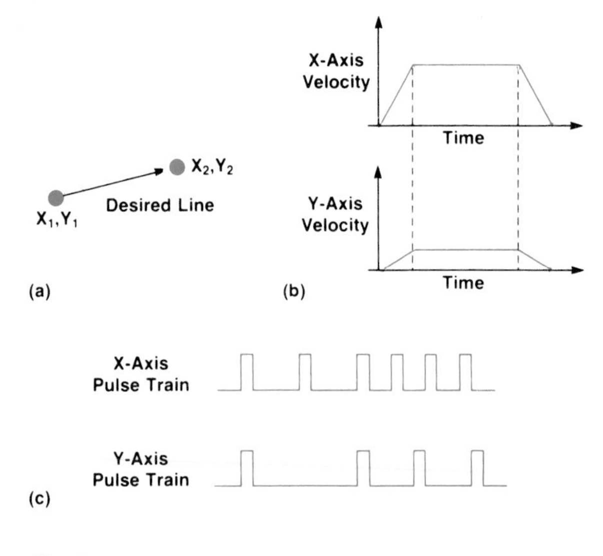

Fig. 2. *To draw a line (a) from one point to another, the axis drive motor requiring the greatest position change is set to its maximum velocity (b) and the velocity of the other axis drive motor is set proportionately. Thus, the drive pulses for each motor (c) will occur at different rates according to the velocities required to draw a straight line.*

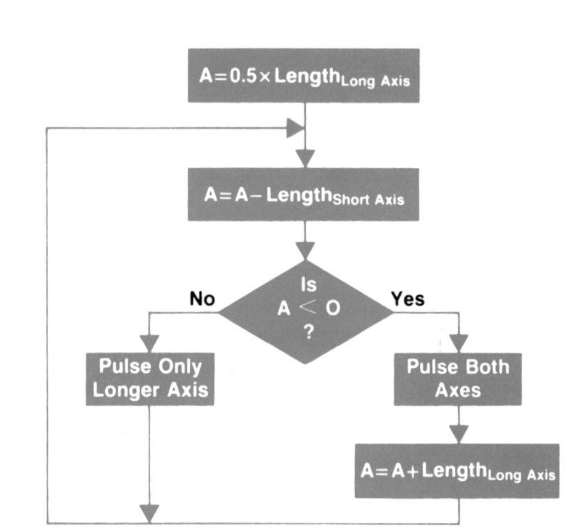

Fig. 3. *Algorithm for controlling pulse rates to each axis drive motor during a move from one point to another.*

The next major block in the architecture is the servo controller. The controller is completely digital to facilitate integration (see box on page 8). The controller's job is greatly simplified by the mechanical structure of the plotter. Since the paper is moved independently along one axis and the pen is moved along the other axis, the two axes are decoupled. Because the position encoders are mounted directly to the back of the drive motors, the problems of resonance are drastically reduced. The dominant inertia of the system is designed to be in the motors. This allows the response of both servos to be matched, which in turn helps keep the lines straight.

The servo controller works much the same as a classical analog servo would. It takes the pulse string from the microprocessor and

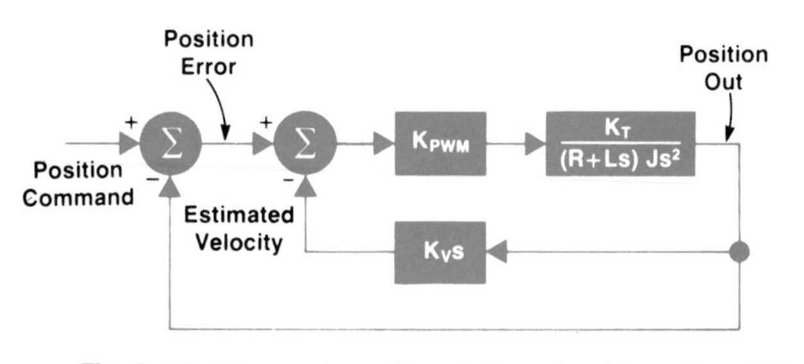

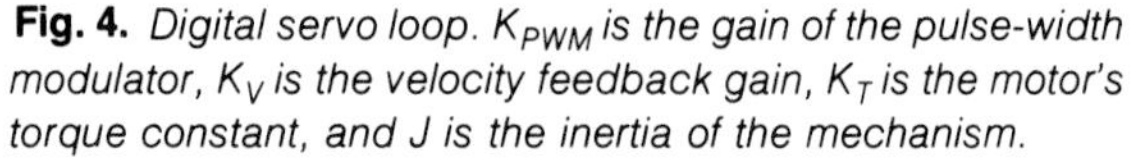

Fig. 4. *Digital servo loop. K_{PWM} is the gain of the pulse-width modulator, K_V is the velocity feedback gain, K_T is the motor's torque constant, and J is the inertia of the mechanism.*

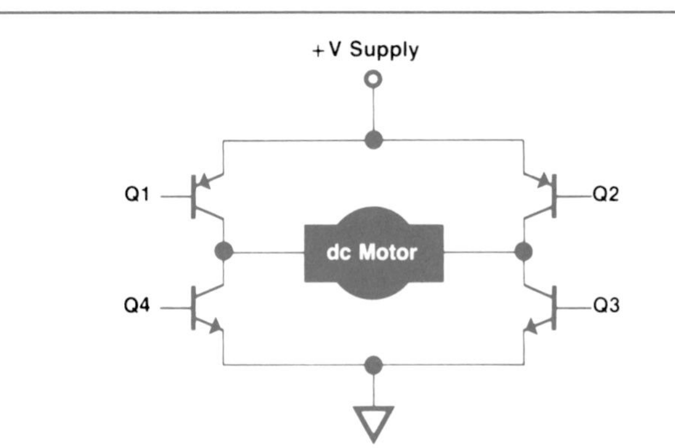

Fig. 5. *Arrangement of power amplifier transistors driving motor for each axis. Appropriate switching of Q1, Q2, Q3, and Q4 allows motor rotation in either direction.*

converts it to an instantaneous position command. At the same time it decodes the quadrature signal from the encoders and determines the current motor position. The difference between these two values is the position error.

For the loop to be stable, it must have compensation (see Fig. 4). In an analog loop this would come from a tachometer. Here a velocity estimate is generated from the encoder signal by using a variation of the reciprocal time method. Thus the need for a second encoder for each axis is eliminated. The velocity estimate is subtracted from the position error to derive the motor command. The motor command is then transformed into a pulse-width modulated signal suitable for driving an H-bridge power amplifier.

The amplifier for each drive motor, in keeping with the all digital theme of the electronics, consists of four transistors which are either turned on or off. By pulsing either Q1 and Q3 or Q2 and Q4, the direction of drive current to the motor can be determined. The average magnitude of the current is controlled by the duty cycle of the pulse.

These currents drive the motors, which in turn move the paper and pen through grit wheels and drive trains respectively, ultimately creating the desired line on the page.

-W.D. Baron

the design of a pen, holder, and guide rod adequate to withstand the new acceleration, we realized an acceleration of 2500 cm/s^2 with good graphic quality, quite beyond anything we had read about or heard of. Fig. 4 shows the new mechanism.

Refinement: Project Sweetheart

This technology needed a name. It was a genuinely satisfying thing to exercise, and it had been wrought for the Andover Division's cardiology team. The name could not be denied: Sweetheart.

We entered a refinement period of repeatedly increasing speed and acceleration until graphic quality suffered, then improving the mechanics, control electronics and software until the quality was regained. Increasingly quick methods for determining the velocity were found, as were better methods of using it to stabilize the servo loop.

New and efficient trajectory control algorithms (see box, page 5) were discovered that allowed high speed and yet could be implemented on a simple Z80 microprocessor. This allowed us to provide greater performance with no increase in electronics cost other than larger motor-driver transistors.

Pen wear became a problem. Higher velocities simply dragged the pen over more paper per unit time, and the much higher accelerations meant many more pen lifts and drops per unit time. These lifts and drops not only hammered the nylon pen tips, but the time they required became an increasingly large fraction of the time to complete a plot containing alphanumeric characters. We needed a much faster, yet gentler, pen-lift technique.

A speaker-coil linear motor was designed with feedback provided by a vane interrupting a light-emitting-diode (LED) optocoupler. This allowed a critically damped design that could drop the pen in less than two milliseconds, yet reach the platen at zero velocity. The system also sampled the platen continuously, allowing manufacturing misalignment of platen and pen guide and freedom from adjustment. A dual-level lift with the initial lift to 0.5 mm followed after 30 ms by a second lift to 2.5 mm meant the pen never lifted above 0.5 mm when printing a line of alphanumerics. This reduced the time devoted to pen lifts even more.

The Sweetheart project had met all its initial objectives. We made numerous prototypes to discover failure modes under various conditions, and they all had an inherent toughness that forgave abuse. Some of these have run for years now with an acceleration of 9400 cm/s^2, speed of 150 cm/s, and 20 characters/s while retaining satisfactory graphic quality.

Andover, having helped us design the missing piece they needed, began to concentrate exclusively on building their digital electrocardiograph. Marco Riera converted the algorithms to a 6801 microprocessor. Marty Mason redesigned the entire mechanism for use as an electrocardiograph component, and added an elegant automatic paper-alignment feature. Marty and Larry Perletz designed a new pen suitable for production. These and all the other contributions that made the HP Model 4700A PageWriter Cardiograph possible are described in the article on page 16.

Could we extend this technology to larger pieces of paper? The condition that the dominant inertia be the motor rotor was still satisfied for much larger paper and pen drive masses. We cut a mechanism in half, lengthened the center until the rollers were just under 56 cm apart, and generated a D-size (56 by 86 cm) plot seven days after beginning the project (Fig. 5).

Marv Patterson of the San Diego Division brought his extensive experience in large-format graphics (and the most insidious diagnostic test routine we had ever seen) to an evaluation of the stretched prototype. He concluded it had real potential for large graphics, and work in San Diego, until then limited to helping us with pens and inks and general technical consultation, began in earnest.

The initial successes with large-format plotting proved deceptive. Plotting on polyester sheets, necessary for many activities such as legally recording subdivision plots in the building industry or master parts files in manufacturing,

Fig. 5. *Engineering team and first large-scale plot made by an expanded version of the basic grit-wheel drive mechanism.*

could not be sustained over long periods without the polyester sheets beginning to slip during a plot. The study, understanding, and correction of this difficulty was led by George Lynch at San Diego and forms a part of a set of articles on large-format plotting in next month's issue.

To design integrable digital control circuitry is one thing, to produce a satisfactory integrated circuit is another. Clement Lo of the Corvallis Division asked numerous questions about our circuit until he understood its operation in the electromechanical control environment. He executed a radical redesign that was essentially the same functionally, but was much more reliable and took maximum advantage of the process he had selected. He and his team then brought in the first working silicon chip ahead of schedule. The box describes this work in more detail.

The promise of combining LSI and digital control has been largely verified, and its commercial realization is just beginning. The new technology provides higher graphic quality and much higher accelerations, yet much lower cost and power consumption than previous art. It can serve a broad range of applications, from a small component within an instrument to a large, stand-alone plotter. Articles in this and next month's issue describe the first products.

These programs have been characterized by conceptual and strategic partnerships of Hewlett-Packard's research laboratories and operating divisions early in the programs, sharing of concerns and suggestions throughout, and the operating divisions discovering and solving some of the more subtle and more difficult problems at the end.

Charles E. Tyler

Chuck Tyler earned the BS degree in physics in 1964 at the Massachusetts Institute of Technology. He then received the MS and PhD degrees in physics from Washington University, St. Louis, Missouri, in 1969. Chuck joined Hewlett-Packard Laboratories as a staff member and now is director of the Applied Physics Lab. He is the author of several papers and an inventor on several patents. A native of Sabinal, Texas, Chuck is married, has two children, and lives in Sunnyvale, California. His interests include ranching, hunting, and fishing.

Lawrence LaBarre

Larry LaBarre joined HP in 1952 with ten years of experience in designing fishing and concrete ships, sawmills, and sawmill machinery. At HP he has designed the 608, 626, and 628 Signal Generators, waveguide and waveguide tooling, and the automatic injection and high pressure pump portions of gas chromatographs. Larry directed special machine work for several years and is now working on low-mass, low-inertia plotter mechanisms. Larry is an inventor on several patents related to an automatic typewriter punch, lapping machinery, automatic injection for gas chromatography, and the grit-wheel paper-drive plotters. He was born in St. Paul, Minnesota and for a time attended the U.S. Naval Academy by presidential appointment. Larry then studied aircraft engineering and shipbuilding through the International Correspondence Schools. He lives in Mountain View, California and is interested in occult philosophy, health, and mechanisms.

Robert G. Younge

Rob Younge joined HP in 1975 after receiving the BSEE (1974) and MSEE (1975) degrees from Stanford University. He was a department manager in Hewlett-Packard Laboratories before he left the company to become one of the founders of Acuson. Rob is the coinventor of one patent (pending) and a member of the IEEE. A native of Grand Junction, Colorado, he now lives in Menlo Park, California and enjoys backpacking and music.

Wayne D. Baron

Wayne Baron was born in Philadelphia, Pennsylvania and attended the Massachusetts Institute of Technology where he received the BSEE and MSEE/Computer Science degrees in 1977. Wayne started at HP as a co-op student in 1974, working on frequency synthesizers. He began full-time work in 1977 as a staff member in Hewlett-Packard Laboratories and now is a project manager. Wayne lives in Cupertino, California and enjoys skiing, folk dancing, and camping.

Plotter Servo Electronics Contained on a Single IC

by Clement C. Lo

The new X-Y plotter technology described in the accompanying article allows the designer to reduce the complexity and cost associated with the servo control electronics of an electromechanical device. Hewlett-Packard Laboratories built numerous prototypes that established the relationship between the digital control circuitry and the mechanism. To achieve the objectives of low cost, small size, low power consumption and high reliability, all of the required digital servo control electronics except the high-current motor drivers were integrated into one NMOS integrated circuit.

The basic plotter servo system for one axis (Fig. 1) consists of a host processor, the servo controller IC, motor drivers, and a dc motor with a quadrature encoder mounted on the shaft. The servo logic can be divided into five major sections:

1. Arithmetic logic unit (ALU) to perform the summing and error-keeping functions (Fig. 2a).
2. Read-only memory (ROM) which stores the predetermined velocity profile of a given electromechanical system.
3. Decoding logic to decode information sent by the host processor or the quadrature encoder signals.
4. Velocity estimation logic.

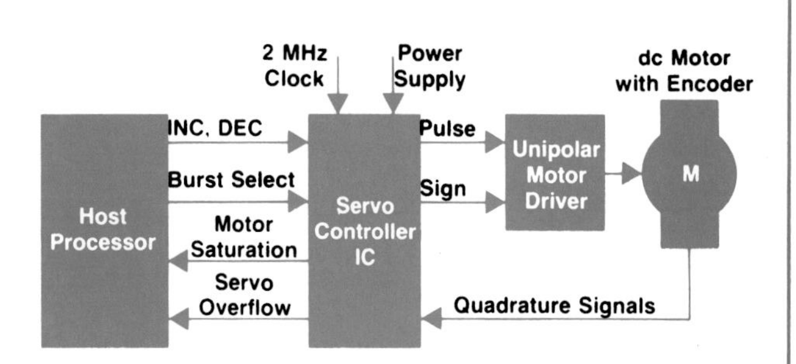

Fig. 1. *Block diagram of servo control system for one axis of a low-mass, low-inertia plotter. One IC contains all of the digital servo control electronics, greatly simplifying the overall system design.*

5. Pulse-width modulation logic.

The architecture of the NMOS servo controller (Fig. 2b) is in many ways a refinement of the initial servo designs. However, the total integration of the whole servo control system into one low-

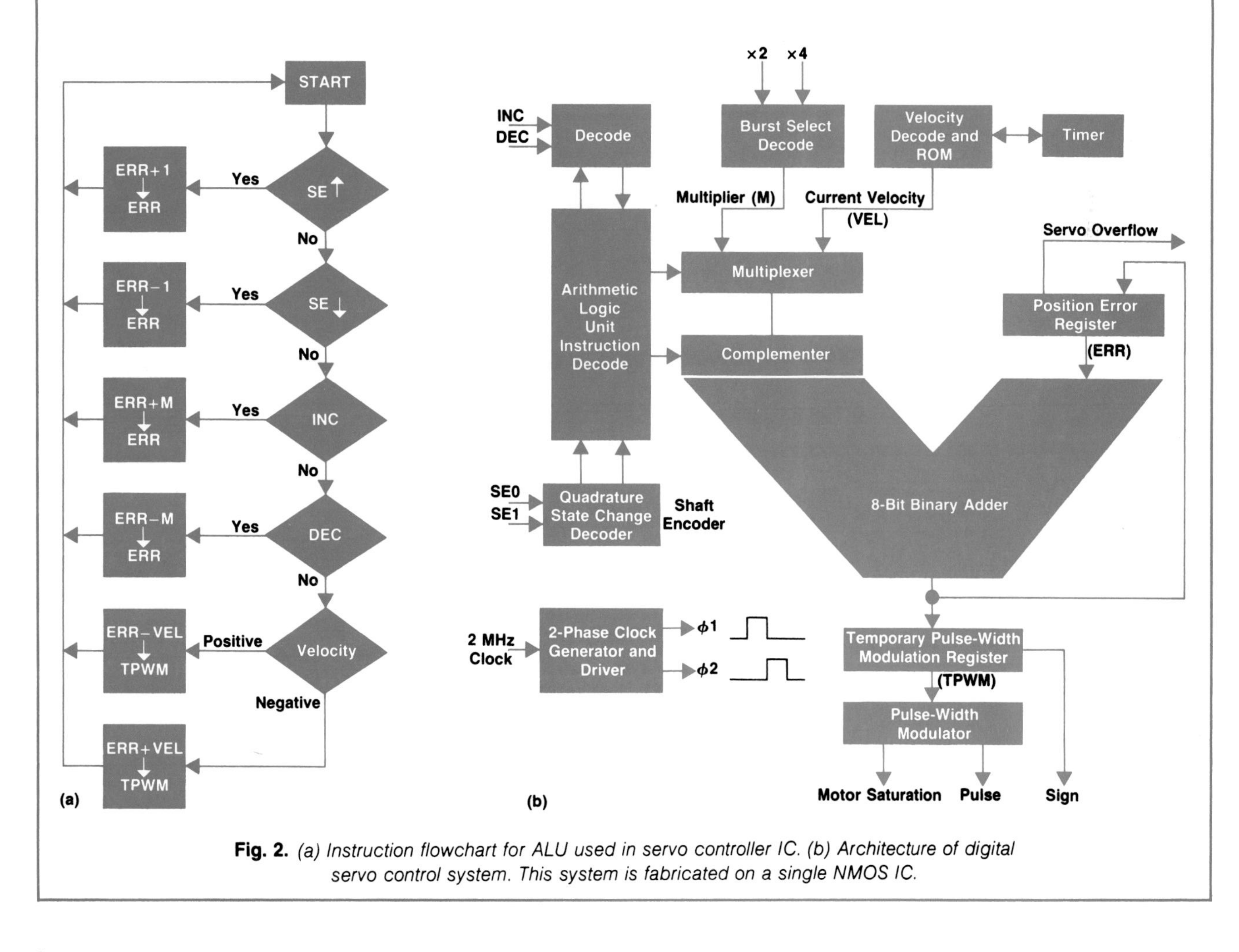

Fig. 2. *(a) Instruction flowchart for ALU used in servo controller IC. (b) Architecture of digital servo control system. This system is fabricated on a single NMOS IC.*

power integrated circuit called for new techniques in both logic and circuit design.

Positioning commands are in the form of INCrement or DECrement. The host processor commands the servo controller with INC, DEC and the burst select inputs (×2, ×4). The direction of motor rotation is determined by either INC or DEC. The number of steps the motor will rotate upon each INC or DEC command is determined by the logic states of the burst select inputs. They provide a multiplication factor of 1, 2, 4, or 8. The servo controller implements the position error summing, velocity feedback summing and error amplification function by using an eight-bit ALU. The velocity decoder obtains the velocity by measuring the time between the shaft encoder quadrature state changes and comparing it to a velocity lookup table stored in an on-chip ROM. The amount of ROM space required for the entire dynamic range is reduced from 512 bytes to 64 bytes through a novel autorange technique.

The servo controller sends two warning signals back to the host processor. Under normal operation both the motor saturation and servo overflow outputs should be at a logic zero state. When the output of the pulse-width modulation logic reaches a 90% duty cycle, the motor saturation output will go to a logic one state. Once it reaches a 100% duty cycle, the servo cannot accelerate the motor any faster and will fall behind. This will eventually cause the controller to lose the shaft position. The position error register can hold ±127 steps. When an overflow situation occurs, the servo overflow output will go to a logic one state and the controller will automatically shut off the drive signals to the motor. The controller will remain in this state until reset by the host processor.

Fig. 3 shows a picture of the integrated circuit. This 3.48-mm-by-3.3-mm IC dissipates about 100 mW at a clock frequency of 2 MHz. Numerous logic cells are used and most of the random logic is implemented by programmable logic arrays (PLA). This greatly reduces the complexity of the layout and was one of the prime factors in the speedy turnaround of the design. The timing of the logic also maximizes the speed/power tradeoff to achieve low power dissipation.

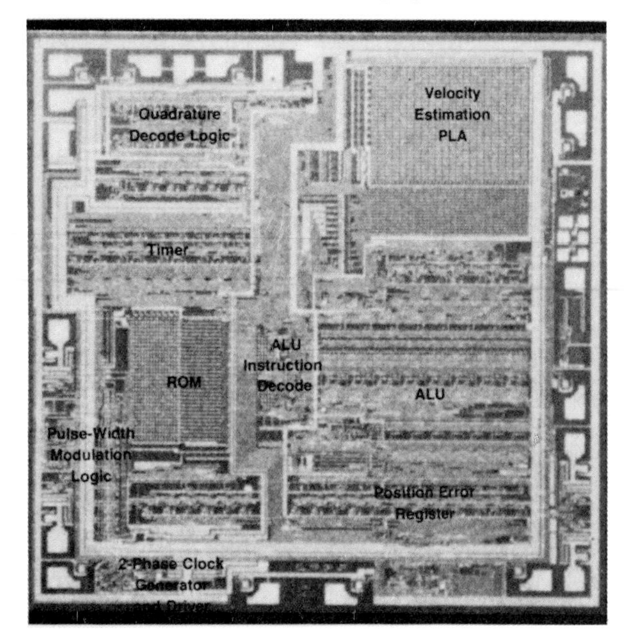

Fig. 3. *Microphotograph of servo controller IC with various functional areas identified. This chip is 3.48 mm wide and 3.3 mm high, roughly this size:* ■

Acknowledgments

I would like to thank Mike Pan for the excellent job he did in circuit design and detail checking of the chip. Don Hale did the layout. Jerry Erickson did the logic simulation of the circuit and Dave Serisky implemented the CMOS breadboard. Special thanks to Mike Lee of HP Laboratories for his advice and encouragement.

Clement C. Lo

Clement Lo was born in Hong Kong and attended the University of California at Berkeley where he was awarded the BSEE degree in 1974. Clement then earned the MSEE degree in 1975 from the University of California at Los Angeles (UCLA). He came to HP in 1975 and has worked on several ICs for the HP-85 Computer and was project leader for the servo-controller chip used in the 4700A. Clement is a project manager in the R&D lab at HP's Corvallis Division and is a co-author of an article on a printer-controller chip. He is married and lives in Corvallis, Oregon. He enjoys woodworking, gardening, and using small computers.